T0207193

Modern Deep Learning for Tabular Data

Novel Approaches to Common Modeling Problems

Andre Ye
Zian Wang

Apress®

Modern Deep Learning for Tabular Data: Novel Approaches to Common Modeling Problems

Andre Ye
Seattle, WA, USA

Zian Wang
Redmond, WA, USA

ISBN-13 (pbk): 978-1-4842-8691-3

ISBN-13 (electronic): 978-1-4842-8692-0

https://doi.org/10.1007/978-1-4842-8692-0

Managing Director, Apress Media LLC: Welmoed Spahr
Acquisitions Editor: Celestin Suresh John
Development Editor: Laura Berendson
Coordinating Editor: Mark Powers

Cover designed by eStudioCalamar

Cover image by Joshua Oluwagbemiga on Unsplash (www.unsplash.com)

Distributed to the book trade worldwide by Apress Media, LLC, 1 New York Plaza, New York, NY 10004, U.S.A. Phone 1-800-SPRINGER, fax (201) 348-4505, e-mail orders-ny@springer-sbm.com, or visit www.springeronline.com. Apress Media, LLC is a California LLC and the sole member (owner) is Springer Science + Business Media Finance Inc (SSBM Finance Inc). SSBM Finance Inc is a **Delaware** corporation.

For information on translations, please e-mail booktranslations@springernature.com; for reprint, paperback, or audio rights, please e-mail bookpermissions@springernature.com.

Apress titles may be purchased in bulk for academic, corporate, or promotional use. eBook versions and licenses are also available for most titles. For more information, reference our Print and eBook Bulk Sales web page at http://www.apress.com/bulk-sales.

Any source code or other supplementary material referenced by the author in this book is available to readers on GitHub (https://github.com/Apress). For more detailed information, please visit http://www.apress.com/source-code.

Printed on acid-free paper

To the skeptics and the adventurers alike

Table of Contents

About the Authors.. xiii

About the Technical Reviewer xv

Acknowledgments xvii

Foreword 1 xix

Foreword 2 xxi

Introduction xxiii

■Part I: Machine Learning and Tabular Data 1

■Chapter 1: Classical Machine Learning Principles and Methods 3

Fundamental Principles of Modeling... 4

    What Is Modeling? 4

    Modes of Learning.. 5

    Quantitative Representations of Data: Regression and Classification................................ 8

    The Machine Learning Data Cycle: Training, Validation, and Test Sets 9

    Bias-Variance Trade-Off... 19

    Feature Space and the Curse of Dimensionality.. 22

    Optimization and Gradient Descent 31

Metrics and Evaluation.. 38

    Mean Absolute Error 38

    Mean Squared Error (MSE) 40

    Confusion Matrix.. 42

    Accuracy 43

Precision 43

Recall 44

F1 Score 45

Area Under the Receiver Operating Characteristics Curve (ROC-AUC) 46

Algorithms 49

K-Nearest Neighbors 50

Linear Regression 57

Logistic Regression 66

Decision Trees 73

Gradient Boosting 83

Summary of Algorithms 90

Thinking Past Classical Machine Learning 90

Key Points 93

■Chapter 2: Data Preparation and Engineering 95

Data Storage and Manipulation 96

TensorFlow Datasets 96

Creating a TensorFlow Dataset 97

TensorFlow Sequence Datasets.. 98

Handling Large Datasets 100

Data Encoding 104

Discrete Data 104

Continuous Data 119

Text Data.. 126

Time Data 141

Geographical Data 144

Feature Extraction 145

Single- and Multi-feature Transformations.. 145

Principal Component Analysis 151

t-SNE 156

Linear Discriminant Analysis 159

Statistics-Based Engineering 161

Feature Selection 163

Information Gain 163

Variance Threshold 165

High-Correlation Method 167

Recursive Feature Elimination 170

Permutation Importance 173

LASSO Coefficient Selection 175

Key Points 177

■Part II: Applied Deep Learning Architectures 181

■Chapter 3: Neural Networks and Tabular Data 183

What Exactly Are Neural Networks? 183

Neural Network Theory 185

Starting with a Single Neuron 185

Feed-Forward Operation 186

Introduction to Keras 189

Modeling with Keras 189

Loss Functions 197

Math Behind Feed-Forward Operation 201

Activation Functions 203

The Math Behind Neural Network Learning 212

Gradient Descent in Neural Networks 212

The Backpropagation Algorithm 213

Optimizers 216

Mini-batch Stochastic Gradient Descent (SGD) and Momentum 216

Nesterov Accelerated Gradient (NAG) 218

Adaptive Moment Estimation (Adam) 219

A Deeper Dive into Keras 221

Training Callbacks and Validation 222

Batch Normalization and Dropout...225

The Keras Functional API...230

The Universal Approximation Theorem... 242

Selected Research 246

Simple Modifications to Improve Tabular Neural Networks.......................246

Wide and Deep Learning 250

Self-Normalizing Neural Networks 253

Regularization Learning Networks 254

Key Points 257

■Chapter 4: Applying Convolutional Structures to Tabular Data 259

Convolutional Neural Network Theory... 260

Why Do We Need Convolutions?...260

The Convolution Operation 268

The Pooling Operation 292

Base CNN Architectures..311

Multimodal Image and Tabular Models 326

1D Convolutions for Tabular Data 339

2D Convolutions for Tabular Data 356

DeepInsight..358

IGTD (Image Generation for Tabular Data) 367

Key Points 378

■Chapter 5: Applying Recurrent Structures to Tabular Data 379

Recurrent Models Theory 379

Why Are Recurrent Models Necessary?...379

Recurrent Neurons and Memory Cells...381

LSTMs and Exploding Gradients 388

Gated Recurrent Units (GRUs)...392

Bidirectionality...396

Introduction to Recurrent Layers in Keras... 397

Return Sequences and Return State 400

Standard Recurrent Model Applications 403

Natural Language 403

Time Series 409

Multimodal Recurrent Modeling 416

Direct Tabular Recurrent Modeling 424

A Novel Modeling Paradigm 424

Optimizing the Sequence 425

Optimizing the Initial Memory State(s) 436

Further Resources 448

Key Points 448

■Chapter 6: Applying Attention to Tabular Data 451

Attention Mechanism Theory 451

The Attention Mechanism 452

The Transformer Architecture 457

BERT and Pretraining Language Models 461

Taking a Step Back 464

Working with Attention 467

Simple Custom Bahdanau Attention 467

Native Keras Attention 472

Attention in Sequence-to-Sequence Tasks 481

Improving Natural Language Models with Attention 485

Direct Tabular Attention Modeling 496

Attention-Based Tabular Modeling Research 500

TabTransformer 500

TabNet 514

SAINT 531

ARM-Net 542

Key Points 546

■Chapter 7: Tree-Based Deep Learning Approaches 549

Tree-Structured Neural Networks 549

Deep Neural Decision Trees..550

Soft Decision Tree Regressors..556

NODE...561

Tree-Based Neural Network Initialization 564

Net-DNF 574

Boosting and Stacking Neural Networks.. 579

GrowNet..579

XBNet...584

Distillation 591

DeepGBM...591

Key Points 597

■Part III: Deep Learning Design and Tools 599

■Chapter 8: Autoencoders 601

The Concept of the Autoencoder 601

Vanilla Autoencoders... 606

Autoencoders for Pretraining 631

Multitask Autoencoders... 640

Sparse Autoencoders 653

Denoising and Reparative Autoencoders... 664

Key Points 680

■Chapter 9: Data Generation 681

Variational Autoencoders.. 681

Theory...682

Implementation...687

Generative Adversarial Networks 697

Theory...697

Simple GAN in TensorFlow..702

CTGAN...706

Key Points 710

■Chapter 10: Meta-optimization...711

Meta-optimization: Concepts and Motivations 711

No-Gradient Optimization 713

Optimizing Model Meta-parameters..725

Optimizing Data Pipelines 738

Neural Architecture Search 747

Key Points 752

■Chapter 11: Multi-model Arrangement..753

Average Weighting 753

Input-Informed Weighting 765

Meta-evaluation 767

Key Points 770

■Chapter 12: Neural Network Interpretability 771

SHAP...771

LIME 784

Activation Maximization 788

Key Points 791

Closing Remarks 792

Appendix: NumPy and Pandas...793

Index..829

About the Authors

Andre Ye is a deep learning (DL) researcher with a focus on building and training robust medical deep computer vision systems for uncertain, ambiguous, and unusual contexts. He has published another book with Apress, *Modern Deep Learning Design and Application Development*, and writes short-form data science articles on his blog. In his spare time, Andre enjoys keeping up with current deep learning research and jamming to hard metal.

Zian "Andy" Wang is a researcher and technical writer passionate about data science and machine learning (ML). With extensive experiences in modern artificial intelligence (AI) tools and applications, he has competed in various professional data science competitions while gaining hundreds and thousands of views across his published articles. His main focus lies in building versatile model pipelines for different problem settings including tabular and computer vision–related tasks. At other times while Andy is not writing or programming, he has a passion for piano and swimming.

About the Technical Reviewer

Bharath Kumar Bolla has over 10 years of experience and is currently working as a senior data science engineer consultant at Verizon, Bengaluru. He has a PG diploma in data science from Praxis Business School and an MS in life sciences from Mississippi State University, USA. He previously worked as a data scientist at the University of Georgia, Emory University, Eurofins LLC, and Happiest Minds. At Happiest Minds, he worked on AI-based digital marketing products and NLP-based solutions in the education domain. Along with his day-to-day responsibilities, Bharath is a mentor and an active researcher. To date, he has published around ten articles in journals and peer-reviewed conferences. He is particularly interested in unsupervised and semi-supervised learning and efficient deep learning architectures in NLP and computer vision.

Acknowledgments

This book would not have been possible without the professional help and support of so many. We want to express our greatest gratitude to Mark Powers, the awesome coordinating editor powering the logistics of the book's development, and all the other amazing staff at Apress whose tireless work allowed us to focus on writing the best book we could. We also would like to thank Bharath Kumar Bolla, our technical reviewer, as well as Kalamendu Das, Andrew Steckley, Santi Adavani, and Aditya Battacharya for serving as our third through seventh pair of eyes. Last but not least, we are honored to have had Tomas Pfister and Alok Sharma each contribute a foreword.

Many have also supported us in our personal lives – we greatly appreciate our friends and family for their unwavering support and motivation. Although these incredible individuals might not be as involved in the technical aspect as those mentioned previously (we have fielded the question "So exactly what is your book about?" at the dinner table many a time), we would not have been able to accomplish what we set out to do – in this book and in life – without their presence.

Foreword 1

Tabular data is the most common data type in real-world AI, but until recently, tabular data problems were almost solely tackled with tree-based models. This is because tree-based models are representationally efficient for tabular data, inherently interpretable, and fast to train. In addition, traditional deep neural network (DNN) architectures are not suitable for tabular data – they are vastly overparameterized, causing them to rarely find optimal solutions for tabular data manifolds.

What, then, does deep learning have to offer tabular data learning? The main attraction is significant expected performance improvements, particularly for large datasets – as has been demonstrated in the last decade in images, text, and speech. In addition, deep learning offers many additional benefits, such as the ability to do multimodal learning, removing the need for feature engineering, transfer learning across different domains, semi-supervised learning, data generation, and multitask learning – all of which have led to significant AI advances in other data types.

This led us to develop TabNet, one of the first DNN architectures designed specifically for tabular data. Similar to the canonical architectures in other data types, TabNet is trained end to end without feature engineering. It uses sequential attention to choose which features to reason from at each decision step, enabling both interpretability and better learning as the learning capacity is used for the most salient features. Importantly, we were able to show that TabNet outperforms or is on par with tree-based models and is able to achieve significant performance improvements with unsupervised pretraining. This has led to an increased amount of work on deep learning for tabular data.

Importantly, the improvement from deep learning on tabular data extends from academic datasets to large-scale problems in the real world. As part of my work at Google Cloud, I have observed TabNet achieving significant performance gains on billion-scale real-world datasets for many organizations.

When Andre reached out to share this book, I was delighted to write a foreword. It is the first book I have come across that is dedicated toward the application of deep learning to tabular data. The book has approachability to anyone who knows only a bit of Python, with helpful extensive code samples and notebooks. It strikes a good balance between covering the basics and discussing more cutting-edge research, including interpretability, data generation, robustness, meta-optimization, and ensembling. Overall, a great book for data scientist–type personas to get started on deep learning for tabular data. I'm hoping it will encourage and enable even more successful real-world applications on deep learning to tabular data.

Tomas Pfister
Head of AI Research, Google Cloud
Author of the TabNet paper (covered in Chapter 6)

Foreword 2

Almost three decades ago, the inception of support vector machines (SVMs) brought a storm of papers to the machine learning paradigm. The impact of SVMs was far-reaching and touched many fields of research. Since then, both the complexity of data and hardware technologies have expanded multifold, demanding the need for more sophisticated algorithms. This need took us to the development of state-of-the-art deep learning networks such as convolutional neural networks (CNNs).

Artificial intelligence (AI) tools like deep learning gained momentum and stretched into industries, such as driverless cars. High-performance graphics processing units (GPUs) have accelerated the computation of deep learning models. Consequently, deeper nets have evolved, pushing both industrial innovation and academic research.

Generally, visual data occupies the space of deep learning technologies, and tabular data (nonvisual) is largely ignored in this domain. Many areas of research, such as biological sciences, material science, medicine, energy, and agriculture, generate large amounts of tabular data.

This book attempts to explore the less examined domain of tabular data analysis using deep learning technologies. Very recently, some exciting research has been done that could drive us into an era of extensively applying deep learning technologies to tabular data.

Andre and Andy have undoubtedly touched this dimension and comprehensively put forward essential and relevant works for both novice and expert readers to apprehend. They described the core models to bridge tabular data and deep learning technologies. Moreover, they have built up detailed and easy-to-follow programming language codes. Links to Python scripts and necessary data are provided for anyone to reimplement the models.

Although the most modern deep learning networks perform very promisingly on many applications, they have their limitations, particularly when handling tabular data. For instance, in biological sciences, tabular data such as multi-omics data have huge dimensions with very few samples. Deep learning models have transfer learning capability, which could be used on a large scale on tabular data to tap hidden information.

Andre and Andy's work is commendable in bringing this crucial information together in a single work. This is a must-read for AI adventurers!

Alok Sharma
RIKEN Center for Integrative Medical Sciences, Japan
Author of the DeepInsight paper (covered in Chapter 4)

Introduction

Deep learning has become the public and private face of artificial intelligence. When one talks casually about artificial intelligence with friends at a party, strangers on the street, and colleagues at work, it is almost always on the exciting models that generate language, create art, synthesize music, and so on. Massive and intricately designed deep learning models power most of these exciting machine capabilities. Many practitioners, however, are rightfully pushing back against the technological sensationalism of deep learning. While deep learning is "what's cool," it certainly is not the be-all and end-all of modeling.

While deep learning has undoubtedly dominated specialized, high-dimensional data forms such as images, text, and audio, the general consensus is that it performs comparatively worse in tabular data. It is therefore tabular data where those with some distaste, or even resentment, toward deep learning stake out their argument. It was and still is fashionable to publish accepted deep learning papers that make seemingly trivial or even scientifically dubious modifications – this being one of the gripes against deep learning research culture – but now it is also fashionable within this minority to bash the "fresh-minted new-generation data scientists" for being too enamored with deep learning and instead to tout the comparatively more classic tree-based methods as instead the "best" model for tabular data. You will find this perspective everywhere – in bold research papers, AI-oriented social media, research forums, and blog posts. Indeed, the counter-culture is often as fashionable as the mainstream culture, whether it is with hippies or deep learning.

This is not to say that there is no good research that points in favor of tree-based methods over deep learning – there certainly is.[1] But too often this nuanced research is mistaken and taken for a general rule, and those with a distaste for deep learning often commit to the same problematic doctrine as many advancing the state of deep learning: taking results obtained within a generally well-defined set of limitations and willfully extrapolating them in ways irrespective of said limitations.

The most obvious shortsightedness of those who advocate for tree-based models over deep learning models is in the problem domain space. A common criticism of tabular deep learning approaches is that they seem like "tricks," one-off methods that work sporadically, as opposed to reliably high-performance tree methods. The intellectual question Wolpert and Macready's classic No Free Lunch Theorem makes us think about whenever we encounter claims of universal superiority, whether this is superiority in performance, consistency, or another metric, is: *Universal across what subset of the problem space?*

---

[1] Two such studies are especially well-articulated and nuanced:

Gorishniy, Y.V., Rubachev, I., Khrulkov, V., & Babenko, A. (2021). Revisiting Deep Learning Models for Tabular Data. *NeurIPS*.

Grinsztajn, L., Oyallon, E., & Varoquaux, G. (2022). Why do tree-based models still outperform deep learning on tabular data?

The datasets used by the well-publicized research surveys and more informal investigations showing the success of deep learning over tabular data are common benchmark datasets – the Forest Cover dataset, the Higgs Boson dataset, the California Housing dataset, the Wine Quality dataset, and so on. These datasets, even when evaluated in the dozens, are undoubtedly limited. It would not be unreasonable to suggest that out of all data forms, tabular data is the most varied. Of course, we must acknowledge that it is much more difficult to perform an evaluative survey study with poorly behaved diverse datasets than more homogenous benchmark datasets. Yet those who tout the findings of such studies as bearing a broad verdict on the capabilities of neural networks on tabular data overlook the sheer breadth of tabular data domains in which machine learning models are applied.

With the increase of data signals acquirable from biological systems, biological datasets have increased significantly in feature richness from their state just one or two decades ago. The richness of these tabular datasets exposes the immense complexity of biological phenomena – intricate patterns across a multitude of scales, ranging from the local to the global, interacting with each other in innumerable ways. Deep neural networks are almost always used to model modern tabular datasets representing complex biological phenomena. Alternatively, content recommendation, an intricate domain requiring careful and high-capacity modeling power, more or less universally employs deep learning solutions. Netflix, for instance, reported "large improvements to our recommendations as measured by both offline and online metrics" when implementing deep learning.[2] Similarly, a Google paper demonstrating the restructuring of deep learning as the paradigm powering YouTube recommendations writes that "In conjugation with other product areas across Google, YouTube has undergone a fundamental paradigm shift towards using deep learning as a general-purpose solution for nearly all learning problems."[3]

We can find many more examples, if only we look. Many tabular datasets contain text attributes, such as an online product reviews dataset that contains a textual review as well as user and product information represented in tabular fashion. Recent house listing datasets contain images associated with standard tabular information such as the square footage, number of bathrooms, and so on. Alternatively, consider stock price data that captures time-series data in addition to company information in tabular form. What if we want to also add the top ten financial headlines in addition to this tabular data and the time-series data to forecast stock prices? Tree-based models, to our knowledge, cannot effectively address any of these multimodal problems. Deep learning, on the other hand, can be used to solve all of them. (All three of these problems, and more, will be explored in the book.)

The fact is that data has changed since the 2000s and early 2010s, which is when many of the benchmark datasets were used in studies that investigated performance discrepancies between deep learning and tree-based models. Tabular data is more fine-grained and complex than ever, capturing a wide range of incredibly complex phenomena. It is decidedly not true that deep learning functions as an unstructured, sparsely and randomly successful method in the context of tabular data.

However, raw supervised learning is not just a singular problem in modeling tabular data. Tabular data is often noisy, and we need methods to denoise noise or to otherwise develop ways to be robust to noise. Tabular data is often also always changing, so we need models that can structurally adapt to new data easily. We often also encounter many different datasets that share a fundamentally similar structure, so we would like to be able to transfer knowledge from one model to another. Sometimes tabular data is lacking, and we need to generate realistic new data. Alternatively, we would also like to be able to develop very robust and well-generalized models with a very limited dataset. Again, as far as we are aware, tree-based models either cannot do these tasks or have difficulty doing them. Neural networks, on the other hand, can do all the following successfully, following adaptations to tabular data from the computer vision and natural language processing domains.

Of course, there are important legitimate general objections to neural networks.

---

[2] https://ojs.aaai.org/index.php/aimagazine/article/view/18140.
[3] Covington, P., Adams, J.K., & Sargin, E. (2016). Deep Neural Networks for YouTube Recommendations. *Proceedings of the 10th ACM Conference on Recommender Systems.*

One such objection is interpretability – the contention that deep neural networks are less interpretable than tree models. Interpretability is a particularly interesting idea because it is more an attribute of the human observer than the model itself. Is a Gradient Boosting model operating on hundreds of features really more intrinsically interpretable than a multilayer perceptron (MLP) trained on the same dataset? Tree models do indeed build easily understandable single-feature split conditions, but this in and of itself is not really quite valuable. Moreover, many or most models in popular tree ensembles like Gradient Boosting systems do not directly model the target, but rather the residual, which makes direct interpretability more difficult. What we care about more is the interaction between features. To effectively grasp at this, tree models and neural networks alike need external interpretability methods to collapse the complexity of decision making into key directions and forces. Thus, it is not clear that tree models are more inherently interpretable than neural network models[4] on complex datasets.[5]

The second primary objection is the laboriousness of tuning the meta-parameters of neural networks. This is an irreconcilable attribute of neural networks. Neural networks are fundamentally closer to ideas than real concrete algorithms, given the sheer diversity of possible configurations and approaches to the architecture, training curriculum, optimization processes, and so on. It should be noted that this is an even more pronounced problem for computer vision and natural language processing domains than tabular data and that approaches have been proposed to mitigate this effect. Moreover, it should be noted that tree-based models have a large number of meta-parameters too, which often require systematic meta-optimization.

A third objection is the inability of neural networks to effectively preprocess data in a way that reflects the effective meaning of the features. Popular tree-based models are able to interpret features in a way that is argued to be more effective for heterogeneous data. However, this does not bar the conjoined application of deep learning with preprocessing schemes, which can put neural networks on equal footing against tree-based models with respect to access to expressive features. We spend a significant amount of space in the book covering different preprocessing schemes.

All of this is to challenge the notion that tree-based models are superior to deep learning models or even that tree-based models are consistently or generally superior to deep learning models. It is not to suggest that deep learning is generally superior to tree-based models, either. To be clear, the following claims were made:

- Deep learning works successfully on a certain well-defined domain of problems, just as tree-based methods work successfully on another well-defined problem domain.

- Deep learning can solve many problems beyond raw tabular supervised learning that tree-based methods cannot, such as modeling multimodal data (image, text, audio, and other data in addition to tabular data), denoising noisy data, transferring knowledge across datasets, successfully training on limited datasets, and generating data.

- Deep learning is indeed prone to difficulties, such as interpretability and meta-optimization. In both cases, however, tree-based models suffer from the same problems to a somewhat similar degree (on the sufficiently complex cases where deep learning would be at least somewhat successful). Moreover, we can attempt to reconcile weaknesses of neural networks with measures such as employing preprocessing pipelines in conjunction with deep models.

---

[4] It is in fact not implausible to suggest that the differentiability/gradient access of neural network models enables improved interpretability of decision-making processes relative to the complexity of the model.
[5] The prerequisite "complex" is important. A decision tree trained on a comparatively simple dataset with a small number of features is of course more interpretable than a neural network, but in these sorts of problems, a decision tree would be preferable to a neural network anyway by virtue of being more lightweight and achieving similar or superior performance.

The objective of this book is to substantiate these claims by providing the theory and tools to apply deep learning to tabular data problems. The approach is tentative and explorative, especially given the novel nature of much of this work. You are not expected to accept the validity or success of everything or even most things presented. This book is rather a treatment of a wide range of ideas, with the primary pedagogical goal of exposure.

This book is written toward two audiences (although you are more than certainly welcome if you feel you do not belong to either): experienced tabular data skeptics and domain specialists seeking to potentially apply deep learning to their field. To the former, we hope that you will find the methods and discussion presented in this book, both original and synthesized from research, at least interesting and worthy of thought, if not convincing. To the latter, we have structured the book in a way that provides sufficient introduction to necessary tools and concepts, and we hope our discussion will assist modeling your domain of specialty.

- - -

Organization of the Book

Across 12 chapters, the book is organized into three parts.

Part 1, "Machine Learning and Tabular Data," contains Chapter 1, "Classical Machine Learning Principles and Methods," and Chapter 2, "Data Preparation and Engineering." This part introduces machine learning and data concepts relevant for success in the remainder of the book.

> Chapter 1, "Classical Machine Learning Principles and Methods," covers important machine learning concepts and algorithms. The chapter demonstrates the theory and implementation of several foundational machine learning models and tabular deep learning competitors, including Gradient Boosting models. It also discusses the bridge between classical and deep learning.

> Chapter 2, "Data Preparation and Engineering," is a wide exposition of methods to manipulate, manage, transform, and store tabular data (as well as other forms of data you may need for multimodal learning). The chapter discusses NumPy, Pandas, and TensorFlow Datasets (native and custom); encoding methods for categorical, text, time, and geographical data; normalization and standardization (and variants); feature transformations, including through dimensionality reduction; and feature selection.

Part 2, "Applied Deep Learning Architectures," contains Chapter 3, "Neural Networks and Tabular Data"; Chapter 4, "Applying Convolutional Structures to Tabular Data"; Chapter 5, "Applying Recurrent Structures to Tabular Data"; Chapter 6, "Applying Attention to Tabular Data"; and Chapter 7, "Tree-Based Deep Learning Approaches." This part constitutes the majority of the book and demonstrates both how various neural network architectures function in their "native application" and how they can be appropriated for tabular data in both intuitive and unintuitive ways. Chapters 3, 4, and 5 each center one of the three well-established (even "traditional") areas of deep learning – artificial neural networks (ANNs), convolutional neural networks, and recurrent neural networks – and their relevance to tabular data. Chapters 6 and 7 collectively cover two of the most prominent modern research directions in applying deep learning to tabular data – attention/transformer methods, which take inspiration from the similarity of modeling cross-word/token relationships and modeling cross-feature relationships, and tree-based neural network methods, which attempt to emulate, in some way or another, the structure or capabilities of tree-based models in neural network form.

> Chapter 3, "Neural Networks and Tabular Data," covers the fundamentals of neural network theory – the multilayer perceptron, the backpropagation derivation, activation functions, loss functions, and optimizers – and the

TensorFlow/Keras API for implementing neural networks. Comparatively advanced neural network methods such as callbacks, batch normalization, dropout, nonlinear architectures, and multi-input/multi-output models are also discussed. The objective of the chapter is to provide both an important theoretical foundation to understand neural networks and the tools to begin implementing functional neural networks to model tabular data.

Chapter 4, "Applying Convolutional Structures to Tabular Data," begins by demonstrating the low-level mechanics of convolution and pooling operations, followed by the construction and application of standard convolutional neural networks for image data. The application of convolutional structures to tabular data is explored in three ways: multimodal image-and-tabular datasets, one-dimensional convolutions, and two-dimensional convolutions. This chapter is especially relevant to biological applications, which often employ methods in this chapter.

Chapter 5, "Applying Recurrent Structures to Tabular Data," like Chapter 4, begins by demonstrating how three variants of recurrent models – the "vanilla" recurrent layer, the Long Short-Term Memory (LSTM) layer, and the Gated Recurrent Unit layer – capture sequential properties in the input. Recurrent models are applied to text, time-series, and multimodal data. Finally, speculative methods for directly applying recurrent layers to tabular data are proposed.

Chapter 6, "Applying Attention to Tabular Data," introduces the attention mechanism and the transformer family of models. The attention mechanism is applied to text, multimodal text-and-tabular data, and tabular contexts. Four research papers – TabTransformer, TabNet, SAINT (Self-Attention and Intersample Attention Transformer), and ARM-Net – are discussed in detail and implemented.

Chapter 7, "Tree-Based Deep Learning Approaches," is dominantly research-focused and focuses on three primary classes of tree-based neural networks: tree-inspired/emulative neural networks, which attempt to replicate the character of tree models in the architecture or structure of a neural network; stacking and boosting neural networks; and distillation, which transfers tree-structured knowledge into a neural network.

Part 3, "Deep Learning Design and Tools," contains Chapter 8, "Autoencoders"; Chapter 9, "Data Generation"; Chapter 10, "Meta-optimization"; Chapter 11, "Multi-model Arrangement"; and Chapter 12, "Neural Network Interpretability." This part demonstrates how neural networks can be used and understood beyond the raw task of supervised modeling in somewhat shorter chapters.

Chapter 8, "Autoencoders," introduces the properties of autoencoder architectures and demonstrates how they can be used for pretraining, multitask learning, sparse/robust learning, and denoising.

Chapter 9, "Data Generation," shows how Variational Autoencoders and Generative Adversarial Networks can be applied to the generation of tabular data in limited-data contexts.

Chapter 10, "Meta-optimization," demonstrates how Bayesian optimization with Hyperopt can be employed to automate the optimization of meta-parameters including the data encoding pipeline and the model architecture, as well as the basics of Neural Architecture Search.

Chapter 11, "Multi-model Arrangement," shows how neural network models can be dynamically ensembled and stacked together to boost performance or to evaluate live/"real-time" model prediction quality.

Chapter 12, "Neural Network Interpretability," presents three methods, both model-agnostic and model-specific, for interpreting the predictions of neural networks.

All the code for the book is available in a repository on Apress's GitHub here: `https://github.com/apress/modern-deep-learning-tabular-data`.

We are more than happy to discuss the book and other topics with you. You can reach Andre at `andreye@uw.edu` and Andy at `andyw0612@gmail.com`.

We hope that this book is thought-provoking and interesting and – most importantly – inspires you to think critically about the relationship between deep learning and tabular data. Happy reading, and thanks for joining us on this adventure!

Best,
Andre and Andy

## PART I

Machine Learning and Tabular Data

CHAPTER 1

■ ■ ■

Classical Machine Learning Principles and Methods

True wisdom is acquired by the person who sees patterns, and comes to understand those patterns in their connection to other patterns – and from these interconnected patterns, learns the code of life – and from the learning of the code of life, becomes a co-creator with God.

—Hendrith Vanlon Smith Jr., Author and Businessman

The focus of this book is on *deep learning* for tabular data (don't worry – you got the right book!), but there are many good reasons to first develop a solid understanding of classical machine learning as a foundation.

Deep learning rests and builds upon the fundamental principles of machine learning. In many cases the axioms and behavior of the modern deep learning regime are different and distorted – these distortions and paradigm shifts will be further discussed and explored. Nevertheless, a strong knowledge of classical machine learning is required to intuit and work with deep learning, in the same way that calculus requires a strong understanding of algebra even though it operates under new, unfamiliar environments of infinity.

Moreover, understanding classical machine learning is even more important to one seeking to apply deep learning for tabular data than one seeking to apply deep learning applications to other sorts of data (image, text, signal, etc.) because much of classical machine learning *was built for and around tabular data*, with later adaptions for other types of data. The core algorithms and principles of classical machine learning often assume the presence of certain data characteristics and attributes that are often present only in tabular data. To understand the modeling of tabular data, one must understand these characteristics and attributes, why classical machine learning algorithms can assume and model them, and how they can potentially be better modeled with deep learning.

Additionally – despite the tremendous hype around it – deep learning is not a universal solution, nor should it be treated as one! In specialized data applications, often the only available viable solutions are based on deep learning. For problems involving tabular data, however, classical machine learning may provide a better solution. Therefore, it's important to know the theory and implementation of that set of classical approaches. Throughout the book, you'll continue to hone your judgment regarding which family of techniques is most applicable to the problems you need to solve.

This chapter will give a wide overview of classical machine learning principles, theory, and implementations. It is organized into four sections: "Fundamental Principles of Modeling," which explores key unifying concepts through intuition, math, visualizations, and code experiments; "Metrics and Evaluation," which provides an overview and implementation for commonly used model assessment methods; "Algorithms," which covers the theory and implementation (from scratch and from existing

© Andre Ye and Zian Wang 2023
A. Ye and Z. Wang, *Modern Deep Learning for Tabular Data*,
https://doi.org/10.1007/978-1-4842-8692-0_1

libraries) of a wide range of foundational and popular classical machine learning algorithms; and "Thinking Past Classical Machine Learning," which introduces the transition into new deep learning territory. The chapter assumes basic working knowledge of Python and NumPy but introduces basic usage of the modeling library scikit-learn. Please see Chapter 2 for an in-depth overview of concepts and implementation in NumPy, as well as Pandas and TensorFlow Datasets.

Fundamental Principles of Modeling

In this section, we will explore several fundamental ideas, concepts, and principles of modeling, especially in relation to the tabular data regime and context. Even if you're familiar or experienced with these topics, you may find discussion of new conceptual approaches or frameworks to solidify and push your understanding.

What Is Modeling?

Thinking generally beyond the field of data science, modeling refers to the process of building appropriate and "smaller" approximations or representations of "larger" phenomena. For instance, a fashion model's role is to demonstrate key elements and trends of a fashion brand. When a designer chooses what clothes a fashion model wears, they cannot have them wear every piece of clothing that the fashion brand carries – they must choose the most representative items, the ones that best capture the spirit, character, and philosophy of the brand. In the humanities and the social sciences (philosophy, sociology, history, psychology, political science, etc.), a model (or, similarly, a "theory") acts as a unifying framework that connects ideas and empirical observations. A psychological theory for love cannot explain or hold *all* the possible reasons and conditions in which love has ever arisen, but it can approximate it by exploring the general contours and dynamics of love.

A scientific model, similarly, articulates a concept that has a general influence on whatever environment it is derived from. Even though the natural world is noisy and scientific models are approximations, models can be used to derive useful approximate predictions. The Copernican model of the solar system suggested that the sun was at the center of the universe, with Earth and other planets orbiting around it in circular paths. This model can be used to understand and predict the motions of planets.

In chemical kinetics, steady-state approximation is a model to obtain the rate law of a multi-step chemical reaction by assuming all state variables are constant; the amount of a reactant produced is equal to the amount of that reactant consumed. This captures another theme of models: because models generalize, they must make certain assumptions.

Both steady-state approximation and the Copernican model, and all models, don't work *perfectly*, but rather "well enough" in "most" cases.

As briefly explored, models are useful in all sorts of domains. In these contexts, the "smallness" of models relative to the "largeness" of the phenomena being modeled is primarily in representation size. For instance, the Copernican model allows us to – under certain assumptions and simplifications – predict the trajectory of planets instead of waiting to let the universe (the phenomena being modeled) play out and observing the result. Restated, they're useful because they are easier to comprehend. One can manipulate and understand a representation of the phenomenon without observing the phenomenon itself. (We don't want to need to actually observe a meteor crashing into Earth to know that a meteor is crashing into Earth – at least, we would hope so.) In the context of data science and computer science, however, "smallness" has another key attribute: *ease of model creation and deployment.*

It takes significant time and effort for designers to curate a fashion model's outfit, for psychologists to collect disparate ideas and unify them into a theory of the human mind, and for scientists to develop models from empirical data. Rather, we would like a way to *autonomously* unify observations into a model to generalize an understanding of the phenomena from which the observations are derived. In this context, *autonomous* means "without significant human intervention or supervision during the creation of the model." In this book, and in contexts of data science, the term *modeling* is usually thus associated with some dimension of automation.

Automated modeling requires a computational/mathematical/symbolic approach to "understanding" and generalizing across a set of given observations, as computers are the most powerful and scalable instrument of automation known to humanity at the writing of this book. Even the most complex-behaving artificial intelligence models – often modeling the dynamics of language or vision – are built upon mathematical units and computation.

Moreover, because a computational/mathematical/symbolic approach to automation must be taken, the observations themselves must be organized in a *quantitative* format. Much of scientific data is quantitative, whereas "data" (perhaps more aptly referenced as "ideas") in the humanities is more often qualitative and textual. Does this mean that automated modeling cannot deal with most data in the humanities or any data that does not appear quantitatively in raw form? No, it merely means that we need another step of converting the data into quantitative form. We'll discuss this more in later subsections.

We can give this concept of automated modeling a name: *learning*. In this context, learning is the process of automatically developing representations of a concept or phenomenon given a set of observations about/derived from it, rather than requiring extensive involvement of another being (usually, a human) to piece the generalization together "for" them.

Modes of Learning

There are two general modes of learning: supervised and unsupervised learning. Note that we will explore more complex learning paradigms, especially with deep learning, since its unique framework and dynamics allow for greater complexity.

Supervised learning is intuitive: given some set of inputs x, the model attempts to predict y. In the supervised learning regime, we seek to capture how various contributing factors (the features) have an effect on changing the output (the label). The data must be correspondingly organized: we must have a *feature set* (referred to in shorthand as the "x") and a *label set* (referred to in shorthand as the "y"). Each "row" or "item" in the feature set is an organization of features derived from the same instance of a *thing* (e.g., pixels from the same image, behavior statistics from the same user, demographic information from the same person) and is associated with a label in the label set (Figure 1-1). The *label* is often also referred to as the "ground truth."

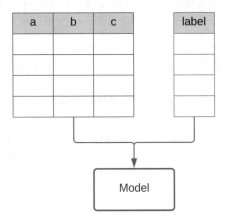

Figure 1-1. *Supervised learning relationship between data and model*

To accomplish the task of modeling an output given a set of inputs, the model must develop an understanding of how the inputted features interact with and relate to each other in producing a certain output. These associations can be simple, like a weighted linear sum of relevant features, or more complex, involving multiple duplications and concatenations to model the relevancy of the provided features with respect to the label (Figure 1-2). The complexity of the associations between the inputs and the outputs depends both on the model and the inherent complexity of the dataset/problem being modeled.

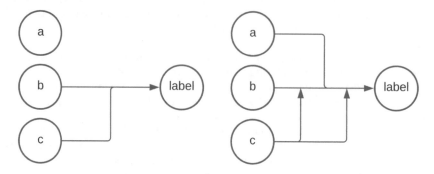

Figure 1-2. *Varying levels of complexity in discovering relationships between features and labels*

Here are some examples of supervised learning, drawn from various domains and levels of complexity:

- *Evaluating diabetes risk*: Provided with a dataset of patient information (age, sex, height, medical history, etc.) and the associated labels (if they have diabetes/severity or type of diabetes), the model automatically learns relationships and can predict a new patient's probability/severity/type of diabetes.

- *Forecasting a stock*: Provided with a recent time window of data (e.g., the stock price for the last n days) and the corresponding label (the stock price on the $n+1$-th day), the model can associate certain patterns in a sequence of data to predict the next value.

- *Facial recognition*: Given a few images taken of a person's face (the features being the pixels of the image) and the corresponding label (the person's identity, which can be any sort of personal identifier, like a unique number), the model can formulate and recognize identifying characteristics of a face and associate it with that person.

■ **Note**   Although in this early-stage introductory context we refer to the algorithms performing the learning and prediction all as "models," the types of models used in each context differ widely. The "Algorithms" section of this chapter engages in more rigorous exploration of different machine learning algorithms.

Unsupervised learning, on the other hand, is a little bit more abstract. The goal of unsupervised learning is to formulate information and trends about a dataset without the presence of a label set (Figure 1-3). Thus, the only input to an unsupervised model or system is the feature set, x.

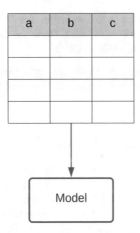

Figure 1-3. *Unsupervised learning relationship between data and model*

Instead of learning the relationship between features with respect to a label, unsupervised algorithms learn interactions between features with respect to the feature set itself (Figure 1-4). This is the essential difference between supervised and unsupervised learning. Like supervised learning, these interactions can range from simple to complex.

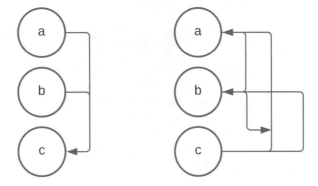

Figure 1-4. *Varying levels of complexity in discovering relationships in between features*

Here are some examples of unsupervised learning:

- *Customer segmentation*: A business has a dataset of their users with features describing their behavior. While they don't have any particular labels/targets in mind, they want to identify key segments or *clusters* of their customer base. They use an unsupervised *clustering algorithm* to separate their customer base into several groups, in which customers in the same group have similar features.

- *Visualizing high-dimensional features*: When there are only one or two features in a dataset, visualizing is trivial. However, the introduction of additional dimensions complicates the task of visualization. As new features are added, additional dimensions of separation must be added (e.g., color, marker shape, time). At a certain dataset dimensionality, it becomes impossible to effectively visualize data. *Dimensionality reduction* algorithms find the projection of high-dimensional data

7

into a lower-dimensional space (two dimensions for visualization) that maximizes the "information" or "structure" of the data. These are defined mathematically and do not require the presence of a label feature.

This book will focus primarily on supervised learning, but we will see many instances in which unsupervised and supervised learning approaches are combined to create systems more powerful than any one of its isolated parts.

Quantitative Representations of Data: Regression and Classification

Earlier, it was discussed that data must be in some sort of *quantitative form* to be used by an automated learning model. Some data appears "naturally" or "raw" in quantitative form, like age, length measurements, or pixel values in images. However, there are many other types of data that we would like to incorporate into models but don't appear in original form as a number. The primary data type here is categorical (class-based or label-based data), like animal type (dog, cat, turtle) or country (United States, Chile, Japan). Class-based data is characterized by a discrete separation of elements with no intrinsic ranking. That is, we cannot place "dog" as "higher than" a "cat" in the same way that 5 is higher than 3 (although many dog owners may disagree).

When there are only two classes present in a feature or label set, it is *binary*. We can simply associate one class with the quantitative representation "0" and the other class with the quantitative representation "1." For instance, consider a feature in a sample dataset "animal" that contains only two unique values: "dog" and "cat." We may attach the numerical representation "0" to "dog" and the numerical representation "1" to "cat" such that the dataset is transformed as follows (Figure 1-5).

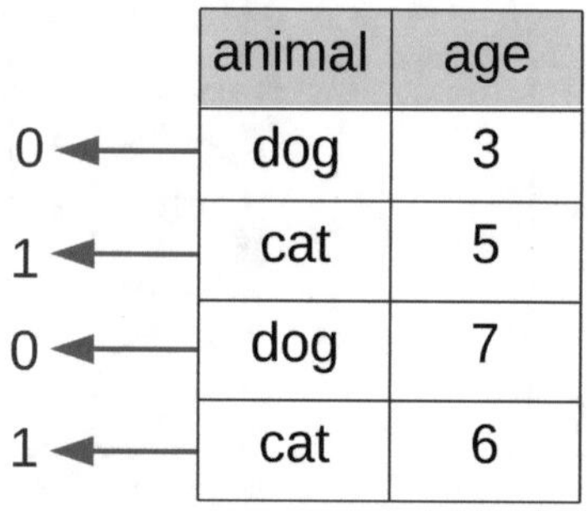

Figure 1-5. Example quantitative conversion/representation of a categorical feature "animal"

You may want to point out that because 1 is higher than 0, we are placing the "cat" label higher than the "dog" label. However, note that when we convert a data form into a quantitative representation, the quantitative measurement does not always (and often doesn't) mean the same thing anymore. In this case, the new feature should be interpreted as "Is the animal a cat?" rather than still as "animal," in which a 0 represents "no" and a 1 represents "yes." We also could have formed a quantitative representation meaning "Is the animal a dog?" in which all "dog" values are replaced with "1" and all "cat" values are replaced with "0." Both approaches hold the same *information* (i.e., the separation between animal types) as the original data (Figure 1-6), but it is represented quantitatively and thus becomes readable/compatible with an automated learning model.

cat?	age
0	3
1	5
0	7
1	6

dog?	age
1	3
0	5
1	7
0	6

Figure 1-6. *Equivalent binary representations of a categorical variable*

However, when there are more than two unique values in a feature or label, things become more complicated. In Chapter 2, we will more deeply explore effective data representation, manipulation, and transformation in the context of modeling tabular data. A challenge we will encounter and explore throughout the book is that of converting features of all forms into a more readable and informative representation for effective modeling.

Supervised problems can be categorized as *regression* or *classification* problems depending on the form of the desired output/label. If the output is continuous (e.g., house price), the problem is a regression task. If the output is binned (e.g., animal type), the problem is a classification task. For *ordinal outputs* – features that are quantized/discrete/binned but are also intrinsically ranked/ordered, like grade (A+, A, A–, B+, B, etc.) or education level (elementary, intermediate, high, undergraduate, graduate, etc. – the separation between regression and classification depends more on the approach you take rather than being inherently attached to the problem itself. For instance, if you convert education level into quantitative form by attaching it to an ordered integer (elementary to 0, intermediate to 1, high to 2, etc.) and the model is to output that ordered integer (e.g., an output of "3" means the model predicted "undergraduate" education level), the model is performing a regression task because the outputs are placed on a continuous output. On the other hand, if you designate each education level as its own class without intrinsic rank (see Chapter 2 ➤ "Data Encoding" ➤ "One-Hot Encoding"), the model is performing a classification task.

Many classification algorithms, including neural networks, the main study of this book, output a probability that an input sample falls into a certain class rather than automatically assigning a label to an input. For instance, a model may return a 0.782 probability an image of an ape belongs to the class "ape," a 0.236 probability it belongs to the class "gibbon," and lower probabilities for all other possible classes. We then interpret the class with the highest probability as the designated label; in this case, the model correctly decides on the label "ape." Although the output is technically continuous in that it is almost always not an integer 0 or 1 in practice, the *ideal* behavior/the expected labels are; thus, the task is classification rather than regression.

Understanding whether your problem and approach perform a classification or a regression task is a fundamental skill in building functioning modeling systems.

The Machine Learning Data Cycle: Training, Validation, and Test Sets

Once we fit our model to the dataset, we want to evaluate how well it performs so we can determine appropriate next steps. If the model performs very poorly, more experimentation is needed before it is deployed. If it performs reasonably well given the domain, it may be deployed in an application. If it performs *too* well (i.e., perfect performance), the results may be too good to be true and require additional investigation. Failing to evaluate a model's true performance can lead to headaches later at best and can be dangerous at worst. Consider, for instance, the impact of a medical model deployed to give diagnoses to real people that incorrectly appears to perfectly model the data.

The purpose of modeling, as discussed previously, is to represent a *phenomenon* or concept. However, because we can only access these phenomena and concepts through observations and data, we must approximate *modeling the phenomenon* by instead *modeling data derived from the phenomenon*. Restated, since we can't directly model the phenomenon, we must understand the phenomenon through the information we record from it. Thus, we need to ensure that the approximation and the end goal are aligned with one another.

Consider a child learning addition. Their teacher gives them a lesson and six flash cards to practice (Figure 1-7) and tells them that there will be a quiz the next day to assess the child's mastery of addition.

1 + 1 = 2	2 + 2 = 4
1 + 2 = 3	2 + 3 = 5
1 + 3 = 4	1 + 4 = 5

Figure 1-7. *Example "training dataset" of flash cards. The feature (input) is the addition prompt (e.g., "1+1," "1+2"); the label (desired output) is the sum (e.g., "2," "3")*

The teacher could use the same questions on the flash cards given to the students on the quiz (Quiz 1), or they could write a set of different problems that weren't given explicitly to the students, but were of the same difficulty and concept (Quiz 2) (Figure 1-8). Which one would more truly assess the students' knowledge?

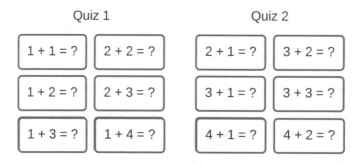

Figure 1-8. *Two possible quizzes: equal to the training dataset (left) or slightly different examples representing the same concept*

If the teacher gave Quiz 1, a student could score perfectly by simply memorizing six associations between a prompt "x + y" and the answer "z." This demonstrates *approximation-goal misalignment*: the *goal* of the quiz was to assess the students' mastery of the concept of addition, but the *approximation* instead assessed whether the students were able to memorize six arbitrary associations.

On the other hand, a student who merely memorized the six associations given to them on the flash cards would perform poorly on Quiz 2, because they are presented with questions that are different in explicit form, despite being very similar in concept. Quiz 2 demonstrates *approximation-goal alignment*: students can score well on the quiz only if they understand the concept of addition.

The differentiating factor between Quiz 1 and Quiz 2 is that the problems from Quiz 2 were not presented to the students for studying. This separation of information allows the teacher to more genuinely evaluate the students' learning. Thus, in machine learning, we always want to separate the data the model is trained on from the data the model is evaluated/tested on.

See the "Bias-Variance Trade-Off" subsection of this chapter for more rigorous exploration on these concepts of "memorization" and "genuine learning."

In (supervised/prediction) machine learning problems, we are often presented with the following data setup (Figure 1-9): a dataset with labels (i.e., a feature set and a label set, associated with one another) and an unlabeled dataset. The latter generally consists of residue data in which features are available but labels have not been attached yet, intentionally or for other purposes.

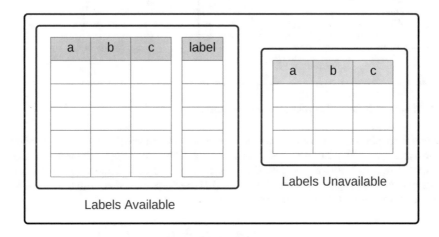

Complete Dataset

Figure 1-9. *Data type hierarchy*

The dataset that the model is exposed to when it is learning ('studying') is the *training set*. The *validation set* is the dataset the model is evaluated on, consisting of the set of data for which there are labels associated with features but that have not been shown to the model. The validation set is used to tune the model at the structural level – for instance, what type of algorithm is used or how complex the algorithm should be.

The *test set* is controversial, and there are two dominant definitions. One is as such: the *test set* is the dataset that the model is applied to in order to obtain predictions; the test set contains features but is not linked with any labels – since the objective is to obtain the labels via model prediction. The other interpretation is that the test set is used to form an "objective evaluation" of the model, since it is not used to inform the development of the model in any way – either through training or through structural tuning.

For instance, consider a machine learning model for an ecommerce platform that predicts whether a customer will purchase a particular product or not (Figure 1-10). The platform has a database of all their previous customers' information (the features) and whether they purchased the product or not (the label) that can be fed into a predictive machine learning model. The ecommerce platform also has data on site visitors who have not purchased anything yet; the platform wants to determine which visitors have a high probability of purchasing a particular product in order to tailor a marketing campaign toward their particular preferences (test set).

Age	Region	Visits/Week	Label
32	US	2	1
18	CAN	8	1
22	MEX	5	0
48	UK	10	1
43	US	1	0

Customer Data

Age	Region	Visits/Week
42	MEX	2
21	US	20
36	UK	4

Visitor Data

Figure 1-10. *Example dataset hierarchy of ecommerce data*

The ecommerce platform randomly splits the customer data into a *train dataset* and a *validation dataset* (in this example, with a 0.6 "train split," meaning that 60% of the original dataset with labels is placed in the training set). They train the model on the training set and evaluate it on the validation data to obtain an honest report of its performance (Figure 1-11). If they're not satisfied with the model's validation performance, they can tweak the modeling system.

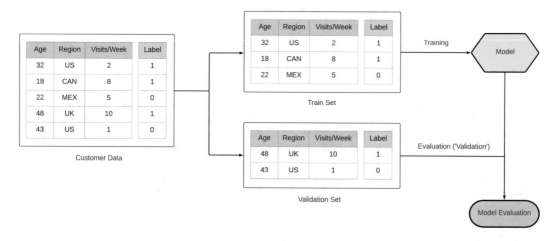

Figure 1-11. *Separation of a complete dataset into training and validation datasets*

When the model obtains satisfactory validation performance, the platform deploys the model on visitor data to identify promising customers (Figure 1-12). The platform launches customized marketing campaigns to those the model identified as having a high probability of purchasing the product, based on previous customer data.

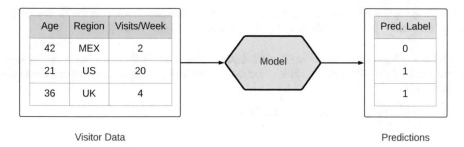

Visitor Data   Predictions

Figure 1-12. *Demonstration of how the training dataset is used in relation to the model*

■ **Note**   You'll often see the validation set referred to as the "test set" from other resources. Others might use "validation" and "test" synonymously. This book uses "train," "validation," and "test" to refer to three separate types of data defined by the presence of an associated label and the use case. It is useful to use this naming method because it provides a meaningful system to reference the primary components of the modern data pipeline – training, evaluation, and deployment.

In Python, one of the fastest and most commonly used methods to separate data into training and validation is with the scikit-learn library. Let's begin by creating a Pandas DataFrame with dummy data (Listing 1-1, Figure 1-13).

Listing 1-1. Loading a dummy dataset as a DataFrame

```
ages = [32, 18, 22, 48, 43]
regions = ['US', 'CAN', 'MEX', 'UK', 'US']
visitsWeek = [2, 8, 5, 10, 1]
labels = [1, 1, 0, 1, 0]
data = pd.DataFrame({'Age':ages,
                    'Region':regions,
                    'Visits/Week':visitsWeek,
                    'Label':labels})
```

	Age	Region	Visits/Week	Label
0	32	US	2	1
1	18	CAN	8	1
2	22	MEX	5	0
3	48	UK	10	1
4	43	US	1	0

Figure 1-13. *Example complete dataset in DataFrame form*

Now, we can separate the data into *features* (referred to in shorthand as *X/x*, or the "independent variables" to the model) and *labels* (the *y*, or the dependent variable(s)) (Listing 1-2).

Listing 1-2. Separating the dataset into a feature set and a label set. Note that the feature set can more efficiently be indexed via data.drop('Label', axis=1) – this returns a collection of all columns except for the "dropped" column

```
features = data[['Age', 'Region', 'Visits/Week']]
label = data['Label']
```

The sklearn.model_selection.train_test_split function (notice that scikit-learn uses "test set" to refer to the validation set) takes in three important parameters: the feature set, the label set, and the size of the training set (Listing 1-3). It returns four sets of data – the training feature set (Figure 1-14), the validation feature set (Figure 1-15), the training label set (Figure 1-16), and the validation label set (Figure 1-17).

Listing 1-3. Splitting into training and validation sets

```
from sklearn.model_selection import train_test_split as
    tts
X_train, X_val, y_train, y_val = tts(features, label,
                        train_size = 0.6)
```

	Age	Region	Visits/Week
3	48	UK	10
1	18	CAN	8
4	43	US	1

Figure 1-14. Example X_train result

	Age	Region	Visits/Week
2	22	MEX	5
0	32	US	2

Figure 1-15. Example X_val result

```
3    1
1    1
4    0
Name: Label, dtype: int64
```

Figure 1-16. Example y_train result

```
2    0
0    1
Name: Label, dtype: int64
```

Figure 1-17. *Example y_val result*

Notice that scikit-learn randomizes the order and split of the dataset to ensure the training and validation sets contain similar "topics" or "types of data." If the original dataset is already randomized, randomization doesn't hurt; if the original dataset is ordered (which is often the case – for instance, ordered alphabetically or by date), the training set likely is not optimally representative of the data in the validation set. Randomization can only help.

The train_test_split function accepts standard Python data forms, including native Python lists and NumPy arrays – as long as they are ordered in a sequential list format where each item is a data instance. Note that the feature and label datasets must be the same length, since they need to correspond with one another.

You can also implement train-validation-split "manually" with NumPy. Consider the following dataset in which the feature is a single integer and the label is 0 or 1, corresponding to even (0) or odd (1) (Listing 1-4).

Listing 1-4. Example dummy dataset construction

```
x = np.array([1, 2, 3, 4, 5, 6, 7, 8, 9])
y = np.array([1, 0, 1, 0, 1, 0, 1, 0, 1])
```

If seeing simple pattern-based data being typed out manually makes you cringe, one can also express the data with logical structures as such (Listing 1-5).

Listing 1-5. Another example of dummy dataset construction

```
x = np.array(list(range(1, 10)))
y = np.array([0 if i % 2 == 0 else 1 for i in x])
```

We'll use the following process to randomly split the dataset (Listing 1-6):

1. Zip the feature and label datasets together. In Python, the zipping operation pairs together elements of the same index from different lists. For instance, zipping the list [1, 2, 3] and [4, 5, 6] yields [(1, 4), (2, 5), (3, 6)]. The purpose of zipping is to associate the individual features and labels together so they do not become "unlinked" during shuffling.

2. Shuffle the feature-label pairs (pairing having been accomplished by the zipping operation).

3. Unzip the feature-label pairs.

4. Obtain the "split index." This is the index at which all previous elements will be part of the training set and all following elements will be part of the validation set. The index is calculated as the ceiling of the length of the train size (a fraction, like 0.8 for 80%) multiplied by the dataset size. One is subtracted from this quantity to account for zero-indexing.

5. Use the "split index" to index *x* train, *x* validation, *y* train, and *y* validation sets.

Listing 1-6. "Manual" implementation of train-validation split with NumPy

```
train_size = 0.8

zipped = np.array(list(zip(x, y)))
np.random.shuffle(zipped)
shuffled_x = zipped[:, 0]
shuffled_y = zipped[:, 1]

split_index = np.ceil(len(x) * train_size) - 1
X_train = shuffled_x[:split_index]
y_train = shuffled_y[:split_index]
X_valid = shuffled_x[split_index:]
y_valid = shuffled_y[split_index:]
```

Note that it is convention for the train set size to be larger than the validation set size, such that the model has access to most of the dataset when it is trained.

A very important source of potential error that is often committed during the train-validation split stage but is only noticed after the model has been trained is *data leakage*. Data leakage occurs when part of the training data is represented in the validation data or vice versa. This undermines a desired complete separation between training and validation sets, which leads to dishonest metric results.

For instance, consider an instance in which you're working with multiple stages of training. This can be the case with large models (e.g., heavy neural networks) that require a long time to train and therefore need to be trained in several sessions or if a trained model is being passed around team members for fine-tuning and inspection. At each session, the data is randomly split into training and validation sets, and the model continues to be fine-tuned on that data. After several sessions, the validation performance of the model is evaluated, and shockingly, it is perfect!

You pump your fist in the air and high-five your teammates. However, your enthusiasm is sadly misguided – you have been the victim of data leakage. Because the training-validation split operation was rerun each session and the operation is random, the training and validation sets were different each time (Figure 1-18). This means that across all sessions, the model was able to see almost all the dataset. Scoring a high validation performance is thus trivial, reduced only to a task of memorization rather than the much more difficult and important one of generalization and "true learning."

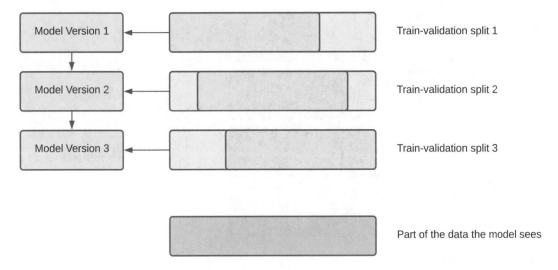

Figure 1-18. *Data leakage problems occurring from repeated train-validation split*

Differences in training-validation set splitting across multiple sessions can also lead to discrepancies in metric outcomes. A model fitted from scratch on one set of training data may perform slightly different than the same model type fitted from scratch on another set of training data, even though the formation of the training data is randomized. This slight difference can introduce errors when comparing model results.

In order to avoid these sorts of problems, we use *random seeding*. By setting a seed in the program, any random processes yield the same result every time they are run. To use random seeding, pass `random_state=101` (or some other value) into scikit-learn's `train_test_split`, or use `np.random.seed(101)` in addition to other `numpy.random` functions that may accept random seeding arguments (see NumPy documentation) at the start of your program for custom implementations.

Additionally, for small datasets, randomized train-validation split may not reflect model performance to the desired precision. When the number of samples is small, the selected validation examples may not be represented accurately in the training set (i.e., represented "too well" or "not well enough"), leading to biased model performance metrics.

In order to address this, one can use *k*-fold evaluation (Figure 1-19). In this evaluation schema, the dataset is randomly split into *k* folds (equally sized sets). For each of the *k* folds, we train a model from scratch on the other *k* – 1 folds and evaluate it on that fold. Then, we average (or aggregate through some other method) the validation performance for each of the *k*-folds.

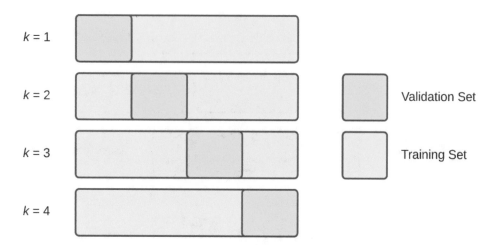

Figure 1-19. *K-fold evaluation*

This method allows us to obtain model validation performance on the "entire" dataset. However, note that it may be expensive to train k models – an alternative is simply to randomly split into training and validation sets without a random seed a given number of times without regard for which specific parts of data are allocated into either set.

This approach is also known as k-fold cross-validation, because it allows us to validate our model across the entire dataset.

Sci-kit learn offers the sklearn.model_selection.KFold object, which assists in implementing k-fold cross-validation by returning the appropriate indices for training and validation sets at each fold (Listing 1-7).

Listing 1-7. Template code to evaluate models on small datasets with scikit-learn's k-fold methods

```
n = 5

from sklearn.model_selection import KFold
folds = KFold(n_splits = n)

performance = 0

for train_indices, valid_indices in kf.split(X):

        X_train = X[train_indices]
        X_valid = X[valid_indices]
        y_train = y[train_indices]
        y_valid = y[valid_indices]

        model = InitializeModel()
        model.fit(X_train, y_train)
        performance += model.evaluate(X_valid, y_valid)

performance /= n
```

There is a balance in choosing the value of k and the train-validation split size generally – we want enough data to properly train the model, but also enough set aside in the validation set to properly evaluate the model. In the section "Metrics and Evaluation," you'll learn further about different metrics and evaluation methods. The point is evaluation is far from an objective operation; there is no "true performance" of the model. Like most methods in machine learning, we need to make different choices to understand our model as holistically as possible given operating circumstances. Any number that you are given as an indicator is showing only one side of a multisided, complex model!

Bias-Variance Trade-Off

Consider the following set of five points (Figure 1-20): how would you draw a smooth curve to model the dataset?

Figure 1-20. *Example set of data points*

There are many possible approaches one could take (Figure 1-21). The simplest approach may be just to draw a straight line in the general direction of the set – in the "middle" of the points. Another approach would be to connect the points with a slight curve such that there are no sudden changes in the derivative than are necessary (i.e., the second derivative remains near zero). A more complex one would be to connect the points, but draw large, extravagant peaks and valleys in between each.

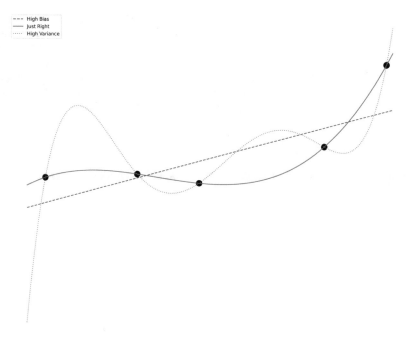

Figure 1-21. *Possible curves to fit the data points presented in Figure 1-20*

Which one best models the given dataset? In this case, most people would agree on the basis of intuition that the second choice (visualized in a continuous green line in Figure 1-21) is the most "natural" model. The second curve models the *given training data* well (i.e., it passes through each point), but its "predictions" for other intermediate locations also seem reasonable. The first curve seems too generally inaccurate, whereas the second passes through all the training data but deviates too much at intermediate values.

It would be helpful to formalize this intuition. In modeling contexts, we work with harder problems than connecting dots and must construct mathematical/computational models rather than drawing curves. The *bias-variance trade-off* can help us compare the behavior of different models in modeling data.

Bias is a model's ability to identify general trends and ideas in the dataset, whereas *variance* is the ability to model data with high precision. Since we think of bias and variance in terms of error, a lower bias or variance is better. The bias-variance relationship is often visualized as a set of dart throws on a bull's-eye (Figure 1-22). The ideal set of throws is a cluster of hits all around the center ring, which is *low bias* (the general "center" of the throws is not shifted off/biased) and *low variance* (the collection of hits are clustered together rather than far out, indicating consistently good performance).

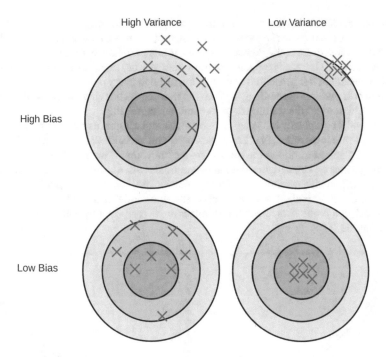

Figure 1-22. *Variance/bias trade-off bull's-eye representation. The ideal model is one that is both low in variance and bias error, in which all the dart throws are concentrated in the ideal central target ring*

On the other hand, a high-bias low-variance set of throws is highly concentrated in a very region far from the desired central ring, and a low-bias high-variance set of throws is centered around the desired ring but very disparate. A high-bias high-variance set of throws is both improperly centered and very spread out.

Given a point estimator $\hat{\theta}$ (in simple terms, an estimator for the value of an unknown parameter of a population) of a parameter θ, the bias can be defined as the difference between the expected value of the parameter and the true value:

$$\text{Bias} = E\left[\hat{\theta}\right] - \theta$$

If the expected value of the parameter is equal to the true value ($E\left[\hat{\theta}\right] = \theta$), the bias error term is 0 (and thus perfect). We can interpret the model as being "unbiased" – the true value of the parameter reflects the value it "should" take on.

Variance is defined as the expected value of the difference between the expected estimated parameter value and the true parameter value, squared:

$$\text{Variance} = E\left[\left(E\left[\hat{\theta}\right] - \hat{\theta}\right)^2\right]$$

If the parameter value difference fluctuates wildly, the variance will be high. On the other hand, if the difference is more consistent, the variance will be low.

Given these definitions, we can show that the Mean Squared Error (MSE) $\left(y - \hat{y}\right)^2$ (where y is the true value and $\hat{y}$ is the predicted value) can be *decomposed* as a relationship between the bias and variance error terms: MSE = Bias^2 + Variance. (The mathematical proof can be easily accessed but is beyond the scope of this book.) The *bias-variance decomposition* tells us that we can understand the error of a model as a relationship between bias and variance errors.

The bias-variance relationship lends itself to the concept of *underfitting* vs. *overfitting*. A model *underfits* when its bias is too high, while its variance is too low; the model doesn't adapt/bend too much at all toward specific data instances of the dataset and thus poorly models the training dataset. On the other hand, a model *overfits* when its bias is too low and the variance is too high; the model is incredibly sensitive to the specific data instances it is presented with, passing through every point.

Models that overfit perform exceedingly well on the training set but very poorly on the validation dataset, since they "memorize" the training dataset without generalizing.

Generally, as model complexity increases (can be measured roughly as number of parameters – although other factors are important), bias error decreases, and variance error increases. This is because the increased number of parameters allows the model greater "movement" or "freedom" to fit to the specifics of the given training dataset. (Thus, higher-degree polynomials with a larger number of coefficients are generally associated with higher-variance overfitting behavior, and lower-degree polynomials with a smaller number of coefficients are generally associated with higher-bias underfitting behavior.)

The ideal model for a problem is one that minimizes the overall error by balancing the bias and variance errors (Figure 1-23).

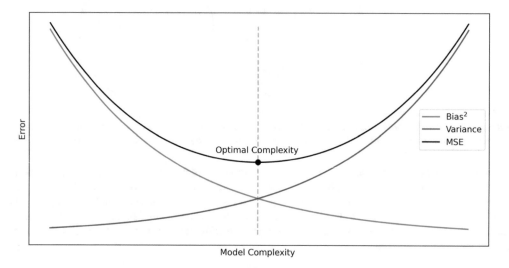

Figure 1-23. *Bias/variance trade-off curve representation. As bias decreases, variance increases, and vice versa. The sum of bias-squared and variance is the total model error. The best model is the one that manages to balance this trade-off*

Feature Space and the Curse of Dimensionality

As we discuss more mathematically involved concepts and representations of modeling, it becomes useful to think of a dataset not only as a collection of rows organized across a set of columns but also as a collection of points situated in the *feature space*.

For instance, the following dataset (Table 1-1) can be situated in three-dimensional space (Figure 1-24).

Table 1-1. *An example dataset with three features corresponding to three dimensions*

x	y	z
0	3	1
3	2	1
9	1	2
8	9	6
5	8	9

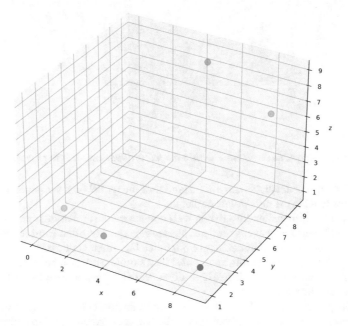

Figure 1-24. *Feature space representation of the data in Table 1-1*

The feature space is incredibly useful in a modeling concept because we can think of models as finding spatial relationships between points in space. For instance, certain binary classification models that attach either a "0" or "1" label to each data point can be thought of as conceptually drawing a hyperplane to separate points in space (Figure 1-25). (Learn more about these types of models in the "Algorithms" section of this chapter.)

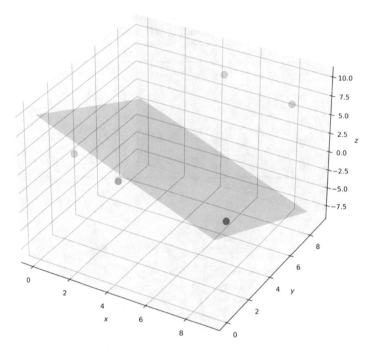

Figure 1-25. *Example hyperplane separation within the feature space containing the data in Table 1-1*

With the feature space concept, we can understand how the model behaves generally with any input, rather than being restricted to the data points available.

Feature spaces in low dimensionality are relatively tame to model and understand. However, the mathematics of high-dimensional spaces becomes bizarre very quickly, often contradicting our intuition – well-honed to the low dimensions of our existence but much less adept at understanding the dynamics of higher dimensions. To demonstrate this, let us consider a simpler variation on a common but quite mathematically inclined illustration of bizarre high-dimensional behavior by exploring how the volume of a hypercube changes with dimensionality.

A hypercube is a generalization of a three-dimensional cube to all dimensions. A two-dimensional hypercube is a square, and a one-dimensional hypercube is a line segment. We can define a hypercube with one corner at the origin and side length l as consisting of the entire set of n-dimensional points of form $(x_1, x_2, ..., x_n)$ that satisfy $0 \leq x_1 \leq l, 0 \leq x_2 \leq l, ..., 0 \leq x_n \leq l$. That is, the point is constrained within a certain range for each dimension (the length of the range being the same), independent of all other dimensions. Working with hypercubes makes for simpler mathematics, compared with other geometric generalizations like hyperspheres, which are defined by interactions between axes/dimensions.

We can draw a "smaller" hypercube within the original hypercube such that the side length of the smaller hypercube is 90% that of the original. We can sample points with uniform randomness throughout the space of the original hypercube and count how many fall in the inner hypercube to obtain an approximation for the volume. Examples of the smaller and larger hypercubes are shown in Figure 1-26 for first, second, and third dimensions.

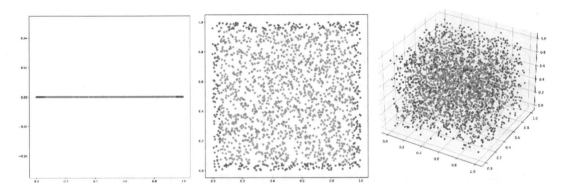

Figure 1-26. *Inner and outer hypercubes, visualized in one-, two-, and three-dimensional feature spaces*

To produce the visualizations, we will need the NumPy and matplotlib libraries, which can be imported via import numpy as np and import matplotlib.pyplot as plt, respectively. This code will become the basis upon which we generalize to hypercubes in higher dimensions (Listings 1-8, 1-9, 1-10).

Listing 1-8. Code to generate a one-dimensional nested hypercube set (left panel in Figure 1-26). Note that we can generate a relatively small number of points because the one-dimensional space is more compact relative to other higher-dimensional spaces

```
# set number of points to generate

n = 500

# generate random points in the space
features = np.random.uniform(0, 1, (500))

# condition for 'inside' or 'outside' point
labels = np.array(['blue' if i < 0.05 or i > 0.95
                   else 'red' for i in features])

# show on a one-dimensional hypercube
plt.figure(figsize=(10, 10), dpi=400)
plt.scatter(features, np.zeros(500), color = labels,
            alpha = 0.5)
plt.show()
```

Listing 1-9. Code to generate a two-dimensional nested hypercube set (center in Figure 1-26)

```
# set number of points to generate

n = 2000

# generate random points in the space
features = np.random.uniform(0, 1, (2, n))

# condition for 'inside' or 'outside' point
labels = np.array(['blue' if (features[0, i] < 0.05
```

```
                            or features[0, i] > 0.95)
                    or (features[1, i] < 0.05
                            or features[1, i] > 0.95)
              else 'red' for i in range(n)])

# show on a two-dimensional hypercube
plt.figure(figsize=(10, 10), dpi=400)
plt.scatter(features[0], features[1], color = labels, alpha = 0.5)
plt.show()
```

Listing 1-10. Code to generate a three-dimensional nested hypercube set (right panel in Figure 1-26)

```
# set number of points to generate

n = 3000

# generate random points in the space
features = np.random.uniform(0, 1, (3, n))

# condition for 'inside' or 'outside' point
labels = np.array(['red' if (features[0, i] < 0.05
                            or features[0, i] > 0.95)
                    or (features[1, i] < 0.05
                            or features[1, i] > 0.95)
                    or (features[2, i] < 0.05
                            or features[2, i] > 0.95)
                else 'blue' for i in range(n)])

# show on a three-dimensional hypercube
fig = plt.figure(figsize=(10, 10), dpi=400)
ax = fig.add_subplot(projection='3d')
ax.scatter(features[0], features[1], features[2],
          color = labels)
plt.show()
```

We want to track the percentage of "outside points" (i.e., sampled points that fall *outside* the smaller hypercube) out of the total points as the dimensionality of the space increases or the difference in volumes between the smaller and larger hypercubes (Figure 1-27).

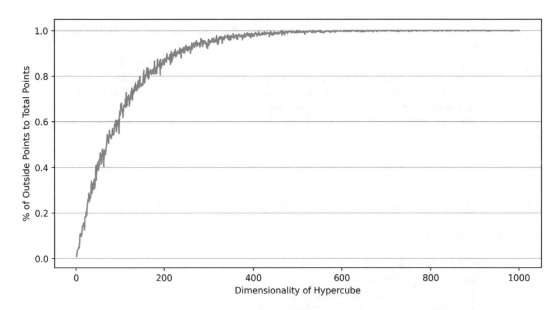

Figure 1-27. *Outside-points-to-total-points ratio, plotted against the dimensionality of the hypercube used in the ratio calculation. Jaggedness or noise is due to randomness in sampling*

At low dimensionality, the ratio seems reasonable and confirms our intuition: 0.01 in one dimension, 0.01 in two dimensions, 0.04 in three dimensions, 0.05 in five dimensions. The inside hypercube has a side length that is 90% that of the outer hypercube – it seems more than reasonable to expect the large majority of points to fall within the inner hypercube and thus for a low outside-points-to-total-points ratio.

However, the ratio for a 200-dimensional hypercube is about 0.85, and at 400 dimensions it is practically 1. This means that in a 400-dimensional hypercube, *essentially every single sampled point falls outside an inner hypercube defined with 90%-long of a side length of that of the outer hypercube.*

A similar result is obtained with hyperspheres. A common interpretation is that practically all of a very high-dimensional orange's mass is in the peel, not the pulp!

■ **Note**   You may be wondering we sample points and count how many land "outside" the inner hypercube rather than using simple mathematical volume formulas to determine the proportion. While the latter is one approach, sampling allows us to generalize findings to more complex shapes, in which defining the *boundary* of the shape is simple but evaluating the *volume* is much more difficult – like hyperspheres.

There are many ways to make sense of oddities and unintuitive behavior in high-dimensional space. For one, the math for this particular example of hypercubes supports the observations. Given dimensionality of the hypercube d, outer hypercube side length s, and the proportionality constant p representing the ratio of the inner hypercube side length to the outer hypercube side length, we can calculate the proportion of volumes as

$$\text{Volume Proportion} = \frac{\text{inner volume}}{\text{outer volume}} = \frac{(p \cdot s)^d}{s^d} = p^d$$

Given that $0 < p < 1$, we have that $\lim_{d \to \infty} p^d = 0$ – any fractional value between 0 and 1 that is repeatedly exponentiated will inevitably decrease toward zero.

A similar understanding can be revealed by working through the implementation for generating the curve (Listing 1-11). At each point, we loop through each dimension of that point and randomly generate a value for that dimension. If that randomly generated value fulfills the condition that it must fall within a certain range, it "passes" for that dimension, and we move on to evaluating the next dimension. If it does not "pass" (i.e., it falls outside the "inside" hypercube side length), we mark the point as being an "outside" point and evaluate the next point.

Listing 1-11. Generalized function to count proportion of "outside" to total points in the nested hypercube example

```
def hyperCube(dims = 1, num = 500, prop = 0.99):

    outsidePoints = 0

    for point in range(num):
        for dim in range(dims):
            randPos = np.random.uniform(0, 1)
            if randPos < (1-prop)/2 or randPos > 1 -
                1-prop)/2:
                    outsidePoints += 1
                    break

    return outsidePoints/num
```

As the dimensionality of sampled points increases, the condition for a point qualifying as an "outside" point gets evaluated more times. This is equivalent to exponentiating a value p between 0 and 1 repeatedly. Thus, more uniformly sampled points satisfy the "outside" point qualifications and thus are distributed near the "outsides" in higher-dimensional spaces. When we define surfaces in high-dimensional spaces, we must take into account this warping effect. The "peel" of the high-dimensional orange covers becomes so large in high dimensions that it outranks the "pulp" in volumes.

This phenomenon, along with other mathematical and empirical observations on high-dimensional spaces, is known collectively as the "Curse of Dimensionality." Simply put, the density/distribution of points in high-dimensional spaces is much different from that of low-dimensional spaces. This has important implications for how distance is measured.

Euclidean distance, or the L2 norm, between two points $(a_1, a_2, ..., a_n)$ and $(b_1, b_2, ..., b_n)$ is defined as $\sqrt{(a_1 - b_1)^2 + (a_2 - b_2)^2 + ... + (a_n - b_n)^2}$. It is our most "intuitive" or native sense of the distance between two points. However, in high-dimensional spaces, Euclidean distance becomes so diluted and "blurry" that it ceases to become useful.

Consider a unit hypercube with n dimensions. We will observe how the difference between the *second longest distance and the longest* distance between two corners of the hypercube changes as $n \to \infty$ (Listing 1-12, Figure 1-28). Beginning with the origin point $(0, 0, ..., 0)$, the farthest point can be found by "flipping" all the dimensions from 0 to 1: $(1, 1, ..., 1)$. The second farthest point can be found by arbitrarily "flipping" all the dimensions from 0 to 1 except one: $(0, 1, ..., 1)$. The distance from the origin to the farthest point is $\sqrt{\underbrace{(0-1)^2 + (0-1)^2 + \cdots + (0-1)^2}_{n}} = \sqrt{n}$, whereas the distance from the origin to the closest point is

$\sqrt{(0-0)^2 + \underbrace{(0-1)^2 + (0-1)^2 + \cdots + (0-1)^2}_{n-1}} = \sqrt{n-1}$. The difference between the two distances is $\sqrt{n} - \sqrt{n-1}$.

Listing 1-12. Plotting sample difference in distance of two selected points on a hypercube as dimensionality increases

```
features = np.linspace(1, 100, 500)
labels = np.sqrt(features) - np.sqrt(features - 1)

plt.figure(figsize=(10, 5), dpi=400)
plt.plot(features, labels, color='red')
axes = plt.gca()
axes.yaxis.grid()
plt.ylabel('Distance ($\sqrt{n} - \sqrt{n-1}$)')
plt.xlabel('Dimensionality ($n$)')
plt.show()
```

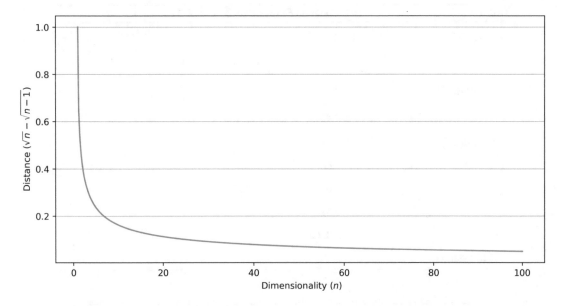

Figure 1-28. *Difference between the distance between the origin and the farthest point on the hypercube and the distance between the origin and the second farthest point on the hypercube, across dimensionality n*

We can find $\lim\limits_{n\to\infty}\sqrt{n}-\sqrt{n-1}=\lim\limits_{n\to\infty}\dfrac{n-(n-1)}{\sqrt{n}+\sqrt{n-1}}=\lim\limits_{n\to\infty}\dfrac{1}{\sqrt{n}+\sqrt{n-1}}\to\dfrac{1}{\infty}\to 0$. Thus, as the dimensionality of the hypercube increases, the two neighboring points on the farthest and second farthest corners of the hypercube become essentially the same distance from the origin. In fact, the distances between the origin and *all other corners* approach the same value in higher dimensions, as the generalization $\lim\limits_{n\to\infty}\sqrt{n}-\sqrt{n-k}=0$ is true in our context ($k < n$, representing the distance between the origin and the $k + 1$th farthest corner). This has the effect of polarizing and warping the distribution of distance in traditionally intuitively defined geometric shapes.

While Euclidean is our most intuitive sense of geometric distance, we can generalize it to different norms (Table 1-2, Listing 1-13, Figure 1-29). The kth norm distance is $\sqrt[k]{\sum_{i=1}^{n}|a_i - b_i|^k}$. *Manhattan distance* uses the L1 norm ($k = 1$) and behaves as a sum of the absolute differences in each dimension. Its name comes from the grid-like layout of Manhattan, New York – we measure distance strictly along paths parallel to the dimensions, just like navigating the blocks of a bustling metropolis. *Chebyshev distance*, on the other extreme, uses the L-infinity norm ($k = \infty$), which returns only the longest distance of any dimension, since raising the largest difference term to the "infinity power" will "outweigh" all other terms. Chebyshev distance is generally used in contexts in which computing distances of other norms is too costly (i.e., only very high-dimensional spaces).

Table 1-2. *Sampled values for the distance between points (0,0) and (2,1) in different norms*

Norm	Distance Between (0, 0) and (2, 1)
$\dfrac{1}{2}$	5.83
1 (Manhattan)	3
$\dfrac{3}{2}$	2.45
2 (Euclidean)	2.24
3	2.08
4	2.03
5	2.01
10	2.0002
∞ (Chebyshev)	2

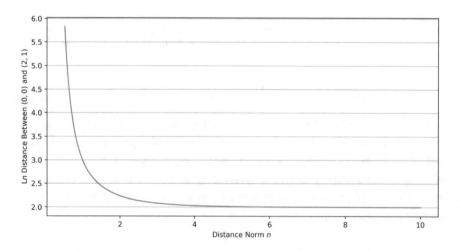

Figure 1-29. *Distance between (0, 0) and (2, 1) calculated with norm n*

Listing 1-13. Plotting sample distance between (0,0) and (2,1) in different norms

```
features = np.linspace(0.5, 10, 500)
labels = np.power((2**features) + 1, 1/features)

plt.figure(figsize=(10, 5), dpi=400)
plt.plot(features, labels, color='red')
axes = plt.gca()
axes.yaxis.grid()
plt.ylabel('L$n$ Distance Between $(0, 0)$ and $(2, 1)$')
plt.xlabel('Distance Norm $n$')
plt.show()
```

Norms other than L2 become useful in high-dimensional spaces since they can better "model" or "work with" the nature of distance in these environments. As you'll see in the later discussion of algorithms, many machine learning models require a defined definition of "distance" – which should not necessarily be Euclidean in the high-dimensional spaces they often operate in (see "Algorithms" ➤ "K-Nearest Neighbors").

The Curse of Dimensionality is a powerful theoretical tool that will be especially relevant when we think about how neural networks – which are indisputably the highest-dimensional machine learning algorithms to be in popular use – generalize and learn.

Optimization and Gradient Descent

Supervised machine learning models (i.e., models that predict an output given an input) "learn" by *optimizing* their *parameters*. Every algorithm has a set of parameters that determine how it works with, or "interprets," the inputs to produce an output. For instance, the parameters of a Simple Linear Regression model $f(x_1, x_2, x_3) = \beta_0 + \beta_1 x_1 + \beta_2 x_2 + \beta_3 x_3$ are $\{\beta_0, \beta_1, \beta_2, \beta_3\}$. The specific values of each of the parameters in this very rudimentary machine learning model determine how heavy a weight each of the inputs $\{x_1, x_2, x_3\}$ has on the final output.

In order to find the "best" set of parameters, we first need to define what "best" means. A *loss function* takes in two inputs – the predicted output (from the model) and the true output (the ideal expectation, the ground truth) – and returns a number indicating how "good" the model's performance is. (More complex loss functions can consider more than these two primary inputs.) Since "loss" is roughly equivalent with "error," smaller is better. We'll further solidify our understanding of loss functions in the following "Metrics and Evaluation" section.

When optimizing, models generally operate by the following sequence of steps:

1. Parameters begin initialized with random values. Depending on what *initialization schema/strategy* is used, parameters can be all set to zero, randomly pulled from a normal distribution, or initialized via some other process. Sometimes statistics or domain knowledge, for instance, can provide initialization values.

2. The model's loss is evaluated via the loss function.

3. Depending on how poorly the model performs (as indicated via the loss), the *optimization algorithm* makes changes to the parameters.

4. Steps 2 and 3 are repeated several times (ranging from about a dozen iterations for simple problems to millions of loops for more complex ones) until a stopping condition is met. This condition can be counter-based (a certain quantity of iterations has been met), time-based (x hours have elapsed since the beginning of training), or performance-based (the model has obtained a certain satisfactory validation score).

Conceptually, we can understand this process of beginning from a bad "guess" and iteratively improving the parameter set in response to feedback given by the loss function via the *loss landscape*. The loss landscape is an $n+1$-dimensional space, in which n is the number of parameters in the model. The landscape associates each possible combination of parameter values a model can take on (the n dimensions) with the loss a model with that set of parameters would incur (the additional dimension).

For instance, the loss landscape visualized in Figure 1-30 shows that a parameter value of 4 in the hypothetical one-parameter model it represents would incur a loss slightly less than 2 (units don't necessarily matter here; what is important is comparing higher/lower losses). It also demonstrates that the optimal parameter value for this model is around –1.5, because it minimizes the loss (this location is marked with an "X").

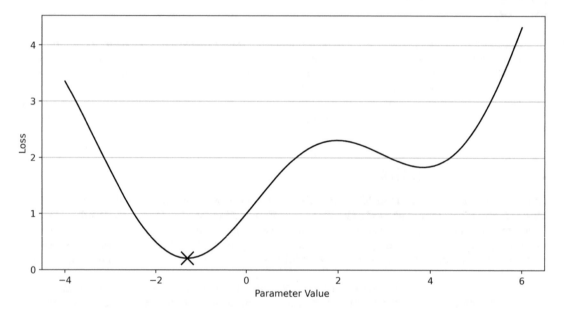

Figure 1-30. *Example loss landscape with global minimum (ideal solution) marked*

During optimization, we can think of the model as "traveling through" the loss landscape, seeking to find the place of lowest elevation (Figure 1-31). Every time it evaluates its error via the loss function (step 2), the model is gauging how "high" in the landscape it is. Every time it makes an update to its parameters (step 3), it is changing its position in the landscape with the goal of eventually ending up in the *global minimum*, or the point in the loss landscape with the lowest loss.

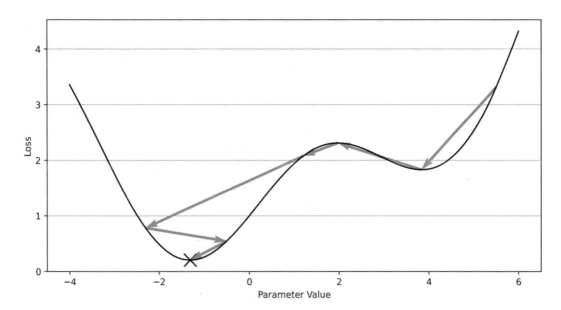

Figure 1-31. *Example descent from an initial parameter value of around 5.5 to the ideal solution of a parameter value around –1.5*

In deep learning, the code and math that determines the update strategy (i.e., how the model changes its position within the landscape) is known as the *optimizer*. Multiple different types of optimizers exist, with different strengths and weaknesses.

The most popular framework for optimization in deep learning is *gradient descent*. The fundamental idea of gradient descent is to use the *gradient*, or the derivative, of the function we're trying to minimize in order to determine which direction to move in. If the derivative at the current location is *positive*, then that means that moving forward will increase the cost function, whereas moving backward will decrease the cost function. On the other hand, if the derivative is negative, then moving forward will decrease the cost function, whereas moving backward will increase it. Since the goal is to decrease the cost function overall, we want to move in the direction opposite to the sign of the derivative. Moreover, we want to take a bigger step if the gradient at some location is larger in magnitude and a smaller step if the gradient is smaller, such that we can more efficiently arrive at a solution and avoid overshooting solutions (since the derivative nears zero at local minima).

We can mathematically represent this idea as follows: at each step instance, we update the current position x by $-\alpha \cdot c'(x)$, or the negative derivative of the cost function $c(x)$ multiplied by a *learning rate parameter* α. α is a positive constant that determines how large the step at each instance is.

Let's consider minimizing the function $c(x) = \sin x + \dfrac{x^2}{10} + 1$. The derivative $c'(x)$ is $\cos x + \dfrac{2x}{10}$. After a few iterations, the position *converges* (i.e., reaches a point at which its position does not change significantly anymore) very close to the true minimum of around $x = -1.037$ (Listing 1-14, Figure 1-32, Figure 1-33, Figure 1-34).

Listing 1-14. Minimizing c(x) with gradient descent

```
cost = lambda x: np.sin(x) + x**2 / 10 + 1
gradient = lambda x: np.cos(x) + 2*x / 10
learn_rate = 0.5
curr_x = -4

for iteration in range(10):
    curr_x += -learn_rate * gradient(curr_x)
```

33

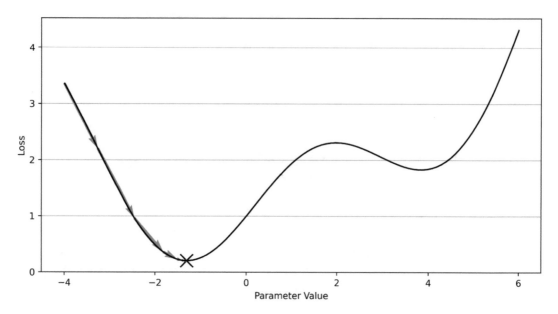

Figure 1-32. *Gradient descent decrease of the provided loss landscape from an initial parameter value of –4. We do not have access to the complete cost function (we wouldn't need to run gradient descent at all if that were the case!) and thus run the risk of poor initialization*

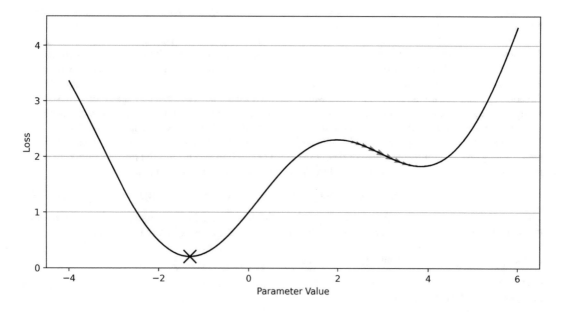

Figure 1-33. *Gradient descent decrease of the provided loss landscape from an initial parameter value of around 2*

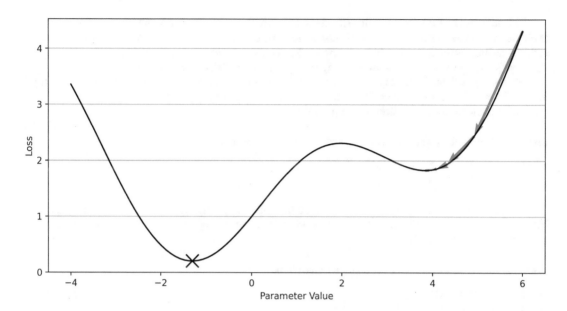

Figure 1-34. *Gradient descent decrease of the provided loss landscape from an initial parameter value of 6*

There are many mechanisms, ranging from simple to incredibly complex, to address the problems of this simple gradient descent algorithm. We'll explore these more in later chapters on neural networks when it becomes more relevant.

It should also be noted that while for the purposes of visualization in these examples the loss landscape is two-dimensional (i.e., a relationship between one parameter and the loss), it is almost always higher in dimensionality and thus more complex. Since successful machine learning models have at least multiple parameters (in nontrivial problems), the loss landscape becomes subject to the Curse of Dimensionality.[1]

Let's explore another demonstration of gradient descent for the purpose of modeling. We have a sequence of numbers [0, 2, 5, 6, 7, 11], in which each number is associated with a timestep: [0, 1, 2, 3, 4, 5]. Since the sequence seems to be linearly spaced with respect to the timestep, we can construct a simple approximating model that begins from the prediction 0 at time 0 and simply adds some constant β at each following timestep. For instance, if $\beta = 3$, the generated sequence is [0, 3, 6, 9, ...]. In other words, we are constructing a linear model with $y = \beta x$ with no y-intercept.

---

[1] There is the theoretical problem of being trapped in a local minimum with gradient descent. However, this may be less of a problem than one might expect due to the Curse of Dimensionality. As Andrew Steckley commented, It's been conjectured that this is because most potential trapping points tend to be saddle points instead of local minima, thus providing an escape route for the gradient descent process. So while high dimensions are a curse for many aspects of DL, it might be a savior when it comes to the problem of gradient descent traps in local minima. This is because it is the same multiplication of probabilities that work against us in making it more probable for a point to fall in the peel of a hypercube (or hypersphere) that works for us to make it more probable that the curvatures of the loss landscape along all the dimensions are not all of the same sign, so they tend to be saddle points rather than full minima. Therefore, when a minimum is found in a high-dimensional space, then it is just more likely to be a true global minimum.

The objective is to minimize the loss function. In this case, we will use the Mean Squared Error (MSE; see the "Metrics and Evaluation" section in this chapter for a more detailed introduction), which is defined as the difference between the *prediction* and the *true label* squared. Thus, $c(\text{params}) = (y_{\text{true}} - \text{prediction}(\text{params}))^2$, or $c(\beta) = (\beta x - y)^2$. Differentiating with respect to β yields $c'(\beta) = 2(\beta x - y) \cdot x$. Since in this case x and y are arrays, we'll need to take the mean of all the gradients across each of the samples in the sequence and update the parameter β from this mean.

Let's define our x and y datasets (Listing 1-15).

Listing 1-15. Initializing sample x and y datasets for Linear Regression fit

```
x = np.array([0, 1, 2, 3, 4, 5])
y = np.array([0, 2, 5, 6, 7, 11])
```

We'll use two helpful functions, predict and gradient, that, respectively, return the prediction given the beta parameter and the input set and the mean gradient given the beta parameter, the input set, and the ground-truth set (Listing 1-16).

Listing 1-16. Defining prediction and gradient functions

```
def predict(beta, x):
    return beta * x

def gradient(beta, x, y):
    diffs = predict(beta, x) - y
    gradients = 2 * diffs * x
    return np.mean(gradients)
```

Then, we'll iteratively optimize the beta parameter with gradient descent (Listing 1-17, Figures 1-35 and 1-36).

Listing 1-17. Iteratively updating the beta parameter

```
learn_rate = 0.005
curr_beta = -1
for iteration in range(100):
    curr_beta += -learn_rate * gradient(curr_beta, x, y)
```

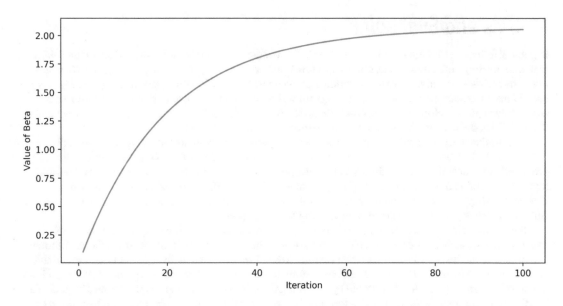

Figure 1-35. *Value of the optimized beta parameter across each iteration via gradient descent*

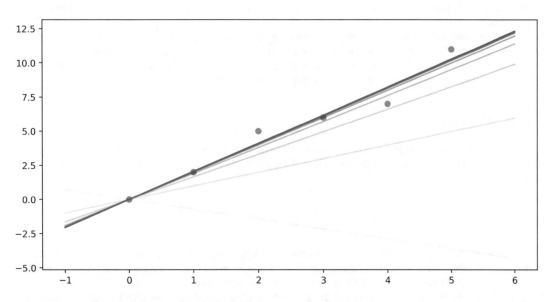

Figure 1-36. *Gradient descent optimization of a Simple Linear Regression over time. Sampled lines become less transparent further in the optimization process*

The beta parameter converges very quickly to around 2.05, which models the sequence quite nicely.

Gradient descent is a powerful idea that underlies many modern machine learning applications, from the simplest statistical model to the most advanced neural network system.

Metrics and Evaluation

In general terms, metrics are tools to set standards and evaluate the performance of models. In the field of machine learning and data science, obtaining a functional model is far from enough. We need to develop methods and assessment metrics to determine how well models perform on tasks that are given. Metrics used in machine learning are formulas and equations that provide specific quantitative measurements as to how well the developed methods perform on the data provided. Evaluation refers to the process of distinguishing between the better and the worse models by comparing metrics.

As introduced in the previous section, there are two general categories of problems that generally describe most datasets: regression and classification. Since for each type of problem the predictions produced by models vary greatly, there's a clear distinction between classification metrics and regression metrics. Regression metrics evaluate continuous values, while classification deals with discrete ones. However, both types of evaluation methods require two types of inputs: the ground-truth value (the desired correct prediction) and the predicted value that the model outputs.

Similar to metrics, there are loss functions. Loss functions accomplish the same goals as those of metrics: they measure the performance or, in some cases, the error for certain model predictions. Although loss functions can be used as metrics, when we refer to loss functions, it usually represents a differentiable function that can be used to train gradient-based models. As mentioned previously, gradient descent allows the model to converge to a local optimum by exploiting the differentiability of a loss function. The goal of a loss function is to provide an informative formula that can describe a model's error that can be utilized by gradient descent. They are generally optimal for gradient descent but not necessarily wholly intuitive from our perspective. On the other hand, metrics prioritize an intuitive explanation of how effective the model performed on the given task instead of designing a formula targeted specifically to the optimization of gradient descent.

As we will see later on in the section, some metrics simply cannot act as loss functions due to their undifferentiability. While sometimes loss functions are easily interpretable on a human scale and some metrics can be considered loss functions due to their differentiability, loss functions and metrics are generally seen as different tools with somewhat similar functionality.

Mean Absolute Error

One of the most common regression metrics is the Mean Absolute Error (MAE). In simple terms, it calculates the difference between the ground-truth labels and the predictions of the model, averaged across all predictions:

$$\text{Mean Absolute Error}\left(\text{MAE}\right) = \frac{\sum_{i=1}^{n}\left|y_i - \hat{y}_i\right|}{n}$$

For example, imagine a class of 50 students taking an exam that is scored out of 100 points. The teacher makes assumptions about each student's expected score based on their classwork in the past. After grading the exams, the teacher realizes that their predictions aren't exactly correct. They want to know how much they're off from the ground-truth values, or the actual scores that the students received. Guided by intuition, they simply calculate the difference in scores between the prediction and the students' results. The teacher finds it helpful to see how off they were for each student, but they also want a summative, aggregate-level idea of how generally wrong the predictions were. They do this by taking the average of each of the errors (Figure 1-37).

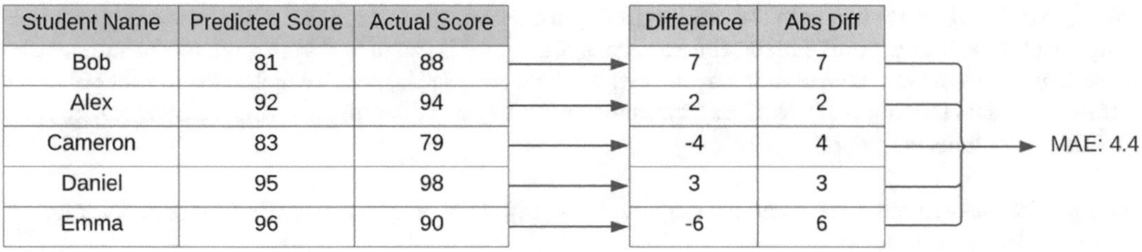

Student Name	Predicted Score	Actual Score		Difference	Abs Diff	
Bob	81	88	→	7	7	
Alex	92	94	→	2	2	
Cameron	83	79	→	-4	4	→ MAE: 4.4
Daniel	95	98	→	3	3	
Emma	96	90	→	-6	6	

Figure 1-37. *Example of MAE in terms of student test scores*

Although it's helpful at times to see by how much you are off for every prediction, the result is more interpretable if it's represented in one value, by taking the average of the differences. This is the Mean Absolute Error.

This method is straightforward and easily understood telling us the average difference between the labels and model predictions. However, it can raise some concerns when dealing with certain types of predictions/data as we will discuss later in the section.

While the Python library scikit-learn provides an easily useable implementation, it's always crucial to understand the nuts and bolts of each as you can only utilize something fully when you intuitively grasp the process of the algorithm/formula. Thus, for each metric introduced in the following, both the scikit-learn implementation and a full walk-through of implementation from scratch only using the NumPy library will be provided.

Before writing any specific code, we will briefly overview NumPy arrays. Simply put, NumPy arrays are Python lists that operate significantly faster as the foundation of the NumPy library is written in the programming language C. One foremost use for NumPy arrays is to perform Linear Algebra operations, framing nested one-dimensional lists as structured matrices. When we define the function, we assume that the inputs, model predictions, and ground-truth labels are one-dimensional NumPy arrays with their shape being the number of entries in the data. For each ground-truth label, there must exist a prediction.

Listing 1-18. The basic definition of a function

```
import numpy as np
def mean_absolute_error(y_true, y_pred):
    # check if y_true and y_pred are the same shape
    assert y_pred.shape == y_true.shape
```

We start by defining the function and its parameters, y_true and y_pred, representing the ground-truth labels and model predictions, respectively. Following up on the definition, we write an `assert` statement to ensure the shape of both inputs is identical since mistakes can happen when preprocessing data and/or retrieving model predictions (Listing 1-18).

Listing 1-19. MAE function in NumPy

```
import numpy as np
def mean_absolute_error(y_true, y_pred):
    # check if y_true and y_pred are the same shape
    assert y_pred.shape == y_true.shape
    return np.sum(np.absolute(y_true-y_pred))/len(y_pred)
```

When understanding these nested return statements, it's extremely helpful to start from the inside and work our way out. Referring back to the equation of MAE, the first step is to calculate each element's difference between the ground-truth labels and model prediction, which is reflected in the code by y_

39

pred-y_true. Then, for each item in the resulting array, we take the absolute value of the result, represented by np.absolute(y_true-y_pred). Finally, the average of the output is calculated by first taking the sum across all elements, np.sum, and then dividing by the length of the array len(y_pred), which is the number of entries in the data (Listing 1-19). Note we imported NumPy as np at the top; thus when referring to np, we are calling functions from the NumPy library.

Listing 1-20. MAE in scikit-learn and its usage, where y_reg_true = ground-truth labels and y_reg_pred = model predictions

```
# implementation in Scikit-Learn
from sklearn.metrics import mean_absolute_error

# usage
mean_absolute_error(y_reg_true, y_reg_pred)
```

As mentioned previously, the Python library sklearn provides a convenient one-liner function for MAE, which can be used the same way as our implementation from scratch (Listing 1-20).

Mean Squared Error (MSE)

The Mean Absolute Error may be simple, but there are a few drawbacks. One major issue regarding MAE is that it does not punish outlier predictions. Consider two samples A_1 and A_2 that the model is 1% and 6% off on (absolute error divided by total possible domain value). Using the Mean Absolute Error, A_2 is weighted 5% worse than A_1. Now, consider two additional samples B_1 and B_2 that the model is 50% and 55% off on. Using the Mean Absolute Error, B_2 is weighted 5% worse than B_1. We should ask ourselves: Is a model degradation from 1% to 6% error the same as a model degradation from 50% to 55% error? A 50–55% error change is a lot worse than a 1–6% error change since in the latter case the model still performs reasonably well.

We can use the Mean Squared Error to disproportionately punish larger mistakes:

$$\text{Mean Squared Error} \left(\text{MSE}\right) = \frac{\sum_{i=1}^{n} \left(\hat{y}_i - y_i\right)^2}{n}$$

As the name suggests, the Mean Squared Error raises the difference between model predictions and the ground-truth label to the second power instead of taking the absolute value.

To further demonstrate, continuing with the student test score example from earlier, the teacher predicts a score of 58 instead of 83 for the student Cameron. Using the formula for MSE, we obtain 107.8 (Figure 1-38). However, we do not have an intuition of how exactly the model is incorrect. Since we squared the result instead of taking the absolute value, the score we obtain doesn't scale to the range of test scores as an average error of 107.8 is impossible for a test that's scored between 0 and 100. To counter such an issue, we can take the square root of the result to "reverse" the operation of raising to the second power, putting the error into contextual scale while still emphasizing any outlier predictions. This metric is the Root Mean Squared Error (RMSE):

$$\text{Root Mean Squared Error} \left(\text{RMSE}\right) = \sqrt{\frac{\sum_{i=1}^{n} \left(y_i - \hat{y}_i\right)^2}{n}}$$

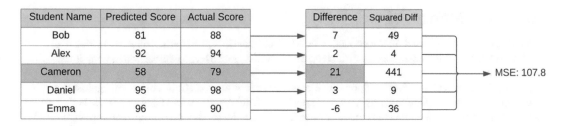

Student Name	Predicted Score	Actual Score		Difference	Squared Diff	
Bob	81	88	→	7	49	
Alex	92	94	→	2	4	
Cameron	58	79	→	21	441	→ MSE: 107.8
Daniel	95	98	→	3	9	
Emma	96	90	→	-6	36	

Figure 1-38. *Example of MSE*

Using the preceding equation, we obtain a 10.38 RMSE for the student test score example. On the other hand, if we use MAE, the resulting score is 7.8. Without looking at any predictions, RMSE gives us more information about the extremely incorrect prediction.

Listing 1-21. MSE and RMSE implementations

```
def mean_squared_error(y_true, y_pred):
    assert y_pred.shape == y_true.shape
    return np.sum((y_pred-y_true)**2)/len(y_pred)

# RMSE
def root_mean_squared_error(y_true, y_pred):
    assert y_pred.shape == y_true.shape
    mse = mean_squared_error(y_true, y_pred)
    return mse**0.5
```

The implementations of MSE and RMSE are extremely like that of MAE. As we see in the function, the difference between the ground-truth labels and model prediction is raised to the second power using the double asterisk symbol. Finally, the average is taken by summing all the elements in the array and then dividing by the total length, or the number of elements in the array, using `np.sum` and `/len(y_pred)`. To obtain the value for RMSE, we simply need to square root the result from MSE as shown previously; we raise the output from MSE to the ½ power or take its square root (Listing 1-21).

Listing 1-22. Scikit-learn implementation of RMSE and MSE, where `y_reg_true = ground-truth labels` and `y_reg_pred = model predictions`

```
# implementation in sklearn
from sklearn.metrics import mean_squared_error

# usage
mean_squared_error(y_reg_true, y_reg_pred)

# RMSE
mean_squared_error(y_reg_true, y_reg_pred)**(1/2)
```

Scikit-learn only provides an implementation for MSE. For RMSE, we simply square root the result like we did previously (Listing 1-22).

Although there are other regression metrics, MAE, MSE, and RMSE are some of the most important ones. Other regression metrics such as the Tweedie Deviance or the Coefficient of Determination revolve around ideas similar to the metrics mentioned previously.

Confusion Matrix

Like regression metrics based on the concept of MAE, most classification metrics rely on the idea of a Confusion Matrix, which describes how much error is made and what type of error with a convenient visualization.

A Confusion Matrix has four components that describe model predictions compared against the ground-truth labels; they are true positives, false positives, true negatives, and false negatives. Let's break down this terminology: the first word indicates whether the model prediction was correct or not ("true" if correct, "false" if not); the second indicates the ground-truth value ("positive" for the label 1 and "negative" for the label 0).

Let's better understand these concepts with real-life examples. Say that a doctor is diagnosing possible cancerous patients based on their screening results. If the doctor deduces that an actual cancerous patient does in fact have cancer, it's an example of a true positive. When the model, or in this case the doctor, predicts the positive class of being cancerous and the actual ground-truth label of whether the patient has cancer or not ends up positive too, it's called a true positive. However, the doctor is not accurate every single time. When the doctor predicts the patient is not cancerous while the patient in fact has cancer, it's an example of a false negative. When the doctor concludes that a patient is cancerous but in truth the patient is healthy, it's an example of a false positive. If a patient is healthy and the doctor predicts healthy too, it's an example of a true negative.

Let's summarize these concepts into numeric language:

- When the model prediction is 1 and the ground truth is 1, it's a true positive.

- When the model prediction is 1 and the ground truth is 0, it's a false positive.

- When the model prediction is 0 and the ground truth is 0, it's a true negative.

- When the model prediction is 0 and the ground truth is 1, it's a false negative.

After we've identified all the cases of this within our model prediction against the ground-truth labels, we can conveniently organize these values into a matrix form, thus the name Confusion Matrix (Figure 1-39).

10 Total Samples	Ground-truth Positive	Ground-truth Negative
Predicted Positive	True Positive 6	False Positive 1
Predicted Negative	False Negative 2	True Negative 1

Figure 1-39. Example of Confusion Matrix

Accuracy

One of the most straightforward ways to evaluate the performance of a classification model is accuracy. Accuracy is simply the percentage of values that the model predicted correctly. In more technical terms, it's the number of true positives plus the number of true negatives divided by the number of all the prediction values:

$$\text{Accuracy} = \frac{TN + TP}{TP + FP + TN + FN}$$

Accuracy may be convenient and intuitive, but it does come with a cost. Imagine if there are 100 samples, where 90 of them belong to the positive class and the other 10 belong to the negative class. Without having any other information about the data, if the model predicts the positive class for all samples, it will receive an accuracy of 0.9, or 90% correct. Consider a well-developed model in which for the 90 positive samples, it predicted 81 samples correctly; on the other hand, for the 10 negative classes, the model predicted 9 samples correctly. Clearly, the second model has a better understanding of the data instead of blindly guessing as seen in the first model, but they all receive the same accuracy score of 0.9. When the data contains significantly more samples of one class than the other, it's referred to as imbalanced data. Thus, accuracy is not a reflective metric when dealing with imbalanced data.

Listing 1-23. Implementation of accuracy

```
def accuracy(y_true, y_pred):
    assert y_true.shape == y_pred.shape
    # returns a boolean array(of 1s and 0s) indicating if element of both arrays match
    return np.average(y_true == y_pred)
```

The NumPy implementation of accuracy is straightforward: y_true == y_pred returns an array of Booleans indicating if elements of the arrays match. Boolean values in Python can be interpreted as integers; thus, we calculate the average across all elements in the array since technically each Boolean represents the accuracy of that element (Listing 1-23).

Listing 1-24. Scikit-learn implementation of accuracy, where y_class_true = ground-truth labels and y_class_pred = model predictions

```
from sklearn.metrics import accuracy_score
accuracy_score(y_class_true, y_class_pred)
```

Again, sklearn provides a simple function for accuracy; it can be used the same way we used our own accuracy score implemented from scratch by inputting model predictions and the ground-truth labels (Listing 1-24).

Precision

On the contrary, precision considers the issue of class imbalance in data. Precision calculates the accuracy only within all predicted positive classes, or the number of true positives divided by the sum of true positives and false positives. This will punish models that perform poorly on the positive class in an imbalanced dataset with a large number of negative values:

$$\text{Precision} = \frac{TP}{TP + FP}$$

Precision gives a sense of how accurate the model is for the positive class while addressing the issue of accuracy on class imbalance. The precision score is extremely useful as a metric as in many real-life datasets such as disease diagnosing, negative samples overwhelm the number of positive samples and precision would produce meaningful intuition of how accurate the model may be when predicting the rare positive class.

Listing 1-25. Implementation of precision

```
# precision
# number of correctly classified positive values/number of all predicted positive values
def precision(y_true, y_pred):
    assert y_true.shape == y_pred.shape
    # y_true * y_pred
    return ((y_pred == 1) & (y_true == 1)).sum() / y_pred.sum()
```

Going from the equation of precision, we first find the number of true positive values. Recall from earlier true positive is when the model predicts correctly for the positive class; in code, (y_pred == 1) & (y_true == 1) returns elements where both y_true and y_pred are equal. Then to find the number of true positives, we simply sum the array using sum() and then divide by all the predicted positive values, which we can obtain by summing the prediction array as all the negative prediction is represented with 0s and will not contribute to the sum (Listing 1-25).

Listing 1-26. Scikit-learn implementation of precision, where y_class_true = ground-truth labels and y_class_pred = model predictions

```
from sklearn.metrics import precision_score
precision_score(y_class_pred, y_class_pred)
```

In sklearn, we can import the precision score function from sklearn.metrics and use it the same way as our own implementation (Listing 1-26).

Recall

Recall score, or the true positive rate, is slightly different from precision as instead of calculating the accuracy of when the model predicts positive, it returns the accuracy of the positive class as a whole. Recall cleverly solves the problem of accuracy by only calculating the correctness in the positive class or the percentage of true positives across all positive labels. The recall metric is useful when we only care about our model predicting the positive class accurately. For example, when diagnosing cancer and most diseases, avoiding having false negatives is much more crucial than avoiding having false positives when resources are limited to develop a perfect model. In cases like such, we would be optimizing for a higher recall as we want the model to be as accurate as possible, especially in predicting the positive class:

$$\text{Recall} = \frac{TP}{TP + FN}$$

The implementation in NumPy is extremely similar to that of precision with only a minor change: instead of y_pred.sum(), we have y_true.sum() as now we are finding all the ground-truth positive values (Listing 1-27).

Listing 1-27. Recall implementation in sklearn and NumPy, where `y_class_true` = ground-truth labels and `y_class_pred` = model predictions

```
# recall
# number of correctly classified positive values/number of all true positive values
def recall(y_true, y_pred):
    assert y_true.shape == y_pred.shape
    return ((y_pred == 1) & (y_true == 1)).sum() / y_true.sum()

# sklearn recall
from sklearn.metrics import recall_score
recall_score(y_class_true, y_class_pred)
```

In sklearn, calling the `recall_score` function from `sklearn.metrics` will accomplish what we've just implemented (Listing 1-27).

F1 Score

The intuition behind the F1 score is creating a universal metric that measures the correctness of the positive class while having both the advantages of precision and recall. Thus, the F1 score is the harmonic mean between precision and recall. Note we used harmonic mean here since precision and recall are expressed as percentages/rates:

$$\text{F1 Score} = 2 \cdot \frac{\text{precision} \cdot \text{recall}}{\text{precision} + \text{recall}} = \frac{\text{TP}}{\text{TP} + \frac{1}{2}(\text{FP} + \text{FN})}$$

The F1 score typically gives a better sense of how correctly the positive class is classified compared with precision and recall. The F1 score can be generalized into the F-beta score where beta represents the amount of weight or attention on precision and recall. When beta equals 1, it's equivalent to the F1 score:

$$\text{F-Beta Score} = \left(1 + \beta^2\right) \cdot \frac{\text{precision} \cdot \text{recall}}{\left(\beta^2 \cdot \text{precision}\right) + \text{recall}}$$

Setting beta lower than 1 gives more weight to precision, meaning that the precision score will be valued more during the harmonic average. On the other hand, setting beta to a value greater than 1 emphasizes the recall score. This gives more flexibility to the combination of precision and recall scores as the amount of attention can be adjusted based on the situation.

Listing 1-28. F1 score implementation

```
# F1 Score
# Harmonic Mean of Precision and Recall

def f1_score(y_true, y_pred):
    num = (2*precision(y_true, y_pred)*recall(y_true, y_pred))
    denom = (precision(y_true, y_pred)+recall(y_true, y_pred))
    return  num / denom
```

Here, we implement the F1 score as most of the time the F1 score is used instead of other F-beta variations. However, this can be easily expanded into the F-beta score. Following the formula for harmonic mean, we can divide the number of observations, 2, by the sum of the reciprocals of each observation, 1 over precision and 1 over recall. Note that the formula and the implementation are different from this formula since we simplified the expression so that it's represented in one fraction. We can utilize the precision and recall functions that were implemented previously: in the numerator, the product of precision and recall is multiplied by 2, while in the denominator they're summed, and dividing them calculates the harmonic mean between the two (Listing 1-28).

Listing 1-29. Sklearn implementation of the F1 score, where y_class_true = ground-truth labels and y_class_pred = model predictions

```
from sklearn.metrics import f1_score
f1_score(y_class_true, y_class_pred)
```

In sklearn, we can use the f1_score function to achieve the same results (Listing 1-29).

Area Under the Receiver Operating Characteristics Curve (ROC-AUC)

Although the F1 score accounts for both the advantages of precision and recall, it only measures the positive class. Another major issue of the F1 score along with precision and recall is that we're inputting binary values for evaluation. Those binary values are determined by a threshold, converting probabilities to binary targets usually during model predictions. In most cases, 0.5 is chosen as the threshold for converting probabilities to binary targets. This not only might not be the optimal threshold, but it also decreases the amount of information that we receive from the output as probabilities also show how confident the model is in its prediction. The Receiver Operating Characteristics (ROC) curve solves these issues cleverly: it plots the true positive rate against the false positive rate for various thresholds:

$$\text{True Positive Rate}\,(\text{TPR})\,/\,\text{Recall} = \frac{TP}{TP+FN},$$

$$\text{False Positive Rate}\,(\text{FPR}) = \frac{FP}{TN+FP}$$

The true positive rate, or recall, gives us how many predictions are correct among all predicted positive labels, or the probability that the model will be correct when the prediction is positive. The false positive rate calculates the number of false negatives within the negative prediction. In sense, it gives a probability that the model will be incorrect when the prediction is negative. The ROC curve visualizes the values of both measurements on different thresholds, generating a plot where we can see what threshold gives the best values for true positive rate/false positive rate (Figure 1-40).

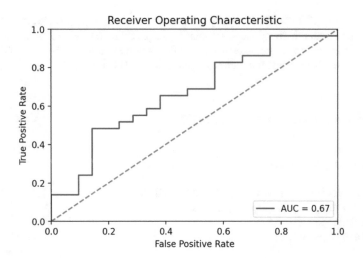

Figure 1-40. *An example of a ROC curve with its area = 0.67*

Furthermore, calculating the area under the curve (AUC) provides a measurement of how well the model distinguishes between classes. An area of 1 indicates a perfect distinction between classes, thus producing correct predictions. An area of 0 indicates complete opposite predictions, meaning that all predictions are reversed between 1 and 0 (labels – 1). An area of 0.5 represents complete random prediction where the model cannot distinguish between classes at all (Figure 1-41).

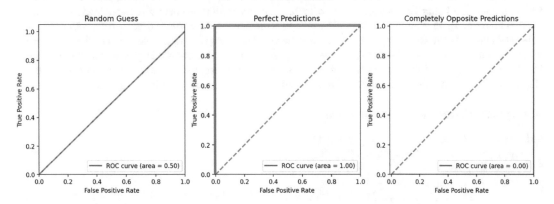

Figure 1-41. *Three extreme cases of a ROC curve*

Listing 1-30. The function that retrieves the TPR and FPR

```
def get_tpr_fpr(y_pred, y_true):
    tp = (y_pred == 1) & (y_true == 1)
    tn = (y_pred == 0) & (y_true == 0)
    fp = (y_pred == 1) & (y_true == 0)
    fn = (y_pred == 0) & (y_true == 1)

    tpr = tp.sum() / (tp.sum() + fn.sum())
    fpr = fp.sum() / (fp.sum() + tn.sum())

    return tpr, fpr
```

The implementation of ROC-AUC is slightly complicated as we first need to plot the ROC curve and then calculate the area under it. The preceding function simply calculates the true positive rate and the false positive rate by first retrieving arrays of true positives, false positives, true negatives, and false negatives and then calculating TPR and FPR, returning them as separate arrays (Listing 1-30).

Listing 1-31. ROC curve function

```
def roc_curve(y_pred, y_true, n_thresholds=100):
    fpr_thresh = []
    tpr_thresh = []
    for i in range(n_thresholds + 1):

        threshold_vector = (y_true >= i/n_thresholds)
        tpr, fpr = get_tpr_fpr(y_pred, y_true)
        fpr_thresh.append(fpr)
        tpr_thresh.append(tpr)

    return tpr_thresh, fpr_thresh
```

In order to plot the ROC curve, we need the TPR and the FPR for various thresholds. In the preceding function, we choose 100 equally spaced thresholds from 0 to 1. For each threshold we choose, the function calculates the TPR and the FPR for that particular threshold based on predicted probabilities provided by some model. Then each value is added to a separate list. Finally, both lists containing TPR and FPR for the amount of threshold specified are returned (Listing 1-31).

Using the points returned from our preceding function, we can plot the ROC curve with the x-axis being FPR and the y-axis being TPR. It's difficult to calculate the area under the curve directly using integrals since we don't have an exact function that would be able to represent the curve plotted every time with different values of TPR and FPR. However, we can estimate the area by slicing rectangular sections under the graph and then summing the area of those rectangles. As we slice more and more rectangles, with each rectangle becoming thinner and thinner, the sum of their area tends to be closer to the exact area under the graph (Figure 1-42). This method of estimating the area under graphs is known as a Riemann sum.

Listing 1-32. The area under the ROC curve

```
def area_under_roc_curve(y_true, y_pred):
    fpr, tpr = roc_curve(y_pred, y_true)
    rectangle_roc = 0
    for k in range(len(fpr) - 1):
            rectangle_roc += (fpr[k]- fpr[k + 1]) * tpr[k]
    return 1 - rectangle_roc
```

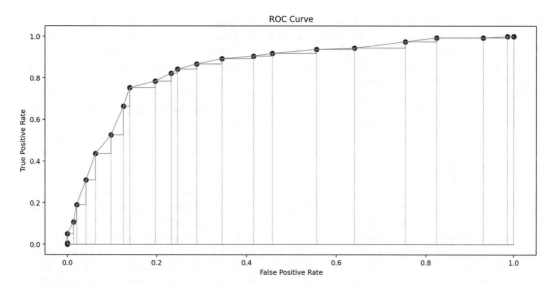

Figure 1-42. *Example of area under the curve calculation*

For calculating the area, we can imagine the gap between two points as the width of a rectangle while letting the height be the point to the bottom of the graph. Although there will be a tiny gap between the top of the rectangle and the line drawn by the graph, this gap will become negligible as the number of points, or rectangles, increases. In the preceding function, after we obtain lists of TPR and FPR containing different values for different thresholds, each value within the list is looped through, and the respective rectangular area is calculated (Listing 1-32). The resulting ROC-AUC score tends to approach that of the sklearn implementation when the number of thresholds is greater than 10,000.

Listing 1-33. Scikit-learn implementation of the ROC-AUC score

```
from sklearn.metrics import roc_auc_score
print(f"Scikit-Learn implementation of ROC-AUC: {roc_auc_score(y_class_true, y_class_
pred)}")
```

The sklearn implementation is as simple as calling the function `roc_auc_score`, inputting model predictions and ground-truth labels (Listing 1-33).

Algorithms

Before diving into any deep learning–related topics, it's crucially important to understand classical machine learning algorithms, many of which are competitors to deep learning models. Some of these classical machine learning algorithms have existed for as long as 70 years, and when used correctly, their performance is considerably high with surprisingly fast computational time compared with various deep learning methods.

To this day, classical machine learning algorithms still play a significant role in modern industries and machine learning–related contests. Not only do these algorithms train faster than deep learning approaches but powerful methods like Gradient Boosting can also outperform standard deep learning methods on tabular data. Moreover, models such as XGBoost (Extreme Gradient Boosting) and LightGBM (Light Gradient Boosting Machines) are still the most popular generic choice for benchmarking tabular datasets. Many advances in classical machine learning set the foundation for modern deep learning techniques. The neural network – the basis of deep learning – relies upon the conceptual and mathematical foundation of Linear Regression and gradient descent. Thus, it's critical to understand and implement these methods before moving on to deep learning. In this section, you will be introduced to six popular and important classical machine learning algorithms, including K-Nearest Neighbors (KNN), Linear and Logistic regressions, Decision Trees, Random Forests, and Gradient Boosting.

K-Nearest Neighbors

K-Nearest Neighbors, or KNN, is one of the simplest and the most intuitive classical machine learning algorithms. KNN was first proposed by Evelyn Fix and Joseph Lawson Hodges Jr. during a technical analysis produced for USAF in 1951. It was unique at its time as the method proposed is nonparametric: the algorithm does not make any assumptions about the statistical properties of the data. Although the paper wasn't ever published due to the confidential nature of the work, it laid out the groundwork for the first-ever nonparametric classification method, K-Nearest Neighbors. The beauty of KNN lies in its simplicity, and unlike most other algorithms, KNN does not contain a training phase. Additional data can be incorporated seamlessly as the algorithm is memory-based, easily adapting to any new data. Over seven decades later, KNN still stands as a popular classification algorithm, and innovations are still constantly being proposed surrounding it.

K-Nearest Neighbors is merely an algorithm concept. Although it was originally proposed for classification tasks, it can also be adapted to regression tasks. One major downfall to KNN is its speed. The time it takes for inference grows significantly as the number of samples increases. The Curse of Dimensionality, as introduced earlier, also weighs down the performance of KNN; as the amount of data features increases, the algorithm struggles to predict samples correctly. However, due to KNN's easy implementation, the convenience of only having one single hyperparameter, and interoperability, it's one of the best algorithms to quickly pick up and perform fast predictions on relatively small-sized data.

Theory and Intuition

The main intuition behind KNN is to group unlabeled data points with existing, labeled data points and classify data based on their distance to the labeled data points. As unlabeled data points are inputted, the distance between them and other training data points is calculated. Then the closest data point's label becomes the model prediction for the new data point. We can visualize this with a simple two-dimensional example (Figure 1-43).

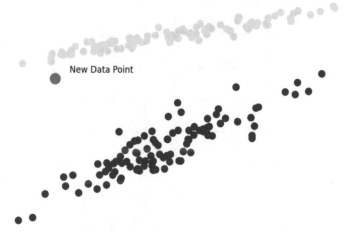

Figure 1-43. *Visualization of intuition on KNN*

Assuming that the labeled dataset has two features that are on the same scale, we graph each data point according to its features, with one being on the x-axis and the other on the y-axis. Here, different labels are distinguished with different colors. When a new, unlabeled data point is inputted, we first graph the data point according to its features. Then we calculate the distance from the new data point to each and every other labeled data point. Once we obtain a list of distances associated with each point in the dataset, the list is sorted ascendingly, and the first K elements are picked. K is a hypermeter that can be adjusted to improve performance. The selection of K will be discussed later. Finally, we perform a majority voting on the labels of the points to determine the final prediction. KNN takes advantage of the fact that data points with the same label most likely contain similar values of its features. This idea can be further extended into higher dimensions or data with more than two features.

Before diving into the implementation of KNN, let's step back and discuss the distance functions, which can be used in the algorithm. A base equation for many distance formulas is Minkowski distance, as shown in the following:

$$D(x,y) = \left(\sum_{i=1}^{n} |x_i - y_i|^p \right)^{\frac{1}{p}}$$

For vectors x and y, n represents the dimension of the vectors with values $(x_1, x_2, x_3, ..., x_n)$ and $(y_1, y_2, y_3, ..., y_n)$ where each value is a feature of each sample. The absolute difference between each value in the two vectors is raised to the pth power. Then these differences between values are summed and taken to the pth root. For different values of p, the function results in different forms of distance calculations. When $p = 1$, the function is known as Manhattan distance; when $p = 2$, the function becomes Euclidian distance; as $p \to \infty$, the function becomes Chebyshev distance. For the result to reflect traditional understandings of distance, p should be larger than or equal to 1.

As mentioned in the previous section, Manhattan distance is inspired by the shape of streets in Manhattan, New York, where each street forms right-angled paths from one point to another. Manhattan distance between two points can be interpreted as the distance of the shortest right-angled path on an n-dimensional grid where the two points are plotted (Figure 1-44).

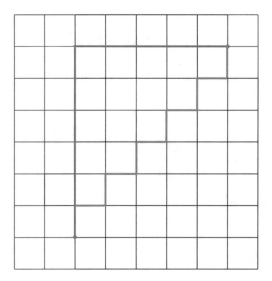

Figure 1-44. *2D Manhattan distance. Both the red line and the green line demonstrate Manhattan distance with the same distance values*

The formula for Manhattan distance is Minkowski distance where $p = 1$, or the sum of the absolute difference of each value between two vectors. Manhattan distance usually describes paths that could be realistically taken within the values described in the features. It is preferred over the more popular Euclidian distance used by KNN in the case of binary/discrete features. However, note that Manhattan distance becomes less understandable when it's calculated in higher dimensions, and the resulting distance value increases significantly as the dimension of space increases:

$$D(x,y) = \sum_{i=1}^{n} |x_i - y_i|$$

Euclidian distance is the shortest path between two points in an n-dimensional space; it's the most common distance formula used in KNN. Euclidian distance is simply an extension of the Pythagorean Theorem in higher dimensions (Figure 1-45).

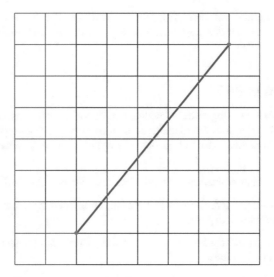

Figure 1-45. *2D Euclidian distance*

The formula for Euclidian distance results when $p = 2$ in Minkowski distance. It can also be seen as the Pythagorean Theorem for finding the hypotenuse in higher dimensions. One major disadvantage of Euclidian distance is the Curse of Dimensionality as described in previous sections: when the dimension of data increases, the less meaningful Euclidian distance becomes:

$$D(x,y) = \sqrt{\sum_{i=1}^{n} |x_i - y_i|^2}$$

As we increase the value of p in Minkowski distance, approaching infinity, we obtain

$\lim_{p \to \infty} \left(\sum_{i=1}^{n} |x_i - y_i|^p \right)^{\frac{1}{p}} = \max_{i} \left(|x_i - y_i| \right)$, which becomes Chebyshev distance. As introduced in previous

sections, Chebyshev is the maxim distance between two points on a single axis. Chebyshev distance in two dimensions can also be understood as the number of moves required for the King chess piece to move from one point to another:

$$D(x,y) = \max_{i} \left(|x_i - y_i| \right)$$

The usage of Chebyshev distance in KNN is rare, as most of the time it is outperformed by other distance functions in various scenarios. In KNN, Chebyshev distance calculates the greatest difference in value across all features between two data points. It raises a crucial issue as two data points could be very similar in every feature except one, thus belonging to the same class but significantly apart on the other feature; based on that single feature, Chebyshev distance will output a large value, which would lead to potentially incorrect prediction by the model.

When choosing KNN, there are some key points to consider. First, all features must be on the same scale. More specifically, during the computation of distances between points, a certain value of distance in one feature must mean the same for all other features. It is recommended to normalize or standardize data before applying KNN. Normalization refers to the process of changing the values of features to lie within a common scale. Standardization is the process of scaling features, accomplishing two things: changing the mean of the data to a value of 0 and ensuring the resulting distribution has a unit standard deviation. Depending on the data, standardization can be used to ensure a Gaussian distribution, while normalization can restrict data to a certain range, reducing the influence of outliers. For the size of the dataset, both the number of features and the number of samples are best kept low due to the Curse of Dimensionality and computational speed, respectively.

Implementation and Usage

Implementing KNN from scratch using NumPy involves two major components, calculating distances and choosing the first K "neighbors" along with making predictions by majority voting. Before writing down any code, it's always a good practice to list out the flow of steps in the algorithm as shown in the following for KNN:

1. Retrieve train data or labeled samples; retrieve test data or unlabeled samples.

2. Calculate distance using the chosen distance function between the test data point and every train data point.

3. Sort the resulting distances from closest to furthest away from the test data point and select the first K values.

4. Perform majority voting to determine the final prediction.

5. Repeat for all test data points.

We can assume that both train and test data are inputted in a NumPy array with shape (num_samples, num_features), while the labels for training data are NumPy arrays with shape (num_samples, 1). Following the preceding steps, we can define a function that calculates the distance between one single test data sample and every train data sample.

Listing 1-34. The basic definition of function

```python
def calculate_distance(train_data, single_test):
    distances = []

    for single_train in train_data:
        single_distance = distance_function(single_train, single_test)
        distances.append(single_distance)

    return distances
```

In the preceding function, we are taking the whole training data array and a single test sample as input. We then iterate through each row of the training data, apply a predefined distance function that we refer to as "distance_function" with inputs x and y for now, and append the resulting value to a list for later. Once we loop through the entirety of the training data, we simply return the list containing all the values produced by the distance function. Now, we have accomplished the first major component of KNN as mentioned previously; for the second part, we'll define another function that sorts the list, selects the first K elements, and performs majority voting for determining prediction (Listing 1-34).

Listing 1-35. Function to predict a single test data sample in KNN

```python
from collections import Counter

def knn_predict_single(distances, y_train, k=7):

    # Convert distances to series with the index being the index of y_train
```

```
distances = pd.Series(distances, index=list(range(len(y_train))))

# Sort values and select first K elements while preserving index
k_neighbors = distances.sort_values()[:k]
# Create Counter object with the lables of each `k_neighbor`
counter = Counter(y_train[k_neighbors.index])

# Most common label in the list by majority voting
prediction = counter.most_common()[0][0]

return prediction
```

Combining the two major components in KNN, the preceding function performs on a single test data sample. We first organize the distance value, which is produced by `calculate_distance`, into a Pandas series with the index being the index of `y_train` using `list(range(len(y_train)))`. We then sort the list ascendingly and select the first K-Nearest "Neighbor" while preserving the indices assigned earlier. For each distance value, we find its respective label by indexing `y_train` using the index in the sorted series. Finally, we use the `Counter` object from the Python library `collections` for the majority vote, deciding the final prediction for the test data sample and returning it as a single value (Listing 1-35).

Listing 1-36. KNN function

```
def knn_pred(X_train, y_train, X_test, distance_function, k):

    predictions = np.array([])

    # Loop through every test sample
    for test in X_test:
        distances = calculate_distance(X_train, test, distance_function)
        single_pred = knn_predict_single(distances, X_train, y_train, k)
        predictions = np.append(predictions, [single_pred])

return predictions
```

The preceding function essentially loops through the whole test data array and transforms KNN on each single test data point. For each test data point, we first retrieve the list of distances using the function `calculate_distance` and then make predictions using the function `knn_predict_single`. Afterward, we append the single prediction to an array that will store all the predictions from the testing data. Finally, we return the array (Listing 1-36).

Listing 1-37. Minkowski distance

```
# p=2, Euclidian Distance
def minkowski_distance(x, y, p=2):
    return (np.abs(x - y)**p).sum(axis=1)**(1/p)
```

For the distance function, we will implement Minkowski distance with an adjustable parameter p. However, this distance function can be any other distance function with inputs x and y (Listing 1-37).

Listing 1-38. Example usage of KNN

```
from sklearn import datasets

iris = datasets.load_iris()

# assign X(features) and y(targets)
X = iris.data
y = iris.target

from sklearn.model_selection import train_test_split
X_train, X_test, y_train, y_test = train_test_split(X, y, test_size=0.3, random_state=42)

# predict using knn
predictions = knn_pred(X_train, y_train, X_test, minkowski_distance, k=5)

from sklearn.metrics import accuracy_score
# returns 1.0
accuracy_score(y_test, predictions)
```

We can test our KNN implementation on the Iris dataset. The dataset aims to classify the type of iris flower – Setosa, Versicolour, or Virginica – based on the physical features of the flower: Sepal Length, Sepal Width, Petal Length, and Petal Width. We can conveniently load the dataset from the library scikit-learn and then split it into training and testing sets using the function `train_test_split`. We then simply need to call the knn_pred function and input the respective parameters (Listing 1-38). To evaluate the performance of our model, we can use the accuracy score from scikit-learn. With K set to 5, our model performs perfectly on the data, reaching an accuracy of 1.0.

Listing 1-39. Scikit-learn implementation of KNN

```
from sklearn.neighbors import KNeighborsClassifier
# create object
neigh = KNeighborsClassifier(n_neighbors=5)
# call fit on training data
neigh.fit(X_train, y_train)
# predict on testing data
sklearn_pred = neigh.predict(X_test)
# returns 1.0
accuracy_score(y_test, sklearn_pred)
```

The library scikit-learn provides an out-of-box implementation for KNN as shown previously: simply create the `KNeighborsClassifier` object and call `fit` and then `predict` on test data (Listing 1-39). We obtain the same result as our own implementation, reaching an accuracy of 1.0 with `n_neighbors`, or k in our implementation, set to 5.

Note that there's no single "method" of finding the best "K" value. Most of the time there must be trials and tuning to find the optimal K value that gives the best performance. Usually, a common technique taken to search for the best K value that performs well both on training data and potential testing data is called the elbow method. The elbow method states that when the performance of KNN is plotted against different K values, the optimal K value for testing data sits at the steepest increase in performance between two K values. Note that usually the elbow method is used along with cross-validation and the performance is measured against the validation dataset. An example of such a selection method is shown in Figure 1-46.

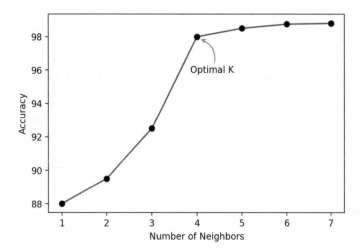

Figure 1-46. *Elbow method for KNN*

In the preceding example on an arbitrary dataset, the optimal K value would be 4 as the steepest increase in performance happens between 3 and 4. This method can be used with any metric that measures the performance of predictions whether the goal is minimization or maximization.

KNN may be simple and easy to use, but keep in mind its limitations such as not being able to scale for larger datasets and high computational time. In later sections, algorithms are introduced to handle this problem better than KNN and generally perform much better. Although KNN sits as one of the simplest ML algorithms, its cons usually outweigh the pros, making it merely a quick and dirty testing algorithm for most data scientists.

Linear Regression

The term *Linear Regression* may be familiar to most as it's a common technique used in statistics to model the relationship between variables. The earliest form of regression analysis, Linear Regression using Least Squares, was first published by Adrien-Marie Legendre in 1805 and then again by Carl Friedrich Gauss in 1809. Both used them to predict astronomical orbitals, specifically bodies that orbit the sun. Later in 1821, Gauss published his continued work on the Least Square Theory; however, the term *regression* wasn't coined until the late nineteenth century by Francis Galton. Galton discovered a linear relationship between the weights of mother and daughter seeds across generations. To Galton, regression was merely a term to describe a biological phenomenon. It was not until Udny Yule and Karl Pearson expanded this method to a more general statistics view.

To this day, Linear Regression has evolved with many variations and solving methods, but the usefulness of the algorithm has remained the same. Linear Regression is not only used in the field of machine learning and data science but also in epidemiology, finance, economics, and so on. Practically any situation where a relationship between continuous variables is presented, Linear Regression can model such relation in one way or another. To generalize, the application of Linear Regression falls under the use of predicting or forecasting certain response variables based on related explanatory features; Linear Regression can also be used to quantify or measure the linear relation between two variables and to identify possible redundant or misleading features in a dataset.

Theory and Intuition

Linear Regression models the relationship between one response, or target variable, and one or more features, or explanatory variables that may be correlated to the target variable in a linear fashion. Linear Regression with two variables, with one feature describing the target, is called Simple Linear Regression. On the other hand, Linear Regression that models multiple feature variables against one target variable is referred to as Multiple Linear Regression. The goal of Linear Regression is to find a function that produces a "line of best fit" that best models the relationship between the features and the target variables. There are many ways to determine the "line of best fit"; however, the most common and successful method is through gradient descent as mentioned earlier in the chapter.

Let's start with a basic example of Simple Linear Regression with only two variables: the explanatory variable, the size of someone's foot, and the target variable, that person's height (Figure 1-47). Our goal in this situation is to predict someone's height based on the length of their foot. There are a few data samples given; we're tasked with writing a function that produces a line that best models the relationship between the two variables.

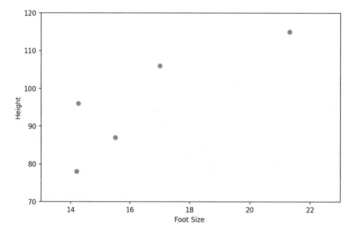

Figure 1-47. *Example data for predicting height based on foot size*

After observing the data, we may hypothesize an estimation of a function that could possibly model the relationship between the two variables. In a two-dimensional graph, the general equation for a line is given in the slope-intercept form:

$$y = \beta x + \varepsilon$$

where β is the slope and ε is the y-intercept. In the field of machine learning and data science, we refer to the slope as the weight and y-intercept as the bias of a model. After estimating an equation for a line that may fit the data, say $y = 5x + 3$, we would want to calculate the performance of our estimation by some method. The most common metric used for optimizing Linear Regression is MSE. The error is calculated as the difference between the y value of the ground-truth data samples (height in our case) and the y value of the line at the ground-truth x value (model predictions) squared for every data point and then summed up together. This method of error calculation is also known as the sum of the squared residuals.

In the preceding figure, the blue line shows the model's estimation, given as a function of the inputs. The green lines in the preceding diagram represent the error between each point and the predicted values. Summing each difference up, we would obtain our final error. Squaring the value would give us the sum of the squared residuals (Figure 1-48). During the process of gradient descent, MSE is referred to as the cost function or the loss function of the model. To reiterate, the formula for MSE is shown in the following:

$$\text{Mean Squared Error}\,(\text{MSE}) = \frac{1}{n}\sum_{i=1}^{n}\left(\hat{y}_i - y_i\right)^2$$

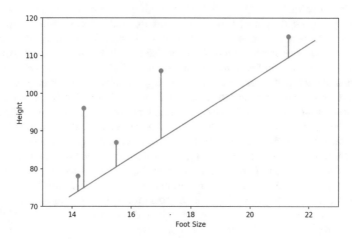

Figure 1-48. *MSE error calculation between estimation line and data point*

Our goal is to minimize this function by finding a more accurate linear function that represents our data, which in turn increases the performance of our "model." We could possibly find the line by brute force, adjusting the values of weight and bias based on errors; however, this quickly fails when the number of samples or features increases. This is where gradient descent comes in. The concept of gradient descent is similar to brute force, adjusting the values of weight and bias based on the error produced by previous values, but gradient descent does so according to formulas in a sophisticated manner.

Imagine if we plot different values of weight and bias against the value of the cost function. We would be able to visualize how changes in weight and bias would affect the value of the cost function. This visualization would be in a three-dimensional space. We can consider a slice through this space (say by fixing the bias as zero) and then visualize a simpler two-dimensional case with only the weight values as shown in Figure 1-49.

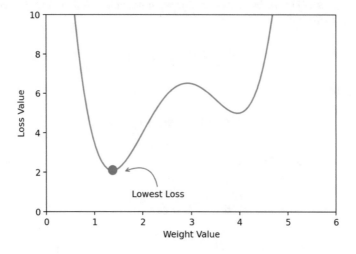

Figure 1-49. *Loss plotted against weight values*

The smallest loss sits at the lowest y value of the function, or when the derivative of the function is 0. One might think that setting the derivative of the function to 0 would give us the optimal weight value. This would work for a convex function like MSE; however, it would not for the example loss function shown previously as there are multiple points where the derivative of the function is 0. As introduced earlier, a common technique in machine learning to find the global minima of a function is gradient descent. Gradient descent aims to find the global minima using the gradient, or its partial derivatives, to take appropriate steps down to the global minima. Gradient descent takes careful steps toward the global minima, and with the right hyperparameter, it can "boost" itself through local minima to reach the desired global minima. Such techniques and parameters will be discussed in Chapter 3. Recall from the section "Optimization and Gradient Descent" where a simple example of how gradient descent aims to search for the global minima of the function was shown. Linear Regression's adaptation of gradient descent does involve some extra steps since we also have the bias to optimize and the derivation of the cost function MSE isn't very straightforward. Note it is not necessary to understand the derivation shown in the following for later sections in the book. The same can apply to Logistic Regression's version of gradient descent.

As mentioned earlier in the chapter, the algorithm uses the negative gradient of the cost function to determine the direction of descent, and the learning rate, α, contributes to how much of a step will the algorithm take. However, different from the equation previously in the chapter, we have a bias term, ε. Thus, instead of multiplying the derivative of the cost function by α, we will multiply it with the partial derivative with respect to the weight by α where c is our cost function, MSE. The value that we obtain with the preceding equation will be added to the current weight and thus become the new weight for another iteration. Note that in MSE, our $\hat{y}$ term is model prediction modeled by the line $y = \beta x + \varepsilon$ and it will replace $\hat{y}$ during differentiation:

$$-\alpha \cdot \frac{\delta}{\delta \beta} c(\beta, \varepsilon)$$

We will only differentiate the weight and calculate the bias term at the end of gradient descent, and during the process of gradient descent, the bias will be set to 0. Let's differentiate β with respect to the cost function c.

We have the equation

$$c(\beta, \varepsilon) = \frac{1}{n} \sum_{i=1}^{n} (\beta x_i + \varepsilon - y_i)^2$$

Let's treat the expression inside brackets as a separate function and differentiate as a composite function, where $f(\beta, \varepsilon)_i = \beta x_i + \varepsilon - y_i$:

$$c(\beta, \varepsilon) = \frac{1}{n} \sum_{i=1}^{n} (f(\beta, \varepsilon)_i)^2$$

Let's apply the chain rule:

$$\frac{\delta}{\delta \beta} c(f(\beta, \varepsilon)_i) = \frac{\delta}{\delta \beta} c(\beta, \varepsilon) \frac{\delta}{\delta \beta} f(\beta, \varepsilon)_i$$

Let's compute the derivatives for each function:

$$\frac{\delta}{\delta \beta} c(\beta, \varepsilon) = \frac{2}{n} \sum_{i=1}^{n} f(\beta, \varepsilon)_i$$

$$\frac{\delta}{\delta \beta} f(\beta, \varepsilon)_i = x_i$$

Our final answer becomes

$$\frac{\delta}{\delta\beta}c\left(f\left(\beta,\varepsilon\right)_i\right)=\frac{\delta}{\delta\beta}c\left(\beta,\varepsilon\right)\frac{\delta}{\delta\beta}f\left(\beta,\varepsilon\right)_i=$$

$$\frac{2}{n}\sum_{i=1}^{n}f\left(\beta,\varepsilon\right)_i x_i=\frac{2}{n}\sum_{i=1}^{n}\left(\beta x_i+\varepsilon-y_i\right)x_i$$

We can then plug in this value for $\frac{\delta}{\delta\beta}c\left(\beta,\varepsilon\right)$ and multiply this with $-\alpha$ to update the weight.

Finally, instead of differentiating the bias, we can estimate this value by setting it to the average difference between our prediction without the bias and the ground-truth values after training the weights with gradient descent:

$$\varepsilon=\frac{1}{n}\sum_{i=1}^{n}\beta x_i-y_i$$

Implementation and Usage

Continuing with our preceding example dataset along with the formulas presented, we can write down the following code (Listing 1-40).

Listing 1-40. Implementing Linear Regression

```
# define our example dataset
foot_size = np.array([14.2, 21.3, 17, 15.5, 14.4])
height = np.array([78, 115, 106, 87, 96])

# foot_size is `x` while height is `y_true`

# MSE cost function
def mse_loss_func(y_true, y_pred):
    return np.sum((y_pred-y_true)**2)/len(y_pred)

# prediction function, namely the equation for the line, y=beta*x+epsilon
def predict(beta, epsilon, x):
    return beta*x + epsilon

# calculates the partial weight derivative
def weight_deriv(beta, epsilon, x, y_true):
    error = predict(beta, epsilon, x) - y_true
    return np.mean(2 * error * x)

# calculates the bias term
def calculate_bias(beta, x, y_true):
    return np.mean(y_true - beta * x)
```

Referring back to the equation of gradient descent introduced in the "Optimization and Gradient Descent" section and covered earlier in this section, we can then write down the following code (Listing 1-41).

Listing 1-41. Gradient descent equation

```
def gradient_descent(beta, epsilon, x, y_true, alpha):

    weight_derivative = weight_deriv(beta, epsilon, x, y_true)
    new_beta = -alpha * weight_derivative
    return new_beta
```

Finally, we iterate through the gradient descent process, updating the weights according to the function. Using our example dataset defined previously, we can calculate the cost using MSE to see how much we're improving with each iteration (Listing 1-42).

Listing 1-42. Linear Regression with gradient descent

```
epsilon = 0
beta = 1

# usually a value between 0.001 and 0.0001 works well
alpha = 0.0001

# Can be adjusted to affect performance, the more #iterations usually means better but
not always
iterations = 200

for i in range(1, iterations + 1):
    beta += gradient_descent(beta, epsilon, foot_size, height, alpha)

    # evaluate every 20 iterations
    if i%20 == 0:
        pred = predict(beta, epsilon, foot_size)
        mse = mse_loss_func(height, pred)
        print(f"Error at iteration {i}: {mse}")

# Calculate the bias
epsilon = calculate_bias(beta, foot_size, height)
final_pred = predict(beta, epsilon, foot_size)
final_mse = mse_loss_func(height, final_pred)
print(f"Final Prediction with bias has error: {final_mse}")
```

The beta or the weight value can be initialized randomly or set to a certain value(s) for producing deterministic results. In this case, we've set the weight to 1, obtaining a final MSE of around 46. Plotting our results out as a "line of best fit," we see that our line fits the trend of the data much better than our initial guess (Figure 1-50).

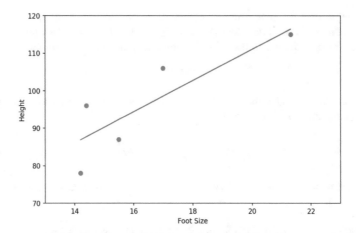

Figure 1-50. *Line of best fit produced by our own implementation of Linear Regression*

The scikit-learn implementation of Linear Regression is simple to use, and due to a slight difference in the method used to calculate the weight and bias, scikit-learn's Linear Regression will obtain a slightly better result than that of our own implementation.

Listing 1-43. Scikit-learn's implementation of Linear Regression

```
from sklearn.linear_model import LinearRegression

lr = LinearRegression()
lr.fit(foot_size.reshape(-1, 1), height)
# get the weight and bias
beta = lr.coef_
epsilon = lr.intercept_

# Predict function
lr_pred = lr.predict(foot_size.reshape(-1, 1))
```

Note that we used reshape(-1, 1) on foot_size because in scikit-learn's implementation when we have a single feature in our case, the data must be in the shape (num_samples, 1), and reshape(-1, 1) does just that (Listing 1-43). Calling the fit function with features and labels inputted performs the training process, and calling predict with features inputted outputs the prediction.

Our own implementation is far from perfect. In situations where the loss function is convex (such as the MSE loss function), using the Ordinary Least Squares (OLS) method may produce slightly better results. Essentially, OLS obtains the optimal result in one step by setting the derivative of the cost function to 0. This ensures parameters obtained from the calculation are values at the global minima of the loss landscape. Another improvement that could be implemented is that we could try to differentiate the bias along with the weight and update it in the same manner; it may produce better results.

Other Variations on Simple Linear Regression

As mentioned before, Multiple Linear Regression refers to regression with multiple features or explanatory variables. Instead of singular values for features, we can treat them as vectors with the length being the number of features and the entirety of training data being a matrix in the shape of *number of samples × number of features.* Following, instead of a single weight value, it would be a vector with the length being the number of features (it's viewed as a column matrix in order to be dotted with X). Thus, our equation for the line, or hyperplane, in this case of dimensions greater than 2, should be

$$y = X\beta + \varepsilon$$

where X is the feature matrix and β is the vector of weights. Instead of differentiating term by term, if we differentiate as a matrix, the final derivative we obtain becomes

$$\frac{\delta}{\delta\beta}c(\beta,\varepsilon) = \frac{2}{n}(X\beta + \varepsilon - y)X^{-1}$$

In order to implement Multiple Linear Regression, only a few minor changes need to be made in the code written previously.

Listing 1-44. Modified gradient descent functions for Multiple Linear Regression

```
# MSE cost function
def mse_loss_func(y_true, y_pred):
    return np.sum((y_pred-y_true)**2)/len(y_pred)

# prediction function, namely the equation for the line, y=beta*x+epsilon
def predict(beta, epsilon, x):
    return np.dot(x, beta) + epsilon

# calculates the partial weight derivative
def weight_deriv(beta, epsilon, x, y_true):
    error = predict(beta, epsilon, x) - y_true
    return np.mean(2 * np.dot(x.T, error))

# calculates the bias term
def calculate_bias(beta, x, y_true):
return np.mean(y_true - np.dot(x, beta))
```

Observe that all multiplication sign has been changed to np.dot to reflect the dot product between matrices (Listing 1-44). In the weight_deriv function, x.T transposes the matrix x. Similarly, the gradient descent process needs to be adjusted if we're still using the example dataset from earlier as we need to reshape the features into the correct shape in vector and matrix form (Listing 1-45).

Listing 1-45. Modified gradient descent process for Multiple Linear Regression

```
epsilon = 0
beta = np.array([1.]).reshape(1, -1)

# usually a value between 0.001 and 0.00001 works well
alpha = 0.00001

iterations = 200

for i in range(1, iterations + 1):
    beta += gradient_descent(beta, epsilon, foot_size.reshape(-1, 1), height, alpha)
```

```
# evaluate every 20 iterations
if i%20 == 0:
    pred = predict(beta, epsilon, foot_size.reshape(-1, 1))
    mse = mse_loss_func(height, pred)
    print(f"{epsilon}, {beta}")
    print(f"Error at iteration {i}: {mse}")

# Calculate the bias
epsilon = calculate_bias(beta, foot_size.reshape(-1, 1), height)
final_pred = predict(beta, epsilon, foot_size.reshape(-1, 1))
final_mse = mse_loss_func(height, final_pred.flatten())
print(f"Final Prediction with bias has error: {final_mse}")
```

When –1 is passed in during reshape, it simply means that the other dimension is set, and the –1 dimension must satisfy the original shape of the array. The function flatten, as the name suggests, flattens the array into a single-dimension vector. Note that the scikit-learn implementation is automatically compatible with Multiple Linear Regression. Nothing needs to be changed except when the number of features is more than one, reshape(-1, 1) is not needed.

In real-life machine learning applications, there rarely are times when one single variable would be able to form a perfect linear relation with the target variable. Most of the time it's tens and hundreds of features, thus making Multiple Linear Regression a default for many data scientists.

Other variants of Linear Regression include LASSO (Least Absolute Shrinkage and Selection Operator), Ridge, and ElasticNet regressions, which introduce a regularization term that will prevent overfitting on train data leading to poor performance on testing data.

Ridge Regression introduces an L2 regularization term on the MSE cost function:

$$c(\beta,\varepsilon) = \frac{1}{n}\|X\beta + \varepsilon - y\|_2^2 + \lambda\|\beta\|_2^2$$

The terms act as a penalty such that if the weight of the model gets too large, the value of the cost function will increase depending on the lambda parameter. Ridge Regression reduces overfitting by lowering the model complexity.

LASSO regression, on the other hand, introduces an L1 regularization term on the MSE cost function:

$$c(\beta,\varepsilon) = \frac{1}{n}\|X\beta + \varepsilon - y\|_2^2 + \lambda\|\beta\|_1$$

The L1 regularization can lead to 0 coefficients depending on the value of lambda, reducing overfitting while also accomplishing feature selection when certain coefficients are 0.

ElasticNet Regression uses both L1 and L2 regularizations, accomplishing both the goals of LASSO and Ridge regressions, further regularizing the model and preventing overfitting:

$$c(\beta,\varepsilon) = \frac{1}{n}\|X\beta + \varepsilon - y\|_2^2 + \lambda_1\|\beta\|_1 + \lambda_2\|\beta\|_2^2$$

The scikit-learn implementation of all three regressions can be used the same as Linear Regression while specifying the parameter lambda at the initialization of the object (Listing 1-46).

Listing 1-46. Regularization Linear Regressions in scikit-learn

```
from sklearn.linear_models import Ridge, Lasso, ElasticNet

# alpha is the lambda value
ridge = Ridge(alpha=1.0)

lasso = Lasso(alpha=1.0)

# alpha is the penalty term for both while the l1_ratio controls how much penalty is
weighted toward l1, the closer to 1 the higher weight to l1
elr = ElasticNet(alpha=1.0, l1_ratio=0.5)

# Fit and predict as normal where X and y are training features and training targets,
respectively
# Replace X and y with your own data
# ridge.fit(X, y)
# lasso.fit(X, y)
# elr.fit(X, y)

# ridge.predict(X_test)
# lasso.predict(X_test)
# elr.predict(X_test)
```

Finally, one other important variation of Linear Regression is Logistic Regression. In simple terms, Logistic Regression changes the cost function of Linear Regression, restraining the output between 1 and 0 to produce binary predictions for classification tasks. Logistic Regression will be covered more in detail during the next section.

Linear Regression serves as the beginner algorithm for hundreds and thousands of data scientists, for a reason. Not only gradient descent is used widely across many machine learning algorithms but modeling the trend of the data based on features is one of the crucial uses of machine learning to this day with applications in forecasting, business, medics, and many more.

Logistic Regression

The logistic function first appeared in Pierre Francois Verhulst's publication "Correspondance mathmematique et physique" in 1838. Then later in 1845, a more detailed version of the logistic function was published by him. However, the first practical application of such a function wasn't apparent until 1943 when Wilson and Worcester used the logistic function in bioassay. In the following years, various advances were made toward the function, but the original logistic function is used for Logistic Regression. The Logistic Regression model found its use not only in areas related to biology but also widely in social science.

Among many variants, the general goal of Logistic Regression remains the same: classification based on explanatory variables or features provided. It utilizes the sample principles of Linear Regression.

Theory and Intuition

The logistic function is a broad term referring to a function with various adjustable parameters. However, Logistic Regression only uses one set of parameters, turning the function into what's referred to as a sigmoid function:

$$S(x) = \frac{1}{1 + e^{-x}}$$

The sigmoid function is known to have an "S"-shaped curve with two horizontal asymptotes at $y = 1$ and $y = 0$ as shown in Figure 1-51. In other words, any value inputted to the function will be "squished" between 1 and 0.

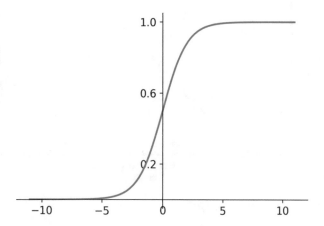

Figure 1-51. *Sigmoid function*

Although there are other variants of the sigmoid function that restrain the output at different values, such as the hyperbolic tangent function in which the output values are restrained between –1 and 1, binary classification on Logistic Regression uses the sigmoid function.

Logistic Regression operates on the same principles as those of Linear Regression, finding an equation that graphs a "line of best fit" of the data. In this case, the "line of best fit" will take the shape of the sigmoid function, while the labels are binary labels sitting either at $y = 0$ or $y = 1$.

The goal of binary classification is to predict outputs of either 1 (positive) or 0 (negative). One example could be to predict whether an animal is a cat based on the characteristics of its fur provided as numerical data. Logistic Regression's predictions will be inputted through the sigmoid function and outputted as probabilities between 1 and 0, which represents how confident the model believes certain data sample belongs to one class (Figure 1-52). Then the probability can be converted to binary outputs by setting a threshold. Usually, the choice is 0.5; however, the performance of the model can vary if the threshold is not carefully selected. See ROC-AUC in the "Metrics and Evaluation" section for more details.

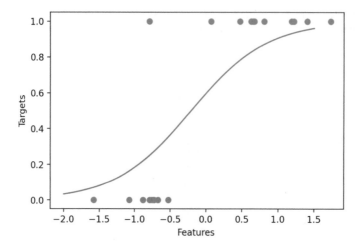

Figure 1-52. *Logistic Regression's "line of best fit"*

Although the process of gradient descent for Logistic Regression is the same as that of Linear Regression, the cost function would be different. MSE calculates the difference between two values in a regression-like situation. While it may be able to learn under classification situations, there are better cost functions where it suits the problem.

Recall from the preceding sigmoid function the x term will be our prediction outputted from the model by the linear function $y = X\beta + \varepsilon$, so the prediction function of Logistic Regression becomes

$$y = \frac{1}{1 + e^{-(X\beta + \varepsilon)}}$$

The output of this function is probability values ranging from 0 to 1, while our labels are discrete binary values. Instead of MSE, our cost function will be the log loss or sometimes called the binary cross-entropy (BCE):

$$\text{Logloss} / \text{BCE} = -\frac{1}{n}\sum_{i=1}^{n} y_i \cdot \log\left(\widehat{y_i}\right) + \left(1 - y_i\right) \cdot \log\left(1 - \left(\widehat{y_i}\right)\right)$$

$$\text{Where } \widehat{y_i} = \frac{1}{1 + e^{-(x_i\beta + \varepsilon)}}$$

Note that in the preceding log loss, the x variable or the features are single values instead of a matrix as written in the sigmoid function. The probability represents how likely the prediction is 1; and vice versa, 1 – probability will be how likely the prediction will be 0. Depending on the label, the cost function calculates the negative log-likelihood of the model prediction for each sample. Then the average of these values is the final cost.

In order to minimize this cost function, we take the exact same approach as Linear Regression, by using gradient descent. Different from Linear Regression, the bias term cannot be calculated after gradient descent. Thus, we will differentiate the weight along with bias and update them the same way. Now that we're familiar with gradient descent, we will dive straight to Multiple Logistic Regression, or Logistic Regression where the training data has more than one feature. As a reminder, the features, X, will be in matrix form with shape number of samples × number of features as mentioned in Multiple Linear Regression; the ground-truth values, or the labels, will be a vector with the length being the number of samples; the weights, or β, will be a vector with the length being the number of features; finally, we will not be including the summation part of the loss as it's simply taking the average of the result and will not affect the differentiation.

Differentiating the weight first, we have

$$c(\beta,\varepsilon)=-\left[y\cdot log(\hat{y})+(1-y)\cdot log\left(1-(\hat{y})\right)\right]$$

where

$$\hat{y}=\frac{1}{1+e^{-x}}$$

We'll refer to this function as $s(\beta, \varepsilon)$ for sigmoid, where $x = X\beta + \varepsilon$.

This we'll refer to as $p(\beta, \varepsilon)$ for prediction. Using the chain rule, the derivative of the cost function becomes

$$\frac{\delta}{\delta\beta}c(\beta,\varepsilon)=\frac{\delta}{\delta s}c(\beta,\varepsilon)\frac{\delta}{\delta p}s(\beta,\varepsilon)\frac{\delta}{\delta\beta}p(\beta,\varepsilon)$$

Let's calculate the derivative of each function:

$$\frac{\delta}{\delta s}c(\beta,\varepsilon)=-\left[\frac{y}{s(\beta,\varepsilon)}+\frac{1-y}{1-s(\beta,\varepsilon)}\right]$$

$$\frac{\delta}{\delta p}s(\beta,\varepsilon)=\frac{e^{-p}}{\left(1+e^{-p}\right)^{2}}=\frac{1}{1+e^{-p}}\left(1-\frac{1}{1+e^{-p}}\right)=s(\beta,\varepsilon)\cdot\left[1-s(\beta,\varepsilon)\right]$$

$$\frac{\delta}{\delta\beta}p(\beta,\varepsilon)=X$$

Let's plug the derivatives back in:

$$\frac{\delta}{\delta\beta}c(\beta,\varepsilon)=-\left[\frac{y}{s(\beta,\varepsilon)}+\frac{1-y}{1-s(\beta,\varepsilon)}\right]\cdot s(\beta,\varepsilon)\cdot\left[1-s(\beta,\varepsilon)\right]\cdot X$$

Here's our final derivative:

$$-\left[s(\beta,\varepsilon)-y\right]\cdot X=-\left[\frac{1}{1+e^{-(X\beta+\varepsilon)}}-y\right]\cdot X$$

Here's the differentiation of the bias term:

$$\frac{\delta}{\delta\varepsilon}c(\beta,\varepsilon)=\frac{\delta}{\delta s}c(\beta,\varepsilon)\frac{\delta}{\delta p}s(\beta,\varepsilon)\frac{\delta}{\delta\varepsilon}p(\beta,\varepsilon)$$

$$\frac{\delta}{\delta s}c(\beta,\varepsilon)=-\left[\frac{y}{s(\beta,\varepsilon)}+\frac{1-y}{1-s(\beta,\varepsilon)}\right]$$

$$\frac{\delta}{\delta p}s(\beta,\varepsilon)=\frac{e^{-p}}{\left(1+e^{-p}\right)^{2}}=\frac{1}{1+e^{-p}}\left(1-\frac{1}{1+e^{-p}}\right)=s(\beta,\varepsilon)\cdot\left[1-s(\beta,\varepsilon)\right]$$

$$\frac{\delta}{\delta\varepsilon}p(\beta,\varepsilon)=1$$

$$\frac{\delta}{\delta\varepsilon}c(\beta,\varepsilon)=-\left[\frac{y}{s(\beta,\varepsilon)}+\frac{1-y}{1-s(\beta,\varepsilon)}\right]\cdot s(\beta,\varepsilon)\cdot\left[1-s(\beta,\varepsilon)\right]\cdot1$$

The derivative for bias becomes

$$-\left[s(\beta,\varepsilon)-y\right]\cdot1=-\left[\frac{1}{1+e^{-(X\beta+\varepsilon)}}-y\right]$$

We can update both the weight and bias based on the equation of gradient descent:

$$\text{new }\beta=-\alpha\cdot\frac{\delta}{\delta\beta}c(\beta,\varepsilon)+\beta$$

$$\text{new }\varepsilon=-\alpha\cdot\frac{\delta}{\delta\varepsilon}c(\beta,\varepsilon)+\varepsilon$$

Implementation and Usage

The code structure of Logistic Regression will be very similar to that of Linear Regression, only with the cost function and the derivatives modified accordingly to the equations shown previously (Listing 1-47).

Listing 1-47. Functions for Logistic Regression

```
# for log loss, we will be using the sklearn implementation
from sklearn.metrics import log_loss

# the sigmoid function
def sigmoid(x):
    return 1 / (1 + np.exp(-x))

# logloss/BCE cost function
def bce_loss_func(y_true, y_pred):
    return log_loss(y_true, y_pred)

# prediction function, namely the equation for the line, y=sigmoid(beta*x+epsilon)
def predict(beta, epsilon, x):
    linear_pred = np.dot(x, beta) + epsilon
    return sigmoid(linear_pred)

# calculates the partial weight derivative
def weight_deriv(beta, epsilon, x, y_true):
    error = predict(beta, epsilon, x) - y_true
    return np.mean(np.dot(x.T, error))
```

```
# calculates the partial bias derivative
def bias_deriv(beta, epsilon, x, y_true):
    error = predict(beta, epsilon, x) - y_true
    return np.mean(np.sum(error))
```

The gradient descent process is very similar to Linear Regression except that we update the bias too (Listing 1-48).

Listing 1-48. Logistic Regression gradient descent

```
def gradient_descent(beta, epsilon, x, y_true, alpha):

    # Update weight
    weight_derivative = weight_deriv(beta, epsilon, x, y_true)
    new_beta = beta - alpha * weight_derivative

    #Update bias
    bias_derivative = bias_deriv(beta, epsilon, x, y_true)
    new_epsilon = epsilon - alpha * bias_derivative

    return new_beta, new_epsilon
```

Combining everything, the implementation for Logistic Regression is shown in the following. We can use scikit-learn's function to create an example dataset for classification with a single feature and 20 samples (Listing 1-49).

Listing 1-49. Implementation of Logistic Regression

```
from sklearn.datasets import make_classification

X, y = make_classification(n_samples = 20, n_features = 1, n_informative = 1, n_redundant = 0, flip_y=0.05, n_clusters_per_class=1)

epsilon = 0
beta = 1

# usually a value between 0.001 and 0.0001 works well
alpha = 0.0001

iterations = 200

for i in range(1, iterations + 1):

    beta, epsilon = gradient_descent(beta, epsilon, X, y, alpha)

    # evaluate every 20 iterations
    if i%20 == 0:
        pred = predict(beta, epsilon, X)
        bce = bce_loss_func(y, pred.flatten())
        print(f"Error at iteration {i}: {bce}")
```

```
final_pred = predict(beta, epsilon, X)
final_bce = bce_loss_func(y, final_pred.flatten())
print(f"Final Prediction has error: {final_bce}")
```

Due to some differences in implementation, scikit-learn's Logistic Regression yields better results than ours. The usage is simply instantiating the object and calling fit (Listing 1-50). One thing to note here, in our implementation, the predict function returns raw probabilities from the sigmoid function, while scikit-learn's predict function returns binary labels. To retrieve the probabilities, we need to call predict_proba.

Listing 1-50. Scikit-learn Logistic Regression

```
from sklearn.linear_model import LogisticRegression

lr = LogisticRegression()
lr.fit(X.reshape(-1, 1), y)

# Retreives the weight and bias
w = lr.coef_
b = lr.intercept_

#Returns probability
pred_prob = lr.predict_proba(X)

#Returns binary labels
pred_binary = lr.predict(X)
```

Other Variations on Logistic Regression

Both the L1 and L2 regularizations can be applied to Logistic Regression and perform the same task as regularizing Linear Regression, to prevent overfitting. Scikit-learn supports L1, L2, and elastic net regularizations. It uses a parameter named C, which controls the inverse regularization strength, or $C = 1/\lambda$ where λ is the hyperparameter in L1 and L2 regularizations. The specific type of regularization can be specified with the penalty parameter (Listing 1-51).

Listing 1-51. Scikit-learn's regularization Logistic Regression

```
regularization_lr = LogisticRegression(penalty="L1", C=0.5)
# Penalty can be either 'L1', 'L2', or 'elasticnet' #or 'none'
```

One other important variation is known as Multinomial Logistic Regression. Normally, binomial, or the concept of Logistic Regression that we've been working with, deals with data that contain two classes. However, in Multinomial Logistic Regression, labels are given in "multiclass form" where there are more than two classes. To illustrate, instead of predicting between cats and dogs, the model might have to predict ten different types of animals.

The labels in multiclass classification are given in the shape of number of samples × number of classes. Following, the weights are in matrices with dimension number of features × number of classes. Normally, we would use the sigmoid function after plugging the weights and biases into $y = X\beta + \varepsilon$. However, due to the nature of multiclass classification, instead of outputting a vector with the length being the number of samples, a matrix with the shape number of samples x number of classes is produced. Thus, sigmoid would not be appropriate. Instead, the softmax function is used. Softmax outputs the prediction matrix in which each class contains its own probability and the final class is the class with the highest probability:

$$\text{Softmax}(x) = \frac{e^x}{\sum_{i=1}^{n} e^{x_i}}$$

In scikit-learn, Multinomial Logistic Regression is defined with the parameter `multi_class="multinomial"` when instantiating the object (Listing 1-52).

Listing 1-52. Scikit-learn regularization LR

```
regularization_lr = LogisticRegression(penalty="L1", C=0.5)
```

Both Linear and Logistic regressions have apparent disadvantages that outweigh themselves compared with more complex deep learning methods. Linear Regression assumes a linear relationship between features and labels, while Logistic Regression assumes features are related to targets by log odds, which sets a major boundary to many real-world datasets as most of them cannot be modeled solely with linear or logarithmic relationships. Especially with Logistic Regression, linearly separable data is rarely found in real-world scenarios. Later, we will look at newer classical machine learning algorithms that introduce completely new concepts that address some of the disadvantages of regression methods.

Decision Trees

The concept of Decision Trees appeared in the 1960s in the field of psychology for modeling the concept of human learning, visually presenting possible outcomes to potential situations that stem from one prompt. It was around that time when people discovered the usefulness of Decision Trees in programming and mathematics. The first paper that was able to develop a concept of "Decision Tree" mathematically was published by William Belson in 1959. In 1977, various professors from Berkley and Stanford developed an algorithm known as Classification and Regression Trees (CART), which, true to its name, consisted of Classification and Regression Trees. To this day, CART still stands as an important algorithm for data analysis. In the field of ML, Decision Tree serves as one of the most popular algorithms for real-world data science problems.

Theory and Intuition

Decision Trees serve as a tool that models the outcomes of observations in a visually interpretable format. It consists of nodes and branches, which stem from a root. Consider the following example of a series of choices. The first question asked "What color is the coin?" is called the root of the tree. Each possible outcome to the question is represented under nodes that branch out from the root. Imagine flipping a coin. There are two possible outcomes, heads or tail. Following the outcome of "Silver" to the question asked, we can continue the splitting process by asking another question: "What size is the coin?" For questions that are not the root, they're considered parent nodes. Once we've answered all possible questions and created every node to their possible outcomes, the final nodes are called leaf nodes (Figure 1-53).

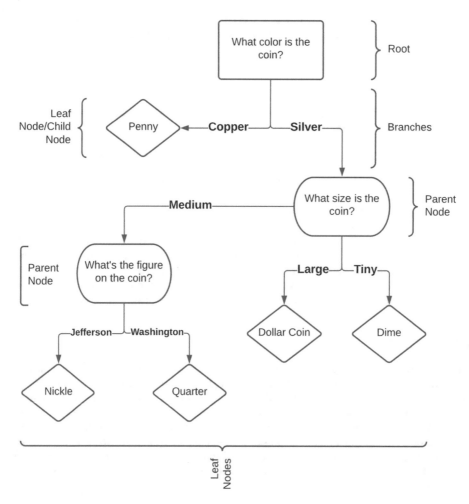

Figure 1-53. *Decision Tree example*

In a sense, Decision Trees can store and organize data based on their attributes. The ML concept of Decision Trees utilizes this intuition to categorize or classify data based on the patterns of their features. By splitting the samples according to their associated features, Decision Trees can be trained to learn different patterns in the data. Using this method, Decision Trees can handle high-dimensional data with ease as their approach differs greatly from KNN and regression methods. One of the major advantages of Decision Trees over the regression method is their interoperability as seen previously. It would be hard to visualize the gradient descent process on high-dimensional data but not for Decision Trees.

With features and targets, Decision Trees seek the best "questions" to ask in the best order to split the data that resembles those assigned by the targets. Take the Titanic dataset for example. It contains features about travelers on the Titanic such as their age, sex, cabin, fare, and so on. The aim is to predict whether someone survived based on their attributes given. A Decision Tree would split the data as follows; the process is demonstrated in Figure 1-54:

1.   Choose a feature that best splits the data.

2.   For that feature, choose a threshold that would best split the data by its labels. A threshold for a categorical feature could be true/false, or if the feature contains more than one category, it would be converted to numbers representing each class and treated as a continuous feature.

3.   For each node that the split produces, repeat the preceding steps until the data is perfectly split, meaning that each node contains exactly one class or the max split, depth, defined by the user is reached.

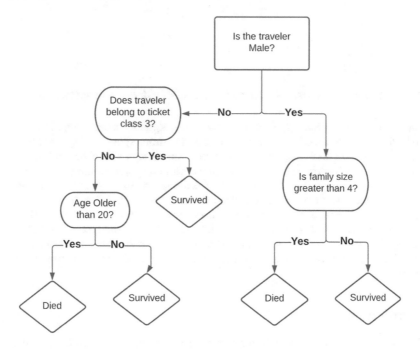

Figure 1-54. *An example Decision Tree on the Titanic dataset*

Simply stating "choosing the best feature to split the data" isn't enough. Decision Trees utilize metrics that measure the "impurity" of data, which determines how well the data is split. If data samples only contain one class, it's pure; on the other hand, if data samples contain half of one class and half of the other class, it's considered impure and anything in between. The two common metrics used to determine impurity in Decision Trees are Gini Impurity and Entropy as shown in the following:

$$\text{Gini Impurity} = 1 - \sum_{i=1}^{n} p_i^2$$

$$\text{Entropy} = -\sum_{i=1}^{n} p_i \cdot \log_2 p_i$$

In both equations, p is the probability of class i, or the percentage of samples that belong to class i in a node after a split. Gini Impurity sits inside the interval $[0, 0.5]$, while vales of Entropy sit inside the interval $[0, 1]$. However, the difference between the two is minimal as seen in the following. When the values of Gini Impurity×2 (to scale with Entropy) and Entropy are plotted against the probability of a class, their output varies little as the input ranges from 0 to 1 (Figure 1-55).

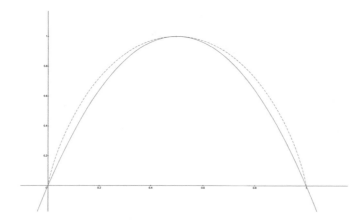

Figure 1-55. *Gini Impurity compared with Entropy where solid is Gini x 2 and dotted is Entropy*

However, computationally Entropy is more complex as it requires the use of logarithms, so Gini Impurity is usually the metric used in the Decision Tree to determine splits. During the split of each node, we treat the unique values of the feature as one threshold, loop through all the thresholds, and calculate its respective Gini Impurity. We perform the preceding process for every feature and then determine the best Gini Impurity, which we will split the data according to its threshold. As mentioned before, the splitting continues until each node is pure or a user-specified hyperparameter "depth" is reached.

The depth hyperparameter decides how many "levels" of nodes should there be in a Decision Tree. Technically, the deeper a Decision Tree is, the higher the performance would be as the splits get more specific. Although increasing the depth may improve performance, most of the time it only applies to training data as the tree overfits easily with a high amount of depth, leading to worse performance on unseen testing data.

Implementation and Usage

Note here that looping through each specific threshold for each feature and computing its Gini Impurity is painfully slow as its time complexity results in $O(N^2)$. Instead, we can sort the features and treat each unique value as a threshold and then compute the Gini Impurity as a weighted average between the left and the right splits while keeping track of the number of samples in each class on both sides of the split. As we move on to the next threshold, we can decrease/increase the class count on both sides. More details are shown in the following in the code (Listing 1-53). In the implementation of Decision Tree, we will be defining classes instead of stand-alone functions as it makes things a lot easier to work with.

Listing 1-53. DecisionTree Class

```
class DecisionTree:
    def __init__(self):
        pass

    def fit(X, y):
        self.num_features = X.shape[1]

    # Returns the index of the best feature and the threshold of the best split of
    that feature
    def find_best_split(self, X, y):
```

```
# cant split if there's only 1 sample
if y.size <= 1:
    return None, None

#keeping track of the best threshold and index
best_index, best_thresh = None, None

# Get the number of positive and negative samples in this node
num_pos, num_neg = (y == 1).sum(),(y == 0).sum()

# Calculate the Gini Impurity of current node, if no other splits exceeds this, no
further split will be performed
best_impurity = 1 - ((num_pos/len(y))**2 + (num_neg/len(y))**2)

for i in range(self.num_features):

    # Sort features values along with their respective labels
    # Threshold associated with their index
    threshold = pd.Series(X[:, 0]).sort_values()
    # Grab the labels in the order of thresholds
    labels = y[threshold.index]
    # Convert threshold to numpy arrays
    threshold = threshold.values

    #Keep track of the number of pos, neg samples on both sides with a matrix,
    [[left_pos, right_pos], [left_neg, right_neg]]
    classes_count = np.array([[0, num_pos], [0, num_neg]])

    #len(X) is the number of samples
    for j in range(1, len(y)):
        # get the current sample's label
        curr_label = labels[j-1]
        # change class count accordingly
        classes_count[curr_label, 0] -= 1
        classes_count[curr_label, 1] += 1

        # Calculate Gini Impurity for both sides, then combine with weighted average
        where the weight is the number of samples on each side,
        # this cleverly gives us the Gini Impurity for the split
        left_gini = 1 - ((classes_count[0, 0]/j)**2 + (classes_count[1, 0]/j)**2)
        right_gini = 1 - ((classes_count[0, 1]/(len(y)-j))**2 + (classes_count[1,
        1]/(len(y)-j))**2)
        gini_combined = left_gini * (j/len(y)) + right_gini * ((len(y)-j)/len(y))

        # Make sure we dont use the same threshold
        if threshold[j] == threshold[j-1]:
            continue

        # Set the best threshold to current threshold if we do better than our best
        gini impurity
        if gini_combined < best_impurity:
```

```
                best_impurity = gini_combined
                best_index = i
                best_thresh = (threshold[j-1] + threshold[j])/2

    return best_index, best_thresh
```

Each time the find_best_split function is called, it's performing a split based on the data of the node, using the samples split by the previous step on the node a level above the current one. We will then define a node class that will keep track of all the data, threshold, and features of the split, and we can build a Decision Tree recursively (Listing 1-54).

Listing 1-54. Node class

```
class Node:
    def __init__(self, best_feature, best_threshold,):
        self.best_feature = best_feature
        self.best_threhold = best_threshold
        # represents the node on the left and the right of the split
        self.left = None
        self.right = None
```

Note that the following methods will be in the DecisionTree class (Listing 1-55).

Listing 1-55. Fit and other methods to build the tree recursively

```
    def __init__(self, max_depth):
        self.max_depth = max_depth

    def fit(self, X, y):
        self.num_features = X.shape[1]
        self.decision_tree = split_tree(X, y, 0)
        return self

    def split_tree(self, X, y, current_depth):
        # Find majority_class first
        num_pos, num_neg = (y == 1).sum(),(y == 0).sum()
        majority_class = np.argmax(np.array([num_neg, num_pos]))

        best_ind, best_thresh = self.find_best_split(X, y)
        curr_node = Node(best_ind, best_thresh, majority_class)

        if current_depth < self.max_depth and best_ind is not None:
            right_split = X[:, best_ind] > best_thresh
            X_right, y_right, X_left, y_left = X[right_split], y[right_split], X[~right_
            split], y[~right_split]
            curr_node.left = self.split_tree(X_left, y_left, current_depth + 1)
            curr_node.right = self.split_tree(X_right, y_right, current_depth + 1)

        return curr_node
```

For the prediction method, we will keep going down the tree on the appropriate side depending on the split until we reach the leaf node where there's no left or right node left. We simply return majority_class as our prediction (Listing 1-56). The predict method will also belong to the DecisionTree class. In the _predict method, we perform splits on the data according to trained thresholds, while in the predict method, we loop through the whole X dataset to perform prediction on every sample.

Listing 1-56. Prediction functions

```
def _predict(self, X):
    curr_node = self.decision_tree
    while curr_node.right:
        if X[curr_node.best_feature] > curr_node.best_threshold:
            curr_node = curr_node.right
        else:
            curr_node = curr_node.left

    return curr_node.majority_class

def predict(self, X):

    return np.array([self._predict(i) for i in X])
```

To use our implemented Decision Tree, we simply need to instantiate the object and call fit with our respective datasets and then call predict for inference (Listing 1-57).

Listing 1-57. Usage of Decision Tree

```
dt = DecisionTree(mac_depth=12)
dt.fit(X, y)
predictions = dt.predict(X)
```

For the scikit-learn implementation (Listing 1-58), the usage is the exact same as our own implementation.

Listing 1-58. Scikit-learn implementation of Decision Tree

```
from sklearn.tree import DecisionTreeClassifier
dt = DecisionTreeClassifier(max_depth = 6)
dt.fit(X, y)
dt.predict(X)
```

The implementation that we wrote earlier does not support multiclass classification. However, the scikit-learn implementation supports multiclass natively without having to specify any extra parameters. Scikit-learn also provides an implementation for Regression Trees that predicts continuous values (Listing 1-59). The intuition of Regression Trees is the exact same as Classification Trees other than that instead of Gini Impurity, we use MSE to evaluate how good is each split while the prediction of each node is determined by the average of the data samples.

Listing 1-59. Scikit-learn implementation of Regression Tree

```
from sklearn.tree import DecisionTreeRegressor
dt = DecisionTreeRegressor(max_depth = 6)
dt.fit(X, y)
dt.predict(X)
```

Random Forest

Random Forest is an algorithm that provides some crucial improvements to the design of Decision Trees. Simply put, Random Forest is an ensemble of many smaller Decision Trees working together. Random Forest uses the simple concept that the wisdom of crowds is always better than one strong individual. By using lowly correlated small Decision Trees, their ensemble of predictions can outperform any single Decision Tree. The technique that Random Forest uses to ensemble smaller Decision Trees is called bagging. Bagging, also known as bootstrap aggregation, is randomly drawing different subsets, with replacements, from the training data, and the final prediction is decided by majority voting.

Random Forest selects subsets of data from the entirety of training data and trains Decision Trees on each of the subsets separately and then combines the results based on majority voting or averaging for classification or regression, respectively (Figure 1-56). When building Random Forest, it's important to notice that not every feature is selected for building each Decision Tree to create diversity while solving the Curse of Dimensionality problem. The ensemble of multiple Decision Trees makes Random Forest very stable in terms of performance. However, Random Forest does come with downfalls as it's fairly difficult to interpret and is substantially slower than Decision Trees. Due to their speed and interoperability, Decision Trees are usually chosen over Random Forest during quick tests on smaller sets of data. Random Forest is suitable when interoperability is not needed along with a large dataset.

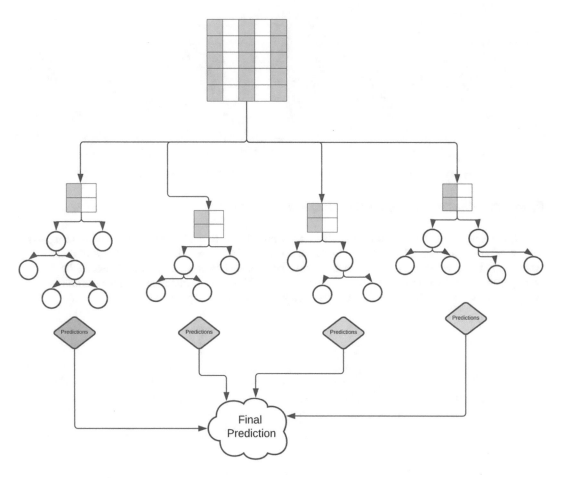

Figure 1-56. *Random Forest concept*

Broadly, this technique is known as bagging, or *b*ootstrap *agg*regat*ing*. It's an interesting idea: without adding any new data, knowledge, or unique learning algorithms, we can increase a model's performance simply by injecting randomness into the system. We train an ensemble of models, where each one is trained on a randomly selected subset of the data; after training, the predictions of this ensemble are aggregated to form the final prediction.

Interestingly, Random Forest reliably performs better than Decision Tree in most cases. Even given the commonly used explanation of "diversity" and "many thinkers are better than one," it can be confusing to understand exactly why bagging increases performance if it doesn't actually introduce unique new learning structures.

Decision Tree algorithms are generally high-variance algorithms (recall the bias-variance trade-off discussed in the "Fundamental Principles of Modeling" section). An unregulated Decision Tree will continue to construct nodes and branches to fit noisy data with little effort toward generalizing or identifying broad trends. We can intuitively understand the behavior of the Decision Tree as something like the very noisy sine wave displayed in Figure 1-57.

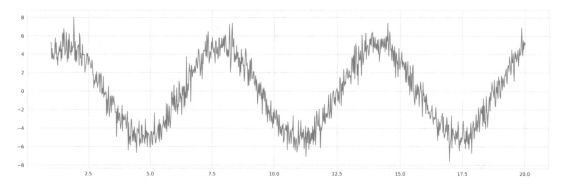

Figure 1-57. *Visualization of high-variance intuition*

Now, we will draw ten "bootstrapped" sine waves. Each bootstrapped sine wave (shown as transparent) consists of a randomly sampled 20% of points from the original sine wave. Then, we "aggregate" these bootstrapped sine waves together to form a "bagged" sine wave, in which each point is informed by averaging nearby values from bootstrapped waves. The resulting wave is visibly smoother, resembling a more generalized and stable curve than before bootstrap aggregating (Figure 1-58).

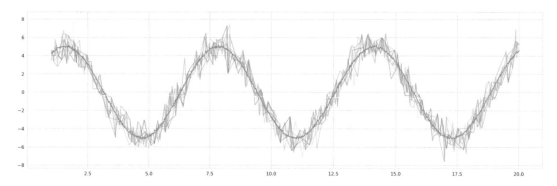

Figure 1-58. *Visualization of bagging to mitigate variance intuition*

Thus, we can understand bagging as *mitigating variance* in a base model by introducing a bias effect. Using this, we can also deduce that bagging does not generally work well on models that are already high bias; this has indeed been empirically observed.

The implementation of Random Forest from scikit-learn is shown in Listing 1-60.

Listing 1-60. Scikit-learn implementation of Random Forest

```
from sklearn.ensemble import RandomForestClassifier

# n_estimators is the number of tree that should be trained
rf = RandomForestClassifier(n_estimators = 30, max_depth = 12)
rf.fit(X, y)
predictions = rf.predict(X)
```

Tree methods like Random Forest and Decision Trees brought much diversity into the classical machine learning field as they utilize completely different techniques while solving many problems in previous algorithms. Tree-based methods are one of the fastest high-performing algorithms, and the concept of them is adapted into better algorithms including Gradient Boosting and some deep learning algorithms. The last chapter of Part 2 is dedicated to deep learning models, which take inspiration from tree-based models.

Gradient Boosting

Gradient Boosting describes a certain technique in modeling with a variety of different algorithms that are based upon it. The first successful algorithm that utilizes Gradient Boosting is AdaBoost (Adaptive Gradient Boosting) in 1998 formulated by Leo Breiman. In 1999, Jerome Friedman composed the generalization of boosting algorithms emerging at this time, such as AdaBoost, into a single method: Gradient Boosting Machines. Quickly, the idea of Gradient Boosting Machines became extremely popular and proved to be high performing in many real-life tabular datasets. To this day, various Gradient Boosting algorithms such as AdaBoost, XGBoost, and LightGBM are the first choice for many data scientists operating on large, difficult datasets.

Theory and Intuition

Gradient Boosting operates on similar ideas to those of Random Forest, using an ensemble of weaker models to build one, strong prediction. What differentiates Gradient Boosting from Random Forest is that instead of having individual, uncorrelated models, Gradient Boosting constructs models that are based on others' errors to continuously "boost" the performance.

Say you're picking a team of five people to represent your school, organization, or country at a trivia tournament in which the team that can answer the most questions about music, history, literature, mathematics, science, etc. wins. What is your strategy? You could pick five people who each have broad and overlapping knowledge of most topics likely to be covered (a Random Forest–style ensembling approach). Perhaps a better approach is to pick a person A who's really good in one area, then pick and train person B on the topics that person A isn't a specialist in, then pick and train person C on the topics that person B isn't a specialist in, and so on. By boosting learners on top of each other, we can build sophisticated and adaptive ensembles.

The concept of Gradient Boosting can be adapted to many different machine learning models such as Linear Regression, Decision Trees, and even deep learning methods. However, the most common use of Gradient Boosting is in conjunction with Decision Trees or tree-based methods in general. Although there have emerged various popular Gradient Boosting models, their initial and core intuition relies on the same algorithm; thus, we will be analyzing the original Gradient Boosting process while briefly going over the differences in each specific type of Boosting model.

Gradient Boosting naively adapts to both regression and classification. We can start by understanding the regression approach first as demonstrated in Figure 1-59.

1. Assuming the target variable is continuous, create a leaf with the average of the targets, which represents our initial guess of the labels.

2. Build a Regression Tree based on the errors of the first leaf. More specifically

   a. Calculate the error between our initial prediction and ground-truth labels for every sample, which we refer to as pseudo residuals.

   b. Construct a Regression Tree to predict the pseudo residuals of samples with a restricted number of leaves. Replace leaves with more than one label with the average of all labels that are in the leaf, which will be the prediction for that leaf.

3. Predict the target variables with a trained Decision Tree as follows:

    a. Start with the initial prediction, which is the average of all labels.

    b. Predict pseudo residuals using the Decision Trees we trained.

    c. Add predictions to the initial guess multiplied by a factor that we refer to as the learning rate, which becomes the final prediction.

4. Calculate new pseudo residuals based on the prediction of the previous model(s).

5. Construct a new tree to predict the new pseudo residuals, repeating steps 2–4. For step 3, we simply add the new tree's prediction and multiply by the same learning rate along with other trees that were created in previous iterations.

6. Repeat steps 2–4 until the maximum specified models are reached or the predictions start to worsen.

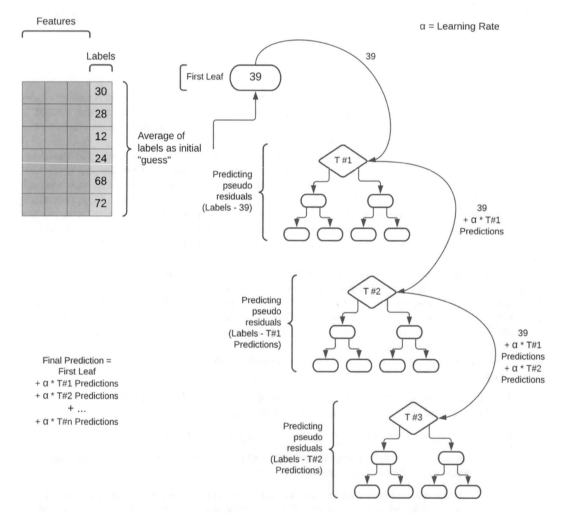

Figure 1-59. *Visualization of Gradient Boosting*

The approach for Gradient Boosting classification is extremely like regression. During our first initial "guess," instead of the average across targets, we compute the log odds on the labels. In order to calculate the pseudo residuals, we convert the log odds to probability using the sigmoid function:

$$\log \text{Odds} = \log\left(\frac{\text{number of positive samples}}{\text{number of negative samples}} \right)$$

In binary labels, we treat positive samples as 1 and negative samples as 0. Then, using the probabilities we calculated, we subtract each of them from their respective labels to obtain pseudo residuals. Just like regression, we construct a tree to predict the pseudo residuals based on the features. Again, with a restricted number of leaves, there will be leaves where more than one value is present. Since our leaf values are derived from probability while our prediction is in log odds, we need to apply transformations to the leaf predictions using a common formula as shown in the following. The summation sign represents the sum across all values in the same leaf, if any. P refers to the previous tree's, or leaf's, predicted probability:

$$\frac{\sum \text{residuals in leaf}_i}{\sum P_i (1 - P_i)}$$

After updating the prediction values in every leaf, we essentially perform the same procedures as those shown in regression, continuously building trees off the previous ones' errors until a user-specified parameter is reached or our predictions stop improving.

The approach described previously is a generalization of many Gradient Boosting–based algorithms. In the sections that follow, we're going to briefly cover some of the most popular ones and their respective implementations.

AdaBoost

Adaptive Gradient Boosting, or AdaBoost, is one of the earliest forms of Gradient Boosting. It was published before the generalization of Gradient Boosting Machines was proposed. The main idea that drives AdaBoost lies in the fact it uses weighted stumps, which combine into the final prediction. A stump is a Decision Tree with a root and only two leaves. The process of training an AdaBoost model is as follows:

1. Assign each sample the same weight, $\frac{1}{\text{number of samples}}$.

2. Train a stump, or a Decision Tree, with only one split as mentioned above, predicting the targets.

3. Calculate the amount of error that the trained stump makes. The amount of error is defined as the sum of the weights across all incorrectly classified samples.

4. Calculate the weight of the trained stump, defined as follows:

$$\text{Weight} = \frac{1}{2} \log\left(\frac{1 - \text{error}}{\text{error}} \right)$$

In practice, a small amount of extra error is added to prevent division by 0.

5. Adjust sample weight on training data accordingly. We increase the sample weights of each incorrectly classified sample by original sample weight $\cdot\ e^{\text{weight of stump}}$. On the other hand, we decrease each correctly classified sample weight by original sample weight $\cdot\ e^{-\text{weight of stump}}$.

6. Adjust, or normalize, sample weights to add up to 1. We divide each sample weight by the sum of all sample weights.

7. Repeat steps 2–6 using weighted Gini Impurity to train stumps until a user-specified parameter is reached or model performance stopped increasing.

8. When samples are passed through for prediction, the final prediction is determined by trees that classified one class, which has a total weight more than the trees that classified the other class.

The implementation in scikit-learn is shown in Listing 1-61.

Listing 1-61. Implementation of AdaBoost in scikit-learn

```
from sklearn.ensemble import AdaBoostClassifier
from sklearn.datasets import make_classification
X, y = make_classification(n_samples=1000, n_features=4, n_informative=2, n_redundant=0,
random_state=42)
# n_estimator is the amount of stumps
clf = AdaBoostClassifier(n_estimators=100, random_state=0)
# Fit and predict
clf.fit(X, y)
clf.predict(X)
```

If the stumps in AdaBoost are replaced with regression stumps, AdaBoost is capable of performing regression tasks too. Implementation for AdaBoost Regressor is shown in Listing 1-62.

Listing 1-62. Implementation of AdaBoost Regressor in scikit-learn

```
from sklearn.ensemble import AdaBoostRegressor
from sklearn.datasets import make_regression
X, y = make_regression(n_samples=1000, n_features=4, n_informative=2, n_redundant=0, random_
state=42)
# n_estimator is the amount of stumps
clf = AdaBoostRegressor(n_estimators=100, random_state=0)
# Fit and predict
clf.fit(X, y)
clf.predict(X)
```

XGBoost

XGBoost, short for Extreme Gradient Boosting, was developed in the early 2010s as a regularized variation of Gradient Boosting Machines. The development of XGBoost began as a research project by Tianqi Chen as part of the Distributed Machine Learning Community. It was then later popularized to the Machine Learning and Data Science Community during the Higgs Boson Machine Learning competition in 2014. Compared with AdaBoost, XGBoost is a lot more optimized for speed and performance and closely resembles the original approach of Gradient Boosting Machines.

The general steps taken to train an XGBoost are listed as follows:

1. Initialize a random prediction. This can be any number regardless of the classification of regression; a typical value is 0.5.

2. Similar to Gradient Boosting, we construct a Decision Tree to predict the pseudo residuals.

   a. Start with a single leaf with all the pseudo residuals and calculate a similarity score across all residuals using the following formula where ri is the ith pseudo residual and λ is a regularization term:

   $$\frac{\left(\sum_{i=1}^{n} r_i\right)^2}{n+\lambda}$$

   b. Find the best split similar to that of the Decision Tree. However, the metric we use to evaluate the effectiveness of a split differs from the Gini Impurity. The split that results in the greatest "gain" is chosen:

   $$\text{Gain} = \left(\text{Leaf Similarity}_{\text{left}} + \text{Leaf Similarity}_{\text{right}}\right) - \text{Root Similarity}$$

   c. Continue building the tree until the user-specified parameter is reached.

   d. A pruning parameter is specified, γ, which will remove certain nodes if the gain of the split is less than γ. This helps with regularization.

3. Calculate the output of prediction in each leaf by using the following formula:

   $$\text{Prediction} = \frac{\sum_{i=1}^{n} r_i}{n+\lambda}$$

4. Similar to Gradient Boosting, our final prediction is the initial leaf plus some learning rate that we define multiplied by the output of the trees. After the first tree is built, its prediction will be used to calculate new residuals and build new trees, until a user-specified parameter is satisfied.

Instead of using scikit-learn for implementation, XGBoost has its own library that provides functionalities and syntax structure similar to that of scikit-learn models (Listing 1-63).

Listing 1-63. XGBoost implementation

```
import xgboost as xgb
X, y = make_regression(n_samples=1000, n_features=4, n_informative=2, n_redundant=0, random_
state=42)

# train test split
from sklearn.model_selection import train_test_split
X_train, X_test, Y_train, Y_test = train_test_split(X, y, test_size=0.2)

# covert data to xgboost data types
dtrain = xgb.DMatrix(X_train, label=y_train)
dtest = xgb.DMatrix(y_test, label=y_test)
# specify parameters via map
```

```
param = {
    'max_depth': 4,
    # learning rate
    'eta': 0.3,
    # gamma, pruning term
    'gamma': 10,
    # lambda, regularization term
    'lambda': 1,
    # subsample, the percentage of samples selected to train each tree
    'subsample': 0.8,
    # tasks types, some common ones including reg:squarederror, regression with
    squared loss,
    # multi:softmax, multiclass classification using softmax, binary:logistic, binary
    classification
    'objective':'binary:logistic',
    # use GPU
    'tree_method': 'gpu_hist',

}

# number of boosted trees
num_round = 200
# train
bst = xgb.train(param, dtrain, num_round)
# make prediction
preds = bst.predict(dtest)
```

LightGBM

LightGBM, short for Light Gradient Boosting Machines, is an optimized Gradient Boosting algorithm that aims for reduced memory usage and speed while keeping the performance high. Developed by Microsoft in 2016, LGBM quickly gained popularity due to the advantages stated above along with its capability of handling large-scale data with parallel computing on GPUs.

One major difference between LGBM and other similar Gradient Boosting algorithms is the fact that Decision Trees grow leaf-wise in LGBM, while in other cases they grow level-wise. Leaf-wise growth handles overfitting better than level-wise and is much faster when operating on large datasets. Additionally, level-wise growth produces many unnecessary leaves and nodes. In contrast, leaf-wise growth only expands on nodes with high performance, thus keeping the number of decision nodes constant.

Furthermore, LGBM samples the datasets using two novel techniques known as Gradient-Based One-Side Sampling (GOSS) and Exclusive Feature Bundling (EFB), which reduce the size of the dataset without affecting performance.

GOSS samples the dataset aiming for the model to focus on data points where there's a larger error. By the concept of gradient descent, samples with lower gradients produce lower training error and vice versa. GOSS selects samples with a greater absolute gradient. Note that GOSS also chooses samples that have relatively lower gradients as it keeps the distribution of the output data like before GOSS.

The goal of EFB is to eliminate, or more appropriately merge, features. The algorithm bundles features that are mutually exclusive, meaning that they could never be of the same value simultaneously. The bundled features are then converted to a single feature, reducing the size and dimensionality of the dataset.

Finally, LGBM performs binning on continuous features to decrease the number of possible splits during Decision Tree building, introducing another major speedup to the algorithm.

Microsoft has its own library for implementing LGBM and provides simple and scikit-learn-like syntax as shown in Listing 1-64.

Listing 1-64. Scikit-learn implementation of LGBM

```
import lightgbm as lgbm

X, y = make_regression(n_samples=1000, n_features=10, n_informative=10, random_state=42)

# train test split
from sklearn.model_selection import train_test_split
X_train, X_test, y_train, y_test = train_test_split(X, y, test_size=0.2)

train_set = lgbm.Dataset(X_train, label=y_train)

params = {
            # number of leafs
            'num_leaves': 15,
            # max depth, counters overfitting when training data is small, but the tree
            still grows leaf wise
            'max_depth': 12,
            # max number of bins to seperate continuous features
            'max_bin':200,
            # task type, some other common ones: binary, multiclass, regression_l1
            'objective': 'regression',
            # change to gpu for training on gpu
            'device_type':'cpu',
            # learning rate
            'learning_rate':0.01,
            # amount of logging
            'verbose':0,
            # set random seed for reproducibility
            'random_seed':0
}

model = lgbm.train(params, train_set,
                   # number of boosted trees
                   num_boost_round=100,)

predictions = model.predict(X_test)
```

Summary of Algorithms

Algorithm Name	Description	Advantages	Disadvantages
KNN	Determines final classification based on the values of K-Nearest Neighbors	Simple, requires no training phase	Performs poorly with large datasets and takes a long time to train and find the optimal K value
Linear Regression	Finds a linear relationship, aka the line of best fit to the given regression dataset	Simple, interpretable	Cannot model complex nonlinear relationships
Logistic Regression	Finds a line of best fit to the given classification dataset	Simple, interpretable	Cannot model complex nonlinear relationships
Decision Tree	Splits data based on attributes to produce the result	Immune to the Curse of Dimensionality, can model nonlinear relationships	Can overfit easily and requires a large amount of hyperparameter tuning
Random Forest	Ensemble of smaller Decision Trees	Immune to the Curse of Dimensionality, can model nonlinear relationships, and generalizes well	Takes longer to train and requires a large amount of hyperparameter tuning
Gradient Boosting	Ensemble of weak learners, each trying to correct the errors of the previous one	Immune to the Curse of Dimensionality, can model nonlinear relationships, and flexible in terms of modeling	Sensitive to outliers, requires a large amount of hyperparameter tuning

Thinking Past Classical Machine Learning

Throughout this chapter, we've taken a comprehensive look at many key classical machine learning principles, concepts, and algorithms. These govern many important applications of automated learning across vast ranges of disciplines and remain an important bedrock for the continually developing vision of self-learning computers.

However, at a certain point, classical machine learning became too limited for ambitious new directions and ideas. As more sophisticated forms of data like images and text became more abundant, there was a need to effectively model high-dimensional *specialized* data forms that classical machine learning could not satisfy.

Deep learning is, on a very general level, the study of *neural networks*. Neural networks are more of an idea or governing concept rather than a concrete algorithm. (Chapter 3 is dedicated to exploring the technical specifics of how standard feed-forward neural networks operate and how you can apply them to tabular data.) Because they are so versatile and manipulatable, deep learning has never been short of applications and new research directions since computational power and data availability in the late 2000s and early 2010s made rapid experimentation possible.

Deep learning powers many important applications, such as the following:

- *Object recognition*: A deep learning model can identify the presence of an object or multiple objects in an image. This has applications in security (automatically detected unwanted entities), self-driving cars and other robotics, biology (recognizing very small biological objects), and more.

- *Image captioning*: Applications like Microsoft Word suggest automatic alternative descriptions for images put into a document. To caption an image, a deep learning model needs to extract visual information from an image, develop an internal representation of what the image "means," and translate that representation into natural language.

- *Image colorization*: Prior to the invention of color-capturing cameras, photographs of the world were taken in black-and-white. Deep learning can bring these grayscale images "back to life" by inferring colors from the image, as well as performing similar tasks like restoring damaged images or performing corrections on corrupted photos.

- *Art generation*: Deep learning can, through several different well-documented methods, create meaningful computer-generated art with little to no human involvement in the creation of each individual artwork.

- *Protein folding prediction*: Proteins are chains of amino acids that serve important biological roles in living organisms. The function of a protein is dependent on its shape, which is dictated by how individual amino acids linked in a sequence repel or attract each other to form intricate protein folding. Understanding the shape of a protein gives biologists a tremendous amount of information about its function and how to manipulate it (e.g., to disable a malicious protein or to design a protein to disable another malicious biological object). Protein folding was an unsolved problem until DeepMind released the AlphaFold model in 2020, which obtained extraordinarily high accuracy in predicting the shape of previously unknown proteins.

- *Text generation*: Neural networks can generate text to inform and entertain. One application of "speaking models" is in conversational chatbots, like AI-based therapy services that provide a natural interface to get counsel and company with the advantage of perhaps being more discreet and carefree than talking with a real human. These can also be used to help developers with coding by inferring the developer's intent and helping autocomplete code, sometimes several lines at a time (see Kite and Github Copilot).

Deep learning is being used to tackle modern, difficult, and incredibly complex problems. Correspondingly, deep learning models are often very large. Tens of thousands of parameters are standard for image recognition models, and state-of-the-art natural language models reach *hundreds* of billions of parameters. In these cases, the number of parameters is often almost always much larger than the number of samples in the training dataset. The (technically inappropriate but somewhat illustrative) analog in much lower-dimensional space is to fit a ten-dimensional polynomial, for instance, on two data points.

Despite being so large, these deep learning models obtain extraordinary validation performance on benchmark datasets. Moreover, it has been observed that increasing the size of a deep learning model almost always increases its validation performance. This conflicts with the bias-variance/overfitting-underfitting paradigm in classical machine learning, which suggests that continuously increasing the number of parameters in a model will lead to overfitting and thus poor generalization. What is happening here?

In order to expand the theory to explain or incorporate these observations, researchers at OpenAI have proposed the *Deep Double Descent* theory, which suggests that the infamous bias-variance U-shaped curve

(Figure 1-23) illustrates the relationship between generalization and model complexity only in the *classical regime* and that it is only part of a more complete curve. If we continue traveling along the *x*-axis far enough, progressively increasing model complexity, we reach a point where model performance becomes *better again* (Figure 1-60).

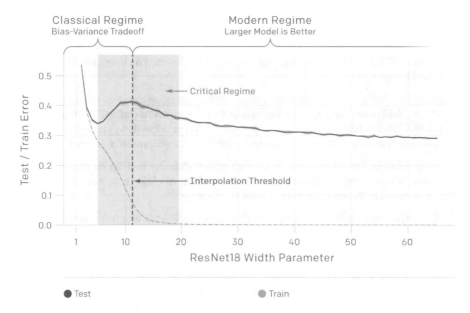

Figure 1-60. *Deep Double Descent, from the Nakkiran "Deep Double Descent" paper*

The Deep Double Descent phenomenon is one of many observed patterns indicating a fundamental shift between the classical and modern regimes. Given how stark the divide is, it may seem inappropriate to cross between the two paradigms. The subject of this book itself – deep learning for tabular data – bridges these two: deep learning, a study of the modern regime, and tabular data, which has always been thought to be modeled "best" by techniques of the classical regime.

This perspective is widespread largely because it makes sense and because, for a significant period of time, it has been empirically validated. Training a neural network on tabular data almost never performed as well as an algorithm like Gradient Boosting. The primary (arguably post hoc) explanation given is that tabular data can simply hold much fewer relationships, both in complexity and quantity, than much higher-dimensional image and text data. Neural networks, with their bulky and overpowering mass of parameters, instinctively by the bias-variance trade-off seem to be overkill and lead to overfitting (i.e., poor generalization).

However, recent developments in deep learning have demonstrated that deep learning has promise as an equal or, in some cases, a superior contender to machine learning methods in tabular data. The Deep Double Descent phenomenon hints that neural networks may not be subject to the same classical machine learning regime paradigms and that if we push far enough and in the right direction, we may end up somewhere better than we could have ever imagined before.

This book is dedicated to exploring an unlikely marriage between tabular data and deep neural networks, with the hope of establishing deep learning as a solid contender in the standard data scientist's toolkit, rather than being reserved for specialized deep learning engineers building image, language, and signal applications. It is an attempt to merge the modern with the classical – to reunderstand specialized developments on the frontier of deep learning with respect to their usefulness in application to structured data. You'll find that working with deep learning is, in many ways, more free and less rigid rinse-and-repeat than the standard classical machine learning workflow. As the state of data science advances, so must the tools we use.

Key Points

In this chapter, we discussed several key principles, concepts, and algorithms of classical machine learning:

- In the context of machine learning, *modeling* is the mostly automated process of developing representations of phenomena from data.

- A model uses supervised learning when it is trained to associate an input with a label. On the other hand, a model uses unsupervised learning when it is trained to find associations within the data itself, without any labels.

- Regression is when the target is continuous, whereas classification is when the target is discretized/binned.

- The training set is the data the model is fitted on; the validation set is the data the model is evaluated on but does not see during training; the test set is the data the model predicts on (with no associated ground-truth labels).

- A model that has too high of variance is incredibly sensitive to specific data instances it is presented with and adapts/"bends" its general prediction capabilities to exactly represent the provided data; this is *overfitting*. On the other hand, a model with too high of a bias doesn't adapt/"bend" its general prediction capabilities much toward specific instances of the dataset; this is *underfitting*. In the classical machine learning regime, an increase in the number of parameters ("degrees of freedom" is generally associated with increased overfitting behavior; the optimal size of a model is somewhere in between underfitting and overfitting.

- The Curse of Dimensionality describes a suite of phenomena involving distances, surfaces, and volumes in high-dimensional space that appears contrary/counter to our intuition in low-dimensional spaces.

- Gradient descent is an important concept in machine learning optimization, in which parameters are updated in the opposite direction of the derivative of the loss landscape with respect to the parameters. This moves in a direction that minimizes the loss. One weakness of gradient descent is that the final convergence result often depends heavily on initialization; this can usually be overcome by more advanced mechanisms.

- The Deep Double Descent phenomenon attempts to explain the success of heavily parametrized deep learning models over traditional machine learning models by extending the classical bias-variance U-shaped curve into a one-ridge descent, with the error decreasing when the model complexity ventures high enough. This phenomenon is one of many indicating operating paradigms and principles in deep learning are fundamentally different from machine learning.

- Modern developments in deep learning suggest promising applications for tabular data.

In Chapter 2, we will cover important tabular data storage, manipulation, and extraction methods for successful modeling pipelines. These skills will prepare you to handle different forms of data required in topics throughout the rest of the book.

CHAPTER 2

■ ■ ■

Data Preparation and Engineering

The goal is to turn data into information, and information into insight.

—Carly Fiorina, Former CEO of Hewlett-Packard

We define data preparation, or data preprocessing, as a transformation or set of transformations applied to raw data collected directly from the data source to better represent information. We do this for the purpose of being able to better model it (Figure 2-1).

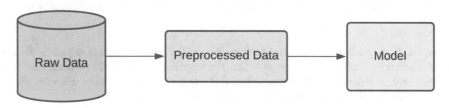

Figure 2-1. *Preprocessing pipeline from raw data to model*

Deep learning has a complex relationship with data preparation and preprocessing. It is well-known that classical machine learning or statistical learning algorithms often require extensive preprocessing to be successful. One of the most exciting advantages of using neural networks for modeling is its purported ability to reliably learn on raw data. It is well-known that neural networks have significantly more predictive power than standard machine learning algorithms. Therefore, theoretically, a neural network can learn the optimal set of preprocessing schemes for data modeling, equal to or better than the scheme a human can select or design.

However, this does not mean that deep learning has eliminated the need for data preparation and preprocessing. Despite the theoretical lack of a need for data preprocessing, applying complex or domain-informed preprocessing schemes can still improve neural network performance in practice. The existence of a theoretical state in which a deep learning model can represent optimal preprocessing schemes does not by itself also show that deep learning models can reliably *converge to such a state* during training. (In Chapter 10 on meta-optimization, we'll see an application of the techniques discussed in this chapter by optimizing the data preprocessing pipeline for a neural network using machine learning.)

Data preparation and preprocessing are still invaluable skills relevant to working with tabular data, regardless of whether you use a classical machine learning or deep learning approach. Machine learning algorithms learn best when they draw associations from data in which the associations are already somewhat clear or at least not obfuscated. The objective of the chapter is to provide you with ideas and tools to accomplish that goal.

© Andre Ye and Zian Wang 2023
A. Ye and Z. Wang, *Modern Deep Learning for Tabular Data*,
https://doi.org/10.1007/978-1-4842-8692-0_2

Roughly, we can identify three primary components of data preprocessing: data encoding, feature extraction, and feature selection (Figure 2-2). Each of these components is somewhat interrelated: the goal of data encoding is to make raw data both readable and "true to its natural characteristics"; the goal of feature extraction is to identify abstracted or more relevant features from within the data space; the goal of feature selection is to identify if and which features are not relevant to the predictive process and can be removed. Generally, data encoding takes precedence over the latter two components because data must be readable and representative of itself before we can attempt to extract features from it or select which features are relevant.

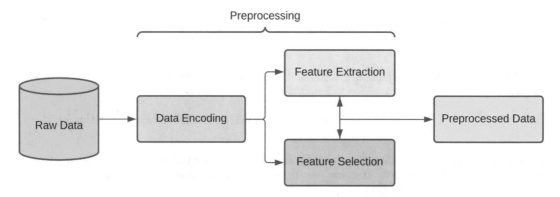

Figure 2-2. *Key components of the preprocessing pipeline: data encoding, feature extraction, and feature selection*

The chapter will begin by discussing various libraries and tools to store and manipulate tabular datasets, with a focus particularly on large and difficult-to-store data. Then, it will cover each of the three discussed components of the data preprocessing pipeline – data encoding, feature extraction, and feature selection – in application to several example tabular datasets.

Data Storage and Manipulation

In this section, we'll explore TensorFlow Datasets, a strong submodule for dataset storage, pipelines, and manipulations, as well as several libraries for loading large tabular datasets.

This chapter and the remainder of the book assume a working knowledge of NumPy and Pandas. If you are not familiar working with these libraries, please see the Appendix for a thorough overview of both.

TensorFlow Datasets

TensorFlow Datasets is especially important for image and text tabular data because they are very high-dimensional specialized data formats that take up a significant amount of space. Tabular data is generally more compact, so in many cases Pandas and NumPy suffice, and it is not necessary to use TensorFlow Datasets. However, TensorFlow Datasets offers significant speedups and space efficiency advantages with higher-frequency tabular data (e.g., high-precision stock market data).

Creating a TensorFlow Dataset

In order to create a TensorFlow dataset, one must first have a data source.

If you already have the data stored in memory, for instance, as a NumPy or Pandas array, you can copy the information into a TensorFlow dataset using the from_tensor_slices function (Listing 2-1).

Listing 2-1. Creating a TensorFlow dataset from a NumPy array

```
arr1 = [1, 2, 3]
data1 = tf.data.Dataset.from_tensor_slices(arr1)

arr2 = np.array([1, 2, 3])
data2 = tf.data.Dataset.from_tensor_slices(arr2)
```

When we call the value of data1 (for instance, by printing it), we obtain the result <TensorSliceDataset shapes: (), types: tf.int32>. However, data2 yields the result <TensorSliceDataset shapes: (), types: tf.int64>. Notice the difference in data types in tensors, which should technically be the same type – what's going on here?

When TensorFlow takes in raw integers, it casts them as 32-bit integers by default. However, NumPy casts raw integers as 64-bit by default; when TensorFlow takes in the NumPy array of 64-bit integers, it retains their representation and stores them under the tf.int64 type. It is important to be conscious of such "hidden" transformations of data when passing through in different containers.

You can control the TensorFlow dataset element representation thus by casting/setting certain element types in the data source, which TensorFlow will faithfully retain, or using TensorFlow's tf.cast.

You may also notice that the displayed shape in all created TensorFlow datasets appears to be empty. Let's create a multidimensional array with shape (2, 3, 4) and pass the data into .from_tensor_slices (Listing 2-2).

Listing 2-2. Creating a TensorFlow dataset from a multidimensional NumPy array

```
arr = np.arange(2*3*4).reshape((2, 3, 4))
data = tf.data.Dataset.from_tensor_slices(arr)
```

The result is <TensorSliceDataset shapes: (3, 4), types: tf.int64>. This is because .from_tensor_slices creates a TensorSliceDataset, which is organized by the storage of *slices*. The shape indicated is the size of each slice. You can think of a slice as a data sample or item arranged in a list; for instance, the .from_tensor_slices constructor will interpret input data with shape (2, 3, 4) as two data samples of shape (3, 4). This data storage mechanism explicitly captures the dimension distinguishing data samples from one another, which NumPy arrays do not.

If you want standard array-like storage or to work with TensorFlow-stored data in a less restrictive format (TensorSliceDatasets are not very modifiable because they are "ready"/"primed" to be fed into a model), you can use a TensorDataset. Rather than storing *slices*, a TensorDataset stores a raw tensor (Listing 2-3).

Listing 2-3. Creating a TensorFlow dataset from a multidimensional NumPy array using "from_tensors" instead of "from_tensor_slices"

```
arr = np.arange(2*3*4).reshape((2, 3, 4))
data = tf.data.Dataset.from_tensors(arr)
```

The code in Listing 2-3 yields `<TensorDataset shapes: (2, 3, 4), types: tf.int64>`.

You'll usually be feeding TensorSliceDatasets into models rather than TensorDatasets, since they are explicitly designed to efficiently capture separation between data samples. TensorFlow models will accept properly set-up datasets like such: `model.fit(data, other parameters...)`.

TensorFlow offers many utilities to convert files (e.g., `.csv`) or data stored in other utilities (e.g., Pandas DataFrames) into TensorFlow-compatible forms. You can explore these on the TensorFlow input/output (`tf.io`) documentation page.

Generally, for the purposes of tabular data modeling, TensorFlow Datasets is only necessary for very large datasets (e.g., fine-precision stock market data collected every few seconds across decades). Throughout the examples used in the book, we will not use TensorFlow Datasets as the default because it isn't necessary for most tabular contexts.

TensorFlow Sequence Datasets

The "traditional" TensorFlow dataset configuration requires you to arrange all your data at once into a compact dataset object that is passed into a model for training. This tends to be quite rigid and inflexible, especially if you are working with large datasets that simply can't be loaded into memory all at once or complex datasets with multiple inputs and/or outputs.

A TensorFlow *Sequence dataset* is a more flexible agreement between TensorFlow and the developer: rather than being forced to assemble and manipulate the entire dataset on an aggregate level at once, the developer agrees to give TensorFlow parts of the dataset on demand when they are requested. These datasets are valuable for their flexible structure and corresponding ease of data pipeline development.

A custom-defined Sequence dataset (Listing 2-4, Figure 2-3) inherits from the `tf.keras.utils.Sequence` class and must have two implemented methods: `__len__()` and `__getitem__(index)`. The former specifies the number of batches in the dataset; this can be calculated as $\frac{\# sample\ points}{batch\ size}$. The latter returns the x and y data subsets corresponding to the index, which indicates which batch is requested by the model. During model training, the model will request a maximum index of `__len__() - 1`. You can also add a method `__on_epoch_end__()`, which executes after every epoch. This feature is very helpful if you want to dynamically alter the dataset parameters throughout training or want to measure model performance in a custom manner. If you are familiar with PyTorch, you may observe that this custom TensorFlow dataset structure reflects the required dataset definition for PyTorch models.

Listing 2-4. The general structure of a TensorFlow Sequence dataset inheriting from tf.keras.utils.Sequence

```
class CustomData(tf.keras.utils.Sequence):

    def __init__(self):
        # set up internal variables

    def __len__(self):
        # return the number of batches

    def __getitem__(self, index):
        # return a request chunk of the data
```

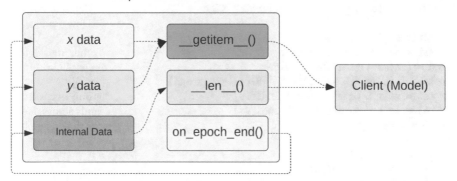

Figure 2-3. *One of several possible arrangements for the relationship between state (x data, y data, and internal data) and behavior (the __getitem__(), __len__(), and on_epoch_end() functions)*

Generally, a custom Sequence dataset contains the following features (see a code example in Listing 2-5):

- Internal data parameters that are set upon initialization and either remain static or change throughout training, for instance, batch size (constant) and image augmentation severity parameters like blurring kernel size or brightness (constant or dynamic).

- A variable to keep track of the current epoch, which is updated after each on_ epoch__end() call.

- Internal x and y fields that store the dataset or references to the dataset. For instance, if our dataset is relatively compact, it may be possible to store the data itself within the custom Sequence dataset and to simply index and return the desired batch of the data on demand when the model requests it. Alternatively, it may store image paths, text files, or references of another form that are loaded in the __getitem__ (index) call.

- Fields to keep track of training and validation indices (assuming this is not taken care of outside of the dataset).

Listing 2-5. Example filler code for building successful TensorFlow Sequence datasets

```
class CustomData(tf.keras.utils.Sequence):

    def __init__(self, param1, param2, batch_size):
        self.param1 = param1
        self.param2 = param2
        self.batch_size = batch_size
        self.x_data = ...
        self.y_data = ...
        self.train_indices = ...
        self.valid_indices = ...
        self.epoch = 1
```

```
def __len__(self):
    return len(self.x_data) // self.batch_size

def __getitem__(self, index):
    start = index*self.batch_size
    end = (index+1)*self.batch_size
    relevant_indices = self.train_indices[start:end]
    x_ret = self.x_data[relevant_indices]
    y_ret = self.y_data[relevant_indices]
    return x_ret, y_ret

def on_epoch_end(self):
    self.param1 = update(param1)
    self.epoch += 1
```

If you want to keep track of model performance over time, define a dataset to initialize with a model object and evaluate the performance of the model on the internal dataset every epoch (or every multiple of epochs). The pseudocode for how this could look is in Listing 2-6.

Listing 2-6. One way to track history internally in the custom dataset

```
class CustomData(tf.keras.utils.Sequence):

    def __init__(self, model, k, ...):
        self.model = model
        self.epochs = 1
        self.k = k
        self.train_hist = []
        self.valid_hist = []
            ...

    def on_epoch_end(self):
        if self.epochs % k == 0:
            self.train_hist.append(model.eval(train))
            self.valid_hist.append(model.eval(valid))
        self.epochs += 1
```

We'll see examples of TensorFlow Sequence datasets helping us a lot in Chapters 4 and 5, when we build models that require both a tabular and an image input to be simultaneously fed into two input heads.

Handling Large Datasets

Biomedical data represents a large portion of modern tabular datasets, such as datasets containing genetic information such as RNA, DNA, and protein expressions. The size of typical biomedical datasets is often very large due to the high precision of data. Thus, it's usually difficult or impossible to manipulate data in memory with Pandas DataFrames.

With TensorFlow datasets, it is simple to convert a dataset that is already loaded into memory into a compact model data feeder. Alternatively, using Sequence datasets, you can load select parts of your data on demand, either from memory or from the disk, as desired. However, there are other tools to deal with large tabular datasets, which may be more convenient than TensorFlow datasets. We will explore the handling of large datasets in a twofold manner – first exploring datasets that fit into memory and then those that do not.

Datasets That Fit in Memory

Datasets that fit into memory can be directly loaded in a session. However, with hardware limitations, further operations such as training with large models can often result in OOM errors. The following methods can be used to reduce the memory footprints of datasets that barely fit into memory. This can allow for complex manipulations of datasets without worrying about memory issues.

Pickle

Converting large, in-memory, Pandas DataFrames to pickle files is a universal approach to reduce the file size of any .csv datasets. Pickle files use the ".pkl" extension, and it is a Python-specific file format. That being said, any Python object can be saved the same way as it is shown for Pandas DataFrames in Listing 2-7. Pickle is usually the default method for saving and loading sklearn-based models (since there's no naïve support of saving and loading functionalities). Due to the efficient storage method adopted by pickle, it can not only reduce the original .csv file size but also able to be loaded in up to 100 times faster than Pandas DataFrames.

Listing 2-7. Saving and loading Pandas DataFrames using pickle files

```
import pickle
# saving
with open("path/to/file", "wb") as f:
      pickle.dump(dataframe, f)
# loading
with open("path/to/file", "rb") as f:
      loaded_dataframe = pickle.load(f)
```

Alternatively, the pickle file can also be loaded with `pd.load_pickle`.

SciPy and TensorFlow Sparse Matrices

If your dataset is especially sparse – a common culprit is the presence of many one-hot encoded columns – you can use either SciPy or TensorFlow sparse matrix storage objects.

SciPy offers several different sparse matrix methods. Compressed Sparse Row (CSR) format supports efficient row slicing and will likely be the go-to data storage object for efficient sample access. You can import the CSR object from **from** `scipy.sparse` **import** `csr_array` and use it as a wrapper against other array objects, such as NumPy arrays, tuples, or lists. Alternatively, `csc_array` supports efficient column indexing. Behind the scenes, nonzero elements are stored by their index, column, and value; sparse matrices support efficient mathematical operations and transformations. There are five more types of sparse arrays/matrices available in `scipy.sparse` with slightly more complicated instantiation processes. Refer to the SciPy sparse documentation for more information. Note that most scikit-learn models accept SciPy sparse matrices as inputs.

While SciPy matrices are certainly efficient and compatible with many other operations, they cannot be directly passed into TensorFlow/Keras models without additional work. TensorFlow supports a *sparse tensor* format, which can be called from `tf.sparse.from_dense`. Similarly to SciPy, sparse tensors are stored as bundles of three standard (dense) tensors: indices (locations), values, and the tensor shape. This provides the necessary information to efficiently store sparse data. You can cast sparse tensors into a TensorFlow dataset with the standard `tf.data.Dataset.from_tensor_slices`, which still preserves the sparsity and can be used for more memory-efficient training.

Datasets That Do Not Fit in Memory

Datasets that do not fit in memory cannot be loaded as a whole in a single session. We can load large image datasets batch by batch since each individual image is stored within its own file. With predefined pipelines in TensorFlow (which we will see in future chapters), images can be easily loaded from the disk in small amounts. However, most tabular datasets are not structured in a sample-by-sample manner; rather, the entire dataset is usually stored in one file. In the following methods, we will explore approaches that can "preload" the entirety of datasets to disk while being able to access the data through typical means.

Pandas Chunker

If you have a CSV file that is too large to be directly loaded into memory, you can instead use an iterator by specifying the iterator=True argument. You can specify the chunk size, which determines how many rows will be loaded on each iteration. We can directly build this into a TensorFlow custom dataset as follows (Listing 2-8), in which the iterator is kept as a dataset internal variable and each call to index the data returns the next chunk. Note that while TensorFlow will use an index, this index is irrelevant to the code written within the function. This means that each chunk can only be called once; the following time it is called, the next chunk will be returned.

Listing 2-8. A custom TensorFlow Sequence dataset using Pandas chunking

```
class CustomData(tf.keras.utils.Sequence):

    def __init__(self, filename, chunksize):
        self.csv_iter = pd.read_csv(filename,
                                    iterator=True,
                                    chunksize=chunksize)

    def __getitem__(self, index):
        for chunk in self.csv_iter:
            return chunk
```

This allows us effectively to stream data into the model with a much smaller memory footprint by directly reading from the file.

h5py

When a dataset is too large to load into memory, we want a way to load parts of it at a time on demand. However, this is easier said than done. Large CSV files that cannot be directly loaded into Pandas DataFrames, for instance, are difficult to chunk on demand. The Pandas Chunker introduced previously can only be used as an iterator where each chunk is returned once and returned in order. The same applies for many other types of large data.

The Python library h5py offers the ability to save and load files in the Hierarchical Data Format (h5). Compared with .csv files, h5 is a much more compressed storage format and cannot be interpreted/opened by typical spreadsheet/document programs. h5py creates a direct reference through variables between the program and the disk, which allows for programs to directly index and access select parts of the dataset stored in disk (which is typically much larger than memory size). Essentially, once the reference link between the defined variable in the program and the dataset stored in the disk is made, the linked variable can be treated like any other NumPy array. To understand and apply operations on h5 files, we're first going to instantiate a h5 dataset with the create_dataset method. The method requires two arguments: (a) the name of the dataset within the file and (b) the shape of the data. Shown in Listing 2-9, we wrote a h5 file with two groups titled "group_name1" and "group_name2", respectively. Both groups contain a two-dimensional matrix with shape (100, 100).

Listing 2-9. Creating a h5 dataset

```
import h5py
import numpy as np
# the hdf5 extension simply means the 5th version of the h5 format
with h5py.File("path/to/file.hdf5", "w") as f:
group1 = f.create_dataset("group_name1", (100, 100),
dtype="i8") # int8
group2 = f.create_dataset("group_name2", (100, 100),
dtype="i8") # int8
# optional argument "data" to create dataset from np arrays
```

To insert data into the empty dataset, we can access certain parts (or the entire dataset with "[:]") of the dataset with NumPy-like index notations.

Listing 2-10. Inserting contents into a h5 dataset

```
group1[10, 10] = 10
group1[2, 50:60] = np.arange(10)
```

We can read a h5 file and access its "dataset keys" in a similar fashion.

Listing 2-11. Reading a h5 file

```
f = h5py.File("path/to/file.hdf5", "r")
```

The "dataset keys" we created previously (group1 and group2) can be seen as key-value pairs in a dictionary. A list of keys present in the h5 file can be accessed by "f.keys()". Note that keys can be nested, meaning that there may be more keys corresponding to data under the top-level key. Values contained under keys (such as the (100, 100) matrix we defined for group1 and group2) can be retrieved in the same way as retrieving a value from a Python dictionary.

Listing 2-12. Retrieving values from keys in a h5 file

```
# access a certain dataset using key-value pairs
f["group1"][:] # retrieves everything from group 1
```

The indexed values will be loaded into memory. We can use this in conjunction with TensorFlow datasets to allow for the training of large tabular datasets.

The h5py library provides convenient and universal solutions to loading in data that does not fit in memory as well as reducing the actual size of the file significantly. Refer to the h5py documentation for further details regarding manipulating h5 files.

NumPy Memory Map

NumPy memory maps can be seen as a quick and dirty alternative to h5 for mapping variables in program to data stored in disk. By wrapping a file path with np.memmap, a link is created between the variable assigned to the return value of np.memmap and the file stored in disk. Two of the popular file paths that np.memmap accepts are .npy and .npz files, which store NumPy arrays and SciPy sparse matrices, respectively. Specifically, a simple usage of NumPy memory maps can be as such: arr = np.memmap("path/to/arr"). The "arr" variable will contain the reference to the file stored on disk as specified by the memmap call. For details on other optional arguments, refer to the documentation of NumPy memory maps.

Data Encoding

We define data encoding as the process of making data merely readable to any algorithm (feature extraction, feature selection, machine learning models, etc.) that may be applied on the dataset afterward (Figure 2-4). Here, *readable* not only means "quantitative" (as discussed in Chapter 1, machine learning models require inputs to somehow be numerically represented) but "representative of the nature of the data." For instance, we could technically make a feature containing countries of the world "readable" in the sense that there would be no code-run errors by randomly associating a country with a unique number – the United States

to 483.23, Canada to –84, India to $e - i \sum_{j=0}^{\infty} j$, and so on. However, this arbitrary numerical conversion does

not represent the nature of the feature: there is no relevant relationship between the representations that reflects the relationships between the represented things in themselves. A numerical representation of a feature, to be "readable," must be faithful to its attributes.

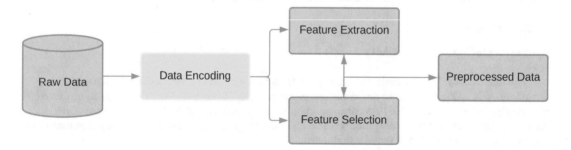

Figure 2-4. *The data encoding component of the data preprocessing pipeline*

Discrete Data

We define "discrete" data as data that can only theoretically take on a limited set of values. These can be binary features (e.g., "yes" or "no" responses), multiclass features (e.g., types of animals), or ordinal data (e.g., education grades or tiered features). There are many different mechanisms to represent information contained within discrete data, known as *categorical encodings*. This section will explore the theory and implementation for these methods.

We'll be using a selection of the Ames Housing dataset (Figure 2-5) to demonstrate these categorical encoding methods. You can load it as such (Listing 2-13).

Listing 2-13. Reading and selecting part of the Ames Housing dataset

```
df = pd.read_csv('https://raw.githubusercontent.com/
hjhuney/Data/master/AmesHousing/train.csv')
df = df.dropna(axis=1, how='any')
df = df[['MSSubClass', 'MSZoning', 'LotArea', 'Street',
        'LotShape', 'OverallCond', 'YearBuilt',
        'YrSold', 'SaleCondition', 'SalePrice']]
```

	MSSubClass	MSZoning	LotArea	Street	LotShape	OverallCond	YearBuilt	YrSold	SaleCondition	SalePrice
0	60	RL	8450	Pave	Reg	5	2003	2008	Normal	208500
1	20	RL	9600	Pave	Reg	8	1976	2007	Normal	181500
2	60	RL	11250	Pave	IR1	5	2001	2008	Normal	223500
3	70	RL	9550	Pave	IR1	5	1915	2006	Abnorml	140000
4	60	RL	14260	Pave	IR1	5	2000	2008	Normal	250000
...	...	...	...	...	...	...	...	...	...	...
1455	60	RL	7917	Pave	Reg	5	1999	2007	Normal	175000
1456	20	RL	13175	Pave	Reg	6	1978	2010	Normal	210000
1457	70	RL	9042	Pave	Reg	9	1941	2010	Normal	266500
1458	20	RL	9717	Pave	Reg	6	1950	2010	Normal	142125
1459	20	RL	9937	Pave	Reg	6	1965	2008	Normal	147500

1460 rows × 10 columns

Figure 2-5. *Visualizing part of the Ames Housing dataset*

The dataset has many categorical features. While we will only apply categorical encoding consistently to one categorical feature, it is encouraged to experiment with each of the different features in the dataset.

Label Encoding

Label encoding is perhaps the simplest and most direct method of encoding discrete data – each unique category is associated with a single integer label (Figure 2-6). This is almost always not the final encoding that you should use for categorical variables, since attaching encodings in this way forces us to make *arbitrary decisions* that lead to *meaningful outcomes*. If we associate a category value "Dog" with 1 but "Snake" with 2, the model has access to the explicitly coded quantitative relationship that "Snake" is two times "Dog" in magnitude or that "Snake" is larger than "Dog." Moreover, there is no good reason "Dog" should be labeled 1 and "Snake" should be labeled 2 instead of vice versa.

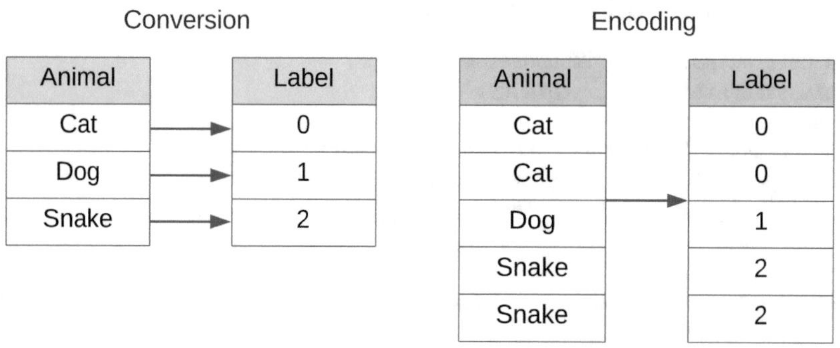

Conversion

Animal		Label
Cat	→	0
Dog	→	1
Snake	→	2

Encoding

Animal		Label
Cat		0
Cat		0
Dog	→	1
Snake		2
Snake		2

Figure 2-6. *Label encoding*

However, label encoding is the basis/primary step upon which many other encodings can be applied. Thus, it is useful to understand how to implement it (Listing 2-14).

We can implement label encoding from scratch by collecting the unique elements with np.unique(), mapping each unique element to an index, and then applying the mapping to each element in the original array. To create the mapping, we use dictionary comprehension, an elegant feature of Python that allows us to quickly construct dictionaries with logical structures. We loop through the enumerate() function, which takes in an array and returns a list of elements "bundled" with their respective indices (e.g., enumerate(['a','b','c']) yields [('a', 0), ('b', 1), ('c', 2)]). This can be unpacked and reorganized into a dictionary to create the desired mapping.

Listing 2-14. Label encoding "from scratch"

```
def label_encoding(arr):
    unique = np.unique(arr)
    mapping = {elem:i for i, elem in enumerate(unique)}
    return np.array([mapping[elem] for elem in arr])
```

Applying from-scratch implementation of label encoding to the LotShape feature of the dataset yields a satisfactory result (Listing 2-15).

Listing 2-15. Applying label encoding

```
lot_shape = np.array(df['LotShape'])
# array(['Reg', 'Reg', 'IR1', ..., 'Reg', 'Reg', 'Reg'])

encoded = label_encoding(lot_shape)
# array([3, 3, 0, ..., 3, 3, 3])
```

We can then update/replace the original DataFrame with the encoded feature with df['LotShape'] = encoded.

The sklearn library offers an implementation of label encoding (Listing 2-16). This is more efficient and convenient than implementing encoding methods by yourself and generally suffices unless your problem necessitates a very specialized encoding method.

In scikit-learn, the sklearn.preprocessing.LabelEncoder() object is used to store information about the encoding procedure. After the object is initialized, it can be fitted on a provided feature with .fit(feature) and then used to transform any given input with .transform(feature); alternatively, the two steps can be merged into one command with .fit_transform(feature).

Listing 2-16. Applying label encoding using sklearn

```
from sklearn.preprocessing import LabelEncoder
encoder = LabelEncoder()
encoded = encoder.fit_transform(df['LotShape'])
```

This object-oriented design is helpful because it allows for additional accessible functionality, like inverse encoding, in which we would like to obtain the original input from the encoded representation: classes = encoder.inverse_transform(encoded). This is helpful when we have a prediction in encoded form (e.g., "2") and want to interpret it by "inverse transforming/encoding" it (e.g., "cat" was encoded as "2," and thus the result of inverse transforming on "2" yields "cat").

A more lightweight function that lacks this additional inversion functionality is pandas. factorize(feature).

One-Hot Encoding

In cases of categorical variables in which no definitive quantitative label can be attached, the simplest satisfactory choice is generally *one-hot encoding*. If there are n unique classes, we create n binary columns, each representing whether the item belongs to that class or not (Figure 2-7). Thus, there will be one "1" across each of the n columns for every row (and all others as "0").

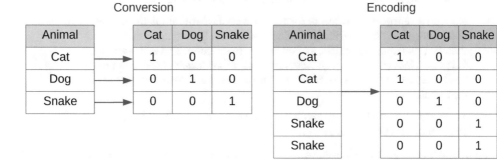

Figure 2-7. *One-hot encoding*

We can think of one-hot encoding as first applying label encoding to obtain some integer label i for each item and then generating a matrix where the ith index of a vector associated with each item is marked as "1" (and all others as "0") (Listing 2-17). This can be implemented by initializing an array of zeros with shape (number of items, number of unique classes). For each item, we index the appropriate value and set it to 1. After processing all items, we return the encoded matrix.

Listing 2-17. One-hot encoding "from scratch"

```
def one_hot_encoding(arr):
    labels = label_encoding(arr)
    encoded = np.zeros((len(arr), len(np.unique(arr))))
    for i in range(len(arr)):
        encoded[i][labels[i]] = 1
    return encoded
```

The encoded result allows us to represent the same information as is present in the original raw column without arbitrarily communicating influential quantitative assumptions to any model or method that may process it:

```
[[0. 0. 0. 1.]
 [0. 0. 0. 1.]
 [1. 0. 0. 0.]
 ...
 [0. 0. 0. 1.]
 [0. 0. 0. 1.]
 [0. 0. 0. 1.]]
```

Because one-hot encoding is so commonly used, a wide variety of libraries offer implementations:

- `pandas.get_dummies(feature)` takes in an array representing a categorical feature and automatically returns the corresponding one-hot encoding. The "dummies" in the function name refer to the dummy variables/features in the encoded result, which are "artificially created" props – in a certain sense – that allow us to faithfully capture data in quantitative form. The Keras deep learning library (install with `pip install keras`) also offers a similar lightweight function: `keras.to_categorical(feature)`. While being quick to use, the disadvantage is that more advanced functionality that may be useful like decoding is not accessible.

- `sklearn.preprocessing.OneHotEncoder()` is scikit-learn's implementation of one-hot encoding. Like the Label Encoder, it must be initialized and can then be fitted upon input data, and encode it in one command with `encoder.fit_transform(features)`. `encoder.inverse_transform(encoded)` can be used to translate one-hot encoded data into its original categorical single-feature representation.

- TensorFlow/Keras also offers one-hot encoding for text. See the subsection "Text Data" of this section to learn more.

One problem that may arise from using one-hot encoding, however, is *multicollinearity*. Multicollinearity occurs when several features are highly correlated such that one can reliably be predicted as a linear relationship of the others. In a one-hot encoding, the sum of feature values across each row is always 1; if we know the values of all other features for some row, we also know the value of the remaining feature.

This can become problematic because each feature is no longer independent, whereas many machine learning algorithms like K-Nearest Neighbors (KNN) and regression assume that each dimension of the dataset is not correlated with any other. While multicollinearity may only have a marginal negative effect on model performance (this can often be addressed with regularization and feature selection – see the "Feature Selection" section of this chapter), the larger problem is the effect on parameter interpretation. If two independent variables in a Linear Regression model are highly correlated, their resulting parameters after training become almost meaningless because the model could have performed just as well with another set of parameters (e.g., switching the two parameters, ambiguous multiplicity of solutions). Highly correlated features act as approximate duplicates, which means that the corresponding coefficients are halved as well.

One simple method to address multicollinearity in one-hot encoding is to randomly drop one (or several) of the features in the encoded feature set. This has the effect of disrupting a uniform sum of 1 across each row while still retaining a unique combination of values for each item (one of the categories will be defined by all zeros, since the feature that would have been marked as "1" was dropped). The disadvantage is that the equality of representation across classes of the encoding is now unbalanced, which may disrupt certain machine learning models – especially ones that utilize regularization. The take-away for best performance: choose either regularization + feature selection or column dropping, but not both.

Column dropping can be implemented in sklearn by passing drop='first' into initializing the OneHotEncoder object. Alternatively, you can index a one-hot encoded array with shape (number_items, number_unique_classes) with encoded[:, 1:]; this removes the first one-hot encoded column.

Another problem is *sparsity*: there is a very low information-to-space ratio. A new column is created for every unique value. Some machine learning models may struggle to learn on such sparse data.

In modern deep learning, one-hot encoding is a widely accepted standard. (Many libraries, like Keras/ TensorFlow – which we will use for neural network modeling – accept label-encoded inputs for the sake of class differentiation and perform one-hot encoding behind the scenes for you.) Neural networks used today are generally deep and powerful enough to effectively work with one-hot encoding. However, in many cases – especially with traditional machine learning models and shallow neural networks – alternative categorical encoding techniques may perform better.

Binary Encoding

Two weaknesses of one-hot encoding – sparsity and multicollinearity – can be addressed, or at least improved, with *binary encoding*. The categorical feature is label encoded (i.e., each unique category is associated with an integer); the labels are converted to binary form and transferred to a set of features in which each column is one place value (Figure 2-8). That is, a column is created for each digit place of the binary representation.

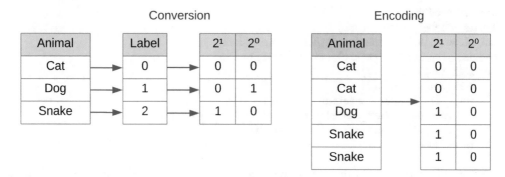

Figure 2-8. *Binary encoding*

Because we use binary representations rather than allocating one column per unique class, we more compactly represent the same information (i.e., the same class has the same combination of "1"s and "0"s across the features) at the cost of decreased interpretability (i.e., it's not clear what each column represents). Moreover, there is no reliable multicollinearity between each of the features used to represent the categorical information.

We can implement binary encoding from scratch by first obtaining the label encodings for a given array and then converting them into binary (Listing 2-18). Here's the pseudocode, given step-by-step:

1. Obtain integer label encodings for the array.

2. Find the "maximum place" of the binary representation – this is the maximum number of places we need to allocate to store a binary representation of the label. For instance, the number "9" is 1001 ($1 \cdot 2^3 + 0 \cdot 2^2 + 0 \cdot 2^1 + 1 \cdot 2^0$); this means that we need four places to represent it. We need to find the maximum place to determine the size of the encoded matrix: (number of elements, maximum place). This can be calculated as $\lfloor v \rfloor$, where v is a list of label-encoded integers.

Taking the floor of the base-2 logarithm of v returns the largest power to which 2 can be raised to yield a value less than v. We add 1 to account for the additional place taken up by $c \cdot 2^0$.

3. Allocate an array of zeros with shape (number of elements, maximum place).

4. For each item in the array

   a. Obtain the associated label encoding.

   b. For each place beginning from the maximum place to 0 (counting backward)

      i. Perform integer division between the current label-encoded value and 2 to the current place. This will either be 0 or 1, indicating whether the current label-encoded value is larger or smaller than 2 to the current place. This value will be marked in the associated encoding matrix.

      ii. Set the current-label encoded value to the remainder yielded by dividing by 2 to the current place. This allows us to "remove" the current place and focus on the "next" place.

Listing 2-18. Binary encoding "from scratch"

```
def binary_encoding(arr):
    labels = label_encoding(arr)
    max_place = int(np.floor(np.math.log(np.max(labels), 2)))
    encoded = np.zeros((len(arr), max_place+1))
    for i in range(len(arr)):
        curr_val = labels[i]
        for curr_place in range(max_place, -1, -1):
            encoded[i][curr_place] = curr_val // (2 ** curr_place)
            curr_val = curr_val % (2 ** curr_place)
    return encoded
```

The result of binary-encoding the sample array `['a', 'b', 'c', 'c', 'd', 'd', 'd', 'e']` is as follows:

```
array([[0., 0., 0.],
       [1., 0., 0.],
       [0., 1., 0.],
       [0., 1., 0.],
       [1., 1., 0.],
       [1., 1., 0.],
       [1., 1., 0.],
       [0., 0., 1.]])
```

Binary encoding is supported by `category_encoders` (Listing 2-19, Figure 2-9).

Listing 2-19. Binary encoding using category_encoders

```
from category_encoders.binary import BinaryEncoder
encoder = BinaryEncoder()
encoded = encoder.fit_transform(df['LotShape'])
```

	LotShape_0	LotShape_1	LotShape_2
0	0	0	1
1	0	0	1
2	0	1	0
3	0	1	0
4	0	1	0
...	...	...	...

Figure 2-9. *Binary encoding on a feature from the Ames Housing dataset*

Frequency Encoding

Label encoding, one-hot encoding, and binary encoding each offer methods of encoding that reflect the "pure identity" of each unique class; that is, we devise quantitative methods to assign a unique identity to each class.

However, we can both assign unique values to each class *and* communicate additional information about each class at once. To frequency-encode a feature, we replace each categorical value with the proportion of how often that class appears in the dataset (Figure 2-10). With frequency encoding, we communicate how often that class appears in the dataset, which may be of value to whatever algorithm is processing it.

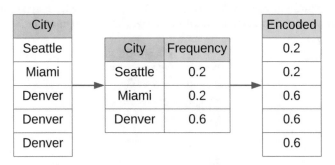

Figure 2-10. *Frequency encoding*

Like one-hot encoding and binary encoding, we attach a unique quantitative representation to each class (although frequency encoding does not guarantee a unique quantitative representation, especially in small datasets). With this encoding scheme, however, the actual value of the representation lies on a continuous scale, and quantitative relationships between the encodings are not arbitrary but instead communicate some piece of information. In the preceding example, "Denver" is "three times" Miami because it appears three times as often in the dataset.

Frequency encoding is the most powerful when the dataset is representative and free of bias. If this is not the case, the actual quantitative encodings may be meaningless in the sense of not providing relevant and truthful/representative information for the purposes of modeling.

To implement frequency encoding (Listing 2-20), we can begin by obtaining the unique labels and how often they appear in the provided feature with np.unique(arr, return_counts=True). Then, we'll scale the frequencies to be between 0 and 1, build a mapping dictionary to map each label to a frequency, create a mapping function to apply the dictionary, and push the array into a vectorized mapping function.

Listing 2-20. Frequency encoding "from scratch"

```
def frequency_encoding(arr):
    labels, counts = np.unique(arr, return_counts=True)
    counts = counts / np.sum(counts)
    mapping_dic = {labels[i]:counts[i] for i in range(len(counts))}
    mapping = lambda label:mapping_dic[label]
    return np.vectorize(mapping)(arr)
```

We can call frequency_encoding() on df['LotShape'], which yields array([0.63, 0.63, 0.33, ..., 0.63, 0.63, 0.63]) (values have been rounded to two decimal places with np.round(arr, 2)).

The category encoders library (pip install category_encoders) contains scikit-learn-style implementations for additional categorical encoding schemes like frequency encoding. The syntax will look familiar (Listing 2-21).

Listing 2-21. Frequency encoding using category_encoders

```
from category_encoders.count import CountEncoder
encoder = CountEncoder()
encoded = encoder.fit_transform(df['LotShape'])
```

The result is a DataFrame with one column (Figure 2-11). This is one advantage of using category_encoders: it is built to accept and return multiple different common data science storage objects. Thus, we can pass in a DataFrame column and receive a DataFrame column in return, rather than needing to integrate the encoded NumPy array with the original DataFrame afterward. If you desire a NumPy array output, set return_df = False in the initialization of the encoder object (this is True by default).

	LotShape
0	925
1	925
2	484
3	484
4	484
...	...

Figure 2-11. Frequency encoding on a feature from the Ames Housing dataset

Note that the category_encoders implementation of frequency encoding does not scale the frequencies between 0 and 1, but rather associates each class with the raw count (number of times it appears in the dataset). You can force scaling by passing normalize = True into the CountEncoder() initialization.

You can also pass an entire DataFrame into `encoder.fit_transform` or `encoder.fit`. If you do not desire for all categorical columns to automatically be encoded via frequency encoding, specify which columns to encode via `cols = ['col_name_1', 'col_name_2', ...]`.

Target Encoding

Frequency encoding is often unsatisfactory because it often doesn't directly reflect information in a class that is directly relevant to a model that uses it. Target encoding is an attempt to model the relationship more directly between the categorical class x and the dependent variable y to be predicted by replacing each class with the mean or median (respectively) value of y for that class. It is assumed that the target class is already in quantitative form, although the target does not necessarily need to be continuous (i.e. a regression problem) to be used in target encoding. For instance, taking the mean of binary classification labels, which are either 0 or 1, gives insight into the proportion of items in the dataset with that class that were associated with class 0 (Figures 2-12 and 2-13). This can be interpreted as the probability the item belongs to a target class given only one independent feature.

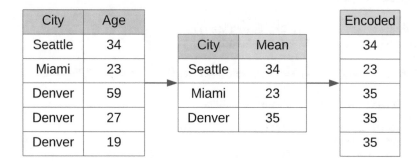

Figure 2-12. *Target encoding using the mean*

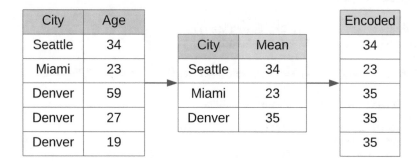

Figure 2-13. *Target encoding using the median*

Note that target encoding can lead to data leakage if encoding is performed before the training and validation sets are split, since the averaging function incorporates information from both the training and validation sets. In order to prevent this, encode the training and validation sets separately after splitting. If the validation dataset is too small, target-encoding the set independently may yield skewed, unrepresentative encodings. In this case, you can use averages per class from the training dataset. This form of "data leakage" is not inherently problematic, since we are using training data to inform operation on the validation set rather than using validation data to inform operation on the training set.

To implement target encoding (Listing 2-22), we need to accept both the categorical feature (we'll call this x) and the target feature (we'll call this y). We'll create a mapping dictionary that maps each unique categorical value to the mean or median of all values in y that corresponds with that categorical value. Then, we will define a function that uses the mapping to return the encoding given the class and apply a vectorized version of it to the given feature x.

Listing 2-22. Target encoding "from scratch"

```
def target_encoding(x, y, mode='mean'):
    labels = np.unique(x)
    func = np.mean if mode=='mean' else np.median
    mapping_dic = {label:func(y[x==label]) for label in labels}
    mapping = lambda label:mapping_dic[label]
    return np.vectorize(mapping)(x)

target_encoding(df['Lot Shape'], df['SalePrice'])
# returns array([164754.82, 164754.82, 206101.67, ...])
```

The `category_encoders` library also supports target encoding (Listing 2-23, Figure 2-14).

Listing 2-23. Target encoding using category_encoders

```
from category_encoders.target_encoder import TargetEncoder
encoder = TargetEncoder()
encoded = encoder.fit_transform(df['LotShape'], df['SalePrice'])
```

	LotShape
0	164754.818378
1	164754.818378
2	206101.665289
3	206101.665289
4	206101.665289
...	...

Figure 2-14. Result of using target encoding with category_encoders

Note that this implementation offers more advanced functionality, including support for multi-categorical targets. The feature to be encoded is replaced with a combination of the *posterior probability* of the target given a certain label and the *prior probability* of the target over all the data. Think of it as a generalization to the idea of representing the relationship between a categorical value and a continuous target by replacing each categorical value with the "expected value" for that label.

Like all `category_encoders` encoding objects, you can choose which columns you want to encode if providing a DataFrame with `cols = [...]` and determine the return type with `return_df = True/False`. This particular encoder offers two additional specifications: `min_samples_leaf`, which is the minimum number of data samples required to consider the average of the category (set this high to ideally remove the negative effects of poorly represented data points), and smoothing, which is a floating-point value larger than zero that controls the balance between the categorical average and the prior for categorical targets. A higher value more strongly regularizes the balance. Generally, there is no need to adjust this.

Leave-One-Out Encoding

Mean-based target encoding can be quite powerful, but it suffers from the presence of outliers. If outliers are present that skew the mean, their effects are imprinted across the entire dataset. Leave-one-out encoding is a variation on the target encoding scheme by leaving the "current" item/row out of consideration when calculating the mean for all items of that class (Figure 2-15). Like target encoding, encoding should be performed separately on training and validation sets to prevent data leakage.

City	Age	Calculation	Encoded
Seattle	34	(23 + 36) / 2	29.5
Seattle	23	(34 + 36) / 2	35
Seattle	36	(34 + 23) / 2	28.5
Denver	59	(27 + 19) / 2	23
Denver	27	(59 + 19) / 2	39
Denver	19	(59 + 27) / 2	43

Figure 2-15. *Leave-one-out-encoding using the mean*

Leave-one-out encoding can be interpreted as an extreme case of the k-fold data splitting scheme, in which k is equal to the dataset length and thus the "model" (in this case a simple averaging function) is applied to all relevant items except one (Listing 2-24).

Listing 2-24. Leave-one-out encoding "from scratch"

```
def leave_one_out_encoding(x, y, mode='mean'):
    labels = np.unique(x)
    func = np.mean if mode=='mean' else np.median
    encoded = []
    for i in range(len(x)):
        leftout = y[np.arange(len(y)) != i]
        encoded.append(np.mean(leftout[x == x[i]]))
    return np.array(encoded)
```

Leave-one-out encoding is supported by `category_encoders` (Listing 2-25, Figure 2-16).

Listing 2-25. Leave-one-out encoding using category_encoders

```
from category_encoders.leave_one_out import LeaveOneOutEncoder
encoder = LeaveOneOutEncoder ()
encoded = encoder.fit_transform(df['LotShape'], df['SalePrice'])
```

	LotShape
0	164707.475108
1	164736.695887
2	206065.643892
3	206238.521739
4	206010.778468
...	...

Figure 2-16. *Leave-one-out encoding on a feature from the Ames Housing dataset*

James-Stein Encoding

Target encoding and leave-one-out encoding assume that each categorical feature is directly and linearly related to the dependent variable. We can take a more sophisticated approach to encoding with James-Stein encoding by incorporating both the *overall mean for a feature* and the *individual mean per class for a feature* into an encoding (Figure 2-17). This is achieved by defining the encoding for a category as a weighted sum of the overall mean $\underline{y}$ and individual mean per class $\underline{y}_i$ via a parameter β, which is bounded by $0 \leq \beta \leq 1$:

$$\text{Encoded for category } i = \beta \cdot \underline{y} + (1 - \beta) \cdot \underline{y}_i$$

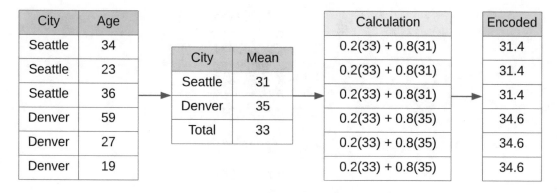

Figure 2-17. *James-Stein encoding*

When $\beta = 0$, James-Stein encoding is the same as mean-based target encoding. On the other hand, when $\beta = 1$, James-Stein encoding replaces all values in a column with the average dependent variable value, regardless of individual class values.

Charles Stein, a statistician and professor of statistics at Stanford University, proposed a formula to find β rather than manually tuning it. Group variance is the variance of the dependent variable within the relevant class, whereas population variance is the overall variance of the dependent variable without regard to class:

$$\beta = \frac{(group\ variance)}{(group\ variance) + (population\ variance)}$$

Variance is important to understanding how relevant the mean is as a representative of the class, or *uncertainty*. Ideally, we would like to set a high β when we are unsure about how representative the mean is for a certain class. This allows the mean of the overall dependent variable to be given higher weight over the mean for the specific class. Likewise, we desire a low β when there is low uncertainty associated with the mean for a certain class. This is represented quantitatively by Stein's formula: if the group variance is significantly lower than the population variance (i.e., the values for a specific class vary less, and thus we are more "sure"/"certain" that the mean for that class is representative), β is close to 0; it is closer to 1 if otherwise.

James-Stein encoding is implemented in the `category_encoders` library (Listing 2-26, Figure 2-18).

Listing 2-26. James-Stein encoding using category_encoders

```
from category_encoders import JamesSteinEncoder
encoder = JamesSteinEncoder()
encoded = encoder.fit_transform(df['LotShape'], df['SalePrice'])
```

117

	LotShape
0	167097.685845
1	167097.685845
2	201579.625847
3	201579.625847
4	201579.625847
...	...

Figure 2-18. James-Stein encoding on a feature from the Ames Housing dataset

Weight of Evidence

The weight of evidence (WoE) technique originated from credit scoring; it was used to measure how "separable" good customers (paid back a loan) were from bad customers (defaulted on a loan) across a group i (this could be something like customer location, history, etc.):

$$WoE \text{ for group } i = \ln \ln \frac{\%good \ customers \ in \ group \ i}{\%bad \ customers \ in \ group \ i}$$

For instance, let's say that within the subset of customers who brushed their teeth at least three times a day (we won't consider how the loaner got this information), 75% paid back the loan (good customers) and 25% defaulted (bad customers). Then, the weight of evidence for the group with respect to the feature/attribute "brushed teeth at least three times a day" is $\ln \ln \frac{75}{25} = \ln \ln 3 \approx 1.09$. This is a moderately high weight of evidence and means that the suggested feature/attribute separates good and bad customers well.

This concept can be used as an encoding for categorical features with a binary-categorical dependent variable:

$$WoE \text{ for class value } i = \ln \ln \frac{\%target = 0 \ and \ feature \ i}{\%target = 1 \ and \ feature \ i}$$

Weight of evidence is often presented as representing how much the *evidence* undermines or supports the *hypothesis*. In the context of categorical encoding, the "hypothesis" is that the selected categorical feature can cleanly divide classes such that we can reliably predict which class an item falls in given only information about its inclusion or exclusion from the group i. The "evidence" is the actual distribution of target values within a certain group i.

We can also generalize this to multiclass problems by finding the WoE for each class, in which "class 0" is "in class" and "class 1" is "not in class"; the weight of evidence of the complete dataset can then be found by somehow aggregating the individual class-specific WoE calculation, for example, by taking the mean.

Using this multiclass WoE logic, we can also apply WoE to continuous target/regression problems by discretizing the target into n categorical buckets. This converts the continuous target into a categorical one. The buckets can be constructed either by equally long ranges (i.e., from $a \leq x < a + b \rightarrow class\ 1$, $a + b \leq x < a + 2b \rightarrow class\ 2$, $a + 2b \leq x < a + 3b \rightarrow class\ 3$, etc.) or by equally sized bins (i.e., 0th–10th percentiles in class 1, 10th–20th percentiles in class 2, etc.).

The weight of evidence encoder is supported by `category_encoders`; however, the target variable must be binary-categorical. If you wish to apply weight of evidence to multiclass or continuous target variable contexts, you must perform preprocessing required by yourself.

Weight of evidence generally doesn't work well when performed after training-validation split for obvious reasons – we can't calculate an accurate weight of evidence for a group without a very sizable number and representative sample of members in that group. For this reason, it is best to perform weight of evidence encoding on the dataset altogether and separate later. To prevent target leakage, the `category_encoders` implementation introduces additional regularization schemes (Listing 2-27, Figure 2-19).

Listing 2-27. Weight of evidence encoding using category_encoders. We use the "Street" column because it is binary-categorical and use mapping to integer-encode it, as required by the function

```
from category_encoders.woe import WOEEncoder
encoder = WOEEncoder()
y = df['Street'].map({'Pave':0, 'Grvl':1})
encoded = encoder.fit_transform(df['LotShape'], y)
```

	LotShape
0	-0.013101
1	-0.013101
2	-0.284931
3	-0.284931
4	-0.284931
...	...

Figure 2-19. *Weight of evidence encoding on a feature from the Ames Housing dataset*

Continuous Data

Continuous quantitative data is often already in a state that is technically readable, but many algorithms assume a certain shape or distribution that we need to satisfy as well.

Min-Max Scaling

Min-max scaling generally refers to the scaling of the range of a dataset such that it is between 0 and 1 – the minimum value of the dataset is 0, and the maximum is 1, but the relative distances between points remain the same.

The scaled version of an array x is $x_{scaled} = \dfrac{x - x}{x - x}$. The numerator, $x - x$, shifts the dataset such that the lowest value is at 0 and the highest is at $x - x$. Dividing by the highest value yields the lowest value (still) at 0 and the highest value at 1. Because scaling only shifts and stretches/shrinks the elements in the array, we do not change the relative distances between points. Since machine learning algorithms work off relative distances between points (i.e., drawing boundaries and landscapes through the feature space), we have retained the dataset's modeling information capacity. However, we do not explicitly center the data and destroy sparsity, but rather control the end points.

Min-max scaling can be implemented as `min_max(arr) = lambda arr: (arr - np.min(arr)) / (np.max(arr) - np.min(arr))` using NumPy.

sklearn also supports an implementation of min-max scaling with the `MinMaxScaling` object; you can pass in custom ranges (the default being between 0 and 1 inclusive) for specialized cases like standard image scaling, which uses the [0, 255] integer range (Listing 2-28). For distributions that are already centered at 0, it may be wiser to scale between [−1, 1] rather than [0, 1] to preserve 0-centering.

Listing 2-28. Min-max scaling using sklearn

```
from sklearn.preprocessing import MinMaxScaler
scaler = MinMaxScaler(feature_range=(lower, higher))
scaled = scaler.fit_transform(data)
orig_data = scaler.inverse_transform(scaled)
```

We can visualize the effect of min-max scaling on a random distribution with mean 5 and standard deviation 1 (Listing 2-29, Figure 2-20).

Listing 2-29. Visualizing the original vs. normalized/min-max scaled distribution

```
arr = np.random.normal(loc=5, scale=1, size=(250,))
adjusted = (arr - arr.min()) / (arr.max() - arr.min())

plt.figure(figsize=(10, 5), dpi=400)
axes = plt.gca()
axes.yaxis.grid()
sns.histplot(arr, color='red', label='Original', alpha = 0.8, binwidth=0.2)
sns.histplot(adjusted, color='blue', label='Normalized', alpha = 0.8, binwidth=0.2)
plt.legend()
plt.show()
```

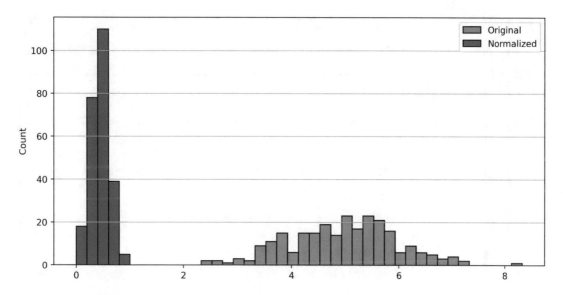

Figure 2-20. *Original distribution vs. normalized/min-max scaled distribution*

Robust Scaling

From the formula for min-max scaling, we see that each scaled value of the dataset is directly impacted by the maximum and minimum values. Hence, outliers significantly impact the scaling operation. We can demonstrate their effect by introducing five instances of the value "30" into our previous example distribution of mean 5 and standard deviation 1 (Listing 2-30, Figure 2-21).

Listing 2-30. Appending outliers to a normal distribution

```
arr = np.random.normal(loc=5, scale=1, size=(250,))
arr = np.append(arr, [30]*5)
```

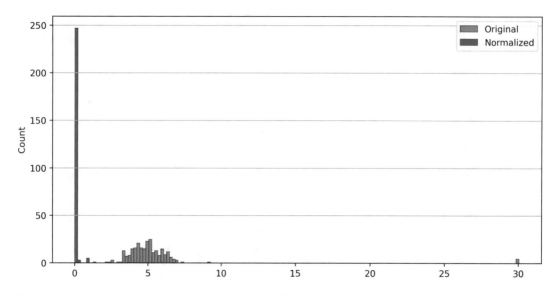

Figure 2-21. *Demonstrating the deleterious effect of outliers on min-max scaling*

Zooming in to further understand the effect of adding a few outlier values, we find that the entire remainder of the distribution is squeezed toward one end just to fit the entire dataset within the set [0, 1] range (Figure 2-22).

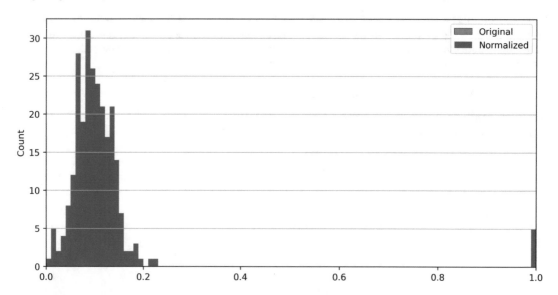

Figure 2-22. *Demonstrating the deleterious effect of outliers on min-max scaling, zoomed in*

Sometimes, however, we don't need values to strictly be between 0 and 1. We just want the dataset to be within a specific local range; to make our scaling method less vulnerable to outliers, we can manipulate the dataset using elements more "within" the dataset rather than on the "outside" with the first, second, and third quartiles:

$$x_{robust\ scaled} = \frac{x - median\ x}{3rd\ quartile\ x - 1st\ quartile\ x}$$

Robust scaling subtracts the median value from all values in the dataset and divides by the interquartile range (Figure 2-23).

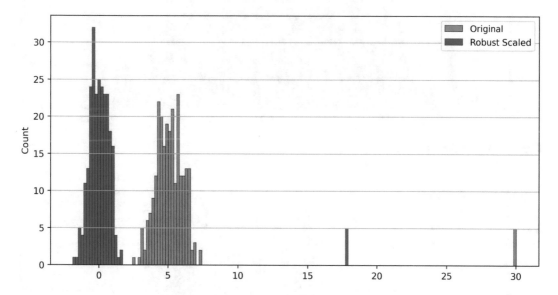

Figure 2-23. *Original distribution vs. robust-scaled distribution*

Zooming in to the primary "bulk" or "mass" of the distribution, we see that even while robust scaling roughly centers the distribution at 0, the values are spread relatively evenly without being affected by the presence of outliers. While the outlier still remains an outlier relative to the primary "mass" of the distribution, the scaled distribution is no longer held dependent on a few outliers (Figures 2-24 and 2-25).

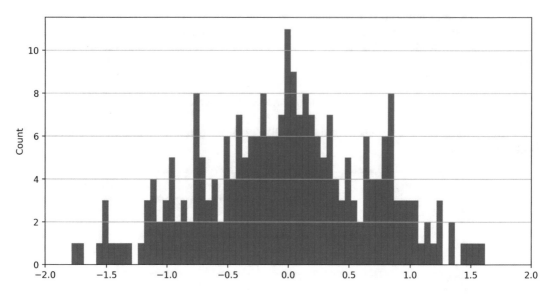

Figure 2-24. *Original distribution*

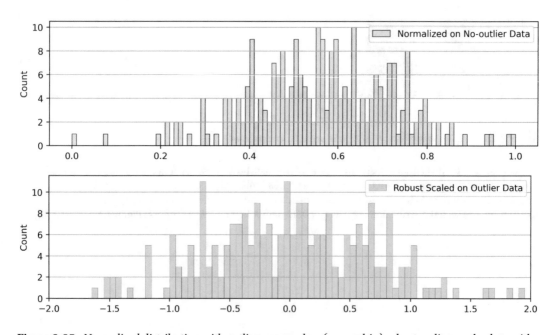

Figure 2-25. *Normalized distribution with outliers removed vs. (zoomed-in) robust scaling on the data with outliers*

Robust scaling can be implemented as robust(arr) = lambda arr: (arr - np.median(arr)) / (np. quantile(arr, 3/4) - np.quantile(arr, 1/4)) using NumPy.

sklearn supports robust scaling with the RobustScaler object (Listing 2-31).

Listing 2-31. Implementing robust scaling using sklearn

```
from sklearn.preprocessing import RobustScaler
scaler = RobustScaler(feature_range=(lower, higher))
scaled = scaler.fit_transform(data)
orig_data = scaler.inverse_transform(scaled)
```

Standardization

More commonly, machine learning algorithms assume that data is *standardized* – that is, in the form of a normal distribution with unit variance (standard deviation of 1) and zero mean (centered at 0). Assuming the input data is already somewhat normally distributed, standardization subtracts the dataset's mean and divides by the dataset's standard deviation. This has the effect of shifting the dataset mean to 0 and scaling the standard deviation to 1 (Figure 2-26).

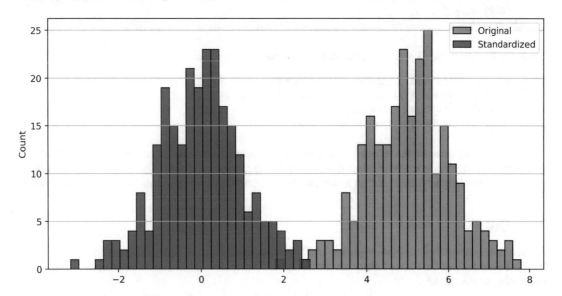

Figure 2-26. *Original distribution vs. standardized distribution*

First, let us prove that subtracting the mean $\mu = \sum_{i=1}^{n} \frac{x_i}{n}$ from a dataset yields a new mean of 0:

$$\mu_{scaled} = \sum_{i=1}^{n} \frac{x_i - \mu}{n} = \sum_{i=1}^{n} \frac{x_i}{n} - \sum_{i=1}^{n} \frac{\mu}{n} = \sum_{i=1}^{n} \frac{x_i}{n} - \mu = \mu - \mu = 0$$

The standard deviation is defined as follows:

$$\sigma = \sqrt{\frac{\sum_{i=1}^{n} (x_i - \mu)^2}{n}}$$

However, when we divide the dataset by the standard deviation, the mean has already been shifted to 0 by subtracting the mean in the previous step. The formula can thus be rewritten as follows, assuming $x_i = x_i - \mu$ (i.e., an already-shifted quantity):

$$\sigma = \sqrt{\frac{\sum_{i=1}^{n} x_i^2}{n}}$$

When we divide each mean-shifted element x_i by the standard deviation σ, the resulting standard deviation (and variance) is 1:

$$\sigma_{scaled} = \sqrt{\frac{\sum_{i=1}^{n}\left(\frac{x_i}{\sigma}\right)^2}{n}} = \sqrt{\frac{1}{\sigma^2} \cdot \frac{\sum_{i=1}^{n} x_i^2}{n}} = \frac{1}{\sigma} \cdot \sqrt{\frac{\sum_{i=1}^{n} x_i^2}{n}} = \frac{1}{\sigma} \cdot \sigma = 1$$

Standardization can be implemented as `standardize(arr) = lambda arr: (arr - np.mean(arr)) / np.std(arr)` using NumPy.

`sklearn` supports standardization with the `StandardScaler` object (Listing 2-32).

Listing 2-32. Implementing standard standardization with the sklearn Standard Scaler

```
from sklearn.preprocessing import StandardScaler
scaler = StandardScaler(feature_range=(lower, higher))
scaled = scaler.fit_transform(data)
orig_data = scaler.inverse_transform(scaled)
```

Note that estimates for the standard deviation and mean can be skewed by the presence of outliers, which makes robust scaling a good alternative to standardization as well. In robust scaling, outlier-robust statistical alternatives are used: median instead of mean, interquartile range instead of standard deviation.

Text Data

Humans interact most naturally through text and language, so it is not surprising that non-categorical text makes up a significant part of many tabular datasets. Text can appear as a customer review, a Twitter biography, or website data.

■ **Note** "Non-categorical text" refers to text that cannot be encoded in quantitative form via a categorical encoding method because text samples are too different from one another (i.e., there are too many unique classes or too few samples per class).

We would like to incorporate information from text data into our models. In future chapters, we will demonstrate how to build advanced multimodal multi-head models that simultaneously consider text inputs along with other forms of data. This section, however, explores various quantitative representation (*vectorization*) methods and how they can be used with classical machine learning models introduced in Chapter 1, the "Algorithms" section.

In this section, we'll work with the well-known SMS Spam Collection Dataset by the University of California Irvine Machine Learning Repository (UCIMLR). It is available on Kaggle at www.kaggle.com/uciml/sms-spam-collection-dataset or on the UCIMLR website at https://archive.ics.uci.edu/ml/datasets/SMS+Spam+Collection.

After elementary cleaning, the dataset should be stored in a Pandas DataFrame called `data` with two columns, `isSpam` and `text` (Figure 2-27).

	isSpam	text
0	0	Go until jurong point, crazy.. Available only ...
1	0	Ok lar... Joking wif u oni...
2	1	Free entry in 2 a wkly comp to win FA Cup fina...
3	0	U dun say so early hor... U c already then say...
4	0	Nah I don't think he goes to usf, he lives aro...
...	...	...

Figure 2-27. *Visualization of part of the SMS Spam Collection Dataset*

Since we'll be evaluating the result of models trained on different text encoding schemes, we'll also need to perform train-validation splitting (Listing 2-33).

Listing 2-33. Train/validation splitting of the spam dataset

```
from sklearn.model_selection import train_test_split as tts
X_train, X_val, y_train, y_val = tts(data['text'], data['isSpam'],
                                     train_size = 0.8)
```

Note that throughout this section, we will present simplified methods for processing text. While one should generally use more thorough approaches like intensive cleaning, lemmatization, stop word removal, etc., deep learning models are generally sophisticated enough to handle these specifics, and therefore such considerations are less necessary. You'll apply deep learning hands-on to text data in Chapters 5 (recurrent neural networks) and 6 (attention and transformers).

Keyword Search

One of the simplest ways to obtain a quantitative representation of textual data is to determine if certain predetermined keywords appear in the text or not (Listing 2-34). This is a very simple, limited, and naïve method. Under the right circumstances, however, it can suffice and serve as a good benchmark model.

Some keywords that often appear in spam messages attempting to advertise a product are "buy," "free," and "win." We can build a simple model that checks if any of these keywords are in a given text sample. If it is present, the text is designated as spam; otherwise, it is designated as ham (safe).

Listing 2-34. A sample simple keyword search function

```
def predict(text):
    keywords = ['buy', 'free', 'win']
    for keyword in keywords:
        if keyword in text.lower():
            return 1
    return 0
```

We can evaluate the accuracy of such a model to understand its performance (Listing 2-35).

Listing 2-35. Evaluating the accuracy of our simple keyword search model

```
from sklearn.metrics import accuracy_score
accuracy_score(data['isSpam'], data['text'].apply(predict))
```

The resulting accuracy is about 88.1%, which may seem pretty good. However, a model that returns "0" all the time (predict = lambda x:0) obtains an accuracy of about 86.6%. Because of the dataset imbalance, a more reflective metric is the F1 score (Listing 2-36).

Listing 2-36. Evaluating the F1 accuracy of our simple keyword search model

```
from sklearn.metrics import f1_score
f1_score(data['isSpam'], data['text'].apply(predict))
```

The F1 score of the keyword search model is about 0.437, whereas a model that predicts the label 0 for any instance obtains a score of 0.0 (as expected).

Raw Vectorization

Raw vectorization can be thought of as "one-hot encoding" for text: it is an explicit quantitative representation of the information contained within text (Figure 2-28). Rather than assigning each text a unique class, texts are generally vectorized as a sequence of *language units*, like characters or words. These are also referred to as *tokens*. Each of these words or characters is considered to be a unique class, which can be one-hot encoded. Then, a passage of text is a sequence of one-hot encodings.

		"the"	"dog"	"jumped"	"over"	"second"
	"the"	1	0	0	0	0
	"dog"	0	1	0	0	0
	"jumped"	0	0	1	0	0
"the dog jumped over the second dog"	"over"	0	0	0	1	0
	"the"	1	0	0	0	0
	"second"	0	0	0	0	1
	"dog"	0	1	0	0	0

Figure 2-28. Raw vectorization

Consider the following small example text dataset, with punctuation stripped and lowercase capitalization (Listing 2-37).

Listing 2-37. Collecting an array of text samples

```
texts = np.array(['the dog jumped over the second dog',
                  'a dog is a dog and nothing else',
                  'a dog is an animal'])
```

One proposed implementation (Listing 2-38) is as follows: we loop through each item and individually one-hot encode a list representation using the one_hot_encoding function defined in the "Discrete Data" section. Then, we apply the raw_vectorize function to each text in the texts array and convert the NumPy array into a list. Using list comprehension, we aggregate the one-hot encodings into a nested list, which can be recast as a NumPy array.

Listing 2-38. Attempting to apply raw vectorization to the text samples

```
def raw_vectorize(text):
    return one_hot_encoding(text.split(' '))

raw_vectorized = np.array([raw_vectorize(text).tolist() for text in texts])
```

When running this code, we receive a VisibleDepreciationWarning, which already should set off alarm bells:

```
/opt/conda/lib/python3.7/site-packages/ipykernel_launcher.py:11: VisibleDeprecationWarning:
Creating an ndarray from ragged nested sequences (which is a list-or-tuple of lists-or-
tuples-or ndarrays with different lengths or shapes) is deprecated. If you meant to do
this, you must specify 'dtype=object' when creating the ndarray # This is added back by
InteractiveShellApp.init_path()
```

This indicates that each element in the array is not the same length; the standard n-d array supports only elements with the same shapes. Thus, NumPy stores the data rather awkwardly as an array of list objects to accommodate for different element sizes.

Investigating further, we find that the first element is represented using five binary features: len(raw_vectorized[0][0]) returns 5. Here, raw_vectorized[0] is the encoding for the first element of texts, and raw_vectorized[0][0] is the encoding for the first token of the first element. The second element is represented using six binary features and the third using five again! It is very problematic if our vector representation does not use the same number of features to represent each sample.

■ **Note**   While it is problematic that this approach represents each sample using a different number of binary features, it is *not an issue* that each sample has a different shape. This is because different text samples inherently have a different number of tokens; while we want to make sure that we are representing tokens the same way across all samples, it is alright that the number of tokens differs. We will discuss models that can handle variable-sized sequence inputs in later chapters (recurrent models) and techniques to address variability in vectorized sample length like *padding*, which adds "blank" tokens to the end of short sequences.

The error here is that we are determining the mapping on a sample-by-sample basis, but each sample's vocabulary is very unlikely to represent the vocabulary of the entire texts. This not only yields a different number of binary features to represent each token (this is the symptom) but – more importantly – causes differences in how each unique token is matched to a unique binary column. In one text sample, the token "dog" may be indicated with a "1" in the third column of a one-hot encoding matrix in one element, but marked in the second column for a different element. This inconsistency in what columns represent completely destroys the broader dataset's informational value.

To address this, we first pool all the texts together to obtain the global vocabulary and create a universal mapping dictionary that applies to all elements in a consistent fashion (Listing 2-39).

Listing 2-39. Obtaining a mapping between each token and an integer

```
complete_text = ' '.join(texts)
unique = np.unique(complete_text.split(' '))
mapping = {i:token for i, token in enumerate(unique)}
```

This method of vectorization is popular in deep learning, because the complexity and power of neural networks are capable of handling and making sense of these very high-dimensional text representations. However, classical machine learning algorithms like the ones introduced in Chapter 1 struggle to produce good results (unless in rare circumstances in which the text samples are short and the vocabulary size/number of unique language units is small).

TensorFlow/Keras offers an implementation of one-hot encoding using the Tokenizer object (Listing 2-40). The Tokenizer automatically removes stop words (semantically "meaningless" words like "a," "an," "the," etc. that contribute more to grammar/convention than content) and performs other preprocessing for you. In this implementation, each token is associated with an integer rather than a one-hot representation of that integer (i.e., an array of 0s with one element marked as a 1). While you can explicitly one-hot encode this representation if desired without too much code, deep learning libraries that you would use to build models that can process raw-vectorized texts generally can perform this conversion automatically, and thus an ordinal representation suffices.

Listing 2-40. Using TensorFlow/Keras's text processing facilities to automatically perform raw label encoding

```
from tf.keras.preprocessing.text import Tokenizer
tk = Tokenizer(num_words=10)
tk.fit_on_texts(texts)
tk.texts_to_sequence(texts)

"'
Returns:
 [[3, 1, 5, 6, 3, 7, 1],
  [2, 1, 4, 2, 1, 8, 9],
  [2, 1, 4]
"'
```

Bag of Words

In order to reduce the sheer dimensionality/size of a raw vectorization text representation, we can use the Bag of Words (BoW) model to "collapse" raw vectorizations. In Bag of Words, we count how many times each language unit appears in a text sample while ignoring the specific order and context in which the language units were used (Figure 2-29).

Unique Tokens

	"the"	"dog"	"jumped"	"over"	"second"
"the"	1	0	0	0	0
"dog"	0	1	0	0	0
"jumped"	0	0	1	0	0
"over"	0	0	0	1	0
"the"	1	0	0	0	0
"second"	0	0	0	0	1
"dog"	0	1	0	0	0
Bag of Words	2	2	1	1	1

"the dog jumped over the second dog"

Figure 2-29. Bag of Words model as a sum of raw vectorization across the sequential axis

One could implement the Bag of Words model by calling np.sum(one_hot, axis=0), assuming the raw vectorization has already been calculated and stored in a variable called one_hot. This takes the sum of token occurrences across the tokens. However, many better options exist.

sklearn implements an easy-to-use CountVectorizer object that performs Bag of Words encoding, with similar syntax to other encoding methods (Listing 2-41). Like the Keras Tokenizer, you can set the maximum vocabulary size/features; if none is specified, scikit-learn will include all detected words as part of the vocabulary. Note that we call .toarray() after transforming the text data because the result of the transformation is a NumPy compressed sparse array format rather than a standard *n*-d array. Additionally, observe that the CountVectorizer() has no inverse_transform() function, unlike many other sklearn encoders, because the Bag of Words transformation is not invertible; multiple different text sequences can still be encoded as the same BoW representation.

Listing 2-41. Applying Bag of Words using sklearn

```
from sklearn.feature_extraction.text import CountVectorizer
vectorizer = CountVectorizer(max_features = n)
encoded = vectorizer.fit_transform(X_train).toarray()
```

The Keras Tokenizer also offers a Bag of Words functionality using the texts_to_matrix() function with the parameter mode = 'count' (Listing 2-42). You can also use mode = 'freq' to reflect word frequency.

Listing 2-42. Applying Bag of Words using Keras

```
from keras.preprocessing.text import Tokenizer
tk = Tokenizer(num_words=3000)
tk.fit_on_texts(X_train)
X_train_vec = tk.texts_to_matrix(X_train, mode='count')
X_val_vec = tk.texts_to_matrix(X_val, mode='count')
```

Let's demonstrate the performance of a model on raw-vectorized data with a maximum vocabulary size of 3000 tokens (there are many more unique words in the dataset). We'll begin by encoding the training and validation text samples (Listing 2-43). Note that we build the tokenizer's vocabulary on the *x*-train dataset and directly apply it to the *x*-validation dataset such that any words part of the validation text corpus that are not represented in the training text corpus are ignored. While this is not ideal (we would hope to have all words represented), we must maintain the same encoding technique when evaluating on the validation set as when fitting on the training set.

Listing 2-43. Applying Bag of Words using sklearn to the spam dataset

```
from sklearn.feature_extraction.text import CountVectorizer
vectorizer = CountVectorizer(max_features = 3000)
X_train_vec = vectorizer.fit_transform(X_train).toarray()
X_val_vec = vectorizer.transform(X_val).toarray()
```

We'll use the Random Forest model to evaluate vectorization methods (Listing 2-44); it builds upon the Decision Tree's high-precision adaptability to nonlinear spaces while reducing overfitting behavior via bagging (as discussed in Chapter 1, "Random Forest").

Listing 2-44. Training a Random Forest classifier on the raw-vectorized dataset

```
model = RandomForestClassifier()
model.fit(X_train_vec, y_train)
pred = model.predict(X_val_vec)
f1_score(pred, y_val)
```

The F1 score of a Random Forest model trained on raw-vectorized text data is about 0.914, which is a significant improvement over keyword search. Note that while this performance certainly isn't bad, neural networks can better make sense of sparse, very-high-dimensional data.

N-Grams

We can be more sophisticated than the Bag of Words model by counting the number of unique *two-word* combinations, or bigrams. This can help reveal context and multiplicity of word meaning; for instance, the Paris in the stripped (no punctuation, no capitalization) text except "paris france" is very different from the Paris in "paris hilton." We can consider each bigram to be its own term and encode it as such (Figure 2-30).

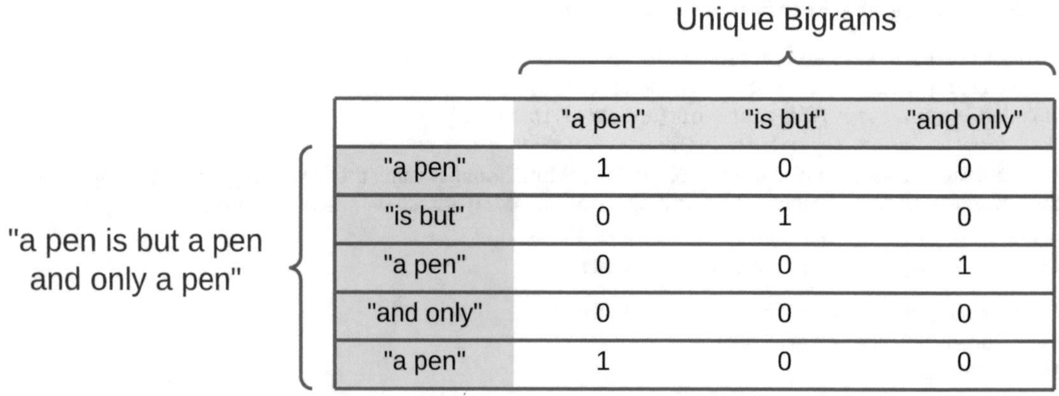

Unique Bigrams

	"a pen"	"is but"	"and only"
"a pen"	1	0	0
"is but"	0	1	0
"a pen"	0	0	1
"and only"	0	0	0
"a pen"	1	0	0

"a pen is but a pen and only a pen"

Figure 2-30. Counting bigrams. There are technically more bigrams involved here, but we're not counting them to simplify the representation

We can generalize bigrams into *n*-grams, with each unit consisting of *n* consecutive words (Listing 2-45). There is an important trade-off one encounters when increasing *n*: precision increases, but the encoding also becomes more sparse. A trigram is able to express more precise, specific ideas than a unigram, but there are fewer instances of that specific trigram than the unigram in the text. If the encoding becomes too sparse, it is difficult for a model to generalize due to the small number of samples available for a unique *n*-gram. You can observe these dynamics in Table 2-1.

Listing 2-45. Training a Random Forest classifier on n-gram data

```
model = RandomForestClassifier()
vectorizer = CountVectorizer(max_features = 3000,
                             ngram_range = (1, 2))
X_train_vec = vectorizer.fit_transform(X_train).toarray()
X_val_vec = vectorizer.transform(X_val).toarray()
```

Table 2-1. *Performance for a Random Forest model trained with n-grams for an n-gram range (row) and upper (column) range. Notice the best performance is with n-grams from lower 1 to upper 4, which is the widest range*

	1	2	3	4
1	0.914	0.903	0.890	0.898
2		0.822	0.822	0.794
3			0.658	0.601
4				0.580

TF-IDF

Another weakness of the Bag of Words model is that the number of times a word appears in the text may not be a good indicator of how important or relevant it is. For instance, the word "the" appears seven times in this paragraph, more than any other. Does this mean that the word "the" is the most significant or holds the most meaning?

No, the word "the" is primarily an artifact of grammar/syntactic structure and reflects little semantic meaning, at least in contexts we are generally concerned with. We usually address the problem of text-saturating syntactic tokens by removing so-called "stop words" from a corpus before encoding it.

However, there are many words left over from stop word screening that hold semantic value but suffer from another problem that the word "the" creates: Because of the structure of the corpus, certain words inherently appear very often throughout the text. This does not mean that they are more important. Consider a corpus of customer reviews for a jacket: naturally, the word "jacket" will appear very often (e.g., "I bought this jacket...," "This jacket arrived at my house..."), but in actuality it is not very relevant to our analysis. We know that the corpus is about the jacket and care instead about words that may occur less but mean more, like "bad" (e.g., "This jacket is bad"), "durable" (e.g., "Such a durable jacket!"), or "good" (e.g., "This was a good buy").

We can formalize this intuition by using TF-IDF, or *Term Frequency–Inverse Document Frequency*, encoding. The logic behind TF-IDF encoding is that we care more about terms that appear often in one document (Term Frequency) but not very often across the entire corpus (Inverse Document Frequency). TF-IDF is calculated by weighting these two effects against each other:

$$TFIDF = TF(t,d) \times IDF(t)$$

Term Frequency $TF(t, d)$ is the number of times a term t appears in a document/text sample d. While there are many methods for calculating Inverse Document Frequency $IDF(t)$, a simple and effective method is $log \dfrac{total \# docs}{number\ of\ docs\ with\ term\ t}$. Consider a word that appears very often in a document but is very rare in the collected corpus. The Term Frequency will be high, since the term appears often in the document; the Inverse Document Frequency will also be high, since there are a small number of documents with the term t (i.e., the denominator is small). Thus, the overall TF-IDF encoding will be high in value, indicating high significance or originality. On the other hand, a word that appears often in a document but also is abundant throughout the collected corpus has a lower TF-IDF weighting.

Scikit-learn supports TF-IDF with a similar maximum-features functionality as the Bag of Words model (Listing 2-46).

Listing 2-46. Applying TF-IDF using sklearn

```
from sklearn.feature_extraction.text import TfidfVectorizer
vectorizer = TfidfVectorizer(max_features = 3000)
X_train_vec = vectorizer.fit_transform(X_train).toarray()
X_val_vec = vectorizer.transform(X_val).toarray()
```

A Random Forest model trained with TF-IDF-vectorized data obtains an F1 score of 0.898 on the validation dataset, which turns out to be slightly worse than the Bag of Words model. This says more about the nature of the dataset than the performance of the encoding method; this spam dataset may not benefit from the introduction of an Inverse Document Frequency term that weights down terms depending on their occurrence over the total corpus.

You can also pass in an `ngrams_range` parameter to consider terms consisting of multiple words, that is, for different values of n.

Sentiment Extraction

In certain problems, it can be beneficial to use more specific text encodings. *Sentiment extraction* is the association of a quantitative label to text representing various qualities of its sentiment, like mood or objectivity. If you are building a classical machine learning model, you can encode text as its quantitative sentiment representation or include semantic extraction as another set of features to other text encoding methods. Note that sentiment extraction is generally unlikely to add much value for deep learning models, which are typically powerful enough to develop internal language understanding mechanisms that are more sophisticated and relevant than a sentiment extraction function defined and selected manually by a human.

The `textblob` library offers a simple sentiment extraction implementation. Begin by initializing a `textblob.TextBlob` object with the phrase or sentence you want to obtain the sentiment of (Listing 2-47).

Listing 2-47. Creating a TextBlob object

```
from textblob import TextBlob
text = TextBlob("Feature encoding is very good")
```

Then, call the `.sentiment` method to obtain the sentiment, which is broken down into *polarity* (negative vs. positive tone, quantified from –1 to +1) and *subjectivity* (whether the phrase is stated as an opinion or more factually, quantified from 0 to 1). For instance, calling `text.sentiment` on the given example in Listing 2-47 yields a polarity of 0.9099 and a subjectivity of 0.7800. This means the `textblob` sentiment analysis implementation associates the sentence "Feature encoding is very good" as being very positive in polarity and moderately subjective. This is a fair assessment.

You can create functions that return the polarity and subjectivity of an inputted string; these can be applied to NumPy arrays, Pandas series, TensorFlow datasets, etc. (Listing 2-48).

Listing 2-48. Extracting the polarity and subjectivity sentiment components from text

```
def get_polarity(string):
    text = TextBlob(string)
    return text.sentiment.polarity

def get_subjectivity(string):
    text = TextBlob(string)
    return text.sentiment.subjectivity
```

textblob uses semantic labels to determine the polarity of a text sample; each relevant word is associated with some inherent polarity (e.g., "bad" with –0.8 polarity, "good" with +0.8 polarity, "awesome" with +1 polarity). The overall polarity is a single aggregation of each of these individual polarities.

Negation words like "not" or "no" flip the sign of affected relevant words. The polarity of "Feature encoding is not very good" is –0.2692.

textblob also considers punctuation and emojis in addition to alphanumeric characters. For instance, "Feature encoding is very good!" yields a polarity of +1.0, over the +0.9 polarity without the exclamation mark.

If a sentence is neutral, like "Feature encoding is a technique.", textblob will yield a polarity and subjectivity near or at 0.

textblob determines subjectivity by the presence of intensity modifiers, like "very."

While textblob works fine in simple cases, the algorithm is quite naïve and is very limited. It does not consider context across multiple words and more complex syntactic and semantic language structures that affect polarity and subjectivity.

VADER (Valence Aware Dictionary for Sentiment Reasoning) is a model that performs similar functions to textblob by associating a given text sample with polarity and subjectivity (referred to as intensity or strength). The VADER model, like textblob, also computes polarity and intensity by aggregating individual sentiments. However, it is more sophisticated; it considers negation contractions (e.g., wasn't vs. was not), advanced punctuation use, capitalization (e.g., ALL CAPS vs. lowercase), slang words, and acronyms ("lmao," "lol," "brb"). VADER was optimized on social media data and thus possesses a wider and more modern vocabulary than most other sentiment analyzers.

You can import the vaderSentiment library with pip install vaderSentiment. Use it as such (Listing 2-49).

Listing 2-49. Getting the polarity of a text using VADER

```
import vaderSentiment
from vaderSentiment.vaderSentiment import SentimentIntensityAnalyzer
analyzer = SentimentIntensityAnalyzer()
sentence = "Feature encoding is very good"
scores = analyzer.polarity_scores(sentence)
```

The sentiment scores (in this example, stored in scores) are represented as a dictionary, with the keys 'neg', 'neu', 'pos', and 'compound'. Respectively, these represent the ratings from 0 to 1 for the text sample's negativity, neutrality, positivity, and overall sentiment (a compound score). The sum of the negativity, neutrality, and positivity scores equals 1. The VADER model is differentiated from other text sentiment models by marking neutrality as its own semantic class, rather than being in between negative and positive.

If you are looking for even more complex features, consider Flair, a powerful natural language processing framework (install with `pip install flair`). Flair contains an implementation for *Task-Aware Representation of Sentences for Generic Text Classification*, or TARS. TARS has an exciting property called *zero-shot learning*, meaning that it can learn to associate text with certain labels and classes that it has had zero exposure to (Figures 2-31 to 2-34).

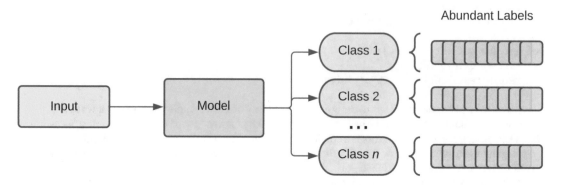

Figure 2-31. *Standard supervised learning scheme*

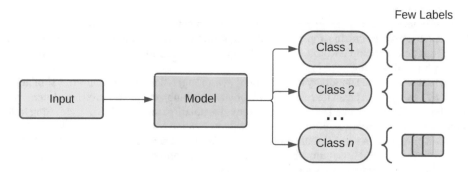

Figure 2-32. *Few-shot learning scheme*

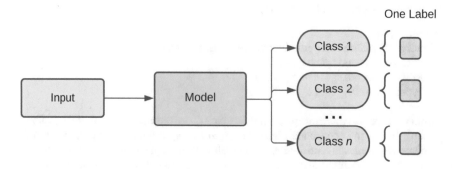

Figure 2-33. *One-shot learning scheme*

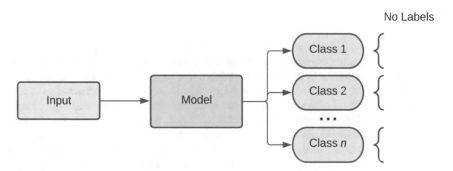

Figure 2-34. *Zero-shot learning scheme. Note that this reflects the ideal behavior of a zero-learning system rather than how it is trained*

This means that you can define your own classes and TARS will automatically assign a probability that a given text falls in any of those classes. Classes need to be defined using a natural language string describing what the class represents. The TARS model is able to "interpret" this definition and use it as a class.

Begin by creating a TARSClassifier model. You'll need to load the model weights, which can take about a minute, with variation depending on environment conditions. Then, create a flair.data.Sentence object and define a list of custom classes. Lastly, run the TARSClassifier's predict_zero_shot() function on the sentence object and the custom classes.

For instance, we can define two classes 'positive' and 'negative' to perform similar functionality to textblob and VADER (Listing 2-50).

Listing 2-50. Using TARS with Flair to obtain deep sentiment analysis extraction

```
import flair
from flair.models import TARSClassifier
from flair.data import Sentence
tars = TARSClassifier.load('tars-base')
sentence = Sentence("Feature encoding is very good")
classes = ['positive', 'negative']
tars.predict_zero_shot(sentence, classes)
```

The model prediction automatically modifies the Sentence object and associates it with sentence labels. You can view these by printing the original sentence object. In this case, Flair assigns the sentence to the class 'positive' with 0.9726 probability.

You can use Flair to define more complex custom classes. For instance, you may want to quantify whether a text sounds "anxious," "nervous," "excited," "ambivalent," "neutral," "empathetic," "pessimistic," or "optimistic." You can also use more descriptive class definitions, like "optimistic but careful." Moreover, you can evaluate the content of text beyond the sentiment; for instance, to evaluate if the text talks about animals in some manner, you can request the TARS model to assign the probability a sentence belongs to the class description 'animals'. It works effectively: for instance, "puppies are so cute" obtains a very high probability, but "plants are so cute" obtains a very low probability.

These are more complex ideas that the TARS model is able to interpret and quantify. Because TARS is a deep learning model rather than a rule-based system (which textblob and VADER use), it is generally more reflective of the text sample's character and content, making it a powerful encoder for text when you have a strong idea of which qualities of the text are relevant to prediction.

Word2Vec

Previous discussion on encoding methods focused on relatively simplistic attempts to capture a text sample's meaning by attempting to extract one "dimension" or perspective. The Bag of Words model, for instance, captures meaning simply by counting how often a word appears in the text. The Term Frequency–Inverse Document Frequency encoding method attempts to improve upon this scheme by defining a slightly more sophisticated level of meaning by balancing the occurrence of a word in a document with its occurrence in the complete corpus. In these encoding schemes, there is always one perspective or dimension of the text that we leave out and simply cannot capture.

With deep neural networks, however, we can capture more complex relationships between text samples – the nuances of word usage (e.g., "Paris," "Hilton," and "Paris Hilton" all mean very different things!), grammatical exceptions, conventions, cultural significance, etc. The Word2Vec family of algorithms associates each word with a fixed-length vector representing *latent* ("hidden", "implicit") features (Figure 2-35).

Token	F1	F2	F3	...	FN
"airplane"	0.4	2.4	9.6	...	-9.2
"car"	1.5	2.3	-0.9	...	5.6
"peace"	-0.1	-8.9	9.7	...	4.2

Figure 2-35. *Hypothetical embeddings/vector associations learned by Word2Vec*

Embeddings are learned by neural networks, which map tokens to embeddings and use the learned embeddings to perform a task. The specific task used varies, but all embedding tasks force the network to understand the internal structure of the text in some way. One commonly used task is to fill in a missing token in a sequence of tokens. For instance, "He was so <masked token> that he threw the deep learning book on the ground and stomped on it" should elicit an output token like "angry" or "upset." In order to complete the task, the network needs to learn the optimal set of latent features associated with each token. The embedding layer is then extracted from the network (Figure 2-36). The learned embeddings can be quite sophisticated, capturing grammatical, cultural, and logical relationships within language. (Read Chapter 6 for a more substantive overview of masked language modeling as a pretraining task.)

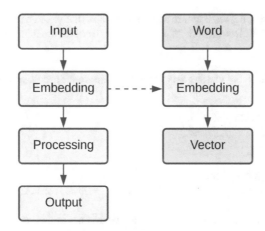

Figure 2-36. *The process of learning embeddings in neural network models and extracting learned embeddings to obtain Word2Vec representations*

A disadvantage of Word2Vec is loss of interpretability. With the Bag of Words representation, we know what each number in the vectorization represents and why it appears so. Even with methods like sentiment extraction or zero-shot classification with TARS, we understand what the vectorization is supposed to represent, even if the process of deriving it is more complex. With Word2Vec, however, we know neither exactly how the vector representations are obtained[1] nor what the numbers themselves mean.

The popular genism library offers a convenient interface to access Word2Vec. Let's begin by installing and importing it, in addition to other relevant libraries for text cleaning and retrieval (Listing 2-51).

Listing 2-51. Installing and importing relevant libraries

```
import gensim
!pip install clean-text
from cleantext import clean
import urllib.request
```

We'll train Word2Vec on *War and Peace* by Leo Tolstoy. Let's load a cleaned version of the full text from Project Gutenberg into a single string.

Listing 2-52. Reading War and Peace from the Project Gutenberg text file

```
NUM_LINES = 25_000
wnp = ""
data = urllib.request.urlopen('https://www.gutenberg.org/files/2600/2600-0.txt')
counter = 0
for line in tqdm(data):
    if counter == NUM_LINES:
        break
    wnp += clean(line, no_line_breaks=True)[1:] + " "
    counter += 1
```

---

[1] Of course, we know that they are obtained through certain technical processes – but we lack a clear intuitive or simple understanding.

Because this text is quite long, we'll create a generator class that yields parts of the text sample, sentence by sentence (Listing 2-53). The alternative is to load all the sentences into a list at once, which is taxing on memory and speed.

Listing 2-53. A generator class that yields parts of the text sample for memory feasibility

```
class Sentences():
    def __init__(self, text):
        self.text = text

    def __iter__(self):
        for sentence in wnp.split('.'):
            yield clean(wnp.split('.')[0], no_punct = True).split(' ')
```

To train the Word2Vec model on the data generator, we first instantiate our data generator, instantiate a Word2Vec model, build the vocabulary, and then train the model on the dataset.

Listing 2-54. Training a Word2Vec model on the data generator

```
sentences = Sentences(wnp)
model = gensim.models.Word2Vec(vector_size = 50,
                               min_count = 50,
                               workers = 4)
model.build_vocab(sentences)
model.train(sentences,
            total_examples=model.corpus_count,
            epochs=5)
```

We can access the word vectors with model.wv. For instance, we can obtain the latent features for the word "war" as such:

```
model.wv['war']
```

```
array([ 0.39098194, -2.5320148 , -1.9733142 ,  0.5213574 , -1.2734774 ,
        1.8427355 ,  1.7073737 ,  0.62725115, -1.4480844 ,  0.3382644 ,
        0.70060515,  2.1146834 , -1.7749621 , -0.06704506, -0.48678803,
        1.1092212 ,  0.4158653 ,  0.8432404 ,  0.68553066, -0.60199624,
        0.6334864 , -2.5865083 ,  1.0051454 ,  2.1787288 , -1.643258  ,
       -0.1480552 ,  0.13485388,  1.7048551 , -1.6034617 ,  0.86792046,
       -0.04222116, -0.55365515, -0.47291237, -3.26655   ,  2.2691224 ,
       -1.2338068 ,  0.40476575, -2.0867212 , -0.30338973,  1.663073  ,
        0.20157905, -0.12529533, -1.8289042 ,  0.38934758,  1.2312702 ,
        2.0223777 ,  0.49417907, -2.7465372 ,  0.67504585, -0.5818529 ],
      dtype=float32)
```

Here are the latent features for the word "peace":

```
model.wv['peace']
Out[141]:
array([ 1.5082303e+00, -2.4013765e+00,  1.8905263e+00,  8.9056486e-01,
       -4.0251561e-02,  1.2571076e+00, -1.0280321e+00, -1.4973698e+00,
       -2.8854045e-01, -1.5057240e+00,  7.9542255e-01,  6.1033070e-01,
        5.5785489e-01,  1.4599910e+00, -2.3478435e-01,  1.3725284e+00,
        1.1054497e+00,  1.8628756e+00,  8.6687636e-01,  2.7426331e+00,
       -9.0635484e-01, -2.1095347e+00, -8.1300849e-01,  7.9262280e-01,
       -3.9320162e-01, -4.6035236e-01, -2.0904967e-01,  2.5718777e+00,
        9.7089779e-01, -5.6960899e-01, -1.8032173e+00, -3.3043328e-01,
       -4.5295760e-01, -2.6447701e+00, -1.0341860e+00, -1.7019720e+00,
        7.6734972e-01, -1.8100220e+00, -8.8125312e-01, -1.6304412e-03,
        1.4674787e-01, -1.4068457e+00,  4.1266233e-01, -2.2529347e+00,
        1.2005507e+00,  1.2053030e+00,  9.5373660e-01, -1.5332963e+00,
        6.0380501e-01, -1.3509953e+00], dtype=float32)
```

Embeddings can also be used to compute the similarity between words. This is done by finding the distance between coordinate points represented by the embeddings associated with each word:

```
model.wv.similarity('war', 'peace') -> 0.35120293
```

You can then use this lookup method to vectorize your text.

We'll see examples of similar embedding techniques in Chapter 4 (section: "Multimodal Image and Tabular Models") and especially in Chapter 5 (section: "Recurrent Models Theory"), where we will deal extensively with text.

Time Data

Time/temporal data often appears in practical tabular datasets. For instance, a tabular dataset of online customer reviews may have a timestamp down to the second indicating exactly when it was posted. Alternatively, a tabular dataset of medical data might be associated with the day it was collected, but not the exact time. A tabular dataset of quarterly company earnings reports will contain temporal data by quarter. Time is a dynamic and complex data type that takes on many different forms and sizes. Luckily, because time is both so rich with information and well-understood, it is relatively easy to encode time or temporal features.

There are several methods to convert time data into a quantitative representation to make it readable to machine learning and deep learning models. The simplest method is simply to assign a time unit as a base unit and represent each time value as a multiple of base units from a starting time. The base unit should generally be the most relevant unit of time to the prediction problem; for instance, if time is stored as a month, date, and year and the prediction task is to predict sales, the base unit is a day, and we would represent each date as the number of days since a starting date (a convenient starting position like January 1, 1900, or simply the earliest date in the dataset). On the other hand, in a physics lab, we may need a base unit of a nanosecond due to high required precision, and time may be represented as the number of nanoseconds since some determined starting time.

Let's look at the Amazon US Software Reviews dataset, which contains customer reviews on software products (this is a subset of the Amazon Reviews dataset). After loading and processing, we find that we want to quantify/encode the date column:

```
0          2015-06-23
1          2014-01-01
2          2015-04-12
3          2013-04-24
4          2013-09-08
              ...
341926     2012-09-11
341927     2013-04-05
341928     2014-02-09
341929     2014-10-06
341930     2008-12-31
Name: data/review_date, Length: 341931, dtype: datetime64[ns]
```

Calling .min() on the date feature yields 1998-09-21, the very first date in the column. dates - dates. min() returns the difference between the date column and the first date in daysw. This associates each date with the number of days that has transpired since the first date:

```
0          6119 days
1          5581 days
2          6047 days
3          5329 days
4          5466 days
              ...
341926     5104 days
341927     5310 days
341928     5620 days
341929     5859 days
341930     3754 days
Name: data/review_date, Length: 341931, dtype: timedelta64[ns]
```

While this approach is explicit, it assumes that any new data will fall within the given range. For instance, if the dataset contains customer reviews from January 1, 2015, to December 31, 2021, we should not expect to apply the model to any customer reviews given before January 1, 2015, or after December 31, 2021. This is because the time representation here limits the model's understanding/interpretation of the feature to the domain provided. If any times outside of this domain are sampled, we should not expect the model to be able to extrapolate to unfamiliar domains (Figure 2-37).

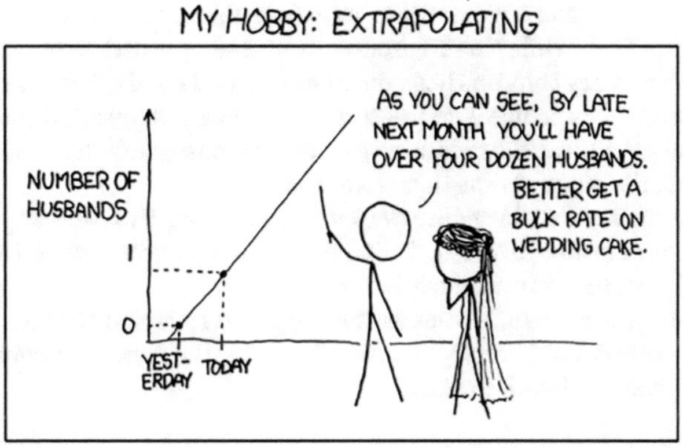

Figure 2-37. "My Hobby: Extrapolating," by Randall Munroe at xkcd.com

Often, we are interested in identifying cyclical patterns in temporal data. Time is full of cycles and units: 60 seconds in a minute, 60 minutes in an hour, 24 hours in a day, 7 days in a week, and so on. A simple way to capture the cyclical quality of time is to represent time as a combination of multiple bounded features.

There are other cyclical methods to keep track of important attributes of time (Listing 2-55, Figure 2-38).

Listing 2-55. Extracting the year, month, and day from a date feature

```
year = dates.apply(lambda x:x.year)
months = dates.apply(lambda x:x.month)
day = dates.apply(lambda x:x.day)
pd.concat([year, months, day], axis=1)
```

	data/review_date	data/review_date	data/review_date
0	2015	6	23
1	2014	1	1
2	2015	4	12
3	2013	4	24
4	2013	9	8
...	...	...	...
341926	2012	9	11
341927	2013	4	5
341928	2014	2	9
341929	2014	10	6
341930	2008	12	31

Figure 2-38. Extracting the year, month, and day as separate features from the original date object

143

When we build the machine learning model to learn from this dataset, we will likely remove the "year" column because it is not cyclical. While we can expect any new sample of data we collect to be reasonably represented by the month and day columns (e.g., the month 12 and the day 7 occurs several times throughout time), we cannot assume this with the year, which does not repeat. If you do include the year column as an input signal for the machine learning model, you must guarantee that any new data the model is evaluated on falls within the range of time data it was trained on.

Another cyclical feature is which day of the week the day falls on. The .weekday attribute of a Python time object returns an integer from 0 to 6 inclusive, where 0 represents Sunday and 6 represents Monday: weekdays = dates.apply(lambda x:x.weekday).

Another possible feature is to include whether the day falls on a holiday or not. Pandas has time-series facilities that contain lists of observed holidays (Listing 2-56). See the Pandas documentation for how to specify custom or more advanced holiday rules.

Listing 2-56. Detecting if a date falls on a holiday or not

```
from pandas.tseries.holiday import USFederalHolidayCalendar
cal = USFederalHolidayCalendar()
allHolidays = cal.holidays(start=dates.min(),
                           end=dates.max()).to_pydatetime()
isHoliday = dates.apply(lambda x:x in allHolidays)
```

There is significant flexibility in time data to encode domain knowledge. Additional techniques include marking the season, relevant commercial events (e.g., single's day, black Friday), day vs. night, working hours, and rush hours. Another approach is to identify or obtain discretized time representations and apply categorical encoding, for instance, on the hour of the day or the day of the week.

Additionally, consider the influence of different time zones across your dataset. In some cases, it is important that the time data uses the same universal system (e.g., UTC) across all its values; in other cases, it is more beneficial to use local times. One technique that exposes the model to both the benefits of a universal and local time – assuming time zone data is available – is to include the local time and the offset from a universal time dependent on the time zone that the local time was collected from.

Geographical Data

Many tabular datasets will contain geographical data, in which a location is somehow specified in the dataset. Similarly to temporal/time data, geographical data can exist in several different levels of scope – by continent, country, state/province, city, zip code, address, or longitude and latitude, to name a few. Because of the information-rich and highly context-dependent nature of geographical data, there aren't well-established, sweeping guidelines on encoding geographical data. However, you can use many of the previously discussed encoding tools and strategies to your advantage here.

If your dataset contains geographical data in categorical form, like by country or state/province, you can use previously discussed categorical encoding methods, like one-hot encoding or target encoding.

Latitude and longitude are precise geospatial location indicators already in quantitative form, so there is no requirement for further encoding. However, you may find it valuable to add relevant abstract information derived from the latitude and longitude to the dataset, like which country the location falls in.

When working with specific addresses, you can extract multiple relevant features, like the country, state/province, zip code, and so on. You can also derive the exact longitude and latitude from the address and append both to the dataset as continuous quantitative representations of the address location.

Feature Extraction

Feature extraction is the derivation of novel features from the existing set of features in an attempt to assist the model by providing potentially useful interpretations of the dataset (Figure 2-39). Although many may argue that feature extraction has been automated or eliminated by neural networks, in practice performing complex feature extraction methods on tabular data often aids performance.

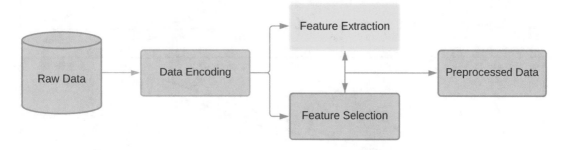

Figure 2-39. *The feature extraction component of the preprocessing pipeline*

Single- and Multi-feature Transformations

A significant component of the statistical learning pipeline is to apply transformations to features to better reflect and amplify their relevance. The simplest of such transformations is a single-feature transformation, in which values from a single feature are mapped to another set of values (Figure 2-40). For instance, if a function has an exponential effect on the target variable as its value increases, we can transform it with the exponential function $f(x) = e^x$, or a modified exponential function informed by domain knowledge. Alternatively, we can apply trigonometric functions to model cyclic relationships, logarithmic or square root functions to model diminishing returns relationships, or quadratics to model two-sided relationships (i.e., values at the two "ends" or "extremes" of the relevant range correspond to one locality of outcomes, whereas values more in the "middle" correspond to a different locality of outcomes). Features can either be transformed individually or appended to the dataset in addition to the raw feature (Figure 2-41).

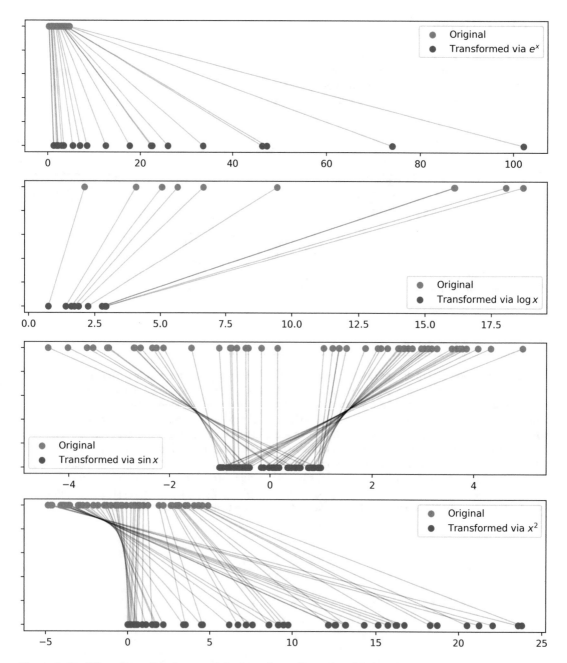

Figure 2-40. *Effect of transforming an original set of one-dimensional data points using various mathematical single-feature transformations*

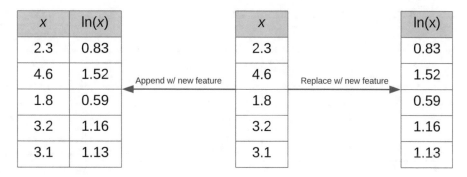

x	ln(x)
2.3	0.83
4.6	1.52
1.8	0.59
3.2	1.16
3.1	1.13

Append w/ new feature

x
2.3
4.6
1.8
3.2
3.1

Replace w/ new feature

ln(x)
0.83
1.52
0.59
1.16
1.13

Figure 2-41. *Two options with feature transformation: replace the original feature with the transformed feature or append it to the original feature*

The Boston Housing Dataset is a well-known benchmark dataset pulled from the US Census Bureau and compiled in 1978 by Harrison and Rubinfeld to use housing data in the Boston area to estimate demand for clean air. The dataset was first published with the paper "Hedonic Prices and the Demand for Clean Air" in volume 5 of the *Economics and Management* journal. The dataset has been included in major data science and machine learning libraries like scikit-learn and TensorFlow.

Some of the features included in the dataset are

- CRIM: Per capita crime rate by town

- INDUS: Proportion of nonretail businesses per town

- PRATIO: Pupil-teacher ratio by school town district

- CHAS: 1 if the town bounds the Charles River, 0 otherwise

- NOX: Nitric oxide concentrations

- PART: Particulate concentrations

- B: $1000(Bk - 0.63)^2$, where Bk is the proportion of African Americans per town

This last feature, B, has been subject to lots of recent discussion. Needless to say, there are many ethical and fairness problems with including a racial feature in a housing dataset. Now, most data science libraries and textbooks that still support or use it include a warning that the dataset is subject to problematic features and suggest the usage of more appropriate housing datasets, like the Ames Housing dataset and the California Housing dataset. The B feature is an interesting case study into how feature transformations can amplify inequitable societal conditions and pose problems for feature correction.

Note that the feature B is a single-feature transformation of another feature, the proportion of African Americans per town. The transformation uses a parabola with a vertex/axis of symmetry at $Bk = 0.63$. At $Bk = 0.63$, the feature "B" is at its lowest, and the feature increases in either direction in quadratic fashion (Figure 2-42). Harrison and Rubinfeld apply this transformation in an attempt to model systemic racism. At low to moderate proportions of African Americans in the town, Harrison and Rubinfeld reason white neighbors will find an increase in Bk as undesirable and hence negatively influencing housing value. At higher values of Bk, Harrison and Rubinfeld note market discrimination yields higher housing values. Thus, Harrison and Rubinfeld assume a parabolic relationship between the proportion of African Americans in the town and the housing value, choosing 0.63 as the "ghetto point" at which an increase in Bk begins to increase rather than decrease house value.

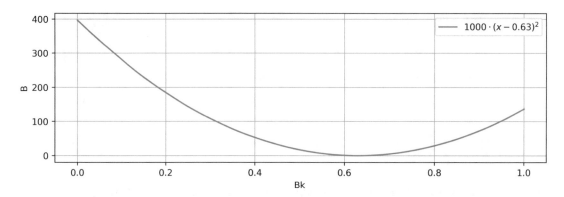

Figure 2-42. *Visualizing the quadratic single-feature transformation used to transform the Bk variable*

Harrison and Rubinfeld's choice of feature transformation was intended to model the effect that societal/institutionalized racism had on how housing was valued with relationship to the proportion of African Americans in the town. Hence, any models that are trained on this dataset take in this transformed data and make decisions assuming the logic Harrison and Rubinfeld attempted to represent.

Hence, there have been recent efforts to examine the feature more and understand the merits for its inclusion in the dataset. For one, those investigating the feature need to know the value of the raw, untransformed feature (i.e., the proportion of African Americans in the town rather than "B"). However, "B" was transformed using a *non-invertible function*, meaning that multiple inputs can be mapped to the same output value. This has the effect of partially destroying the original data (Figure 2-43).

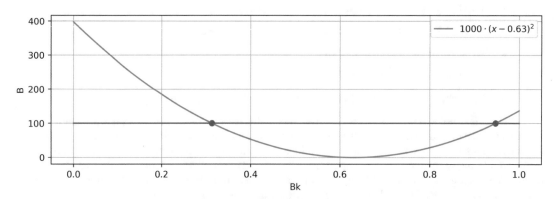

Figure 2-43. *Visualizing the effect of partial data destruction by using a non-one-to-one function as a transformation*

This demonstrates an important property of feature transformations that replace the raw feature: it may not be possible to recover the data if only the non-invertible transformed feature is published. Thus, it is strongly recommended convention for dataset creators in all disciplines to publish the raw dataset separately from any feature transformations that were applied afterward.

Researchers and independent investigators were later able to obtain the raw feature and map it to the dataset using the original US Census data, although with some difficulty (as mapping requires reverse-engineering Harrison and Rubinfeld's data aggregation procedure).

Although single-feature transformations can amplify relevant aspects of a feature in statistical modeling, we can also apply transformations to multiple sets of features to amplify relevant aspects of feature *interaction*.

Many multi-feature transformations utilize elementary operations like adding, subtracting, multiplying, and dividing sets of features (Figure 2-44). For instance, it may be useful to include a feature that holds the average scores or values from a set of comparable columns, like income returns for each year. For multi-feature transformations to be effective in statistical learning and classical machine learning pipelines, it is important to understand the purpose of the multi-feature transformation. If you are adding two columns together for the sake of adding two columns together, you shouldn't expect the model to make sense out of the new feature. Simple models are often more "confused"/distorted by complex, difficult-to-interpret features.

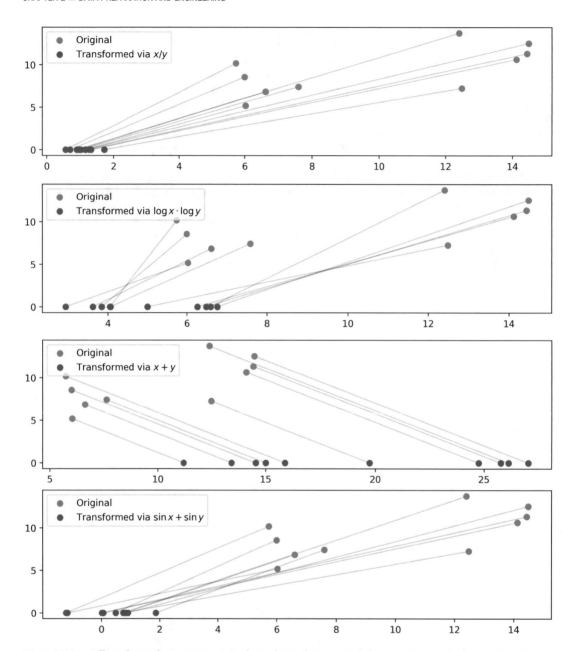

Figure 2-44. *Effect of transforming an original set of two-dimensional data points to a single new feature using various mathematical multi-feature transformations*

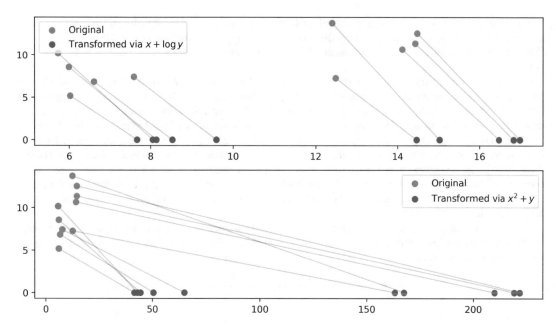

Figure 2-44. *(continued)*

Both simple single- and multi-feature transformations are an important part of statistical learning and classical machine learning, but they are generally superfluous in the context of deep learning. Neural networks are more than capable of learning these transformations themselves; moreover, they can usually learn better and more complex transformations than you can manually design. This is part of the beauty and power of deep learning.

However, this does not mean human-operated feature engineering has no place in deep learning if the methods used are sufficiently introduced and provide nontrivial value, especially in the context of tabular data. The following sections will explore various techniques to introduce helpful input signals into a deep learning dataset.

Principal Component Analysis

Feature engineering and extraction can also come in the form of retrieving information that reconstructs the original data in a lower dimension. Dimensionality reduction techniques reduce the amount of data and training time while synthesizing insights that are not easily accessible in the original data.

There are many advantages to dimensionality reduction techniques. When applied to training data, it can reduce the effects of the Curse of Dimensionality on the model's performance. During preprocessing and analysis, it's crucial to understand and visualize the data as it can better help us discover patterns in which we can build models accordingly. However, many modern datasets consist of hundreds if not thousands of features and are thus impossible to visualize for human eyes. Even the simplest dataset such as the Iris Flower dataset contains four features that cannot be viewed fully at once. By reducing the dimension of our data, we can view the complex structure of the dataset with humanly interpretable visualizations. In addition, we can keep crucial information that the dataset possesses at higher dimensions in most cases.

One of the most popular and effective methods of dimensionality reduction is Principal Component Analysis (PCA). PCA has evolved to many uses and variations from its original design as a convenient tool for exploratory data analysis.

Performing Principal Component Analysis is like writing a summary for a long book. We need to understand what we are reading and grasp the most critical components before we can summarize. Similarly, PCA can mathematically determine which part of the dataset contributes the most to its final "summary."

It is easier to understand this with a real-world example. Let's say we bought three books with different page numbers, and our goal is to distinguish them by just observing their thickness. It would be simple if the first book had 50 pages, the second had 200 pages, and the third had 500 pages. Their thickness will appear dramatically different; thus, it'll be easy for us to determine which book is which. On the other hand, if the first book has 100 pages, the second has 105 pages, and the third has 110 pages, it becomes difficult for us to separate one book from another as their thickness is similar. In our scenario, when the page numbers of these three books are more spread out, it provides us with more information and thus more variance. On the contrary, when the data is closer together, it contains less information, translating to less variance. A more technical overview of variance will be introduced later in the "Feature Selection" section as we'll be using this concept to select features.

Building off the preceding example, we can understand variance as how much information, or spread, the dataset provides. The process of PCA preserves variables with the most variance. We can illustrate such a concept with a simple dataset.

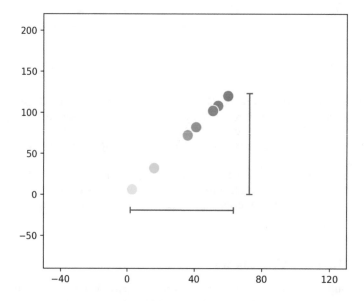

Figure 2-45. *Dummy dataset where both variables have similar variance value*

PCA doesn't simply remove or select features. As presented previously, both features have similar variance (Figure 2-45). Removing either will significantly lessen the information compared to the original data. However, PCA doesn't only consider variance in the variables itself. Instead of looking at the vertical and horizontal axes, we notice that the diagonal axis contains as much variance if not more.

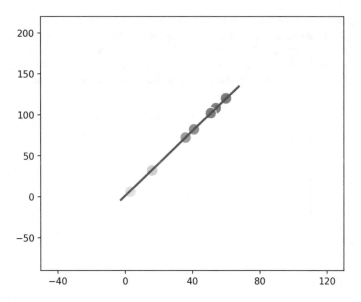

Figure 2-46. *Variance in the diagonal axis*

PCA then creates two distinct variables based on the combination of the original two. In our example, we transform the data with respect to the diagonal axis and the line perpendicular to it as our two new axes (Figure 2-46).

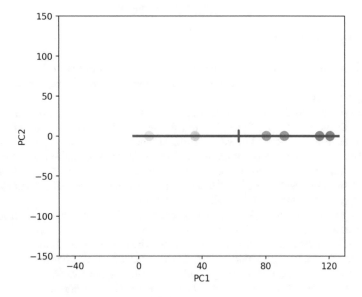

Figure 2-47. *Rotated data relative to the diagonal axis*

We refer to the two new variables as principal components. From the graph, we can clearly see that PC1 has much more variance than PC2 (Figure 2-47). Keeping PC1 will retain the most variance, or the most information from the original data, and that's exactly what PCA does. If we dig deeper, from our example, we discover that PC1 completely explains all the variance in the original data. However, that won't always be the case. We can visually represent how much variance each component explains using a Scree Plot (Figure 2-48).

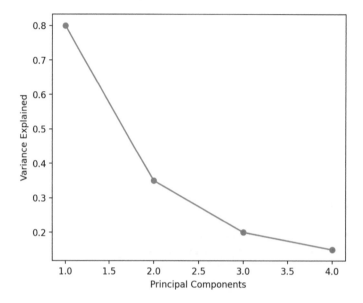

Figure 2-48. *Scree Plot with dummy dataset*

The plot displays variance explained by each principal component in proportion to the total variance. In our preceding example, four principal components are shown; the first one explains 80% of total variance relative to the original dataset.

There is one major drawback to PCA as a feature selection technique surrounding the scale or the distance between each individual data point after transformation. When searching for principal components and transforming them into new variables, we are simply rotating axes and changing the direction of data points. However, when we start to remove dimensions, the restriction of space across one dimension will influence how each data point relates to each other by their Euclidian distance. This problem can influence model performance in some cases, while it can be ignored during other times. Whether PCA will be helpful for the use case can only be determined through experience and testing.

There are two approaches to using PCA for feature extraction. The first is that we can set a certain number of "components" that the algorithm will reduce the data to, thus having a smaller dataset that retains the most information from the original. Second, PCA can also be used for the addition of new features. Instead of solely using the features provided by PCA, in some cases it would be more beneficial to keep the original features and add a set number of "components" from PCA as new features. In some cases, using PCA to create more features can give the model more information that it may or may not consider as essential. To the model, the principal components are a compact way of storing essential information, and as the model sees fit, it can also use the original features as those are the ones that contain the most accurate values.

Both uses for PCA on feature engineering/extraction are demonstrated in the following with scikit-learn (Listing 2-57). It's crucial to standardize the data before performing PCA since the new projection of data and new axes will be based on the standard deviation of original variables. When one's standard deviation is higher than the other, it can cause uneven weights assigned to different features.

Listing 2-57. Example of using PCA to select components

```
# Dummy Dataset where the goal to predict diagnosis of breast cancer
from sklearn.datasets import load_breast_cancer
# PCA
from sklearn.decomposition import PCA
# standardization
from sklearn.preprocessing import StandardScaler

breast_cancer = load_breast_cancer()
breast_cancer = pd.DataFrame(data=np.concatenate([breast_cancer["data"], breast_
cancer["target"].reshape(-1, 1)], axis=1),
                            columns=np.append(breast_cancer["feature_names"],"diagnostic"))

# standardize
scaler = StandardScaler()
breast_cancer_scaled = scaler.fit_transform(breast_cancer.drop("diagnostic", axis=1))
# the data originally have 30 features, for visualization purposes later on,
# we're only going to keep 2 principal components
pca = PCA(n_components=2)
# transform on features
new_data = pca.fit_transform(breast_cancer_scaled)
# reconstruct dataframe
new_data = pd.DataFrame(new_data, columns=[f"PCA{i+1}" for i in range(new_data.shape[1])])
new_data["diagnostic"] = breast_cancer["diagnostic"]
```

If it's better to add those calculated principal components, we can do such as shown in the following (Listing 2-58).

Listing 2-58. Combined data with features extracted from PCA

```
# we're dropping the target column in new_data and not
# the original to keep the ordering of columns

combined_data = pd.concat([new_data.drop("diagnostic", axis=1), breast_cancer], axis=1)
```

Finally, PCA is crucial to visualizing and analyzing patterns beneath high-dimensional datasets. We can do such by graphing two or three principal components in either two or three dimensions, respectively. Furthermore, for each data point, we can color them differently based on their label for better visual analysis (Listing 2-59).

Listing 2-59. Code for displaying principal components

```
plt.figure(figsize=(6, 5), dpi=200)
plt.xlabel("PCA1")
plt.ylabel("PCA2")
plt.scatter(new_data.PCA1, new_data.PCA2, c=new_data.diagnostic, cmap='autumn_r')
plt.show()
```

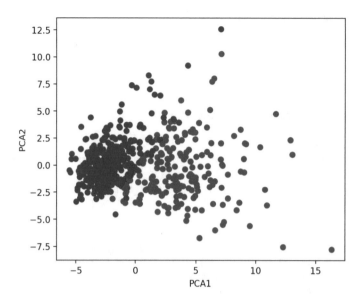

Figure 2-49. *Visualization of principal components*

From observing the graph, notice that the labels are separated based on color (Figure 2-49). It's almost as if two labels are in two separable clusters. Such a phenomenon shows that a significant amount of variance is already explained with only two principal components or two features. From this visualization, we can conclude that reducing the dimensionality of the dataset by a significant amount would still preserve a huge ton of information while decreasing model complexity. It's also possible to obtain the exact value of explained variance for components by calling .explained_variance_ratio_ on a fitted PCA object.

Principal Component Analysis can be useful at times, reducing model complexity while possibly improving performance, but like many other feature extraction techniques, it has its drawbacks. Thus, the data scientist must decide how to apply the algorithm that suits the situation while minimizing computational expenses.

t-SNE

Many modern datasets exist in high dimensions as mentioned previously; we can utilize algorithms such as PCA to reduce the dataset to a lower dimension. However, PCA only performs well on linearly separable data. It's difficult for PCA to project nonlinearly separable data to a lower dimension while preserving most information that would help the model distinguish between labels.

Manifold is a family of dimensionality techniques that focuses on separating nonlinearly separable data (Figure 2-50). Specifically, t-SNE (t-distributed Stochastic Neighbor Embedding) is among one of the most used and effective algorithms.

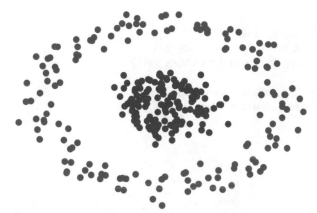

Figure 2-50. *Nonlinearly separable data*

t-SNE, or t-distributed Stochastic Neighbor Embedding, is one of the most popular unsupervised algorithms for high-dimensional visualization. Although t-SNE is not necessarily used for feature selection, it stands as an important step for many complex deep learning pipelines and is commonly used for neural network interpretation. Unlike PCA, t-SNE focuses on the local structure, preserving the local distances between points rather than prioritizing the global structure. The algorithm converts relationships in original space into t-distributions, or normal distributions with small sample sizes and relatively unknown standard deviations.

Instead of relying on variance to determine useful information, t-SNE focuses on grouping local groups of clustered data. The algorithm uses the Kullback-Leibler divergence, a metric for a statistical distance of joint probabilities as a measurement of separation between data. Gradient descent is applied to optimize the metric.

In t-SNE, the perplexity of the transformed data is adjustable by a user-specified hyperparameter. This parameter affects the ending visualization significantly and thus should be tuned and considered when using t-SNE. Perplexity can be interpreted as the number of nearest neighbors that t-SNE will consider for projecting the data. We would expect to see sparser visualization with smaller perplexity and vice versa.

Scikit-learn provides an implementation for t-SNE, as shown in the following. We can generate an example three-dimensional Swiss roll dataset; this is a classic example of nonlinearly separable data (Listing 2-60). The dataset generator also returns an array of values representing each sample's univariate position in the main dimension. We can color each point according to this returned array and evaluate the performance of t-SNE based on it (Figure 2-51).

Listing 2-60. Code for generating Swiss roll data and training t-SNE

```
# generate a nonlinearly separable data in the shape of a swiss roll
# 3-dimensions
swiss_roll, color = datasets.make_swiss_roll(n_samples=3000, noise=0.2, random_state=42)
fig = plt.figure(figsize=(8, 8))

# visualize original data, results shown below
ax = fig.add_subplot(projection="3d")
ax.scatter(swiss_roll[:, 0], swiss_roll[:, 1], swiss_roll[:, 2], c=color, cmap=plt.
cm.Spectral)
```

```
# t-SNE training
from sklearn.manifold import TSNE
embedding = TSNE(n_components=2, perplexity=40)
X_transformed = embedding.fit_transform(swiss_roll)

# visualize t-SNE results, graph shown below
plt.scatter(X_transformed[:, 0], X_transformed[:, 1], c=color)
```

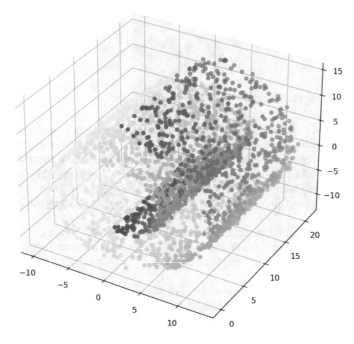

Figure 2-51. *Swiss roll data*

In comparison, t-SNE reduced the data to two dimensions while preserving most structures that were present in the original data, separating different colors well (Figure 2-52).

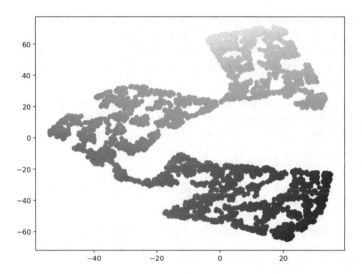

Figure 2-52. *Swiss roll data reduced to two dimensions*

Although t-SNE handles nonlinear data better than PCA, the algorithm has some major disadvantages that should be considered. Due to the randomized initialization in gradient descent used by t-SNE, seed selection can affect results. In addition, t-SNE has an extremely high computational cost. When running on a dataset with millions of samples, t-SNE may take substantially longer to finish than PCA.

If used for feature selection, it can be used in the same way as PCA. Generally, t-SNE plays a pivotal role in model interpretation and feature extraction but not in feature selection. Some common uses include converting image data to tabular data and visualizing various deep learning algorithms.

Linear Discriminant Analysis

Principal Component Analysis is an unsupervised algorithm, meaning that when transforming data, it does not consider the labels. PCA projects higher-dimensional data onto a lower-dimensional space based on the assumption that large variance contains more information and thus will better represent the original data when transformed into a lower dimension. On the other hand, Linear Discriminant Analysis (LDA) aims to maximize separation between label clusters. It may seem that LDA can only be used for classification, but it may be able to better transform higher-dimensional data to humanly understandable visualizations than PCA due to the fact that it is a supervised learning algorithm.

Note while LDA can also be used as an algorithm for classification datasets, this is out of scope for our purposes. We will only focus on the dimensionality reduction and feature selection portion of LDA. For binary classification tasks, LDA attempts to maximize the distance between each label cluster's central points. On the contrary, in multiclass classification, LDA maximizes the distance from clusters to an overall central point.

LDA assumes that the features are normally distributed and each feature must have a similar variance. Specifically, the process of LDA can be summarized into three steps.

1. Calculate the between-class variance for each label. Between-class variance quantifies the distance between the means of each label cluster.

2. Calculate the within-class variance for each label. Within-class variance is the distance between the mean of each class and each sample in that label cluster.

3. According to the number of components to keep, which is specified as a hyperparameter, the data is projected to that number of dimensions. The projection should maximize between-class variance while minimizing within-class variance. This can be accomplished by either singular value decomposition or using eigenvalues.

Scikit-learn has an implementation of LDA, which operates on a similar syntax compared to PCA (Listing 2-61). Note that during dimensionality reduction, the maximum number of components that can be kept is limited to $(num\_classes - 1, num\_features)$.

Listing 2-61. Code for dimensionality reduction in LDA

```
# Dummy Dataset where the goal to predict diagnosis of breast cancer
from sklearn.datasets import load_breast_cancer
# LDA
from sklearn.discriminant_analysis import LinearDiscriminantAnalysis
# standardization
from sklearn.preprocessing import StandardScaler

breast_cancer = load_breast_cancer()
breast_cancer = pd.DataFrame(data=np.concatenate([breast_cancer["data"], breast_
cancer["target"].reshape(-1, 1)], axis=1),
                             columns=np.append(breast_cancer["feature_names"],"diagnostic"))

# standardize
scaler = StandardScaler()
breast_cancer_scaled = scaler.fit_transform(breast_cancer.drop("diagnostic", axis=1))
# the data originally have 30 features, for visualization purposes later on,
# we're only going to keep 2 principal components
lda = LinearDiscriminantAnalysis(n_components=1)
# transform on features
new_data = lda.fit_transform(breast_cancer_scaled, breast_cancer["diagnostic"])
# reconstruct dataframe
new_data = pd.DataFrame(new_data, columns=[f"LDA{i+1}" for i in range(new_data.shape[1])])
new_data["diagnostic"] = breast_cancer["diagnostic"]
```

If we plot the transformed data and separate class labels by colors as shown in the following, LDA does a good job of separating classes in just one single dimension (Figure 2-53).

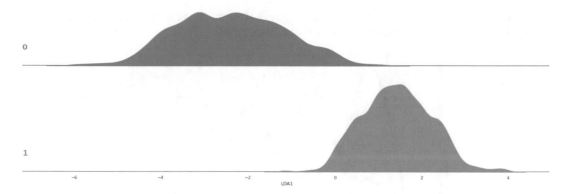

Figure 2-53. *Dimensionality reduction from LDA into one dimension*

LDA isn't commonly used in modern machine learning fields, either as a classification algorithm or a dimensionality reduction technique. As mentioned previously, one major downfall in LDA is that it assumes that all features have a normal distribution and that they have similar variances. Moreover, both PCA and LDA are linear dimensionality reduction algorithms, meaning they can only operate on linearly separable data.

Compared with PCA, LDA is harder to use, and it is a much less popular approach to dimensionality reduction. LDA is better suited for multiclass classification problems as the labels have more than two classes, meaning that the reduced data will still have more than two dimensions, thus retaining more information from the original data. Furthermore, LDA presumably performs better than PCA due to the fact that it is a supervised learning algorithm.

Statistics-Based Engineering

The core purpose of feature engineering boils down to extracting useful information from the set of original features that provides value to the model. This can be achieved by applying transformation as discussed in the "Single- and Multi-feature Transformations" section. Feature extraction can also come in the form of deriving the statistical properties of samples. We can obtain statistical measures of each sample across the entirety of the row, creating new features. However, this method only suits for sets of features where their values are in the same scale hence the statistical measures produced from those features can make practical sense and provide useful information to the model.

Additionally, utilizing this technique provides a different perspective for the model to understand the data. Instead of providing features that extend the relevancy from features to the target or increase their correlation, statistical measures of each sample let the model treat each sample as its own rather than being a part of a larger dataset.

We can calculate simple measures such as the mean, median, or mode of each sample. However, calculating any statistical information related to distribution, such as standard deviation or skewness, has a greater positive influence on the model. We can demonstrate such a concept on a genetics dataset that contains about 250 genomic sequences as features while predicting the bacterial type based on its genetic sequences.

Listing 2-62. Loading data

```
# only take first 2000 rows as the whole data is too large
gene_data = pd.read_csv("../input/cleaned-genomics-data/cleaned.csv", nrows=2000)
```

After loading the data (Listing 2-62), we can visualize the distribution of each sample based on the values of its features at that row as seen in the following (Figure 2-54).

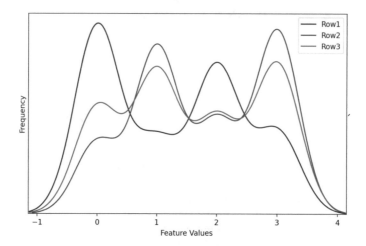

Figure 2-54. *Each row's feature distribution for the first three samples*

Based on each row's distribution, we can then calculate the standard deviation, skewness, kurtosis, mean, and median for each row (Listing 2-63).

Listing 2-63. Calculating the statistics for each row

```
gene_feature = gene_data.drop(["Unnamed: 0", "label"], axis=1).columns.to_list()

for stats in ["mean", "std", "kurt", "skew", "median"]:
gene_data[f"{stats}_feat"] = getattr(gene_data[gene_feature], stats)(axis = 1)
```

After applying transformations, we can observe differences in these measures between different classes, while the model may discover a pattern within and correlate it to the targets, thus improving the performance.

Remember that the features that we group together and whose properties we calculate must be on the same scale; they must make practical sense when combined and used together. That is, the features must be *homogenous*. We'll see in later chapters that certain techniques require or assume homogeneity within features. In our case, it's natural that we can combine each single nitrogen bond since they form a sequence of DNA. In cases where not every feature would fit well into each other and their distribution together would not be useful (i.e., feature heterogeneity), we can try grouping similar features and calculate the statistical measures of each group.

This idea of treating each sample as a group rather than a data point within the feature can be extended such as calculating the sum across each row or the product across rows depending on domain knowledge or simply trial and error. The sum or product of each sample can also be utilized when there are a smaller number of features, and calculating measures related to the distribution of all features on a single sample wouldn't demonstrate much use as extra information.

Depending on the size of the dataset, when there are a larger number of features, using deep learning methods is preferred over classical machine learning algorithms, which is the essence of this book and which this method found most of its success upon. The usage of both will be demonstrated in later chapters.

Feature Selection

Feature selection refers to the process of filtering out or removing features from a dataset (Figure 2-55). There are two main reasons to perform feature selection: removing redundant (very similar information content) features and filtering out irrelevant (information content not valuable w.r.t. the target) features that may worsen model performance. The difference between feature extraction and feature selection lies in that selection reduces the number of features while extraction creates new features or modifies existing ones. A universal approach to feature selection usually consists of obtaining a measure of "usefulness" for each feature and then eliminating those that do not meet a threshold. Note that no matter which method for feature selection is used, the best result will likely come from trial and error since the optimal techniques and tools vary for datasets.

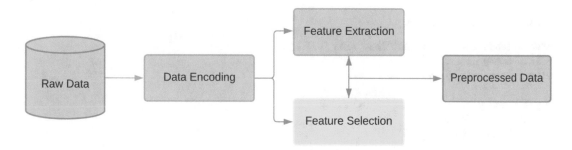

Figure 2-55. *The feature selection component of the preprocessing pipeline*

Information Gain

Information Gain, synonymous with Kullback-Leiber divergence, can be defined as a measure of how much a certain feature tells us about a target class. Before discussing its use in feature selection, note that Information Gain could act as another "metric" for finding the best split in the Decision Tree along with Gini Impurity and Entropy. One major drawback of Information Gain being used in Decision Trees is that it tends to select features with more unique values. One example is if the dataset contains a certain attribute like Date, in usual cases, it would not be useful for Decision Trees to utilize such feature as its values are independent of targets. However, Information Gain would output a score for the Date feature potentially higher than other more useful features. Additionally, when dealing with categorical features, Information Gain favors features with more categories, which might not be ideal. Although Information Gain may be useful in serving as a "metric" during Decision Tree splits in rare cases, most of the time, it's not considered due to its major disadvantage.

In technical terms, Information Gain produces the difference in Entropy before and after a transformation. When applied to classification feature selection, it calculates the statistical dependence between two variables, or how much information the two share; it's sometimes referred to as Mutual Information. In statistics, the term *information* refers to how surprising a certain event is. An event is considered more surprising than another when it has a more balanced probability distribution and thus more Entropy. Entropy measures the "purity" of the dataset in terms of the probability distribution of samples belonging to classes. For example, a dataset with perfectly balanced targets (50-50 split) would result in Entropy of 1, while a dataset with imbalanced targets (90-10 split) would produce a low Entropy. Information Gain evaluates the impact on the purity by splitting the dataset by each unique value in the dataset. Essentially, it calculates one feature's usefulness in relation to the target based on how well the feature splits the target. The equation for Information Gain is shown in the following:

$$Information\ Gain(D,X) = Entropy(D) - Entropy(X)$$

The second calculation of Entropy is a conditional Entropy on the feature X in dataset D defined as

$$Entropy(X) = \sum_{v \in X} \frac{D_v}{D} \cdot Entropy(D_v)$$

For each term in the sum, we split the dataset by that unique value in the feature, v; calculate the Entropy for that subset, D_v; and then multiply by the ratio of the subset to the whole dataset (the number of samples in D_v divided by the total number of samples in the dataset D) and the sum of every one of these subsets. Finally, we compute the difference between the beginning and the ending Entropy. The higher the resulting value, the more information the feature provides us with the target.

We can demonstrate this feature selection technique with the following example on a dummy dataset aiming to classify the type of wine based on various attributes.

Listing 2-64. Loading dataset

```
from sklearn.datasets import load_wine
# load dummy dataset
wine_data = load_wine()

# get X and y
X = wine_data["data"]
y = wine_data["target"]
```

After loading the dataset (Listing 2-64), we can first train a Decision Tree without feature selection for comparison later (Listing 2-65).

Listing 2-65. Training baseline model with Decision Tree

```
# Train a simple Decision Tree on the Dataset without feature selection
from sklearn.tree import DecisionTreeClassifier
from sklearn.model_selection import train_test_split as tts

# train test split so we can evaluate our performance on unseen test data
X_train, X_test, y_train, y_test = tts(X, y, test_size=0.3, random_state=42)

# Decision Tree
dt = DecisionTreeClassifier(max_depth=5)
dt.fit(X_train, y_train)

# Predict on unseen test dataset
predictions = dt.predict(X_test)

# evaluate performance
from sklearn.metrics import classification_report
print(classification_report(y_test, predictions))
```

We obtain a total accuracy of 0.94 with a slight caveat: class 1's performance is substantially worse than others. We can further improve the results using Information Gain to dispose of features that do not serve a positive influence on the model (Listing 2-66).

Listing 2-66. Feature selection with Information Gain

```
# already good enough performance, but can it be better with feature selection?
from sklearn.feature_selection import SelectKBest
from sklearn.feature_selection import mutual_info_classif

# mutual info classif calculates the mutual information between two variables, aka
Information Gain
# select K best chooses the k top features based on the feature selection method provided
# select top 11 features
X_new = SelectKBest(mutual_info_classif, k=8).fit_transform(X, y)
```

We obtain an X_new array with the top eight features selected by Information Gain. Now we train a new Decision Tree model with the same train-test split and the same parameters (Listing 2-67).

Listing 2-67. Retrain Decision Tree and evaluate performance

```
# Decision Tree with Feature selection

# train test split so we can evaluate our performance on unseen test data
X_train, X_test, y_train, y_test = tts(X_new, y, test_size=0.3, random_state=42)

# Decision Tree
dt = DecisionTreeClassifier(max_depth=5)
dt.fit(X_train, y_train)

# Predict on unseen test dataset
predictions = dt.predict(X_test)

# evaluate performance
from sklearn.metrics import classification_report
print(classification_report(y_test, predictions))

# BETTER PERFORMANCE!
```

We see that our accuracy increases to an astounding 0.98 and the performance for class 2 is comparatively much better than the previous model trained without feature selection.

Information Gain is extremely useful as a feature selection technique for relatively smaller datasets since the computational cost increases tremendously for larger datasets with more unique valued features.

Variance Threshold

One key goal that feature selection accomplishes is that it removes excessive information that doesn't provide any practical usage for predicting the target; thus, dropping them reduces the model size while also improving model performance. The statistical measure, variance, tells us about the variability in the distribution of a feature. In simpler terms, it measures how far the data is spread out. Usually when there are more unique values or when the data contains values different from the mean, the feature contains more useful information for the target. For example, a feature with a constant value would have a standard deviation of 0 and thus 0 variances: there's no variation in the data. Thus, our goal is to remove features with low variance. However, measuring the variance of certain features does not take the correlation of a feature to the target into account; it assumes that more unique values tend to perform better than features with fewer variabilities, as this is commonly the case. The variance of a dataset is defined as follows where $\underline{x}$ is the mean of all observations or values and n is the number of values:

$$Variance = \frac{\Sigma(x_i - \underline{x})^2}{n-1}$$

Compared with Information Gain, the Variance Threshold provides a significantly faster and simpler method of feature selection with decent improvements on models. The Variance Threshold is usually used as a baseline feature selector to filter out inadequate features without the significant computational cost. A simple demonstration of using the Variance Threshold to select features is shown in the following. Note that when using the Variance Threshold to select features by comparing and removing data columns over a certain value, all columns' values must be on the same scale. Differently scaled values produce variance that is comparable only to their own scale. In our example, we used MinMaxScaler to scale all features before calculating their variance (Listing 2-68).

Listing 2-68. Loading the Breast Cancer dataset and performing scaling

```
# we can perform Variance Threshold solely using Pandas

# example dataset, using patient's data to predict their breast cancer diagnostic
from sklearn.datasets import load_breast_cancer
breast_cancer = load_breast_cancer()
breast_cancer = pd.DataFrame(data=np.concatenate([breast_cancer["data"], breast_
cancer["target"].reshape(-1, 1)], axis=1), columns=np.append(breast_cancer["feature_
names"],"diagnostic"))
# diagnostic is our target column

# Scale the data before calculation
from sklearn.preprocessing import MinMaxScaler

# Get the name of all the features
features = load_breast_cancer()["feature_names"]

scaler = MinMaxScaler()
breast_cancer[features] = scaler.fit_transform(breast_cancer[features])
```

After scaling the data, a Decision Tree with a max depth set to 7 is trained as a baseline comparison (Listing 2-69). The classifier reaches an accuracy of 0.94, but the accuracy for the negative class can be as low as 0.90.

Listing 2-69. Baseline model

```
from sklearn.model_selection import train_test_split as tts
from sklearn.metrics import classification_report
from sklearn.tree import DecisionTreeClassifier

X_train, X_test, y_train, y_test = tts(breast_cancer[features], breast_cancer["diagnostic"],
random_state=42, test_size=0.3)
```

```
rf = LogisticRegression()
rf.fit(X_train, y_train)
predictions = rf.predict(X_test)
print(classification_report(y_test, predictions))
```

With the help of removing features with low variance, we see an improvement in our model performance as shown in the following (Listing 2-70).

Listing 2-70. Retrain model with removed features

```
# returns the variance of each column that's more than 0.015
var_list = breast_cancer[features].var() >= 0.015
var_list = var_list[var_list == True]

# Select those features from the dataset
features = var_list.index.to_list()

from sklearn.model_selection import train_test_split as tts
from sklearn.metrics import classification_report
from sklearn.tree import DecisionTreeClassifier

X_train, X_test, y_train, y_test = tts(breast_cancer[features], breast_cancer["diagnostic"],
random_state=42, test_size=0.3)

rf = LogisticRegression()
rf.fit(X_train, y_train)
predictions = rf.predict(X_test)
print(classification_report(y_test, predictions))
```

We observe a slight improvement in the accuracy, reaching 0.95 while the precision for true negative predictions increases to 0.92. Depending on the threshold value, worse or better results may be produced. The best value can only be determined through trial and error.

However, there are times when the Variance Threshold provides unsatisfactory results as it doesn't take into account the correlation between features and targets. Categorical and binary features tend to have extremely low variance due to the nature of their feature representation; the entirety of the dataset may only contain a handful of unique values, but it provides crucial clues to predicting the target. It's recommended to exclude categorical or binary features while performing the Variance Threshold. Additionally, some datasets consist of features with high variability but are not necessarily useful for predicting the target, which drives back to the fact that the Variance Threshold does not consider relations between the target and the feature, thus making it an unsupervised feature selection technique.

Finally, choosing the threshold that determines the "cut-off" depends on the dataset used. Some datasets may have features with high variance across the board; in that case, the Variance Threshold would be futile. There's no universal rule of determining the threshold; the best value comes from trial and error.

High-Correlation Method

One of the most straightforward ways to determine whether some features will be adequate indicators of the target is by correlation. In statistics, correlation defines the relevancy between two variables; it usually produces a measure that specifies how well the two variables are related. The relationship between features and targets is arguably the most important factor that determines whether the trained model will predict the target well or not. Those features with low correlation to the target will present as noise and possibly reduce the performance of the trained model. Linear correlation between two variables calculated using Pearson's Correlation Coefficient is frequently used to measure how closely two variables align with each other or their correlation. The equation for it is presented in the following, where $\underline{x}$ and $\underline{y}$ represent the mean of the x and y variables, respectively:

$$Pearson's\ Correlation\ Coeffcient = \frac{\Sigma(x-\underline{x})(y-\underline{y})}{\sqrt{\Sigma(x-\underline{x})^2\ \Sigma(y-\underline{y})^2}}$$

Pearson's Correlation Coefficient produces a value between –1 and 1; –1 illustrates a negative correlation between two variables, while 1 illustrates a perfect positive correlation. A value of 0 indicates no correlation at all between variables. Generally, when the value lies above 0.5 or below –0.5, the two variables are considered to have a strong positive/negative correlation.

Take the Boston Housing Dataset introduced earlier as an example. We can visualize the correlation between each feature and the target as shown in the following, plotted as a heatmap (Figure 2-56).

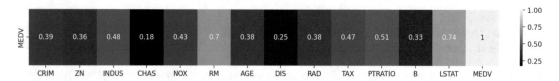

Figure 2-56. *The feature-to-target correlation of the Boston Housing Dataset*

The target column, "MEDV," represents the median value of owner-occupied homes in thousands of dollars. We see that LSTAT has an extremely high correlation value with the target at 0.74. Following, columns "INDUS," "RM," and "PTRAIO" all have a correlation value of 0.5 or greater. Reasonably, the feature LSTAT, representing the lower status of the population in percentage, would have the highest correlation to the target as your social status translates to your financial situation in most cases.

Utilizing our correlation values, we can drop lowly correlated features with the target. To systematically remove such features, we select a threshold of 0.35, discarding any features with correlation values less than the threshold. Let's compare model performance with and without those features using a simple KNN regressor model with a neighbor of 3 (Listing 2-71).

Listing 2-71. Code example for using the high-correlation method to filter features

```
# KNNRegressor without removeval of lowly correlated features

from sklearn.neighbors import KNeighborsRegressor
from sklearn.model_selection import train_test_split as tts
from sklearn.metrics import mean_absolute_error

features = ["CRIM", "ZN", "INDUS", "CHAS", "NOX", "RM", "AGE", "DIS", "RAD", "TAX",
"PTRATIO", "B", "LSTAT"]
target = ["MEDV"]

X_train, X_test, y_train, y_test = tts(boston_data[features], boston_data[target], random_
state=42, test_size=0.3)
knn = KNeighborsRegressor(n_neighbors=3)
knn.fit(X_train, y_train)
no_corr_pred = knn.predict(X_test)
```

```
print(mean_absolute_error(y_test, no_corr_pred))
# 3.9462719298245608

# Remove feartures with correlation less than to 0.35
lowly_corr_feat = ["CHAS", "DIS", "B"]

# Remove those features
features = list(set(features) - set(lowly_corr_feat))
X_train, X_test, y_train, y_test = tts(boston_data[features], boston_data[target], random_
state=42, test_size=0.3)
knn = KNeighborsRegressor(n_neighbors=3)
knn.fit(X_train, y_train)
no_corr_pred = knn.predict(X_test)
print(mean_absolute_error(y_test, no_corr_pred))
# 3.7339912280701753
```

With the three of the lowest correlated features removed, we observe an improvement in the MAE by about 0.212. However, suppose we start raising the threshold to remove more features; model performance will decrease dramatically, indicating that we took away crucial features that the model utilizes in its learning. There's no perfect way to select the best universal threshold; choices are made through trials and errors.

Although linear correlation can be a useful tool to measure features' effectiveness, the complexity it presents compares lowly with other methods. Adding on, Pearson's Correlation Coefficient can only be used when the data is Gaussian distributed. When the data does not satisfy the distribution, Rank Correlation should be used. Instead of using the actual values of the features, Rank Correlation methods compute the ordinal association between two variables: each set of similar values is replaced with a "rank" or "order" that does not assume any distribution of data. These methods produce values with a similar trend as Pearson's Correlation Coefficient but can be used on data with any distribution, thus sometimes referred to as nonparametric correlation. The correlation values of the Boston Housing Dataset are presented in the following using Spearman's Rank Coefficient (Figure 2-57).

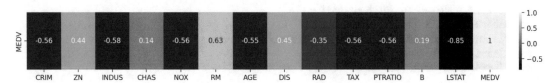

Figure 2-57. *The feature-to-target correlation of the Boston Housing Dataset using Spearman's Rank Coefficient*

Finally, no matter what correlation method is being used, high correlation doesn't always translate to the causation of one feature to the target. Measuring correlation and filtering out features based on their values is useful. However, many large modern datasets model relationships much more complex than linear or even quadratic; thus, correlation filtration methods may be deemed useless in those cases.

Calculating these values may still present a nice visual and a basic idea of direct features related to the target, but whether it'll improve model performance or not depends on what the model produces.

■ **Note** The Boston Housing Dataset contains a problematic feature, B. See the preceding exploration of the significance of B. Using models trained on the Boston Housing Dataset in real-world applications is not recommended.

Recursive Feature Elimination

In previous sections, all feature selection techniques introduced are in the form of measuring some individual properties relative to each feature and then determining the removal of features based on their "measures." These methods are universal and can be applied to any dataset using the same pipeline and process. But at the end of the day, feature selection is aimed at improving model performance, and therefore it is crucial to observe how well each feature specifically contributes to model performance. Recursive Feature Elimination (RFE) is a process in which features are removed (eliminated) based on how much they contribute to a trained model.

Due to its effectiveness and flexibility, RFE is one of the most used feature selection algorithms. RFE is not a single method or tool; it's a wrapper that can adapt to any model depending on the use case. In the following example, Random Forest will be used as the model for feature selection; however, it can be replaced with any other model to improve performance.

Like the Boston Housing Dataset introduced earlier, the Forest Cover dataset is another popular dataset for benchmarking tabular classification models. The data is collected from four wilderness areas in the Roosevelt National Forest in Colorado. Cartographic data that defines an area of 30 meters by 30 meters is used to predict the type of forest for each observation. The data includes 54 features and seven classes with 581 thousand samples, so it is framed as a multiclass classification problem.

We will first establish a baseline model using Random Forest and measure our performance using ROC-AUC. Then, we will use RFE to remove features to reduce model complexity while improving model predictions (Listing 2-72).

Listing 2-72. The baseline model using Random Forest

```
# internet is required to fetch dataset
from sklearn.datasets import fetch_covtype
from sklearn.ensemble import RandomForestClassifier
from sklearn.metrics import roc_auc_score
from sklearn.model_selection import train_test_split as tts

# load data
forest_cover = fetch_covtype()
forest_cover = pd.DataFrame(data=np.concatenate([forest_cover["data"], forest_
cover["target"].reshape(-1, 1)], axis=1))
# rename target column
forest_cover = forest_cover.rename(columns={54:"cover_type"})
# feature name is from 0 to 53
features = range(54)
X_train, X_test, y_train, y_test = tts(forest_cover[features], forest_cover["cover_type"],
random_state=42, test_size=0.3)

rf = RandomForestClassifier(max_depth=7, n_estimators=50, random_state=42)
```

```
rf.fit(X_train, y_train)
predictions = rf.predict_proba(X_test)
print(roc_auc_score(y_test.values, predictions, multi_class="ovr"))
```

After initial modeling, the ROC-AUC score is sitting around 0.939 using all 54 features. From here, we can recursively remove features by iteratively modeling and comparing performances. For each feature, the trained model has a weight or a value of importance assigned to it based on the training results. The determination of this feature importance value works differently for different algorithms. For example, in regression algorithms, the feature importance is simply the weight multiplied by the feature associated with the weight. Coefficients in regression algorithms act as weights that tell how much each feature should contribute to the final prediction. On the other hand, feature importance in tree-based methods such as Decision Tree and Random Forest is computed as the total reduction in the criterion used that one feature brought.

In the scikit-learn API, feature importance for the model can be obtained by calling either the coef_ or the feature_importances_ attribute on a fitted model. By sorting the importance values and visualizing them through bar graphs (Figure 2-58), we can remove features that contribute to the model poorly (Listing 2-73). However, each time one feature is removed, the coefficient or importance values change. So we need to remove features one at a time and recompute feature importance by retraining the model.

Listing 2-73. Code to obtain and graph feature importance

```
# top 15 important features
feat_import = pd.DataFrame(zip(rf.feature_importances_, features), columns=["importance",
"feature"]).sort_values("importance", ascending=False)
feat_import_top = feat_import[:10].reset_index(drop=True)
# plot as bar graph
plt.figure(figsize=(8, 6), dpi=200)
sns.barplot(x=feat_import_top.importance,y=feat_import_top.feature, data=feat_import_top,
orient="h", order=feat_import_top["feature"])
```

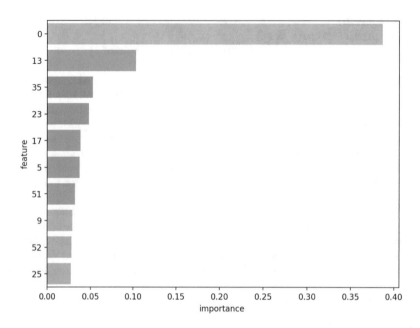

Figure 2-58. *Top ten most important features*

To perform RFE, scikit-learn implemented a wrapper class called RFE (Listing 2-74). The object instantiates with a scikit-learn-like model and various hyperparameters. One major downfall for RFE and the reason that it can't be used in any dataset is its computational expenses. For each iteration of RFE, a new model is trained with the full dataset, which results in heavy computational time and costs. RFE does slightly improve with the parameter step: it defines how many features to remove at each iteration. Instead of removing feature one by one, if necessary, more than one can be removed to decrease training time.

Listing 2-74. Retraining RandomForest with RFE

```
from sklearn.feature_selection import RFE
# select top 20 features by 3
rfe = RFE(estimator=RandomForestClassifier(max_depth=7, n_estimators=50, random_state=42),
n_features_to_select=20, step=3)
rfe.fit(X_train, y_train)

# all the kept features
X_train.columns[rfe.support_]

# evaluate performance with removed feature
X_train_rfe = rfe.transform(X_train)
X_test_rfe = rfe.transform(X_test)

rf = RandomForestClassifier(max_depth=7, n_estimators=50, random_state=42)
rf.fit(X_train_rfe, y_train)
predictions = rf.predict_proba(X_test_rfe)
print(roc_auc_score(y_test.values, predictions, multi_class="ovr"))
```

By keeping the top 20 features and removing the other 34, our ROC-AUC score did not decrease; it even increased to 0.94! But the problem with RFE is apparent: for a dataset with only 50 features and a few hundred thousand samples, the training time is already extremely high. It's difficult to scale RFE to larger datasets containing thousands of features and millions of samples. Even with modern GPUs, one iteration of training can take up to hours.

The trade-off between performance and training time should be considered and kept balanced. Sometimes, RFE would consume a significant amount of computational power and time, while using a faster algorithm can provide slightly decreased accuracy but at a much faster time. It's up to the data scientist to make the decision.

Permutation Importance

Permutation Importance can be seen as another way of calculating feature importance. Both Permutation Importance and feature importance measure how much one feature contributes to the overall prediction. However, the calculation of Permutation Importance is independent of the model, meaning that the algorithm remains the same no matter what machine learning model is used. Permutation Importance's speed depends on the model prediction rate, but it's still relatively faster than other feature selection algorithms such as RFE.

Permutation Importance produces a measure of relevancy from the feature to the target. Logically, features with low Permutation Importance are potentially unnecessary to the model, while features with higher Permutation Importance may be deemed more useful to the model.

The algorithm starts by shuffling the rows of one feature in the validation dataset. After the shuffling, we predict using the trained model and observe the effect that the shuffling has on performance. Theoretically, if one feature is crucial to the model, it would significantly decrease the accuracy of the model prediction. On the other hand, if the feature shuffled does not contribute to model prediction as much, then it wouldn't affect the model performance as much. By computing the loss function compared to the ground-truth values, we can obtain a measure of feature importance by the performance deterioration from shuffled features.

Still using the Forest Cover dataset from RFE as a comparison for feature importance, we can use the following code for the baseline model (Listing 2-75).

Listing 2-75. Code for baseline model

```
# internet is required to fetch dataset
from sklearn.datasets import fetch_covtype
from sklearn.ensemble import RandomForestClassifier
from sklearn.metrics import roc_auc_score
from sklearn.model_selection import train_test_split as tts

# load data
forest_cover = fetch_covtype()
forest_cover = pd.DataFrame(data=np.concatenate([forest_cover["data"], forest_
cover["target"].reshape(-1, 1)], axis=1))
# rename target column
forest_cover = forest_cover.rename(columns={54:"cover_type"})
# feature name is from 0 to 53
features = range(54)
X_train, X_test, y_train, y_test = tts(forest_cover[features], forest_cover["cover_type"],
random_state=42, test_size=0.3)
```

```
rf = RandomForestClassifier(max_depth=7, n_estimators=50, random_state=42)
rf.fit(X_train, y_train)
predictions = rf.predict_proba(X_test)
print(roc_auc_score(y_test.values, predictions, multi_class="ovr"))
```

Scikit-learn does not provide a naïve implementation for Permutation Importance; instead, we can use the eli5 library, which is compatible with scikit-learn models.

To calculate and display Permutation Importance, we can simply call `fit` and `feature_importances_`, respectively (Listing 2-76).

Listing 2-76. Permutation Importance with eli5

```
import eli5
from eli5.sklearn import PermutationImportance
from sklearn.metrics import make_scorer

# convert metric to scorer for eli5
scocer_roc_auc = make_scorer(roc_auc_score, needs_proba=True, multi_class="ovr")
perm = PermutationImportance(rf, scoring=scocer_roc_auc,random_state=42).fit(X_test, y_test)

# top 15 important features
feat_import = pd.DataFrame(zip(perm.feature_importances_, features), columns=["importance",
"feature"]).sort_values("importance", ascending=False)
feat_import_top = feat_import[:10].reset_index(drop=True)
# plot as bar graph
plt.figure(figsize=(8, 6), dpi=200)
sns.barplot(x=feat_import_top.importance,y=feat_import_top.feature, data=feat_import_top,
orient="h", order=feat_import_top["feature"])
```

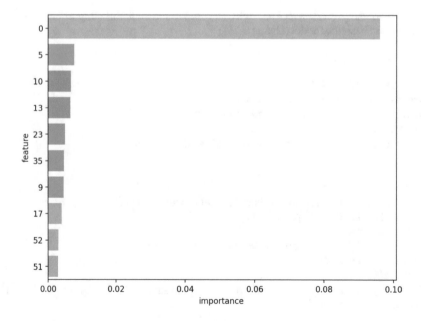

Figure 2-59. *Top ten most important features using Permutation Importance*

Comparing the calculated feature importances with those from Random Forest in the "Recursive Feature Elimination" section, we observe that the order of importance follows a similar pattern but is not exactly the same (Figure 2-59). For comparison against RFE and its calculation of feature importance, we will take the top 20 features and retrain our Random Forest model (Listing 2-77).

Listing 2-77. Retraining Random Forest with top 20 features selected

```
# retrain model with top 20 features
rf = RandomForestClassifier(max_depth=7, n_estimators=50, random_state=42)
X_train_permu = X_train[feat_import[:20]["feature"].values]
X_test_permu = X_test[feat_import[:20]["feature"].values]

rf.fit(X_train_permu, y_train)
predictions = rf.predict_proba(X_test_permu)
print(roc_auc_score(y_test.values, predictions, multi_class="ovr"))
```

We obtain a ROC-AUC score of around 0.9416, slightly higher than using RFE. RFE and Permutation Importance are extremely alike in terms of algorithmic approaches, but their proper use cases are quite different. Permutation Importance requires the testing data to be labeled, and the shuffling of features is random. Sometimes deterministic results may be more important than computational expenses. On the contrary, sometimes where computational time is not easily available, Permutation Importance provides a faster approach to feature selection that produces competitive results.

LASSO Coefficient Selection

Recall that during Linear Regression, a coefficient is assigned to each feature, acting as a weight that decides how much that feature will contribute to the final prediction. Ideally, a perfectly trained regression model would also have perfect coefficients and thus perfect feature importance. As RFE and Permutation Importance demonstrate the concept, we could select and remove features based on their feature importance. Those features with low or zero weight are unimportant or do not contribute to the prediction, so we do not need them for training as they only increase training time and possibly even reduce the performance of our models. Luckily, LASSO regression does this exactly. Depending on the adjustable hyperparameter, the weights of unimportant features will shrink to zero.

As mentioned in the previous chapter, the Least Absolute Shrinkage and Selection Operator (LASSO) regression adds a penalty term to Linear Regression for regularization:

$$c(\beta, \varepsilon) = \frac{1}{n} \|X\beta + \varepsilon - y\|_2^2 + \lambda \|\beta\|_1$$

We can utilize the fact that LASSO regression shrinks the weights of some features to zero as a feature selection technique, removing those with zero weight. The λ parameter controls how much shrinkage there should be. The greater λ is, the more likely regularization is to lead to zero weights.

LASSO regression itself might not be fit for prediction on datasets as it's too simple to model the complex relationship present in every intricate dataset. However, we can utilize its coefficient to select useful features and train these features with another model that might fit the dataset better. The process is demonstrated in the following using the same process as in RFE and Permutation Importance: train a baseline model, perform feature selection, and then retrain the model and observe for improvements (Listing 2-78).

Listing 2-78. Baseline model

```python
# internet is required to fetch dataset
from sklearn.datasets import fetch_covtype
from sklearn.ensemble import RandomForestClassifier
from sklearn.metrics import roc_auc_score
from sklearn.model_selection import train_test_split as tts

# load data
forest_cover = fetch_covtype()
forest_cover = pd.DataFrame(data=np.concatenate([forest_cover["data"], forest_
cover["target"].reshape(-1, 1)], axis=1))
# rename target column
forest_cover = forest_cover.rename(columns={54:"cover_type"})

# feature name is from 0 to 53
features = range(54)
X_train, X_test, y_train, y_test = tts(forest_cover[features], forest_cover["cover_type"],
random_state=42, test_size=0.4)

rf = RandomForestClassifier(max_depth=7, n_estimators=50, random_state=42)
rf.fit(X_train, y_train)
predictions = rf.predict_proba(X_test)
print(roc_auc_score(y_test.values, predictions, multi_class="ovr"))
```

After training a baseline model, we train a LASSO regression model on the data with some value as λ. As the value of λ increases, more features tend to get a weight of zero. There's no optimal way of finding λ other than trial and error. In the following example, λ is set to 0.005, removing 26 features after training (Listing 2-79).

Listing 2-79. Feature selection using LASSO and retraining after selection

```python
from sklearn.linear_model import Lasso

lasso = Lasso(alpha=0.005)
lasso.fit(X_train, y_train)

# convert all weights to positive values,
# feature importance
feat_import_lasso = abs(lasso.coef_)

# select features with coeffcient greater than 0
features = np.array(features)[feat_import_lasso > 0]

# retrain model with selected features
X_train, X_test, y_train, y_test = tts(forest_cover[features], forest_cover["cover_type"],
random_state=42, test_size=0.4)

rf = RandomForestClassifier(max_depth=7, n_estimators=50, random_state=42)
rf.fit(X_train, y_train)
predictions = rf.predict_proba(X_test)
print(roc_auc_score(y_test.values, predictions, multi_class="ovr"))
```

The ROC-AUC score of our model improved to about 0.9406 with only 28 features out of the 54 original ones. Remember that this result can be improved upon by tuning the hyperparameter alpha, or λ in our preceding equation. Lastly, we can visualize some of the most important features calculated by LASSO using a bar graph.

Notice that in our feature importance graph in Figure 2-60, our top features are somewhat different compared with those produced by Random Forest in RFE and Permutation Importance. One major difference between Random Forest and LASSO regression is their ability to model relationships. LASSO regression can only model linear relationships, while Random Forest can handle data in which the feature-target relationship is nonlinear. LASSO regression may not explain the convoluted relationships in modern datasets, but its speed is extremely fast compared with other algorithms including Information Gain, RFE, and Permutation Importance. LASSO coefficient selection can commonly serve as quick insight into feature selection or act as a baseline selection tool. However, it should be used cautiously since it may remove important features to models that can model nonlinear relationships.

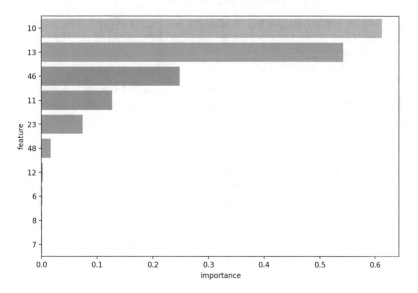

Figure 2-60. *Top ten most important features using LASSO regression*

Key Points

In this chapter, we discussed several key components of data preparation and engineering: TensorFlow datasets, data encoding, feature extraction, and feature selection.

- TensorFlow datasets are used for large datasets intended for neural network models to make them memory-feasible. TensorFlow Sequence datasets are user-defined data loading classes that offer flexibility in how data is loaded and passed into the model.

- Not all tabular datasets are small and convenient to manipulate, especially modern tabular datasets related to the biomedical field. Five methods are introduced to either reduce the dataset size or avoid loading in the entirety of the dataset file into memory:

- Pickle files are Python-specific file formats. Saving Pandas DataFrames to pickle files can decrease loading time and decrease file size.

- SciPy and TensorFlow sparse matrices are both approaches to compress sparse data. This allows for easier manipulation on the dataset without worrying about OOM errors.

- Pandas Chunker allows for Pandas DataFrames to be loaded as an iterator with a user-specified chunk size. This can be used in conjunction with TensorFlow datasets to load in only one batch of data at once.

- Storing data in h5 files compresses the data in binary format. The Python library h5py can create a "link" between a variable defined within program and the file stored on disk. This is a lot more flexible compared with Pandas Chunker as any number or portion of the data can be accessed.

- A NumPy memory map provides similar functionality as pyh5 in terms of creating a reference between the program and data stored on disk. However, NumPy memory maps can be used without all the hassle of using h5py syntax and can be done in one line.

- Often, the raw data that we collect is not in a form suitable for model training. The primary prerequisite for data is that it must be quantitative, but we also want the data's quantitative form to be representative of its nature or properties (i.e., we want to *encode* the attributes of the data into how the model sees it).

  - *Strategies for encoding discrete data*: Label encoding (each class is arbitrarily associated with an integer), one-hot encoding (one position corresponding to the class in a one-hot vector is marked with a 1, whereas all others are marked as 0), binary encoding (the binary representation of the label encoding for each class is used), frequency encoding (the class is associated with the frequency of that class in the dataset), target encoding (the class is associated with the aggregated target value for items of that class), leave-one-out encoding (target encoding but the current row is not considered in the aggregation calculation), James-Stein encoding (like target encoding, but considers both the overall mean and the individual mean per class), weight of evidence encoding (uses the WoE formula to determine how much a class helps distinguish the target).

  - *Strategies for encoding continuous data*: Min-max scaling (the data is scaled such that the lowest value is 0 and the highest is 1 or some other set of bounds), robust scaling (similar to min-max scaling, but the first and third quartiles are used as relevant distribution markers instead of the min and max values), standardization (data is divided by the standard deviation after subtracting the mean).

  - *Strategies for encoding text data*: keyword search, raw vectorization (treat each word/token as a class and run one-hot encoding), a bag of words (count the number of words/tokens in a sequence), *n*-grams (count the number of sequential combinations of *n* words/tokens in a sequence), TF-IDF (balances how often a term appears in a document vs. overall to evaluate its relevance), sentiment extraction (quantitative markers of a text's sentiment qualities), Word2Vec (embeddings learned by a neural network as optimal to performing a language task).

- *Strategies for encoding time/temporal data*: Represent each time as the number of base units after a starting time (only use for interpolation), extract cyclical features of time (season, month, weekday, hour, etc.), and detect if it is a significant day/time (holiday, rush hour, etc.).

- *Strategies for encoding geographical data*: Obtain quantitative abstract subcomponents of location like country or state/province and obtain latitude and longitude.

- Feature extraction/engineering methods are crucial techniques to consider before applying any sort of model. Feature extraction algorithms aim to find a representation of the original data in a lower dimension that preserves the most information.

  - Algorithms like PCA and t-SNE are unsupervised dimensionality reduction techniques that project data onto a lower dimension while trying to preserve the overall structure of data as much as possible.

  - PCA is fast compared with t-SNE but does not focus on local distances between points. While t-SNE does put more attention to spaces between local points, the algorithm is extremely computationally heavy and has rare use cases for feature extraction, although, as we will see later on, t-SNE plays a major role in model interpretation during deep learning pipelines.

  - On the other hand, LDA is supervised and assumes that the data is normally distributed and has a similar variance. LDA is a classification algorithm, but it can also be used as a dimensionality reduction technique with the number of components kept being equal to or lower than the number of classes. Again, LDA isn't as popular as PCA when used for feature extraction, but it does perform fairly well in most situations.

  - Finally, statistics-based engineering creates new features by extracting higher-order statistics of each row in the dataset. This can help the model look at how each single feature associates with the target and how each sample's features as a whole affect the target.

- Feature selection methods can be utilized to not only improve model performance but decrease training time by reducing the size of the dataset. Most if not all feature selection methods follow the two basic steps: obtain value for a certain metric and select features based on the threshold of the metric.

  - Algorithms such as the Variance Threshold, high-correlation method, and LASSO coefficient selection are fast to execute even operating on large datasets. However, none of these methods are model specific, meaning that they apply universally to any dataset regardless of what model will be used to train.

  - Feature selection techniques such as RFE and Permutation Importance are model specific as they rely on the output of trained models. They tend to perform better than those methods mentioned previously, but they are significantly more computationally expensive as they require multiple iterations of model training.

  - Information Gain provides better results than the Variance Threshold and high-correlation method, but like those that are model specific, it's computationally expensive.

In the next chapter, we will begin our exploration of deep learning with neural networks.

PART II

Applied Deep Learning Architectures

■ ■ ■

## CHAPTER 3

Neural Networks and Tabular Data

The foundation of all technology is fire.

—Isaac Asimov, Writer and Professor at Boston University

Deep learning methods operate on entirely different concepts than classical machine learning algorithms introduced earlier. Neural networks are the core and foundation of deep learning and address many of the weaknesses of classical machine learning discussed toward the end of Chapter 1. This chapter introduces you to the core theory and mathematics of deep learning, which operates on different paradigms than much of classical machine learning, as well as corresponding implementation in the popular and easy-to-approach deep learning framework Keras.

This chapter will introduce you to neural networks from a mathematical perspective as well as corresponding implementation in the popular and easy-to-approach deep learning framework Keras. We will begin by exploring what neural networks are and the structure that enables their representation power. You'll learn about defining and training Keras models but also discover that their performance is nowhere near their potential. Afterward, the chapter will dive deeper into the theory and mathematics behind the feed-forward and backpropagation processes. Along with that, you will learn the details of activation functions and why they play such a crucial role in unleashing the full power of neural networks. In the second part of the chapter, we will dive into more advanced neural network usage and manipulation, including training callbacks, the Keras functional API, and model weight sharing. Lastly, we will review several research papers demonstrating simple mechanisms to improve neural network performance on tabular data.

This chapter sets the foundation of deep learning for the rest of the book! The following chapters rely upon a strong knowledge of the theory and tools discussed in this chapter.

What Exactly Are Neural Networks?

Machine learning surrounds the idea of generalization. The ability to adapt to and learn from similar but not identical situations is what differentiates machine learning from hard-coded algorithms.

Take ourselves as an example: if we're presented with two images that both resemble cats, even if the two images slightly differ in appearance, we'll still be able to say that they are both cats confidently. Our minds can differentiate between the two images but also determine that they both are cats – this is a generalization. The brain achieves this by learning patterns that it is presented over our lifetime. Machine learning mimics this concept: algorithms learn from data and recognize patterns in order to generalize for data that they have never seen.

© Andre Ye and Zian Wang 2023
A. Ye and Z. Wang, *Modern Deep Learning for Tabular Data*,
https://doi.org/10.1007/978-1-4842-8692-0_3

The human brain consists of billions of neurons connected with each other by synapses, forming a very large network that controls our thinking and guides our actions. Our senses receive information and pass it to the brain; neurons process and transfer information between each other through electrical pulses and chemical signals. Afterward, the data is passed throughout our body to the nervous systems, which act upon the outputted information.

Each neuron in our body receives input and outputs its processed information. Perceptrons, mathematical models inspired by the neuron model of information processing, were proposed by Frank Rosenblatt in 1958 (long before the modern age of supercomputers). However, due to the technological limitations at that time, the full potential of perceptrons was not discovered. It was not until the 1980s, when more research was put into AI and machine learning, that the idea of a network of perceptions emerged. This is now known as an artificial neural network (Figure 3-1).

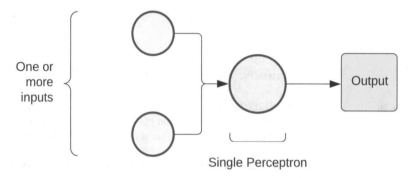

Figure 3-1. *The perceptron model*

The core concept of a neural network mimics how human neurons connect and process information. Instead of electrical pulses, imagine each neuron stores a value that represents its ability and strength to communicate its information to other neighboring neurons (Figure 3-2).

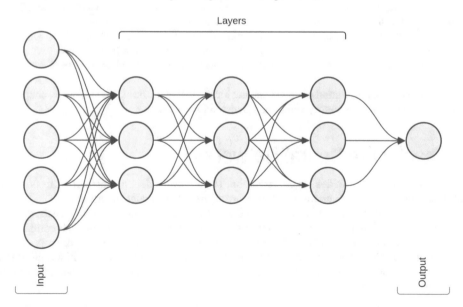

Figure 3-2. *Simple illustration of a neural network*

The network is structured into layers, with the first layer receiving inputs and the final one outputting results. Typically, each neuron in each layer is connected to every neuron in the previous and the next layer. Information flows forward from the first layer to the nth layer. Every single connection differs from the other; some neurons may contribute more to the final prediction, while others may only affect a little bit. Information is passed through from layer to layer, from the input to the output. The training of neural networks happens through a process called backpropagation, which we will discuss in detail in later sections.

Neural Network Theory

There are an absurd number of variations on the standard or so-called "classic" neural network. Some are going to be covered in this book, while others are left for you to explore on your own. The foundation that supports all these variations is known as the artificial neural network (ANN). Alternate names for ANN include multilayer perceptron (MLP) and fully connected network (FCN); these terms will be used interchangeably throughout the book. Understanding and learning to utilize the power of ANNs serve as a prerequisite to the endless land of deep learning.

Starting with a Single Neuron

We start our understanding of neural networks with the simplest mathematical model of a neuron, a single perceptron with two inputs and one output. The neuron represents some function that combines the information from the two inputs and outputs a meaningful result (Figure 3-3).

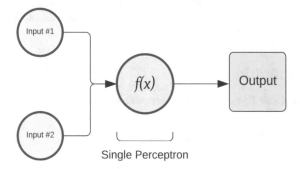

Figure 3-3. *Demonstration of a perceptron model*

Realistically, there would be some method for the perceptron model to learn from its mistakes and correct the output value. We can introduce adjustable weights that multiply with each value of the inputs, and the final output will be determined by the sum of each input multiplied by its weight. The weights of our "model" can be updated iteratively to produce the correct values for predictions. The process in which the weights update is currently unimportant to our current context. (We will discuss how weights are updated later in the chapter.) One additional improvement that we can implement is a trainable bias value. By shifting the weighted sum output by an optimal amount, we can ensure the network will be able to reach a wide variety of values, thus having the ability to model more complex functions (Figure 3-4).

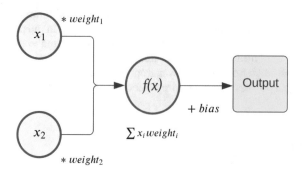

Figure 3-4. *Perceptron model with weight and bias*

We can extend our generalization of producing the output through a single perceptron into a mathematical formula where x_i is the feature where n is the number of features, w_i is the weight, and b is the bias:

$$Output = b + \sum_{i=1}^{n} x_i w_i$$

Some may recognize that Linear Regression operates on similar concepts compared to the perceptron model described previously. The capability of Linear Regression is extremely limited as it can only model and understand linear relationships between variables, just like our simple perceptron model described previously. This is where the "network" in the neural network comes into play. With hundreds of neurons stacked into tens of layers, each layer processes different portions of the data, recognizing local patterns and combining information obtained by each layer or neuron.

Feed-Forward Operation

Expanding from the idea of a single neuron, we move onto a multilayer perceptron model or an ANN (Figure 3-5).

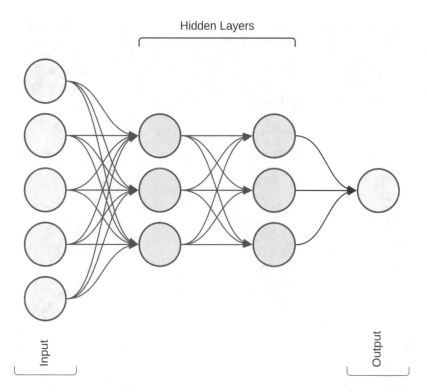

Figure 3-5. *A simple neural network example*

We consider each column of neurons to be a layer within the network, where the first layer receives input and the last outputs predictions. The number of features, or the dimension of the dataset, translates to the number of neurons in the input layer. In the case of regression, the output layer would contain a single neuron that produces a predicted value. Similarly, for binary classification, the output layer will only have one neuron, but this time the value of it will be limited to between 1 and 0 by an activation function. Activation functions will be explored further in-depth during subsequent sections. They can be temporarily seen as tools to help neurons adjust their values to fit the output range. For example, classification tasks require the output to be in the range of [0, 1], represented as a probability. Hence, we can use the sigmoid activation function to transform the raw output values to be between 0 and 1.

The layers between the input and the output layers are referred to as "hidden layers." In the diagram shown previously, we have two hidden layers, each with three neurons. The number of hidden layers and neurons are hyperparameters that can be adjusted to improve a neural network's performance.

We can imagine the neural network analyzes the data by breaking it into different portions. Moreover, we can interpret each weight value as a helper that manipulates the inputting data/intermediate result from the previous layer to fit the task that each neuron is trained to recognize. For instance, every neuron in the first hidden layer could be trained to discover certain underlying statistical distributions within the data, while the second hidden layer processes the information passed down from the first and produces intermediate results for the final hidden layer to compute predictions.

Depending on the input layer weights, each neuron in the first hidden layer could receive a "modified" version of the original input, thus being able to interpret the dataset from different perspectives (Figure 3-6). If we count each weight, or connection, as a trainable parameter, our simple four-layer network with five inputs and one output will have $5 \times 3 + 3 \times 3 + 3 \times 1 = 27$ parameters. However, remember there's also a bias added after the summation of weights from the previous layer for each neuron. Thus, our total parameter from the preceding example network will be $27 + 3 + 3 + 1 = 34$.

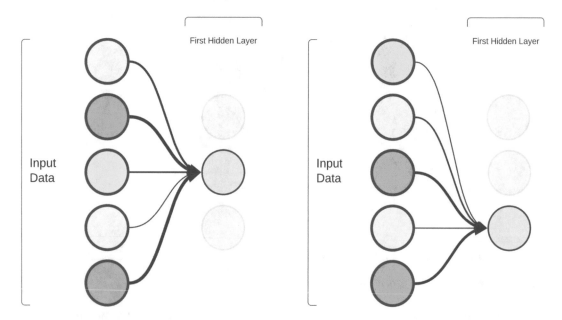

Figure 3-6. *How different weights change input data*

The process of passing information from the first layer forward to the last layer is referred to as a feed-forward operation. Not only is it the starting point for neural network training but it's also how predictions are made. A general process for training a neural network can be outlined in five steps:

1. Initialize random weights and biases.

2. Calculate an initial prediction by feed-forward.

3. Calculate the network's error based on a differentiable metric, essentially telling us how well the network is doing on the data.

4. Adjust the values of weights and biases based on the error made using backpropagation.

5. Repeat until the desired accuracy is reached.

Before diving into the complex yet fascinating math behind how neural networks learn using feed-forward and backpropagation, we're going to construct a simple neural network in Python, familiarizing ourselves with the concepts introduced previously by a concrete example. For now, treat the backpropagation process as an algorithm that adjusts network parameters to increase performance based on the error.

Introduction to Keras

The popular deep learning library Keras, created by Francois Chollet, allows straightforward implementation of neural networks from simple to sophisticated and convoluted models of any kind. Due to its remarkable usability and performance, we chose Keras as the framework in this book to develop and demonstrate deep learning concepts.

Meaning "Horn" in French, the name *Keras* is derived from the literary image presented in the *Odyssey*. Initially developed as a research project for ONEIROS (Open-ended Neuroelectronic Intelligent Robot Operating System), Keras quickly expanded to provide general uses and support across the field of deep learning. Keras offers performance and strength comparable to any modern standard by focusing on the idea of "progressive disclosure of complexity." The Keras framework is widely employed around the world, with renowned companies like NASA and Google using it. While advanced modeling and workflow can be achieved through a clear process, simple and quick ideas can be implemented with minimal effort.

Keras itself is a high-level library designed to run on top of many low-level deep learning packages such as TensorFlow, Theano, and Cognitive Toolkit. The core motivation behind the library is to provide an easy interface that can better bridge ideas and implementation. By default, Keras is built upon the deep learning platform TensorFlow developed by Google. During its early releases, TensorFlow provided detailed but complex systems and classes for developing deep learning models. By version 2.0, Keras's popularity made it the official API for TensorFlow. Although Keras remains a separate library, TensorFlow can fill in the gaps for any low-level training control Keras lacks. It is recommended to install TensorFlow instead of Keras's standalone package to take full advantage of TensorFlow's endless customization abilities along with Keras (Listing 3-1).

> TensorFlow can be installed via pip from the command line as well as directly from the Jupyter Notebook by adding an exclamation mark before the command. Note that inserting an exclamation mark (!) before any line is equivalent to running the command from the command line in Jupyter Notebooks.

Listing 3-1. TensorFlow installation

```
!pip install tensorflow
import tensorflow as tf # import tensorflow as a whole
from tensorflow import keras # only import Keras
```

Modeling with Keras

Before we start, remember that, at this point, not all mathematical concepts and neural network components are explained yet. However, you do not need to understand everything to build a working network in Keras. Proceeding this brief introduction to Keras will be an in-depth explanation of how neural networks function behind the curtains.

Consider the Fashion MNIST dataset. The dataset contains 70,000 grayscale images at 28-by-28-pixel resolution. Fashion MNIST is a multiclass classification task, sorting various images of clothing articles into ten categories (Figure 3-7).

Class Label	Class Description
1	T-shirt/Top
2	Trouser
3	Pullover
4	Dress
5	Coat
6	Sandal
7	Shirt
8	Sneaker
9	Bag
10	Ankle Boot

Figure 3-7. *Class description for Fashion MNIST*

The dataset is automatically installed with TensorFlow and can be imported through the Keras API via `tf.keras.datasets.fashion_mnist.load_data()`, which returns the train images, train labels, test images, and test labels all as NumPy arrays (Listing 3-2). Fashion MNIST provides variety and a relatively challenging task compared with other commonly used benchmarking datasets.

Listing 3-2. Retrieving the Fashion MNIST dataset

```
# retrieving Fashion MNIST
# needs internet
(X_train, y_train), (X_test, y_test) = tf.keras.datasets.fashion_mnist.load_data()
```

By calling `imshow()` from matplotlib, we can display our data as images, with each value representing the brightness of that pixel (Listing 3-3).

Listing 3-3. Using matplotlib to visualize data

```
# description of each target
targets = ['T-shirt/top', 'Trouser', 'Pullover', 'Dress', 'Coat', 'Sandal', 'Shirt',
'Sneaker', 'Bag', 'Ankle boot']

# Display images on a 3x3 grid
plt.figure(figsize=(8,7), dpi=130)
# 9=3x3
for i in range(9):
    # placing the image in the ith position of a 3 by 3 grid
    plt.subplot(3,3,i+1)
    plt.xticks([])
    plt.yticks([])
    plt.grid(False)
```

```
# extract data from training images, shown in grayscale
plt.imshow(X_train[i], cmap="gray")
# get the respective target for the image
plt.xlabel(targets[y_train[i]])
plt.show()
```

Note that images are stored in a three-dimensional array, with each sample having 28 columns and 28 rows of pixels and having 60,000 of 28-by-28 training images in total (Figure 3-8) (Listing 3-4).

Figure 3-8. *Visualization of the first nine images from the training data*

Listing 3-4. The shape of training data

```
X_train.shape
# (60000, 28, 28)
```

Artificial neural networks are only capable of processing one-dimensional input. The input layer treats each row as a separate sample. By flattening each two-dimensional array and reshaping it into one single row, we can obtain the desired shape for the input of a fully connected neural network (Listing 3-5). By doing this, we do sacrifice any structural information that was present in a two-dimensional image, but for our context of ANNs, this is a viable solution. Models that are specifically designed to perform image recognition tasks will be discussed in Chapter 4, along with how to utilize them for tabular data. But for now, treat Fashion MNIST as a tabular dataset with 784 features.

Listing 3-5. Flattening 2D array into 1D

```
# shape into (num_samples, 784)
X_train = X_train.reshape(X_train.shape[0], 28*28)
# shape into (num_samples, 784)
X_test = X_test.reshape(X_test.shape[0], 28*28)
```

Each pixel in the image is a number between 0 and 255, with 0 being the darkest black and 255 being the brightest white. A common practice is to normalize these values between 1 and 0, speeding up model convergence and stabilizing training (Listing 3-6).

Listing 3-6. Normalize data

```
X_train = X_train / 255.
X_test = X_test / 255.
```

The modeling process in Keras follows three basic steps: define the architecture, compile with additional parameters, and finally train with provided data (Figure 3-9).

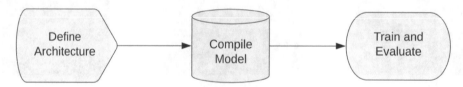

Figure 3-9. *The Keras workflow*

Defining the Architecture

Since the flattened image data contains 784 input features, our first input layer will have 784 neurons, one for each feature. We can start by initializing the model with the Sequential class from keras.models. Keras's sequential workflow allows us to stack layers linearly on top of each other in an ordered manner (Listing 3-7). This method of constructing neural networks is deemed adequate and convenient for simple prediction tasks like Fashion MNIST.

Listing 3-7. Initialize the Keras sequential model

```
# import the dense and input layer
from keras.layers import Dense, Input
# the sequential model object
from keras.models import Sequential
# initialize
fashion_model = Sequential()
```

Layers are added by calling the add() method on the sequential model object after initialization (Listing 3-8). We start by adding our input layer, specifying the shape of the input data.

Listing 3-8. Adding the input layer

```
# add input layer, specify the input shape of one sample as a tuple
# in our case it would be (784, ) as it's one-dimensional
fashion_model.add(Input((784,)))
```

Following, we will add our hidden layers. For the simplicity of architecture, our model will have two hidden layers, each with 64 neurons (Listing 3-9). You can experiment with these parameters and observe possible improvements to the model. Recall from earlier that in an ANN, each neuron in every hidden layer is connected to every neuron in the previous and the next layer. Such layers are categorized as fully connected layers and can be imported through keras.layers as Dense layers.

Listing 3-9. Adding the dense layers in the network by specifying the number of neurons in Dense calls

```
fashion_model.add(Dense(64))
fashion_model.add(Dense(64))
```

Finally, the output layer contains ten neurons as there are ten classes, and similarly to hidden layers, the output layer will also be a Dense layer. The model would ideally be able to output an integer between 1 and 10, with each number corresponding to one class from the Fashion MNIST dataset. However, all neural networks and modern machine learning models output continuous values. Hence, we can insert an activation function that normalizes values to be within our output range. In multiclass classification tasks, the activation function softmax is used. With ten neurons in the output layer, logically, our network will output an array of ten values during prediction. Softmax will then convert these values into probabilities between 1 and 0, which represent the likelihood that the predicted image belongs to each of the ten classes while all the probabilities sum up to 1 (unlike sigmoid, where each output neuron would be interpreted as a single probability with no relations to other outputs). The final prediction can be determined by finding the max of these values, and the position of the max value will be our result, a number between 1 and 10 (Figure 3-10). Again, details of activation functions will be explored in later sections.

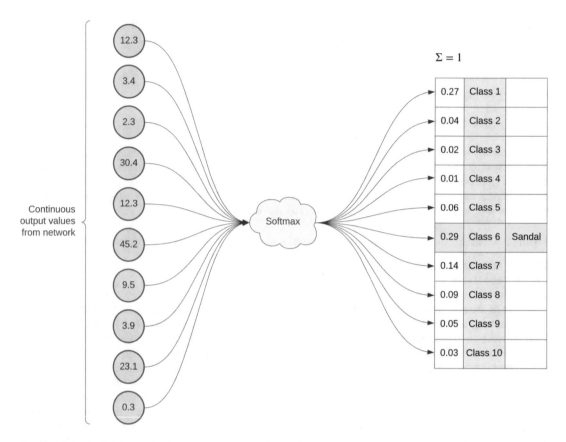

Figure 3-10. *An intuition of softmax*

Thus, the complete code for defining the architecture of our simple network is written as such in Listing 3-10.

Listing 3-10. Complete code for defining the architecture

```
# import the dense and input layer
from keras.layers import Dense, Input
# the sequential model object
from keras.models import Sequential
# initialize
fashion_model = Sequential()
# add input layer, specify the input shape of one sample # as a tuple
# in our case it would be (784, ) as it's one-dimensional
fashion_model.add(Input((784,)))

# add dense layers, the only parameter that we need to
# worry about right now is the number of neurons, which
# we set to 64
fashion_model.add(Dense(64))
fashion_model.add(Dense(64))
```

```
# add output layer
# softmax activation can be specified by the "activation"
# parameter
fashion_model.add(Dense(10, activation="softmax"))
```

Compiling the Model

Before training, our model needs a few more settings that specify how training should be done. There are three key parameters that should be defined during compilation:

- *Optimizer*: Controls the method of "learning" by telling backpropagation how to adjust the values of weight and bias. Different optimizers can affect the speed and the results of training. By default, the Adam (Adaptive Moment Estimation) optimizers are used. The intuition behind optimizers is best understood after learning about backpropagation.

- *Loss function*: Differentiable functions that measure the performance of the model. As explained in Chapter 1, the key difference between metrics and loss functions is that loss functions must be differentiable to be compatible with gradient descent. However, metrics don't necessarily satisfy differentiability; it's simply a measure of correctness used for evaluation to assess model performance. For multiclass classification, categorical cross-entropy is generally used as a loss function, while the Mean Squared Error is typically used for regression tasks. Different loss functions suited for the same task can produce different training results.

- *Metrics*: Different from loss functions, the training of neural networks doesn't rely on the metric; it simply acts as another tool to better monitor the model performance. Metrics can usually provide an understanding of the model from a different aspect than the loss function. In some cases, metrics and loss functions can be the same.

The following code in Listing 3-11 compiles the model with the Adam optimizer, categorical cross-entropy loss, and accuracy as the metric. The categorical cross-entropy loss is simply a modified version of the classic binary cross-entropy loss suited for multiclass classification tasks. More will be covered in the "Loss Functions" section.

Listing 3-11. Compiling the model

```
# import loss function
from keras.losses import SparseCategoricalCrossentropy
# sparse categorical cross entropy simply converts one-
# hot encoded vector(results from the network)
# to the index position at which the probability is the highest (actual target)
# we need to do this since the y_train is not one-hot
# encoded but instead the targets are represented by
# numbers 1-10
# if the target it one-hot encoded (i.e. in our case it
# would be in the shape of (60000,10))
# then just using "CategoricalCrossentropy" would be fine

fashion_model.compile(optimizer="adam",loss=SparseCategoricalCrossentropy(),
metrics=["accuracy"])
```

Training and Evaluation

After compiling, our model is ready to be trained. Again, there are a few key parameters to consider for the training process:

- *Training data*: It's passed into the fit method as x and y. Similar to scikit-learn models, the data in the place of x and y can be NumPy arrays or Pandas DataFrames.

- *Epochs*: This is the number of times that the model will loop through the training data. More specifically, this represents the number of times the network will be evaluated as each loop through the dataset counts as one epoch, and at the end of each epoch, the selected metric will be calculated on the network to track performance. The parameter is passed in as epoch=num_epochs.

- *Batch size*: This parameter controls how many samples are processed at once in each training step. For the model's batch size equal to 1, it means that in order to loop through the whole dataset once, it'd take 60,000 steps of learning as there are 60,000 samples in the training dataset. For a larger batch size, training speed will speed up significantly, and typically a large batch size will not only increase training speed but also improve performance compared with tiny batch sizes. This pattern varies for different datasets, and there's no definite method to calculate the perfect batch size for training. Note that having large batch sizes may result in memory overflows. The parameter is passed in as batch_size=num_batch.

Although there are many more parameters that can affect training or reduce overfitting, these three are the most relevant to consider, while the rest will be discussed in later sections. Shown in Listings 3-12 and 3-13 are the code and results for training, along with the results displayed as progress bars. Remember that each time .fit() is called, the training result may be slightly different since each time, the neural network initializes with random weight.

Listing 3-12. Model training code

```
# batch_size is randomly chosen at 1024 here,
# reader can change it and observe changes in training results
fashion_model.fit(X_train, y_train, epochs=15, batch_size=1024)
```

Listing 3-13. Model training results

```
Epoch 1/15
10/10 [==============================] - 0s 6ms/step - loss: 2.0185 - accuracy: 0.3140
Epoch 2/15
10/10 [==============================] - 0s 6ms/step - loss: 1.1346 - accuracy: 0.6298
Epoch 3/15
10/10 [==============================] - 0s 6ms/step - loss: 0.8539 - accuracy: 0.6997
Epoch 4/15
10/10 [==============================] - 0s 6ms/step - loss: 0.7443 - accuracy: 0.7387
Epoch 5/15
10/10 [==============================] - 0s 6ms/step - loss: 0.6778 - accuracy: 0.7676
Epoch 6/15
10/10 [==============================] - 0s 6ms/step - loss: 0.6344 - accuracy: 0.7867
Epoch 7/15
10/10 [==============================] - 0s 6ms/step - loss: 0.6000 - accuracy: 0.7967
Epoch 8/15
```

```
10/10 [==============================] - 0s 6ms/step - loss: 0.5769 - accuracy: 0.8060
Epoch 9/15
10/10 [==============================] - 0s 6ms/step - loss: 0.5625 - accuracy: 0.8083
Epoch 10/15
10/10 [==============================] - 0s 6ms/step - loss: 0.5420 - accuracy: 0.8153
Epoch 11/15
10/10 [==============================] - 0s 6ms/step - loss: 0.5271 - accuracy: 0.8226
Epoch 12/15
10/10 [==============================] - 0s 5ms/step - loss: 0.5145 - accuracy: 0.8246
Epoch 13/15
10/10 [==============================] - 0s 6ms/step - loss: 0.5060 - accuracy: 0.8268
Epoch 14/15
10/10 [==============================] - 0s 8ms/step - loss: 0.5015 - accuracy: 0.8253
Epoch 15/15
10/10 [==============================] - 0s 6ms/step - loss: 0.4942 - accuracy: 0.8291
```

We observe that the accuracy of the model slowly converges to somewhere around 0.82 at epoch 15, while the loss is at 0.49. Model predictions can be done by fashion_model.predict(X_testdata). Performance measures on the validation or the test data, as the variables are named X_test and y_test in our example, can be done by fashion_model.evaluate(X_test, y_test). Performance is calculated by both the loss and metric(s) passed in at compilation.

As we expand on the knowledge of neural networks, we'll improve upon our model by adjusting learning rates, adding activation functions, and more to improve not only training performance but also validation results.

Before exploring deeper into Keras and its capabilities in building neural networks and developing deep learning pipelines, let's step back and understand a bit more about the math and intuition of neural networks. Prior to Keras, neural networks are introduced as these "components," which multiply and add numbers together to form predictions, and they somehow adjust those "weights and biases" to improve performance through "backpropagation." Along with the actual learning process, loss functions, activations, and optimizers will be introduced intuitively and through mathematical lenses to fully understand how they contribute to the network.

Loss Functions

Loss functions are briefly defined and compared with metrics in Chapter 1. Both terms define a function that measures the performance of model predictions against ground-truth data. The difference that separates loss functions and metrics stems from the function's differentiability.

Loss functions, speaking strictly in the context of deep learning, are differentiable, or their derivative is defined at any point in the domain of the function. We can use its differentiability for gradient descent as it helps us search through the loss landscape effectively and in an orderly manner.

To reiterate, a loss function requires two inputs, the model's prediction and the ground-truth value, and outputs a measure that acts as an indicator of how "good" the predictions are compared with the ground-truth values. In the same way as regression methods, the core concept behind neural network learning relies on gradient descent. The goal of backpropagation is to minimize a high-dimensional non-convex loss function through the iterative process of gradient descent.

Theoretically, the loss landscape can be visualized in dimensions where the number of parameters and the added dimension represent the actual loss value. But our senses and visual abilities limit us to three-dimensional visualizations compared with the millions of dimensions in the context of a neural network. It's helpful to see the actual "landscape" that the network is traversing through during training for a few reasons. First, some networks are known to produce loss functions that are "smoother" or easier to train in general. Being able to have a visual perception of the loss landscape can enable us to better understand the

197

relationship between neural network structure and training results. Second, comparing loss landscapes can be another tool to assess model performance and its actual ability to fit the data based on the complexity of the loss landscape. Along with assessing the ability of the model, visualizations also help a lot in model interpretation and understanding. Interpreting models as complex as neural networks is rewarding as you get an insight into how training progresses and how the model "learns."

There are a few methods through which the "landscape" of loss functions in a neural network can be visualized. Among them, the technique proposed by Hao Li, Zheng Xu, Gavin Taylor, Christoph Studer, and Tom Goldstein in their paper "Visualizing the Loss Landscape of Neural Nets"[1] proves to be effective and visually appealing. Using a method referred to as "Filter-Wise Normalization," plots of loss functions are generated by a random Gaussian direction vector that has corresponding norms to that of the neural network parameters. We can apply this method to the network we previously trained with the Fashion MNIST dataset.

Using the library Landscapeviz, we can implement the graphing method proposed in the paper with three simple lines of code. The process is shown in Listing 3-14.

Listing 3-14. Model training results

```
# https://www.kaggle.com/datasets/andy1010/landscapeviz
# package used, link above, original code by
# by Artur Back de Luca on github
# (https://github.com/artur-deluca/landscapeviz),
# modified to speed up calculations

import landscapeviz
landscapeviz.build_mesh(fashion_model, (X_train, y_train), grid_length=40, verbose=True,
eval_batch_size=1024)
landscapeviz.plot_3d(key="sparse_categorical_crossentropy", dpi=150, figsize=(12, 12))
```

Note that due to time and memory limitations, the graph presented in the following (Figure 3-11) is only trained on the first 10,000 samples of the training data. The actual loss landscape for the entirety of training data can vary from the one shown here.

---

[1] Li, H., Xu, Z., Taylor, G., & Goldstein, T. (2018). Visualizing the Loss Landscape of Neural Nets. *NeurIPS*.

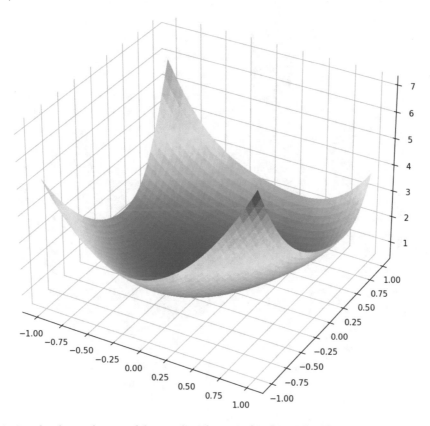

Figure 3-11. Loss landscape for a model trained on the part of Fashion MNIST

The gradient of color represents different values on the graph, with blue being the lowest and red being the highest. The optimal value for the parameters of the network is given at the lowest point in the graph. Remember that the figure plotted is simply a projection of the actual landscape in three dimensions, and the numbers don't correlate to the actual optimal values in the neural network. The only meaningful axis here is the z-axis, which represents the loss value, while the x- and y- axes are arbitrary parameter values. Visualizations like this are meant for analysis and interpretation; by modifying the architecture of the network, the mesh of the loss landscape changes. Shown in Figure 3-12 is the loss landscape for the same data, but instead of two layers with 64 neurons each, the structure of the network is changed to one layer with 512 neurons.

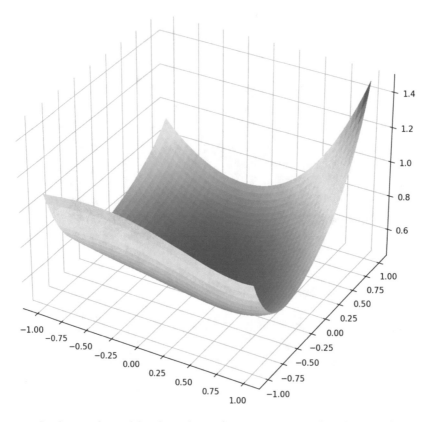

Figure 3-12. *Loss landscape of a model with one layer of 512 neurons trained on the part of Fashion MNIST*

Model interpretation is as important as understanding the mechanics behind the model. Being able to not only learn and understand how the model works but also see how pieces interact and change the results can provide key insights as well. Notice that the network with one hidden layer of 512 neurons produces a relatively flatter and milder landscape compared with the network with two hidden layers of 64 neurons each. This can indicate that model convergence is faster and possibly more optimal on the network with two hidden layers. Remember that visualization is merely a tool for us to briefly understand and interpret the general training path that the model may take, but not a detailed guide on where the global minima are.

Other than the categorical cross-entropy loss that was used in the Fashion MNIST example, here are a couple of more commonly used losses in training neural networks.

- *Binary cross-entropy*: Can be used in Keras with the string "`binary_cross_entropy`" passed in place of the loss parameter. The loss function is used for binary classification; it calculates the log loss between the prediction and the labels. The math behind binary cross-entropy is the exact function used from Logistic Regression introduced in Chapter 1.

- *Categorical cross-entropy/sparse categorical cross-entropy*: Can be used in Keras with string "`categorical_cross_entropy`"/"`sparse_categorical_cross_entropy`" passed in place of the loss parameter. Both loss functions are used for multiclass classification; they calculate the cross-entropy loss or the log loss between the prediction and the labels. The difference between sparse and non-sparse categorical entropy is that the sparse version is only used when the targets are not one-hot encoded. In cases where the classes are one-hot encoded into single columns that represent every single class, categorical cross-entropy should be used.

- *Mean Absolute Error/Mean Squared Error*: Can be used in Keras with the string "mean_squared_error"/"mean_absolute_error" passed in place of the loss parameter. MAE and MSE are both common loss functions used for regression tasks. For their simplicity and easy interpretation, MSE and MAE perform surprisingly well in most cases regardless of tabular or image data.

Math Behind Feed-Forward Operation

The explanation provided previously regarding feed-forward is more intuition based rather than technical. We can formulate the calculation between input, weights, and biases in the language of Linear Algebra.

Recall that the number of neurons in the first layer, referred to as the input layer, equals the number of features. Each feature feeds into one neuron. We organize the values from the input layer into a column vector of shape 1 × n _ features. The weights are then arranged into a matrix where the number of rows is the number of neurons in the first hidden layer and the number of columns is the number of neurons in the input layer. Note that each row of the weight matrix corresponds to the weights of each neuron from the input layer to a particular neuron in the next layer. From the perceptron model, the output is calculated by a "feed-forward" process in which each neuron in the input layer is multiplied by its weight and then summed with every other neuron multiplied by its own weight. The sum, along with the addition of the bias, becomes the input for one neuron in the next layer. Thus, with our feature vector and weight matrices, the calculation becomes as simple as dotting the weight matrix with the feature vector and adding a vector of bias for each neuron in the next layer (Figure 3-13).

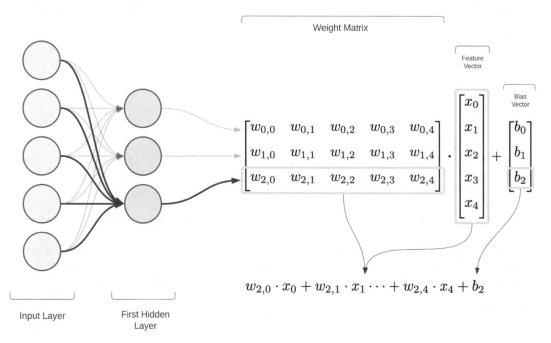

Figure 3-13. Feed-forward operation demonstrated in the input and the first hidden layer

The operation continues between the first hidden layer and the second, the second and the third, and so on until reaching the final layer. For a more concise equation and readability, we can denote the weight matrix as W, the feature vector as X, and the bias vector as b. But the feature vector only exists in the first layer when it's passed through the input layer. For layer outputs, we'll denote it as $L^{(n)}$, where n represents the layer number counting from left to right. Thus, the values of the first layer are defined as a vector of shape $n_{neurons} \times one$; the output can be obtained by $L^{(1)} = W^{(0)} \cdot X + b^{(0)}$. Each layer contains different weights and biases connecting each neuron; thus, we denote the different sets of weights and biases the same way we distinguish layer outputs.

For each subsequent layer, the output of it will be calculated by $L^{(n)} = W^{(n-1)} \cdot L^{(n-1)} + b^{(n-1)}$. The feed-forward operation continues this calculation until it has reached the end of the network, producing the final output (Figure 3-14).

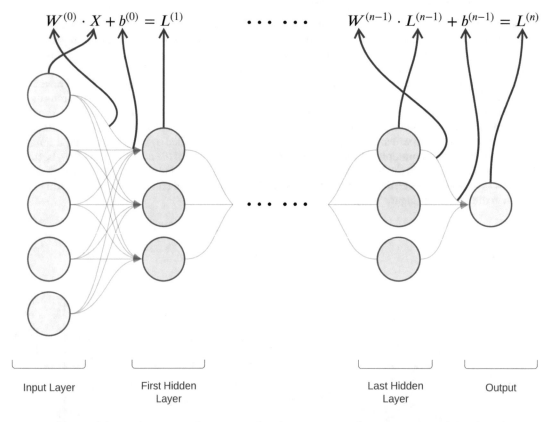

Figure 3-14. *Feed-forward operation demonstrated in the entire network using notations defined previously*

However, in most modern neural networks, we're still missing an important piece in the feed-forward operation as well as in the structure of the entire network. Nonlinearity in data is common. Even though in a neural network, tens if not hundreds of neurons are stacked together in multiple layers, when it comes down to a single neuron's calculations, it's still a linear equation. Activation functions are introduced to add nonlinearity to the model.

Activation Functions

Activation functions are applied at the end output of each neuron in the network. By modifying or restricting output values at the end, the model becomes better at predicting nonlinearly separable data.

Intuitively, activation functions can be seen as switches that control how "active" each neuron is regardless of its weights and biases. In a biological context, the neuron only passes information forward when the signal from the dendrites reaches a certain threshold. Technically speaking, the neural networks that we built previously all contain activation functions at each layer in each neuron. But instead of any fancy nonlinear functions, the activation is simply $y = x$. Thus, its name is "linear activation."

Nonlinear activation functions provide several key improvements to the network:

1. *Add nonlinearity*: This improves the model's performance on nontrivial datasets. Stacking neurons and neurons in layers doesn't change the fundamental fact that each neuron is an affine transformation itself. The composition of affine transformations still results in an affine transformation, unable to model any complex relationship between variables. With a basic example using scalars, we can show that compositions of linear functions still result in a linear function. Let $f(x) = ax + b$ and $g(x) = cx + b$; then $(f \circ g)(x) = a(cx + b) + b = (ac)x + ab + b$. We know that a, b, and c are all scalars; thus, in the end, the equation still results in the form $y = ax + b$. The same concept applies to matrix operation too. Let $f(\vec{x}) = A\vec{x} + \vec{b}$ and $g(\vec{x}) = C\vec{x} + \vec{b}$; then $(f \circ g)(\vec{x}) = (AC)\vec{x} + \vec{b}$. Since both A and C are invertible matrices, AC must be invertible as well. Thus $(f \circ g)(\vec{x})$ is still an affine transformation. Without activation functions, multilayer networks essentially collapse into a network with one single layer as, in the end, hundreds of affine transformations combine into one single affine transformation. On the other hand, having nonlinearity in neural networks improves representation capability.

2. *Address gradient problems*: Certain activation functions that restrict values to certain boundaries address the problems of vanishing gradients and exploding gradients. At each iteration, backpropagation changes the parameters of the network based on the indicators obtained by gradient descent. In some cases, due to calculations that will be discussed in later sections, the indicators from gradient descent will be either too small or too large. In the first case, the network parameters barely change, and it'll be impossible for the network to reach values even close to the optimal ones. In the second case, the parameter values can grow rapidly, leading to overflows or "infinite" losses. Some specifically designed activation functions can help with both exploding and vanishing gradients while adding nonlinearity to the network.

3. *Restrict output values*: As shown in the first example, during classification tasks, the output value of neurons must be in the range of 0 and 1. Sigmoid activation restricts any input to be 0 and 1, while softmax restricts all output values to 0 and 1 while keeping their sum at 1 (useful for multiclass classification). The hyperbolic tangent activation function restricts values between –1 and 1 and is used in recurrent models (see Chapter 5). In most regression tasks, a linear activation function is used as the output value is not restricted.

With the addition of activation functions, the feed-forward operation is slightly modified. An activation function is applied to each layer after multiplying every neuron with its weights and adding the bias term. Thus, the equation for the nth layer of a neural network counting from the input becomes $L^{(n)} = \sigma\left(W^{(n-1)} \cdot L^{(n-1)} + b^{(n-1)}\right)$ where $\sigma(x)$ represents some activation function.

In the following subsections, we will introduce five commonly employed activation functions.

Sigmoid and Hyperbolic Tangent

The sigmoid activation function is mostly used for restricting output values instead of adding nonlinearity between layers of neurons. The sigmoid activation is the logistic function used by Logistic Regression, as mentioned in Chapter 1. Sigmoid activation is not recommended to be used between hidden layers due to the vanishing gradient problem, which is much more common than exploding gradients. Again, vanishing gradient leads to neural networks barely updating their parameters, essentially not learning anything from the data.

Similarly to sigmoid, the hyperbolic tangent (tanh) activation function, displayed in Figure 3-15, is not recommended to be used for activation between hidden layers but instead to restrict output values.

$$\text{Tanh}(x) = \frac{\left(e^x - e^{-x}\right)}{\left(e^x + e^{-x}\right)}$$

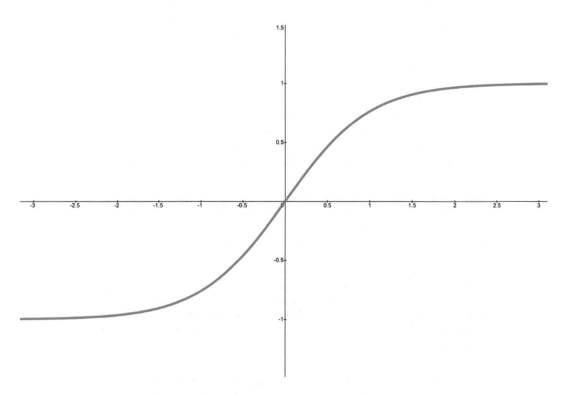

Figure 3-15. *Tanh activation function*

Compared with sigmoid, the tanh activation function will perform better when used between hidden layers as it's a zero-centric function while sigmoid's output value is restricted between 0 and 1. But tanh and sigmoid both suffer greatly from the vanishing gradient problem, and in almost all cases, activations such as ReLU (Rectified Linear Unit) should be chosen over them.

Rectified Linear Unit

Rectified Linear Unit, commonly referred to as ReLU, is the simplest activation function that introduces nonlinearity to the network. It aims to add complexity to the network as well as solve the vanishing gradient problem. It's defined as follows and graphed in Figure 3-16.

$$\text{ReLU}(x) = \max(0, x)$$

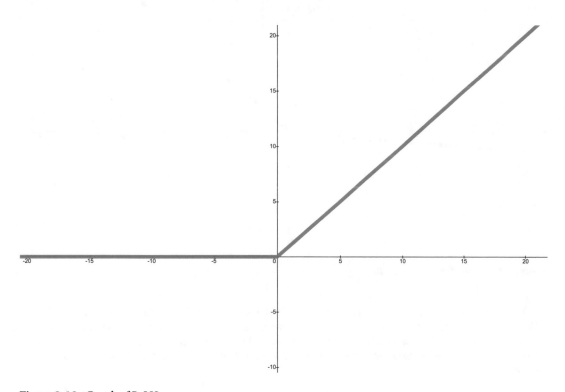

Figure 3-16. *Graph of ReLU*

Despite its simplicity, ReLU avoids the vanishing gradient problem while adding nonlinearity. Compared with other activation functions, ReLU uses less space and has relatively little time complexity. However, with the flattened tail for negative inputs, sometimes too many weights and biases are unchanged, leading to the "dead" ReLU problem. Although it has been proven that sparsity (when a lot of activations are equal to 0) in neural networks caused by ReLU can improve performance, in some cases, this does become a problem leading to convergence issues. Finally, ReLU does not address the exploding gradient problem.

LeakyReLU

LeakyReLU is a modification from ReLU with an adjustable parameter α, which creates a downward tail compared with the flattened line from ReLU. This function addresses some problems in ReLU but also has some disadvantages (Figure 3-17).

$$\text{LeakyReLU}(x) = \begin{cases} x & \text{if } x > 0 \\ \alpha x & \text{if } x \leq 0 \end{cases}$$

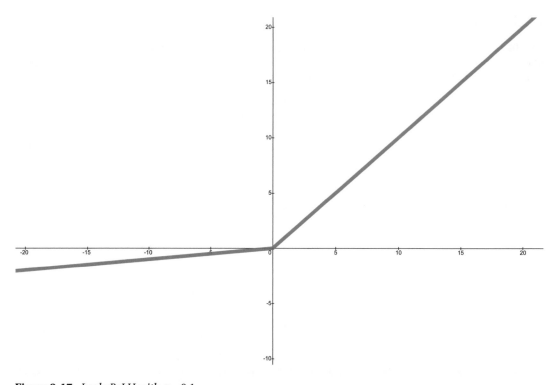

Figure 3-17. *LeakyReLU with α = 0.1*

When α is greater than zero, the dead ReLU problem is solved since for any input less than zero, its value is adjusted so that it does not remain constant at zero, thus being able to update network parameters by a tiny amount. The hyperparameter α is typically chosen between 0 and 0.3. However, LeakyReLU doesn't solve the exploding gradient problem. Besides, to achieve optimal performance, we need to hand-tune the parameter .

Swish

Swish presents itself as a newer but much more effective activation function (Figure 3-18). Developed by researchers at Google Brain,[2] its remarkable performance was seen in image recognition tasks but later also in tabular data predictions.

$$\text{swish}(x) = x \times \text{sigmoid}(\beta x) = \frac{x}{1 + e^{-\beta x}}$$

---

[2] Ramachandran, P., Zoph, B., & Le, Q.V. (2017). Swish: A Self-Gated Activation Function. *arXiv: Neural and Evolutionary Computing.*

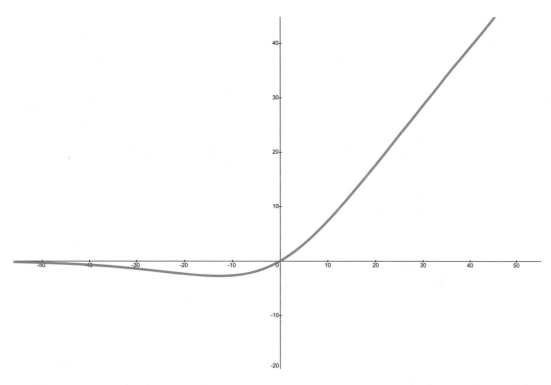

Figure 3-18. *Swish activation function*

Like ReLU, the swish activation function is unbounded above as $\lim\limits_{x\to\infty} f(x)=\infty$ but bounded below, meaning that as the domain of the function approaches negative infinity, there's a definite value that $f(x)$ approaches. However, unlike ReLU, with the combination of sigmoid, the swish function is smooth without any sudden changes or vertices, which prevents unwanted jumps in output values. More importantly, being unbounded avoids slow training time during shrinking gradients produced by backpropagation.

Visually, swish is essentially a "smoother" version of ReLU with a "bump," as the original paper showed, during training, most output values before plugging into the activation fall in the domain range of the "bump," indicating the importance of the addition of the "bump" compared with other similarly shaped activation functions such as ReLU. The importance of this downward "bump" lies in the differentiation of the function. The swish function is non-monotonic, meaning that there are no continually negative or positive values in the derivation. This property of swish addresses the vanishing gradient problem as the values won't ever be restricted or bounded by a certain limit during backpropagation, which causes the gradient to result in extremely small values.

By adjusting the hyperparameter, β, we can see that when $\beta = 0$, the function becomes a scaled linear activation $f(x)=\dfrac{x}{2}$, while as $\beta \to \infty$, the sigmoid part of the function becomes somewhat a 0-1 function, making the entire swish function resemble the shape of ReLU (Figure 3-19).

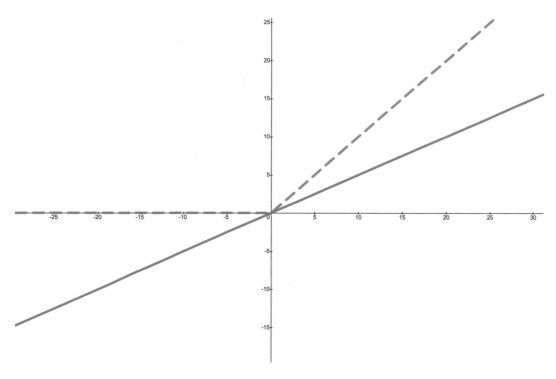

Figure 3-19. *Comparing the values of β. The dotted green line represents β = 10000, and the solid blue line represents swish when β = 0*

Thus, swish can be viewed as a nonlinear interpolation between a linear activation and the ReLU activation where the amount is controlled by the hyperparameter β. Usually, β is set to a certain value for the entirety of the training. However, it can be adjusted throughout training as a trainable parameter.

Finally, the swish activation function has also been proven to generalize well due to its smoothness and ability to optimize networks with larger layer numbers and batch sizes compared with ReLU.

The Nonlinearity and Variability of Activation Functions

The nonlinearity and differences between activation functions can be further proven by a simple demonstration. Imagine the simplest neural network possible: one input neuron, two neurons in one hidden layer, and one output neuron. We will approximate the quadratic function $f(x) = x^2$ using a neural network. A linear function would not be able to accomplish this.

The network is going to be trained on the swish activation function, currently one of the best-performing functions available for structured datasets. Afterward, we're going to change the activation function of the network from swish to ReLU, sigmoid, linear (without activation functions), and more. The predictions of the networks over the range of the input dataset are plotted against the x^2 function for a visual comparison. Here's the basic code to get started shown in Listing 3-15.

Listing 3-15. Defining the dataset

```
# demonstration of activation's non linearity on NNs
# simple dataset that models a quadratic function

demo_x = np.array([i for i in range(-10, 11)])
# smooth out the data points for later graphing
demo_x = np.linspace(demo_x.min(), demo_x.max(), 300)

demo_function = lambda x: ((1/2)*x)**2
demo_y = np.array([demo_function(i) for i in demo_x])
```

Like the network in the previous section, we first start off by defining the Sequential object, then adding Dense layers, specifying the number of neurons, and setting the activation to swish. After compiling the model, it's trained for 20 epochs with the Mean Absolute Error or the MAE as its loss function as it presents a straightforward intuition of how much error the model is producing compared with the targets (Listing 3-16).

Listing 3-16. Defining the network with the swish activation function

```
# construct a simple one hidden layer with two neuron network with activations
# import the dense and input layer
from keras.layers import Dense, Input
# the sequential model object
from keras.models import Sequential
# optimizer, don't worry about it now
from tensorflow.keras.optimizers import Adam

# model with activation
nonlinear_model = Sequential()
nonlinear_model.add(Input((1,)))
# beta parameter is learned in this case
nonlinear_model.add(Dense(2, activation="swish"))
nonlinear_model.add(Dense(1, activation="linear"))
# learning rates will be discussed later
nonlinear_model.compile(optimizer=Adam(learning_rate=0.9), loss="mean_absolute_error",)

nonlinear_model.fit(demo_x, demo_y, epochs=20, verbose=0)
print(f"Nonlinear model with swish activation function results: {nonlinear_model.
evaluate(demo_x, demo_y)}")
```

The final MAE sits around 1.1287; considering the range of our dataset combined with how simple our network is, the model's performance is decently well. Next, we can plot the neural network as a function where the input is the range of values across our input and the output is the predictions produced (Listing 3-17). We can compare this plot to the plot of our function to approximate, $f(x) = x^2$, to visually evaluate the model's performance (Figure 3-20).

Listing 3-17. Plotting $f(x) = x^2$ against the network trained with the swish activation function

```
nonlinear_y = nonlinear_model.predict(demo_x)
plt.figure(figsize=(9, 6), dpi=170)
plt.plot(demo_x, demo_y, lw=2.5, c=sns.color_palette('pastel')[0], label="x²")
```

```
plt.plot(demo_x, nonlinear_y, lw=2.5, c=sns.color_palette('pastel')[1], label=f"network
trained with swish")
plt.title('swish activation')
plt.legend(loc=2)
```

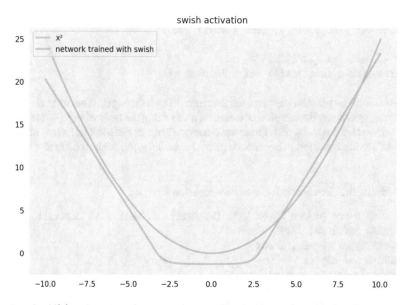

Figure 3-20. *Graph of f(x) = x² against the network trained with the swish activation function*

The trained network was able to replicate the general parabola shape of $f(x) = x^2$ relatively well with only two neurons and only seven trainable parameters (excluding those in the swish activation function). With the same weights in the network, we can replace the activation function with other ones or even remove it and observe the results (Figure 3-21).

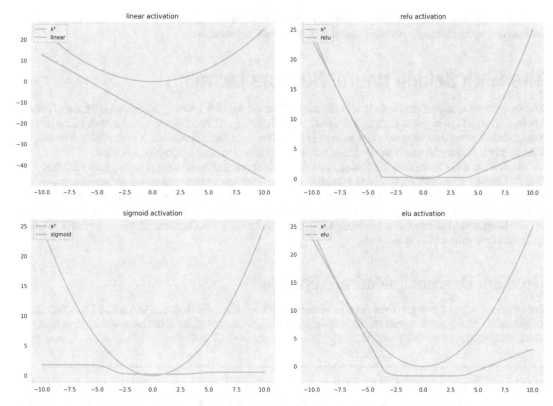

Figure 3-21. *Plots with networks trained on ReLU, sigmoid, linear (without activation functions), and ELU*

■ **Note**   ELU stands for the exponential linear unit, another type of activation function that is not as common as ReLU and swish. It's defined as $ELU(x) = x$ if $x > 0$ and $ELU(x) = \alpha(e^x - 1)$ if $x \leq 0$, where α is a hyperparameter usually between 0.1 and 0.3. ELU contains the advantages of ReLU for $x > 0$ and solves the dead ReLU problem for $x \leq 0$. However, ELU is computationally expensive compared with other activation functions, and the α value is hand-adjusted based on training results.

From a visual perspective, ReLU seems to have the best performance, followed by ELU, then sigmoid, and, lastly, the network trained without activation functions. As proven in the previous section and shown in the graph, no matter how many neurons there are, a network without activation functions collapses into a linear function. Most, if not all, real-life datasets can't be modeled by a simple linear relationship.

Observe that both swish and ELU's graphs contain smooth curves without sharp turns in the plot, while ReLU has vertices and angles that produce a sudden change in output values. Swish and ELU's smoothness gives them an upper hand in approximating functions that contain curves and soft turns. Given that many real-world datasets have gradually changing values instead of sudden jumps, swish and ELU tend to perform better in those cases.

The pool of activation functions is way broader than what was presented here; in most situations, using ReLU or swish will be sufficient, but there's no strict rule for choosing a certain activation function, and the choice will be left for the data scientist.

The curious reader may also find the SELU activation, which is discussed in the "Selected Research" section of this chapter, an interesting addition to this family of functions.

The Math Behind Neural Network Learning

The feed-forward operation comes first in the steps of neural network training. Random weights and biases are initialized and then passed through the network with the input data. Since the parameters of the network are randomly generated, they won't be a suitable fit for the input data. Therefore, we must tell the network how poorly it performed and how we can adjust the parameters to improve its performance.

Loss functions accomplish the first part, while gradient descent does the second. To calculate the "loss" of the network, we simply take the results from the feed-forward operation, which most likely will be gibberish upon initialization, considering the randomly generated parameters. The algorithm then uses it as an initial "prediction," computing the loss against the targets. After the loss is computed, gradient descent is used to optimize the loss function, attempting to search for the global minima with respect to the perhaps millions of parameters in the network.

Gradient Descent in Neural Networks

Recall from Chapter 1 that gradient descent aims to search for the global minima in a non-convex function. By iteratively taking smaller and smaller steps toward the direction of greatest descent, we can eventually reach the global minima or a point relatively close to the theoretical "global minima" in the loss landscape.

■ **Note**   For the global minima, or the minima of a convex function, we can simply set the derivative of that function to zero and solve for its parameters. This method of finding the minima is often referred to as Ordinary Least Squares (OLS), and many Linear Regression algorithms utilize this method over gradient descent as it's less computationally expensive and produces more accurate results. However, this method is intractable for neural network problems.

In the context of neural networks, since they usually contain up to millions of parameters that can be tweaked and optimized, instead of searching for the global minima in a 2D or 3D graph, the goal shifts to finding the lowest point on a hypersurface (Figure 3-22).

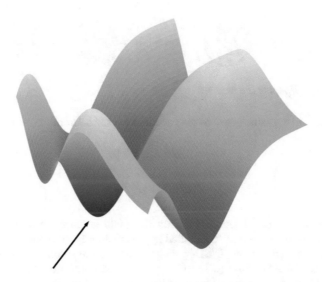

Figure 3-22. *Gradient descent searches for the lowest point on a hypersurface demonstrated in 3D*

The algorithm aims to search for a certain set of parameters that reduces the value of the cost function to the lowest. During the "descent" down the sloping hills of the loss landscape, the algorithm seeks to adjust the parameter, which can move the value of the function in a "downward" direction. Calculus tells us that taking the gradient of a function results in the direction of the steepest increase. Intuitively, negating the gradient gives us the direction of "decrease" or the direction of downhill in the loss landscape.

Taking the negative gradient of the function simply results in a gradient vector that tells us how to change the parameters of the network based on how sensitive the cost function is to changes in the parameters of the network, which, in terms, will take us "downward" in the loss landscape. Backpropagation refers to the algorithm that computes the gradient of the cost function with respect to millions of the parameters that the network contains.

The Backpropagation Algorithm

We can start off by demonstrating backpropagation on a simple model, a network with one input neuron, two hidden layers with one neuron in each, and one output neuron. There are, in total, six parameters in this tiny neural network, three weights and three biases (Figure 3-23).

Figure 3-23. *Simple four-layer neural network*

As previously described, before adjusting the parameters of the network, we need to produce an initial prediction and calculate its cost to know how "bad" the network is performing, thus giving us a starting point for optimization. We can denote each neuron, or essentially each layer in our case, as $a^{(L-n)}$ where $a^{(L)}$ is the last layer, $a^{(L-1)}$ is the second to last layer, and so on (Figure 3-24). Backpropagation starts from the end, or the output of the network, and works its way forward. After producing the initial "guess" from randomized weights, we can calculate the loss between the prediction and the ground-truth labels. In the example here, MSE is used as the cost function.

Figure 3-24. *Notation for layers in the network*

We can further expand the value of $a^{(L)}$, or the output of the last neuron:

$$a^{(L)} = \sigma\left(z^{(L)}\right)$$

$$z^{(L)} = w^{(L)}a^{(L-1)} + b^{(L)}$$

In the preceding equation, $\sigma(z^{(L)})$ denotes the activation of the last layer. $z^{(L)}$ is the weighted sum produced by the current layer's weight $w^{(L)}$ multiplied by the previous layer's activation output $a^{(L-1)}$, then adding the current layer or neuron's bias, $b^{(L)}$. Denoting the loss/cost function as C, the loss of the network after the forward pass can be defined as $C(\ldots) = (a^{(L)} - y)^2$. Recall that the input of the loss function is all the parameters in the network, while the output is the actual loss or the "cost" between the ground-truth labels and the predictions.

To obtain the "indicators" that tell us how to adjust the weights and biases of the network, we need to compute the derivative of the cost function with respect to that layer's weight. Starting from $w^{(L)}$, we know that changing the value of the weight also changes the value of $z^{(L)}$. A slight change in $z^{(L)}$ then influences the value of $a^{(L)}$, which directly changes the value of C, or the cost function (Figure 3-25).

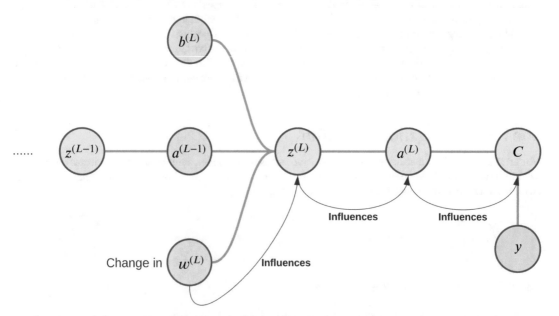

Figure 3-25. *The influence of changing the value of the weight in the last layer with respect to the cost function*

Expanding the derivative of the cost function with respect to the last layer's weight using the chain rule, we obtain $\frac{\partial C}{\partial w^{(L)}} = \frac{\partial z^{(L)}}{\partial w^{(L)}} \frac{\partial a^{(L)}}{\partial z^{(L)}} \frac{\partial C}{\partial a^{(L)}}$. Note that in our example, the derivative calculated applies only to one specific training example. To obtain the full derivative across all training examples, the average of each individual derivative is computed. The same concept is applied to differentiating the bias with respect to the cost function, resulting in $\frac{\partial C}{\partial b^{(L)}} = \frac{\partial z^{(L)}}{\partial b^{(L)}} \frac{\partial a^{(L)}}{\partial z^{(L)}} \frac{\partial C}{\partial a^{(L)}}$. This process can be iteratively applied to every layer in the network. For each "step" or layer that we take backward through the network, one more "chain" is added to the derivative.

For example, the derivative of the weight in the second to the last layer, $L - 1$, can be represented as $\frac{\partial C}{\partial w^{(L-1)}} = \frac{\partial a^{(L-1)}}{\partial w^{(L-1)}} \frac{\partial C}{\partial a^{(L-1)}}$; it essentially tells us that a change in $w^{(L-1)}$ influences the value in $a^{(L-1)}$, while a change in $a^{(L-1)}$ directly influences the cost function. We can use the chain rule to expand the derivative like earlier to obtain the full equation.

We can expand our example to a network with multiple neurons in each layer, as shown in Figure 3-26.

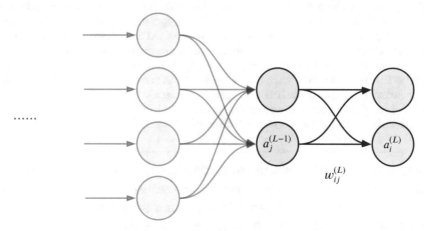

Figure 3-26. *Network with multiple layers and neurons*

For explanatory purposes, let's only focus on the L and $L - 1$ layers; for each subsequent $L - n$ layer, the same concept of computation is applied. Note that for each layer, the neuron is denoted with a subscript, with i representing neurons in the L layer and j representing neurons in the $L - 1$ layer.

The derivative of one specific weight, w_{ij}, turns out to be the same compared to the simple single-neuron network described previously: $\frac{\partial C}{\partial w_{ij}^{(L)}} = \frac{\partial z^{(L)}}{\partial w_{ij}^{(L)}} \frac{\partial a^{(L)}}{\partial z^{(L)}} \frac{\partial C}{\partial a^{(L)}}$. However, the derivative of $a_j^{(L-1)}$ becomes the sum of the derivative of each neuron across the layer L: $\frac{\partial C}{\partial a_j^{(L-1)}} = \sum_{i=0}^{n-1} \frac{\partial z^{(L)}}{\partial w_{ij}^{(L)}} \frac{\partial a^{(L)}}{\partial z^{(L)}} \frac{\partial C}{\partial a^{(L)}}$ where n is the number of neurons in layer L. This applies not only to multiple output neurons but also to hidden layers where neurons are connected to more than one neuron in the next layer (this will be the case for most network architectures encountered) since a change in the activation will influence both outputs of the neurons in the next layer.

The algorithm is applied to compute the derivative iteratively, traveling back to the beginning of the network, hence the name backpropagation. Once the gradient of the cost function is obtained – a vector of every derivative of every parameter in the network with respect to the cost function – the update rule for each parameter is the same as the gradient descent in regression models. A learning rate parameter is chosen, and it's multiplied by the derivative of that parameter; the resulting value becomes the new weight/bias for the network.

Optimizers

Optimizers refer to the algorithm that is used to update the parameters of the network. Recall from earlier, and in Chapter 1, the update rule for gradient descent is as follows:

1. For each training sample, differentiate the cost function with respect to every parameter in the network.

2. Average each derivative across all training samples.

3. Update each parameter in the network by $\theta := \theta - \alpha \cdot \nabla_\theta C(\theta)$ where θ is the parameter being updated, while α is the learning rate and $C(\theta)$ is the cost function.

4. Repeat for the desired number of epochs.

In a real-world context, it's extremely computationally expensive to sum up all the "influences" of each sample on every step. Not to mention that each parameter is only updated once after the model has seen the entire dataset, which proves to be unproductive and exceedingly slow. One solution to this could be updating the parameters of every sample by randomly shuffling the dataset. For every gradient descent step, the gradient is calculated only for that data point. However, taking one sample at a time would produce a very unstable descent. This can occur because the variation between each sample will lead to the model changing too specific for each sample, therefore ignoring the global trend, which to us will appear as unstable training. Unstable training can lead to missing the global minima or swaying away from it as this method focuses too much on the "local optimization" instead of looking at the "big picture" of the entire dataset, like gradient descent.

■ **Note** For the clarity of terms used here, backpropagation refers to the algorithm that computes the gradient of the cost function that informs the model about how to change each parameter to decrease the cost function. On the other hand, one gradient descent step refers to the process of backpropagation over all samples of training data and using the gradient to update the parameters of the network.

Mini-batch Stochastic Gradient Descent (SGD) and Momentum

Mini-batch stochastic gradient descent walks a middle path between the traditional gradient descent and updating sample by sample, compromising both algorithms' advantages and disadvantages. Instead of just focusing on how much impact one sample has on the loss or summing up the influence of every sample, mini-batch SGD divides the dataset into relatively small batches and computes the gradient with respect to each batch. The update rule for mini-batch SGD requires two more steps than gradient descent:

1. Randomly divide the dataset into subsections or mini-batches.

2. For each mini-batch, differentiate the cost function with respect to every parameter in the network.

3. Average the derivatives across the entire mini-batch.

4. Update each parameter in the network by $\theta := \theta - \alpha \cdot \nabla_\theta C(\theta)$ where θ is the parameter being updated, while α is the learning rate and $C(\theta)$ is the cost function.

5. Repeat for all mini-batches.

6. Repeat for the desired number of epochs.

Picture the training of a neural network as a person traveling downhill in a valley while blindfolded. Using traditional gradient descent, the person considers every parameter that may affect their next step. After careful calculation of all the possible factors that may influence their step downhill, they take a conservative step and does this until they slowly reach the bottom of the hill (Figure 3-27).

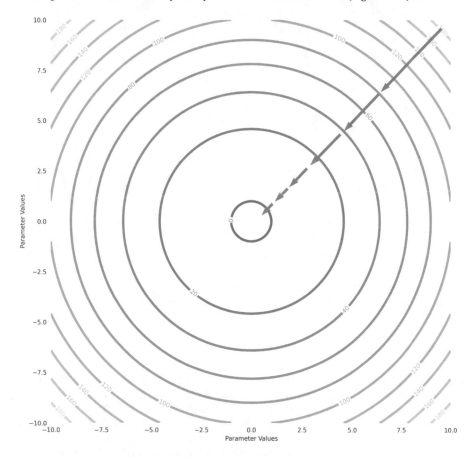

Figure 3-27. *The possible path taken by traditional gradient descent*

On the contrary, mini-batch SGD acts like a drunk man, swaying left and right in the correct direction while keeping with the overall path downward (Figure 3-28). Since in mini-batch SGD the parameter change in the network does not reflect the overall trend of the data but rather a subset, although in most cases it provides the correct direction of descent, the process can be noisy but is way more time and computationally efficient compared with traditional gradient descent. Moreover, escaping saddle points – areas where there's a local maximum in one direction but a local minimum in the other – is near impossible for mini-batch SGD since most of the gradients surrounding the terrain are close to zero.

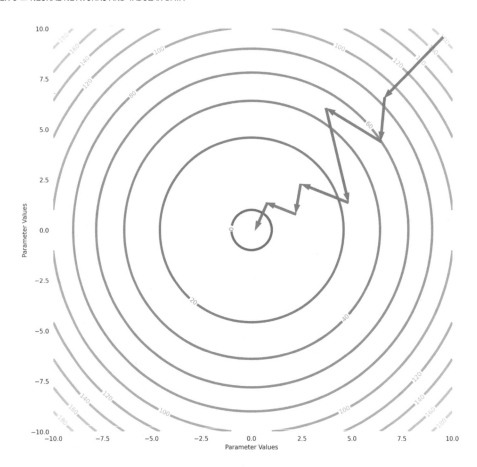

Figure 3-28. *The possible path taken by mini-batch SGD*

Another challenge that SGD usually faces is overcoming "ravines" that have curving dimensions steeper in one direction than the other. SGD tends to oscillate around the slopes of the ravine without making any actual progress. Introducing a technique known as "momentum" can assist with "pushing" SGD in the relevant direction without oscillation. Momentum modifies the update rule to include a partial update vector from the previous step into the current update vector by the following equation: $\nu_t = \gamma\nu_{t-1} - \alpha\nabla_\theta C(\theta)$ where γ is usually set to 0.9 and ν_{t-1} is the previous update vector. We then update the current parameter by $\theta = \theta + \nu_t$. Momentum increases the step size when both update vectors' gradients point in the same direction while decreasing the update size when gradients point in a different direction.

Nesterov Accelerated Gradient (NAG)

We can compare SGD with momentum to the analogy of a ball rolling downhill, accelerating when the slope steepens while slowing down when the steepness decreases or changes to an upward direction. Nesterov accelerated gradient can increase the "precision" that the ball has, letting it know when to slow down when the "terrain" ahead starts sloping up. In the equation involving momentum, we know that the update vector or the gradient of the cost function is "nudged" by an added momentum term, $\gamma\nu_{t-1}$, to "push" the gradient into the direction of the minima, dampening any oscillation. Hence, by adding $\gamma\nu_{t-1}$ to the current value of our parameter, we can approximate its next position; the estimate is not precise as the gradient term,

$-\alpha\nabla_\theta C(\theta)$, is excluded (we don't have the term as we haven't calculated it in the current step yet), but the value is precise enough to provide a point in the vicinity of where we will eventually end up (Figure 3-29). We then can utilize this "look-ahead" technique, computing the gradient with respect to the approximated parameter rather than the current one: $\nu_t = \gamma\nu_{t-1} - \alpha\nabla_\theta C(\theta + \gamma\nu_{t-1})$. The parameters are updated the same way as before: $\theta = \theta + \nu_t$.

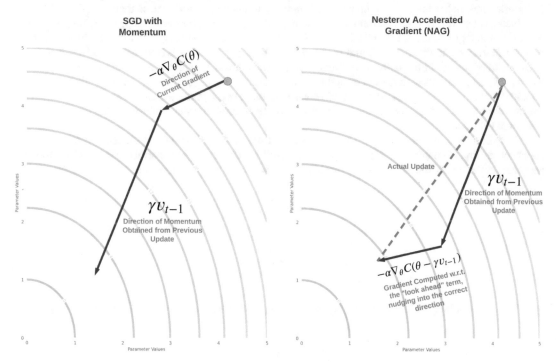

Figure 3-29. *NAG's update movement compared with the momentum*

Nesterov accelerated gradient update can be interpreted in two phases. During the first phase, the accumulated gradient from the previous step acts as momentum, taking a big step in that direction. Then during the second phase, the gradient is computed w.r.t. to the "look-ahead" term; it then nudges or corrects the direction taken by momentum, resulting in the final update. Essentially, this lets the "ball" look ahead to where it's rolling in the loss landscape before taking the next step, thus becoming more efficient and accurate than solely based on momentum.

By incorporating SGD along with a mini-batch into the traditional gradient descent, we were able to speed up the descent while reducing computational workload. Adding momentum and NAG to SGD increases the precision of descent while adjusting the step size adaptively to the slope of the loss landscape. Depending on the values of parameters, some are deemed more important than others; thus, updates can be done according to the importance of parameters.

Adaptive Moment Estimation (Adam)

The Adam optimizer computes the adaptive learning rate based on the importance of each parameter that it serves to the network. The learning rate of each parameter is adjusted based on its computed gradient during the step. In some older optimizers, such as AdaGrad, the gradients of previous steps are accumulated and summed with the current gradient, and the term is then squared to be used as the divisor for the learning rate. Following the trend of past gradients, AdaGrad performs smaller updates for parameters that are associated with frequently occurring features by reducing the learning rate. In contrast, AdaGrad performs larger updates for parameters that are associated with less frequently occurring features by increasing the learning rate. However, one major flaw emerges with AdaGrad. This is because the accumulated gradient grows to extremely large values during training. This shrinks the learning rate to an infinitesimally small value to the point where the model can't acquire additional changes.

Instead of storing all past accumulated gradients, Adam imposes an exponentially decaying average of all past squared gradients. For the conciseness of notation, denote $\nabla_\theta C(\theta_t)$ as g_t, and the decaying average of past squared gradients can be defined as $v_t = \beta_2 v_{t-1} + (1-\beta_2) g_t^2$ where β_2 is usually set to 0.999, as stated in the authors' paper. Additionally, the gradient computed at the current step is replaced by a decaying average of past gradients without squaring the gradient term: $m_t = \beta_1 m_{t-1} + (1-\beta_1) g_t$. β_1 is usually set to 0.9, as stated in the paper. Both m_t and v_t estimate the nth moment of the gradient; specifically m_t is the mean or the first moment, while v_t is the uncentered variance, or the second moment of the gradient, hence the name "Moment Estimation."

Since both m_t and v_t are initialized as zero vectors, the values of both tend to be biased toward 0, especially during the beginning stage of training. The authors computed bias-corrected first and second moments of the gradients, which slightly modifies the value of both after computing the decaying average.

$$\widehat{m_t} = \frac{m_t}{1-\beta_1^t}$$

$$\widehat{v_t} = \frac{v_t}{1-\beta_2^t}$$

Then these values are incorporated into the Adam update rule, which is defined in the following:

$$\theta_{t-1} = \theta_t - \frac{\alpha}{\sqrt{\widehat{v_t}} + \epsilon} \widehat{m_t}$$

To avoid division by 0, a buffer term ϵ is added, usually set to a value around 10^{-9}.

If SGD with momentum represents a ball rolling down a slope, Adam can be seen as a heavier ball rolling down with friction. Adam is one of the most used optimizers in modern deep learning, regardless of the task at hand, due to its significantly improved performance compared with other optimizers such as SGD with NAG and AdaGrad measured on state-of-the-art (SOTA) models.

There are numerous other optimizers in the Adam family, such as AdaMax, which generalizes the squared gradient term to l_∞ norm, or Nadam (Nesterov Accelerated Momentum Estimation), which combines the concept of NAG and Adam, or AdaBelief, an optimizer that considers curvature information in the loss landscape and adjusts its step size by the "belief" of the current gradient direction.

Adaptive methods result in faster convergence but may lead to poorer generalization, while the SGD family may converge slower, but it's typically steadier and generalizes better. Although there are optimizers that try to combine the benefit of both, such as AdaBound or AMSBound, currently, Adam still is one of the most used and best-performing optimizers.

A Deeper Dive into Keras

In the section "Modeling with Keras," only basic operations and techniques involving constructing a functioning neural network for classification were presented. Now, we're going to dive deeper into advanced Keras modeling techniques, covering some of the most useful functionalities of Keras. We attempt to improve upon our initial model, which categorizes images of common clothing articles from the Fashion MNIST dataset.

Recall from earlier that our model is defined as such in Listing 3-18.

Listing 3-18. The model architecture defined in the "Modeling with Keras" section

```
# import the dense and input layer
from keras.layers import Dense, Input
# the sequential model object
from keras.models import Sequential
# initialize
fashion_model = Sequential()
# add input layer, specify the input shape of one sample # as a tuple
# in our case it would be (784, ) as it's one-dimensional
fashion_model.add(Input((784,)))

# add dense layers, the only parameter that we need to
# worry about right now is the number of neurons, which
# we set to 64
fashion_model.add(Dense(64))
fashion_model.add(Dense(64))

# add output layer
# softmax activation can be specified by the "activation"
# parameter
fashion_model.add(Dense(10, activation="softmax"))
```

Notice that there are no activation functions in the hidden layers that we previously constructed. In Keras syntax, there are two methods of adding activation functions; if none are used, the default activation becomes linear.

The first method is to specify the name of the activation as a string input into the "activation" parameter of the Dense call (Listing 3-19). Activation functions available by a string in Keras include elu, exponential, gelu, hard_sigmoid, relu, selu, sigmoid, softmax, softplus, softsign, swish, and tanh.

Listing 3-19. Inserting activation as a string passed into the Dense call

```
fashion_model.add(Dense(64, activation="swish"))
fashion_model.add(Dense(64, activation="swish"))
```

The second method is inserting an activation layer between Dense calls. By importing activation layers as objects, they can be called and added to the sequential model the same way as dense layers are added. For most activations, there's no clear advantage between using strings in dense layers and using separate activation layers other than code readability and conciseness. However, for activations with user-defined hyperparameters such as LeakyReLU, they must be added to the model as separate layers as their hyperparameter is defined during the layer call, as shown in Listing 3-20.

Listing 3-20. Example activation layer using LeakyReLU

```
from keras.layers import LeakyReLU
fashion_model.add(LeakyReLU(alpha=0.2))
```

We can modify the initial model to have swish activations in both hidden layers and possibly observe performance. The complete code for model definition and training is shown in Listing 3-21.

Listing 3-21. Complete code for training the Fashion MNIST network with added swish activation (without comments to reduce space usage)

```
from keras.layers import Dense, Input
from keras.models import Sequential
fashion_model = Sequential()
fashion_model.add(Input((784,)))
fashion_model.add(Dense(64, activation="swish"))
fashion_model.add(Dense(64, activation="swish"))
fashion_model.add(Dense(10, activation="softmax"))

from keras.losses import SparseCategoricalCrossentropy
from tensorflow.keras.optimizers import Adam
fashion_model.compile(optimizer=Adam(learning_rate=1e-3),loss="sparse_categorical_
crossentropy", metrics=["accuracy"])

fashion_model.fit(X_train, y_train, epochs=25, batch_size=1024)

fashion_model.evaluate(X_test, y_test, batch_size=2048)
```

Compared with training results earlier, the accuracy improved from 0.82 to 0.90 after the addition of activation functions, proving the importance of their presence in any neural network. Note that results may be slightly different over different runs as weights are initialized randomly.

Next, we're going to discuss built-in validation methods and techniques to improve overfitting.

Training Callbacks and Validation

During the fit call, Keras has built-in validation functionality by including some extra parameters in the call. During training, it's useful to have the model evaluated on a validation set for every epoch. Instead of adjusting the number of epochs and learning rate for every training cycle based on the results of model. evaluate, we can simply set the epoch number to the epoch in which the validation set has the highest accuracy or whichever metric is used. Specifically, the validation features and targets can be passed into the fit call under the validation_data parameter as a tuple. It's recommended to use cross-validation than simply splitting the data into a training and a validation chunk; the validation_data parameter is usually used for quick tests (Listing 3-22).

Listing 3-22. Fit example with validation data passed in

```
fashion_model.fit(X_train, y_train, epochs=25, batch_size=1024, validation_data=(X_test,
y_test))
```

Although not required for training, Keras callbacks are processes performed after each epoch, allowing users for more in-depth and detailed control over training as well as model adjustments. Here are three common callbacks used:

- *Model Checkpoint: Imported as* `tensorflow.keras.callbacks.ModelCheckpoint`: Saves the model according to user specifications. Options include saving the entire model or solely the model weights every epoch that the training or validation metric improves. The callback can also be set to save every epoch.

- *Early Stopping: Imported as* `tensorflow.keras.callbacks.EarlyStopping`: Terminates the training process prior to reaching the desired number of epochs based on some criteria such as training or validation metric not improving for a certain number of epochs.

- *Reduce Learning Rate on Plateau: Imported as* `tensorflow.keras.callbacks.ReduceLROnPlateau`: Reduces the learning rate by a user-specified factor when training or validation loss stops improving for a certain number of epochs.

The Early Stopping callback is usually used in conjunction with Model Checkpoint – automatically stopping extensive training sessions while saving the best version of the model. It avoids potential progress loss during long, unmonitored training, which could last up to hours to days. The basic syntax of both callbacks is shown in the following, based off our `fashion_model` as an example (Listing 3-23).

Listing 3-23. Early Stopping and Model Checkpoint example

```
from tensorflow.keras.callbacks import EarlyStopping, ModelCheckpoint

checkpoint = ModelCheckpoint(filepath="path_to_weights", monitor="val_accuracy",
                             save_weights_only=True, save_best_only=True)
early_stop = EarlyStopping(patience=3, monitor="val_accuracy",
                           min_delta=1e-7, restore_best_weight=True)
```

Model Checkpoint will save the model weights, as set by "`save_weights_only=True`", to the specified file path in the parameter "`filepath`". The model is saved every epoch unless "`save_best_only`" is set to true as shown previously, in which case only epochs that show improvement to the metric monitored set by the "`monitor`" parameter are saved.

Early Stopping only ends training when the metric set by the "`monitor`" parameter stops improving by some specified epochs in the "`patience`" parameter. Additionally, using `min_delta`, a quantity can be set in which an improvement only counts when the improvement amount exceeds the set value. Finally, by setting the "`restore_best_weight`" parameter to true, Early Stopping will restore model weights from the epoch in which the monitored metric performed the best.

In Keras fit calls, an object is returned, and its attribute "history" returns a detailed training log, including values of all losses and metrics over all epochs. A dictionary containing all the information is returned where each key is the loss/metric monitored, while each value within the key is the value of that specific loss/metric at one epoch. We can utilize the data and visualize the trend in training or validation loss by plotting metric/loss values against the number of epochs (Listing 3-24).

Listing 3-24. Plotting training history

```
# history and plotting
history = fashion_model.fit(X_train, y_train, epochs=40, batch_size=1024, validation_
data=(X_test, y_test))
plt.plot(history.history['val_accuracy'], label="val_acc")
plt.plot(history.history['accuracy'], label="train_acc")
```

```
plt.xlabel('Epochs')
plt.ylabel('accuracy')
plt.title("Training and Validation Accuracy")
plt.legend()
plt.show()
```

By analyzing trends and patterns in training history, valuable information can be retrieved, which can assist future model adjustments and improvements. In the preceding example, the validation accuracy starts to plateau if not decrease at around epoch 20, an indicator that the learning rate should be reduced around the point of increase (Figure 3-30). It's likely that doing so will yield a slight performance boost in validation accuracy.

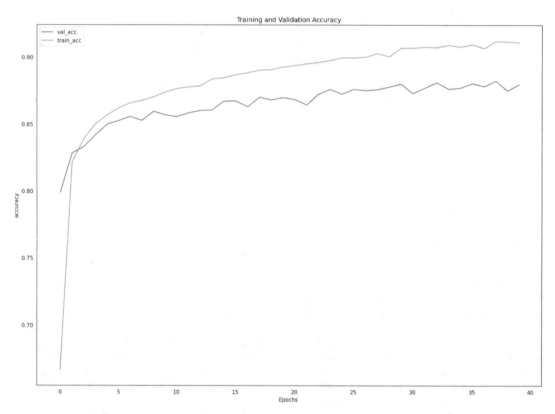

Figure 3-30. *Training performance plotted against the number of epochs*

Reduce Learning Rate on Plateau reduces the learning rate if the monitored metric specified by the "monitor" parameter does not improve for the number of epochs specified in the "patience" parameter. Each time the learning rate is reduced, it's multiplied by the factor set in the "factor" parameter.

By incorporating the callbacks described previously, validation results may be able to improve from previous training (Listing 3-25).

Listing 3-25. Retraining with callbacks

```
# callbacks
from tensorflow.keras.callbacks import EarlyStopping, ReduceLROnPlateau, ModelCheckpoint

checkpoint = ModelCheckpoint(filepath="path_to_weights", monitor="val_accuracy",
                             save_weights_only=True, save_best_only=True, verbose=1)
early_stop = EarlyStopping(patience=6, monitor="val_accuracy",
                           min_delta=1e-7, restore_best_weights=True, verbose=1)
reduce_lr = ReduceLROnPlateau(monitor='val_loss', factor=0.2,
                              patience=3, min_lr=1e-7, verbose=1)


# redefine the model
fashion_model = Sequential()
fashion_model.add(Input((784,)))
fashion_model.add(Dense(64, activation="swish"))
fashion_model.add(Dense(64, activation="swish"))
fashion_model.add(Dense(10, activation="softmax"))
# compile again so the model restarts training progress from above
fashion_model.compile(optimizer=Adam(learning_rate=1e-3),loss="sparse_categorical_
crossentropy", metrics=["accuracy"])

history = fashion_model.fit(X_train, y_train, epochs=100, batch_size=1024, validation_
data=(X_test, y_test), callbacks=[checkpoint, early_stop, reduce_lr])

plt.figure(dpi=175)
plt.plot(history.history['val_accuracy'], label="val_acc")
plt.plot(history.history['accuracy'], label="train_acc")
plt.xlabel('Epochs')
plt.ylabel('accuracy')
plt.title("Training and Validation Accuracy")
plt.legend()
plt.show()
```

Observe that the validation accuracy improved by around 0.01, from 0.87 to around 0.88. You can adjust callback parameters to further improve the results. However, based on the problem setting and the limited modification power that callbacks possess, a 0.01 accuracy improvement is considered a success.

Batch Normalization and Dropout

Previously, different optimizers were explored; some lead to fast convergence, while others can stabilize training while improving the model's ability to generalize on unseen datasets. Although by updating the gradient differently, the entirety of training can be stabilized, there are still issues regarding training stability within the model itself. As the layers of the model receive raw signals from the previous layer's activation, the distribution of the activation could change drastically across different batches. This can lead to the algorithm always trying to fit a moving target. This problem is named internal covariate shift. The network relies on high interdependency between distributions, which can slow down training as well as decrease the stability (Figure 3-31).

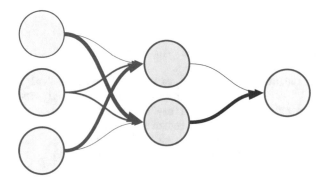

Figure 3-31. *Raw, unnormalized signals from the network can lead to training instability*

Batch normalization produces a normalized signal, which mitigates interdependency between distributions. This not only can increase training stability and speed, but batch normalization can also insert regularization into training or the ability to reduce overfitting and increase generalization.

Batch normalization uses the first and second statistical moments of the current batch, normalizing activation vectors from hidden layers. Usually, batch normalization is applied after the activation function; however, having it before the nonlinear transformation is acceptable too.

We denote the activation vector/output as z. Then the first and second moments of the batch, or the mean and uncentered variance, are defined in the following as such:

$$\mu = \frac{1}{n}\sum_{i}^{n} z_i$$

$$\sigma^2 = \frac{1}{n}\sum_{i}^{n}\left(z_i - \mu\right)^2$$

Then, based on the values obtained, the activation vector is normalized so that each output vector from every neuron follows a normal or Gaussian distribution across the entire batch. Note that ϵ is added for numerical stability:

$$\hat{z}_i = \frac{z_i - \mu}{\sqrt{\sigma^2 - \epsilon}}$$

Then a linear transformation is applied to the normalized output with two learnable parameters, allowing each layer to choose its own optimal distribution. Specifically, γ tunes the standard deviation, while β controls the bias:

$$\text{new } z_i = \gamma * \hat{z}_i + \beta$$

The values of γ and β are trained along with gradient descent using an Exponential Moving Average, similar to the mechanics of the Adam optimizer.

During the evaluation phase, when the data fed in is not sufficient for an entire batch and thus not enough information to calculate the first and second moments, the mean and uncentered variance are replaced by a precomputed estimation from the training phase.

In practice, Keras implemented batch normalization as a layer call and used it just like any Dense call. We can add batch normalization layers to our model as shown in Listing 3-26.

Listing 3-26. Adding batch norm layers to our model

```
from tensorflow.keras.layers import BatchNormalization

fashion_model = Sequential()
fashion_model.add(Input((784,)))
fashion_model.add(Dense(64, activation="swish"))
fashion_model.add(BatchNormalization())
fashion_model.add(Dense(64, activation="swish"))
fashion_model.add(BatchNormalization())
fashion_model.add(Dense(10, activation="softmax"))
```

By retraining the model with batch normalization, a slight improvement to the validation accuracy is observed, while the training accuracy improved by around 0.02. If we analyze the graph produced by the batch normalization model, notice that the accuracy of both training and validation data is a lot more stable than the model without. What's more is that in the graph of the model with batch normalization, both validation and training metrics converge a lot faster than the model without, plateauing at around epoch 15, while the model without stops improving around epoch 50 with similar performance (Figure 3-32).

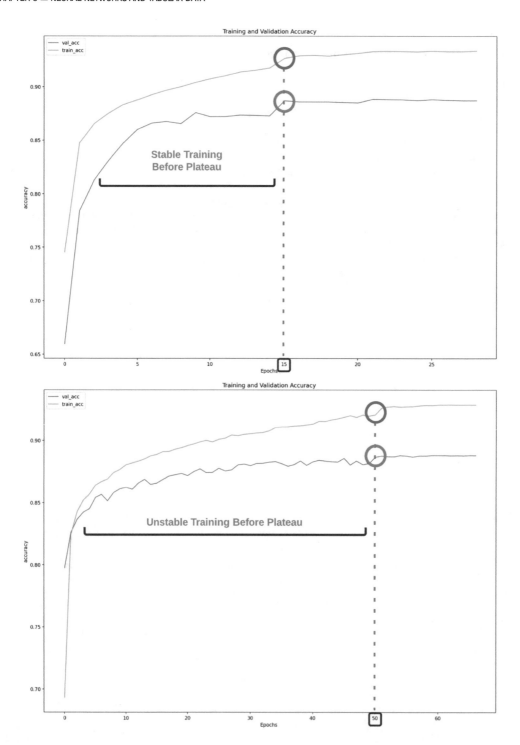

Figure 3-32. *Model with batch normalization compared with the model without*

Note that batch normalization does indeed assist with overfitting (the larger the batch size, the less regularization that it poses). Its most prominent use is to speed up and stabilize training. In modern networks, whether large or small, batch normalization layers are almost a default. However, one major downside to adding batch normalization layers after every hidden layer is the computational cost. With each addition of batch normalization, the convergence or the number of epochs may be reduced, but the amount of time that each epoch takes will increase significantly. It's best to find a balance between adding batch normalization for every layer and having no batch normalization at all.

One major reason that neural networks are so easily overfitted (and, in return, the absurd number of methods that can improve neural network generalization) is the sheer size and the insane number of parameters that one simple network can contain.

An analogy to the superiority of sexual reproduction over asexual reproduction can be described as follows: The role of sexual reproduction is not just to let new and diverse genes spread through the population but also to reduce complex co-adaptions that would decrease the chance of a new gene improving the fitness of an individual. It displays the importance of sexual production by training the genes to not always rely on the vast pool of genes available but rather leading them to be able to work with a small number of genes too. The same concept is applied to neural networks by adding dropouts to layers. Each hidden unit in the neural network can be trained to work with all other parameters but also to work with a randomly chosen small set of parameters. This can, in turn, improve the robustness of the network.

Dropouts, as the name suggests, drop out neurons at random during the training phase (Figure 3-33). The main purpose behind dropout is to decrease the codependency between neurons and increase generalization abilities against unseen datasets.

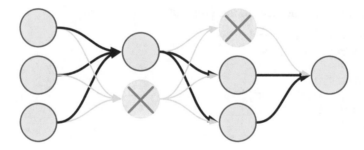

Figure 3-33. *Example of dropouts in neural networks*

In practice, like batch norm layers, Keras provides a dropout layer with a user-specified parameter/value. According to the Keras documentation, the dropout layer "randomly sets input units to 0 with a frequency of rate at each step during training time, which helps prevent overfitting." Here, "rate" is a preset hyperparameter.

The basic syntax for adding any dropout layer to a hidden layer is shown in the following (Listing 3-27). Note that during evaluation, all neurons are turned on.

Listing 3-27. Adding dropouts to hidden layers

```
# dropouts
from tensorflow.keras.layers import Dropout

fashion_model = Sequential()
fashion_model.add(Input((784,)))

# add dropout after dense and batch norm layers
fashion_model.add(Dense(64, activation="swish"))
```

```
fashion_model.add(BatchNormalization())
fashion_model.add(Dropout(0.25))

fashion_model.add(Dense(64, activation="swish"))
fashion_model.add(BatchNormalization())
fashion_model.add(Dropout(0.25))

fashion_model.add(Dense(10, activation="softmax"))
```

You can experiment with different values and amounts of dropout layers on the fashion_model and possibly further improve model performance. In most cases, dropouts will slightly reduce the performance of the model on training data but significantly improve the results on validation sets or unseen data.

The Keras Functional API

In the preceding examples, modeling with Keras is done through the Sequential object, adding layers in an orderly fashion. However, this type of functionality and process of constructing a model is extremely limited, and many complex model architectures are built using the Keras functional API. Before adding any fancy connections and structures to the network, in Listing 3-28 is an example of fashion_model defined previously as an example in the preceding sections built with the Keras functional API.

Listing 3-28. Simple model built using the functional API

```
import tensorflow.keras.layers as L
from tensorflow.keras import Model


inp = L.Input((784, ))
x = L.Dense(64, activation="swish")(inp)
x = L.BatchNormalization()(x)
x = L.Dropout(0.2)(x)

x = L.Dense(64, activation="swish")(x)
x = L.BatchNormalization()(x)
x = L.Dropout(0.2)(x)

out = L.Dense(10, activation="softmax")(x)

new_fashion_model = Model(inputs=inp, outputs=out)

# compile and train as normal
```

As the name "functional API" suggests, layers are defined as functions of the previous one. Each layer can be stored in its own variable while it's created in relation to the layer or layers, as we will see later. Unlike the Sequential model, in which each layer becomes a "part" of the model object, the functional API lets each layer become its own structure with limitless connections and structural possibilities. With the potential of each layer becoming its own unique variable, we can reference it later and create nonlinear network connections or even skip connections. However, if this is not needed, most conventions define every layer as one variable being redefined over and over for conciseness and readability.

The notation of redefining different layers over the same variable can be confusing at times. It's best to pay attention to where each function is directing to, whether it directs to the output layer or another separate branch of layers.

We establish a connection between the current layer and the previous layer by calling the current layer object (e.g., the Dense layer) as a function where its parameter is the previous layer's variable. Then the output of the function call is assigned to a variable (e.g., x is a commonly used variable name). Then the variable that we assigned the current layer to contains information about the current layer and its connection to the previous layer (Figure 3-34).

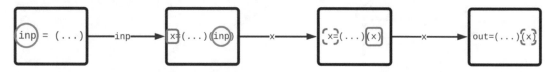

Figure 3-34. *The intuition behind functional API*

A simple analogy to the common data structure of linked lists can be made. Each variable, or layer, defined in the functional API acts as a node in a linked list. The single variable does not contain the full information about the entire model, just as how the user can't access every value in the linked list through one node. However, each variable contains a "pointer" to the next layer in its connection. When "stringing" the model together using the keras.Model object, Keras then internally connects the input layer and the output layer together by looking for these "pointers" and retrieving the layer information. Note that in the actual Keras library, the process of constructing a Model object from the functional API doesn't necessarily work like a linked list, but it's a decent analogy that can help with understanding.

Nonlinear Topologies

The gem of the Keras functional API lies in its ability to create nonlinear topology models and models with multiple inputs and outputs. These model architectures cannot be sequentially defined and may contain complex structures in which one layer's output can be copied and fed into multiple other layers. These nonlinear topology models also utilize merging techniques such as concatenation – one layer consists of the combination of two or more different layers' output.

Having the ability to construct nonlinear topology models is important since they allow for a deeper and possibly more meaningful analysis of the given data, hence being able to produce better results. In a sequentially defined model, the data would be limited to one set of parameters that encodes information from the input or the previous layer's outputs. With nonlinear topology, inputs can be passed through and split into multiple different "branches" of the network, each with different settings and types of connections. Then at one point, the insights of all these different "branches" can be merged back together by concatenation or some other form of operation specified by the user. Although it's been said that sequential models can adapt to data and create different parameterized neurons based on the data, it's still beneficial to use nonlinear topology as the ability of sequential models is limited. Most, if not all, modern state-of-the-art (SOTA) models use nonlinear topology in one way or another, whether trained on tabular data or other forms of data.

Building nonlinear topology in a functional manner is extremely intuitive. In the following, we're going to build a rather simple nonlinear topology network as an example. Then, more sophisticated concepts such as multi-input, multi-output, and weight sharing will be built upon and explored. As an example, we aim to build a network starting off with one input block and then splitting off into two separate branches with a different number of hidden layers and number of neurons; then, afterward, they concatenate back into one branch and finally output the prediction. The concept is illustrated in Figure 3-35.

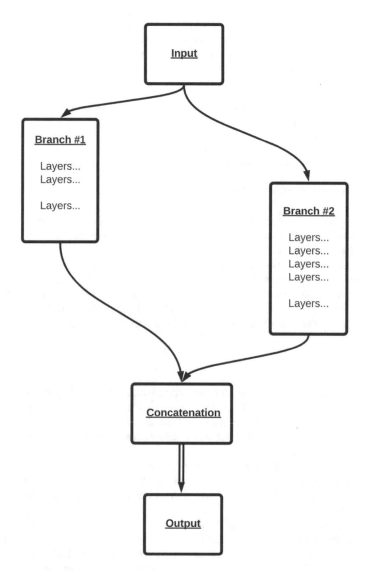

Figure 3-35. *Nonlinear topology network*

Following a similar fashion to the simple, functional model defined previously, we start out by defining the input layer shown in Listing 3-29.

Listing 3-29. Defining the input layer

```
import tensorflow.keras.layers as L
inp = L.Input((784, ))
```

Then, the two branches are defined separately (Listing 3-30). An important note is that the layers within the branch can use the same variable name, but layers of different branches cannot be defined with the same variable name. Doing so can mess up layer-to-layer relations, and it simply doesn't make sense.

Listing 3-30. Definition of two separate branches

```
# use the variable name "x" for 1st branch
x = L.Dense(128, activation="swish")(inp)
x = L.BatchNormalization()(x)

x = L.Dense(32, activation="swish")(inp)
x = L.BatchNormalization()(x)

# use the variable name "y" for 2nd branch
y = L.Dense(64, activation="relu")(inp)
y = L.BatchNormalization()(x)
```

After the creation of these two separate "branches" (the number of layers and neurons are chosen at random – this model only serves as an example), a concatenation is imposed to merge the outputs of the layers. Finally, the output layer is defined in relation to the concatenated layer. Notice that in the "parameter" of the L.Concatenate "function call" is a list consisting of the layer outputs that should be concatenated. In our case, it's the output of x and y (Listing 3-31). Concatenate simply joins the output of the layers on the specified axis. Usually, the axis of concatenation is assumed to be along the feature column (e.g., an array of shape (100,3) concatenated with an array of shape (100,2) would result in an array of shape (100,5)). The concatenate layer is used much more than any other merging method because of its ability to preserve outputs from all layers. Other merging layers include Average (averages value on the specified axis), Dot (performs dot product between the two vectors/matrices), Maximum (applies the max function on the specified axis), and many more. In most cases, Concatenate serves more than enough to the model; it's better for the model to "figure" out what operations to perform on the merged layers than assigning it a strict operation to perform between the outputs of different branches.

Listing 3-31. Concatenation between layers and the output

```
from tensorflow.keras import Model
# use the variable name "concat" for
# concatenated layer
# combine the output from
concat = L.Concatenate()([x, y])
concat = L.BatchNormalization()(concat)

out = L.Dense(10, activation="softmax")(concat)

# combining into one single Model object
non_linear_fashion_model = Model(inputs=inp, outputs=out)

# compile and train as normal
```

Then, the entire model is composed of one keras.Model object where the input layer's variable is passed into the "input" parameter, while the variable that contains the output layer is passed into the parameter "output".

When building complex and intricate networks, it's easy to use the Keras functional API and loose track of variable names, connections, and layers. Keras provides a set of simple functions that display information about the model as well as visualize the model architecture.

By calling `summary()` on the created model object (or a compiled model), Keras will output a detailed summary of the parameters, shapes, and size of each layer, the connection of each layer, as well as the total number of parameters in the network. The output of `summary()` called on the nonlinear model created previously is shown in Listing 3-32 and Figure 3-36.

Listing 3-32. Model summary example

```
non_linear_fashion_model.summary()
```

```
Model: "model_1"
```

Layer (type)	Output Shape	Param #	Connected to
input_2 (InputLayer)	[(None, 784)]	0	
dense_4 (Dense)	(None, 32)	25120	input_2[0][0]
batch_normalization_3 (BatchNor	(None, 32)	128	dense_4[0][0]
batch_normalization_4 (BatchNor	(None, 32)	128	batch_normalization_3[0][0]
concatenate (Concatenate)	(None, 64)	0	batch_normalization_3[0][0] batch_normalization_4[0][0]
batch_normalization_5 (BatchNor	(None, 64)	256	concatenate[0][0]
dense_6 (Dense)	(None, 10)	650	batch_normalization_5[0][0]

```
Total params: 26,282
Trainable params: 26,026
Non-trainable params: 256
```

Figure 3-36. *Output produced by model.summary()*

For a more intuitive and straightforward representation, Keras has a function that can plot the architecture of the model while including information about the layer shapes and types. The function is imported from `keras.utils` as `plot_model`. There are several parameters associated with the function. The model to be plotted is passed in as the first parameter. Then the `to_file` parameter creates the name of the visualization as well as the file path that it's saved to. There are two other minor parameters, `show_shapes` and `show_layer_names`, which control whether the plot displays the input and the output shape of the layer and the name of the layer, respectively (Listing 3-33 and Figure 3-37). The name of each layer can be customized by passing in the name as a string to any layer call under the `name` parameter.

Listing 3-33. Keras function for plotting the model

```
from keras.utils import plot_model
plot_model(non_linear_fashion_model, to_file="model.png", show_shapes=True, show_layer_
names=True)
```

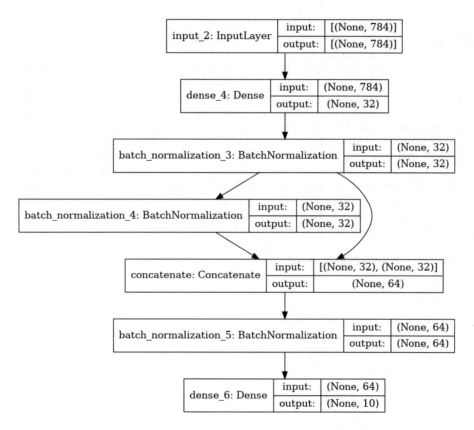

Figure 3-37. *The plotted model diagram*

Especially for nonlinear topologized models, being able to visualize how data is passed through between intertwined branches of a layer is extremely helpful to fix mistakes in model definition or simply to have a better grasp of the general model architecture. We will use this extensively in the remainder of the book as we grapple with new architectures.

Multi-input and Multi-output Models

In certain problem settings, multiple components of data are presented in different formats, or sometimes they fit into completely different categories that don't make sense to be combined and trained together. The obvious solution would be to train a separate model for different types of data. One common example is in the medical field. Imagine classification on radiographs is oftentimes paired with metadata about the image's patient given in the form of tabular data. Both image and tabular data provide crucial information that may be useful to the classification task. One might train a model handling the tabular data part, while another model processes the images. Then after each model predicts, the prediction is averaged to produce the final output. Although viable, the solution does encounter a few problems. Since each model only possesses certain portions of the "whole picture" and attempts to make a prediction based on the partial information, the results produced may be poorly informed. With multi-input models, however, we can jointly consider how all parts of the picture fit together. Luckily, there is a way to have one model receive two inputs and then, after processing both individually, combine the outputs of the layer into one single, merged layer. Then the entire model is trained together with backpropagation splitting its way back into each input "branch." A model trained this way performs much better than multiple singular models as the model learns its own "language" to combine insights from two or more different types of data, producing results that cannot be replicated with separate models.

Similarly, there are situations where one set of data is given to predict multiple different types of outputs. An example could be predicting house prices and whether it will be sold in 5 years using the same set of features about the house. Again, having two models with the same feature predict different types of outputs can be done. However, it's more beneficial to combine the two output "branches" into one single model. Backpropagation can relate the patterns of both branches into the model, acquiring the conjoined knowledge of both parts of the data that can't be learned with two separate models.

Building these types of models with the Sequential object would be impossible, but with the functional API, it becomes both possible and entirely intuitive. For multi-input models, we simply define two input heads with different variables. Then once they're processed through layers in their own "branches," concatenation is used, or any other form of merging between layers can combine the input branches into one single hidden layer. An example code for constructing a multi-input model is shown in Listing 3-34, along with the visualization of model architecture in Figure 3-38.

Listing 3-34. Code for a multi-input model

```
# example data of zeros
X_a = np.zeros((100, 4))
X_b = np.zeros((100, 8))

y = np.zeros((100,))

# first branch of input
inp1 = L.Input((4, ))
x = L.Dense(64, activation="relu")(inp1)
x = L.BatchNormalization()(x)
x = L.Dense(64, activation="relu")(x)
x = L.BatchNormalization()(x)

# second branch of input
inp2 = L.Input((8, ))
z = L.Dense(128, activation="relu")(inp2)
z = L.BatchNormalization()(z)

# concatenation
concat = L.Concatenate()([x, z])
out = L.Dense(1)(concat)

# build into one model
multi_in = Model(inputs=[inp1, inp2], outputs=out)
```

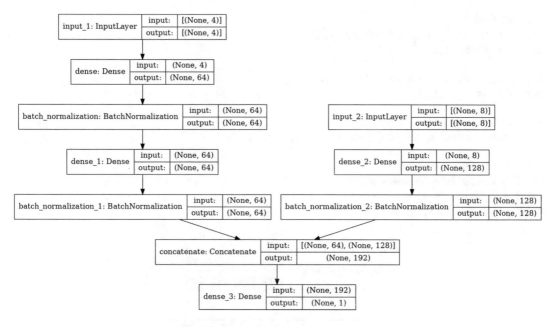

Figure 3-38. *The model architecture of the multi-input model*

During training, the different types of input data are passed in as a list within the (x, y) tuple in the same order that the layers are inputted into the list when creating the Model object (Listing 3-35). The same concept applies to evaluation.

Listing 3-35. Example training code for multi-input model

```
multi_in.compile(optimizer="adam",loss="mse")
multi_in.fit([X_a, X_b], y, epochs=10)
multi_in.evaluate([X_a, X_b], y)
```

Multi-output models can be defined in a similar manner. The different output layers are passed in as a list into the keras.Model object, and during training, it's passed in as a list within the (x, y) tuple. Shown in Listing 3-36 is the basic code for defining a multi-output model, and its plotted model architecture is in Figure 3-39.

Listing 3-36. Example code for multi-output models

```
# example data of zeros and ones
X = np.zeros((100, 12))

y_a = np.zeros((100, 1))
y_b = np.ones((100, 1))

inp = L.Input((12, ))
x = L.Dense(64, activation="relu")(inp)
x = L.BatchNormalization()(x)
x = L.Dense(64, activation="relu")(x)
x = L.BatchNormalization()(x)
```

```
# seperation
out1 = L.Dense(28, activation="relu")(x)
out1 = L.Dense(1)(out1)

out2 = L.Dense(14, activation="relu")(x)
out2 = L.Dense(1)(out2)

# build into one model
multi_out = Model(inputs=inp, outputs=[out1, out2])

# training and evaluation
multi_out.compile(optimizer="adam",loss="mae")
multi_out.fit(X, [y_a, y_b], epochs=10, batch_size=100)
multi_out.evaluate(X, [y_a, y_b])
```

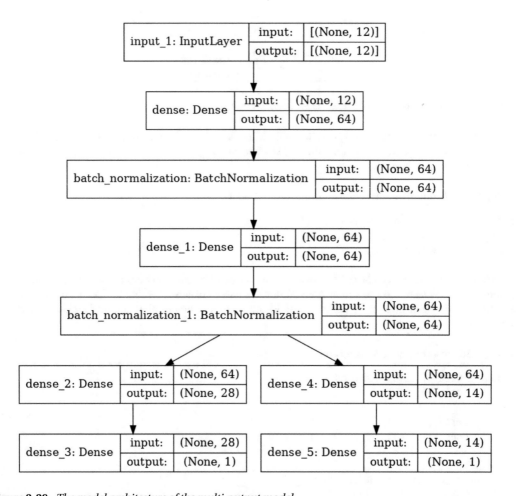

Figure 3-39. *The model architecture of the multi-output model*

Embeddings

Tabular datasets often have continuous features in addition to categorical features. One can use the categorical encoding methods discussed in Chapter 2 to encode the categorical features, which may be successful. (See Chapter 10 to see how you can automate the selection of the best data preprocessing operations.) Another way to encode categorical features is to have the network automatically learn to associate an n-dimensional vector with each unique categorical value – that is, to optimally *embed* unique values into n-dimensional space.

Embeddings are useful when there are a high number of unique values in a categorical feature and it is too difficult to capture their intricacies with alternative "classical" categorical encoding methods.

Embeddings are implemented as matrices applied to one-hot encoded categorical features, which are optimized just like any other parameter in the network. However, to use the Keras embedding layer, your feature must be ordinally encoded (beginning from zero). Moreover, to use embeddings on a categorical feature or multiple categorical features in conjunction with continuous features, you need to designate one head per categorical feature such that it can be passed into a unique embedding. After embedding the feature, the result can be concatenated with other vectors or processed as desired (Listing 3-37).

Listing 3-37. Using an embedding layer

```
embed_inp = L.Input((1,))
embedded = L.Embedding(num_classes, dim)(embed_inp)
flatten = L.Flatten()(embedded)
process = L.Dense(32)(flatten)
```

Note that we flatten the result of the embedding layer because its output is of shape (1, dim); flattening produces a dim-length vector, which can be processed with fully connected layers and the like. The rank-two shape is because the embedding layer is built to work primarily with text, which is functionally a large bundle of categorical features each with the same number of classes (the vocabulary size). A 100-token-long sequence would have an embedding shape (100, dim).

You will see examples of embeddings applied to categorical features and text data later in this chapter ("Selected Research" > "Wide and Deep Learning"), Chapter 4 (Multimodal Image and Tabular Models), Chapter 5 (Multimodal Recurrent Modeling), and Chapter 10, among others.

Model Weight Sharing

Computational costs are one major downside to training multi-input, multi-output, or any other complex nonlinear topologized models. As seen earlier in the chapter, the number of trainable parameters that a network can contain is absurd, up to millions in a simple model; with the addition of different branches of the network, the risk of overfitting increases as well as the training time. One technique that improves upon this disadvantage is weight sharing (Figure 3-40).

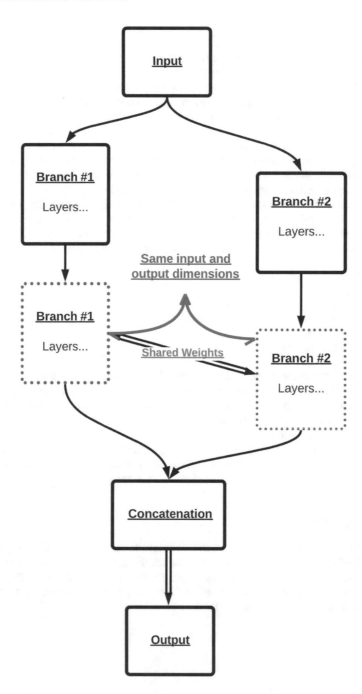

Figure 3-40. *Weight sharing intuition*

It does exactly what it sounds like: sharing weights between different layers of the same shape. By doing so, the sets of weights in two separate layers are identical, meaning that backpropagation only needs to run through it once. However, by reducing the training time, we do come at the cost of model flexibility since the algorithm needs to find the same set of weights that fits both layers. But note that by doing so, we introduce regularization into the model, possibly improving validation performance.

Using the Keras functional API, we can create a single layer that we want to share the weights off and assign to a variable. Following, each time that we want to reuse the layers, we simply call the layers in the functional API style, connecting them to whichever previous layer we want to connect. Shown in Listing 3-38 is a basic example of weight sharing, and the plotted model architecture is in Figure 3-41. The example model built in the following is a multi-input model that shares the same layer weights in two different branches before merging into one. An important point to note, the input dimension from the previous layers that use the same shared layer must be the same.

Listing 3-38. Weight sharing example

```
# example data
X_share_a = np.zeros((100, 10))
# same shape
X_share_b = np.ones((100, 10))

y_share = np.zeros((100, ))

# create shared layer
shared_layer = L.Dense(128, activation="swish")

inp1 = L.Input((10, ))
x = shared_layer(inp1)
x = L.BatchNormalization()(x)

inp2 = L.Input((10, ))
y = shared_layer(inp2)
y = L.BatchNormalization()(y)

out = L.Concatenate()([x, y])
out = L.Dense(1)(out)

shared_model = Model(inputs=[inp1, inp2], outputs=out)
```

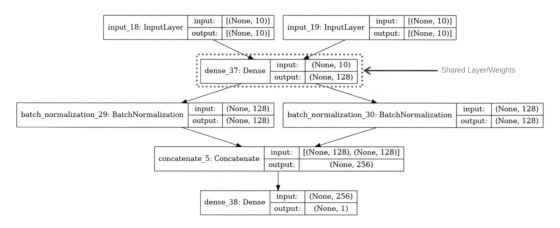

Figure 3-41. *Weight sharing model architecture*

The Universal Approximation Theorem

George Cybenko published the paper "Approximation by Superpositions of a Sigmoidal Function"[3] in a 1989 edition of the *Mathematics of Control, Signals, and Systems* journal (Figure 3-42). This paper proposed the theoretical foundations of the Universal Approximation Theorem, which has been subsequently built on by a large body of expanding and generalizing work. Fundamentally, Cybenko's paper and the Universal Approximation Theorem make statements about the ability of neural networks to theoretically fit any function to an arbitrary degree of correctness, given sufficient size.

Approximation by Superpositions of a Sigmoidal Function*

G. Cybenko†

Abstract. In this paper we demonstrate that finite linear combinations of compositions of a fixed, univariate function and a set of affine functionals can uniformly approximate any continuous function of *n* real variables with support in the unit hypercube; only mild conditions are imposed on the univariate function. Our results settle an open question about representability in the class of single hidden layer neural networks. In particular, we show that arbitrary decision regions can be arbitrarily well approximated by continuous feedforward neural networks with only a single internal, hidden layer and any continuous sigmoidal nonlinearity. The paper discusses approximation properties of other possible types of nonlinearities that might be implemented by artificial neural networks.

Key words. Neural networks, Approximation, Completeness.

Figure 3-42. *Title and abstract of Cybenko's original paper*

---

[3] Cybenko, G. Approximation by Superpositions of a Sigmoidal Function. *Math. Control Signal Systems* **2,** 303–314 (1989). https://doi.org/10.1007/BF02551274.

Cybenko's paper considered a function to be "sigmoidal" if it approached 1 as its argument approached infinity and 0 as its argument approached negative infinity. Note that the "standard" sigmoid function $\frac{1}{1+e^{-x}}$ satisfies this property, but so do others, like a rescaled hyperbolic tangent ($\frac{\tanh(x)+1}{2}$) and the Heaviside step function ($x = 0$ if $x < 0$, else 1). Cybenko was concerned primarily with activation functions that, at root, allowed arguments to manipulate an output near two "binary" states. Under these conditions, a single-hidden-layer neural network with a sufficient number of neurons can be shown to approximate an arbitrary function. In this case, we can interpret the network as a linear sum of sigmoidal functions.

Subsequently, similar results have been demonstrated for other activation functions like ReLU, multilayer networks, and other variants.

The original conditions outlined in Cybenko's paper are difficult to get working. This neural network has a one-dimensional input, a very large hidden layer, and a one-dimensional output. With your experience in Keras, this should be simple to implement. However, it is difficult to obtain a functional approximation using this network. This stems from the difference between proving the existence of a set of weights that can approximate an arbitrary function and finding a method that reliably leads to the discovery of said weights.

However, it is simpler to demonstrate universal approximation capabilities with two modifications: using multilayer networks, which are more expressive than single-layer networks, and the ReLU activation, which is unbounded and therefore easier to manipulate and optimize for in the context of function approximation. Listing 3-39 demonstrates implementation of a "Universal Approximation Theorem model generator," which takes in the number of hidden layers, the number of nodes in each layer, and the activation function to use.

Listing 3-39. A function to create a scalar-input scalar-output architecture with a specified number of layers, number of neurons in each layer, and activation

```
def UAT_generator(n, layers, activation):
    model = tensorflow.keras.models.Sequential()
    model.add(L.Input(1,))
    for i in range(layers):
        model.add(L.Dense(n, activation=activation))
    model.add(L.Dense(1, activation='linear'))
    return model
```

Our arbitrary function of choice will be $\sin^2 x - e^{-\cos x}$ within the domain $[-20, 20]$, which is highly nonlinear (Listing 3-40, Figure 3-43).

Listing 3-40. Title and abstract of Cybenko's original paper

```
def function(x):
    return np.sin(x)**2 - np.exp(-np.cos(x))

x = np.linspace(-20, 20, 4000)
y = function(x)
plt.figure(figsize=(10, 5), dpi=400)
plt.plot(x, y, color='red')
plt.show()
```

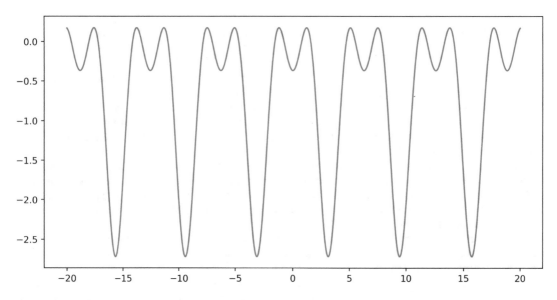

Figure 3-43. *Graph of $sin^2 x - e^{-\cos x}$ within the domain $[-20, 20]$*

Figures 3-44, 3-45, and 3-46 demonstrate the fit for a neural network trained with 1024 nodes in each layer and one, two, and three layers (respectively). Note the angular fit of the ReLU activation functions and the successively improving fits as the number of layers and expressivity increase.

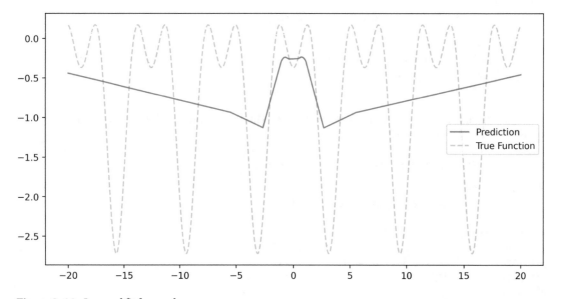

Figure 3-44. *Learned fit for one layer*

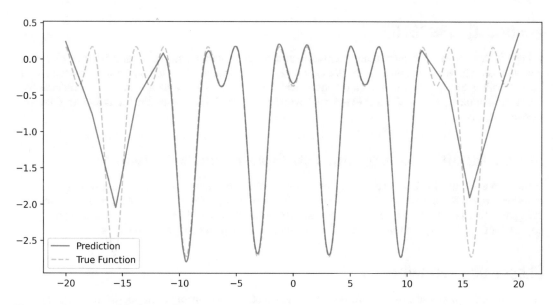

Figure 3-45. *Learned fit for two layers*

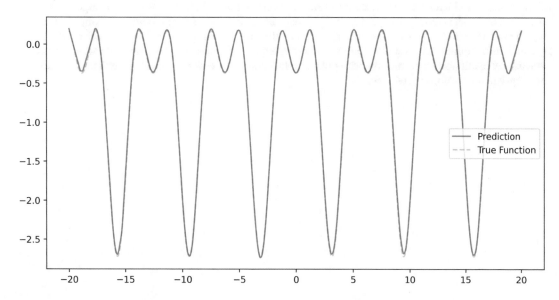

Figure 3-46. *Near-perfect learned fit for three layers*

This is one demonstration of how you can observe interesting theoretical properties of neural networks with the skills you've acquired in constructing and fitting neural networks.

Selected Research

We will further explore how more complex models and designs are being used in future chapters. However, one who has learned the material in just this chapter is already prepared to understand and work with a host of modern tabular deep learning techniques, which modify feed-forward networks to more effectively model tabular data. This section provides a brief overview and implementation directions for four selected research papers.

Simple Modifications to Improve Tabular Neural Networks

James Fiedler synthesizes a set of relatively low-labor modifications to standard neural networks from other research in the area that can be used to possibly improve performance on tabular problems in the 2021 paper "Simple Modifications to Improve Tabular Neural Networks."[4] Two modifications that are approachable at this stage in the book and discussed in Fiedler's paper are presented here.

Ghost Batch Normalization

Previously, batch normalization was introduced as an effective mechanism to smooth the loss landscape and improve training performance by standardizing the inputs (Figure 3-47). However, note that batch normalization performs differently for different batch sizes. The "generalization gap" is an observed phenomenon in which neural networks perform worse on validation data when trained on large batch sizes than on small batch sizes. Large batch sizes are constrained by a larger population of samples, so we would expect it to move "slower" and less sharply than the updates computed with small batch sizes. This, researchers hypothesize, prevents models trained with large batch sizes and batch normalization layers from moving out of attractive local minima with poor generalization. Moreover, computing batch normalization on large batches is subpar in efficiency.

---

[4] Fiedler, J. (2021). Simple Modifications to Improve Tabular Neural Networks. *ArXiv, abs/2108.03214.*

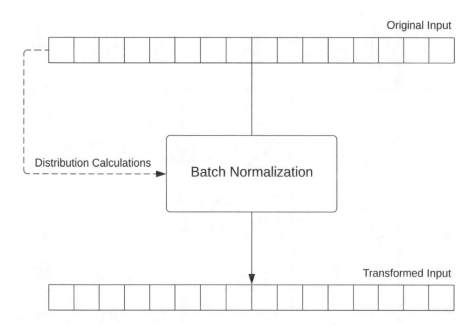

Figure 3-47. *A diagram of batch normalization*

To address this, we can use *ghost batch normalization*[5] (Figure 3-48) in place of batch normalization. Rather than computing batch normalization statistics across the entire batch, we split the batch into virtual "ghost" batches and calculate normalization across these smaller groups of samples. This allows for training models with large batch sizes without experiencing generalization gap problems in which said large batches constrain learning activity.

---

[5] Hoffer, E., Hubara, I., & Soudry, D. (2017). Train longer, generalize better: closing the generalization gap in large batch training of neural networks. *ArXiv, abs/1705.08741.*

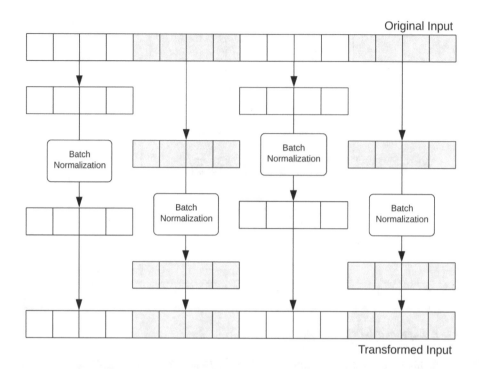

Figure 3-48. *A diagram of ghost batch normalization. The distribution calculation arrows are not shown.*

You can pass the `virtual_batch_size` argument into Keras's `BatchNormalization` layer to enable ghost batch normalization. The virtual batch size must divide into the batch size; for instance, 50 would not be a valid virtual batch size assuming a batch size of 256, but 32 would be.

This should not be confused with the "virtual batch normalization" method, used by the authors of the Generative Adversarial Network paper (see Chapter 9), in which a single batch's calculations are used across an entire dataset to ensure greater stability.

Leaky Gates

A leaky gate is a simple "gating" mechanism in which the network learns simple element-wise linear transformations to determine whether each element "passes" the gate or not. Each element in a vector is multiplied by a vector and added to a bias. If the resulting value is larger than zero, the value passes through unchanged; otherwise, zero (or a value very close to zero) is returned. It is functionally an element-wise linear transformation followed by a ReLU. The leaky gate is defined as follows for a vector input x and an index i denoting the ith element of x:

$$g_i(x_i) = \begin{cases} w_i x_i + b_i, & w_i x_i + b_i \geq 0 \\ \approx 0, & w_i x_i + b_i < 0 \end{cases}$$

Consider the following system:

$$x = \langle 3, 2, 1 \rangle$$
$$w = \langle -4, 2, 3 \rangle$$
$$b = \langle 2, 0, -1 \rangle$$

Then, we have the following element-wise linear transformation (where ⊗ denotes element-wise multiplication and ⊕ denotes element-wise addition):

$$(x \otimes w) \oplus b = \langle 3 \cdot -4 + 2, 2 \cdot 2 + 0, 1 \cdot 3 - 1 \rangle = \langle -10, 4, 2 \rangle$$

After passing through the ReLU condition, we obtain $\langle 0, 4, 2 \rangle$. The first input did not pass, but the other two did.

Such gating mechanisms enable a form of explicit "feature selection" similar to how tree-based models select a subset of features to reason from. We will see more advanced neural network feature selection analogs in future research paper discussions.

We can implement a leaky gate with a custom Keras layer (Listing 3-41). In the __init__ method, we "hook" the layer with the previous layer in the functional API style. Two learnable parameters, the weight and the bias, are each added in the build method. The call method orchestrates the transformation of the input by the layer's internal parameters, which in this case is a simple element-wise multiplication (we use broadcasting to enable the same multiplication over the batch dimension) and addition.

Listing 3-41. A custom leaky gate layer

```
class LeakyGate(keras.layers.Layer):

    def __init__(self):
        super(LeakyGate, self).__init__()

    def build(self, input_shape):
        self.w = self.add_weight(shape=(1,input_shape[-1],),
                            initializer='random_normal',
                            trainable=True)
        self.b = self.add_weight(shape=(1,input_shape[-1],),
                            initializer='random_normal',
                            trainable=True)
        self.mult = keras.layers.Multiply()

    def call(self, inputs):
        return self.mult([inputs, self.w]) + self.b
```

To verify the functionality of this layer, let us construct a simple model and gating task to train on (Listing 3-42).

Listing 3-42. Using the custom leaky gate on a synthetic task

```
inp = L.Input((128,))
gate = LeakyGate()
gated = gate(inp)

model = keras.models.Model(inputs=inp, outputs=gated)
```

249

```
x = np.random.normal(size=(512, 128))
mask = np.random.choice([0,1], size=(1, 128))
y = x * mask

model.compile(optimizer='adam', loss='mse')
model.fit(x, y, epochs=200)
```

One can verify with (np.round(gate.w) == mask).all() that the model exactly learns the mask.

This gating mechanism can be added throughout your network designs to encourage dynamic feature selection.

Wide and Deep Learning

Introduced by Heng-Tze Cheng et al. in the paper "Wide & Deep Learning for Recommender Systems,"[6] Wide and Deep is a relatively simple tabular deep learning paradigm that nevertheless has shown extraordinary success in many domains, including in recommendation systems.

A "wide" model is a simple linear model (Logistic Regression in the context of classification and functionally a zero-hidden-layer neural network). The inputs to such a model include both original features and *cross-product transformations*, in which selected features are multiplied by each other. This can capture when relevant phenomena occur simultaneously across multiple columns – for instance, if the cross-product of two columns, "isVideo?" and "isTrending?", in a hypothetical social media content dataset indicates whether the sample both is a video and is trending. Computing cross-product transformations is necessary because simple linear models are not capable of developing these intermediate features on their own. Even if we add cross-product features, however, these products cannot generalize to new feature pairs because they are hard-coded as inputs to the model.

On the other hand, a "deep" model is a standard neural network that can develop meaningful embedding representations that can be internally combined and generalize well to new data. However, it is difficult for deep neural networks to learn effective low-dimensional embeddings when the dataset itself is complex and contains many intricacies.

The Wide and Deep paradigm (Figure 3-49) is to take the best of both worlds by jointly training a model with both a wide and a deep component. In the resulting model, some features (original and cross-product computed) are passed into a "wide" linear model, and others are passed into a "deep" neural network. The outputs of both models are added to produce the final output, jointly informed by the generalizing influence of the wide component and the specificity power of the deep component.

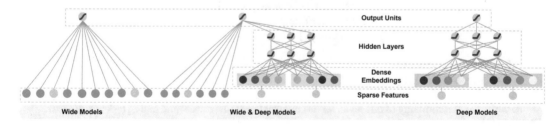

Figure 3-49. Merging a wide linear model and a deep model to form a Wide and Deep model, from Cheng et al.

---

[6] Cheng, H., Koc, L., Harmsen, J., Shaked, T., Chandra, T., Aradhye, H.B., Anderson, G., Corrado, G.S., Chai, W., Ispir, M., Anil, R., Haque, Z., Hong, L., Jain, V., Liu, X., & Shah, H. (2016). Wide & Deep Learning for Recommender Systems. *Proceedings of the 1st Workshop on Deep Learning for Recommender Systems*.

Let us demonstrate the Wide and Deep paradigm on the Forest Cover dataset (Listing 3-43). After loading the dataset into data, we create the inputs to the wide and deep models. There are two relevant categorical features: soil type and wilderness area. We will generate a feature cross between these two features by multiplying every one-hot column from one feature by every one-hot column in the other. The inputs to the deep model are all the continuous inputs and the categorical features, which are isolated and will be passed into their own respective embedding layers.

Listing 3-43. Collecting the data to apply a Wide and Deep approach to the Forest Cover dataset

```python
# initiate data
wide_data = data.drop('Cover_Type', axis=1)
deep_cont_data = data[['Elevation', 'Aspect', 'Slope',
                       'Horizontal_Distance_To_Hydrology',
                       'Vertical_Distance_To_Hydrology',
                       'Horizontal_Distance_To_Roadways',
                       'Horizontal_Distance_To_Fire_Points',
                       'Hillshade_9am', 'Hillshade_Noon',
                       'Hillshade_3pm']]
deep_embed_data = {}

# obtain categorical features
soil_types = [col for col in data.columns if 'Soil_Type' in col]
wild_areas = [col for col in data.columns if 'Wilderness_Area' in col]

# cross soil types and wild areas
for soil_type in soil_types:
    for wild_area in wild_areas:
        crossed = wide_data[soil_type] * wide_data[wild_area]
        wide_data[f'{soil_type}X{wild_area}'] = crossed

# get ordinal representations of categorical features
deep_embed_data['soil_type'] = np.argmax(data[soil_types].values, axis=1)
deep_embed_data['wild_area'] = np.argmax(data[wild_areas].values, axis=1)
```

The wide model is a simple linear model (Listing 3-44).

Listing 3-44. Constructing the wide linear model.

```python
wide_inp = L.Input((len(wide_data.columns)))
wide_out = L.Dense(7)(wide_inp)
wide_model = keras.models.Model(inputs=wide_inp,
                                outputs=wide_out)
```

The deep model is more complex (Listing 3-45). We need to create three inputs – one for the continuous features, one for the soil type feature, and one for the wilderness area feature. Both categorical features are passed through an embedding layer with 16 dimensions, indicating that each unique class in each feature is associated with a unique 16-dimensional vector. These embeddings are concatenated with the continuous features and passed through a series of fully connected layers into an output.

Listing 3-45. Constructing the deep model

```python
deep_inp = L.Input((len(deep_cont_data.columns)))
```

251

```
deep_soil_inp = L.Input((1,))
deep_soil_embed = L.Embedding(np.max(deep_embed_data['soil_type']) + 1,
                              16)(deep_soil_inp)
deep_wild_inp = L.Input((1,))
deep_wild_embed = L.Embedding(np.max(deep_embed_data['wild_area']) + 1,
                              16)(deep_wild_inp)
deep_concat = L.Concatenate()([deep_inp,
                               L.Flatten()(deep_soil_embed),
                               L.Flatten()(deep_wild_embed)])
deep_dense1 = L.Dense(32, activation='relu')(deep_concat)
deep_dense2 = L.Dense(32, activation='relu')(deep_dense1)
deep_dense3 = L.Dense(32, activation='relu')(deep_dense2)
deep_out = L.Dense(7)(deep_dense3)
deep_model = keras.models.Model(inputs={'cont_feats': deep_inp,
                                        'soil': deep_soil_inp,
                                        'wild': deep_wild_inp},
                                outputs=deep_out)
```

We can combine these two models using the WideDeepModel provided in Keras's experimental module (Listing 3-46). The model accepts both a wide model and a deep model, as well as the final activation, and allows for their joint training. Note that we pass the wide model's data and the deep model's data together bundled in the same list, per multi-input model syntax.

Listing 3-46. Combining a wide and a deep model into a Wide and Deep model

```
from tensorflow.keras.experimental import WideDeepModel
model = WideDeepModel(wide_model, deep_model, activation='softmax')
model.compile(optimizer='adam', loss='sparse_categorical_crossentropy')
model.fit([wide_data, {'cont_feats':deep_cont_data,
                       'soil':deep_embed_data['soil_type'],
                       'wild':deep_embed_data['wild_area']}],
          data['Cover_Type'] - 1,
          epochs=10)
```

This method of combining models with different strengths can be adopted as a general modeling paradigm. Have a look at Chapter 11 for more examples of how different models can be combined into effective ensembles.

See the following similar papers that make use of feature crosses and other explicit feature interaction modeling methods. Also, see Chapter 6 to see how feature interactions can be learned (rather than manually set into the data):

Huang, T., Zhang, Z., & Zhang, J. (2019). FiBiNET: Combining feature importance and bilinear feature interaction for click-through rate prediction. *Proceedings of the 13th ACM Conference on Recommender Systems.*

Lian, J., Zhou, X., Zhang, F., Chen, Z., Xie, X., & Sun, G. (2018). xDeepFM: Combining Explicit and Implicit Feature Interactions for Recommender Systems. *Proceedings of the 24th ACM SIGKDD International Conference on Knowledge Discovery & Data Mining.*

Qu, Y., Fang, B., Zhang, W., Tang, R., Niu, M., Guo, H., Yu, Y., & He, X. (2019). Product-Based Neural Networks for User Response Prediction over Multi-Field Categorical Data. *ACM Transactions on Information Systems (TOIS), 37*, 1 –35.

Wang, R., Fu, B., Fu, G., & Wang, M. (2017). Deep & Cross Network for Ad Click Predictions. *Proceedings of the ADKDD'17.*

Wang, R., Shivanna, R., Cheng, D.Z., Jain, S., Lin, D., Hong, L., & Chi, E.H. (2021). DCN V2: Improved Deep & Cross Network and Practical Lessons for Web-scale Learning to Rank Systems. *Proceedings of the Web Conference 2021.*

Self-Normalizing Neural Networks

Günter Klambauer et al. introduce the scaled exponential linear unit (SELU) activation function in the paper "Self-Normalizing Neural Networks"[7] as an alternative to ReLUs. The authors find that simple substitution of SELU over ReLU in standard fully connected neural networks can significantly improve performance on a large number of tasks. While the paper discusses applications to vision and text, we will focus on the implications for modeling tabular data.

Klambauer et al. assert that normalization mechanisms like batch normalization are perturbed by stochastic processes in the network, such as stochastic gradient descent and dropout (i.e., stochastic regularization), which makes it difficult to train deep fully connected neural networks on tabular data.

A *self-normalizing neural network* is roughly defined as a neural network for which the mean and variance of activations stably approach/converge toward a fixed point as the number of layers information is passed through increases. For instance, say that our ideal fixed point is a mean of 0 and a variance of 1. In the following examples of hypothetical means and variances of activations for each layer in a nine-layer network, progression A would be self-normalizing, whereas progression B would not:

```
[Progression A]
    Means: 2.2, 2.1, 1.8, 1.4, 0.8, 0.3, 0.2, 0.1, 0.0
    Variances: 4.9, 4.5, 4.2, 3.4, 3.1, 2.9, 1.5, 1.1, 1.0

[Progression B]
    Means: 2.2, 2.1, -3.4, -2.9, -4.2, -1.2, 0.4, 2.5, 1.3
    Variances: 4.9, 4.5, 3.4, 2.4, 0.1, 1.6, 2.3, 2.1
```

Note how this differs from a normalization scheme like batch normalization, where activations are immediately (perhaps even "abruptly") normalized but don't necessarily stay that way through the network. A network that satisfies the self-normalizing constraint can be thought of as adopting a more "sustainable" normalization trajectory.

The SELU activation, when used by a neural network, is self-normalizing (Figure 3-50). It is defined as follows, given two parameters $\lambda > 1$ and $\alpha > 0$:

$$\mathrm{SELU}(x) = \lambda \begin{cases} x, & x > 0 \\ \alpha e^x - \alpha, & x \leq 0 \end{cases}$$

---

[7] Klambauer, G., Unterthiner, T., Mayr, A., & Hochreiter, S. (2017). Self-Normalizing Neural Networks. *ArXiv, abs/1706.02515.*

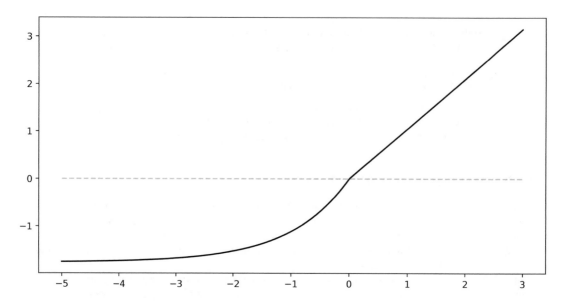

Figure 3-50. *The SELU activation function*

This activation function possesses the following key properties:

1. The capability to represent both negative and positive values to control the mean, which ReLU and sigmoid lack

2. Regions with derivatives approaching zero (as $x \to -\infty$) to dampen the variance if it is too large (this maps high-variance inputs to low-variance activations, that is, to a "flat landscape")

3. Regions with larger derivatives (a slope larger than 1 for $x > 0$) to increase the variance if it is too small (this maps low-variance data to high-variance activations, that is, to a "steep landscape")

4. A continuous curve upon which the previous three properties can be nudged and tuned toward self-normalization as information is propagated throughout the network

These properties endow the SELU function with theoretically and practically demonstrable self-normalizing behavior throughout the network.

SELU comes implemented in Keras with `activation='selu'` in any layer that accepts an activation function.

The SELU function has been empirically observed to either perform about the same as ReLU or better than ReLU; moreover, when SELU performs better than ReLU, it often performs much better (as opposed to marginally better). Thus, it is not ill-advised to use SELU as a "default" activation function.

Regularization Learning Networks

Regularization is a technique to combat overfitting by penalizing large weight values. We can generically express a regularized loss L_r given P that returns model predictions on an input x given the weights w and the default loss L between the predicted labels and the true labels as follows:

$$L_R(x,y,w) = L(P(x,w),y) + \lambda \|w\|_n$$

Note that the double-pipe symbol represents the norm of the weights and λ the weight of the penalty term relative to the default prediction loss. It is most common to either use the L1 or L2 norm.

This modifies the loss such that the model can decrease regularized loss either by updating its weights in the direction that minimizes default prediction loss or by decreasing the magnitude of its weights. Weights are therefore "small" by default and do not exert heavy influence over the internal network information flow. If a model trained with regularized loss does have large weight values, these weights are so important to prediction that their contribution to default prediction loss outweighs their magnitude (penalized by the regularization term).

One can conceptualize the regularization penalty term as having the same effect as gravity does on our movement. Gravity is an always-present force that shapes how we move and expend our energy.

In Keras, users can apply L1 or L2 regularization to a given network layer's parameters and/or activity (Listing 3-47). Applying regularization to parameters is more commonly used to avoid overfitting, whereas applying regularization to activity (i.e., the output of the layer) is used to encourage sparsity. See the "Sparse Autoencoders" section of Chapter 8 for an example of applying regularization for the development of robust sparse learning representations.

Listing 3-47. Using regularization in Keras

```
from keras import regularizers as R
dense = L.Dense(32,
                kernel_regularizer = R.L1(),
                activity_regularizer = R.L2())
```

These regularizers apply a uniform penalty strength across all weights in a layer (this penalty strength being λ). It "equally pushes all weights down," so to speak. This works for homogenous input data forms like images and text in which each feature resides in the same possible range of values as every other feature. However, tabular data often consists of heterogenous data in which features operate on many different scales. It therefore does not make sense to apply an equal weighting across all weights.

Ira Shavitt and Eran Segal propose the *Regularization Learning Network* in the 2018 paper "Regularization Learning Networks: Deep Learning for Tabular Datasets,"[8] which addresses this problem by learning a different penalty strength for every weight. In order to optimize individual λ-values effectively, one needs to use gradient-based methods; however, there is no clear differentiable relation between the λ-values and the loss function as there is between the weights and the loss function. The λ-values do not directly affect the model's predictions; they affect how the weights are learned. Therefore, λ-values enact change *temporally* (i.e., through time).

Exploiting this, Shavitt and Segal put forth *counterfactual loss*, which is the loss that a network updated with the current set of λ-values *will* obtain. Out of this clever reformulation emerges a clear differentiable relation between the λ-values and the future loss. Rather than using a penalty term that augments the model's loss, the model updates in two steps: updating the penalty strengths λ to minimize the counterfactual loss and then updating the weights themselves to minimize the default predictive loss (Figure 3-51).

---

[8] Shavitt, I., & Segal, E. (2018). Regularization Learning Networks: Deep Learning for Tabular Datasets. *NeurIPS*.

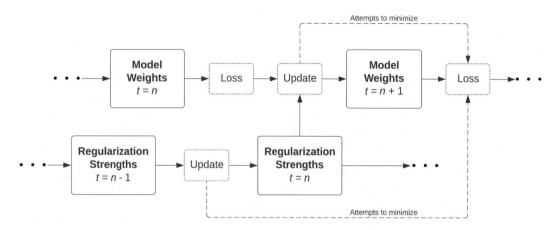

Figure 3-51. *Approximate schematic illustrating how regularization penalty strengths can be optimized along with the model weights themselves by optimizing for the loss incurred by a model updated using the given model weights*

Regularization Learning Networks (RLNs) are implemented as a Keras callback (Listing 3-48).

Listing 3-48. Importing the RLN callback by pulling directly from GitHub from a Jupyter Notebooks cell. Pulling from git also works

```
!wget -O rln.py https://raw.githubusercontent.com/irashavitt/regularization_learning_
networks/master/Implementations/Keras_implementation.py
import rln
import importlib
importlib.reload(rln)
from rln import RLNCallback
```

To use the callback, simply pass a regularized layer into the callback constructor and pass the callback(s) into the .fit function (Listing 3-49).

Listing 3-49. Training a model with optimizer regularization parameters

```
from keras import regularizers as R

NUM_LAYERS = 4

inp = L.Input((X.shape[-1],))
x = inp
for i in range(NUM_LAYERS):
    x = L.Dense(32, activation='selu',
                kernel_regularizer=R.L1())(x)
out = L.Dense(7, activation='softmax')(x)
model = keras.models.Model(inputs=inp, outputs=out)

callbacks = [RLNCallback(model.layers[i]) for i in range(1, 1 + NUM_LAYERS)]

model.compile(optimizer='adam', loss='sparse_categorical_crossentropy')
model.fit(X, y, epochs=10, callbacks=callbacks)
```

Applying regularization learning works well on heterogenous datasets that, for one reason or another, the network is likely to overfit on (the network is overparametrized with respect to the dataset size, the dataset is not very complex, etc.).

Key Points

In this chapter, the theory and foundation of neural networks, their applications to tabular data, and Keras were thoroughly discussed and visually explained.

- Neural networks are machine learning models linked by numerous neurons with each of them assigned a trainable weight and bias. In an artificial neural network (ANN), each neuron is connected to every neuron in the previous and the next layer.

  - Feed-forward operation is needed to make predictions and is also involved in backpropagation. It takes in data in the form of vectors, or matrices when dealing with more than one sample at a time. As the data is passed through the input layer, the result of multiplication and addition with the neuron that's associated is passed to every single neuron in the next layer. This process continues until the data reaches the output layer. Calculations are done by dot products for efficiency and conciseness. The produced output is treated as predictions or an intermediate step during backpropagation

  - Backpropagation is the heart of neural network learning; it uses gradient descent to adjust each parameter in the network. Prior to training, the network is typically initialized with random weights or calculated according to some weight initialization algorithm. Then the data, generally batched by a user-specified parameter, is passed through the network to generate an initial prediction. The cost function acts as an objective or task that the gradient descent algorithm is trying to optimize. The number of parameters in the cost function is the number of parameters in the network as changes in any of these values would influence the final cost or loss value. Then the derivative of the cost function is taken with respect to every parameter in the network, or the gradient of the cost function is calculated by working backward to the input layers. The parameter update rule is determined by the optimizer used. Gradient descent–related algorithms such as NAG and acceleration tend to have stabilizing performance, but training speed is slow. On the contrary, modern optimizers such as Adam have fast convergence, but training results may be unstable when used on large models.

- Keras, the high-level API built on TensorFlow, is a common framework or library chosen for building neural networks with easy-to-understand syntax. One downfall of Keras is that it does not have as much low-level control of the actual training process compared with its competitors such as PyTorch. But in most cases what Keras has to offer is enough.

  - One of the most straightforward methods of constructing a neural network is through the Keras Sequential object. Although only limited to sequentially connected layers with each layer preceding the next without any skip connections or branches, mistakes are less likely to happen, and the readability of the code is high compared with the functional API.

- The Keras functional API provides endless possibilities for neural network architectures from multi-input to multi-output models, skip connections and weight sharing, or a combination of all. By using the functional API, the code readability aspect will be sacrificed for complex structures and creating models to any constraint. Layers are defined in relation with the previous layer functionally, and all layers are stored in a variable.

- The most common ordering of layers in a neural network is as follows: dense/fully connected layer, activation, dropout, and then finally batch normalization if needed.

- Keras has built-in functions for monitoring training progress and retrieving the data for training after it's done. Callbacks are used to monitor and gather certain information regarding the training process at each epoch. Commonly used ones include ModelCheckpoint, EarlyStopping, and ReduceLROnPlateau. By calling history on the object returned after fit, a DataFrame containing the specified metric and loss at each epoch will be generated. This can be plotted and used to analyze the overarching trend of training. Finally, by using the function plot_model imported from keras.utils, a diagram can be produced to display the model architecture.

— Research into tabular deep learning demonstrates that several modifications of fully connected layers can yield successful models.

- Using ghost batch normalization rather than standard batch normalization can improve convergence speed and performance.

- Using a self-gating mechanism allows for implicit feature selection, which somewhat replicates tree-based logic. (See Chapter 7 on specific tree-inspired/replicative deep learning models.)

- Merging different-depth models, such as a "wide" linear model and a deep neural network model, can allow for the simultaneous utilization of each of their respective strengths. Moreover, manually computing feature crosses can provide helpful input signals to models, wide or deep.

- Using the SELU activation can help automatically normalize activations throughout a network.

- Using weight regularization can help reduce overfitting, but tabular networks often consist of heterogenous data for which using a universal regularization penalty strength is unreasonable. Regularization Learning Networks can be utilized as a callback in Keras training code and learn the optimal penalty strength for each weight.

In the next chapter, you'll learn about convolutional neural networks and how you can apply them to effectively solve computer vision, multimodal data, and tabular data problems.

CHAPTER 4

■ ■ ■

Applying Convolutional Structures to Tabular Data

There are things known and there are things unknown, and in between are the doors of perception.

—Aldous Huxley, Writer and Philosopher

In the previous chapter, you explored the application of standard feed-forward/artificial neural network models to tabular data. In this chapter, we will take a significant jump from the well-documented and "traditional" into the new and comparatively uncharted by exploring the application of *convolutional structures* to tabular data. While having been traditionally applied to image data, convolutional layers and convolutional neural networks can offer a unique perspective to tabular data that classical machine learning algorithms lack and artificial neural networks cannot reliably substitute for.

This chapter will begin with an overview of convolutional neural network theory, exploring the intuition and theory behind key convolution and pooling operations, how convolutional and pooling layers are organized among other layers into convolutional neural networks (CNNs), and a brief survey of accessible and successful CNN model architectures. The goal of this first half is to set the theoretical and practical foundation for CNN models in the "natural" image context. The second half demonstrates their applications to tabular data: First, you'll learn about multimodal network design – building models that incorporate both images *and* tabular data to produce a decision, using convolutional layers to process image inputs and feed-forward layers to process tabular inputs. Afterward, the chapter will demonstrate techniques to apply both one-dimensional and two-dimensional convolutions directly and indirectly to tabular data.

© Andre Ye and Zian Wang 2023
A. Ye and Z. Wang, *Modern Deep Learning for Tabular Data*,
https://doi.org/10.1007/978-1-4842-8692-0_4

Convolutional Neural Network Theory

In this section, we'll understand the justification for the existence of convolutions, how convolution operations work, and how pooling operations work. You'll put this knowledge together and implement convolutional neural networks for standard image classification tasks.

Why Do We Need Convolutions?

Let's imagine that convolutional layers have not been invented and all we have available to us as neural network "building blocks" are fully connected layers. We would like to build a neural network that is able to work with images. As an example task, let us consider the famous MNIST digits dataset, which contains several dozen thousand 28-by-28-pixel grayscale images of handwritten digits from 0 to 9.

MNIST is a relatively small dataset that can be easily loaded through the Keras datasets submodule (Listing 4-1).

Listing 4-1. Loading the MNIST dataset

```
from keras.datasets.mnist import load_data as load_mnist
(x_train, y_val), (x_test, y_val) = load_mnist()
```

Let's visualize 25 digits from the dataset (Listing 4-2, Figure 4-1). We'll use a seaborn heatmap(array) instead of the more standard plt.imshow(img) to explicitly show the pixel values.

Listing 4-2. Displaying heatmap representations of sample digits from the MNIST dataset

```
plt.figure(figsize=(25, 20), dpi=400)
for i in range(5):
    for j in range(5):
        plt.subplot(5, 5, i*5 + j + 1)
        sns.heatmap(x_train[i*5 + j], cmap='gray')
        plt.yticks(rotation=0)
plt.show()
```

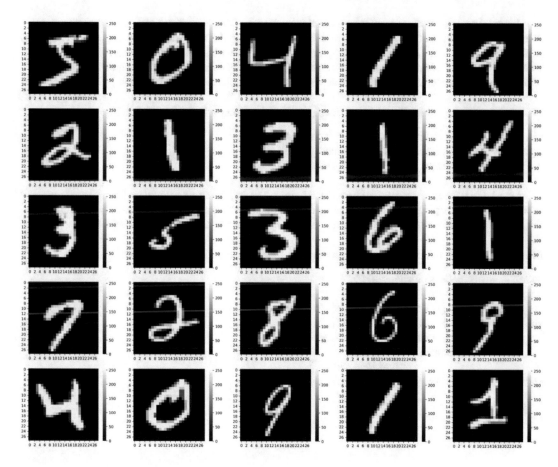

Figure 4-1. *Sample items from the MNIST dataset, shown as heatmaps*

Each 28-by-28-pixel image contains 784 values. We need to flatten x_train and x_val, which have shapes (60000, 28, 28) and (10000, 28, 28), respectively, in order to make them acceptable to a standard fully connected neural network. The desired shapes are (60000, 784) and (10000, 784). We can do this by reshaping the arrays (Listing 4-3).

Listing 4-3. Reshaping the MNIST dataset into "flattened" form

```
x_train = x_train.reshape(60000, 784)
x_val = x_val.reshape(10000, 784)
```

It's good practice to directly reference variable attributes of the dataset rather than hard-coding the value of those attributes. If someone who had 60,001 elements in their dataset wanted to use your code, they would get an error. A more robust reshaping method that works with a variable dataset and image size is shown in Listing 4-4. Note that setting –1 as an axis dimension value is equivalent to requesting NumPy to infer the remaining dimension. (If you are unfamiliar with this, please see the Appendix.)

Listing 4-4. A more robust alternative to the reshaping operation performed in Listing 4-3

```
flattened_shape = x_train.shape[1] * x_train.shape[2])
x_train = x_train.reshape(-1, flattened_shape)
x_val = x_val.reshape(-1, flattened_shape)
```

Let's build a feed-forward neural network (Figure 4-2) using the simple sequential model with the following logic: we begin with an input layer with 784 nodes; each dense layer contains half of the nodes as the one before it; we stop adding layers when the number of neurons is 10 or less (Listing 4-5).

Listing 4-5. Programmatically generating a model to fit the "flattened" MNIST dataset

```
model = keras.models.Sequential()
curr_nodes = 28 * 28
model.add(L.Input((curr_nodes,)))
while curr_nodes > 10:
    curr_nodes = round(1/2 * curr_nodes)
    model.add(L.Dense(curr_nodes,
                        activation='relu'))
model.add(L.Dense(10, activation='softmax'))
```

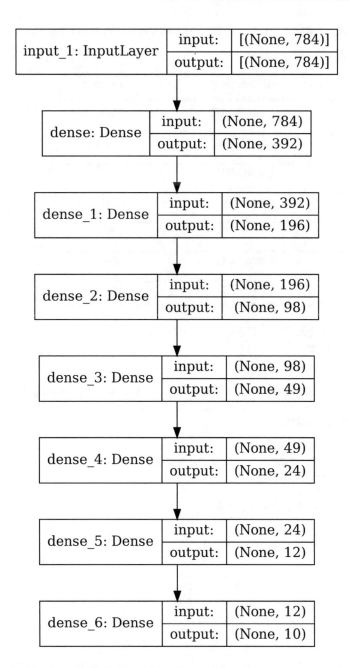

Figure 4-2. *Result of calling keras.utils.plot_model on the model architecture constructed in Listing 4-5*

Calling `model.summary()` prints out information about the model architecture and parameters (Listing 4-6). We see that the first layer learns the mapping between a 784-dimensional vector and a 392-dimensional vector, requiring $784 \cdot 392 = 307,720$ parameters. The second layer learns the mapping between a 392-dimensional vector and a 196-dimensional vector, requiring $392 \cdot 196 = 77,028$ parameters. In total, across six layers, the model architecture requires 410.5k parameters.

Listing 4-6. Parameter and shape summary for the architecture written in Listing 4-5

```
Model: "sequential"
```

Layer (type)	Output Shape	Param #
dense (Dense)	(None, 392)	307720
dense_1 (Dense)	(None, 196)	77028
dense_2 (Dense)	(None, 98)	19306
dense_3 (Dense)	(None, 49)	4851
dense_4 (Dense)	(None, 24)	1200
dense_5 (Dense)	(None, 12)	300
dense_6 (Dense)	(None, 10)	130

```
Total params: 410,535
Trainable params: 410,535
Non-trainable params: 0
```

We can train the model with standard meta-parameters (Listing 4-7, Figure 4-3).

Listing 4-7. Compiling and fitting

```
model.compile(optimizer='adam',
              loss='sparse_categorical_crossentropy',
              metrics=['accuracy'])
model.fit(x_train, y_train,
          validation_data=(x_val, y_val),
          epochs=10)
```

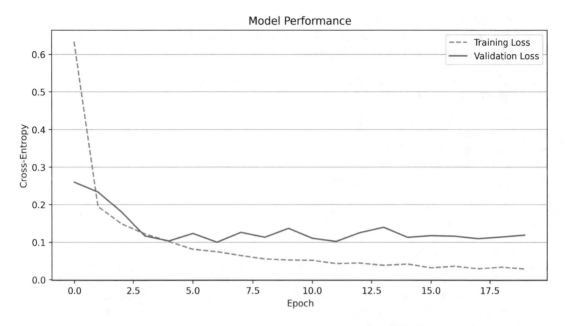

Figure 4-3. *Model training and validation performance modeling the MNIST dataset over 20 epochs*

Our model obtains nearly 0.975 validation accuracy – pretty good performance.

However, blurry 28-by-28 grayscale images of handwritten digits aren't representative of the images used in today's age of high-resolution images. Let's suppose that a "modern" MNIST dataset consists of 200-by-200-pixel images of handwritten digits and we want to scale our model to adapt to this image size. (You can make changes to the dataset using `cv2.resize(img, (200, 200))` and sharpening filters to address digit blurriness if desired. We'll just assume that such a dataset exists to analyze how the parametrization scales.)

For reference, Figure 4-4 is 1000-by-750 pixels in original picture resolution. 200-by-200 pixels is a comparatively small image size to today's image resolution standards.

Figure 4-4. *Sample 1000-by-750-pixel-resolution image, by Error 420 from Unsplash*

Let's build an architecture to process this sort of image, using the same logic as before: each layer should have half as many nodes as the layer before it, and we stop and add the final ten-class output when the number of nodes drops below 20 (Listing 4-8).

Listing 4-8. Programmatically generating an architecture (like in Listing 4-5) for a sample dataset with an input of shape (200, 200, 1)

```
model = keras.models.Sequential()
curr_nodes = 200 * 200
model.add(L.Input((curr_nodes,)))
while curr_nodes > 20:
    curr_nodes = round(1/2 * curr_nodes)
    model.add(L.Dense(curr_nodes,
                      activation='relu'))
model.add(L.Dense(10, activation='softmax'))
```

Let's look at how many parameters are involved here: *more than one trillion parameters,* an almost 2.5k-fold increase in parameter count for about a sevenfold increase in image resolution dimension (compared with the previous network built for 28-by-28-pixel images) (Listing 4-9). The model size here resembles the parametrization of massive modern natural language processing applications that translate languages and generate text (you'll read about these in the next two chapters) – a far cry from the intended application for simple image recognition.

Listing 4-9. Parameter and shape summary for the architecture written in Listing 4-8

```
Model: "sequential"
```

Layer (type)	Output Shape	Param #
dense (Dense)	(None, 20000)	800020000
dense_1 (Dense)	(None, 10000)	200010000
dense_2 (Dense)	(None, 5000)	50005000
dense_3 (Dense)	(None, 2500)	12502500
dense_4 (Dense)	(None, 1250)	3126250
dense_5 (Dense)	(None, 625)	781875
dense_6 (Dense)	(None, 312)	195312
dense_7 (Dense)	(None, 156)	48828
dense_8 (Dense)	(None, 78)	12246
dense_9 (Dense)	(None, 39)	3081
dense_10 (Dense)	(None, 20)	800
dense_11 (Dense)	(None, 10)	210

```
Total params: 1,066,706,102
Trainable params: 1,066,706,102
Non-trainable params: 0
```

The problem with using dense, feed-forward neural network architectures on image data, as one can tell from this experiment, is that they do not scale feasibly with image size. The difference between a 300-by-300 image and a 350-by-350 image isn't very significant to the human eye (Figure 4-5), but to a neural network, the input space of the 350-by-350 image contains 32,500 more dimensions than that of the 300-by-300 image. This increase is compounded in each subsequent layer.

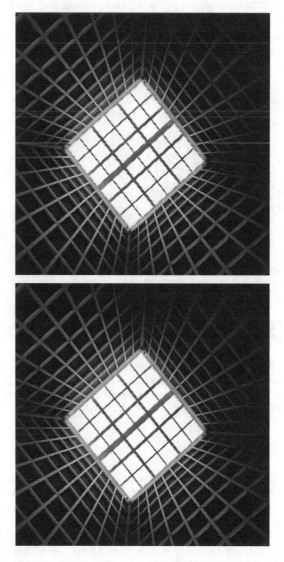

Figure 4-5. *Two sample images, one 300 × 300 and the other 350 × 350 (in an undisclosed order). The visual difference is negligible to humans, but the representation difference is massive for neural network architectures. Image from Unsplash and modified*

In this case, we didn't even consider color. If you considered the color channel and set `curr_nodes = 200 * 200 * 3` into the neural network architecture construction logic outlined in Listing 4-8 (or otherwise increase the resolution of the image), you might even run into an error: "`Allocator ran out of memory trying to allocate [storage size].`" The architecture is so large that Keras physically cannot allocate enough space in memory to store all the parameters!

We need a feasible way to scale neural network architectures to handle images that reflect, to some extent, how vision should work. The number of additional parameters or space needed to model a slightly higher-resolution image should not be significant, because having a slightly more high-resolution image doesn't affect the actual content meaning/semantics in the image.

Moreover, philosophically speaking, we cannot treat each individual pixel as always representing the same concept – the same pixel location on two pictures of dogs, for example, might represent two entirely different values and meanings even though the aggregated pixels contribute to the same label (Figure 4-6). Images should be processed in a way that is somehow consistent across the image yet capable of capturing deep and useful information. Using a standard ANN does not guarantee any of this consistency.

Figure 4-6. *Two images representing the same semantic information but with very different pixel values. Note that a pixel coordinate could be part of the dog in one image but part of the ocean or sky in another image. A neural network processing image data must be invariant to these transformations that change the pixel values but not the semantic content of the image. Image by Oscar Sutton from Unsplash*

The Convolution Operation

The convolutional layer changed the game in image recognition and general deep learning image tasks and applications. It addresses all the concerns previously described well and is still the foundation of almost all deep learning computer vision models – despite having been in use for several decades (a relatively long time in the context of deep learning history).

The concept of a convolution itself for image processing has been known for quite a long time. Given a special filter describing how certain pixels are to be modified in relationship to surrounding pixels, we can apply the filter to an image to obtain a modified image. We can design kernels that force pixels into certain relationships with others, creating effects like blurring, sharpening, or amplifying edges (sharp changes) of an image.

Consider a hypothetical 3-by-3 kernel k:

$$k = \begin{bmatrix} 0 & 0.5 & 0 \\ 0.5 & 1 & 0.5 \\ 0 & 0.5 & 0 \end{bmatrix}$$

Let's construct a sample 4-by-4 matrix I to apply the kernel to:

$$i = \begin{bmatrix} 1 & 2 & 3 & 4 \\ 5 & 6 & 7 & 8 \\ 9 & 10 & 11 & 12 \\ 13 & 14 & 15 & 16 \end{bmatrix}$$

Lastly, let R be the result of the convolution. As we'll see, it has shape (2, 2):

$$R = \begin{bmatrix} ? & ? \\ ? & ? \end{bmatrix}$$

We will begin by filling in the upper-left element of R. This corresponds to the upper-left 3-by-3 window in I (bolded):

$$\begin{bmatrix} \mathbf{1} & \mathbf{2} & \mathbf{3} & 4 \\ \mathbf{5} & \mathbf{6} & \mathbf{7} & 8 \\ \mathbf{9} & \mathbf{10} & \mathbf{11} & 12 \\ 13 & 14 & 15 & 16 \end{bmatrix}$$

We perform an element-wise multiplication between each element of the kernel k and each element in the relevant 3-by-3 window in I:

$$\begin{bmatrix} 0 \cdot \mathbf{1} & 0.5 \cdot \mathbf{2} & 0 \cdot \mathbf{3} \\ 0.5 \cdot \mathbf{5} & 1 \cdot \mathbf{6} & 0.5 \cdot \mathbf{7} \\ 0 \cdot \mathbf{9} & 0.5 \cdot \mathbf{10} & 0 \cdot \mathbf{11} \end{bmatrix} = \begin{bmatrix} 0 & 1 & 0 \\ 2.5 & 6 & 3.5 \\ 0 & 5 & 0 \end{bmatrix}$$

The final result is the sum of the elements in the resulting product matrix: $0 + 1 + 0 + 2.5 + 6 + 3.5 + 0 + 5 + 0 = 18$. The first value of R has been derived:

$$R = \begin{bmatrix} 18 & ? \\ ? & ? \end{bmatrix}$$

We can apply a similar operation to obtain the value for the top-right value of R. The relevant subregion of I is (bolded)

$$\begin{bmatrix} 1 & \mathbf{2} & \mathbf{3} & \mathbf{4} \\ 5 & \mathbf{6} & \mathbf{7} & \mathbf{8} \\ 9 & \mathbf{10} & \mathbf{11} & \mathbf{12} \\ 13 & 14 & 15 & 16 \end{bmatrix}$$

The calculation can be performed as follows:

$$\begin{bmatrix} 0{\cdot}2 & 0.5{\cdot}3 & 0{\cdot}4 \\ 0.5{\cdot}6 & 1{\cdot}7 & 0.5{\cdot}8 \\ 0{\cdot}10 & 0.5{\cdot}11 & 0{\cdot}12 \end{bmatrix} \begin{bmatrix} 0 & 1.5 & 0 \\ 3 & 7 & 4 \\ 0 & 5.5 & 0 \end{bmatrix}$$

$$\rightarrow 0 + 1.5 + 0 + 3 + 7 + 4 + 0 + 5.5 + 0 = 21$$

$$R = \begin{bmatrix} 18 & 21 \\ ? & ? \end{bmatrix}$$

If you calculate through the other two values of R, you should get the matrix

$$R = \begin{bmatrix} 18 & 21 \\ 30 & 33 \end{bmatrix}$$

Thus, given an original matrix of shape $a \times b$ and a kernel of shape $x \times y$ (convolutions can be performed on non-square matrices and with non-square kernels!), the resulting convolved matrix has shape $(a - x + 1, b - y + 1)$. Each value represents the number of "slots" the kernel can occupy across that spatial dimension.

Alright, exactly what does this convolved result mean? In order to interpret the result of a convolution, we need to first understand the design of the kernel. This particular kernel weights the value in the middle the highest, with decaying influence further from the center. We can therefore expect the kernel slightly "averages" values near each pixel; the resulting convolved feature reflects the *general*/"averaged" nature of elements in the original matrix.

For instance, we see that the order of elements follows $A < B < C < D$ where

$$R = \begin{bmatrix} A & B \\ C & D \end{bmatrix}$$

This reflects the general organization of elements across the original matrix I. We can also observe that $B - A = D - C$, which reflects the property that items in the bottom-right region of the original matrix I are roughly the same distance from items in the bottom-left region as items in the top right are from those in the top left.

Let's see a slightly more complex example using a 10-by-10 image (Listing 4-10, Figure 4-7). This image will feature a plus sign consisting of "1"s on a background gradient.

Listing 4-10. Generating a sample low-dimensional "image"

```
# initialize 'canvas' of zeros
img = np.zeros((10, 10))

# draw background gradient
for i in range(10):
    for j in range(10):
        img[i][j] = i * j / 100

# draw vertical stripe
for i in range(2, 8):
    for j in range(4, 6):
        img[i][j] = 1
```

```
# draw horizontal stripe
for i in range(4, 6):
    for j in range(2, 8):
        img[i][j] = 1

# display figure with values
plt.figure(figsize=(10, 8), dpi=400)
sns.heatmap(img, cmap='gray', annot = True)
plt.show()
```

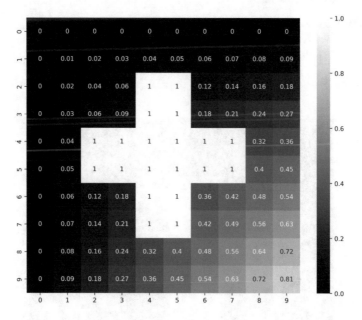

Figure 4-7. *Heatmap representation of a custom "image" – a plus sign overlayed against a gradient background*

We can use cv2's filter2D function to apply kernels to an image. Because cv2.filter2D is adapted for images, cv2 incorporates *padding* on the side such that the convolved image has the same shape as the original image. This entails adding buffer values (most commonly, 0) to the side of the matrix and applying a convolution to the padded matrix. Let's write a function that takes in a kernel, applies it to a matrix, and displays the matrix as a heatmap (Listing 4-11).

Listing 4-11. A function to apply an inputted kernel on the inputted image

```
def applyKernel(kernel, img):
    altered = cv2.filter2D(img, -1, kernel)
    plt.figure(figsize=(10, 8), dpi=400)
    sns.heatmap(altered, cmap='gray', annot = True)
    plt.show()
```

The *identity kernel* is defined as a 1 in the center of the matrix and 0s everywhere else (Listing 4-12, Figure 4-8). Each pixel in the convolved matrix is directly impacted by only one pixel in the previous original matrix, yielding a convolved matrix identical to the original:

$$identity\ kernel = \begin{bmatrix} 0 & 0 & 0 \\ 0 & 1 & 0 \\ 0 & 0 & 0 \end{bmatrix}$$

Listing 4-12. Using the identity kernel

```
kernel = np.array([[0, 0, 0],
                   [0, 1, 0],
                   [0, 0, 0]])

applyKernel(kernel, img)
```

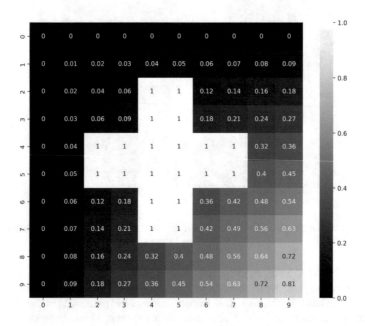

Figure 4-8. *Result of applying a convolution with the identity kernel to the "image" in Figure 4-7 (Note that there is no difference.)*

To apply a blurring effect to an image, we can define a kernel that weights all neighboring pixels identically (Listing 4-13, Figure 4-9):

$$3\times3\ uniform\ blurring\ kernel = \begin{bmatrix} 1 & 1 & 1 \\ 1 & 1 & 1 \\ 1 & 1 & 1 \end{bmatrix}$$

Listing 4-13. Using the 3 × 3 blurring kernel

```
kernel = np.ones((3, 3))
applyKernel(kernel, img)
```

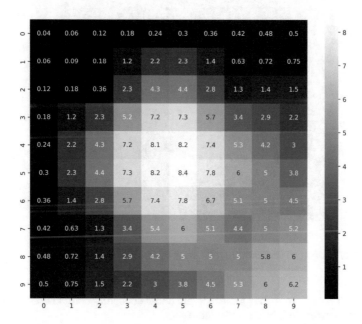

Figure 4-9. *Result of applying a convolution with the 3-by-3 uniform blurring kernel to the "image" in Figure 4-7*

cv2 applies *padding* (adding additional zeros) around the outside of the image before a convolution is applied such that the resulting matrix has the same size. We will further discuss padding in the context of convolutions later.

We can adjust the intensity and range of blurring by changing the size of the kernel, which impacts how many neighboring pixels are considered when calculating a pixel in the convolved image. Consider the result of a 2-by-2 blurring kernel (Listing 4-14, Figure 4-10):

$$2 \times 2\ uniform\ blurring\ kernel = \begin{bmatrix} 1 & 1 \\ 1 & 1 \end{bmatrix}$$

Listing 4-14. Using the 2-by-2 uniform blurring kernel

```
kernel = np.ones((2, 2))
applyKernel(kernel, img)
```

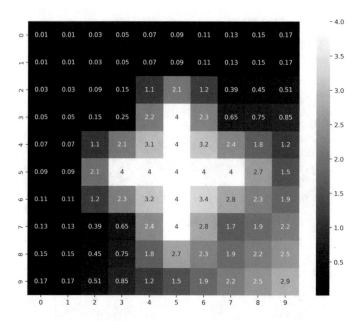

Figure 4-10. *Result of applying a convolution with the 2-by-2 uniform blurring kernel to the "image" in Figure 4-7*

With an 8-by-8 blurring kernel, the plus sign in the middle becomes completely integrated with the background and is unnoticeable (Listing 4-15, Figure 4-11):

$$8 \times 8 \text{ uniform blurring kernel} = \begin{bmatrix} 1 & 1 & 1 & 1 & 1 & 1 & 1 & 1 \\ 1 & 1 & 1 & 1 & 1 & 1 & 1 & 1 \\ 1 & 1 & 1 & 1 & 1 & 1 & 1 & 1 \\ 1 & 1 & 1 & 1 & 1 & 1 & 1 & 1 \\ 1 & 1 & 1 & 1 & 1 & 1 & 1 & 1 \\ 1 & 1 & 1 & 1 & 1 & 1 & 1 & 1 \\ 1 & 1 & 1 & 1 & 1 & 1 & 1 & 1 \\ 1 & 1 & 1 & 1 & 1 & 1 & 1 & 1 \end{bmatrix}$$

Listing 4-15. Using the 8-by-8 uniform blurring kernel

```
kernel = np.ones((8, 8))
applyKernel(kernel, img)
```

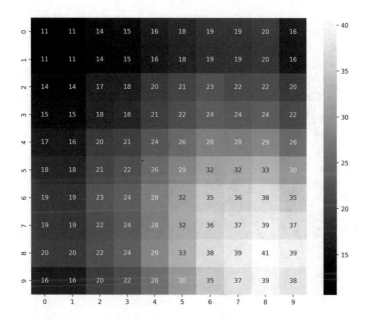

Figure 4-11. *Result of applying a convolution with the 8-by-8 uniform blurring kernel to the "image" in Figure 4-7*

Another operation is the sharpening effect, which makes edges and values contrast more sharply with one another (Listing 4-16, Figure 4-12). The kernel weights the center pixel very highly and the surrounding neighbors lowly; this has the effect of increasing the difference between adjacent pixels in the convolved feature:

$$sharpening\ kernel = \begin{bmatrix} 0 & -1 & 0 \\ -1 & 5 & -1 \\ 0 & -1 & 0 \end{bmatrix}$$

Listing 4-16. Using the sharpening kernel

```
kernel = np.array([[0, -1, 0],
                   [-1, 5, -1],
                   [0, -1, 0]])
applyKernel(kernel, img)
```

275

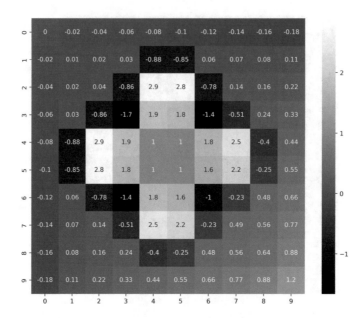

Figure 4-12. *Result of applying a convolution with the 3-by-3 sharpening kernel to the "image" in Figure 4-7*

There are many other kernels that can accomplish a wide variety of effects on image matrices. It's a good exercise to play around with your own custom kernels and view the resulting convolved image.

Consider the following *pixel-by-pixel* image of a dog, derived as such (Listing 4-17, Figure 4-13).

Listing 4-17. Loading and displaying an image of a dog from Unsplash

```
from skimage import io
url = 'https://images.unsplash.com/photo-1530281700549-e82e7bf110d6?ixlib=rb-1.2.1&ixid=
MnwxMjA3fDB8MHxwaG90by1wYWdlfHx8fGVufDB8fHx8&auto=format&fit=crop&w=988&q=80'
image = io.imread(url)
image = cv2.cvtColor(image, cv2.COLOR_BGR2GRAY)
image = image[450:450+400, 250:250+400]

plt.figure(figsize=(10, 10), dpi=400)
plt.imshow(image, cmap='gray')
plt.axis('off')
plt.show()
```

Figure 4-13. *Sample image of a dog*

Let's modify our applyKernel function to display the convolved matrix as an image rather than as a heatmap (Listing 4-18).

Listing 4-18. A function to apply and display a kernel to an image

```
def applyKernel(kernel, img):
    altered = cv2.filter2D(img, -1, kernel)
    plt.figure(figsize=(8, 8), dpi=400)
    plt.imshow(altered, cmap='gray')
    plt.axis('off')
    plt.show()
```

Applying a 3-by-3 blurring kernel discussed before results in the following image (Listing 4-19, Figure 4-14).

Listing 4-19. Applying an (erroneous) 3-by-3 uniform blurring kernel

```
kernel = np.ones((3, 3))
applyKernel(kernel, image)
```

Figure 4-14. *Result of incorrectly applying a convolution with a 3-by-3 uniform blurring kernel to the sample dog image in Figure 4-13*

That's odd! What happened here?

Let's plot the distribution of pixel values across the image (Listing 4-20, Figure 4-15).

Listing 4-20. *Displaying the distribution of pixel values across the (erroneously) convolved/blurred image*

```
plt.figure(figsize=(40, 7.5), dpi = 400)
sns.countplot(x=cv2.filter2D(image, -1, kernel).flatten(), color='red')
plt.xticks(rotation=90)
plt.show()
```

Figure 4-15. *Distribution of pixel values after incorrectly applying a convolution with a 3-by-3 uniform blurring kernel to the sample dog image in Figure 4-13*

It seems that almost all the values in the image have been pushed up to 255. This is because convolving the feature has the effect of changing the domain of possible values. If all nine values in a convolution region have value 255, then the convolved result is $255 \cdot 9 = 2295$ – which is way out of the valid domain for unsigned int-8 image pixel values. In these cases, `cv2.filter2D` caps the maximum value as 255. In fact, the convolved result for any region with average value larger than 255/9 will be capped at 255.

Thus, we need to modify the kernel such that the domain of values in the convolved result does not extend beyond the original domain. We can solve this problem by defining the blur kernel as

$$3 \times 3 \, uniform \, blurring \, kernel = \begin{bmatrix} 1/9 & 1/9 & 1/9 \\ 1/9 & 1/9 & 1/9 \\ 1/9 & 1/9 & 1/9 \end{bmatrix}$$

Let M represent the maximum pixel value (in this context it is 255). The maximum result of applying this kernel across even a region in which all pixels are populated with Ms is $9 \cdot \left(\frac{1}{9} \cdot M \right) = M$. Thus, we preserve the scale of pixel values.

After applying this modified kernel, we see that the resulting convolved distribution is very similar to that of the original distribution and remains within a valid range (Figures 4-16 and 4-17).

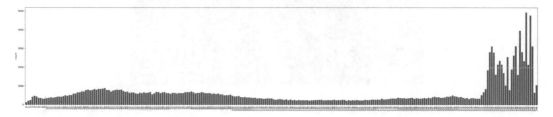

Figure 4-16. *Original distribution of pixel values from the sample dog image in Figure 4-13*

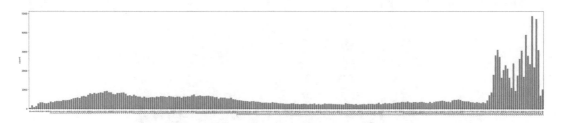

Figure 4-17. *Distribution of pixel values after correctly applying a convolution with a 3-by-3 uniform blurring kernel to the sample dog image in Figure 4-13*

The resulting image is slightly blurred and displays properly, as desired (Figure 4-18).

Figure 4-18. *Result of correctly applying a convolution with a 3-by-3 uniform blurring kernel to the sample dog image in Figure 4-13*

Applying an 8-by-8 blurring kernel, defined as an 8×8 matrix filled with the value 1/64, yields an even more blurred image (Figure 4-19).

Figure 4-19. *Result of correctly applying a convolution with an 8-by-8 uniform blurring kernel to the sample dog image in Figure 4-13*

Similarly, applying the sharpening kernel has the following effect (Figure 4-20).

Figure 4-20. *Result of correctly applying a convolution with a 3-by-3 sharpening kernel to the sample dog image in Figure 4-13*

Convolutions can be constructed to extract meaningful features from images and matrices. For instance, it may be helpful to blur an image in order to minimize the distance between adjacent pixels to minimize variance in the image and perform general, broad analyses of the image minimally affected by noisy variation. Alternatively, sharpening may assist with amplification of important features and edges that act as landmarks and features for the content of the image.

The fundamental idea of a *convolutional neural network* (CNN) is similar to that of a standard feed-forward fully connected neural network introduced previously in Chapter 3. Standard artificial neural networks can demonstrate sophisticated behavior by arranging large quantities of a simple extractive unit – the perceptron. By providing the architecture/arrangement of parameters, the network can learn the optimal values to extract desirable information to perform the intended task. Similarly, CNNs can model image data quite well by stacking convolution operations upon each other; the neural network learns the values of each of the kernels in the convolutions to extract the optimal features from the image. Optimization is still performed via gradient descent.

A *convolutional layer* is a collection of convolutions, like a fully connected layer is a collection of perceptrons. There are a few important attributes of a convolutional layer that define its specific implementation:

- *The number of filters, n*: This is the number of "convolution operations" that exist in the layer. The network will learn n different kernels for this layer.

- *The kernel shape* (a, b): This defines how large the learned kernel will be.

- *Input or valid padding*: This determines whether padding is used or not. If input padding is used, then any incoming matrices will be padded such that the resulting convolved matrices have the same shape as the incoming matrices before padding. Otherwise, no padding will be applied, and the convolved matrix will have shape $(x - a + 1, y - b + 1)$, where (x, y) is the shape of the original matrix.

A layer must learn one filter per connection between each feature map in the previous layer and each feature map in the current layer. For instance, if layer A produces 8 feature maps (because it has 8 filters) and layer B produces 16 feature maps (because it has 16 filters), layer B learns $8 \times 16 = 128$ filters. If all the filters are 3-by-3 matrices, then layer B uses $128 \times (3 \times 3) = 1152$ parameters.

Let's start building a simple convolutional neural network using the sequential syntax for simplicity. (We'll use the functional API in more complex situations when using the sequential API becomes difficult or impossible.) We will build a convolutional neural network that processes 28-by-28 images from the MNIST dataset and classifies them into one of ten digits.

We begin with an input layer (Listing 4-21). All image data must have three specified spatial dimensions: width, height, and depth. Grayscale images have a depth of 1, whereas color images generally have a depth of 3 (where depth-wise layers correspond to red, green, and blue). In this case, our input data has shape (28, 28, 1).

Listing 4-21. Building the base and input to a convolutional network

```
import keras.layers as L
from keras.models import Sequential
model = Sequential()
model.add(L.Input((28, 28, 1)))
```

After the input, we should add convolutional layers to process the image (Listing 4-22). In Keras, convolutional layers can be instantiated via `keras.layers.Conv2D(num_filters, kernel_size = (a, b), activation = 'activation_name', padding = 'padding_type')`. The default activation is linear (i.e., $y = x$, applies no nonlinearity to the data), and the default padding type is valid.

Listing 4-22. Stacking convolutional layers

```
model.add(L.Conv2D(8, (5, 5), activation='relu'))
model.add(L.Conv2D(8, (3, 3), activation='relu'))
model.add(L.Conv2D(16, (3, 3), activation='relu'))
model.add(L.Conv2D(16, (2, 2), activation='relu'))
```

It's worth understanding how each layer changes the shape of the input:

1. The original input layer takes in data of shape (28, 28, 1).

2. The first convolutional layer uses 8 filters and applies a 5-by-5 kernel, yielding an output of shape (24, 24, 8).

3. The second convolutional layer uses 8 filters and applies a 3-by-3 kernel, yielding an output of shape (22, 22, 8).

4. The third convolutional layer uses 16 filters and applies a 5-by-5 kernel, yielding an output of shape (20, 20, 16).

5. The fourth convolutional layer uses 16 filters and applies a 1-by-1 kernel, yielding an output of shape (19, 19, 16).

To confirm this, we can plot the model to understand how each layer transforms the shape of incoming data with `keras.utils.plot_model(model, show_shapes=True)` (Figure 4-21).

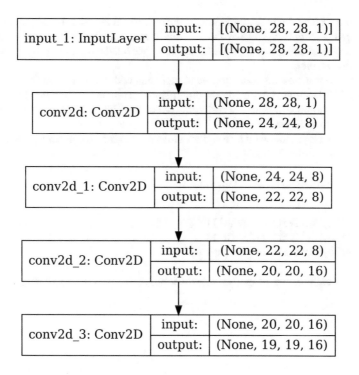

Figure 4-21. *Visual representation of the model architecture defined in Listings 4-21 and 4-22*

Moreover, we can see that the convolutions require very few parameters (Listing 4-23).

Listing 4-23. Parameter counts for the convolutional neural network

```
Model: "sequential"
```

Layer (type)	Output Shape	Param #
conv2d (Conv2D)	(None, 24, 24, 8)	208
conv2d_1 (Conv2D)	(None, 22, 22, 8)	584
conv2d_2 (Conv2D)	(None, 20, 20, 16)	1168
conv2d_3 (Conv2D)	(None, 19, 19, 16)	1040

```
Total params: 3,000
Trainable params: 3,000
Non-trainable params: 0
```

However, our model isn't done! The objective of the classification task is to map an input image of shape (28, 28, 1) to an output vector of length 10. No matter how many convolutional layers we add, we're always going to have data arranged along three spatial dimensions. We need a method to force the data from three spatial dimensions to just one.

Flattening might be the most obvious method to map data with three spatial dimensions to one: we simply unpack individual elements in the higher-dimensional arrangement and lay them out along a one-dimensional axis. This is similar to standard reshaping operations, like `arr.reshape` in NumPy: all values are retained, just arranged in a different format.

Let's add a flattening layer, followed by a series of fully connected layers that eventually map to the desired ten-class output (Listing 4-24).

Listing 4-24. Adding a flattening layer and an output

```
model.add(L.Flatten())
model.add(L.Dense(32, activation='relu'))
model.add(L.Dense(16, activation='relu'))
model.add(L.Dense(10, activation='softmax'))
```

Now, the network architecture properly maps the input to the desired output shape (Figure 4-22).

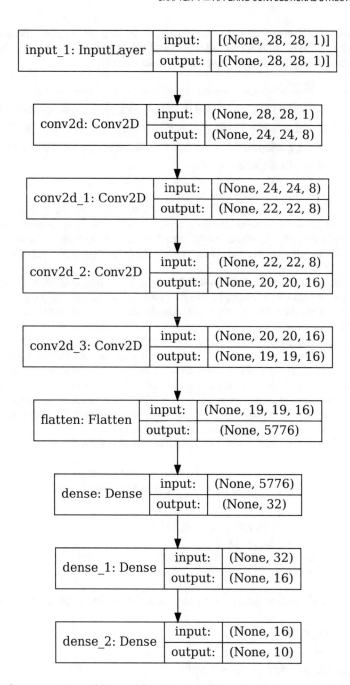

Figure 4-22. Visual representation of the model architecture defined in Listings 4-21, 4-22, and 4-24

Our finalized network has a few hundred thousand parameters, about half the number of parameters in our previous network architecture design using only fully connected layers.

When we scale the image size to 200-by-200 while using the same architecture, the number of model parameters used is 18,682,002 – compare this with the 1,066,706,102 parameters used in the previously discussed hypothetical 200-by-200 image fully-connected network.

■ **Note** The parameter count for the convolutional neural network, you'll probably notice, isn't great. The parameter count especially balloons at the flattening layer. We'll explore another family of layers used in convolutional neural networks, *pooling layers*, which help us address this problem and further improve the parameter scaling of a convolutional neural network.

For convenience of reference, the part of the convolutional neural network before conversion from three spatial dimensions to one spatial dimension is generally called the "convolutional component" and the part afterward the "fully connected component." Another name is the "bottom" and "top," respectively – which can be confusing because the last part of the model (the fully connected component) is referred to as the "top."

The convolutional component can be thought of as serving an *extractive* role by identifying and amplifying the most relevant/important qualities of the input. In contrast, the fully connected component performs an *aggregative/compiling/interpretive* role by processing all the extracted features and interpreting how they relate to the output.

For instance, the convolutional component of our example network built for MNIST data might detect and amplify corners, like the top-left part of a "5" or the left and top part of a "4." The fully connected component might be able to aggregate the various detected corner properties and interpret them as support for belonging to a certain class. If there are many sharp corners, the image could be a "4" or "5." If there are a low number of sharp corners, the image could be "1," "2," "3," or "7." If there are no sharp corners, the image could be a "0," "6," "8," or "9." Combining this information with other extracted features allows the fully connected component to pin down exactly which digit the image is.

Let's compile and train the model, using familiar syntax (Listing 4-25).

Listing 4-25. Compilation and training of a sample convolutional neural network. Assumes that the training and validation sets from the MNIST dataset have already been loaded into X_train, y_train, X_val, and y_val

```
model.compile(optimizer = 'adam',
              loss = 'sparse_categorical_crossentropy',
              metrics = ['accuracy'])
history = model.fit(X_train, y_train, epochs = 100,
              validation_data = (X_val, y_val))
```

The model very quickly obtains good training and validation performance (Listing 4-26, Figure 4-23).

Listing 4-26. Plotting model training history

```
plt.figure(figsize=(20, 7.5), dpi=400)
plt.plot(history.history['loss'], color='red',
         label='Train')
plt.plot(history.history['val_loss'],
         color='blue', linestyle='--',
         label='Validation')
plt.xlabel('Epoch')
plt.ylabel('Loss')
plt.title('Loss over Epochs')
```

```
plt.legend()
plt.grid()
plt.show()

plt.figure(figsize=(20, 7.5), dpi=400)
plt.plot(history.history['accuracy'], color='red',
        label='Train')
plt.plot(history.history['val_accuracy'],
        color='blue', linestyle='--',
        label='Validation')
plt.xlabel('Epoch')
plt.ylabel('Accuracy')
plt.title('Accuracy over Epochs')
plt.legend()
plt.grid()
plt.show()
```

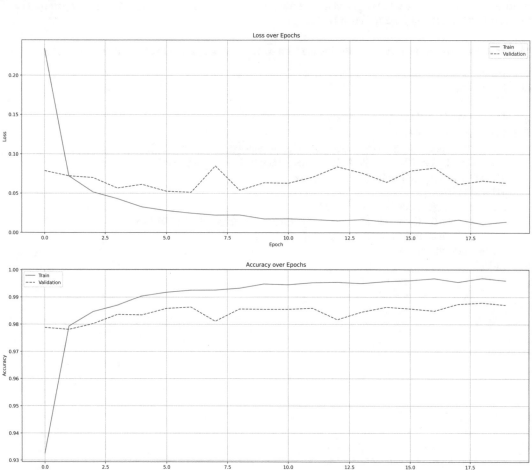

Figure 4-23. *Loss and accuracy history of the previously defined convolutional neural network architecture as it trains on the MNIST dataset for 20 epochs*

To gain a better idea of what the convolutional neural network is really doing in terms of spatially extracting features, let us pass an input image through only the *first convolutional layer* of the trained network. This allows us to visually identify what types of transformations the network has learned as being optimal for classification.

In order to "peek" into the input-output flow of a layer within the network, we can create a new model constructed from the same layer objects as in the original network. This allows us to isolate the weights within that specific layer (Listing 4-27).

Listing 4-27. Building a model to "peek" into the learned weights within each layer of the model

```
peek = Sequential()
peek.add(L.Input((28, 28, 1)))
peek.add(model.layers[0])
```

We can use `peek.predict` to get the result after passing in a sample input. Recall that the first layer maps the input of shape (28, 28, 1) to an output of shape (24, 24, 8) – meaning that the first layer outputs 8 feature maps of shape (24, 24). We can visualize a sample input and a sample of the feature map outputs from the first layer (Listing 4-28, Figure 4-24).

Listing 4-28. Plotting the result of applying learned convolutions through the first convolutional layer

```
NUM_IMAGES = 8
GRAPHIC_WIDTH = 8

plt.figure(figsize=(40, 40), dpi=400)
for index in range(NUM_IMAGES):
    peek_out = peek.predict(x_train[index].reshape((1, 28, 28, 1)))[0]
    plt.subplot(NUM_IMAGES, GRAPHIC_WIDTH, index*GRAPHIC_WIDTH+1)
    plt.imshow(x_train[index], cmap='gray')
    plt.axis('off')
    for i in range(GRAPHIC_WIDTH - 1):
        plt.subplot(NUM_IMAGES, GRAPHIC_WIDTH, index*GRAPHIC_WIDTH+i+2)
        plt.imshow(peek_out[:,:,i], cmap='gray')
        plt.axis('off')
plt.show()
```

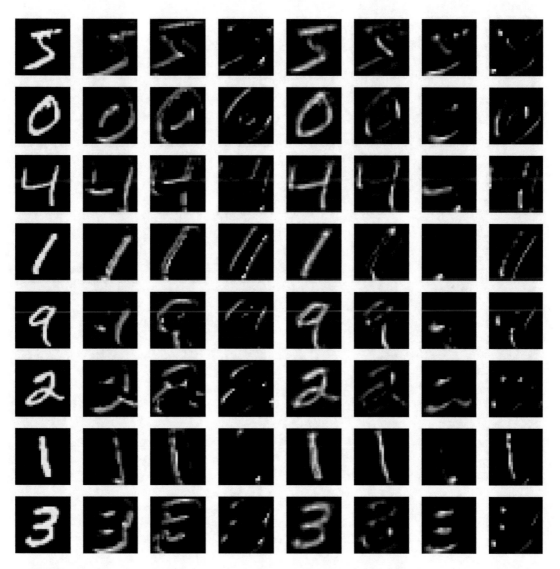

Figure 4-24. *Representing the effect of the learned first-layer convolution kernels (all columns except the left one) on an input image (left column)*

The left column contains the original input image to the layer; each of the images to the right displays one feature map output. You can begin to observe various transformations performed just by the first layer: inversion, shifting, edge detection, corner detection, and line detection, to name a few.

We can modify the peek model to include the second layer and view the feature maps obtained by passing an input through both the first and second layers (Listing 4-29, Figure 4-25).

Listing 4-29. Plotting the result of applying learned convolutions through the second convolutional layer

```
peek = Sequential()
peek.add(L.Input((28, 28, 1)))
peek.add(model.layers[0])
peek.add(model.layers[1])
```

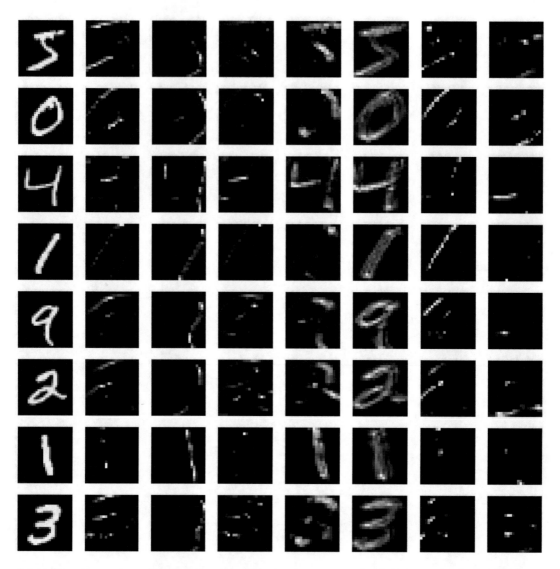

Figure 4-25. *Representing the effect of the learned first- and second-layer convolution kernels (all columns except the left one) on an input image (left column)*

The second layer is able to capture more specific components of each digit. If you look closely, you can see that each feature map "looks for" increasingly specialized features. The first displayed feature map (second column from the left in Figure 4-26), for instance, appears to "look for" horizontal lines in digits: there are three horizontal lines for the digit 3, two for the digit 2, two for the digit 5, etc. The outputs of the third layer become more difficult to interpret due to increasing distance from the input.

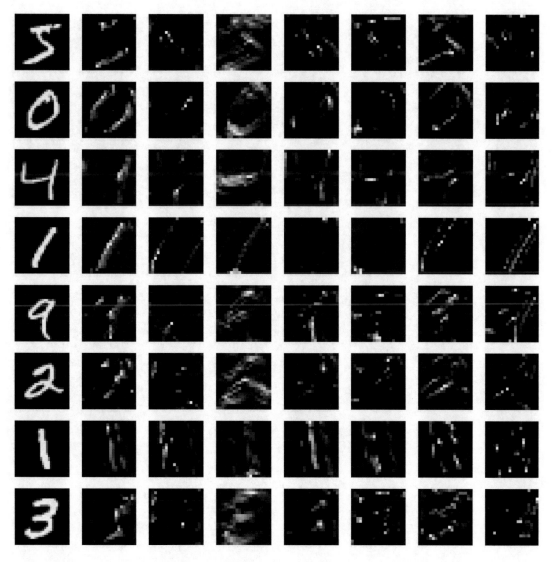

Figure 4-26. *Representing the effect of the learned first-, second-, and third-layer convolution kernels (all columns except the left one) on an input image (left column)*

The fourth layer is the last convolutional layer before the outputs are flattened and passed into the fully connected component (Figure 4-27).

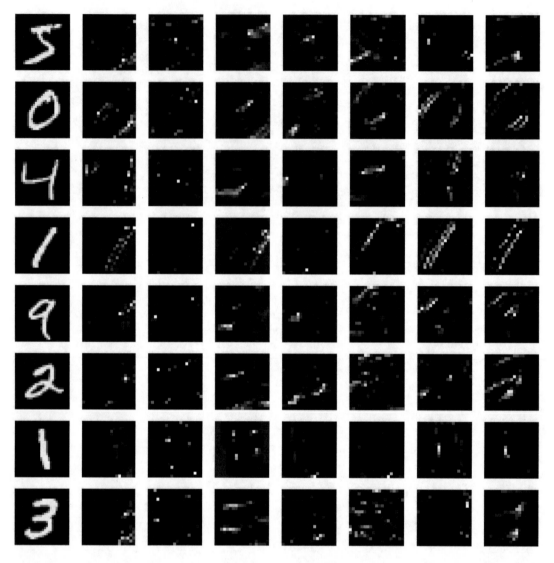

Figure 4-27. *Representing the effect of the learned first- through fourth-layer convolution kernels (all columns except the left one) on an input image (left column)*

The Pooling Operation

Convolutions are incredibly helpful – they allow neural networks to extract features from images in a systematic, low-parameter, and effective manner. However, they're relatively insignificant with respect to changing the shape of the image. We might refer to this as the "information compression factor" – the convolution doesn't do much "squeezing." As explored before, performing a convolution with kernel size n on an image of shape (a, a) produces an output of shape $(a - n + 1, a - n + 1)$ – this is quite marginal for standard single-digit sizes of n.

Because of the convolution's relatively low information compression factor, the fully connected component after the flattening layer has to process a very large number of parameters. The number of parameters was forgivable for a small-resolution dataset like MNIST, but is unfeasible for larger (and more practical) datasets and applications.

Let's use the same architecture as in the previous section and explore how the number of parameters (which can be accessed via model.count_params()) scales for a wide range of image sizes (Listing 4-30, Figure 4-28).

Listing 4-30. Plotting the scaling capability of a model with only convolutions

```
def build_model(img_size):
    model = Sequential()
    model.add(L.Input((img_size, img_size, 1)))
    model.add(L.Conv2D(8, (5, 5), activation='relu'))
    model.add(L.Conv2D(8, (3, 3), activation='relu'))
    model.add(L.Conv2D(16, (3, 3), activation='relu'))
    model.add(L.Conv2D(16, (2, 2), activation='relu'))
    model.add(L.Flatten())
    model.add(L.Dense(32, activation='relu'))
    model.add(L.Dense(16, activation='relu'))
    model.add(L.Dense(10, activation='softmax'))
    paramCount = model.count_params()
    del model
    return paramCount


x = [10, 15, 20, 25, 30, 35, 40, 45, 50,
     60, 70, 80, 90, 100, 120, 130, 140, 150,
     170, 180, 190, 200, 225, 250, 275, 300,
     350, 400, 450, 500, 600, 700, 800]
y = [build_model(i) for i in tqdm(x)]

plt.figure(figsize=(15, 7.5), dpi=400)
plt.plot(x, y, color='red')
plt.xlabel('Image Size (one spatial dimension)')
plt.ylabel('# Parameters')
plt.grid()
plt.show()
```

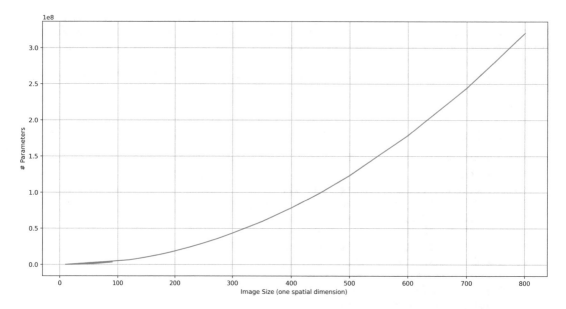

Figure 4-28. *Demonstrating the parametrization scaling of a convolutional neural network as the image input dimension increases*

The parameter count scales much better than a fully connected network, but still is quite poor. A 300-by-300 image input requires almost 50,000,000 parameters; a 600-by-600 image input requires almost 175,000,000 parameters; an 800-by-800 image input requires over 800,000,000 parameters.

We need a way to address this problem. You'll notice that the main source of parameters is in the fully connected component after flattening, because convolutions don't reduce the feature map representation size sufficiently quickly. We can simplify the problem of building a network with a feasible number of parameters as a problem of building a network that efficiently reduces the feature map size.

How can we address this? A convolution is the only current operation we have in our toolbelt to reduce the dimensionality of an image with a feasible number of parameters. Let's build an architecture-defining program that just keeps on stacking convolutional layers until the feature maps collectively contain 2048 elements or less (Listing 4-31, Figure 4-29). We keep track of one spatial dimension of the feature map size (we only need to keep track of one rather than both because we assume feature maps are square) and calculate the representation size as $s^2 \cdot 16$, where s is the feature map size and 16 is derived from the number of feature maps.

Listing 4-31. Plotting the scaling capability of a model with "continual convolution stacking"

```
def build_model(img_size):

    model = Sequential()
    model.add(L.Input((img_size, img_size, 1)))
    model.add(L.Conv2D(16, (1, 1), activation='relu'))
    featureMapSize = img_size
    while (featureMapSize ** 2) * 16 > 2048:
        model.add(L.Conv2D(16, (3, 3),
                  activation='relu'))
        featureMapSize -= 2
    model.add(L.Flatten())
```

```
    model.add(L.Dense(32, activation='relu'))
    model.add(L.Dense(16, activation='relu'))
    model.add(L.Dense(10, activation='softmax'))

    paramCount = model.count_params()
    del model
    return paramCount

x = [10, 15, 20, 25, 30, 35, 40, 45, 50,
     60, 70, 80, 90, 100, 120, 130, 140, 150,
     170, 180, 190, 200, 225, 250, 275, 300,
     350, 400, 450, 500, 600, 700, 800]
y = [build_model(i) for i in tqdm(x)]

plt.figure(figsize=(15, 7.5), dpi=400)
plt.plot(x, y, color='red')
plt.xlabel('Image Size (one spatial dimension)')
plt.ylabel('# Parameters')
plt.grid()
plt.show()
```

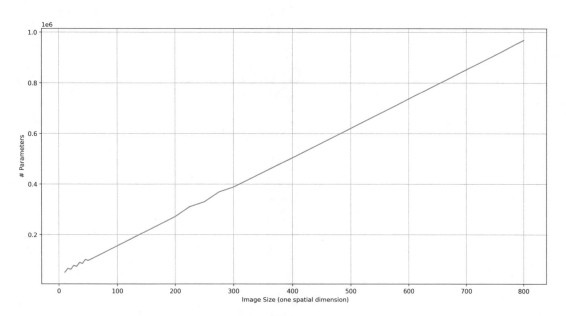

Figure 4-29. *Demonstrating the parametrization scaling of a specialized "self-extending" convolutional neural network design as the image input dimension increases*

This technique scales better: it is linear rather than exponential, resulting in parameter counts two orders of magnitude less than the previous convolutional neural network design.

However, we run into another problem: as the image size increases, the *length* of the network increases significantly as well. Look at the length of the architecture just for a 75-by-75 image input, for instance (Figure 4-30).

Figure 4-30. *The (very long) generated architecture just for a 75-by-75-pixel input image using the "self-extending" convolutional neural network design*

While this is a *valid* and *feasible* neural network, it's not a good design. Having so many layers stacked together like this causes problems for signal propagation throughout the network and – most importantly – isn't necessary nor effective. It's the most manual, brute-force approach.

We can replace the return in previous code plotting the scaling of parameter counts by image size instead with the number of parameters (given by len(model.layers)) to view how the layer count by this method scales with the image size (Figure 4-31).

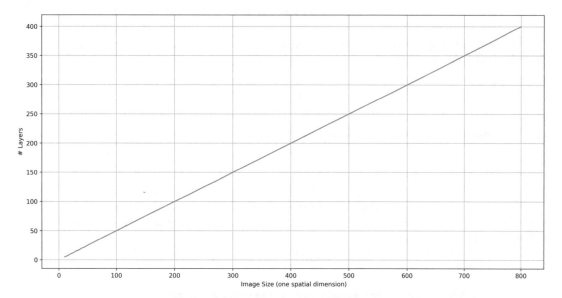

Figure 4-31. *The layer requirement scaling of the "self-extending" convolutional neural network architecture design as the image dimension increases*

In order to reduce the dimensionality of the image more effectively (and correspondingly improve parameter count scaling), we need to use a more efficient mechanism: pooling. To apply pooling to a matrix i with size (a, b), we divide i into non-overlapping blocks of shape (a, b), aggregate all values in each block, and fill in the aggregated values into a pooled matrix j corresponding to the locations of those blocks. For instance, consider pooling on the following matrix with a pooling shape of $(2, 2)$:

$$i = \begin{bmatrix} 1 & 2 & 3 & 4 \\ 5 & 6 & 7 & 8 \\ 9 & 10 & 11 & 12 \\ 13 & 14 & 15 & 16 \end{bmatrix}$$

There are four non-overlapping blocks of shape $(2, 2)$ in i. The pooled matrix j has shape $\left(\frac{4}{2}, \frac{4}{2} \right) = (2,2)$.

Generally, for a matrix of shape (m, n) and a pooling size of (a, b), the resulting pooled matrix has shape $\left(\frac{m}{a}, \frac{n}{b} \right)$.

$$j = \begin{bmatrix} ? & ? \\ ? & ? \end{bmatrix}$$

Let's fill in the upper-left corner of *j*. Because we are using pooling with shape $(2, 2)$, this corresponds to the following bolded subregion of *i*:

$$i = \begin{bmatrix} 1 & 2 & 3 & 4 \\ 5 & 6 & 7 & 8 \\ 9 & 10 & 11 & 12 \\ 13 & 14 & 15 & 16 \end{bmatrix}$$

To keep things simple, we will employ max pooling instead of average pooling. The maximum value in the bolded subregion is 6. We can fill this into the corresponding element slot in *j*:

$$j = \begin{bmatrix} 6 & ? \\ ? & ? \end{bmatrix}$$

We can fill in the remaining elements in a similar fashion:

$$j = \begin{bmatrix} 6 & 8 \\ 14 & 16 \end{bmatrix}$$

Similarly to convolutions, the resulting pooled matrix reflects the "gist" or "main ideas" from the raw matrix. However, there is a key difference between convolutions and pooling that can only be realized with a higher-dimensional matrix.

Let's return to the previous "image" we created in Listing 4-10, a plus sign on top of a gradient background (Figure 4-32).

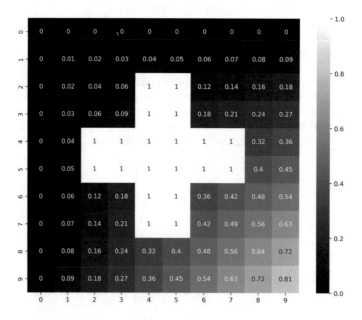

Figure 4-32. *Revisiting the synthetic figure created in Listing 4-10*

We can use `skimage.measure`'s `block_reduce` function to simulate pooling on an image. The function takes in an array representing the input, the pooling shape, and the function to apply to all elements in each pooling subregion.

Listing 4-32 produces Figure 4-33, showing the result of pooling the matrix shown in Figure 4-32 with average pooling.

Listing 4-32. Applying two-dimensional average pooling to an image and displaying the result

```
pooled = skimage.measure.block_reduce(img, (2,2), np.mean)

plt.figure(figsize=(10, 8), dpi=400)
sns.heatmap(pooled, cmap='gray', annot = True)
plt.show()
```

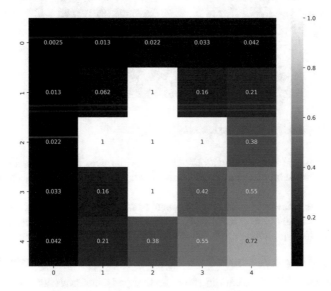

Figure 4-33. The result of applying 2 × 2 mean pooling on Figure 4-32

Listing 4-33 and Figure 4-34 show the effect of pooling using max pooling.

Listing 4-33. Applying two-dimensional max pooling to an image and displaying the result

```
pooled = skimage.measure.block_reduce(img, (2,2), np.max)
plt.figure(figsize=(10, 8), dpi=400)
sns.heatmap(pooled, cmap='gray', annot = True)
plt.show()
```

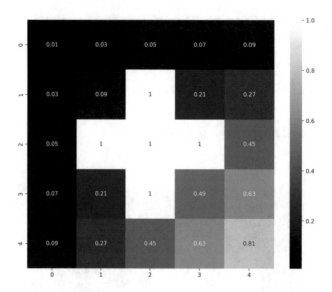

Figure 4-34. *The result of applying 2 × 2 max pooling on Figure 4-32*

The results of average and max pooling look more or less identical. In this case, it is because the plus in the middle is cleanly arranged into blocks of 2 that will yield a pooled result of 1 regardless of the aggregating function used.

Let's perform another round of pooling on the image to see the effect that multiple pooling operations have on a matrix. Note that the shape of this current matrix is $(5, 5)$, but the pooling size is $(2, 2)$: our previous formula describing the effect of pooling on shape would suggest that the resulting array has a fractional shape of $\left(\dfrac{5}{2}, \dfrac{5}{2}\right)$, which is clearly inaccurate. To deal with array sizes that don't cleanly divide by the pooling size, pooling layers automatically use a padding function that pads the array with a "default value" (usually 0 or some sort of average) until it is a valid size.

Consider the following matrix with shape $(3, 3)$ when we desire to apply pooling with shape $(2, 2)$:

$$bad\ matrix = \begin{bmatrix} 1 & 1 & 1 \\ 1 & 1 & 1 \\ 1 & 1 & 1 \end{bmatrix}$$

We can pad the matrix as such:

$$valid\ matrix = \begin{bmatrix} 1 & 1 & 1 & 0 \\ 1 & 1 & 1 & 0 \\ 1 & 1 & 1 & 0 \\ 0 & 0 & 0 & 0 \end{bmatrix}$$

If we apply max pooling, the zero padding is more or less irrelevant because it is always the smallest value in the region. The padding approach is sort of a cheat trick to have fractionally sized pooling regions (if max pooling is used) because there is no possibility that all outputs are less than zero (unless the network is behaving very oddly or is using a bizarre activation function). The three relevant pooling regions in the previous example are

$$
\begin{bmatrix} 1 & 1 & - & - \\ 1 & 1 & - & - \\ - & - & - & - \\ - & - & - & - \end{bmatrix}, \begin{bmatrix} - & - & 1 & - \\ - & - & 1 & - \\ - & - & - & - \\ - & - & - & - \end{bmatrix}, \begin{bmatrix} - & - & - & - \\ - & - & - & - \\ 1 & 1 & - & - \\ - & - & - & - \end{bmatrix}
$$

Let's apply pooling again, using both mean (Listing 4-34, Figure 4-35) and max pooling (Listing 4-35, Figure 4-36).

Listing 4-34. Applying mean pooling two times

```
pooled = skimage.measure.block_reduce(img, (2,2), np.mean)
pooled2 = skimage.measure.block_reduce(pooled, (2,2), np.mean)

plt.figure(figsize=(10, 8), dpi=400)
sns.heatmap(pooled2, cmap='gray', annot = True)
plt.show()
```

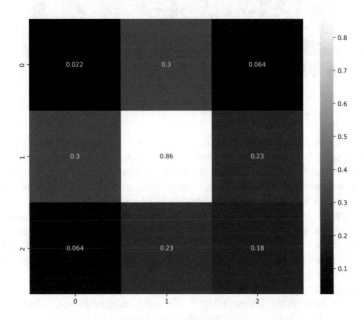

Figure 4-35. *The result of applying 2 × 2 mean pooling twice on Figure 4-32*

Listing 4-35. Applying max pooling two times

```
pooled = skimage.measure.block_reduce(img, (2,2), np.max)
pooled2 = skimage.measure.block_reduce(pooled, (2,2), np.max)

plt.figure(figsize=(10, 8), dpi=400)
sns.heatmap(pooled2, cmap='gray', annot = True)
plt.show()
```

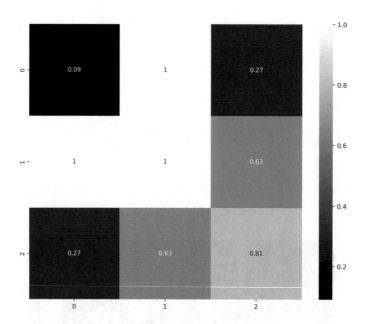

Figure 4-36. *The result of applying 2 × 2 max pooling on Figure 4-32*

Here, the difference between average and max pooling becomes clearer. We see that the strongest signal is the only signal propagated to the next component of the network in max pooling, whereas average pooling considers all signals. Generally, max pooling is more commonly used because it allows the neural network to easily form if/else-style switch points instead of manipulating a comparatively complex averaging game to propagate important signals to the next component of the network.

Convolutions perform an *extractive role* in a neural network, whereas pooling serves an *aggregative role*. Note that while the neural network needs to learn and optimize the parameter values for the kernels in convolutional layers, pooling layers operate without any trainable parameters. Pooling significantly reduces the representation size of a set of feature maps, which allows us to build more efficient and sustainable neural network architectures.

Let's construct an example neural network architecture, improving upon our previous network design employing only convolutional layers (Listing 4-36). We'll use the same architecture, but insert just one max pooling layer in between the second and third convolutional layers.

Listing 4-36. Building a convolutional neural network with a pooling layer

```
import keras.layers as L
from keras.models import Sequential
model = Sequential()
model.add(L.Input((28, 28, 1)))
```

```
model.add(L.Conv2D(8, (5, 5), activation='relu'))
model.add(L.Conv2D(8, (3, 3), activation='relu'))
model.add(L.MaxPooling((2, 2)))
model.add(L.Conv2D(16, (3, 3), activation='relu'))
model.add(L.Conv2D(16, (2, 2), activation='relu'))
model.add(L.Flatten())
model.add(L.Dense(32, activation='relu'))
model.add(L.Dense(16, activation='relu'))
model.add(L.Dense(10, activation='softmax'))
```

Let's compare the effect of pooling on neural network parametrization across different image sizes to see how its benefits scale (Listing 4-37).

Listing 4-37. Comparing parameter scaling of an architecture with and without pooling

```
def build_model(img_size):
    inp = L.Input((img_size, img_size, 1))
    x = L.Conv2D(8, (5, 5), activation='relu')(inp)
    prev = L.Conv2D(8, (3, 3), activation='relu')(x)
    pool = L.MaxPooling2D((2, 2))(prev)
    after = L.Conv2D(16, (3, 3), activation='relu')(pool)
    x = L.Conv2D(16, (2, 2), activation='relu')(after)
    x = L.Flatten()(x)
    x = L.Dense(32, activation='relu')(x)
    x = L.Dense(16, activation='relu')(x)
    x = L.Dense(10, activation='softmax')(x)
    model = keras.models.Model(inputs = inp, outputs = x)

    yesPoolingParamCount = model.count_params()

    after = L.Conv2D(16, (3, 3), activation='relu')(prev)
    x = L.Conv2D(16, (2, 2), activation='relu')(after)
    x = L.Flatten()(x)
    x = L.Dense(32, activation='relu')(x)
    x = L.Dense(16, activation='relu')(x)
    x = L.Dense(10, activation='softmax')(x)
    model = keras.models.Model(inputs = inp, outputs = x)

    noPoolingParamCount = model.count_params()

    del model

    return noPoolingParamCount, yesPoolingParamCount
```

Now, let's plot out the difference (Listing 4-38, Figure 4-37).

Listing 4-38. Plotting parameter scaling of an architecture with and without pooling

```
x = [20, 25, 30, 35, 40, 45, 50,
    60, 70, 80, 90, 100, 120, 130, 140, 150,
    170, 180, 190, 200, 225, 250, 275, 300,
    350, 400, 450, 500, 600, 700, 800]
```

```
y = [build_model(i) for i in tqdm(x)]

plt.figure(figsize=(7.5, 3.25), dpi=400)
plt.plot(x, [i for i, j in y], color='red', label='No Pooling')
plt.plot(x, [j for i, j in y], color='blue', label='Yes Pooling', linestyle='--')
plt.xlabel('Image Size (one spatial dimension)')
plt.ylabel('# Layers')
plt.grid()
plt.legend()
plt.show()
```

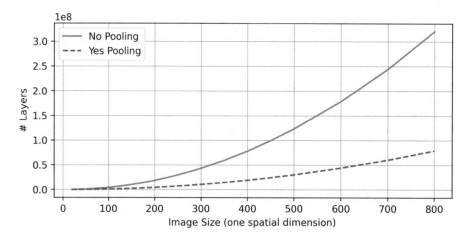

Figure 4-37. *Comparing the parametrization scaling of a convolutional neural network with and without pooling layers*

Note that while the addition of the pooling layer doesn't add any parameters, it forces a decrease in representation size that has downstream effects, since *every layer* after pooling works with a smaller layer.

In fact, we can observe even larger improvements in parametrization by stacking even more pooling layers upon each other.

■ **Note** We usually use padding = 'same' for convolutions, which automatically calculates and applies the padding needed to retain the same image shape as the input, because the effect of pooling is trivial and in fact may help detection of edge features. Moreover, in many cases, we don't want to mix un-padded convolutions and pooling layers for shape management convenience: we might have a nicely sized image (e.g., 256-by-256 pixels), which we only want to cut down by divisible factors (e.g., 2-by-2 max pooling operations).

Many deep learning figures are not fans of the pooling mechanism. Despite its assistance in parametrization scaling, it itself is unparametrized (unlearnable) and therefore can be interpreted as a messy, brute-force way to cut down information size. Opponents of pooling often argue instead for the usage of *strides* in convolutions, which we will discuss later. Strides can accomplish the same feature map size reduction effect that pooling does.

We've seen that pooling layers can help significantly reduce the size of learned feature maps – but pooling can help us out again, in a different form.

Note that the flattening layer, despite being intuitive and retaining all information, is a bottleneck for feasibility in parametrization. "Small" data packed neatly across three spatial dimensions can become very large when flattened out into one spatial dimension. Using pooling, we can derive a more efficient mechanism to collapse a set of feature maps into a vector. It comes from the simple realization that applying a pooling kernel with a size equal to the feature maps it is applied on yields a singular aggregative value. For instance, a set of feature maps with shape $(5, 5, 32)$ (i.e., has 32 "versions" of a 5-by-5 feature representation) passed through pooling with size $(5, 5)$ will yield one value for each of the outputs, or aggregated data with shape (32).

The name given to this special case of pooling in which the pooling shape is equal to the shape of the incoming feature maps is *global pooling*. Like standard pooling, it comes in two common flavors: global average pooling and global max pooling. Global average pooling averages all values in each feature map, whereas global max pooling finds the maximum of all values in each feature map. Whereas max pooling is generally preferred over average pooling in the extractive component of a neural network, *global average pooling is generally preferred over global max pooling* to perform collapsing from three to one spatial dimension. Global max pooling is a sharply reductive operation in that only one of the n^2 elements in a $n \times n$ feature map "counts"; that is, very little of the learned signals are propagated to the next part of the network with max pooling. On the other hand, in global average pooling, all n^2 elements "have a say" in determining the output signal. Because global pooling performs the crucial switch-off between the extractive (convolutional) and the interpretative (fully connected) component of the network, we generally do not want to introduce any new signal bottlenecks.

Let's demonstrate how replacing flattening with global max pooling can further help us improve how the parametrization scales with input size. Listing 4-39 demonstrates a function we might build to calculate the parameter count for a convolutional neural network using both convolutions and global pooling, and the parametrization scaling of this model compared with the previous two (no pooling or yes pooling but no global pooling) is shown in Figure 4-38.

Listing 4-39. Parameter scaling of an architecture with global max pooling in replacement of flattening

```
def build_pooling_model(img_size):
    inp = L.Input((img_size, img_size, 1))
    x = L.Conv2D(8, (5, 5), activation='relu')(inp)
    prev = L.Conv2D(8, (3, 3), activation='relu')(x)
    pool = L.MaxPooling2D((2, 2))(prev)
    after = L.Conv2D(16, (3, 3), activation='relu')(pool)
    x = L.Conv2D(16, (2, 2), activation='relu')(after)
    x = L.GlobalAveragePooling2D()(x)
    x = L.Dense(32, activation='relu')(x)
    x = L.Dense(16, activation='relu')(x)
    x = L.Dense(10, activation='softmax')(x)
    model = keras.models.Model(inputs = inp, outputs = x)
    paramCounts = model.count_params()
    del model
    return paramCounts
```

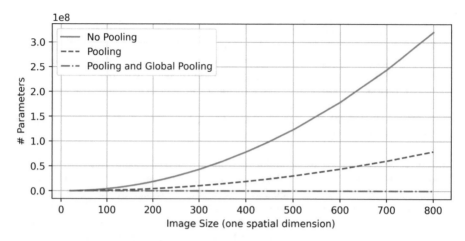

Figure 4-38. *Comparing the parametrization scaling of a convolutional neural network without pooling layers, with pooling but without global pooling, and with pooling and global pooling*

The difference is incredibly significant. The parametrization of a convolutional neural network with both pooling and global pooling appears functionally flat. Upon obtaining a closer look by only comparing CNN designs with pooling but variable global pooling (Figure 4-39), we observe that the parametrization scaling of a model with pooling and global pooling really is flat. If you observe the raw values, you'll find that for this particular model, the number of parameters is always 4242 – we've achieved *constant parameter scaling*, which is the best sort of scaling one can obtain (since it usually doesn't make sense for the number of parameters to decrease as input complexity increases).

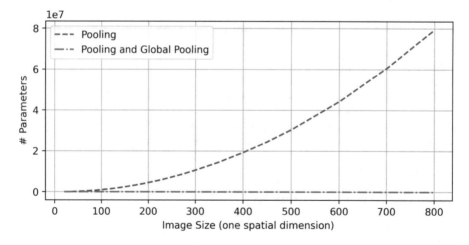

Figure 4-39. *"Zooming in" on Figure 4-38 with a focus only on the parametrization scaling of a convolutional neural network with pooling but without global pooling and one with pooling and global pooling*

It's worth thinking how the global pooling mechanism allows us to obtain constant parameter scaling. The convolutions themselves have constant parametrization, because they simply learn the values for a convolutional kernel of a certain size regardless of the input size. Recall that convolutions are a good choice for neural networks that process images because they have an inherently scalable sliding-window/kernel design. However, it is the crucial switch-off from feature map to vector (i.e., between the extractive/convolutional and interpretive/fully connected components) that has a variable number of parameters. If the last layer in the convolutional component outputs k feature maps of shape (a, b) and the first layer of the fully connected component contains d nodes, then the number of parameters using the flattening layer is $(k \cdot a \cdot b) \times d$. While k and d are constant in that in our experiment they are the same regardless of input size, note that the values of a and b are variable. Thus, as the input size grows, we expect the parameter scaling to grow roughly quadratically using the flattening layer.

However, if we use max pooling, the k feature maps of shape (a, b) are pooled into a vector of shape k. The number of parameters required is $k \cdot d$, which both consist of constants! Thus, regardless of the size of the input, a convolutional neural network employing both pooling and max pooling will always have a constant number of parameters (assuming the architecture stays constant). Note that this does not necessarily mean that the model will still perform as well as we scale the input complexity up; an inherently more complex task likely requires more parameters to represent and model it, which may require modifications to the architecture (e.g., more layers, a large number of nodes/filters in each layer, etc.).

One can think of flattening and global pooling as complementary methods: where one lacks, the other provides. When using flattening, we gain retention of all information at the cost of losing recognition of feature map separation (it does not matter which feature map any value is part of; all values are indiscriminately placed side-by-side on the same vector). When using global pooling, we gain recognition of feature map separation at the cost of losing information. Generally, larger convolutional neural networks favor global pooling, whereas smaller ones favor flattening. Large CNNs produce large output feature maps that, if flattened, would be unscalable to work with. Global pooling is generally a sufficient mechanism to capture the "main ideas" of the extracted features. Small CNNs, on the other hand, generally produce smaller feature maps in which each element holds a higher proportion of information. Using flattening over global pooling can help explicitly preserve the raw extracted features and is generally feasible. In most "standard" modeling problems, however, sufficiently sized neural networks should obtain roughly similar performance using either flattening or global pooling (albeit under different training conditions and requirements).

Now that we have the two key building blocks of convolutional neural networks under our belt, let's implement a slightly modified *AlexNet architecture* layer-by-layer in Keras. AlexNet was an instrumental model in computer vision. Released in 2012 by Alex Krizhevsky in collaboration with Ilya Sutskever and Geoffrey Hinton in the paper "ImageNet classification with deep convolutional neural networks,"[1] AlexNet set the foundation for rapid research development in convolutional neural network architectures in the following years. As of the writing of this book, the paper has been cited over 80,000 times.

AlexNet follows a relatively simple architecture (Figure 4-40).

---

[1] Krizhevsky, A., Sutskever, I., & Hinton, G.E. (2012). ImageNet classification with deep convolutional neural networks. *Communications of the ACM, 60*, 84–90.

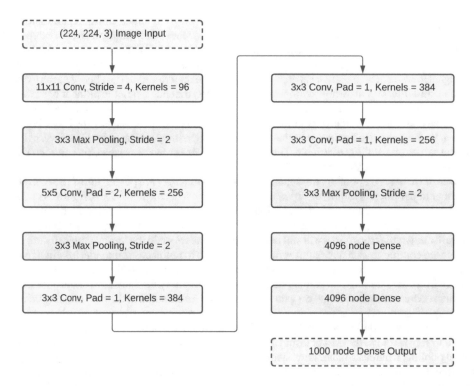

Figure 4-40. *Visual representation of the AlexNet architecture*

Note that this architecture utilizes *strides*. Strides specify the number of elements to "jump over" every time a kernel is shifted during convolution or pooling. Convolutions are usually introduced with stride 1, but applying a convolution with stride 2 to a sample (5, 5) matrix affects the relevant bolded subsections:

$$\begin{bmatrix} \mathbf{1} & \mathbf{1} & 1 & \mathbf{1} & \mathbf{1} \\ \mathbf{1} & \mathbf{1} & 1 & \mathbf{1} & \mathbf{1} \\ 1 & 1 & 1 & 1 & 1 \\ \mathbf{1} & \mathbf{1} & 1 & \mathbf{1} & \mathbf{1} \\ \mathbf{1} & \mathbf{1} & 1 & \mathbf{1} & \mathbf{1} \end{bmatrix}$$

Strides are a useful tool to give feature extraction layers more "leeway" by allocating regions in the feature maps that are not propagated into the remainder of the network. Note that while strides do not decrease the number of parameters in the individual convolutional layer they are applied to, they do decrease the size of the output feature maps, which has downstream parametrization implications (i.e., that the rest of the network uses fewer parameters). Strides are another important method to reduce parametrization for image processing neural networks.

The code to build the layers of AlexNet is fairly straightforward (Listing 4-40).

Listing 4-40. Building a sample AlexNet architecture

```
model = Sequential()
model.add(L.Input((224, 224, 3)))
model.add(L.Conv2D(96, (11, 11), strides=4))
model.add(L.MaxPooling2D((3, 3), strides=2))
model.add(L.Conv2D(256, (5, 5)))
model.add(L.MaxPooling2D((3, 3), strides=2))
model.add(L.Conv2D(384, (3, 3)))
model.add(L.Conv2D(384, (3, 3)))
model.add(L.Conv2D(256, (3, 3)))
model.add(L.MaxPooling2D((3, 3), strides=2))
model.add(L.Flatten())
model.add(L.Dense(4096, activation='relu'))
model.add(L.Dense(4096, activation='relu'))
model.add(L.Dense(1000, activation='softmax'))
```

Plotting the model confirms our desired architecture (Figure 4-41). Note something interesting – the specific kernel shape and arrangement of pooling and convolutional layers are designed such that the last layer in the extractive/convolutional component outputs feature maps of shape $(1, 1, 256)$: meaning that each of the 256 feature maps has been compressed into only one value! In this unique case, note that flattening and global pooling are functionally no different.

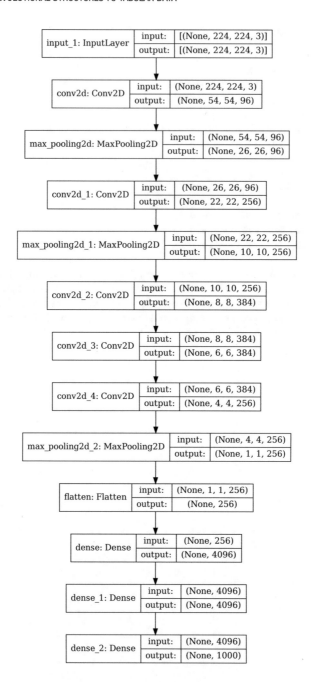

Figure 4-41. *Keras-style visual representation of the AlexNet architecture after implementation*

The AlexNet model, we can observe, uses a carefully hand-crafted flow of information. When we build convolutional neural networks (and neural networks generally), we want to carefully monitor how the *representation* size changes across each the length of the network. The representation size is simply the total number of elements a layer holds; a fully connected layer with n nodes has a representation size of n elements, and a convolutional layer that outputs feature maps with shape (a, b, c) has a representation size of $a \cdot b \cdot c$. Suppose the change in representation size drastically decreases at any point in the architecture (i.e., a bottleneck). In that case, the network is forced to compress a large quantity of information into a small amount of space. On the other hand, if the change in representation size drastically increases at any point in the architecture (i.e., *inflation*), then the network is forced to expand a small quantity of information into a large amount of space. Depending on the "inherent complexity" of the data, bottlenecks could be restrictive, and inflation could lead to redundant computation.

Analyzing the design of successful convolutional neural network architectures gives us tremendous insight into how convolutions and pooling layers operate. In the next section, we'll continue an analysis of other more modern architectures.

Base CNN Architectures

All content discussed before this section has focused on the low-level building blocks of convolutional neural networks. However, very few people build their own convolutional neural network architectures "by hand" by arranging together layers for now-standard problems like image classification. Instead, researchers have discovered a set of effective general architectures that can serve as the *base* of a convolutional neural network, upon which minor customizations can be made to "specialize" the base model for your desired task. In this section, we'll discuss three key base CNN architectures and demonstrate how to instantiate and use them for quick custom modeling.

ResNet

As suggested by its name, the ResNet architecture features residual connections as a primary element of its topological design.

Residual connections are the "first step" toward architectural nonlinearity – these are simple connections placed between nonadjacent layers. They're often presented as "skipping" over a layer or multiple layers, which is why they are also often referred to as "skip connections" (Figure 4-42).

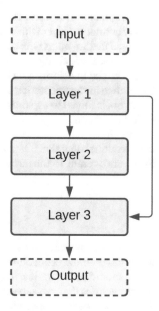

Figure 4-42. *Residual connection*

Note that in implementation the connections are first merged through a method like adding or concatenation. The merged components are then passed into the next layer (Figure 4-43). This is the implicit assumption of all residual connection diagrams that do not explicitly demonstrate merging.

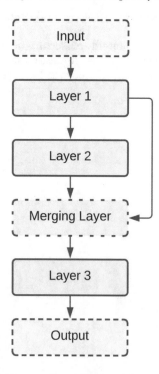

Figure 4-43. *Technically correct residual connection with merge layer*

Adding residual connections decreases *degradation* of the input signal, in which the original input gets saturated or lost in a long heap of network layers. Residual connections enable greater flow of information, which allows the network to reason more nonlinearly by combining information from different stages or areas of reasoning.

"ResNet-style" residual connection desgins employ a series of short residual connections that are repeated routinely throughout the network (Figure 4-44).

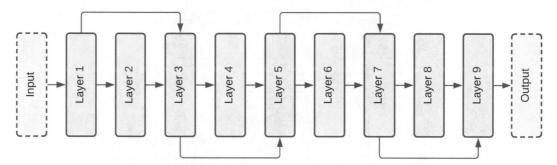

***Figure 4-44.** ResNet-style residual connection design*

The ResNet architecture, originally introduced by Kaiming He, Xiangyu Zhang, Shaoqing Ren, and Jian Sun at Microsoft Research in the 2015 paper "Deep Residual Learning for Image Recognition,"[2] comprises 34 layers with residual connections that skip over every two layers. Figure 4-45 compares ResNet with a plain "equivalent" architecture and the more classic VGG-19 architecture.

---

[2] He, K., Zhang, X., Ren, S., & Sun, J. (2016). Deep Residual Learning for Image Recognition. *2016 IEEE Conference on Computer Vision and Pattern Recognition (CVPR)*, 770–778.

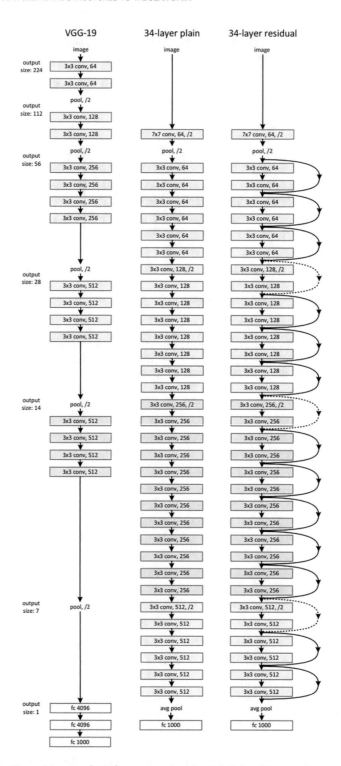

Figure 4-45. *The ResNet architecture (right), compared with a "plain" architectural equivalent (i.e., without residual connections, middle) and the VGG-19 architecture*

Note that there are other architectural interpretations of what a residual connection is. Rather than relying upon a linear backbone, you can interpret a residual connection as splitting the layer before it into two branches, which each process the previous layer in their own unique ways. One branch (layer 1 to layer 2 to layer 3 in the following diagram) processes the output of the previous layer with a specialized function, whereas the other branch (layer 1 to identity to layer 3) processes the output of the layer with the identity function – that is, it simply allows the output of the previous layer to pass through, the "simplest" form of processing (Figure 4-46).

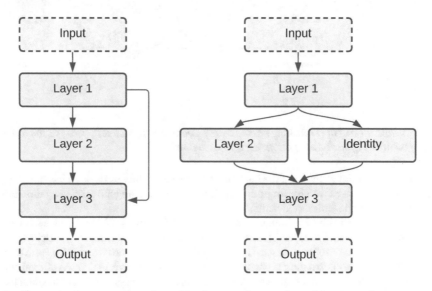

Figure 4-46. *Alternative interpretation of a residual connection as a branching operation*

This method of conceptually understanding residual connections allows you to categorize them as a subclass of general nonlinear architectures, which can be understood as a series of branching structures.

Residual connections are often presented as a solution for the *vanishing gradient problem* (Figure 4-47): in order to access some layer, we need to travel through several other layers first, diluting the information signal. In the vanishing gradient problem, the backpropagation signal within very deep neural networks used to update the weights gets progressively weaker such that the front layers are barely utilized at all. (Note that in many cases using ReLU activations instead of bounded functions like sigmoid can address this problem – but residual connections are another method.)

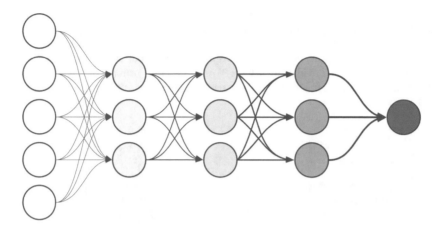

Figure 4-47. *Visualization of the vanishing gradient problem: when the network becomes too long, information signals during backpropagation "die out" or "vanish"*

With residual connections, however, the backpropagation signal travels through fewer average layers to reach some particular layer's weights for updating. This enables a stronger backpropagation signal that is better able to make use of the entire model architecture.

Residual connections can also be thought of as a "fail-safe" for poor-performing layers. If we add a residual connection from layer A to layer C (assuming layer A is connected to layer B, which is connected to layer C), the network can "choose" to disregard layer B by learning near-zero weights for connections from A to B and from B to C while information is channeled directly from layers A to C via a residual connection. In practice, however, residual connections act more as an additional representation of the data for consideration than as fail-safe mechanisms.

The ResNet architecture is implemented in several flavors in Keras: ResNet50, ResNet101, and ResNet152 (each with two versions). The number attached to each ResNet architecture is a rough indicator of how many layers deep the network is (although if you count, you won't get an exact number because of certain technicalities). ResNet50 is the smallest offered version of ResNet, and ResNet 152 is the deepest.

You can instantiate and train a ResNet model simply by calling the model object and specifying the input shape and number of classes (Listing 4-41). Most model architectures in Keras come with pretrained weights on the ImageNet dataset, but the number of output classes must be 1000 because the ImageNet dataset contains 1000 output classes.

Listing 4-41. Boilerplate code to train a ResNet50 model on an image classification task with input shape (a, b, 3) and c output classes

```
from tensorflow.keras.applications import ResNet50
model = ResNet50(input_shape = (a, b, 3),
                 classes = c,
                 weights = None)
model.compile(...)
model.fit(...)
```

You can also treat any model (including ResNets, other Keras application models, and your own sequential or functional models) as a submodel or component of a larger overarching model. For instance, consider a hypothetical architecture in which an input is independently passed through a ResNet50 and a ResNet121 architecture and then joined and processed into an output implemented in Listing 4-42 (visualized in Figure 4-48). We begin by instantiating the ResNet50 and ResNet121 models and then use the `result = model(inp_layer)` syntax.

Listing 4-42. Constructing a "hybrid" ResNet architecture by instantiating ResNet50 and ResNet121 architectures as components/submodels

```
from tensorflow.keras.applications import ResNet50, ResNet152
inp = L.Input((a, b, 3))
resnet50 = ResNet50(input_shape = (a, b, 3),
                    classes = c,
                    weights = None)
resnet50out = resnet50(inp)
resnet121 = ResNet152(input_shape = (a, b, 3),
                      classes = c,
                      weights = None)
resnet121out = resnet121(inp)
concat = L.Concatenate()([resnet50out, resnet121out])
out = L.Dense(c, activation='softmax')(concat)
model = keras.models.Model(inputs = inp,
                           outputs = out)
```

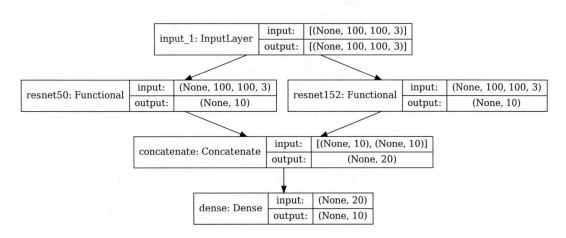

Figure 4-48. *Keras visualization of the architecture produced in Listing 4-42*

We'll see an example usage of this model compartmentalization in the "Multimodal Image and Tabular Models" section.

Another architecture, DenseNet, introduced by Gao Huang, Zhuang Liu, and Killian Q. Weinberger in the 2016 paper "Densely Connected Convolutional Networks,"[3] uses residual connections in a more extreme or "dense" fashion. DenseNet architectures feature uniformly spaced "anchor points"; residual connections are placed between every set of anchor points (Figure 4-49). Like ResNet, DenseNet comes implemented in Keras with multiple flavors: DenseNet121, DenseNet169, and DenseNet201. All these models are housed under keras.applications.DenseNetx.

---

[3] Huang, G., Liu, Z., & Weinberger, K.Q. (2017). Densely Connected Convolutional Networks. *2017 IEEE Conference on Computer Vision and Pattern Recognition (CVPR)*, 2261–2269.

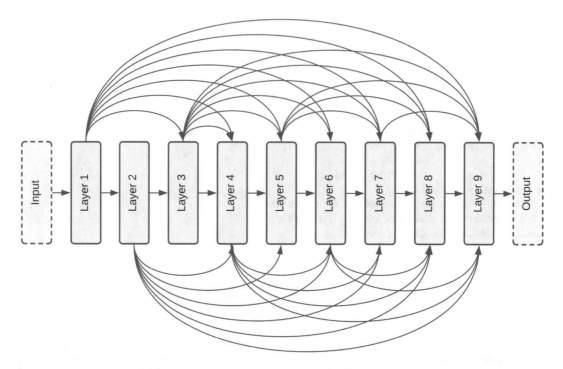

Figure 4-49. *Example of DenseNet-style residual connections in which every layer has an anchor point*

Inception v3

Christian Szegedy, Vincent Vanhoucke, Sergey Ioffe, Jonathon Shlens, and Zbigniew Wojna introduced the Inception v3 architecture, an improved member of the Inception family of models that has become a pillar of image recognition, in their 2015 paper "Rethinking the Inception Architecture for Computer Vision."[4] In many ways, the Inception v3 architecture laid out the key principles of convolutional neural network design for the following years to come. The aspect most relevant for this context is its cell-based design.

The Inception v3 model attempted to improve upon the designs of the previous Inception v2 and original Inception models. The original Inception model employed a series of repeated cells (referred to in the paper as "modules") that followed a multi-branch nonlinear architecture (Figure 4-50). Four branches stem from the input to the output of the module; two branches consist of a 1 × 1 convolution followed by a larger convolution, one branch is defined as a pooling operation followed by a 1 × 1 convolution, and another is just a 1 × 1 convolution. Padding is provided on all operations in these modules such that the size of the filters is kept the same such that the results of the parallel branch representations can be concatenated depth-wise back together.

---

[4] Szegedy, C., Vanhoucke, V., Ioffe, S., Shlens, J., & Wojna, Z. (2016). Rethinking the Inception Architecture for Computer Vision. *2016 IEEE Conference on Computer Vision and Pattern Recognition (CVPR)*, 2818–2826.

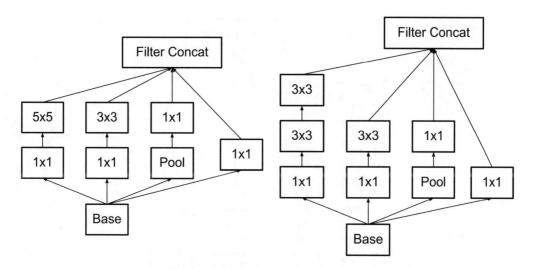

Figure 4-50. *Left: Original Inception cell. Right: One of the Inception v3 cell architectures. From Szegedy et al.*

A key architectural change in the Inception v3 module designs is the *factorization* of large filter sizes like 5 × 5 into a combination of smaller filter sizes. For instance, the shape effect of a 5 × 5 filter can be "factored" into a series of two 3 × 3 filters; a 5 × 5 filter applied on a feature map (with no padding) yields the same output shape as two 3 × 3 filters: (w-4, h-4, d). Similarly, a 7 × 7 filter can be "factored" into three 3 × 3 filters. Szegedy et al. note that this factorization promotes faster learning while causing no hindrance to representative power. This module will be termed the *symmetric factorization module*, although in implementation within the context of the Inception v3 architecture, it is referred to as *Module A*.

In fact, even 3 × 3 and 2 × 2 filters can be factorized into sequences of convolutions with smaller filter sizes. An *n*-by-*n* convolution can be represented as a 1-by-*n* convolution followed by an *n*-by-1 convolution (or vice versa). Convolutions with kernel heights and widths that are different lengths are known as *asymmetric convolutions* and can be valuable fine-grained feature detectors (Figure 4-51). In the Inception v3 module architecture, *n* was chosen to be 7. This module will be termed the *asymmetric factorization module* (also known as *Module B*). Szegedy et al. find that this module performs poorly on early layers but works well on medium-sized feature maps. Correspondingly, it is placed after symmetric factorization modules in the Inception v3 cell stack.

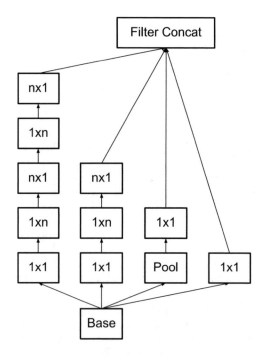

Figure 4-51. *Factorizing n-by-n filters as operations of smaller filters. From Szegedy et al.*

For extremely *coarse* (i.e., small-sized) inputs, a different module with *expanded filter bank outputs* is used. This model architecture assists in highly specialized processing by using a tree-like topology – the two left branches in the symmetric factorization module are further "split" into "child nodes," which are concatenated along with the outputs of the other branches at the end of the filter (Figure 4-52). This type of module is placed at the end of the Inception v3 architecture to handle feature maps when they have become spatially small. This module will be termed the *expanded filter bank module* (or *Module C*).

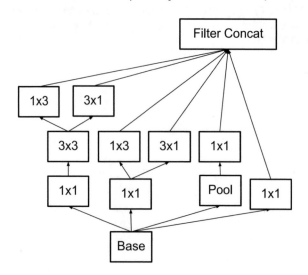

Figure 4-52. *Factorizing n-by-n filters as operations of smaller filters. From Szegedy et al.*

Another reduction-style Inception module is designed to efficiently reduce the size of the filters (Figure 4-53). The reduction-style module uses three parallel branches; two use convolutions with a stride of 2, and the other uses a pooling operation. These three branches produce the same output shapes, which can be concatenated depth-wise. Note that Inception modules are designed such that a decrease in size is correspondingly counteracted with an increase in the number of filters.

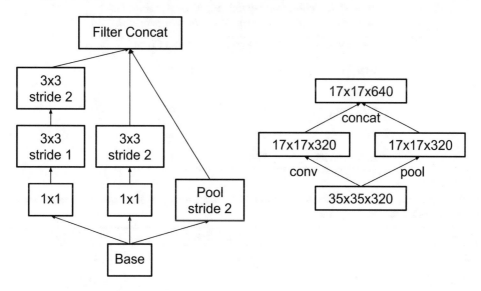

Figure 4-53. *Design of an Inception v3 reduction cell. From Szegedy et al.*

The Inception v3 architecture is formed by stacking these module types in a linear fashion, ordered such that each module is placed in a location where it will receive a feature map input shape that it succeeds in processing. The following sequence of modules is used:

1. A series of convolutional and pooling layers to perform initial feature extraction (these are not part of any module)

2. Three repeats of the symmetric convolution module/Module A

3. Reduction module

4. Four repeats of the asymmetric convolution module/Module B

5. Reduction module

6. Two repeats of the expanded filter bank module/Module C

7. Pooling, dense layer, and softmax output

Another often unnoticed but important feature of the Inception family of architectures is the 1 × 1 convolution, which is present in every Inception cell design – often as the most frequently occurring element in the cell architecture. In terms of model performance, 1 × 1 convolutions serve a key purpose in the Inception architecture: computing cheap filter reductions before expensive, larger kernels are applied to feature map representations. For instance, suppose at some location in the architecture 256 filters are passed into a 1 × 1 convolutional layer; the 1 × 1 convolutional layer can reduce the number of filters to 64 or even 16 by learning the optional combination of values for each pixel from all 256 filters. Because the 1 × 1 kernel does not incorporate any spatial information (i.e., it doesn't consider pixels next to one another), it is cheap to compute. Moreover, it isolates the most important features for the following larger (and thus more expensive) convolution operations that incorporate spatial information.

The Inception v3 architecture performed very well in that the 2015 ILSVRC (ImageNet competition) and has become a staple in image recognition architectures (Tables 4-1 and 4-2).

Table 4-1. *Performance of the Inception v3 architecture against other models in ImageNet. From Szegedy et al.*

Architecture	Top-5 Error	Top-1 Error
GoogLeNet	-	9.15%
VGG	-	7.89%
Inception	22%	5.82%
PReLU	24.27%	7.38%
Inception v3	18.77%	4.2%

Table 4-2. *Performance of an ensemble of Inception v3 architectures compared against ensembles of other architecture models. From Szegedy et al.*

Architecture	# Models	Top-5 Error	Top-1 Error
VGGNet	2	23.7%	6.8%
GoogLeNet	7	-	6.67%
PReLU	-	-	4.94%
Inception	6	20.1%	4.9%
Inception v3	4	17.2%	3.58%

The full Inception v3 architecture is available at `keras.applications.InceptionV3` with available ImageNet weights for transfer learning or just as a powerful architecture (used with random weight initialization) for image recognition and modeling.

Building an Inception v3 module itself is relatively simple with Keras (and a good exercise). We can build four branches in parallel to one another, which are concatenated. Note that we specify `strides=(1,1)` in addition to `padding='same'` in the max pooling layer to keep the input and output layers the same. If we only specify the padding, the strides argument is set to the entered pool size. These cells can then be stacked alongside other cells in a sequential format to form an Inception v3–style architecture (Listing 4-43, Figure 4-54).

Listing 4-43. Building a simple Inception v3 Module A architecture

```
def build_iv3_module_a(inp, shape):

    w, h, d = shape

    branch1a = L.Conv2D(d, (1,1))(inp)
    branch1b = L.Conv2D(d, (3,3), padding='same')(branch1a)
    branch1c = L.Conv2D(d, (3,3), padding='same')(branch1b)
```

```
branch2a = L.Conv2D(d, (1,1))(inp)
branch2b = L.Conv2D(d, (3,3), padding='same')(branch2a)

branch3a = L.MaxPooling2D((2,2), strides=(1, 1),
                          padding='same')(inp)
branch3b = L.Conv2D(d, (1,1), padding='same')(branch3a)

branch4a = L.Conv2D(d, (1,1))(inp)

concat = L.Concatenate()([branch1c, branch2b,
                          branch3b, branch4a])
return concat, shape
```

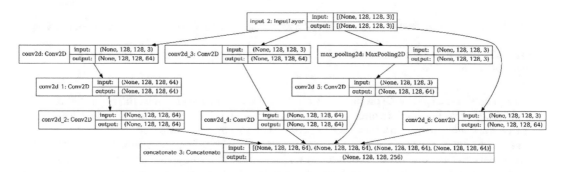

Figure 4-54. *Visualization of the Keras Inception v3 cell built in Listing 4-43*

Besides getting to work with large neural network architectures directly, another benefit of implementing these sorts of architectures from scratch is customizability. You can insert your own cell designs, add nonlinearities across cells (i.e., ResNet-/DenseNet-style connections between cells), or increase or decrease how many cells you stack to adjust the network depth. Moreover, cell-based structures are incredibly simple and quick to implement, so this comes at little cost.

EfficientNet

Convolutional neural networks have historically been scaled relatively arbitrarily. "Arbitrary" scaling entails adjusting these dimensions of a network without much of a justification for how the adjusting is performed; there is ambiguity in how large to scale the dimensions of a neural network to equip it for more complex tasks. For instance, the family of ResNet models (ResNet50, ResNet 101, etc.) are examples of scaling primarily by network *depth,* or the number of layers in the architecture. However, to address the arbitrariness of network scaling, we need a *systematic method of scaling neural network architectures across several architectural dimensions* for the highest expected success (Figure 4-55).

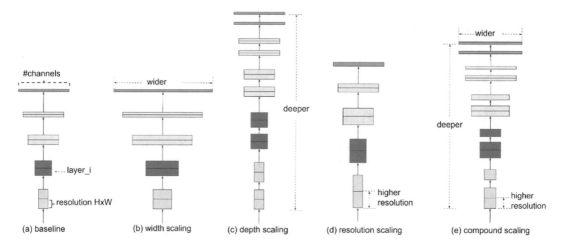

Figure 4-55. *Dimensions of a neural network that can be scaled, compared with the compound scaling method. From Tan and Le*

Mingxing Tan and Quoc V. Le proposed the *compound scaling method* in their 2019 paper "EfficientNet: Rethinking Model Scaling for Convolutional Neural Networks."[5] The compound scaling method is a simple but successful scaling method in which each dimension is scaled by a constant ratio.

A set of fixed scaling constants is used to uniform-scale the width, depth, and resolution used by a neural network architecture. These constants – α, β, γ – are scaled by a compound coefficient, ϕ, such that the depth is $d = \alpha^{\phi}$, the width is $w = \beta^{\phi}$, and the resolution is $r = \gamma^{\phi}$. ϕ is defined by the user, depending on how many computational resources/predictive power they are willing to allocate toward a particular problem.

The value of the constants can be found through a simple grid search. This is both feasible and successful given the small search space. Two constraints on the constants are imposed:

- $a \geq 1, \beta \geq 1, \gamma \geq 1$. This ensures that the constants do not decrease in value when they are raised to the power of the compound coefficient, such that a larger compound coefficient value yields in a larger depth, width, and resolution size.

- $a \cdot \beta^2 \cdot \gamma^2 \approx 2$. The FLOPS (floating-point operations per second) of a series of convolution operations are proportional to the depth, the width squared, and the resolution squared. This is because depth operates linearly by stacking more layers, whereas the width and the resolution act upon two-dimensional filter representations. To ensure computational interpretability, this constraint ensures that any value will raise the total number of FLOPS by approximately $(\alpha \cdot \beta^2 \cdot \gamma^2)^{\phi} = 2^{\phi}$.

This scaling method is very successful in application to previously successful architectures like MobileNet and ResNet (Table 4-3). Through the compound scaling method, we can expand the network's size and computational power in a structured and non-arbitrary way that optimizes the resulting performance of the scaled model.

[5] Tan, M., & Le, Q.V. (2019). EfficientNet: Rethinking Model Scaling for Convolutional Neural Networks. *ArXiv, abs/1905.11946.*

Table 4-3. *Performance of the compound scaling method on MobileNetV1, MobileNetV2, and ResNet50 architectures. From Tan and Le*

Model	FLOPS	Top-1 Acc.
Baseline MobileNetV1	0.6B	70.6%
Scale MobileNetV1 by width (w=2)	2.2B	74.2%
Scale MobileNetV1 by resolution (r=2)	2.2B	74.2%
Scale MobileNetV1 by compound scaling	**2.3B**	**75.6%**
Baseline MobileNetV2	0.3B	72.0%
Scale MobileNetV2 by depth (d=4)	1.2B	76.8%
Scale MobileNetV2 by width (w=2)	1.1B	76.4%
Scale MobileNetV2 by resolution (r=2)	1.2B	74.8%
Scale MobileNetV2 by compound scaling	**1.3B**	**77.4%**
Baseline ResNet50	4.1B	76.0%
Scale ResNet50 by depth (d=4)	16.2B	76.0%
Scale ResNet50 by width (w=2)	14.7B	77.7%
Scale ResNet50 by resolution (r=2)	16.4B	77.5%
Scale ResNet50 by compound scaling	**16.7B**	**78.8%**

By intuition, when the input image is larger, all dimensions – not just one – need to be correspondingly increased to accommodate the increase in information. Greater depth is required to process the increased layers of complexity, and greater width is needed to capture the greater quantity of information. Tan and Le's work is novel in quantitatively expressing the relationship between the scaling of network dimensions.

Tan and Le's paper proposes the *EfficientNet* family of models, which is a family of differently sized models built from the compound scaling method. There are eight models in the EfficientNet family – EfficientNetB0, EfficientNetB1, ..., to EfficientNetB7, ordered from smallest to largest. The EfficientNetB0 architecture was discovered via Neural Architecture Search, a subfield beyond the scope of this book in which the optimal architecture of a neural network is derived via a "meta" or "controller" machine learning model. (We do, however, briefly talk about Neural Architecture Search in Chapter 10.) In order to ensure that the derived model optimized both performance and FLOPS, the objective of the search was not merely to maximize the accuracy, but to maximize a combination of performance and the FLOPS. The resulting architecture was then scaled using different scaling values to form the other seven EfficientNet models.

■ **Note**   The actual open source EfficientNet models are slightly different from the models obtained by "pure" compound scaling. As you may imagine, compound scaling is a successful but approximate method, as is to be expected with most scaling techniques. To maximize performance more fully, some fine-tuning of the architecture is still needed afterward. The publicly available versions of the EfficientNet model family in Keras applications contain some additional architectural changes after scaling via compound scaling to further improve performance.

The EfficientNet family of models impressively obtains higher performance on benchmark datasets like ImageNet, CIFAR-100, Flowers, and others than similarly sized models – both manually designed and NAS-discovered architectures (Figure 4-56). While the core EfficientNetB0 model was created as a product of Neural Architecture Search, the remaining members of the EfficientNet family were constructed via the relatively simple compound scaling paradigm.

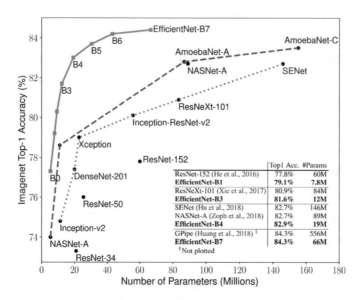

Figure 4-56. Plot of various EfficientNet models against other important model architectures in number of parameters and ImageNet Top-1 accuracy. From Tan and Le

The EfficientNet model family is available in Keras applications at `keras.applications.EfficientNetBx` (substituting x for any number from 0 to 7). The EfficientNet implementation in Keras applications ranges from 29 MB (B0) to 256 MB (B7) in size and from 5,330,571 parameters (B0) to 66,658,687 parameters (B7). Note that the desired input shape for different members of the EfficientNet family is different. EfficientNetB0 expects images of spatial dimension (224,224); B4 expects (380,380); B7 expects (600,600). Note that these are recommendations for input size – you could use an EfficientNetB0 model on a 600-by-600 image if you deemed it necessary. You can find the full list of the desired input shapes listed in the Keras/TensorFlow application documentation.

Multimodal Image and Tabular Models

In this section, we'll explore an application of convolutional neural networks to *multimodal models*, which consider both image and tabular data to produce a prediction. "Multimodal" refers to many (*multi-*) data forms (*-modal*). Image-based multimodal models can be applied to datasets in which images are associated with individual rows in a tabular/structured dataset. (In the next chapter, we'll see another application of sequential multimodal models, which process both a sequence – like text – and tabular data to produce a jointly informed output).

Multimodal applications are particularly interesting, because – at least in principle – they somewhat reflect how humans perceive the world. Rather than making a judgment or decision based on only one modality of input, we consider a variety of data input types and combine them to form more robust and well-informed conclusions. Rephrased, we do not process different input modalities (e.g., sight, hearing, touch, etc.) independently but rather jointly.

These sorts of models must be *multi-head*; that is, they take in more than one input. Each of these inputs is processed independently and molded into the "universal neural network computation form" (i.e., a vector). An image input head would be processed using convolutional layers (convolutions, pooling layers, etc.) and cast into a vector either through flattening or global pooling. A standard tabular/vector input head is already in vector form, but it can be further processed with a series of fully connected layers to extract and amplify relevant features. Once all the images have been processed and cast into vector form, we can apply merging techniques like concatenation, adding, or multiplying. This has the effect of aggregating the combined observations from all the input heads. The aggregation can then be processed through a series of fully connected layers into an output. This generalized blueprint structure for a multimodal model is visualized in Figure 4-57.

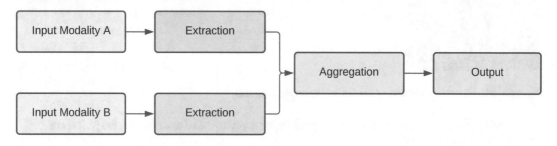

Figure 4-57. *General structure of a multimodal model*

In this sort of network design, the different data modalities are first independently processed to extract relevant features and express the information in vector representation; afterward, these representations are combined and jointly considered to produce an output. Following this, we'll show how we can build more advanced network topologies to capture more complex knowledge flows.

Multimodal models are a powerful extension of convolutional neural network techniques.

We will be building a multimodal model to predict house prices given both tabular and image data. This will be done using the SoCal House Prices and Images dataset on Kaggle, assembled and maintained by Kaggle user ted8080.[6] The dataset comprises a .csv file storing tabular data; each row represents a house with its street, the city it is located in, the number of bedrooms it has, the number of baths it has, the square footage, the house price, and an image identifier. Each image identifier is associated with an image in the image directory; the image corresponding to the row with image ID 0 is titled "0.jpg," the image corresponding to image ID 123 is "123.jpg," and so on (Figure 4-58).

---

[6]`www.kaggle.com/datasets/ted8080/house-prices-and-images-socal`. The dataset was used with permission by the user.

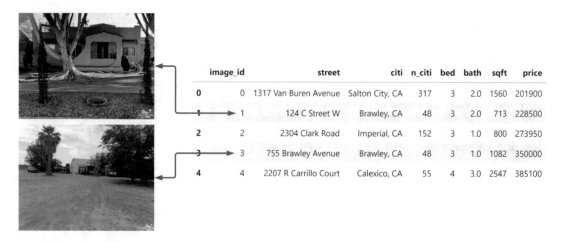

	image_id	street	citi	n_citi	bed	bath	sqft	price
0	0	1317 Van Buren Avenue	Salton City, CA	317	3	2.0	1560	201900
1	1	124 C Street W	Brawley, CA	48	3	2.0	713	228500
2	2	2304 Clark Road	Imperial, CA	152	3	1.0	800	273950
3	3	755 Brawley Avenue	Brawley, CA	48	3	1.0	1082	350000
4	4	2207 R Carrillo Court	Calexico, CA	55	4	3.0	2547	385100

Figure 4-58. *Visual exploration of the SoCal House Prices dataset*

We will begin from a baseline (i.e., poorly performing approach) and demonstrate the process of moving toward a better model.

In a baseline approach, we may consider the following multimodal model arrangement: we feed the image identified by the image_id column into the image head, feed the ['n_citi', 'bed', 'bath', 'sqft'] columns into the tabular head, and use the 'price' column as the desired target value.

In order to manage the data, we will use a TensorFlow Sequence dataset. Refer back to Chapter 2 for introductory information on how TensorFlow Sequence datasets are constructed and used by the model. Simply put, it allows us to define a data flow any way we like – as long as we give it the inputs and the desired targets when the model requests them. This gives us a lot of flexibility and makes life generally easier: we don't need to chase down TensorFlow exceptions or warnings trying to load the entire dataset up front into a TensorFlow dataset. This comes at the cost of potential slight inefficiency (caused by reloading the data when it is requested), but it's a decision we'll make in this case for ease of implementation.

Our multimodal dataset will contain three sets of internal data: an array of image IDs, a DataFrame containing relevant features, and an array of target house prices. We will store training and validation indices that indicate which indices in each of these internal datasets correspond to the training set and which correspond to the validation set. When the model requests data by making a .__getitem__(index) call, we execute the following steps:

1. Identify a batch interval of indices from the training set to use.

2. Load the images using IDs from the selected indices.

3. Obtain the tabular features from the selected indices.

4. Obtain the target from the selected indices.

5. Bundle the image and tabular feature inputs together into a list.

6. Return the bundled inputs and the target.

The Keras model will then read, process, and utilize the given dataset (Figure 4-59).

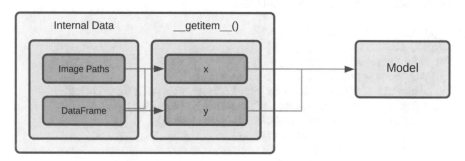

Figure 4-59. *Structure of our TensorFlow Sequence multimodal dataset*

Although we are using this dataset for our specific multimodal house price modeling task, it can be used for any dataset in which there are associated image-tabular pairs used to predict a regression or classification output. One implementation of this dataset is presented in Listing 4-44.

Listing 4-44. Implementing our custom TensorFlow Sequence dataset to handle multimodal data flows

```python
class MultiModalData(tf.keras.utils.Sequence):

    def __init__(self,
                 imageCol, targetCol,
                 tabularFeatures, oneHotFeatures,
                 imageDir, csvDir,
                 batchSize = 8, train_size = 0.8,
                 targetScale = 1000):

        self.batchSize = batchSize
        self.imageDir = imageDir

        df = pd.read_csv(csvDir)
        self.imagePaths = df[imageCol]
        self.targetCol = df[targetCol] / targetScale
        self.tabular = df.drop([imageCol, targetCol],
                               axis = 1)[tabularFeatures]
        for feature in oneHotFeatures:
            self.tabular = self.tabular.join(pd.get_dummies(self.tabular[feature]))
            self.tabular.drop(feature, axis=1, inplace=True)

        self.dataSize = len(df)
        self.trainSize = round(self.dataSize * train_size)

        dataIndices = np.array(df.index)
        self.trainInd = np.random.choice(dataIndices,
                                         size = self.trainSize)
        self.validInd = np.array([i for i in dataIndices if i not in self.trainInd])
```

```
def __len__(self):
    return self.trainSize // self.batchSize

def __getitem__(self, index):
    images, tabulars, y = [], [], []
    for i in range(self.batchSize):
        currIndex = index * self.batchSize + i
        imagePath = f'{self.imageDir}/{self.imagePaths[currIndex]}.jpg'
        image = cv2.resize(cv2.imread(imagePath), (400, 400))
        tabular = np.array(self.tabular.loc[currIndex])
        target = self.targetCol[currIndex]
        images.append(image)
        tabulars.append(tabular)
        y.append(target)
    return [np.stack(images), np.stack(tabulars)], np.stack(y)
```

Note that the dataset takes in a parameter target_scale that determines what constant to divide the target by. We do this to reduce the scale of the output from individual dollars to thousands of dollars. Although neural networks can theoretically work with outputs of any scale, typically regression targets should be kept close to 0 for faster training, especially in which the scaling process does not eliminate any crucial precision. We don't expect our model to accurately model house prices down to the dollar because many other factors not captured by the model inputs influence the final pricing of a house property. In this case, such a scaling is reasonable.

Listing 4-45 demonstrates the instantiation of the multimodal dataset object, which provides custom parameters for our house prices dataset.

Listing 4-45. Instantiating the multimodal dataset with relevant information from the dataset

```
data = MultiModalData(imageCol = 'image_id',
                      targetCol = 'price',
                      tabularFeatures = ['n_citi', 'bed', 'bath', 'sqft'],
                      oneHotFeatures = ['n_citi'],
                      imageDir = '../input/house-prices-and-images-socal/socal2/socal_pics',
                      csvDir = '../input/house-prices-and-images-socal/socal2.csv',)
```

To verify that we have implemented data feeding process correctly, we can call x, y = data.__getitem__(0) to test the data feeding pattern. Recall that x is a two-element list bundling the image and tabular inputs together; we have x[0].shape as (8, 400, 400, 3) and x[1].shape as (8, 418). The target shape y.shape is (8,). Our dataset behaves as expected.

Now, we can design the neural network. At a high level, the model must have two heads to take in the two modalities – image and tabular – with independent processing, followed by a merger and joint processing into a single-node output with a rectified linear unit output.

■ **Note** For regression problems, it is generally convention to use a linear output activation, but in this particular domain no target value will be negative. Therefore, using a rectified linear unit is functionally the same, with the additional benefit of imposing a reasonable bound. If you wanted, you could also use a rectified linear unit with both the standard lower bound at $y = 0$ and at $y = \alpha$, where α is some value set using domain knowledge representing the maximum possible output. This can be implemented by defining a custom ReLU object: `crelu = lambda x: keras.backend.relu(x, max_value = alpha)`. Another alternative choice is to use a sigmoid multiplied by α (and perhaps horizontally stretched $\sigma(\alpha x)$ as opposed to the standard sigmoid function $\sigma(x)$) such that the upper bound is $y = \alpha$.

Note that the tabular dataset does not contain many features – a few dozen and less than a dozen if you considered all the features created by the one-hot expansion of the `n_citi` column as only one feature. Thus, a few fully connected layers with a small number of nodes per layer should suffice. In this particular implementation, we apply three dense layers with 16 nodes each.

The convolutional head, on the other hand, requires more intensive processing. We can build a repeating nonlinear block-style design, in which different branches process the input with convolutions of varying kernel sizes, followed by a merging and a pooling. As the spatial dimensionality is reduced with max pooling, we increase the number of filters. After several iterations of this simple nonlinear topology, we flatten the output and use three dense layers to squeeze the extracted feature maps into a 16-element vector. The 16-element vectors from the image input and the tabular input are then merged via concatenation and further processed into a single one-node prediction output. The full architecture is implemented in Listing 4-46 and visualized in Figure 4-60. Note that this architecture follows the standard components of a multimodal model in Figure 4-59 (refer to the previous figure).

Listing 4-46. Defining a custom two-head architecture to process our multimodal data

```
imgInput = L.Input((400, 400, 3))
x = L.Conv2D(8, (3, 3), activation='relu', padding='valid')(imgInput)
for filters in [8, 8, 16, 16, 32, 32]:
    x1 = L.Conv2D(filters, (1, 1), activation='relu', padding='same')(x)
    x2 = L.Conv2D(filters, (3, 3), activation='relu', padding='same')(x)
    x3a = L.Conv2D(filters, (5, 5), activation='relu', padding='same')(x)
    x3b = L.Conv2D(filters, (3, 3), activation='relu', padding='same')(x)
    add = L.Add()([x1, x2, x3b])
    x = L.MaxPooling2D((2, 2))(add)
flatten = L.Flatten()(x)
imgDense1 = L.Dense(64, activation='relu')(flatten)
imgDense2 = L.Dense(32, activation='relu')(imgDense1)
imgOut = L.Dense(16, activation='relu')(imgDense2)

tabularInput = L.Input((418,))
tabDense1 = L.Dense(16, activation='relu')(tabularInput)
tabDense2 = L.Dense(16, activation='relu')(tabDense1)
tabOut = L.Dense(16, activation='relu')(tabDense2)

concat = L.Concatenate()([imgOut, tabOut])
dense1 = L.Dense(16, activation='relu')(concat)
dense2 = L.Dense(16, activation='relu')(dense1)
out = L.Dense(1, activation='relu')(dense2)

model = keras.models.Model(inputs = [imgInput, tabularInput],
                           outputs = out)
```

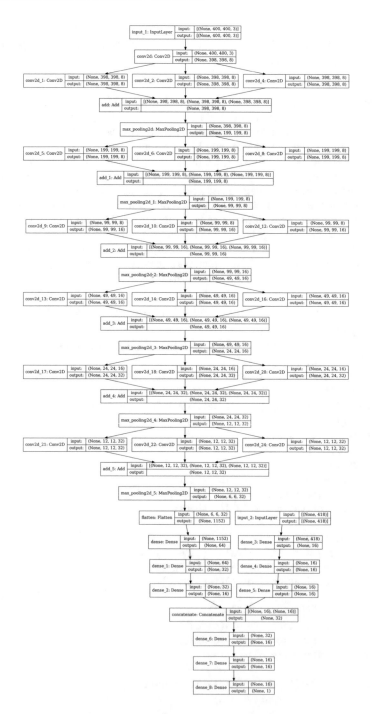

Figure 4-60. *Keras visualization of the custom multimodal architecture built in Listing 4-46*

This particular custom model has 123,937 parameters – which isn't bad. As one may expect, the bulk of these parameters come from processing the high-dimensionality image input. We can compile and fit the model with standard hyperparameters (Listing 4-47, Figure 4-61).

Listing 4-47. Compiling and fitting our custom two-head architecture.

```
model.compile(optimizer='adam',
              loss='mse',
              metrics=['mae'])

history = model.fit(data, epochs = 100)

plt.figure(figsize=(10, 5), dpi=400)
plt.plot(history.history['loss'], color='red', label='Train')
plt.xlabel('Epochs')
plt.ylabel('Loss')
plt.legend()
plt.grid()
plt.show()
```

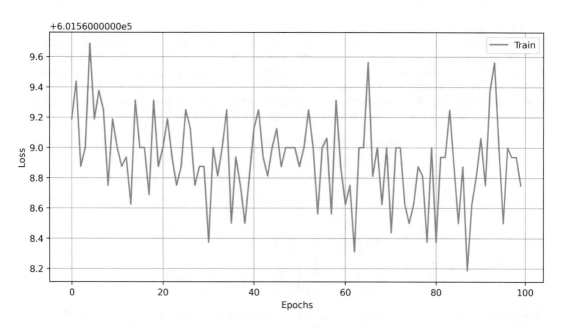

Figure 4-61. *Training history of our first multimodal modeling attempt*

The model performance is very poor. Throughout the 100 epochs of training, the model has improved the training loss only marginally and exhibited incredibly volatile behavior. This volatility and lack of any progress are a sign that our model is simply not equipped to solve the problem at all.

We can make two improvements to our model:

- *Use a better image processing component.* Earlier in this chapter, we discussed various "base model" successful architectures that are easily accessible via Keras applications or other platforms. Researchers have spent a tremendous amount of effort to develop a wide range of successful model architectures – instead of attempting to build our own, we can simply use a "prebuilt" architecture. We will replace our custom convolutional component with an EfficientNet model, which has proven a generally robust, high-performing architecture across a wide domain of problems.

- *Explicitly structure the desired data interpretation.* Despite theoretical universal approximation properties, neural networks cannot successfully model anything thrown at them in practice. Sometimes, we can nudge the model in the right direction by building an interpretation of the data or part of the data directly into the design of the model or the system. In this case, we observe that the n_citi feature contains incredibly valuable data – the city a house is in is probably one of the most reliable predictors of house price – but the standard one-hot expansion form may not be conducive to meaningfully utilizing this data. We can use the *embedding* mechanism to more explicitly communicate how to interpret and make use of the city feature data.

Recall the embedding mechanism discussed in Chapter 3. An embedding takes in a certain token – for instance, a specific word or symbol – and maps it to a set of learned features or attributes. This is treated like a standard parametrized layer (the features associated with each token are optimized) (Figure 4-62). We can use this logic to reflect how we want the neural network to interpret and make use of the city data: for each city, we want the embedding to learn a particular set of optimally useful attributes. Although we don't care how the network derives the associated features or what they represent, we do want the network to interpret the city feature that way; using the embedding mechanism accomplishes our goal.

Token	F1	F2	...	F3
1	0.63	2.93	...	-3.45
2	0.98	-5.47	...	-2.69
3	1.23	4.32	...	-1.90
...	...	...	...	...
n	-0.02	0.03	...	0.52

Figure 4-62. *Visualization of embedding. Note that when implementing embeddings in Keras, the first token should start from 0 – you may receive errors if otherwise*

Let's begin by rewriting the dataset class for our second attempt at modeling. Because we want to use embeddings to process the specific city that a house is in, we need *three heads*: an image head that accepts an image input, a tabular head that accepts a tabular input, and an embedding head that accepts a single integer representing the city of the sample (i.e., the n_citi column). As such, our dataset needs to bundle together three inputs (Figure 4-63).

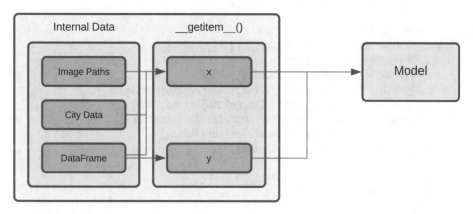

Figure 4-63. *Structure of our updated multimodal dataset*

The modified dataset is implemented in Listing 4-48.

Listing 4-48. Implementing an updated TensorFlow Sequence dataset to separate the embedding data from the tabular dataset component

```
class MultiModalData(tf.keras.utils.Sequence):

    def __init__(self,
                 imageCol, targetCol,
                 tabularFeatures, embeddingFeature,
                 imageDir, csvDir,
                 batchSize = 8, train_size = 0.8,
                 targetScale = 1000):

        self.batchSize = batchSize
        self.imageDir = imageDir

        df = pd.read_csv(csvDir)
        self.imagePaths = df[imageCol]
        self.targetCol = df[targetCol] / targetScale
        self.tabular = df.drop([imageCol, targetCol],
                               axis = 1)[tabularFeatures]
        self.onehotData = self.tabular[embeddingFeature]
        self.tabular.drop(embeddingFeature, axis=1, inplace=True)

        self.dataSize = len(df)
        self.trainSize = round(self.dataSize * train_size)

        dataIndices = np.array(df.index)
        self.trainInd = np.random.choice(dataIndices,
                                         size = self.trainSize)
        self.validInd = np.array([i for i in dataIndices if i not in self.trainInd])
```

```
    def __len__(self):
        return self.trainSize // self.batchSize

    def __getitem__(self, index):
        images, embeddingInps, tabulars, y = [], [], [], []
        for i in range(self.batchSize):
            currIndex = index * self.batchSize + i
            imagePath = f'{self.imageDir}/{self.imagePaths[currIndex]}.jpg'
            image = cv2.resize(cv2.imread(imagePath), (400, 400))
            embeddingInp = np.array(self.onehotData.loc[currIndex])
            tabular = np.array(self.tabular.loc[currIndex])
            target = self.targetCol[currIndex]
            images.append(image)
            embeddingInps.append(embeddingInp)
            tabulars.append(tabular)
            y.append(target)
        return [np.stack(images), np.stack(embeddingInps), np.stack(tabulars)], np.stack(y)
```

We also need to adjust our model (Listing 4-49). Firstly, instead of building a custom convolutional (sub)network to process the image head, we can use the EfficientNetB1 model, which is the second smallest architecture in a family of eight. Using the functional API, models in Keras can be treated like layers; to "link" them with other layers, we use the syntax after_layer = buildModel(params)(prev_layer). Secondly, we construct an additional embedding head that takes in a single integer representing one of 415 cities (here, 415 is the "vocabulary size") and maps it to a vector with eight elements. The result of embedding is concatenated with the results of image and tabular input processing. Then the result is passed through several dense layers into a single-node regression output.

Listing 4-49. Defining a novel multimodal architecture, using the EfficientNetB1 model architecture to process the image component

```
from keras.applications import EfficientNetB1

imgInput = L.Input((400, 400, 3))
effnet = EfficientNetB1(input_shape = (400, 400, 3),
                        weights = None,
                        classes = 16)(imgInput)
imgOut = L.Dense(16, activation='relu')(effnet)

embeddingInput = L.Input((1,))
embedding = L.Embedding(415, 8)(embeddingInput)
reshape = L.Reshape((8,))(embedding)

tabularInput = L.Input((3,))
tabDense1 = L.Dense(4, activation='relu')(tabularInput)
tabDense2 = L.Dense(4, activation='relu')(tabDense1)
tabOut = L.Dense(4, activation='relu')(tabDense2)

concat = L.Concatenate()([imgOut, tabOut, reshape])
dense1 = L.Dense(16, activation='relu')(concat)
dense2 = L.Dense(16, activation='relu')(dense1)
out = L.Dense(1, activation='relu')(dense2)

model = keras.models.Model(inputs = [imgInput, embeddingInput, tabularInput], outputs = out)
```

336

Our three-head model is visualized in Figure 4-64.

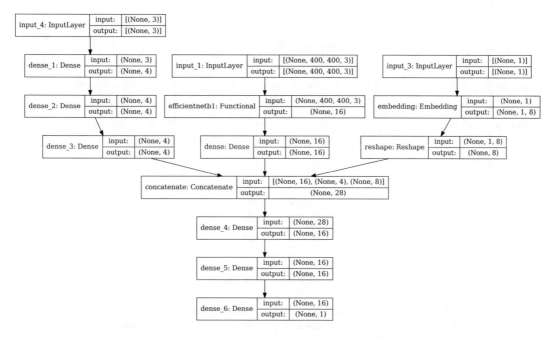

Figure 4-64. *Keras's visual representation of our redesigned three-head network*

This model has 6,538,081 trainable parameters, the majority of these parameters coming from the EfficientNetB1 model. When training this updated model on the updated dataset (Figure 4-65), we obtain significantly improved performance. After almost 60 epochs of training, the model converges to a Mean Squared Error of 83584 and a Mean Absolute Error of 192 – a significant improvement over the previous model's performance. Recall that our target is in units of thousands of dollars, which means that the model is, on average, $192k off on the house price estimation. While this is certainly not incredible performance, it is understandable given the many other house pricing factors not included in this multimodal model.

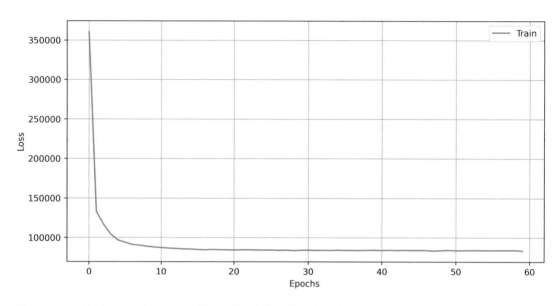

Figure 4-65. *Training performance of the updated three-head model*

If we replace the EfficientNetB1 model with the larger and more powerful EfficientNetB3 model (Listing 4-50), we can obtain even better performance (Figure 4-66). The expanded model has 10,725,225 parameters and converges to a 38739.26 MSE and 124.4208 MAE.

Listing 4-50. Defining a novel multimodal architecture, using the EfficientNetB3 model architecture to process the image component

```
imgInput = L.Input((400, 400, 3))
effnet = EfficientNetB3(input_shape = (400, 400, 3),
                        weights = None,
                        classes = 16)(imgInput)
imgOut = L.Dense(16, activation='relu')(effnet)

embeddingInput = L.Input((1,))
embedding = L.Embedding(415, 8)(embeddingInput)
reshape = L.Reshape((8,))(embedding)

tabularInput = L.Input((3,))
tabDense1 = L.Dense(4, activation='relu')(tabularInput)
tabDense2 = L.Dense(4, activation='relu')(tabDense1)
tabOut = L.Dense(4, activation='relu')(tabDense2)

concat = L.Concatenate()([imgOut, tabOut, reshape])
dense1 = L.Dense(16, activation='relu')(concat)
dense2 = L.Dense(16, activation='relu')(dense1)
out = L.Dense(1, activation='relu')(dense2)

model = keras.models.Model(inputs = [imgInput, embeddingInput, tabularInput],
                           outputs = out)
```

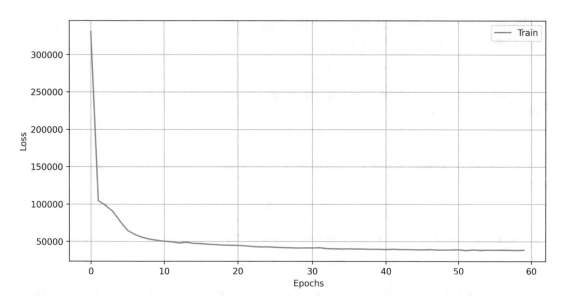

Figure 4-66. *Training history for the three-head model using EfficientNetB3 instead of EfficientNetB1*

Still, there are many areas for system improvement, like optimizing the exact nonlinear architecture, the dimensionality of the input, the image processing architecture (e.g., trying out NASNet, ResNet, or another design), the training meta-parameters, and so on. Improvement of this model is left as an open exercise for you.

1D Convolutions for Tabular Data

We've done a lot of work setting up the usage of convolutional neural networks for image data, both for pure image datasets and multimodal datasets with images and an associated tabular/structured data component. Now, we'll demonstrate the usage of convolutions directly on tabular datasets.

The most "natural" application of convolutions to tabular data is a *one-dimensional convolution*. Images have two spatial dimensions (ignoring the depth/color channel), so we can apply two-dimensional convolutions across them. Standard tabular data, on the other hand, has one "spatial dimension," so we can apply a one-dimensional convolution.

One-dimensional convolutions operate by the same logic as two-dimensional convolutions, except with a one-dimensional kernel that slides along the axis rather than a two-dimensional kernel sliding along two spatial axes.

Consider a hypothetical kernel k with length 3:

$$k = \begin{bmatrix} 1, & 3, & -1 \end{bmatrix}$$

We will attempt to apply k to a "single row of data," or a one-dimensional matrix/vector i:

$$i = \begin{bmatrix} 1, & 5, & 6, & 3, & 9, & 2, & 3, & 8, & 20, & 3 \end{bmatrix}$$

The feature i contains ten features; we can fit *feature size – kernel size* + 1 = 10 – 3 + 1 = 8 convolutions, meaning the resulting convolved feature R contains eight elements:

$$R = \begin{bmatrix} ?, ?, ?, ?, ?, ?, ?, ? \end{bmatrix}$$

To determine the value of the first element of R, we apply the convolution kernel k to the first set of contiguous values along the feature i (bolded):

$$i = \begin{bmatrix} \mathbf{1}, & \mathbf{5}, & \mathbf{6}, & 3, & 9, & 2, & 3, & 8, & 20, & 3 \end{bmatrix}$$

Applying the kernel to these three elements yields the output $1 \cdot 1 + 3 \cdot 5 + (-1) \cdot 6 = 1 + 15 - 6 = 10$. The first value of R is thus 10:

$$R = \begin{bmatrix} 10, & ?, & ?, & ?, & ?, & ?, & ?, & ? \end{bmatrix}$$

We can find the second value of R by applying the same kernel to the second set of contiguous values along the feature (bolded):

$$i = \begin{bmatrix} 1, & \mathbf{5}, & \mathbf{6}, & \mathbf{3}, & 9, & 2, & 3, & 8, & 20, & 3 \end{bmatrix}$$

This yields $1 \cdot 5 + 3 \cdot 6 + (-1) \cdot 3 = 5 + 18 - 3 = 20$. The second value of R is thus 20:

$$R = \begin{bmatrix} 10, & 20, & ?, & ?, & ?, & ?, & ?, & ? \end{bmatrix}$$

The remainder of the convolved matrix can be filled out by continuing this procedure. One-dimensional convolutions serve a similar purpose to two-dimensional convolutions: they can act as "filters" that amplify or dim the presence of certain attributes or characteristics in the data, depending on the kernel values.

Neural networks can learn the optimal kernel values for a one-dimensional convolutional neural network for a certain task.

To demonstrate this, let's consider a sequence modeling task: given sequentially sampled points $[f(x_1), f(x_2), ..., f(x_n)]$ where $x_n - x_{n-1} = x_{n-1} - x_{n-2}$ (i.e., the inputs to the function are equally spaced) from a noisy function f, the one-dimensional convolutional neural network must classify f as either being a linear, quadratic, or cyclic function.

Listing 4-51 generates such a dataset, given numElements (the value of n, determining how many points are sampled from f) and numTriSamples (the number of times a sample for each of the three classes is generated). baseRange represents the set x; it is equally spaced from −5 to 5 with numElements elements. Each time we generate a linear, quadratic, or cyclic function, we choose random parameters (for instance, $\{a, b, c, d\}$ in $a \, sin \, sin \, (bx - c) + d$ for a cyclic function). The limits of the uniform random distribution have been chosen such that the functions generally occupy the same rectangular region such that the class of a function ideally cannot be predicted by how high it is.

Listing 4-51. Generating our custom function identification synthetic dataset

```
numElements = 400
numTriSamples = 2000

x, y = [], []
baseRange = np.linspace(-5, 5, numElements)

for i in range(numTriSamples):

    # get random linear sample
    slope = np.random.uniform(-3, 3)
    intercept = np.random.uniform(-10, 10)
    x.append(baseRange * slope + intercept)
    y.append(0)
```

```
# get random quadratic sample
a = np.random.choice([np.random.uniform(0.2, 1),
                      np.random.uniform(-0.2, -1)])
b = np.random.uniform(-1, 1)
c = np.random.uniform(-1, 1)
x.append(a * baseRange**2 + b * baseRange + c)
y.append(1)

# get random sinusoidal sample
a = np.random.uniform(1, 10)
b = np.random.uniform(1, 3)
c = np.random.uniform(-np.pi, np.pi)
d = np.random.uniform(-20, 20)
x.append(a * np.sin(b * baseRange - c) + d)
y.append(2)

x = np.array(x)
x += np.random.normal(loc = 0, scale = 1,
                      size = x.shape)
y = np.array(y)
```

Additionally, note that we add random noise to make the problem more interesting. In this case, we add normally distributed noise with mean 0 and standard deviation 1 to x.

We need to split the dataset into training and validation datasets to evaluate genuine model performance (Listing 4-52).

Listing 4-52. Splitting the dataset into training and validation sets

```
import sklearn
from sklearn.model_selection import train_test_split as tts
X_train, X_val, y_train, y_val = tts(x, y, train_size = 0.8)
```

Figure 4-67 displays three samples from each class. While there is moderate noise, the overall trajectory of each function is clearly identifiable.

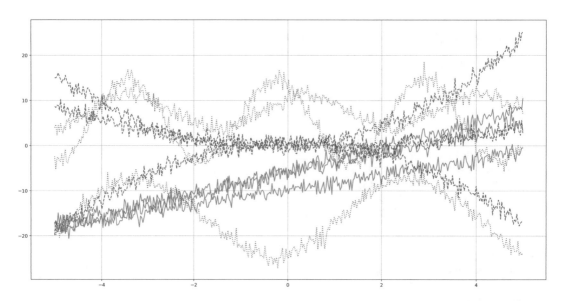

Figure 4-67. *Three sampled functions from each class in our custom dataset – linear, quadratic, and cyclic*

We can construct a simple model that employs one-dimensional convolutions and one-dimensional pooling layers to process the inputs (Listing 4-53). In Keras, one-dimensional convolutions are instantiated via L.Conv1D(...). Note that while the input is a single vector with shape (a), we need to reshape it into shape $(a, 1)$ to apply one-dimensional convolutions for the same reason that a grayscale image of shape (a, b) needs to be reshaped into shape $(a, b, 1)$ to be processed using two-dimensional convolutions. After three iterations of a convolution-convolution-pooling block, the input is flattened back to one spatial dimension and processed with two more full connected layers to an output of shape 3 (such that each class is associated with a probability). Because this is a multiclass problem in which all output probabilities should sum to 1, we use a softmax activation output.

Listing 4-53. Constructing a 1D CNN model architecture for our function identification synthetic dataset

```
model = Sequential()
model.add(L.Input(numElements))
model.add(L.Reshape((numElements, 1)))
for i in range(3):
    model.add(L.Conv1D(8, 3, padding='same',
                       activation='relu'))
    model.add(L.Conv1D(8, 3, padding='same',
                       activation='relu'))
    model.add(L.MaxPooling1D(2))
model.add(L.Flatten())
model.add(L.Dense(16, activation='relu'))
model.add(L.Dense(3, activation='softmax'))
```

We can compile and train the model using standard meta-parameters (Listing 4-54, Figure 4-68).

Listing 4-54. Compiling, training, and plotting performance history

```
model.compile(loss='sparse_categorical_crossentropy',
              optimizer='adam',
              metrics=['accuracy'])
history = model.fit(X_train, y_train,
                    epochs = 20,
                    validation_data = (X_val, y_val))

plt.figure(figsize=(10, 5), dpi=400)
plt.plot(history.history['loss'], color='red', label='Train')
plt.plot(history.history['val_loss'], color='blue', label='Validation')
plt.xlabel('Epochs')
plt.ylabel('Loss')
plt.legend()
plt.grid()
plt.show()

plt.figure(figsize=(10, 5), dpi=400)
plt.plot(history.history['accuracy'], color='red', label='Train')
plt.plot(history.history['val_accuracy'], color='blue', label='Validation')
plt.xlabel('Epochs')
plt.ylabel('Accuracy')
plt.legend()
plt.grid()
plt.show()
```

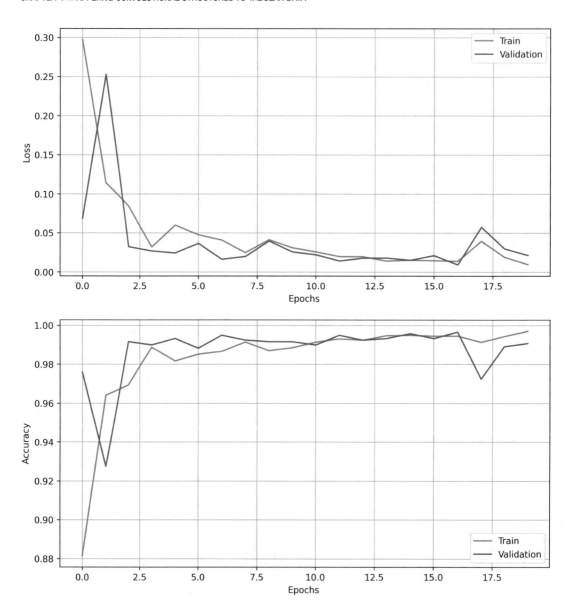

Figure 4-68. *Loss and accuracy of the 1D CNN model's performance on the function identification task*

The model obtains good performance very quickly: 0.0098 training loss, 0.9971 training accuracy, 0.0215 validation loss, and 0.9908 validation accuracy. Let's make the problem more difficult by increasing the standard deviation of noise from 1 to 2. Now, the dataset looks like the following (Figure 4-69).

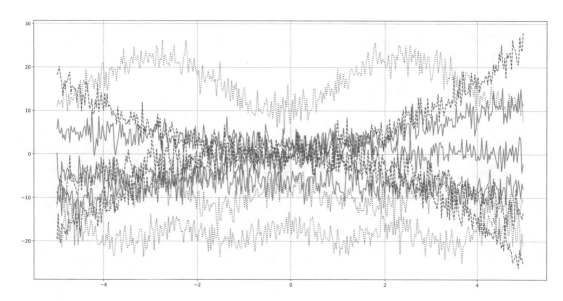

Figure 4-69. *Three sampled functions from each class in our custom dataset, with an increased noise standard deviation of 2*

After two epochs, the model reaches 0.0140 validation loss and 0.9958 validation accuracy – functionally equal to (and, in fact, slightly better than) the performance on a dataset with noise drawn from a distribution with standard deviation 1.

Let's increase the noise even more, to a standard deviation of 3 (Figure 4-70). The neural network obtains 0.0934 validation loss and 0.9717 validation accuracy, which is a relative significant decrease in performance – but still isn't bad.

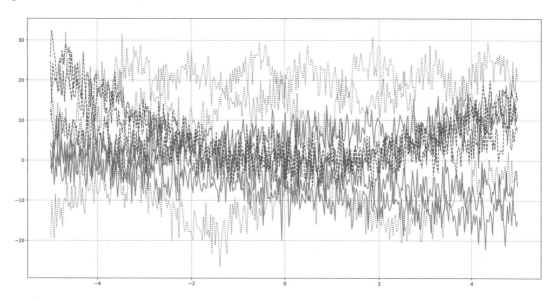

Figure 4-70. *Three sampled functions from each class in our custom dataset, with an increased noise standard deviation of 3*

When we increase the noise standard deviation to 10, the dataset becomes quite tricky to separate (Figure 4-71) – but the model still does reasonably well (Figure 4-72), scoring a validation loss of 0.2230 and a validation accuracy of 0.9300.

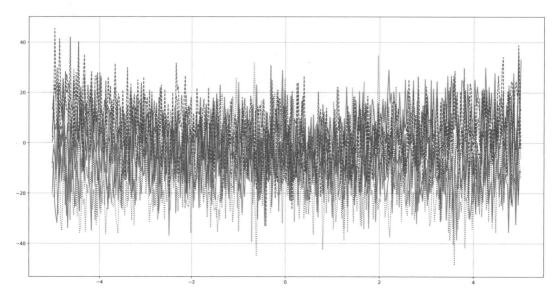

Figure 4-71. *Three sampled functions from each class in our custom dataset, with an increased noise standard deviation of 10*

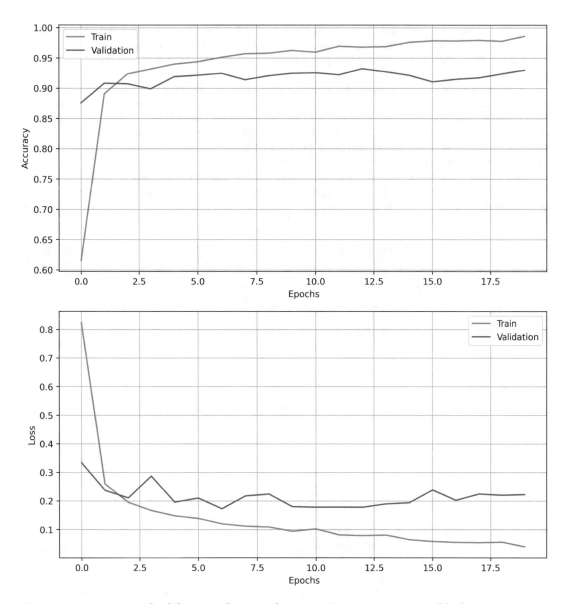

Figure 4-72. *Accuracy and validation performance for a 1D CNN on a noisy version of the function identification synthetic dataset*

The main idea of these experiments is this: we were able to build a robust and powerful signal processing model using a simple neural network comprising of one-dimensional convolutions and pooling operations. To further improve the network's modeling power, you could add more sophisticated convolutional structures, like residual connections and other topological nonlinearities.

Note that you can use one-dimensional convolutions to process data like audio signals or time series in their raw or natural form. For instance, consider one-dimensional convolutions to perform speaker diarization – categorizing which speaker is speaking at some point in a time series. We can do this by processing an input signal with one-dimensional convolutions, "flattening" into one vector, and using dense layers to form the learned features into an output vector (a very similar pattern to two-dimensional

347

CNNs). If you have a dataset in which signals are associated with tabular data, you can build multimodal models as discussed in the previous section, but with one-dimensional convolutional heads rather than two-dimensional ones. (We'll see an example of applying convolutions, in addition to recurrent layers, to model audio signals in the next chapter.)

However, we're unlikely to get success by directly applying a one-dimensional convolutional neural network to *tabular data*. Our example task and all natural applications of one-dimensional CNNs like audio and time series contain an essential attribute: they are ordered along a sequential axis. That is, there is a clear relationship between x_i and x_{i+1}. Tabular datasets generally contain independent features that are not linked to each other in order. There is no reason one column should necessarily be placed "before," "after," or "next to" another – these relational concepts don't apply to tabular data.

We can give this attribute a name: *contiguous semantics*. *Contiguous* means "next to" or "adjacent" (e.g. a contiguous block of memory), and *semantics* refers to the "meaning" or "concept" implied by a symbolic or syntactic representation. If a feature or dataset possesses contiguous semantics, then it makes sense to apply one-dimensional convolutions directly on the dataset (Figure 4-73).

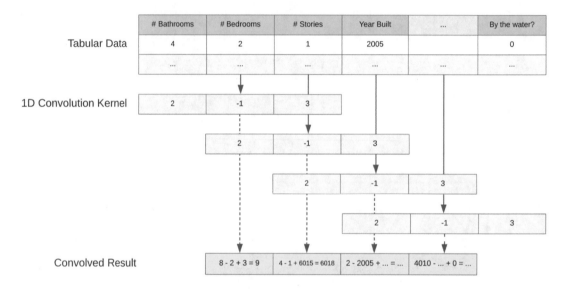

Figure 4-73. *Example demonstration of applying an operation requiring contiguous semantics (a convolution) to a dataset that does not possess contiguous semantics and the bizarre result we get by trying to do so*

We can attempt to get around this problem using a clever trick: *soft ordering*. Even though each sample *x* contains unordered columns, we can imagine that some *ordered representation q* exists containing the same information as *x*; that is, it "rearranges" and "molds" the semantics (information content, meaning) of the raw unordered feature into a contiguous semantics. This is not necessarily hard to believe: we deform and reform data spaces into possessing new attributes often in machine learning (e.g., different distances, different dimensionalities, different modalities, different statistical properties) while retaining their "information content."

It's difficult for us to imagine or design a transformation from *x* to *q*, but if *q* exists – which we argue it does – then the mapping $x \rightarrow q$ should exist too. We can task a neural network, which has been proven to be a theoretical "universal function approximator," to learn this mapping for us by extracting and rearranging our tabular data in an ordered sequence optimally readable by a one-dimensional convolution (Figure 4-74).

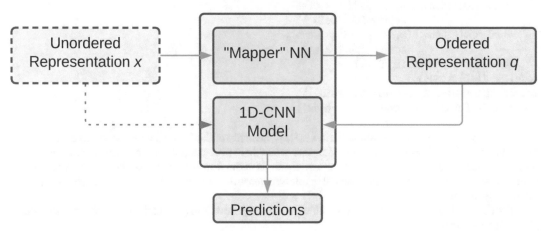

Figure 4-74. *Blueprint for applying 1D CNNs to tabular data using soft ordering*

A simple approach is to introduce a fully connected component, or an *encoding component, before* convolutions. The hope is that the fully connected component learns the mapping from raw input data x to optimal ordered representation q, which is then further processed in convolutional fashion by the remainder of the network.

Thus, in practice, the "mapper NN" and the 1D CNN model can be merged into one single neural network process, in which the output of the mapping component directly feeds into the input of the 1D CNN model.

This concept emerged from a Kaggle competition, popularized by the user "tmp" for using this approach in the second place solution for the Harvard Laboratory for Innovation Science Mechanisms of Action Prediction competition. As tmp writes on a forum post[7] outlining their solution

> Using such a structure in tabular data is based on the idea that [a] CNN structure
> performs well in feature extraction, but it is rarely used in tabular data because
> the correct features ordering is unknown. A simple idea is to reshape the data
> directly into a multi-channel image format, and the correct sorting can be
> learned by using FC layer through back-propagation.

Let's take the architecture used for our previous function classification task and slightly modify it to include a soft ordering component (Listing 4-55). A simple example of such a component would be a series of dense layers.

Listing 4-55. Defining a 1D CNN model with a fully connected soft ordering component

```
model = Sequential()
model.add(L.Input(numElements))
model.add(L.Dense(numElements, activation='relu'))
model.add(L.Dense(numElements, activation='relu'))
model.add(L.Dense(numElements, activation='relu'))
model.add(L.Reshape((numElements, 1)))
```

---

[7] Available here: www.kaggle.com/competitions/lish-moa/discussion/202256#1106810.

```
for i in range(5):
    model.add(L.Conv1D(8, 3, padding='same',
                        activation='relu'))
    model.add(L.Conv1D(8, 3, padding='same',
                        activation='relu'))
    model.add(L.MaxPooling1D(2))
model.add(L.Flatten())
model.add(L.Dense(16, activation='relu'))
model.add(L.Dense(3, activation='softmax'))
```

Let's apply this on a sample tabular dataset. The University of California Irvine Forest Cover dataset – one of many well-established benchmark datasets available in the University of California Irvine dataset repository – contains tree observations from several areas of the Roosevelt National Forest. This dataset contains several dozen features and over half a million measurements, making it a good dataset to apply neural networks to.

The dataset features are shown in the following. The objective is to predict the cover type of the region given measurements:

```
['Elevation', 'Aspect', 'Slope', 'Horizontal_Distance_To_Hydrology',
 'Vertical_Distance_To_Hydrology', 'Horizontal_Distance_To_Roadways',
 'Hillshade_9am', 'Hillshade_Noon', 'Hillshade_3pm',
 'Horizontal_Distance_To_Fire_Points', 'Wilderness_Area1', ... ,
 'Wilderness_Area4',
 'Soil_Type1', 'Soil_Type2', ... 'Soil_Type39', 'Soil_Type40',
 'Cover_Type'])
```

When we instantiate the previously discussed architecture on this dataset by customizing the number of input nodes to match the number of features in the dataset, we obtain the following model (Listing 4-56).

Listing 4-56. Layers and parameter summary of a 1D soft ordering CNN

```
Model: "sequential"
```

Layer (type)	Output Shape	Param #
dense (Dense)	(None, 108)	5940
dense_1 (Dense)	(None, 864)	94176
reshape (Reshape)	(None, 54, 16)	0
conv1d (Conv1D)	(None, 54, 16)	784
conv1d_1 (Conv1D)	(None, 54, 16)	784
average_pooling1d (AveragePo	(None, 27, 16)	0
conv1d_2 (Conv1D)	(None, 27, 16)	784
conv1d_3 (Conv1D)	(None, 27, 16)	784

```
average_pooling1d_1 (Average (None, 13, 16)          0
_____
conv1d_4 (Conv1D)            (None, 13, 16)          784
_____
conv1d_5 (Conv1D)            (None, 13, 16)          784
_____
average_pooling1d_2 (Average (None, 6, 16)           0
_____
conv1d_6 (Conv1D)            (None, 6, 16)           784
_____
conv1d_7 (Conv1D)            (None, 6, 16)           784
_____
average_pooling1d_3 (Average (None, 3, 16)           0
_____
conv1d_8 (Conv1D)            (None, 3, 16)           784
_____
conv1d_9 (Conv1D)            (None, 3, 16)           784
_____
average_pooling1d_4 (Average (None, 1, 16)           0
_____
flatten (Flatten)            (None, 16)              0
_____
dense_2 (Dense)              (None, 16)              272
_____
dense_3 (Dense)              (None, 16)              272
_____
dense_4 (Dense)              (None, 7)               119
=================================================================
Total params: 108,619
Trainable params: 108,619
Non-trainable params: 0
```

The model can be trained with standard meta-parameters on the dataset (which has been loaded into X_train and y_train datasets) (Listing 4-57).

Listing 4-57. Compiling and fitting

```
model.compile(optimizer='adam',
              loss='sparse_categorical_crossentropy',
              metrics=['accuracy'])

history = model.fit(X_train, y_train, epochs=100,
                    validation_data = (X_val, y_val),
                    batch_size = 256)
```

The model converges to a train loss slightly under 0.2 and obtains a validation accuracy of 0.90 (Figure 4-75).

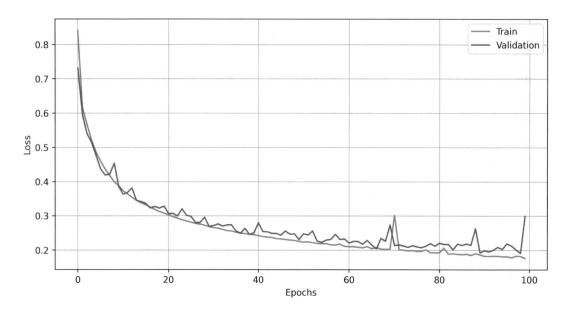

Figure 4-75. *Performance history of the initial soft ordering 1D CNN model on the Forest Cover dataset*

While this is reasonable performance, it's not great – we can very easily obtain comparable or higher performance using purely fully connected networks or traditional machine learning applications. We can try building a more sophisticated neural network structure (implemented in Listing 4-58). The key architectural components are as follows:

- *Soft ordering expansion*: Instead of merely transforming one vector to another (as in the previous architecture), we develop multiple "vector feature maps" during the soft ordering component. This can be thought of as the number of depth-wise channels in an image.

- *Nonlinear convolutional cells*: Three difference branches with different kernel sizes are applied to the input before merging and pooling.

- *Use of the SELU activation*: Instead of using the standard ReLU activation function, we allow for the alternative SELU activation. Recall the SELU function discussed in Chapter 3.

Listing 4-58. Designing a more powerful/sophisticated soft ordering 1D CNN

```
numElements = len(data.columns) - 1

inp = L.Input(numElements)
d1 = L.Dense(numElements*4, activation='selu')(inp)
d2 = L.Dense(numElements*8, activation='selu')(d1)
d3 = L.Dense(numElements*16, activation='selu')(d2)
x = L.Reshape((numElements, 16))(d3)
for i in [16, 8, 4]:
    x1a = L.Conv1D(i, 3, padding='same', activation='selu')(x)
    x1b = L.Conv1D(i, 3, padding='same', activation='selu')(x1a)
```

```
    x2a = L.Conv1D(i, 5, padding='same', activation='selu')(x)
    x2b = L.Conv1D(i, 3, padding='same', activation='selu')(x2a)
    x3 = L.Conv1D(i, 2, padding='same', activation='selu')(x)
    add = L.Add()([x1b, x2b, x3])
    x = L.AveragePooling1D(2)(add)
    x = L.Conv1D(i, 3, padding='same', activation='selu')(x)
flatten = L.Flatten()(x)
d3 = L.Dense(16, activation='selu')(flatten)
d4 = L.Dense(16, activation='selu')(d3)
out = L.Dense(7, activation='softmax')(d4)


model = keras.models.Model(inputs = inp, outputs = out)
```

This model uses 487,879 parameters (Figure 4-76).

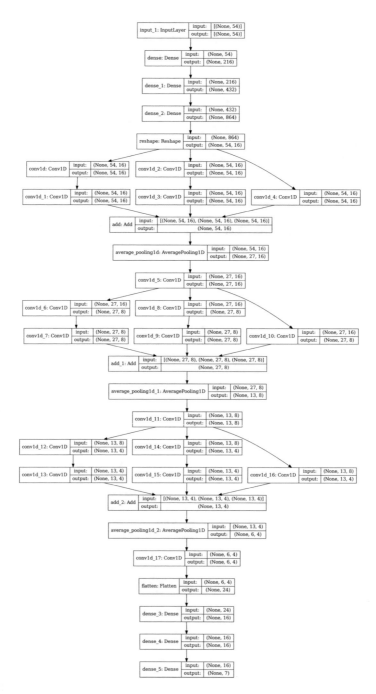

Figure 4-76. *Keras's architecture visualization of our updated soft ordering 1D CNN*

This adapted architecture performs better, obtaining near 96% validation accuracy on the dataset – which is a great result given only two architectural attempts and relatively simple training techniques (Figure 4-77).

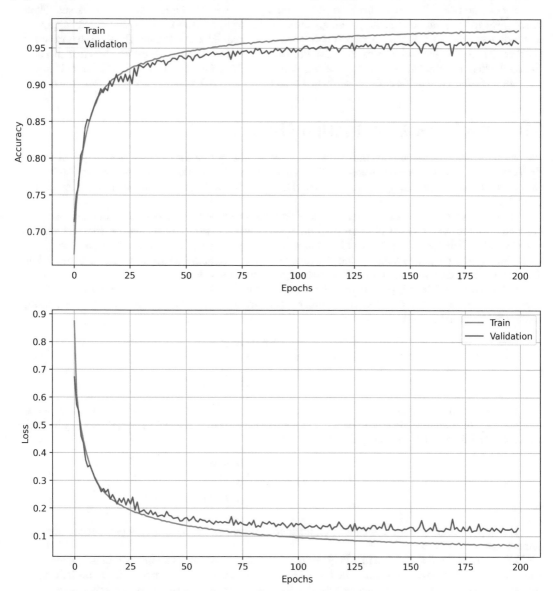

Figure 4-77. *Performance history (accuracy and loss) of our updated soft ordering 1D CNN design on the Forest Cover dataset*

There are several areas in which you could attempt to improve the model. You may notice that there are two one-hot encoded columns in the dataset that can be processed via a multi-head embedding approach discussed in the "Multimodal Image and Tabular Models" section, in which each possible unique value is associated with a set of learned embeddings/features. Another path of inquiry is the exploration of more complex training procedures, like learning rate adjustment and choosing another optimizer. It also may be fruitful to experiment with the specific size or scale of the architecture. This last inquiry can be pursued using meta-optimization, which is discussed in Chapter 6 on advanced neural network methods.

For additional reading on applications of one-dimensional convolutional neural networks, see "Flow-based malware detection using convolutional neural network"[8] by Minsoo Yeo et al. (2018), who apply a one-dimensional CNN directly to cybersecurity domain tabular data, and the Kaggle user tmp's forum post[9] on the second place solution to the Laboratory for Innovation Science at Harvard's Mechanisms of Action Prediction competition, which uses a one-dimensional CNN in addition to a TabNet model[10] and a conventional DNN. Out of the entire ensemble, the 1D CNN is given the highest weight (65% of the output).

2D Convolutions for Tabular Data

We saw in the previous section that one-dimensional convolutions made sense for a certain type of data – data with sequential or ordered features, that is – and how neural networks can be used to learn both the convolutions and the mapping from unordered to ordered data (i.e., satisfying the prerequisite for convolutions).

Using a similar logic, we can identify the data prerequisite for feasibly applying two-dimensional convolutions: the input must be of shape $(width, height, depth)$, and pixels must have a spatial relationship to one another. If this latter requirement is not met, then applying convolutions doesn't make sense. Convolutions group together neighboring pixels, so any input to a standard two-dimensional convolution should be arranged such that a spatial relationship between values exists.

Just like we *could* directly pass an unordered tabular dataset into a series of one-dimensional convolutions, we *could* pass an unordered tabular dataset reshaped into form $(width, height, depth)$ through a series of two-dimensional convolutions:

$$[a,b,c,d,e,f,g,h,i] \rightarrow \begin{bmatrix} a & b & c \\ d & e & f \\ g & h & i \end{bmatrix}$$

However, it's not going to work well because this raw dataset doesn't fulfill the data prerequisite for two-dimensional convolutions: there is no contiguous semantics; the same data could be reshaped/expressed in a completely different ordering like such:

$$[a,b,c,d,e,f,g,h,i] \rightarrow \begin{bmatrix} b & g & a \\ d & h & e \\ c & f & i \end{bmatrix}$$

This sort of arbitrariness of contiguous organization indicates that data does not possess contiguous semantics in its raw form.

However, we can *imagine* that for every raw tabular input x, there exists an image representation p that contains the same information but has the crucial property of *contiguous semantics* – that is, each data point (pixel) is somehow related to spatially neighboring data points (adjacent pixels). As previously asserted, if both x and p exist, then there must be some mapping $x \rightarrow p$ that we can task the neural network to discover and approximate.

---

[8] Yeo, M., Koo, Y., Yoon, Y., Hwang, T., Ryu, J., Song, J., & Park, C. (2018). Flow-based malware detection using convolutional neural network. *2018 International Conference on Information Networking (ICOIN)*, 910–913.
[9] www.kaggle.com/competitions/lish-moa/discussion/202256#1106810.
[10] The TabNet model and its implementation are discussed extensively in Chapter 6, on the attention mechanism.

Many successful and sophisticated convolutional neural network architectures exist that take an input in image form. Some of these have been discussed previously in this chapter, like Inception and EfficientNet. When we obtain another neural "subnetwork" that can learn the crucial mapping $x \to p$, we can train the CNN architecture directly on the "image representation" p (Figure 4-78).

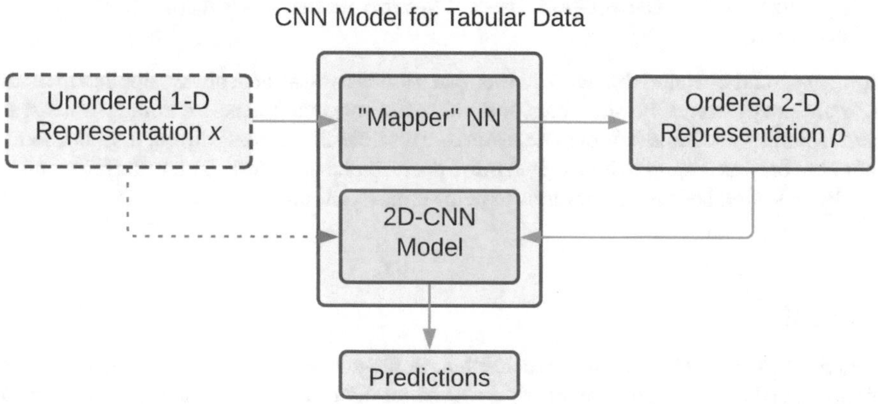

CNN Model for Tabular Data

Figure 4-78. *Blueprint for applying standard two-dimensional CNNs to tabular data using a "mapper" from unordered 1D representation to ordered 2D representation*

Like in the previous section, we can use "vanilla soft ordering" simply by stacking several fully connected layers together, reshaping the produced vector into an image form, and passing the formed image into a standard convolutional neural network (Listing 4-59).

Listing 4-59. Boilerplate code to use standard soft ordering for two-dimensional convolutional components

```
inp = L.Input((q,))
x = L.Dense(A, activation='relu')(inp)
x = L.Dense(B, activation='relu')(x)
...
x = L.Dense(X, activation='relu')(x)
x = EfficientNetB0(params, classes=n)(x)
model = keras.models.Model(inputs = inp, outputs = x)
```

However, in many cases it's difficult to learn the crucial mapping from unordered representation x to ordered representation p, especially when p is two-dimensional (although it's generally easy to set up and thus worth a try). The task may be too complex to feasibly learn with the "vanilla" soft ordering approach. In these cases, it may be more successful to use human-guided machine learning mappings from $x \to p$ – that is, mappings that employ machine learning techniques like Principal Component Analysis within a human-built pipeline/framework but do not use generic fully connected layers to approximate the mapping.

We will cover two papers that propose similar novel methods to transform tabular data into an image for the application of conventional convolutional neural networks: DeepInsight and IGTD (Image Generation for Tabular Data). These are not the only works in the space; for additional reading, consult the following works as examples:

Bazgir, O., Zhang, R., Dhruba, S.R., Rahman, R., Ghosh, S., & Pal, R. (2020). Representation of features as images with neighborhood dependencies for compatibility with convolutional neural networks. *Nature Communications, 11.*

Ma, S., & Zhang, Z. (2018). OmicsMapNet: Transforming omics data to take advantage of Deep Convolutional Neural Network for discovery. *ArXiv, abs/1804.05283.*

Moreover, it should be noted that work in this space tends to be almost always applied to medical, biological, or physics problems, because convolutions really only make sense on homogenous-scale, standardized features – like hundreds of gene features. This does not necessarily bar application to other contexts, but note that such applications may require the artificial insertion of a reordering component (like several dense layers) before convolutions to help support automatic transformation into said per-pixel homogeneity.

DeepInsight

Alok Sharma, Edwin Vans, Daichi Schigemizu, Keith A. Boroevich, and Tatsuhiko Tsunoda describe such an $x \rightarrow p$ mapping in the 2019 paper "DeepInsight: A methodology to transform a non-image data to an image for convolution neural network architecture."[11] The DeepInsight method is a pipeline to transform structured/tabular data (this does not necessarily exclude sequential or text-based data, as long as it is framed in a structured data format) into image form data, which a standard convolutional neural network can be trained on.

The first step of DeepInsight is to acquire a feature matrix that is used to map individual features in structured data to spatial coordinates in the corresponding image. This feature density matrix is used as a "template" to generate individual images for each individual vector. Each feature is associated with a pixel in the "template" matrix. This is essentially learning the optimal "coordinate" correspondence to a feature in the structured dataset.

Consider the following example optimal translation of nine features into a 4×4 template matrix. This scheme will produce 4×4 "images":

$$\begin{bmatrix} a & b & c \\ d & e & f \\ g & h & i \end{bmatrix} \rightarrow \begin{bmatrix} a & - & b & h \\ - & c & - & e \\ d & e & i & g \\ - & f & - & - \end{bmatrix}$$

This association is performed via a clever trick in which the data is *transposed* and used in dimensionality reduction via methods like kernel-PCA or t-distributed Stochastic Neighbor Embedding (t-SNE) (recall from Chapter 2). Traditionally, in a dataset of n samples and d features, dimensionality reduction to two spatial dimensions yields a dataset of n samples and two features. However, if we apply dimensionality reduction to the *transpose* of such a dataset – that is, we treat each of the d features as a sample and each of the n samples as a feature – this yields a reduced dataset with d samples and two features. Thus, each of the d features has been mapped to a two-dimensional point in the "template" matrix.

Using a transformation method like kernel-PCA and t-SNE that preserves local relationships, we can map features that behave similarly to one another to physically closer locations in the feature matrix (Figure 4-79). This allows similar features in the generated images to be processed more efficiently by convolutions.

---

[11] Sharma A., Vans E., Shigemizu D., Boroevich K.A., & Tsunoda T. DeepInsight: A methodology to transform a non-image data to an image for convolution neural architecture. 2019. Paper link: www.nature.com/articles/s41598-019-47765-6.pdf.

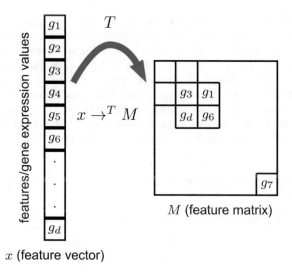

Figure 4-79. *Visualization of mapping features in a tabular dataset to pixel coordinates in an ordered 2D representation (the template matrix) using the DeepInsight method. From Sharma et al.*

Once the "template" feature matrix has been established, we can create an image for an input vector by establishing a point in the image corresponding to the location allocated for that specific feature (Figure 4-80). (You can observe from this that DeepInsight is designed for high-dimensional data; a high number of features are needed to populate the image since each point in the image is one feature.) In order to prevent redundancy in image representation, the Convex Hull Algorithm is utilized to select the smallest rectangle consisting of all the data, cropping out unnecessary blank sides. The data is correspondingly rotated, and the space is mapped into pixel, image-based format, which can then be passed through a standard convolutional neural network.

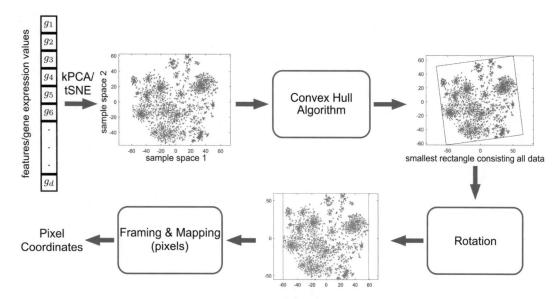

Figure 4-80. *The DeepInsight pipeline: mapping vectors to pixel coordinates. From the DeepInsight paper (see footnote #2). From Sharma et al.*

The resulting DeepInsight pipeline performs very well on both genetic datasets, which the model was originally designed for, and other high-dimensional data contexts. Sharma et al. evaluate the method on five benchmark datasets: RNA-seq, a biological RNA sequence dataset from the NIH TCGA dataset; a subset from the TIMIT corpus, a speech dataset; the Relathe dataset, derived from news documents; the Madelon dataset, which is a synthetically constructed binary classification problem; and ringnorm-DELVE, another synthetically constructed binary classification problem. These five datasets represent a wide array of problem contexts and data spaces; the DeepInsight method performs much better than other algorithms that have become successful staple methods in modeling structured/tabular datasets (Table 4-4). See Figure 4-81 for a visualization on how DeepInsight generates meaningful visual representations across these datasets.

Table 4-4. *Performance of DeepInsight against other common methods for structured data with various datasets. From Sharma et al*

Dataset	Decision Tree	AdaBoost	Random Forest	DeepInsight
RNA-seq	85%	84%	96%	99%
Vowels	75%	45%	90%	97%
Text	87%	85%	90%	92%
Madelon	65%	60%	62%	88%
Ringnorm-DELVE	90%	93%	94%	98%

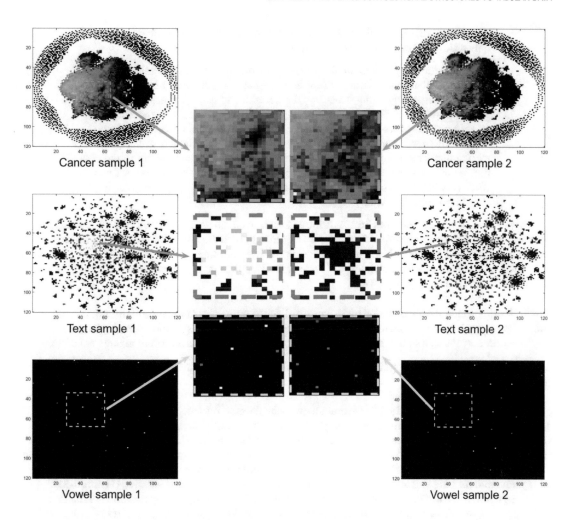

Figure 4-81. *Visualization of image patterns mapped by DeepInsight. Differences in samples from Cancer, Text, and Vowel datasets are shown as the small patches in the middle columns. These differences are extracted by convolutional filters to perform more effective classification than another method. From Sharma et al.*

We can reflect on key advantages of DeepInsight:

- CNNs (theoretically and generally in practice) do not require any additional feature extraction techniques, which are often used for structured data. They automatically derive advanced and information-rich features from the raw input data without need for preprocessing via a series of convolutions and pooling. The nonlinear architecture of the model used to process the image aids in the development of advanced, rich representations.

- Convolutions process image data locally in a restricted subarea. This allows for greater network depth with a relatively small quantity of parameters, which spurs healthy network understanding and generalization. Such performance would be more difficult if using a fully connected network, in which building more depth to increase modeling power leads to a faster increase in parameters, risking overfitting, which neural networks trained on structured data are especially prone to.

- The unique structure of the CNN allows it to run very efficiently, given recent hardware advancements like GPU utilization.

- CNNs and the DeepInsight pipeline broadly are much more customizable/ optimizable than standard algorithms like tree-based methods that have traditionally shown success in modeling structured data. Besides adjusting hyperparameters like the model architecture, the vector-to-template matrix mapping, and the learning rate, among countless others, one can also easily use image augmentation methods to generate "new" image data. Such data augmentation is difficult to accomplish with tabular data because the structured format does not contain an inherent robustness due to the low dimensionality of its representation space relative to images; that is, while rotating an image should not affect the phenomenon it represents, altering structured data likely will.

In practice, DeepInsight should be a contributing member in an ensemble of other decision-making models. Combining the locality-specific nature of DeepInsight with the more global approach of other modeling methods will likely yield a more informed predictive ensemble.

Sharma et al. have provided prepackaged code to use DeepInsight in Python, which can be installed from the GitHub repository (Listing 4-60).

Listing 4-60. Installing code provided by Sharma et al. for DeepInsight. At the time of this book's writing, the authors of pyDeepInsight are making active changes that make this installation command erroneous. If you encounter errors, check the GitHub repository for the most up-to-date information on installation

```
!python3 -m pip -q install git+https://github.com/alok-ai-lab/pyDeepInsight.
git#egg=pyDeepInsight
```

The dataset we will use is the Mice Protein Expression dataset from the infamous University of California Irvine Machine Learning Repository, which is a classification dataset with 1080 instances and 80 features modeling the expression of 77 proteins in the cerebral cortex of mice exposed to contextual fear conditioning. A cleaned version of the dataset is available in the source code for this book to be downloaded.

Assuming that the data has been loaded as a Pandas DataFrame in the variable data, the first step is to separate into training and testing datasets, a standard procedure in machine learning (Listing 4-61). We'll also need to convert the labels to one-hot format, which in their original organization are integers corresponding to a class. This can be accomplished easily using keras.util's to_categorical function.

Listing 4-61. Selecting a subset of data and converting to one-hot form as necessary

```
import pandas as pd
# download csv from online source files
data = pd.read_csv('mouse-protein-expression.csv')
from sklearn.model_selection import train_test_split
X_train, X_test, y_train, y_test = train_test_split(data.drop('class',axis=1),
                data['class'],
                train_size=0.8)
y_train = keras.utils.to_categorical(y_train)
y_test = keras.utils.to_categorical(y_test)
```

We will need to use the LogScaler object from the DeepInsight library to scale the data between 0 and 1 using the L2 norm (Listing 4-62). We transform both the training dataset and the testing dataset, fitting the scaler on the training dataset only. All new data used for prediction by the DeepInsight model should pass through this scaler first.

Listing 4-62. Scaling data

```
from pyDeepInsight import LogScaler
ln = LogScaler()
X_train_norm = ln.fit_transform(X_train)
X_test_norm = ln.transform(X_test)
```

The `ImageTransformer` object performs the image transformation by first generating the "template" matrix via a dimensionality reduction method passed into `feature_extractor`, which accepts either `'tsne'`, `'pca'`, or `'kpca'`. This method is used to determine a mapping of features in the input vector to an image of pixels dimensions. We can instantiate an ImageTransformer with the kernel-PCA dimensionality reduction method to generate 32-by-32 images (`feature_extractor='kpca'`, `pixels=32`) (Listing 4-63).

Listing 4-63. Training and transforming with the ImageTransformer

```
from pyDeepInsight import ImageTransformer
it = ImageTransformer(feature_extractor='kpca',
                      pixels=32)
tf_train_x = it.fit_transform(X_train_norm)
tf_test_x = it.transform(X_test_norm)
```

Kernel-PCA is used rather than t-SNE because of the relatively low dimensionality and quantity of the data. PCA is not employed because its linearity limits the nuance it captures. An image length of 32 pixels is chosen as a balance between making the generated images too sparse (too high an image length) and too small (too small an image length) to meaningfully and accurately represent spatial relationships between features. As image size decreases, the conception of distance in the DeepInsight pipeline – that is, the placement of features as pixels farther or closer to one another dependent on their similarity – becomes more approximated to the point of being arbitrary.

We can visualize the generated images of the ImageTransformer easily using `matplotlib.pyplot.imshow()` to get a feel for how the dimensionality reduction method and image size influence the arrangement of the features and the likelihood of success (Figure 4-82). The differences between images are subtle, but the distinguishing factors are identified and amplified by a series of convolutional operations. Note that the 32-by-32-pixel space allows for a clustering of similar features and for less related features to be distanced farther away in a corner.

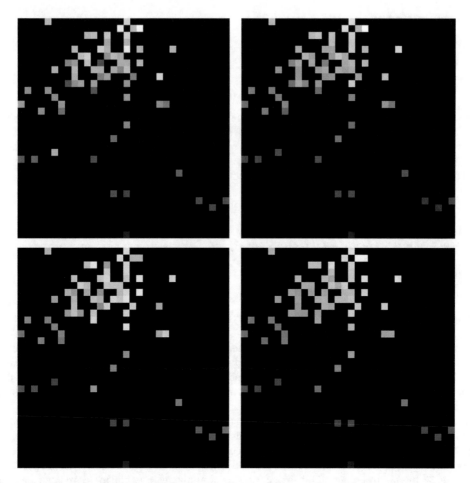

Figure 4-82. *Four example images generated from our Mice Protein Expression dataset using the DeepInsight method*

We'll build an architecture in a similar style to the two-branch cell design used in the DeepInsight paper, with three key adaptations: Inception v3–style factorization/expansion of filters, dropout in cells, and a longer fully connected component (Listing 4-64, Figure 4-83). These aid to develop more specific filters with smaller areas to better parse densely packed features, further aid generalization by preventing overfitting, and better process derived features, respectively. One branch processes images with a kernel of size (2,2), and the other uses a kernel of size (5,5) (with additional factorization, e.g., 5×1 and 1×5).

Listing 4-64. Sample implemented architecture in Keras

```
# input
inp = L.Input((32,32,3))

# branch 1
x = inp
for i in range(3):
    x = L.Conv2D(2**(i+3), (2,1), padding='same')(x)
    x = L.Conv2D(2**(i+3), (1,2), padding='same')(x)
```

```
    x = L.Conv2D(2**(i+3), (2,2), padding='same')(x)
    x = L.BatchNormalization()(x)
    x = L.Activation('relu')(x)
    x = L.MaxPooling2D((2,2))(x)
    x = L.Dropout(0.3)(x)
x = L.Conv2D(64, (2,2), padding='same')(x)
x = L.BatchNormalization()(x)
branch_1 = L.Activation('relu')(x)

# branch 2
x = inp
for i in range(3):
    x = L.Conv2D(2**(i+3), (5,1), padding='same')(x)
    x = L.Conv2D(2**(i+3), (1,5), padding='same')(x)
    x = L.Conv2D(2**(i+3), (5,5), padding='same')(x)
    x = L.BatchNormalization()(x)
    x = L.Activation('relu')(x)
    x = L.MaxPooling2D((2,2))(x)
    x = L.Dropout(0.3)(x)
x = L.Conv2D(64, (5,5), padding='same')(x)
x = L.BatchNormalization()(x)
branch_2 = L.Activation('relu')(x)

# concatenate + output
concat = L.Concatenate()([branch_1, branch_2])
global_pool = L.GlobalAveragePooling2D()(concat)
fc1 = L.Dense(32, activation='relu')(global_pool)
fc2 = L.Dense(32, activation='relu')(fc1)
fc3 = L.Dense(32, activation='relu')(fc2)
out = L.Dense(9, activation='softmax')(fc3)

# aggregate into model
model = keras.models.Model(inputs=inp, outputs=out)
```

Figure 4-83. *Visualization of the DeepInsight model architecture*

The model, when compiled and trained on the data for several dozen epochs, yields almost perfect training accuracy *and* validation accuracy (Listing 4-65).

Listing 4-65. Compiling and fitting the model

```
model.compile(optimizer='adam',
              loss='categorical_crossentropy',
              metrics=['accuracy'])
model.fit(tf_train_x, y_train, epochs=100,
          validation_data=(tf_test_x, y_test))
```

DeepInsight can also be modified in other ways. Because DeepInsight tends to produce sparse mappings – that is, most pixels in the image are empty and not associated with features – one recently proposed technique that has shown to be exceptionally successful is to blur the images after they have been generated.[12] This can help enhance the effects of localization by precomputing interactions/interpolations between adjacent pixels and helps spread the "influence" of filled pixels to nearby empty pixels.

DeepInsight has successfully been applied to many other domains and is one example of a successful $x \rightarrow q$ 1D unordered to 2D ordered mapping.

IGTD (Image Generation for Tabular Data)

Introduced by Yitan Zhu, Thomas Brettin, Fangfang Xia, Alexander Partin, Maulik Shukla, Hyunseung Yoo, Yvonne A. Evrard, James H. Doroshow, and Rick L.Stevens in the paper "Converting tabular data into images for deep learning with convolutional neural networks,"[13] the IGTD (Image Generation for Tabular Data) is another method to transform standard tabular data into image form.

The IGTD paper was written after the DeepInsight paper. The authors argue that the primary advantage of IGTD over the DeepInsight method is decreased sparsity (i.e., improved compactness); a large number of pixels are left blank in images generated by DeepInsight and may be comparatively inefficient to train.

Let c represent the number of columns in a dataset and X represent the dataset matrix. Our objective is to transform each of the c features in X into an image we will denote I, which is of shape $n \times n$ (we will assume it is square for simplification here, but the method also generalizes to rectangular images). To ensure feature density, we want every feature to be mapped to a pixel and every pixel to a feature – there are no redundant pixels. As such, $n^2 = c$; the number of pixels in the image is equivalent to the number of features.[14]

We can generate a matrix of shape $c \times c$ denoted ***R***. ***R***$_{i,j}$ – that is, the ith row and jth column of ***R*** – is the Euclidean distance between the ith and jth features of the dataset X. (Other distance metrics can also be used; the point is to measure the similarity between feature pairs.) This structure systematically organizes the similarity between every feature pair. Figure 4-84 visualizes such a matrix on a 2,500-feature ($c = 2500$) gene expression dataset.

[12] Castillo-Cara, M., Talla-Chumpitaz, R., Orozco-Barbosa, L., & García-Castro, R. (2022). A Deep Learning Approach Using Blurring Image Techniques for Bluetooth-Based Indoor Localisation. *SSRN Electronic Journal.*

[13] Zhu, Y., Brettin, T.S., Xia, F., Partin, A., Shukla, M., Yoo, H.S., Evrard, Y.A., Doroshow, J.H., & Stevens, R.L. (2021). Converting tabular data into images for deep learning with convolutional neural networks. *Scientific Reports, 11.*

[14] One possible remedy to having a dataset with a non-factorable number of features c is to add redundant columns of zeros or noise.

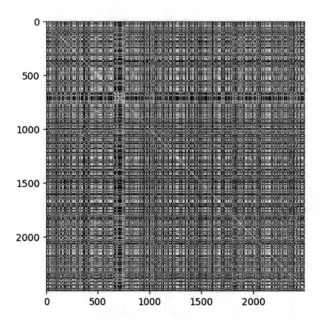

Figure 4-84. *Visualization of the **R** matrix. From Zhu et al.*

Let $\boldsymbol{Q}$ also be a matrix of shape $c \times c$. This represents the pairwise spatial distance between each pixel of I – all $n^2 = c$ of them. The distance between the pixel at $(3, 9)$ and the pixel at $(6, 13)$ is $\sqrt{3^2 + 4^2} = 5$. The main diagonal of $\boldsymbol{Q}$ is all zeros, since this represents the distance between a pixel and itself.

Figure 4-85 demonstrates such a matrix on the same 2,500-feature gene expression dataset. Since $n = 50$, we observe a "mosaic" pattern in which small 50-by-50 tiles are repeated (50 times in each dimension) due to the end-to-end concatenation/flattening of the pixels in I. The bottom-left and top-right corners are darkest, indicating that the 2,500th feature (which occupies the pixel at location $(50,50)$) is farthest from the 1st feature (which occupies the pixel at location $(1,1)$).

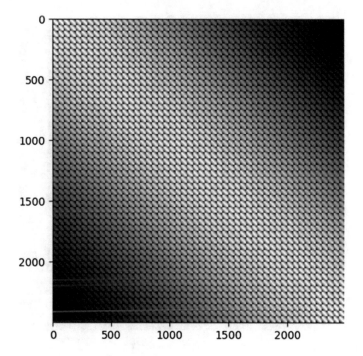

Figure 4-85. *Visualization of the **Q** matrix. From Zhu et al.*

Here is the clever part of the IGTD paradigm: we want to map the dataset to the image, so we find a way to directly map **R** – which contains the pairwise feature distances computed from the dataset – to **Q**, which contains the pairwise distances computed from the image.

To do so, we rearrange the columns in **R**. The initial ordering of the columns was arbitrary, but we can arrange it such that the pairwise distances of the columns match the pairwise distances of the pixels. That is, we *calibrate the columns with the pixels such that more similar features are mapped to spatially closer pixels.*

We guide this process by attempting to minimize the error between a rearrangement of **R** and the pairwise pixel difference matrix **Q** for some distance metric (L1 or L2) diff(*a*, *b*):

$$err(\boldsymbol{R},\boldsymbol{Q}) = \sum_{i=1}^{c}\sum_{j=1}^{c} \mathrm{diff}\left(\boldsymbol{R}_{i,j},\boldsymbol{Q}_{i,j}\right)$$

For practical computation purposes, we only need to compute the difference between the bottom-left 90-degree "half-triangles" of **R** and **Q**, since both matrices are symmetrical across the major diagonal:

$$err(\boldsymbol{R},\boldsymbol{Q}) = \sum_{i=2}^{c}\sum_{j=1}^{i-1} \mathrm{diff}\left(\boldsymbol{R}_{i,j},\boldsymbol{Q}_{i,j}\right)$$

Indeed, after optimally arranging the columns of **R** to minimize the error function, the resulting matrix appears very similar visually to **Q** (Figure 4-86).

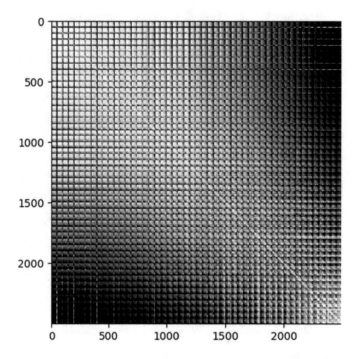

Figure 4-86. *Visualization of the **R** matrix after rearrangement. From Zhu et al*

Note that this calibration between **R** and **Q** has mapped every feature in the dataset X (from **R**) to every pixel in the image I (from **Q**).

The algorithm runs for several iterations and attempts different swap operations to minimize the error. The algorithm will not be described here but is viewable in the original paper.

This can be tricky to grasp, especially in high dimensions and symbolic abstraction. Let us demonstrate this logic on a small scale using a sample nine-feature dataset (Figure 4-87).

a	b	c	d	e	f	g	h	i
1	0	3	2	5	0	4	1	2
0	4	5	5	2	4	9	1	1
4	3	2	4	5	5	5	2	3

Figure 4-87. *A sample dataset*

If we compute the distance between each of these features, we get the following pairwise distance matrix **Q** (Figure 4-88).

	a	b	c	d	e	f	g	h	i
a	0	4.24	5.74	5.09	4.58	4.24	9.54	2.24	1.73
b	4.425	0	3.32	2.45	5.74	2.00	6.71	3.17	3.61
c	5.74	3.32	0	2.23	4.69	4.36	5.11	4.47	4.24
d	5.09	2.45	2.23	0	4.36	2.45	4.58	4.58	4.12
e	4.58	5.74	4.69	4.36	0	5.39	7.07	5.11	3.74
f	4.24	2.00	4.36	2.45	5.39	0	6.40	4.36	4.12
g	9.54	6.71	5.11	4.58	7.07	6.40	0	9.06	8.49
h	2.24	3.17	4.47	4.58	5.11	4.36	9.06	0	1.41
i	1.73	3.61	4.24	4.12	3.74	4.12	8.49	1.41	0

Figure 4-88. *The pairwise feature distance matrix calculated across features using Euclidean distance*

Since $c = 9$, we have $n = 3$; we will have a 3×3 image. We can compute the pairwise pixel distances between each of the pixels to produce R (Figure 4-89).

	(1,1)	(1,2)	(1,3)	(2,1)	(2,2)	(2,3)	(3,1)	(3,2)	(3,3)
(1,1)	0	1.00	2.00	1.00	1.41	2.24	2.00	2.24	2.83
(1,2)	1.00	0	1.00	1.41	1.00	1.41	2.24	2.00	2.24
(1,3)	2.00	1.00	0	1.24	1.41	1.00	2.83	2.24	2.00
(2,1)	1.00	1.41	2.24	0	1.00	2.00	1.00	1.41	2.24
(2,2)	1.41	1.00	1.41	1.00	0	1.00	1.41	1.00	1.41
(2,3)	2.24	1.41	1.00	2.00	1.00	0	2.24	1.41	1.00
(3,1)	2.00	2.24	2.83	1.00	1.41	2.24	0	1.00	2.00
(3,2)	2.24	2.00	2.24	1.41	1.00	1.41	1.00	0	1.00
(3,3)	2.83	2.24	2.00	2.24	1.41	1.00	2.00	1.00	0

Figure 4-89. *The pairwise feature distance matrix calculated across pixels using Euclidean distance*

Now, we want to map Q to R by repeatedly swapping different features and finding the configuration that minimizes the difference between the two matrices. Figures 4-90 and 4-91 demonstrate how we might swap features d and g.

	a	b	c	d	e	f	g	h	i
a	0	4.24	5.74	5.09	4.58	4.24	9.54	2.24	1.73
b	4.425	0	3.32	2.45	5.74	2.00	6.71	3.17	3.61
c	5.74	3.32	0	2.23	4.69	4.36	5.11	4.47	4.24
d	5.09	2.45	2.23	0	4.36	2.45	4.58	4.58	4.12
e	4.58	5.74	4.69	4.36	0	5.39	7.07	5.11	3.74
f	4.24	2.00	4.36	2.45	5.39	0	6.40	4.36	4.12
g	9.54	6.71	5.11	4.58	7.07	6.40	0	9.06	8.49
h	2.24	3.17	4.47	4.58	5.11	4.36	9.06	0	1.41
i	1.73	3.61	4.24	4.12	3.74	4.12	8.49	1.41	0

Figure 4-90. *Identification of two features to be swapped*

	a	b	c	g	e	f	d	h	i
a	0	4.24	5.74	9.54	4.58	4.24	5.09	2.24	1.73
b	4.425	0	3.32	6.71	5.74	2.00	2.45	3.17	3.61
c	5.74	3.32	0	5.11	4.69	4.36	2.23	4.47	4.24
g	9.54	6.71	5.11	4.58	7.07	6.40	0	9.06	8.49
e	4.58	5.74	4.69	7.07	0	5.39	4.36	5.11	3.74
f	4.24	2.00	4.36	6.40	5.39	0	2.45	4.36	4.12
d	5.09	2.45	2.23	0	4.36	2.45	4.58	4.58	4.12
h	2.24	3.17	4.47	9.06	5.11	4.36	4.58	0	1.41
i	1.73	3.61	4.24	8.49	3.74	4.12	4.12	1.41	0

Figure 4-91. *Result of two features to be swapped*

Say that after we run the algorithm, we obtain the following modified Q (Figure 4-92).

	c	i	f	d	b	a	e	h	g
c	...	...	...	...	...	...	...	...	...
i	...	...	...	...	...	...	...	...	...
f	...	...	...	...	...	...	...	...	...
d	...	...	...	...	...	...	...	...	...
b	...	...	...	...	...	...	...	...	...
a	...	...	...	...	...	...	...	...	...
e	...	...	...	...	...	...	...	...	...
h	...	...	...	...	...	...	...	...	...
g	...	...	...	...	...	...	...	...	...

Figure 4-92. *Hypothetical resulting matrix after optimal swapping*

We can map each feature to the corresponding pixel in R:

- $c \rightarrow (1,1)$
- $i \rightarrow (1,2)$
- $f \rightarrow (1,3)$
- $d \rightarrow (2,1)$
- $b \rightarrow (2,2)$
- $a \rightarrow (2,3)$
- $e \rightarrow (3,1)$
- $h \rightarrow (3,2)$
- $g \rightarrow (3,3)$

From this result, we can infer that columns c and g are very different from each other, since (a) c was mapped to the pixel $(1,1)$, (b) g was mapped to the pixel $(3,3)$, (c) the distance between these two pixels is very large relative to other intra-pixel distances, and (d) the IGTD algorithm places high-distance (high-dissimilarity) features into high-distance pixels.

Figure 4-93 demonstrates the images derived from a 2,500-feature genomics dataset using IGTD (left), another tabular-to-image method REFINED[15] (center), and DeepInsight (right). Note that IGTD generates much more compact image representations compared with DeepInsight.

---

[15] Bazgir, O., Zhang, R., Dhruba, S.R., Rahman, R., Ghosh, S., & Pal, R. (2020). Representation of features as images with neighborhood dependencies for compatibility with convolutional neural networks. *Nature Communications, 11.*

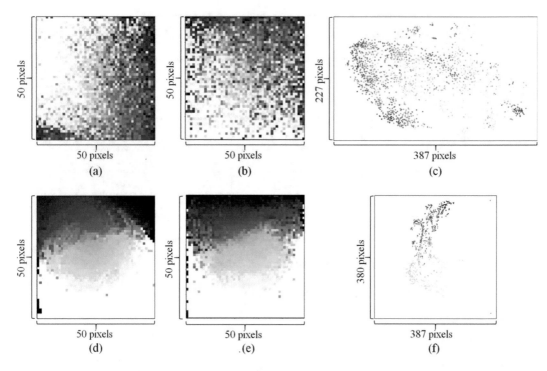

Figure 4-93. *Sample results of images generated with different methods. From Zhu et al*

Table 4-5 reports performance of IGTD with a CNN compared with other models and tabular-to-image generation methods on two genomics datasets, the Cancer Therapeutics Response Portal (CTRP) and Genomics of Drug Sensitivity in Cancer (GDSC).

Table 4-5. *Performance of IGTD against other methods on two datasets. From Zhu et al*

Dataset	Prediction model	Data representation	R²	P-value
CTRP	LightGBM	Tabular data	0.825 (0.003)	8.19E−20
	Random forest		0.786 (0.003)	5.97E−26
	tDNN		0.834 (0.004)	7.90E−18
	sDNN		0.832 (0.005)	1.09E−16
	CNN	IGTD images	**0.856** (0.003)	
		REFINED images	0.855 (0.003)	8.77E−01
		DeepInsight images	0.846 (0.004)	7.02E−10
GDSC	LightGBM	Tabular data	0.718 (0.006)	2.06E−13
	Random forest		0.682 (0.006)	4.53E−19
	tDNN		0.734 (0.009)	1.79E−03
	sDNN		0.723 (0.008)	6.04E−10
	CNN	IGTD images	**0.74** (0.006)	
		REFINED images	0.739 (0.007)	5.93E−01
		DeepInsight images	0.731 (0.008)	2.96E−06

The authors provide a software package to use IGTD. It can be loaded from the IGTD_Functions.py file in the paper repository (Listing 4-66).

Listing 4-66. Loading the IGTD library from GitHub

```
!wget -O IGTD_Functions.py https://raw.githubusercontent.com/zhuyitan/IGTD/With_CNN_
Prediction/Scripts/IGTD_Functions.py
import IGTD_Functions
import importlib
importlib.reload(IGTD_Functions)
```

We need to scale the dataset such that it resides within a constant scale and demonstrates homogeneity across features (Listing 4-67). Only homogenous-domain features make sense in the context of a transformation to images.

Listing 4-67. Scaling the dataset to ensure feature homogeneity

```
from IGTD_Functions import min_max_transform
norm_data = min_max_transform(data.drop('class', axis=1).values)
```

To generate the images, specify the number of rows and columns in the generated images (their product being the number of features), the width of sample images generated (just samples and informative plots; the actual data can be collected in raw form), the maximum number of steps before IGTD algorithm termination, the number of validation steps to run through to determine convergence, the distance method used to determine the similarity/distance between features, the distance method used to compute distance between pixels, the metric used to determine error between Q and R (squared or absolute), and the directory to store the data results in.

Listing 4-68 demonstrates such a configuration, generating 8-by-10-pixel images from the 80-feature dataset with Pearson's Correlation Coefficient used to calculate feature distance and Euclidean distance to calculate pairwise pixel differences.

Listing 4-68. Running the tabular-to-image conversion

```
num_row = 8
num_col = 10
num = num_row * num_col
save_image_size = 10
max_step = 10000
val_step = 300

from IGTD_Functions import table_to_image
fea_dist_method = 'Pearson'
image_dist_method = 'Euclidean'
error = 'squared'
result_dir = 'images'
os.makedirs(name=result_dir, exist_ok=True)
table_to_image(norm_data,
               [num_row, num_col],
               fea_dist_method,
               image_dist_method,
               save_image_size,
```

```
            max_step,
            val_step,
            result_dir,
            error)
```

After running the table_to_image function – which takes a few minutes, depending on the size of the dataset – the results are stored in the provided results directory.

In our sample run on the Mice Protein Expression dataset, the original feature similarity matrix R looks as follows in Figure 4-94. Note that there is no similarity with the pattern of Q in Figure 4-86.

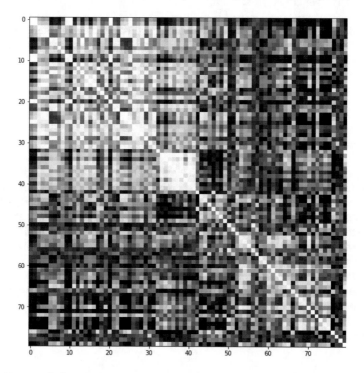

Figure 4-94. *The R matrix before sorting*

After the features are sorted with the IGTD algorithm, R (Figure 4-95) mimics the mosaic-like pixel distance grid, with "cells" (recall – caused by end-to-end row concatenation) overlayed on an overall gradient with larger distances toward the bottom left and top right and shorter distances toward the major diagonal.

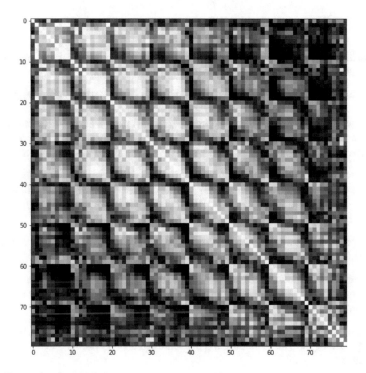

Figure 4-95. *The R matrix after sorting*

Figure 4-96 visualizes two samples generated using the IGTD algorithm.

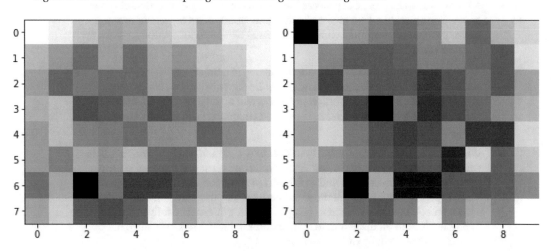

Figure 4-96. *Two sample images generated with IGTD*

IGTD and DeepInsight, as well as other proposed tabular-to-image methods, are all strong contenders for high-feature datasets with high homogeneity.

Key Points

In this chapter, we discussed the theory and implementation of convolutional neural networks and how to apply them to tabular datasets.

- Convolutions consist of a set of kernels that "sweep" across the spatial axes of the image, extracting and amplifying relevant features. In convolutional neural networks, the convolutional/extractive component learns the optimal kernels to identify visual information, while the fully connected/interpretative component learns how to arrange and understand the extracted features in relation to the prediction task.

- In terms of parametrization, the rough relationship is true: *pooling and global pooling < pooling and no global pooling < convolutional < only FC*. Using both pooling and global pooling in a convolutional neural network design yields the optimal parametrization scaling as input size increases.

- Base or foundation models can be directly used for training or as compartmentalized submodels in larger overarching models.

  - The ResNet architecture extensively uses residual connections, which skip over certain layers. The DenseNet architecture uses a denser arrangement of residual connections.

  - The Inception v3 model uses a cell-based structure with an emphasis on optimizing filter sizes and the highly nonlinear arrangement of layers within cell designs.

  - The EfficientNet model was designed to scale a NAS-optimized "small model" across all dimensions of the network (width, height, depth) uniformly.

- Multimodal models that can process image and tabular inputs together have two heads. The image head is processed using convolutions and outputted into a vector that can be combined with the processed tabular input to produce a jointly informed output.

- Applying one-dimensional convolutions directly to tabular data is possible but unlikely to work because tabular data generally does not possess contiguous semantics. We can add a soft ordering component before the one-dimensional convolutional component to learn the optimal mapping from unordered to ordered representation.

- While soft ordering techniques may suffice for two-dimensional convolutions (in which the input is processed using dense layers and reshaped into image form), the complexity of image data makes this sort of approach generally unsuccessful. The DeepInsight and IGTD methods are both human-designed, machine learning–assisted mappings from tabular to image data that work successfully in a wide array of problems. Both methods aim to map features to images in a way that places similar features spatially closer together.

In the next chapter, we'll take a similar approach to understanding the application of specialized neural network structures designed for non-tabular inputs to tabular inputs with recurrent layers.

CHAPTER 5

■ ■ ■

Applying Recurrent Structures to Tabular Data

I have to write in sequence and only in sequence.

—Andrew Scott, Actor

Chapter 4 demonstrated the application of convolutional neural networks both to images and signals (their "natural" data domain), as well as to tabular data through clever tricks – soft ordering, DeepInsight, and IGTD. This chapter will pursue a similar path of exploration: exploring the application of *recurrent networks*, traditionally applied to sequences like text and signals, to tabular data.

This chapter opens with a discussion of the theory underlying recurrent neural networks, with a focus on understanding the paradigm mechanics of the recurrent operation and the three primary recurrent model designs: "vanilla" recurrent neural networks (RNNs), Long Short-Term Memory (LSTM) networks, and Gated Recurrent Unit (GRU) networks. Then, you'll use these recurrent models in their "natural" data domain with three key applications – text modeling, audio modeling, and time-series modeling. Finally, similarly to the previous chapter, we will demonstrate multimodal usage and methods of directly applying recurrent layers to tabular data.

The last section in particular may seem foreign or controversial, just as the last section of Chapter 4 may have seemed counterintuitive. We encourage you to approach it with an open mind.

Recurrent Models Theory

In the upcoming sections, three different types of recurrent-based models are introduced through visualization along with their mathematical theory. You will acquire adequate foundation knowledge to apply the theories into modeling tabular and sequence data using recurrent models in Keras.

Why Are Recurrent Models Necessary?

In the previous chapter, we learned that standard ANNs have difficulties handling flattened image data because of the sheer number of trainable parameters that the network may contain. As a result, CNNs are introduced to scale down parameter size by replacing fully connected layers with 2D convolutions; this also allows the network to search for 2D patterns and structures that feed-forward networks cannot recognize.

© Andre Ye and Zian Wang 2023
A. Ye and Z. Wang, *Modern Deep Learning for Tabular Data*,
https://doi.org/10.1007/978-1-4842-8692-0_5

Another common type of tabular data is sequence-based data. Generally, this refers to datasets that are in a particular order, with each sample having a specific label indicating its position in the data. One common example is the time-series dataset.

Say that we are given the task of predicting an arbitrary company's stock prices over time. Each sample contains various features containing information about the stock each day, such as its high, low, open, and close prices. Then, each sample is assigned a timestamp representing the day the data is obtained. We're asked to predict the stock's opening price at the next future timestamp (Figure 5-1).

Date	High	Low	......	Volume
1/2/2022	20	14	......	20345
1/3/2022	22	12	......	21222
1/4/2022	19.5	15	......	23435
⋮	⋮	⋮	⋮	⋮
⋮	⋮	⋮	⋮	⋮
12/31/2023	12.5	9.8	......	19890

Figure 5-1. *Example of time-series/sequential data*

With a normal, fully connected network, one might propose the solution of having the next day's opening price as the target for the previous day's data and dropping the last data row (since it wouldn't be possible for the previous data row to have a target as there are no future data points assuming that the dataset is arranged chronologically). Then, treating each row as one single training sample, we can pass in the dataset to a fully connected network as a regression task.

Despite this seemingly viable workaround to the problem, using a standard ANN for time-series prediction in this fashion does raise a few issues. When each timestamp and associated data are treated as a single sample, the entire data is shuffled during training or cross-validation across multiple folds. Shuffles are applied to the dataset prior to cross-validation for more robust results. However, shuffling can result in look-ahead in the case of time-series data where data ordering is not arbitrary. The model may see future data for training and past data for validation during one fold while seeing past data in training and future data for validation in another fold. Due to the inherent chronological relationship between samples in time-series datasets, it is most likely that future data points can hint at the trend in the past. After all, time-series data is ordered chronologically, and each row of data is built from or in relation to the previous sample. This is commonly referred to as look-ahead or data leakage, causing surprisingly outstanding but false validation scores (Figure 5-2).

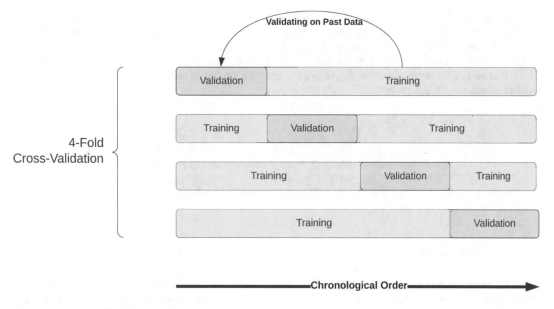

Figure 5-2. *Example of data leakage*

Furthermore, time-series data is usually structured where each new sample in the time span is affected by the sample or samples in past timestamps. Simply training and making predictions from one row of data can be insufficient since data in previous rows can contain valuable information for predicting the current sample. Instead of processing and treating each sample individually, recurrent networks process each sample at the current timestamp along with signals learned from past timestamps.

Although RNNs may be best suited for sequence and time-related data, we can also take advantage of the unique structure and improve its performance over standard ANNs on tabular modeling tasks.

Recurrent Neurons and Memory Cells

RNNs typically solve problems that involve sequence modeling or predictions in time-series datasets. The stock prediction example mentioned previously can be categorized into time-series modeling. The autocomplete feature in most document and writing software attempts to predict the next word that the user will type based on their previous behaviors. Tasks like these are referred to as sequence modeling. RNN relies on the concept that it can obtain and store information not only from the current sample but data from the previous sample relative to the ordering of the dataset.

One of the simplest examples of sequence modeling can be a literal sequence prediction. Given a sequence set, the goal is to predict the next term according to what was presented (Figure 5-3).

```
Given [1, 3, 5, 7]
Next Term(s)? [3, 5, 7, ?, ?,...]
```

Figure 5-3. *Example of sequence data*

For readers, it's simple to recognize the pattern: each preceding term is 2 less than the current term. We can obtain the next term by adding 2 to the current number. We can discover the pattern because we have access to historical data or numbers before the current term. RNN operates on a similar concept, utilizing "memory cells" that can essentially remember previously fed-in information stored for later predictions.

A normal ANN neuron, shown in Figure 5-4, receives the input and outputs a value processed by the weights associated with the neuron.

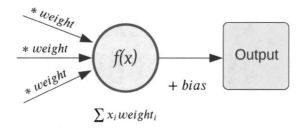

Figure 5-4. *ANN neuron example*

RNN, however, introduces a loop within the neuron to achieve a "memory" effect that can include data from previous timestamps during the prediction or training of the current sample. Using the example from earlier, the entire sequence, from elements 0 to 4, would be inputted into the network as a sequence. The first element is processed, and an output is produced. Since there's no previous timestamp on which the model can base its prediction, the calculation and process would be the same as a standard ANN. Next, the second element would be passed into the same neuron, retaining its state from the previous element's pass-through (meaning that the parameters are not updated). The output from the first element would be inputted along with the second element. Reasonably, the third element would be inputted along with the prediction or the output from the second element, whose prediction was based on the output of the first element. In a sense, the looping process mimics a chain connecting every element's influence in past timestamps (Figure 5-5).

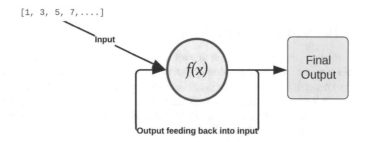

Figure 5-5. *Recurrent neuron example*

Once the neuron reaches the last element to make its final prediction for the next unknown term, not only samples at the current timestamp will be given, but also an aggregation of all past timestamps' influence.

For a more visual and intuitive representation, we can attempt to "unwrap" a recurrent neuron into a link of multiple aggregations or calculations done on a single neuron (Figure 5-6).

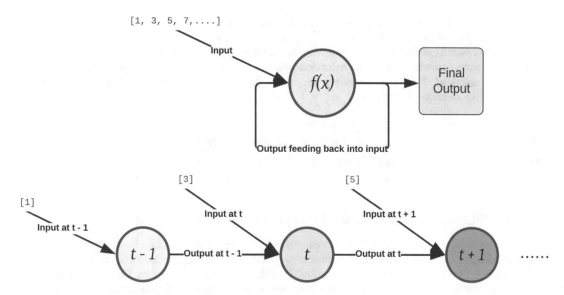

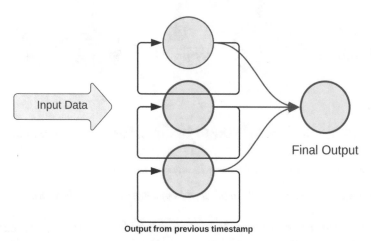

Figure 5-6. *An example of an "unrolled" recurrent neuron*

We will denote the current timestamp as *t*, the previous timestamp as *t* − 1, and the next timestamp (the term that the model is tasked to predict) as *t* + 1. When the sequence of terms is fed into the model, the term at *t* − 1 is inputted, and an output is produced from it. Then, an aggregation between the term at timestamp t and the output from *t* − 1 is fed into the same neuron. We vaguely use "aggregation" here, but the details of how to "combine" hidden states and inputs from the current timestamp will be explained in the next section. The output of that calculation using features *t* and memory from term *t* − 1 would then become the final prediction for the next term in the sequence. This process continues for sequences with more than two terms as historical data in the same fashion.

The idea of a "memory neuron" or "memory cell" by utilizing a loop predict sequence can be easily expanded into layers stacked with multiple of these "memory cells" (Figure 5-7).

Figure 5-7. *Recurrent neurons in layers*

These work the same way as single neurons; instead of one output, layer outputs from previous timestamps are passed back into the layer along with the following input.

The feed-forward operation in RNNs is exactly like what was shown previously since making a prediction requires a forward pass through the network. However, backpropagation in RNNs is slightly different from standard ANN backpropagation since multiple inputs and outputs are processed through a neuron retaining the same parameters.

■ **Note** For the clarity of terms here, the "memory" outputted from the model based on the previous timestamp is usually referred to as the "hidden state" of the recurrent neuron or the memory cell.

Backpropagation Through Time (BPTT) and Vanishing Gradients

Backpropagation Through Time, or BPTT, as the name suggests, is the algorithm for backpropagation in RNN dealing with sequential data such as time series. RNNs cannot adapt to the standard ANN backpropagation algorithm since different timestamps within one neuron are not considered. BPTT not only adjusts parameters based on the influence from the current timestamp but also takes account of what had happened before the timestamp. Remember that the output (whether it's a new hidden state or the final output) computed at the current timestamp does not entirely rely on the past results; new features at that time are also passed into the model. Thus, BPTT only partially considers the "influence" that past hidden states have on the final output. Consider the following representation of an unrolled RNN neuron in Figure 5-8. To simplify matters, the model is tasked with predicting only one output based on a string of sequences, commonly referred to as a many-to-one prediction task.

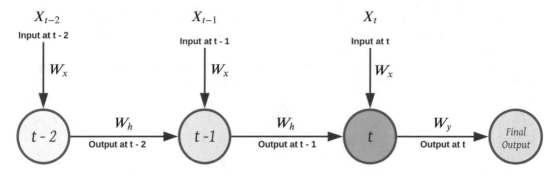

Complete Pass-through of ONE Recurrent Neuron

Figure 5-8. *An unrolled RNN neuron with an input of a sequence containing three terms*

The hidden state of the recurrent neuron is determined by two factors: the output from the last timestamp (the hidden state) and the current timestamp's input. Each input is associated with a different weight matrix. We can denote the weights assigned to the current timestamp's input as Wx and the weights assigned to process the hidden states as Wh. The aggregation between the hidden state and X_{t-n} is defined by the equation $h_{t-1} = \sigma(W_x X_{t-1} + W_h h_{t-2})$ where h represents the hidden state produced by the aggregation and σ is the activation function. Activation functions will be ignored in later calculations for explanatory purposes and simplicity. The final output can be calculated by Predictions $= Y = \sigma(W_y h_t)$.

Suppose the sequence has three total timestamps, $t-2$, $t-1$, and t as shown in Figure 5-8. Backpropagation computes the gradient of the loss function with respect to all the parameters in the network or, in our example, all the parameters in the recurrent neuron. In a many-to-one sequence prediction task, typically, only the loss at the last timestamp would be calculated instead of computing the loss separately for every timestamp. Denote the loss/cost function as L where for each neuron, the loss is accumulated across all timestamps: $L = \sum_{t=1}^{T} L_t$.

Differentiating W_y is straightforward since changing the value of it only influences the final output: this differential is not associated with the previous intertwined loops of the recurrent neuron. Its differentiation is as such: $\frac{\partial L}{\partial W_y} = \frac{\partial L}{\partial Y} \frac{\partial Y}{\partial W_y}$. Here, Y represents the final output.

As their influence stretches across the entire sequence span, things get tricky with W_h and W_x. We'd not only have to consider the changes in the current timestamp but also those that occurred in previous timestamps. Consider the following diagram where the inputs and their weights are truncated from the picture for clarity (Figure 5-9).

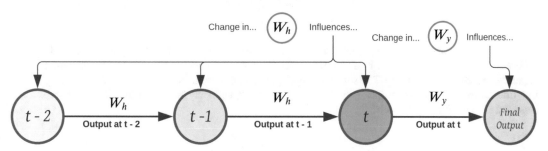

Figure 5-9. *Unrolled RNN excluding the inputs*

Recall from Chapter 3 that to "sum" up all the influence that one parameter has, we need to find the affected values if the parameter that we're differentiating changes. For a normal ANN, the backpropagation process for Wh would be as follows:

$$\frac{\partial L}{\partial W_h} = \frac{\partial L}{\partial Y} \frac{\partial Y}{\partial h_t} \frac{\partial h_t}{\partial W_h}$$

But remember changing the value of Wh not only alters the value of the current hidden state but also all previous "hidden states" or timestamps. Hence, we can partially differentiate the loss function to previous hidden states. For $t-1$ the differentiation is

$$\frac{\partial L}{\partial Y} \frac{\partial Y}{\partial h_t} \frac{\partial h_t}{\partial h_{t-1}} \frac{\partial h_{t-1}}{\partial W_h}$$

Similarly, for timestamp $t-2$, the differentiation will be the following:

$$\frac{\partial L}{\partial Y} \frac{\partial Y}{\partial h_t} \frac{\partial h_t}{\partial h_{t-1}} \frac{\partial h_{t-1}}{\partial h_{t-2}} \frac{\partial h_{t-2}}{\partial W_h}$$

Finally, these are summed up to "add up" the influence that the weight Wh has on the neuron:

$$\frac{\partial L}{\partial W_h} = \left(\frac{\partial L}{\partial Y} \frac{\partial Y}{\partial h_t} \frac{\partial h_t}{\partial W_h} \right) + \left(\frac{\partial L}{\partial Y} \frac{\partial Y}{\partial h_t} \frac{\partial h_t}{\partial h_{t-1}} \frac{\partial h_{t-1}}{\partial W_h} \right) + \left(\frac{\partial L}{\partial Y} \frac{\partial Y}{\partial h_t} \frac{\partial h_t}{\partial h_{t-1}} \frac{\partial h_{t-1}}{\partial h_{t-2}} \frac{\partial h_{t-2}}{\partial W_h} \right).$$

The preceding equation can be generalized to any neuron with any sequence length as input, and it can be written compactly as

$$\frac{\partial L}{\partial W_h} = \frac{\partial L}{\partial Y} \frac{\partial Y}{\partial h_t} \left(\sum_{i=1}^{t} \frac{\partial h_t}{\partial h_i} \frac{\partial h_i}{\partial W_h} \right)$$

where the term $\dfrac{\partial h_t}{\partial h_i}$ equals to the product between all adjacent timestamps:

$$\frac{\partial h_t}{\partial h_i} = \prod_{j=i+1}^{t} \frac{\partial h_j}{\partial h_{j-1}}$$

The differentiation for Wx is identical to Wh as changing the parameter Wx influences all the values that Wh influences when it changes. Simply replace the Wh term in the formula, and we can obtain the differentiation rule for Wx:

$$\frac{\partial L}{\partial W_x} = \frac{\partial L}{\partial Y} \frac{\partial Y}{\partial h_t} \left(\sum_{i=1}^{t} \frac{\partial h_t}{\partial h_i} \frac{\partial h_i}{\partial W_x} \right)$$

In cases of many-to-many prediction tasks (Figure 5-10), the process of backpropagation is slightly different. Instead of retrieving the loss after the final timestamp, for each output, the loss is calculated separately up to that timestamp. The same concept applies to differentiation: we only differentiate up to the timestamp of the current output with the loss calculated specifically for that output. Then the steps are repeated for each output "branch."

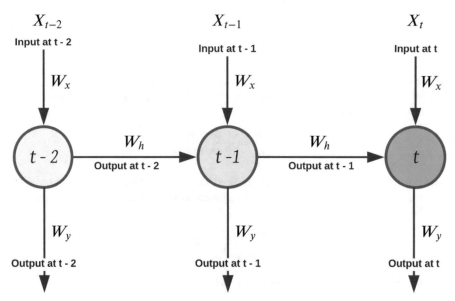

Figure 5-10. *A "many-to-many" prediction task represented in the "unrolled" RNN neuron*

Standard RNN architecture and neurons are only suitable for relatively short sequences due to vanishing gradients. In standard ANNs, vanishing gradients occur when the number of layers in a network becomes too deep. During backpropagation, when derivatives are calculated, the closer we get to the front, the smaller the gradient tends to be. This is due to the fact that each parameter is a function of the updates made previously during backpropagation. In other words, one tiny change in a parameter toward the front of the network can affect much more values than changing a parameter toward the back of the network. As we accumulate the gradient, the multiplication can likely reduce its value significantly to the point where when the algorithm gets toward the front of the network, each parameter would only update by a little or even close to none. This can lead to slow or even impossible convergence. Usually, for networks with large numbers of hidden layers, skip connections are implemented into the architecture, as seen in Figure 5-11.

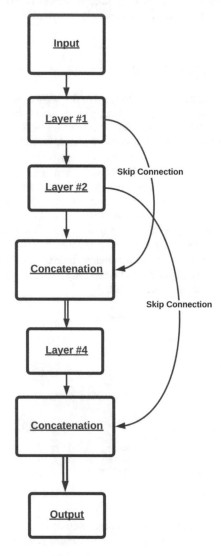

Figure 5-11. *A network with skip connections*

Skip connections allow the network to pass gradients from the back directly to layers ahead during backpropagation. Essentially, skip connections act as "transporters" that transport gradients with more considerable influences into those layers where the size of the gradient is extremely small, thus reducing the effect of vanishing gradients. Also mentioned in Chapter 3, batch normalization can reduce the effect of vanishing gradients and activation functions such as LeakyReLU.

However, in RNNs, the vanishing gradient problem is much more apparent and severe, since not only does the number of hidden layers affect the size of the gradient but also the length of the inputted sequence. As the size of sequences increases, so does the number of terms in $\sum_{i=1}^{t} \frac{\partial h_t}{\partial h_i} \frac{\partial h_i}{\partial W_h}$. Again, this can lead to gradients becoming extremely small. Adding on to the large number of hidden layers that can also cause vanishing gradients, it makes training standard RNNs with longer sequence data extremely difficult to impossible.

LSTMs and Exploding Gradients

Long Short-Term Memory (LSTM) networks effectively solve the issue as they use a gated structure that allows them to directly access data stored in memory units instead of traveling back through each hidden state. The key difference between LSTMs and standard RNNs is that LSTMs have the ability to retain past information for longer periods of time and they can more holistically understand sequences rather than treating each value in the sequence as an independent point. Furthermore, LSTMs keep track of larger patterns over longer periods of time; they can retrieve information further back than simply the previous timestamp, thus the name "Long Short-Term Memory." On the other hand, past information stored in RNN tends to be "forgotten" after processing longer sequences as information is lost at each step. Compared with RNNs, LSTMs are generally better at discovering patterns that span across long periods of time since, as mentioned before, they have access to "long-term memories" – this allows them to base their predictions not only on short-term data but also on overarching patterns discovered in the long run.

Recall that a standard RNN neuron receives the previous hidden state of the neuron along with the current timestamp's input. Then, an aggregation of the two is applied and passed through an activation function to produce the next hidden state (Figure 5-12).

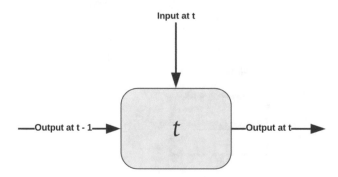

Figure 5-12. *Standard RNN neuron*

However, in an LSTM cell, an additional storage component that acts as long-term memory – preserving information from further back in time – is fed along with the hidden state. Shown in Figure 5-13 is a representation of an LSTM cell.

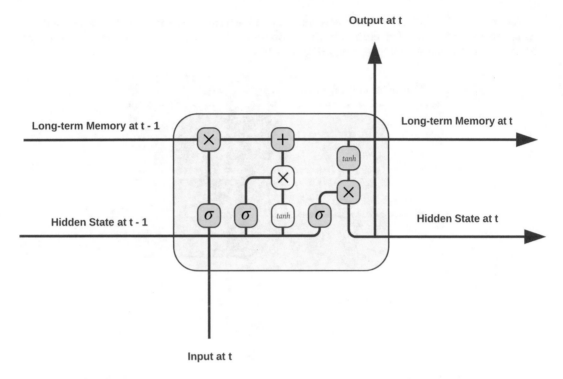

Figure 5-13. *LSTM cell*

In the context of RNNs, the "short-term memory" in LSTM is the hidden state of the neuron or information from the immediate past. The LSTM cell comprises four components: the forget gate, the input gate, the output gate, and the update gate. The forget gate and the update gate deal with the "memory" of the network, such as deciding whether to forget or replace a piece of information stored in long-term memory. The input and output gates control what is fed into the network and what is outputted. For clarity, we can denote the long-term memory (the *cell state*) as C, the short-term memory (the *hidden state*) as h, and the input as x.

The long-term memory runs through the entire neuron like a conveyor belt. Unlike hidden states, it carries information without fading as the timestamp progresses.

LSTM utilizes gates that control which information to be added to or removed from the "conveyor belt" of the long-term memory. Gates in LSTMs are composed of a sigmoid layer that restricts the amount of information that passes through them. A value of zero means that nothing is passed through, while a value of one means everything is passed through (Figure 5-14).

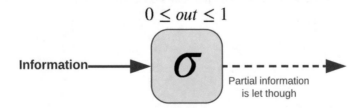

Figure 5-14. *Representation of a gate where the amount of information that's let pass is decided by a number between 1 and 0*

The first step of an LSTM is to decide what information will be thrown away from the cell state or the long-term memory C_{t-1}, based on the inputs. The previously hidden state, h_{t-1}, and the current timestamp's input, x_t, are used to train a forget gate layer f_t (Figure 5-15).

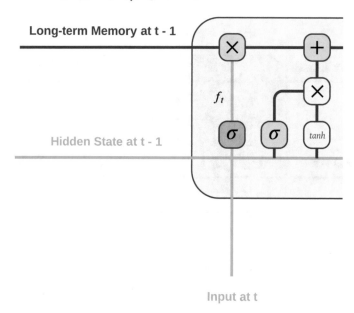

Figure 5-15. *Forget gate and long-term memory*

The forget gate outputs a value between 0 and 1 squished by the sigmoid activation function; these are multiplied by the cell state. Intuitively, a value closer to 0 will tell the network to "forget" most parts of the current cell state and vice versa. In the forget gate layer, both h_{t-1} and x_t are associated with the same weight, W_f. The output of f_t is determined from the following equation: $f_t = \sigma(W_f \cdot [h_{t-1}, x_t] + b_f)$. Here, b_f is the bias of the layer.

The next step is creating new information to add to the long-term memory. Again, the amount of information and its values are calculated based on the hidden state and the current timestamp's inputs. Before producing any new long-term memories, the LSTM cell computes how many of these "memories" will be updated. This is accomplished using the exact mechanism as the forget gate by introducing an input gate (Figure 5-16). Values between zero and one are produced with the equation $i_t = \sigma(W_i \cdot [h_{t-1}, x_t] + b_i)$.

Then a vector of possible candidate values, $\tilde{C}_t$, is generated. Thus, another layer of weights and biases are trained to produce these possible "new memories": $\tilde{C}_t = tanh\left(W_c \cdot [h_{t-1}, x_c] + b_c\right)$. Note that the hyperbolic tangent activation function is used to restrict output values between –1 and 1.

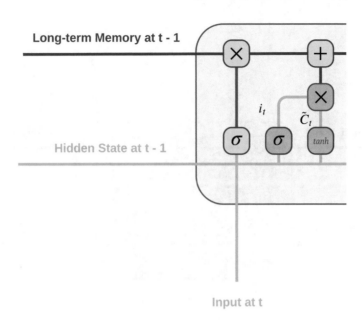

Long-term Memory at t - 1

Hidden State at t - 1

Input at t

Figure 5-16. *Input gate and new memories*

With all this work done on creating and calculating long-term memories, it's finally time to update the cell state C_{t-1} into the new cell state C_t. This represents the new long-term memory of the neuron (Figure 5-17). The new cell state is simply a linear combination of the terms we've calculated before: $C_t = f_t * C_{t-1} + i_t * \tilde{C}_t$. For a better understanding of this equation, treat it as a weighted average between the old memory C_{t-1} and the newly calculated memory $\tilde{C}_t$ where its weights are f_t and i_t, respectively.

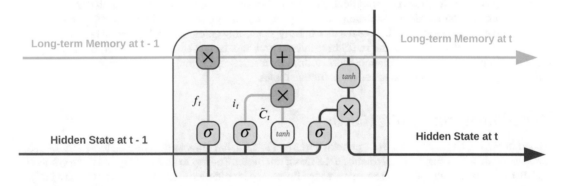

Long-term Memory at t - 1   **Long-term Memory at t**

Hidden State at t - 1   **Hidden State at t**

Figure 5-17. *Updating the long-term memory*

The final step in the process is generating a prediction along with the next hidden state (not the cell state since that was already done previously). It will be based on a partial version of the current, updated cell state. Again, a gate is implemented to decide which part of the cell state to use for prediction and which are thrown out. An output gate layer is trained on h_{t-1} and x_t, fed through the gate with a sigmoid function: $o_t = \sigma(W_o \cdot [h_{t-1}, x_t] + b_o)$. The current cell state is put through the tanh activation function and finally multiplied with the output from the output gate: $h_t = o_t * \tanh(C_t)$ (Figure 5-18).

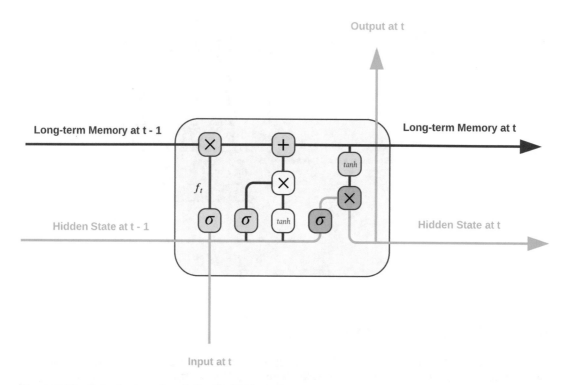

Figure 5-18. *Output gate and updating the hidden state*

The hidden state h_t will be the final output if the sequence reaches its end but a hidden state if the sequence passed in does not reach its end. It can also be both in a many-to-many prediction task.

Although LSTMs solved the vanishing gradient problem, the opposite can also happen – gradients become exceedingly large, and the loss reaches infinity. Exploding gradients can happen in both RNNs and LSTMs due to similar reasons as the vanishing gradient problem. It can generally be reliably addressed by gradient clipping during the gradient descent update. By restricting the gradient to be always between two values, large gradient values will never cause infinite losses.

Gated Recurrent Units (GRUs)

Gated Recurrent Units, or GRUs, are a popular variant of LSTMs that reduces the complexity of various gates and layers while keeping the performance of LSTMs if not better. The key to eliminating vanishing gradients in GRU lies in its update gate and reset gate, which function a bit differently compared with standard LSTMs. The forget and input gates in LSTM are combined into one single component known as the update gate. Following, the reset gate acts similar to the operation done on the long-term memory in LSTM – deciding which piece or pieces of information to forget and which gets carried on to the next timestamp. Shown in Figure 5-19 is a representation of a single GRU.

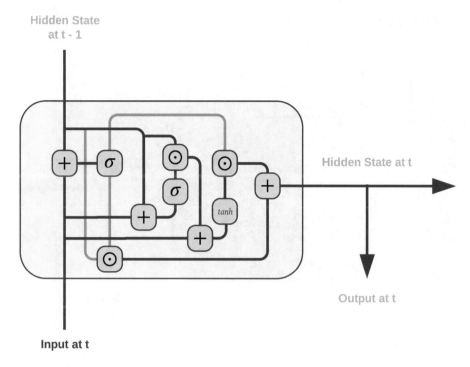

Figure 5-19. *Representation of a GRU cell*

Entering the GRU are the current timestamp's inputs and the hidden state obtained from the previous timestamp. Note that GRU removed the long-term memory part of LSTM. The update gate handles how much information the network is going to keep from the previous hidden state. Instead of having the same parameters as the LSTM, the input and the hidden state are both multiplied by their own individual trained weights. The result of both is added up and then passed through the sigmoid activation function: $z_t = \sigma(W^{(z)}x_t + U^{(z)}h_{t-1})$ where $W^{(z)}$ is the weight matrix for the input x_t and $U^{(z)}$ is the weight matrix for the hidden state h_{t-1}. By implementing a gate here, we can decide how much information are we going to keep and how much are we going to pass on to the future (Figure 5-20). The cleverness of GRU lies in its ability to simply carry entire chunks of information from the past without having the risk of vanishing gradients.

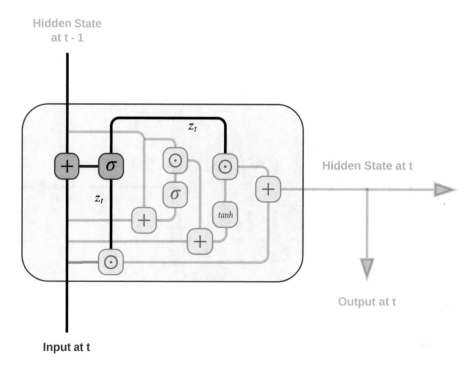

Figure 5-20. *Update gate*

The reset gate decides how much information to forget. This essentially performs the opposite job compared with the update gate (Figure 5-21). The computation of the reset gate value r_t is the exact same compared to the update gate simply trained with different weights: $r_t = \sigma(W^{(r)}x_t + U^{(r)}h_{t-1})$.

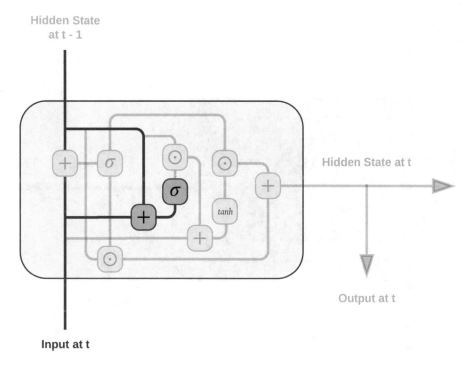

Figure 5-21. *Reset gate*

That was all that we needed in terms of gates in GRU. The structure and complexity of GRUs compared with LSTMs is much simpler. Now we can piece together the information gathered through the update and reset gates, keeping some part of h_{t-1} while forgetting other parts. This creates a new memory content specific to the current timestamp, which will serve as an intermediate result to computing the new hidden state. Again, like the preceding reset and update gates, we train two distinct sets of weights for the current timestamp's feature and the hidden state or memory from the past timestamp. By taking the Hadamard product (element-wise matrix multiplication) between the reset gate value and the hidden state, parts of the hidden state are "forgotten":

$$\widetilde{h}_t = \tanh\left(Wx_t + U\left(r_t \odot h_{t-1}\right)\right)$$

The result of $U(r_t \odot h_{t-1})$ is then summed with Wx_t to produce $\widetilde{h}_t$. Finally, the sum is passed through the hyperbolic tangent activation function (Figure 5-22).

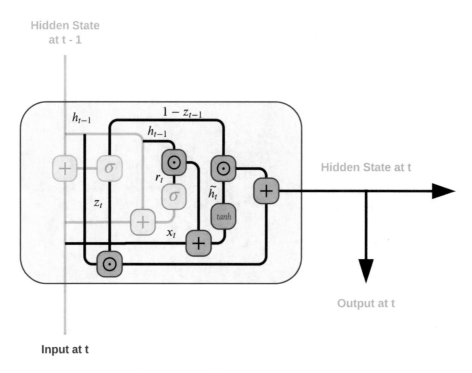

Figure 5-22. *Components to produce the new hidden states and output*

Lastly, the model needs to compute the hidden state that will be passed onto the next timestamp, h_t. This is simply calculated as a weighted average between the current memory content $\widetilde{h}_t$ and the previous hidden state h_{t-1} passed in as input. Utilizing the update gate that is trained to decide how much information is going to be kept from h_{t-1}, we can piece together h_t as such: $h_t = z_t \odot h_{t-1} + (1 - z_t) \odot \widetilde{h}_t$.

The hidden state that's passed onto the network also serves as an output. For example, on the last timestamp when h_t is produced, usually a nonlinear activation function is applied and then a linear layer, which matches the shape of the output. Unlike LSTM, there's no specific step to separately calculate an output. The hidden state and output in GRU are synonymous. GRU does not provide as much memory manipulation as LSTM does with its long-term memory. However, in most cases GRU performs similarly compared to LSTM, if not better, due to its reduced training complexity.

Bidirectionality

The standard recurrent, LSTM, and GRU layers all allow future timesteps in a sequence to be informed by previous timesteps, to differing degrees of effectiveness at differing levels of sophistication. However, in many cases we also want the *past* to be informed by the *future*. For instance, consider the sentence "John Doe – shaken and emotional – cried tears of joy." If we are to process this in the standard sequential paradigm of recurrent modeling, the model – up to the last word – reasons that this sentence is associated with sadness, negative sentiment, and so on. It will be very difficult for the last word to completely change the character of the hidden (and cell, if using an LSTM) state. In reality, however, that last word has a profound effect on how we interpret the beginning of the sentence.

To address this, we can use bidirectionality. A bidirectional recurrent layer is really a stack of two recurrent layers; one is applied in the forward direction and the other in the backward direction. The hidden states of each are then combined, either through addition or concatenation. The output of the bidirectional recurrent layer at any timestep, therefore, is informed by the entire sequence.

It is generally not necessary to stack multiple bidirectional recurrent layers on top of each other in a stack. A bidirectional recurrent layer is often placed as the first recurrent layer, which already produces a sequence informed by all elements of the sequence, not just the timesteps before it. We can continue to process the resulting sequence in a standard sequential manner in later layers of the recurrent stack.

Introduction to Recurrent Layers in Keras

Before diving into real-world applications of recurrent modeling, this section serves as a brief introduction to the syntax of recurrent and sequence modeling. Due to the special nature of recurrent networks and the way they handle data, the syntax for modeling these situations is going to be a bit different from tabular data models.

Preprocessing data for a recurrent model can be done in various ways depending on the data type. Tabular data or time-series data are typically simpler to preprocess, while language data is more difficult to deal with. This introductory section will only cover the major components surrounding recurrent modeling, including input shape and recurrent batch sizes. More advanced preprocessing techniques will have to depend on the application. Such usages will be demonstrated to you in later sections where recurrent models are applied in various scenarios.

We can start with one of the simplest representations of a time-series dataset: a list of numbers following a pattern. Specifically, we'll use the sequence shown in Figure 5-23.

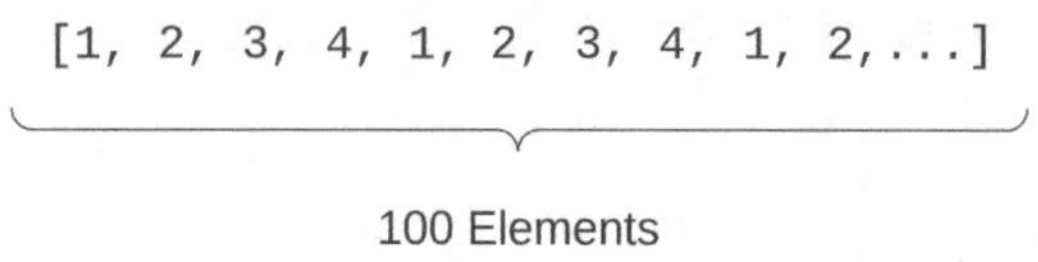

```
[1, 2, 3, 4, 1, 2, 3, 4, 1, 2, ...]
```

100 Elements

Figure 5-23. *Example sequence with length 100*

The pattern present in the sequence can easily be identified as it is plainly the numbers 1, 2, 3, and 4 repeating in that order. The major difference between a typical tabular data prediction task and time-series prediction is that in tabular models sets of features are given corresponding to a different set of targets, while this is not the case in time-series prediction. In a time-series prediction, features are assigned to timestamps. When building the training set, chunks of the data are split into past and future based on its timestamps where the past is trained to predict the future. In a time-series prediction task, features and labels come from the same variable/column of data.

There is a lot of flexibility in time-series prediction tasks in terms of how many "labels" should be predicted or how much training data should be used. There's typically no set amount on how much training data should be portioned from the dataset. However, one general rule of thumb is that the training data should be large enough to capture any periodic trends relevant to predicting future trends. This also applies to how much training data should be fed into the network at one batch. In our preceding example, we might decide that the first 80 terms will be used for training, while the last 20 terms will be testing or validation data (Figure 5-24). We can arbitrarily decide to feed in eight data points during model training while only letting the model predict one output into the future. The length of the sequence we feed in is referred to as the "window size." We can modify this value based on training results or domain knowledge to improve model accuracy.

Timestamps

Figure 5-24. *Representation of time-series training and testing data*

There are multiple ways to set up a time-series dataset with Keras. One of the easiest methods is using the TimeseriesGenerator class.

We can start off by defining our preceding toy dataset in NumPy arrays. Then we import the TimeseriesGenerator class from tensorflow.keras.preprocessing.sequence as shown in Listing 5-1.

Listing 5-1. Creating the example dataset and importing TimeseriesGenerator

```
from tensorflow.keras.preprocessing.sequence import TimeseriesGenerator

example_data = np.array([1, 2, 3, 4]*25)

# 80-20 train test split
train = example_data[:80]
test = example_data[80:]
```

When the class is instantiated, there are a few important parameters that should be set. Like any other machine learning model or dataset class, we need to pass in the features and targets. In our case, they would be the same NumPy array, since our features and targets come from the same column of data. Next, the length parameter defines the number of samples that the model will use as features to predict values in the next timestamp. Finally, the batch_size represents the number of time-series samples per batch. Usually for smaller datasets like ours, a batch size of 1 will do. Listing 5-2 defines a TimeseriesGenerator with a length of 8 and a batch size of 1.

Listing 5-2. Instantiating the generator

```
generator = TimeseriesGenerator(train, train, length=8, batch_size=1)
```

Remember that the length of the generator will be 8 less than our 80-element training data since the last eight values do not have a corresponding target. In the following code snippet, we imported the SimpleRNN, LSTM, and GRU layers corresponding to RNN layers, LSTM layers, and GRU layers (Listing 5-3).

Listing 5-3. Importing recurrent model layers

```
# we can also use L.LSTM, but these are imported
# separately for clarity
from tensorflow.keras.layers import SimpleRNN, LSTM, GRU
```

We can first start off by constructing a basic RNN model with the Keras functional API (Listing 5-4). Our dataset is a univariate time series, or in other words, there's only one feature. Thus, our input shape would be (length, 1). Notice here that we're skipping the extra dimension created by batches; it's not necessary for the argument input_shape. The actual input shape would be (batch_size, length, n_features).

Listing 5-4. Code for creating a model with one SimpleRNN layer

```
from tensorflow.keras.layers import SimpleRNN, LSTM, GRU

inp = L.Input(shape=(8, 1)) # NOT THE SAME as (8,)
x = SimpleRNN(10, input_shape=(8, 1))(inp)
# extra layer to process info before output
x = L.Dense(4)(x)
# output layer
out = L.Dense(1)(x)

model = Model(inputs=inp, outputs=out)
```

By default, we compile the model with the Mean Squared Error loss function while using the Adam optimizer. Note that our dataset is a generator instead of a plain NumPy array. We need to call `fit_generator` instead of `fit` to initiate the training process (Listing 5-5).

Listing 5-5. Fitting the model using fit_generator

```
model.fit_generator(generator, epochs=40)
```

For prediction on testing data, the process again is slightly different from standard tabular model predictions. To achieve time-series prediction, we need to use data from the immediate past. To put it more concisely, some of our future prediction will be based on past predictions in previous timestamps. Shown in the following diagram is the logic behind time-series forecasting (Figure 5-25).

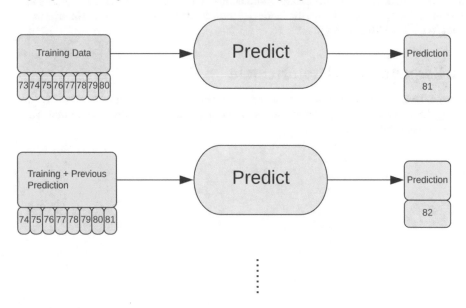

Figure 5-25. *Logic for time-series prediction*

In our example, we have a portioned test data, meaning that we can just create a generator the same way as we instantiated the training one. Then we can obtain predictions by simply passing in the test generator to `predict_generator`. However, during real-time forecasting, we don't have a "test dataset," which can feed in the correct values that precede the timestamp we're going to predict. Our trained model predicts one future value based on eight values from past timestamps. During real-time forecasting, we need to get the last eight samples of data that precede the timestamp of the sample that we're going to predict. Then for predicting the next sample, we will need seven samples from the past along with the sample that we just predicted. This process continues until we have predicted our desired number of timestamps. One drawback that comes with time-series forecasting is that predictions tend to get more inaccurate as the timestamp moves forward. This occurs since predictions in the far future will likely be based on predictions in the near future. No model is perfect, and there may be slight errors to our prediction for the near future, even if it might not make a difference at that point. But as our prediction continues, the error will be amplified, thus making predictions into the far future less accurate than one might expect from their models.

The usage of LSTM and GRU layers is like that of `SimpleRNN`; they can be used the same way as `SimpleRNN` by replacing the layer call with LSTM or GRU (Listing 5-6).

Listing 5-6. Example of integrating GRU layers

```
inp = L.Input(shape=(8, 1))
x = GRU(10, input_shape=(8, 1))(inp)
# extra layer to process info before output
x = L.Dense(4)(x)
# output layer
out = L.Dense(1)(x)

model_gru = Model(inputs=inp, outputs=out)
# compile and train as normal
```

Additionally, to make any recurrent layer bidirectional, add a `keras.layers.Bidirectional` wrapper around the layer. For instance, the following code creates a bidirectional LSTM layer: `x = L.Bidirectional(L.LSTM(...))(x)`.

Return Sequences and Return State

In all recurrent layers, there are two additional arguments that are crucially important: `return_sequences` and `return_state`. For a better explanation, we can refer to the LSTM diagram as an example shown in the previous section in Figure 5-26.

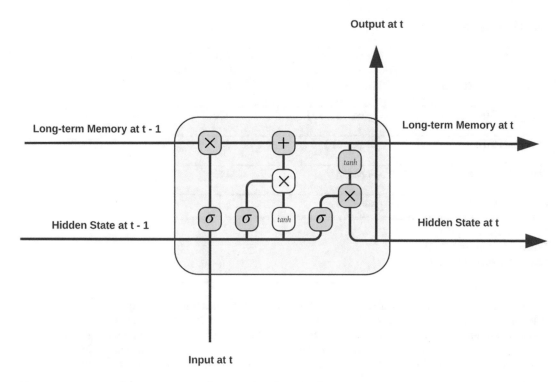

Figure 5-26. *A visual representation of LSTM taken from the "LSTMs and Exploding Gradients" section*

By default, both return_sequences and return_state are set to false.

Return sequences output all hidden states at every timestamp. Currently, with the default parameter in the LSTM layer, only the hidden state at the last timestamp is returned as the final output. Return sequences will return all hidden states from every timestamp in the "unrolled" LSTM cell. Setting this parameter to true allows for stacking LSTMs as it can "start off from before" by receiving the result from the previous LSTM in the correct data and shape format. The following code shows stacking LSTMs as well as the model.summary for a clearer understanding of the LSTM output shape (Listing 5-7) . The second dimension of the LSTM layer output is where the hidden state from every timestamp is stored, and it serves as an input to the next LSTM layer.

Listing 5-7. Stacking LSTMs with return_sequences set to true

```
inp = L.Input(shape=(8, 1))

# stacking lstms
x = LSTM(10, input_shape=(8, 1), return_sequences=True)(inp)
# no additional params needed
x = LSTM(10)(x)
x = L.Dense(4)(x)
# output layer
out = L.Dense(1)(x)

model_lstm_stack = Model(inputs=inp, outputs=out)
model_lstm_stack.summary()
```

Model: "model_3"

Layer (type)	Output Shape	Param #
input_5 (InputLayer)	[(None, 8, 1)]	0
lstm_5 (LSTM)	(None, 8, 10)	480
lstm_6 (LSTM)	(None, 10)	840
dense_6 (Dense)	(None, 4)	44
dense_7 (Dense)	(None, 1)	5

Total params: 1,369
Trainable params: 1,369
Non-trainable params: 0

On the other hand, return state outputs the hidden state of the last timestamp, equivalent to the output, twice, as two separate NumPy arrays. Then the long-term memory, or the cell state of the last timestamp, is also outputted as a separate NumPy array. In total, by setting the return state argument to true, three separate NumPy arrays (two for RNN and GRUs since there's no cell state) will be outputted. Notice in model.summary there are three output shapes corresponding to three output values (Listing 5-8).

Listing 5-8. LSTM with return_state set to true

```
inp = L.Input(shape=(8, 1))

lstm_out, hidden_state, cell_state = LSTM(10, input_shape=(8, 1), return_state=True)(inp)
x = L.Dense(4)(lstm_out)
# output layer
out = L.Dense(1)(x)

model_return_state = Model(inputs=inp, outputs=out)
model_return_state.summary()

# extra print for cell_state's shape
# since it was cut off in the summary
print(cell_state.shape)
```

Model: "model_5"

Layer (type)	Output Shape	Param #
input_8 (InputLayer)	[(None, 8, 1)]	0
lstm_9 (LSTM)	[(None, 10), (None, 10),	480
dense_11 (Dense)	(None, 4)	44

```
dense_12 (Dense)              (None, 1)                    5
=================================================================
Total params: 529
Trainable params: 529
Non-trainable params: 0
```

(None, 10)

Both `return_sequences` and `return_state` come in handy during complex RNN manipulation as they can retrieve intermediate results also allowing RNN stacking, which, when used correctly without stacking too many layers as it can cause exploding gradients, can be very powerful. Such examples will be demonstrated in later sections.

Standard Recurrent Model Applications

In this section, we will look at two important applications of recurrent models: natural language and time series. Recall that recurrent layers can be reasonably applied to most sequential data; natural language and time-series data are some of the most common forms of data ordered along a sequential or temporal axis.

Natural Language

In this subsection, we'll consider the example of building recurrent models for text classification/regression, a commonly encountered problem in a variety of contexts. We will use a software product reviews subset of the US Amazon Reviews dataset (Figure 5-27), which is a large corpus of Amazon product reviews and associated data like the star rating, the date of the review, the number of upvotes, whether the review was associated with a verified purchase, etc. We will try to build a model to predict the star rating of a review given the review text. Such a model could be used to automatically extract a quantity metric of customer satisfaction given customer input in natural language form, which could be used to gauge customer satisfaction in contexts where natural language is available but a concrete star rating isn't (e.g., discussions about a product on social media).

	data/review_body	data/star_rating
0	I haven't worked for a long time with Final Dr...	3
1	This was a gift for my husband. He does a lot...	5
2	Good condition, and a nice set'	5
3	I learned 3D design using Autodesk Inventor. ...	5
4	This is a great way to help your child with ma...	5
...	...	...
341926	There is no custom card size so you are forced...	1
341927	Nothing wrong with this purchase. As expected...	5
341928	In the past I used H&R block. I was very sati...	2
341929	This seems to be a well designed set of tool f...	4
341930	Do not buy this product. Nortons does not even...	1

Figure 5-27. *The Amazon US Software Reviews dataset*

First, we will vectorize the data by using ordinal encoding (Listing 5-9, Figure 5-28). After a sequence is converted to lowercase and punctuation is removed, each token is associated with an integer, and a passage of text is represented as a sequence of tokens. To ensure that all sequences are of the same length, we add padding tokens at the end (Figure 5-29). This is all taken care for us with TensorFlow's TextVectorization layer. After instantiating the vectorizer with our desired parameters, we adapt it to our dataset so it learns a mapping from tokens to integers. Once adapted, we can call the vectorizer on our text to obtain a set of tensors that can be passed directly for training into our model.

Listing 5-9. Vectorizing the text

```
SEQ_LEN, MAX_TOKENS = 128, 2048

From tensorflow.keras.layers import TextVectorization
vectorize = TextVectorization(max_tokens=MAX_TOKENS,
                        output_sequence_length=SEQ_LEN)
vectorize.adapt(data['data/review_body'])
vectorized = vectorize(data['data/review_body'])
```

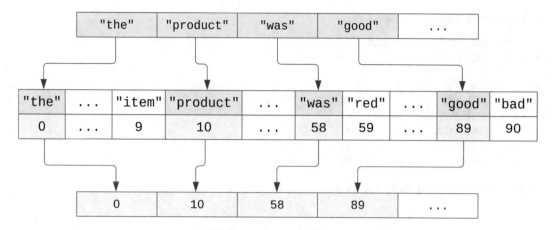

Figure 5-28. *Mapping each token in the input to a word-specific integer*

0	1	5	6				
0	1	5	6	7	7	7	7

5	2	4	3	1	0	0	
5	2	4	3	1	0	0	7

1	5	3	2	6	5	2	4
1	5	3	2	6	5	2	4

Figure 5-29. *Adding an additional token to serve as a padding token such that all sequences are the same length*

Recall that in Chapter 2 we discussed different ways of encoding data. We don't need that here because we have learnable embeddings. When we construct a natural language model (Listing 5-10, Figure 5-30), the first step is to build an embedding layer. The embedding layer, as previously discussed in Chapters 3 and 4, learns to associate separate tokens represented by integers with fixed vectors. The embedding layer requires us to specify the vocabulary size and the embedding dimension, or the dimensionality of the vector each token will be associated with. After embedding, the data will have shape (SEQ_LEN, EMBEDDING_DIM). We can pass this through a "vanilla" recurrent layer (keras.layers.SimpleRNN) with 32 recurrent units. The output vector will thus be 32-dimensional. We can pass this through several dense layers to process or "interpret" the result and then map it to a softmax output for classification.

405

Listing 5-10. Building a text model

```
SEQ_LEN, MAX_TOKENS = 64, 2048
EMBEDDING_DIM = 16

inp = L.Input((SEQ_LEN,))
embed = L.Embedding(MAX_TOKENS, EMBEDDING_DIM)(inp)
rnn = L.SimpleRNN(32)(embed)
dense = L.Dense(32, activation='relu')(rnn)
dense2 = L.Dense(32, activation='relu')(dense)
out = L.Dense(5, activation='softmax')(dense2)

model = keras.models.Model(inputs=inp, outputs=out)
```

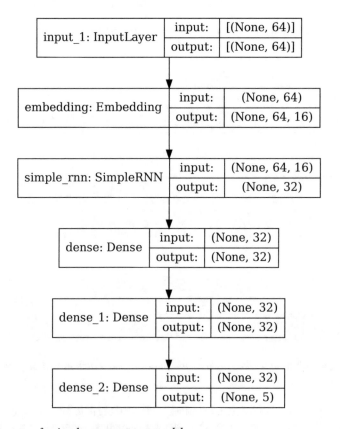

Figure 5-30. *Architecture of a simple recurrent text model*

To be completely clear about what's going on, let's track how a sample sequence of tokens [0, 1, 2, 3, 4] gets processed by this model. First, each of the tokens is mapped to a learned embedding vector $\vec{e}_n$ (Figure 5-31). This embedding vector contains important latent features that capture the meaning/essence of each word among many dimensions relative to the problem at hand.

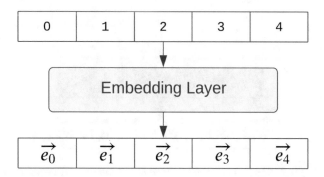

Figure 5-31. *Example of sequence data*

Then, each embedding vector in the embedding layer is processed sequentially by a recurrent cell, which begins with an initialized hidden state and takes in the first element (Figure 5-32). The hidden state produced by this recurrent cell is fed back into the recurrent cell, which then takes the second element. This proceeds for each element in the sequence until the final output $\vec{h}$ is produced.

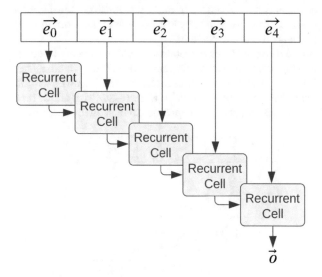

Figure 5-32. *Recurrent cell processing each embedding vector*

This vector $\vec{o}$ contains relevant information that has theoretically been informed by all elements of the sequence in a sequential manner. We can then pass $\vec{o}$ through several fully connected layers to further interpret this information with respect to optimizing performance on the prediction task and then finally apply a softmax layer such that each output indicates the probability prediction for each class (Figure 5-33).

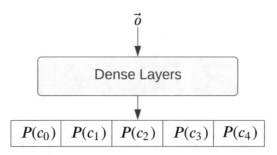

Figure 5-33. *Probability prediction for each class derived from a recurrently generated information vector, or hidden state*

After the model has been defined, we can compile and fit the model on our dataset. This is fairly straightforward and similar to previous models we have built in Keras.

Note that Keras's syntax for recurrent layers is especially convenient (especially in comparison with other popular deep learning frameworks, such as PyTorch). To use an LSTM or GRU layer instead of the vanilla recurrent layer, replace L.SimpleRNN with L.LSTM and L.GRU.

However, a recurrent layer can only capture so many relationships. Ideally, we would want to be able to capture multiple layers of depth and complexity in language. To make our model more complex, we can stack multiple recurrent layers on top of each other.

Recall that in a standard recurrent layer, we retain the hidden state of the current timestep for consideration in the following timestep, but the outputs of the cell at each timestep are ignored except at the last timestep. If we instead collect the outputs at each timestep, however, we obtain another sequence (Figure 5-34), which we can recurrently process again (Figure 5-35). This allows for the model to learn multiple layers of complexity that may require deeper recurrent sequence processing to uncover.

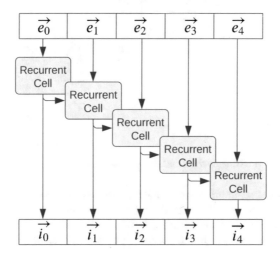

Figure 5-34. *A recurrent layer in which all states are collected (as opposed to just the last)*

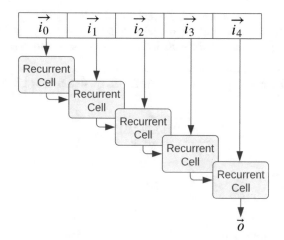

Figure 5-35. *Accepting the sequence of states from a previous hidden layer and passing it through another layer*

Recall from earlier that to collect the outputs at each timestep, we set `return_sequences=True` and continue to stack additional layers (Listing 5-11).

Listing 5-11. Using a double-RNN stack

```
inp = L.Input((SEQ_LEN,))
embed = L.Embedding(MAX_TOKENS, EMBEDDING_DIM)(inp)
rnn1 = L.SimpleRNN(32, return_sequences=True)(embed)
rnn2 = L.SimpleRNN(32)(rnn1)
dense = L.Dense(32, activation='relu')(rnn2)
dense2 = L.Dense(32, activation='relu')(dense)
out = L.Dense(5, activation='softmax')(dense2)

model = keras.models.Model(inputs=inp, outputs=out)
```

If you observe your network is overfitting, one approach is to increase recurrent dropout. In recurrent dropout, a certain proportion of the hidden vector is dropped at each timestep. Note that this is different from applying dropout as a separate layer after processing with a recurrent layer; recurrent dropout is applied "within" the recurrent layer at every timestep, whereas standard dropout is applied only to its final output. This can be set by passing `recurrent_dropout=...` into a recurrent layer's instantiation.

Chapter 10 will discuss the Neural Architecture Search library AutoKeras. AutoKeras supports working with high-level blocks that make modeling text data simpler to work with.

Time Series

Time-series data can come in many forms. In general, it is data that is taken sequentially at equal time intervals. The objective is usually either next-timestep forecasting (predict t_n from $\{t_{n-w}, t_{t-w+1}, ..., t_{n-1}\}$ for some window length w, Figure 5-36), time-dependent target prediction (predict some time-dependent target y_n from $\{t_{n-w+1}, t_{t-w+2}, ..., t_n\}$ for some window length w, Figure 5-37), or time-independent target prediction (predict some time-independent target y from $\{t_{n-l+1}, t_{t-l+2}, ..., t_n\}$ for some sequence interval length *l*, Figures 5-38 and 5-39). The difference between the tasks is largely a difference in the data rather than the model: you can still define a recurrent model with the same structure (adjusting the input and output sizes as appropriate), and most of the work will be in preparing the format of the data.

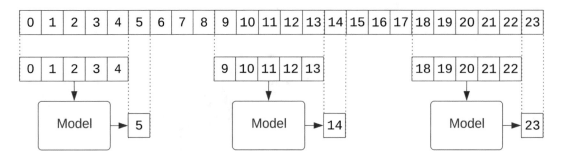

Figure 5-36. *Next-timestep forecasting*

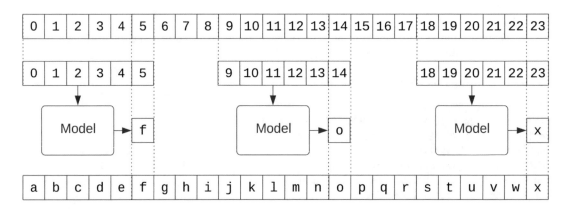

Figure 5-37. *Time-dependent prediction*

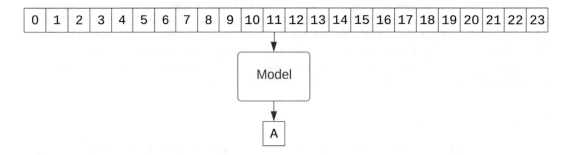

Figure 5-38. *One mode of time-independent prediction*

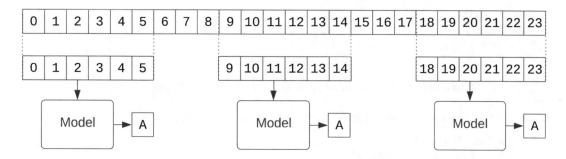

Figure 5-39. *Another mode of time-independent prediction*

Some examples of time-series problems include the following:

- *Stock forecasting*: Predict the next day's stock given the previous w days' worth of stock data (next-timestep forecasting).

- *Epidemic forecasting*: Predict the next day's number of infections given the previous w days' worth of infection data (next-timestep forecasting).

- *Political sentiment forecasting*: Predict the mean sentiment of a political community (social platforms, parties, figures, etc.) at some timestep given a window of previous political activity (legislative activity, electoral results, etc.) (time-dependent target prediction).

- *Voice accent classification*: Predict the accent (British, American, Australian, etc.) of the audio file (time-independent target prediction).

Recurrent layers do not always do well by their own on time-series data. Often, time-series data has a very large sampling frequency (consider: an audio file, high-frequency stock data), yet recurrent layers still process this information on a timestep-by-timestep basis. This is equivalent to listening to music by hearing its constituent elements on a 1/16000th-of-a-second basis (assuming a sampling rate of 16000 elements per second). Even advanced memory-equipped recurrent models fail to retain information over just a few seconds because there are so many elements per second of audio. The same applies for other forms of high-frequency data.

For this reason, it is often a successful strategy to use one-dimensional convolutional layers in conjunction with recurrent layers. These layers can, if well-designed, systematically extract relevant sequentially ordered features, which can be more effectively processed by recurrent layers.

The Speech Accent Archive dataset (`www.kaggle.com/datasets/rtatman/speech-accent-archive`) contains many audio files of the same phrase being spoken by people with different accents. We will attempt to build a time-independent target prediction model that predicts the accent of the speaker from the audio file. The directory contains several files organized as such:

```
['spanish47.mp3',
 'english220.mp3',
 'arabic64.mp3',
 'russian7.mp3',
 'dutch36.mp3',
 'english518.mp3',
 'bengali5.mp3',
 'english52.mp3',
```

```
'arabic11.mp3',
'farsi11.mp3',
'khmer7.mp3',
...
```

Listing 5-12 is a helper function to extract the accent class from the filename.

Listing 5-12. Helper function extracting class information from the filename

```
def clean_name(filename):
    for i, v in enumerate(filename):
        if v in '0123456789':
            break
    return filename[:i]
```

Listing 5-13 identifies the top five accents by frequency (we want to predict on classes with sufficient training data) and stores the corresponding audio files and labels (ordinal-encoded).

Listing 5-13. Obtaining relevant classes and ordinal encodings

```
directory_path = '../input/speech-accent-archive/recordings/recordings/'
filenames = os.listdir(directory_path)
classes = [clean_name(name) for name in filenames]
i, j = np.unique(classes, return_counts=True)
top_5_accents = [x for _, x in sorted(zip(j, i))][::-1][:5]

top_5_files = [file for file in filenames if clean_name(file) in top_5_accents]
top_5_classes = [clean_name(file) for file in top_5_files]
ordinal_encoding = {val:i for i, val in enumerate(np.unique(top_5_classes))}
top_5_classes = [ordinal_encoding[class_] for class_ in top_5_classes]
```

Audio files are long strings of floats (Figure 5-40). We will use the librosa library to read the .wav files into a NumPy array. When loading, we need to provide a sampling rate – the number of data points sampled each second. If you use a sampling rate of 1000, then a 5-second audio clip will have 5000 elements in array form. Choosing the sampling rate is a balancing problem: if the sampling rate is too large, the audio quality is optimal, but it may be too long and cause training issues; if the sampling rate is too small, it may be a feasible size, but the audio quality is too degraded to perform the task. A sampling rate in between 3,000 and 10,000 is a good range to look for when dealing with human voice audio. (To play around with the optimal rate, try loading with a certain sampling rate, save the audio file with that sampling rate, and then listen to the altered audio.) In this case, we choose a sampling rate of 6,000.

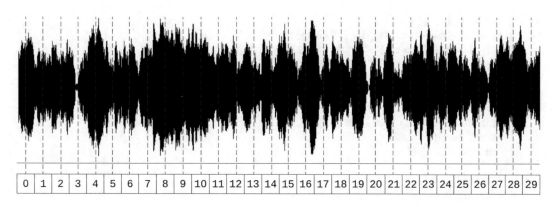

Figure 5-40. *Visualization of mapping audio to a sequence of values*

The audio files vary in length, but we need a uniform interval length to feed into the model. We will choose a 5-second window length, which should be enough context to classify the accent of the speaker (Figure 5-41, Listing 5-14). Moreover, we will choose a 5-second window shift length, which means that the start times of adjacent windows are 5 seconds apart. This means that there is no overlap in audio, which is fine in this case because we have sufficient training data. (One strategy to address limited audio data is to decrease the shift size such that samples overlap with each other.) We store each of the windows and their associated targets in audio and target, respectively.

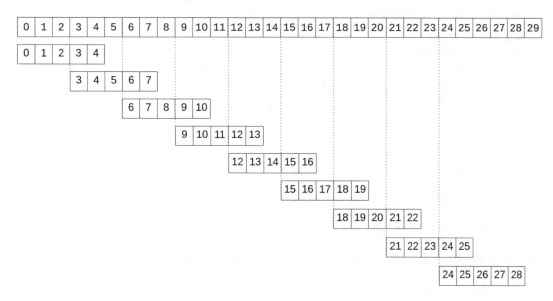

Figure 5-41. *Windowing data across a sequence with an overlap of two elements*

Listing 5-14. Obtaining windows from the dataset

```
SAMPLE_RATE = 6_000
WINDOW_SEC = 5
WINDOW_LEN = WINDOW_SEC * SAMPLE_RATE
SHIFT_SEC = 5
SHIFT_LEN = SHIFT_SEC * SAMPLE_RATE
```

```
audio, target = [], []
for i, file in tqdm(enumerate(top_5_files)):
    y, sr = librosa.load(os.path.join(directory_path, file),
                         sr=SAMPLE_RATE)
    start, end = 0, WINDOW_LEN
    while (end < len(y)):
        audio.append(y[start:end])
        target.append(top_5_classes[i])
        start += SHIFT_LEN
        end += SHIFT_LEN

audio = np.array(audio)
target = np.array(target)
```

Let us construct the model (Listing 5-15, Figure 5-42). We will begin by reshaping the input into a two-dimensional array, which should be interpreted as a sequence with one feature map representation. Then, we apply a series of one-dimensional convolutions; these increase the number of feature map representations and decrease the length of sequence by making large strides. Since the four convolutions use strides of 8, 8, 4, and 4, the original sequence length is cut down by a factor of $8 \times 8 \times 4 \times 4 = 1024$ (in actuality, a little bit more, due to the kernel size; we don't use padding here). This helps reduce the 30,000-length input to an information-dense 25-length sequence of size-8 vectors, which are processed recurrently.

Listing 5-15. Constructing the audio model

```
inp = L.Input((WINDOW_LEN,))
reshape = L.Reshape((WINDOW_LEN,1))(inp)
conv1 = L.Conv1D(4, 16, strides=8, activation='relu')(reshape)
conv2 = L.Conv1D(4, 16, strides=8, activation='relu')(conv1)
conv3 = L.Conv1D(8, 16, strides=4, activation='relu')(conv2)
conv4 = L.Conv1D(8, 16, strides=4, activation='relu')(conv3)

lstm1 = L.LSTM(16, return_sequences=True)(conv4)
lstm2 = L.LSTM(16)(lstm1)
dense1 = L.Dense(16, activation='relu')(lstm2)
dense2 = L.Dense(16, activation='relu')(dense1)
out = L.Dense(5, activation='softmax')(dense2)

model = keras.models.Model(inputs=inp, outputs=out)

model.compile(optimizer='adam', loss='sparse_categorical_crossentropy',
              metrics=['accuracy'])
model.fit(audio, target, epochs=100)
```

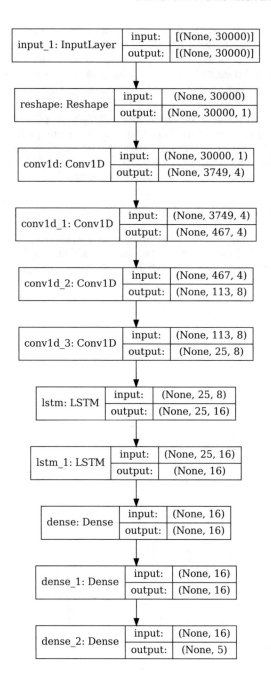

Figure 5-42. *Architecture of the audio model*

An advantage of using deep learning to model time-series data is the ability to jointly model multiple time series. For instance, rather than modeling a single stock price, a neural network can simultaneously model dozens or even hundreds of stock prices throughout time: each additional stock is just another channel. This allows for jointly informed modeling – for instance, the prices of stocks in a similar industry likely have strong relationships with each other, and joint modeling is likely to yield superior performance to independent modeling.

While recurrent models are good for high-frequency and high-complexity time-series data (audio, high-frequency stock market data), they can be overkill for simpler time-series problems. Domain-specific time-series problems often have a long history of research that has yielded tried-and-true modeling techniques that should not be overlooked. Moreover, more "classical" or "manual" approaches in signal processing may be useful, with or without the involvement of deep learning models.

Multimodal Recurrent Modeling

Another application of recurrent layers is in multimodal recurrent modeling. We can process sequences in recurrent fashion and tabular data in standard feed-forward fashion and then combine the extracted information in the universal form of the vector such that a jointly informed prediction can be made (Figure 5-43).

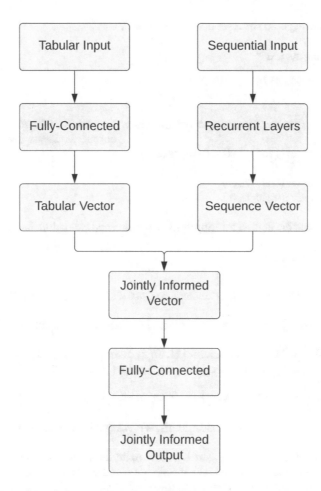

Figure 5-43. *Example multimodal usage of recurrent models*

The sequential input, in this case, could be one of the two natural applications discussed in the previous section (natural language or a signal). With a multimodal architecture, you can make predictions informed by both the sequential input and the tabular input, as opposed to isolated models that only predict on the basis of one or the other.

Recall the software reviews dataset, which contains a tabular component and two text components. Consider the design for a model that takes in all three components to make a jointly informed prediction (Listing 5-16).

Listing 5-16. Selecting relevant components of the data

```
tabular = data[['data/helpful_votes', 'data/total_votes', 'data/star_rating']]
body_text = data['data/review_body']
head_text = data['data/review_headline']
target = data['data/verified_purchase']
```

We begin by vectorizing both corpuses of text (Listing 5-17).

Listing 5-17. Vectorizing the text

```
SEQ_LEN, MAX_TOKENS = 64, 1024
EMBEDDING_DIM = 32

vectorize = tensorflow.keras.layers.TextVectorization(max_tokens=MAX_TOKENS,
                                        output_sequence_length=SEQ_LEN)
vectorize.adapt(pd.concat([body_text, head_text]))

vec_body_text = vectorize(body_text)
vec_head_text = vectorize(head_text)
```

Afterward, we can split the dataset into appropriate training and validation datasets (Listing 5-18).

Listing 5-18. Train-validation split

```
TRAIN_SIZE = 0.8
train_indices = np.random.choice(data.index, replace=False, size=round(TRAIN_SIZE *
len(data)))
valid_indices = np.array([i for i in data.index if i not in train_indices])

tabular_train, tabular_valid = tabular.loc[train_indices], tabular.loc[valid_indices]
body_text_train, body_text_valid = vec_body_text.numpy()[train_indices], vec_body_text.
numpy()[valid_indices]
head_text_train, head_text_valid = vec_head_text.numpy()[train_indices], vec_head_text.
numpy()[valid_indices]
target_train, target_valid = target[train_indices], target[valid_indices]
```

To construct the model (Listing 5-19, Figure 5-44), we build several heads to process each of the components independently, followed by vector concatenation and continued processing. We use a shared embedding between the review body text input and the review heading text input for simplicity. While the two are embedded with the same embeddings, they are processed independently.

Listing 5-19. Constructing a multimodal recurrent model

```
body_inp = L.Input((SEQ_LEN,), name='body_inp')
head_inp = L.Input((SEQ_LEN,), name='head_inp')

embed = L.Embedding(MAX_TOKENS, EMBEDDING_DIM)

body_embed = embed(body_inp)
head_embed = embed(head_inp)

body_lstm1 = L.GRU(16, return_sequences=True)(body_embed)
body_lstm2 = L.GRU(16)(body_lstm1)

head_lstm = L.GRU(16)(head_embed)

tab_inp = L.Input((3,), name='tab_inp')
tab_dense1 = L.Dense(8, activation='relu')(tab_inp)
```

```
tab_dense2 = L.Dense(8, activation='relu')(tab_dense1)

concat = L.Concatenate()([body_lstm2, head_lstm, tab_dense2])
outdense1 = L.Dense(16, activation='relu')(concat)
outdense2 = L.Dense(16, activation='relu')(outdense1)
out = L.Dense(1, activation='sigmoid')(outdense2)

model = keras.models.Model(inputs=[body_inp, head_inp, tab_inp], outputs=out)
```

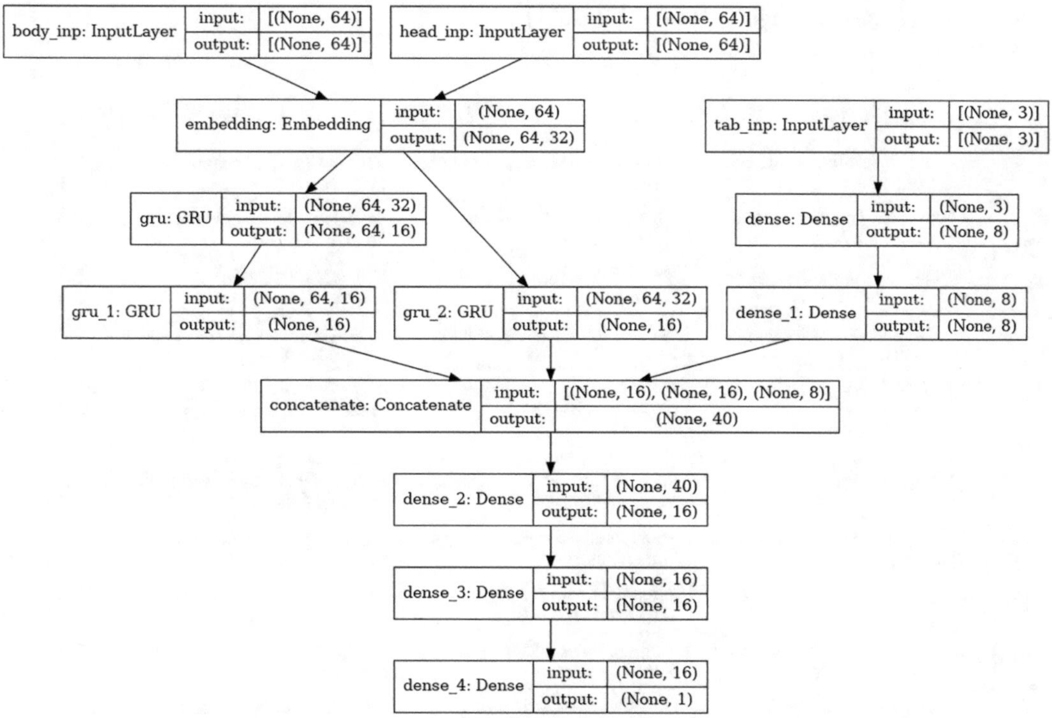

Figure 5-44. *Architecture of the multimodal recurrent model*

Let's consider another example: news-informed stock prediction. Stock prediction is often treated as a direct forecasting problem, in which the model attempts to predict the value at t_n given $\{t_{n-w}, t_{t-w+1}, ..., t_{n-1}\}$ for some window size w. However, predicting time series in this manner is difficult. More recent research shows that incorporating other data sources describing the state of consumer sentiment and related factors significantly helps improve stock prediction (as one would expect it to, since it captures external influential factors). Thus, many stock models incorporate consumer sentiment indices and other painstakingly acquired measurements. However, with deep learning, we can construct models that directly interpret and understand text data in relation to stock prices.

The Daily News for Stock Market Prediction dataset on Kaggle (www.kaggle.com/datasets/aaron7sun/stocknews) provides the top headlines in the r/worldnews subreddit with the associated Dow Jones Industrial Average (DJIA) that day. We will provide the model not only the DJIA at the previous w timesteps $\{t_{n-w}, t_{t-w+1}, ..., t_{n-1}\}$ but also the top three headlines at timestep t_n. The objective is to predict the DJIA at t_n. In this particular example, all our inputs are sequential – but the sequences are not uniform in type or context.

To prepare the dataset, we will read the datasets, select only the top three headlines for that day, and merge (Listing 5-20, Figure 5-45). Because the DJIA value ranges across a very wide range of values, we will scale the target by 100.

Listing 5-20. Reading multimodal stock data

```
news = pd.read_csv('../input/stocknews/Combined_News_DJIA.csv')
news = news[['Top1', 'Top2', 'Top3', 'Date']]
stock = pd.read_csv('../input/stocknews/upload_DJIA_table.csv')
data = news.merge(stock, how='inner', left_on='Date', right_on='Date')
stock = data[['Open', 'High', 'Low', 'Close']]
stock /= 100
```

	Top1	Top2	Top3	Date
0	b"Georgia 'downs two Russian warplanes' as cou...	b'BREAKING: Musharraf to be impeached.'	b'Russia Today: Columns of troops roll into So...	2008-08-08
1	b'Why wont America and Nato help us? If they w...	b'Bush puts foot down on Georgian conflict'	b"Jewish Georgian minister: Thanks to Israeli ...	2008-08-11
2	b'Remember that adorable 9-year-old who sang a...	b"Russia 'ends Georgia operation'"	b"'If we had no sexual harassment we would hav...	2008-08-12
3	b' U.S. refuses Israel weapons to attack Iran:...	b"When the president ordered to attack Tskhinv...	b' Israel clears troops who killed Reuters cam...	2008-08-13
4	b'All the experts admit that we should legalis...	b'War in South Osetia - 89 pictures made by a ...	b'Swedish wrestler Ara Abrahamian throws away ...	2008-08-14
...	...	...	...	...
1984	Barclays and RBS shares suspended from trading...	Pope says Church should ask forgiveness from g...	Poland 'shocked' by xenophobic abuse of Poles ...	2016-06-27
1985	2,500 Scientists To Australia: If You Want To ...	The personal details of 112,000 French police ...	S&P cuts United Kingdom sovereign credit r...	2016-06-28
1986	Explosion At Airport In Istanbul	Yemeni former president: Terrorism is the offs...	UK must accept freedom of movement to access E...	2016-06-29
1987	Jamaica proposes marijuana dispensers for tour...	Stephen Hawking says pollution and 'stupidity'...	Boris Johnson says he will not run for Tory pa...	2016-06-30
1988	A 117-year-old woman in Mexico City finally re...	IMF chief backs Athens as permanent Olympic host	The president of France says if Brexit won, so...	2016-07-01

Figure 5-45. *A sample of the top news headlines*

Next, we will prepare the stock history component (Listing 5-21). Assuming a window length of 20, we will store 20 timesteps' worth of values in x_stock and the 21st in y_stock. We repeat this at every valid starting timestep in the dataset. This is the standard "next-timestep" modeling paradigm. We'll also correspondingly select the associated relevant headlines.

Listing 5-21. Windowing the stock data in correspondence with the top text data

```
WINDOW_LENGTH = 20

x_stock = np.zeros((len(stock) - WINDOW_LENGTH,
                    WINDOW_LENGTH,
                    len(stock.columns)))
```

```
y_stock = np.zeros((len(stock) - WINDOW_LENGTH,
                    len(stock.columns)))

for i in range(len(stock) - WINDOW_LENGTH):
    x_stock[i] = np.array(stock.loc[i:i+WINDOW_LENGTH-1])
    y_stock[i] = np.array(stock.loc[i+WINDOW_LENGTH])

data = data.loc[WINDOW_LENGTH:]
top1_text, top2_text, top3_text = data['Top1'], data['Top2'], data['Top3']
```

Additionally, we need to vectorize the top one, two, and three headlines like before (Listing 5-22).

Listing 5-22. Vectorizing the top headline texts

```
SEQ_LEN, MAX_TOKENS = 64, 1024
EMBEDDING_DIM = 32

vectorize = tensorflow.keras.layers.TextVectorization(max_tokens=MAX_TOKENS,
                                                      output_sequence_length=SEQ_LEN)
vectorize.adapt(pd.concat([top1_text, top2_text, top3_text]))

top1_text = vectorize(top1_text)
top2_text = vectorize(top2_text)
top3_text = vectorize(top3_text)
```

We'll need to separate the dataset into training and validation (Listing 5-23). Because this is a forecasting problem, we will not use standard random train-validation split to prevent data leakage. Instead, we will fit on the first 80% worth of data and evaluate on the last 20%. Training and validation sets will be generated for each relevant variable. (There are other ways of doing this – exec is one cheap trick for doing so, which runs strings as Python code to avoid manual variable assignment.)

Listing 5-23. Train-validation split

```
variables = ['x_stock', 'y_stock',
             'top1_text', 'top2_text', 'top3_text']

train_prop = 0.8
train_index = round(train_prop * len(data))
for variable in variables:
    exec(f'{variable}_train = {variable}[:{train_index}]')
    exec(f'{variable}_valid = {variable}[{train_index}:]')
```

Our model will have three text inputs and one time-series input (Listing 5-24, Figure 5-46).

Listing 5-24. Constructing the multimodal text, time-series, and tabular data model

```
top1_inp = L.Input((SEQ_LEN,), name='top1')
top2_inp = L.Input((SEQ_LEN,), name='top2')
top3_inp = L.Input((SEQ_LEN,), name='top3')

embed = L.Embedding(MAX_TOKENS, EMBEDDING_DIM)
top1_embed = embed(top1_inp)
```

```
top2_embed = embed(top2_inp)
top3_embed = embed(top3_inp)

lstm1 = L.LSTM(32, return_sequences=True)
top1_lstm1 = lstm1(top1_embed)
top2_lstm1 = lstm1(top2_embed)
top3_lstm1 = lstm1(top3_embed)

top1_lstm2 = L.LSTM(32)(top1_lstm1)
top2_lstm2 = L.LSTM(32)(top2_lstm1)
top3_lstm2 = L.LSTM(32)(top3_lstm1)

concat = L.Concatenate()([top1_lstm2, top2_lstm2, top3_lstm2])
concat_dense = L.Dense(16, activation='relu')(concat)

stock_inp = L.Input((WINDOW_LENGTH, 4), name='stock')
stock_cnn1 = L.Conv1D(8, 5, activation='relu')(stock_inp)
stock_lstm1 = L.LSTM(8, return_sequences=True)(stock_cnn1)
stock_lstm2 = L.LSTM(8)(stock_lstm1)

joint_concat = L.Concatenate()([concat_dense, stock_lstm2])
joint_dense1 = L.Dense(16, activation='relu')(joint_concat)
joint_dense2 = L.Dense(16, activation='relu')(joint_dense1)
out = L.Dense(4, activation='relu')(joint_dense2)

model = keras.models.Model(inputs={'top1': top1_inp,
                                   'top2': top2_inp,
                                   'top3': top3_inp,
                                   'stock': stock_inp},
                           outputs=out)
```

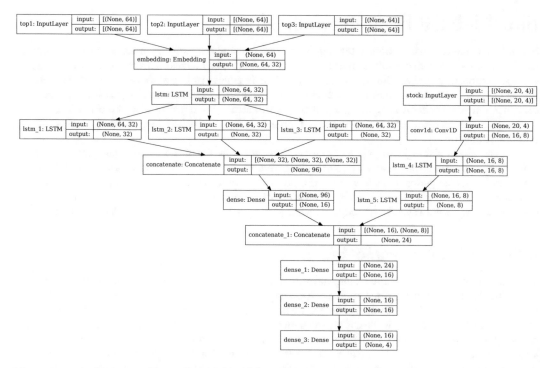

Figure 5-46. *Architecture of the multimodal stock forecasting model*

The model can be accordingly compiled and fitted (Listing 5-25).

Listing 5-25. Compiling and fitting the model

```
model.compile(optimizer='adam', loss='mse', metrics=['mae'])

model.fit(x={'top1': top1_text_train,
             'top2': top2_text_train,
             'top3': top3_text_train,
             'stock': x_stock_train},
          y=y_stock_train,
          validation_data=({'top1': top1_text_valid,
                            'top2': top2_text_valid,
                            'top3': top3_text_valid,
                            'stock': x_stock_valid},
                           y_stock_valid),
          batch_size=128,
          epochs=20)
```

At this point in the book, you have seen enough examples of multimodal modeling to be able to construct valid architectures for a variety of inputs: tabular, image, text, sequence. Multimodal compatibility is one of the powerful features of using deep learning to model problems with tabular data.

Direct Tabular Recurrent Modeling

Recall from Chapter 4 that we were able to apply one-dimensional and two-dimensional convolutions to tabular data, even though convolutions were designed to operate on images or, more generally speaking, value structures with spatially contiguous semantics. That is, pixels in an image have some sort of inherent spatial relationship with one another, or values in a sequence have some sort of inherent ordering. Even though tabular data generally does not possess contiguous semantics in raw form, we were able to get the model to learn a mapping from an unordered representation to an ordered representation using soft ordering strategies.

We can use similar logic to apply recurrent models directly for the purpose of modeling tabular data. Similarly to one-dimensional convolutions, recurrent models are designed to work on sequence data, or data possessing contiguous semantics. There should be some ordered quality to the data, such that it can be modeled in a recurrent fashion. However, because recurrent models generally have more components than convolutions, there is an even larger range of techniques we can use to apply recurrent layers to tabular data.

A Novel Modeling Paradigm

The methods introduced in Chapter 4 may have already seemed somewhat dubious upon initial impression. How could applying a tool intended for images possibly work on data that possessed near none of the properties of image data? In response, note that we *have* demonstrated that applying convolutional processing techniques to data while tasking the network to learn representations with contiguous semantics opens up new meta-nonlinearities in learning that can match or improve performance of standard feed-forward modeling techniques. However, we will even further stretch and manipulate the capacities of recurrent models in ways that may seem immediately sacrilegious or otherwise objectionable.

To pad this instinct, consider the following theoretical model (Figure 5-47). The key principle to understand here, which is hopefully approachable given the empirical work and demonstration of this sort of nontraditional usage of established methods in the previous chapter, is a subversion/reversion of the traditional modeling application paradigm. Traditionally, we abide by if the data has [property], then apply [model built for data with this property]. If the data is comprised of images, then apply convolutional layers. If the data is comprised of text, then apply recurrent/attention layers.

However, the novel insight is that we can also go boldly in the other direction: if we apply [model built for data with this property] to [data without this property in raw form], then the data will realize [property]. That is, we will be able to extract representations or information projections of that original data possessing some property (e.g., contiguous semantics) that can then be processed using a host of novel and powerful tools that would have otherwise been closed off. In this perspective of modeling, the purpose of a model is not passive (i.e., to be applied in line with the nature of the raw data itself) but rather active (i.e., to be applied to push/project/transform the data into new spaces with different properties to access novel processing methods).

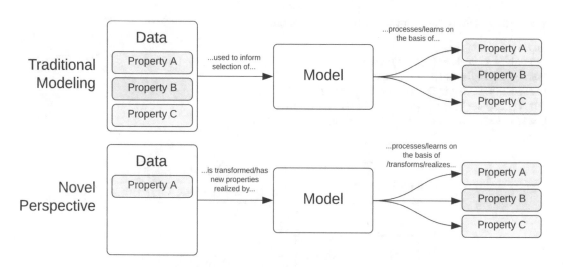

Figure 5-47. *A novel perspective of operations as transformative rather than just passive*

Optimizing the Sequence

One naïve approach is to directly treat the tabular data like it possesses contiguous semantics and to process it as if it were a sequence (Figure 5-48). That is, we treat each element of the feature vector as an element associated with a particular timestep, which is accordingly fed into the recurrent layer in a temporal fashion.

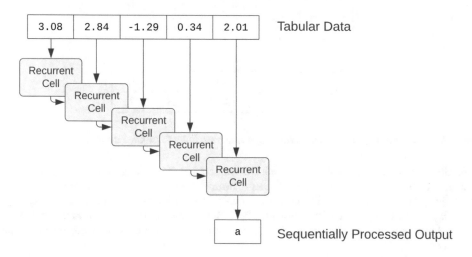

Figure 5-48. *Applying a recurrent layer to tabular data. Variables represent an arbitrary output*

We might arbitrarily implement such a model (Figure 5-49) for the 54-feature Forest Cover dataset previously used in Chapter 4 for pragmatic demonstration purposes, as in Listing 5-26. Note that we need to reshape our input into the shape (number of timesteps, elements in vector associated with each timestep) to use the recurrent layer. In this particular example, we use 32 hidden units, which means that the recurrent network will use and return a 32-length vector containing information informed from throughout the sequence. The 32-dimensional output of the recurrent layer can then be interpreted by a fully connected layer and mapped to a seven-class softmax output.

Listing 5-26. Directly applying a recurrent layer to tabular data

```
inp = L.Input((54,))
reshape = L.Reshape((54,1))(inp)
rnn1 = L.SimpleRNN(32)(reshape)
predense = L.Dense(32, activation='relu')(rnn1)
out = L.Dense(7, activation='softmax')(predense)
model = keras.models.Model(inputs=inp, outputs=out)
```

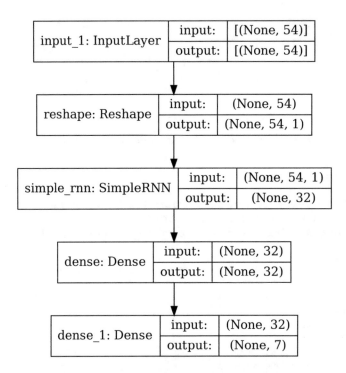

Figure 5-49. *Architecture of a direct application of a recurrent layer to tabular data*

This model, expectedly, doesn't do very well. Such an approach is analogous to directly applying convolutional kernels to a feature vector in Chapter 4. Instead, we can use a similar approach as employed to address this problem in the context of applying convolutions to tabular data and add additional fully connected layers at the beginning of the network (Figure 5-50). The hope is that these layers will be able to transform the input into a form with contiguous semantics, which is more readily readable by a recurrent layer.

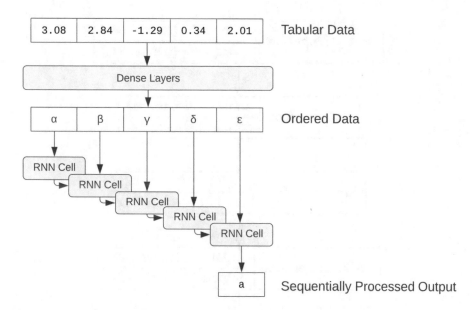

Figure 5-50. Applying a soft reordering component with a dense layer before applying a recurrent layer. "Recurrent cell" abbreviated as "RNN cell," but this does not necessarily mean the vanilla recurrent cell, but includes GRU and LSTM cells, which are also fundamentally recurrent

We can implement this in Keras by simply adding some more fully connected layers at the beginning of the network before reshaping and passing into the recurrent layer (Listing 5-27, Figure 5-51).

Listing 5-27. Using a soft ordering component consisting of fully connected layers before applying a recurrent layer

```
inp = L.Input((54,))
dense1 = L.Dense(32, activation='relu')(inp)
dense2 = L.Dense(32, activation='relu')(dense1)
reshape = L.Reshape((32,1))(dense2)
rnn1 = L.SimpleRNN(32)(reshape)
predense = L.Dense(32, activation='relu')(rnn1)
out = L.Dense(7, activation='softmax')(predense)
model = keras.models.Model(inputs=inp, outputs=out)
```

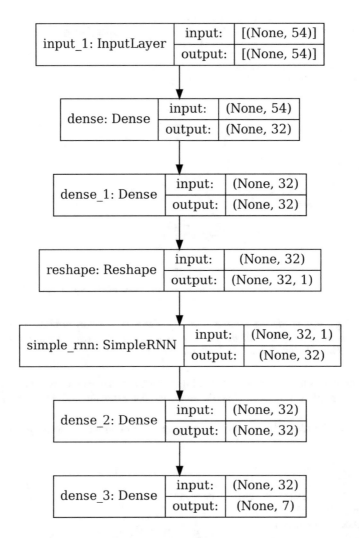

Figure 5-51. *Architecture of a soft ordering recurrent model*

Moreover, observe that we do not need to reshape the processed vectors into a sequence of one-element vectors. If we so desired, we could reshape the output of the second fully connected layer into shape (8, 4), which would represent a sequence of eight timesteps with a four-element vector being associated with each timestep. This is preferable, because it provides more information to the recurrent cell at each timestep to develop better-informed internal data representations. In general, if one must make a trade-off, it is better to spread information "vertically" (i.e., across the vectors associated with each timestep, by adding new elements) rather than "horizontally"/"temporally" (i.e., across timesteps, by adding additional timestep vectors). This is because recurrent models – even the more sophisticated ones – suffer from a decaying signal along the temporal axis. If we can provide it more information within each timestep, it will be able to extract and propagate more relevant information than if it were distributed sparsely across timesteps. (Of course, this is as at a limit – if we distributed all the information across one timestep or an otherwise very small number of timesteps, there would be little point in using the structure of the recurrent layer in the first place.) For simplicity, however, for the remainder of this chapter, we will use "simple" reshaping (i.e., from shape (a,) to (a,1)) to highlight other moving pieces and to minimize confusion.

The full -connected layer provides a soft ordering–like pre-manipulation of the data. We can visualize how various tabular inputs are transformed by the two-layer fully connected head into inputs to the recurrent layers by creating a submodel (Listing 5-28; recall from Chapter 4). Figure 5-52 demonstrates the sequences fed into the recurrent layers, shown in batches of ten. Note that high-magnitude values are dispersed sparsely at different locations in the sequence, which will result in differing readings by the recurrent layers.

Listing 5-28. Visualizing learned sequential representations

```
inp = L.Input((54,))
dense1 = model.layers[1](inp)
dense2 = model.layers[2](dense1)

submodel = keras.models.Model(inputs=inp,
                              outputs=dense2)

i = 0

plt.figure(figsize=(10/2.5, 33/2.5), dpi=400)
batch_prediction = submodel.predict(X_train[10*i:10*i + 10])
sns.heatmap(batch_prediction.reshape((32, 10)), cbar=False)
plt.xlabel('Sample')
plt.ylabel('Sequence Index')
plt.show()
```

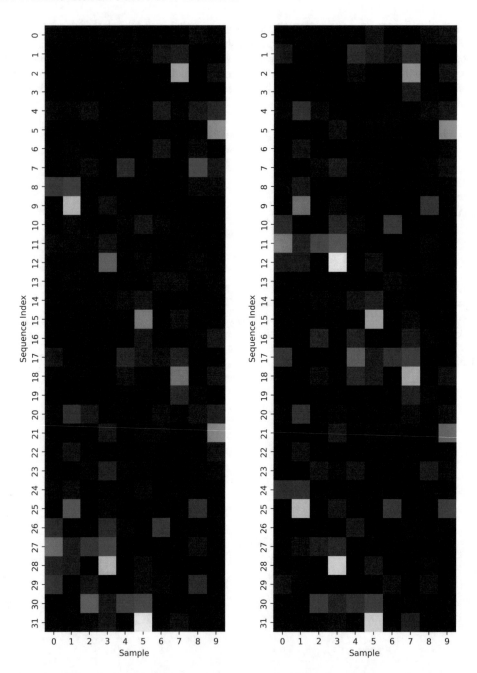

Figure 5-52. *Visualization of generated sequence mappings in batches of ten samples. Each column contains the sequence that some tabular sample has been projected into*

However, we can do one better – we know that convolutional layers can help process the input in a sequential way (Figure 5-53). Because convolutions are applied sequentially, they can help "pull out" contiguous semantics properties from our feature vector before being passed into the recurrent layer. This addition increases the expressivity of the input sequence with respect to its sequential/temporal quality. You can think of this as adding recurrent layers to the previous direct modeling design discussed in the previous chapter (in which we applied convolutions to fully-connected-layer-processed feature vectors).

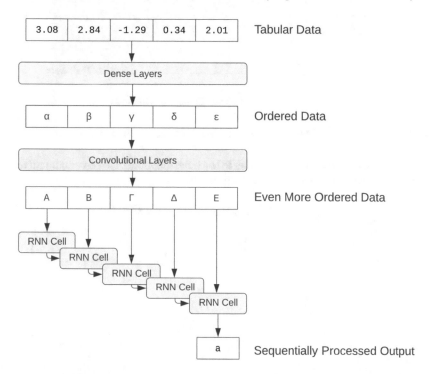

Figure 5-53. *Applying a dense soft ordering component and a 1D convolution component before using recurrent layers*

Moreover, an advantage of this sort of design is that we are able to pass in richer sequence inputs to the recurrent layer in a more natural way. Ideally, as previously discussed, the recurrent layer would have a substantially sized vector associated with each timestep to inform its hidden/internal representations. We could simply reshape a fully connected vector into such a shape, but it is unnatural in the sense that the fully connected layers are being tasked to learn complex spatial relationships between different outputs with linear mappings. When we pass in a "grid of values" to the recurrent layer, we assume certain elements with certain spatial properties have specific relationships to other elements, each with its own spatial properties. For instance, when we pass in a grid of elements to a recurrent model, we understand that there is a *temporal relationship* between a vector at some timestep t and $t + 1$. However, there is a *nontemporal relationship* between elements within each vector. We cannot "compare" or "quantify" the relationship between the 0th–1st vector index pair and the 0th–2nd vector index pair in the same way that we can with the vector pairs at the 0th–1st timesteps and the 0th–2nd timesteps (i.e., that the "duration" of the latter is twice that of the first). These are complex relationships that exist implicitly in the recurrent layer's treatment of such data (Figure 5-54).

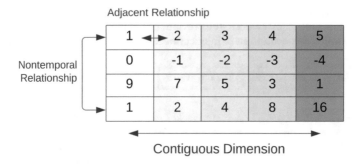

Figure 5-54. Assumed relationships between elements in an array as treated by a recurrent model

You can probably see the difficulty with trying to learn the mapping between a standard feature vector and this complex arrangement of relationships (Figure 5-55).

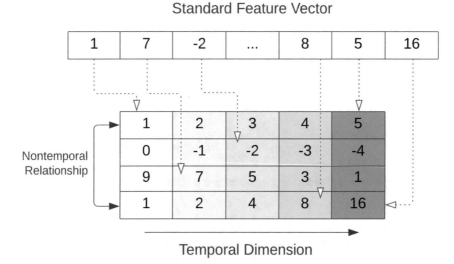

Figure 5-55. High precision and nonlinearity of mapping elements from a standard tabular feature vector to elements in an array with recurrent relationships

However, the assumptions in the data that one-dimensional convolutions operate on are very similar to those of the data that recurrent layers operate on. Each "row" in this representation is a sequence across which the convolution windows are "slid," generated and read by a different filter (a different "lens" or "perspective" to feature extraction) (Figure 5-56).

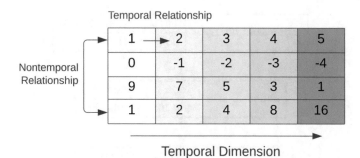

Figure 5-56. *Assumed relationships between elements in an array as treated by a convolutional model*

Thus, we can more "naturally" give the recurrent layer richer and more "readable" data (on a timestep-by-timestep basis) by processing the information with convolutions that can expand the depth with different filters (Listing 2-29, Figure 5-57).

Listing 5-29. Implementing a model with dense, 1D convolution, and recurrent components

```
inp = L.Input((54,))
dense1 = L.Dense(32, activation='relu')(inp)
dense2 = L.Dense(32, activation='relu')(dense1)
reshape = L.Reshape((32,1))(dense2)
conv1 = L.Conv1D(16, 3)(reshape)
conv2 = L.Conv1D(16, 3)(conv1)
rnn1 = L.LSTM(16, return_sequences=True)(conv2)
rnn2 = L.LSTM(16)(rnn1)
predense = L.Dense(16, activation='relu')(rnn2)
out = L.Dense(7, activation='softmax')(predense)
model = keras.models.Model(inputs=inp, outputs=out)
```

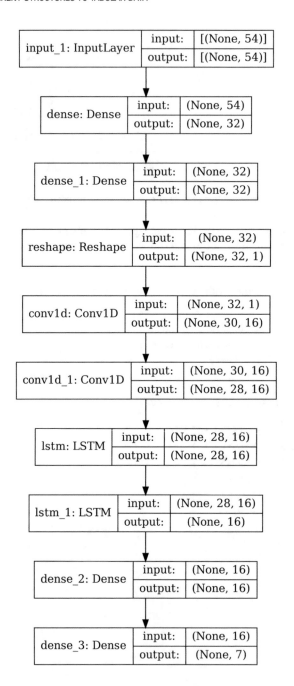

Figure 5-57. *Architecture of the dense-convolution-recurrent model*

When visualizing the result of transforming the original raw tabular inputs to the inputs to the recurrent layers (i.e., outputs of the last convolutional layer), we notice much higher activity and information-rich signals. Indeed, empirically adding convolutional layers generally yields superior performance to a similarly parametrized fully-connected-only head (Figure 5-58).

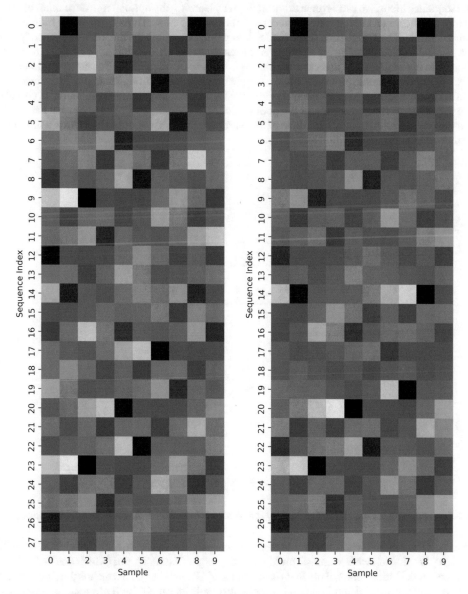

Figure 5-58. *Visualization of generated sequence mappings in batches of ten samples. Each column contains the sequence that some tabular sample has been projected into*

Optimizing the Initial Memory State(s)

However, we can exploit a different input "inlet" of the recurrent mechanism: the initial hidden state. The initial hidden state is initialized to zeros in TensorFlow, but changes as it becomes informed by elements from the sequence input. We can reverse this paradigm by learning the optimal transformation of the raw tabular input to the initial hidden state of a recurrent unit (Figure 5-59). Think of the initial hidden state as the "canvas" and the input sequence as the "brush and paint." Traditionally, the input sequence "paints" a "blank canvas," applying new layers and details at each timestep. The result is a "painting" informed by all timesteps in the sequence. In this case, the canvas is not initialized as blank, but rather as already quite complex. We can use a simple input sequence ("simple brushes and painting strategies") to provide a stimulus to modify the canvas sequentially. For simplicity, this "dummy stimulus sequence" is visualized as a zero vector; in practice, a zero vector is a poor choice for a stimulus sequence because it has minimal effect on the original hidden state, regardless of the learned weights. Two better choices are a one vector and transformer-style sinusoidal position encodings.

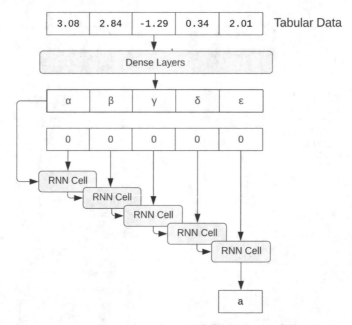

Figure 5-59. *Learning the initial hidden state of the recurrent layer rather than its initial sequence*

Moreover, if you desire to stack multiple recurrent layers, you can also specify the initial state of the second, third, fourth, etc. layers as transformations of the original tabular input (which can be linked to the learned initial hidden state of the first recurrent layer) (Figure 5-60). This means that the learned transformation of the original tabular input is now being dually interpreted and produces tremendously complex and expressive topological nonlinearity without many parameters.

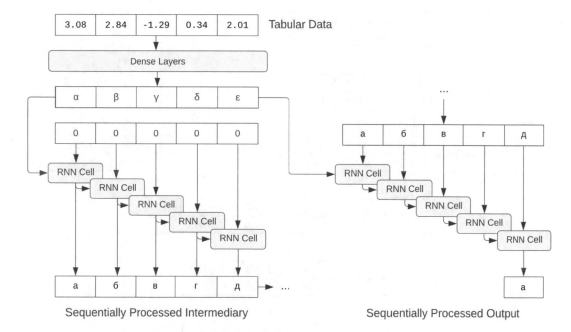

Figure 5-60. *Learning initial states across recurrent layer stacks*

Listing 5-30 and Figure 5-61 demonstrate a multi-stack recurrent model with only the first recurrent layer's initial hidden state learned as a function of the original tabular input.

Listing 5-30. Implementing a model that maps a tabular input to the initial state of a recurrent layer

```
init_hidden_vec = L.Input((54,), name='Init Hidden Vec')
init_inp_vec = L.Input((32, 1), name='Init Inp Vec')
dense1 = L.Dense(16, activation='relu')(init_hidden_vec)
dense2 = L.Dense(16, activation='relu')(dense1)
rnn1 = L.GRU(16, return_sequences=True)(init_inp_vec,
                                        initial_state=dense2)
rnn2 = L.GRU(16)(rnn1)
predense = L.Dense(16, activation='relu')(rnn2)
out = L.Dense(7, activation='softmax')(predense)
model = keras.models.Model(inputs=[init_hidden_vec, init_inp_vec],
                           outputs=out)
```

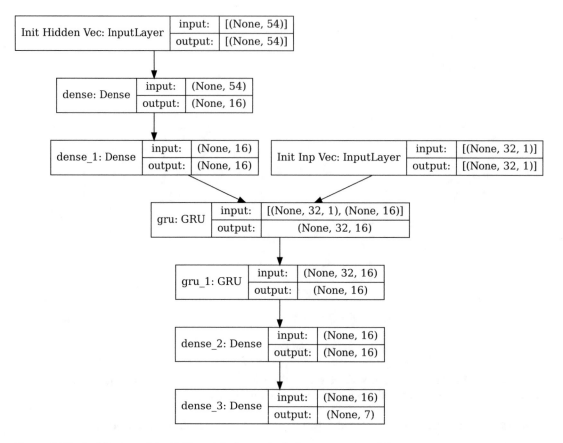

Figure 5-61. *Architecture of the hidden-state-learned tabular recurrent model*

We can modify `initial_state=dense2` into the second recurrent layer to create a dual linkage (Figure 5-62). This increases expressivity and connection, generally yielding superior empirical results during training faster.

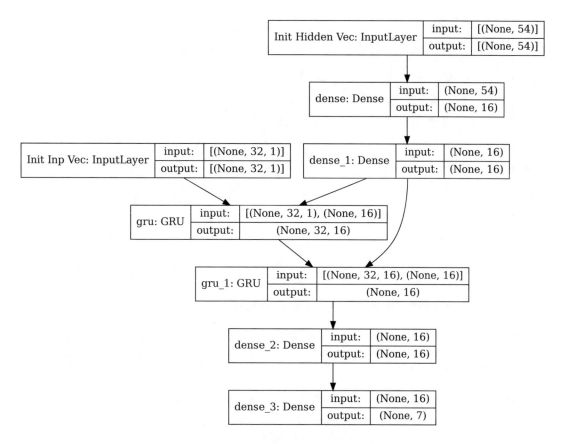

Figure 5-62. *Architecture of the dually connected hidden-state-learned tabular recurrent model*

The elegance of such a design is that it does not require the transformed tabular vector to possess contiguous semantics, but still produces a result that is informed/generated sequentially. It can be "easier" for a fully connected head to learn such a transformation.

We can compile and fit the model with one vectors (Listing 5-31).

Listing 5-31. Compiling and fitting

```
model.compile(optimizer='adam', loss='sparse_categorical_crossentropy',
              metrics=['accuracy'])
model.fit([X_train, np.ones((len(X_train), 16, 1))],
          y_train, epochs=20,
          validation_data=([X_valid, np.ones((len(X_valid), 16, 1))],
                            y_valid),
          batch_size = BATCH_SIZE)
```

Alternatively, we could opt for a more complex type of encoding. Transformer-style positional encoding generates a set of sinusoidal curves such that the values of the curves at any timestep are enough to inform a model which approximate timestep it is at while remaining bounded (Figure 5-63). See Chapter 6 for more context on the usage of positional encoding in a transformer model.

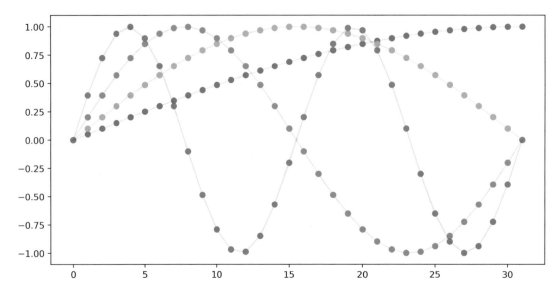

Figure 5-63. *A visualization of sinusoidal positional encoding vectors*

In a simple hypothetical implementation here (Listing 5-32), we generate four sinusoidal curves with differing periods and store this stimulus sequence as a 32-timestep sequence of length-four vectors. We need to correspondingly adjust the model architecture such that init_inp_vec = L.Input((32,4)).

Listing 5-32. Training with transformer-style sinusoidal position encodings

```
individ_seq = np.stack([np.sin(np.linspace(0, 1/2 * np.pi, 32)),
                        np.sin(np.linspace(0, np.pi, 32)),
                        np.sin(np.linspace(0, 2*np.pi, 32)),
                        np.sin(np.linspace(0, 4*np.pi, 32))],
                        axis=1)
train_pos_encoding = np.stack([individ_seq] * len(X_train))
valid_pos_encoding = np.stack([individ_seq] * len(X_valid))

model.compile(optimizer='adam', loss='sparse_categorical_crossentropy',
              metrics=['accuracy'])

model.fit([X_train, train_pos_encoding],
           y_train, epochs=20,
           validation_data=([X_valid, valid_pos_encoding],
                            y_valid),
          batch_size = BATCH_SIZE)
```

Putting these methods together, we can simultaneously learn both the optimal initial hidden state and the optimal input sequence to recurrent layers (Listing 5-33, Figures 5-64 and 5-65).

Listing 5-33. Implementing a model that learns both the sequence and hidden state inputs

```
init_vec = L.Input((54,))

dense1 = L.Dense(32, activation='relu')(init_vec)
dense2 = L.Dense(32, activation='relu')(dense1)
reshape = L.Reshape((32,1))(dense2)
conv1 = L.Conv1D(16, 3)(reshape)
conv2 = L.Conv1D(16, 3)(conv1)

hidden_dense1 = L.Dense(16, activation='relu')(init_vec)
hidden_dense2 = L.Dense(16, activation='relu')(hidden_dense1)

rnn1 = L.GRU(16, return_sequences=True)(conv2,
                                   initial_state=hidden_dense2)
rnn2 = L.GRU(16)(rnn1)

predense = L.Dense(16, activation='relu')(rnn2)
out = L.Dense(7, activation='softmax')(predense)
model = keras.models.Model(inputs=init_vec, outputs=out)

model.compile(optimizer='adam', loss='sparse_categorical_crossentropy',
              metrics=['accuracy'])

model.fit(X_train, y_train, epochs=20,
          validation_data=(X_valid, y_valid),
          batch_size = BATCH_SIZE)
```

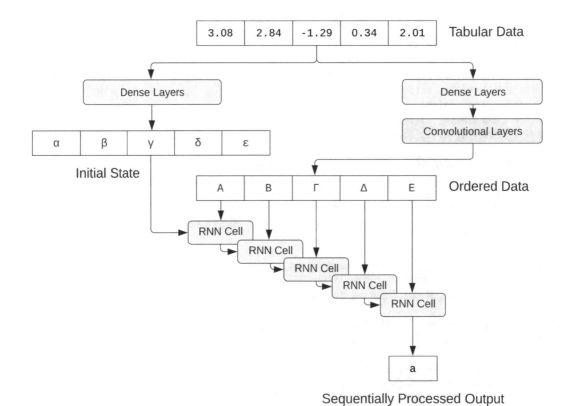

Figure 5-64. *Schematic diagram of the model*

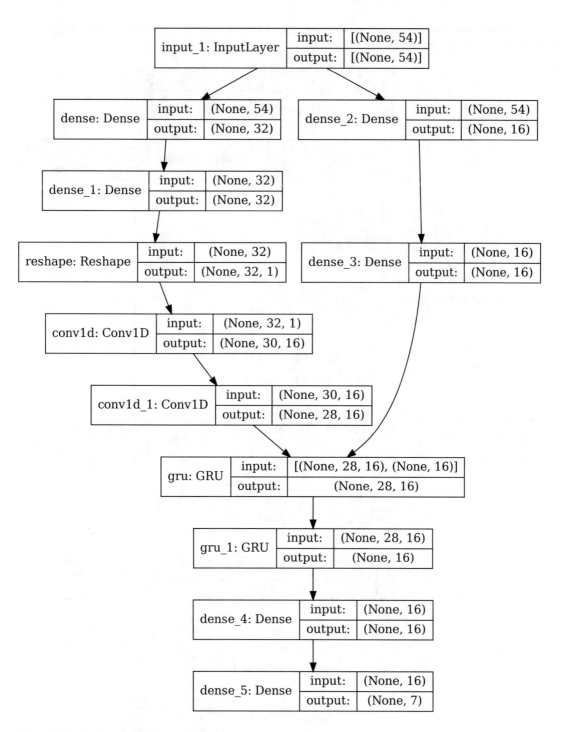

Figure 5-65. *Architecture of the model*

As before, we can connect the learned initial hidden state for the first recurrent layer to the second recurrent layer too (Figure 5-66).

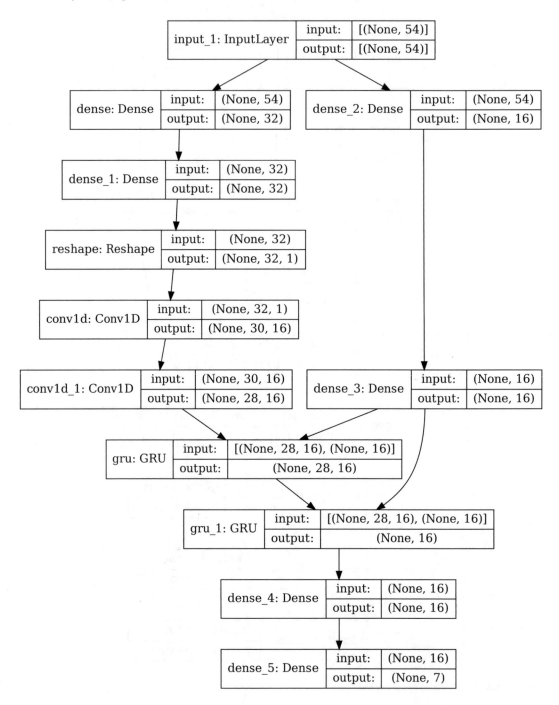

Figure 5-66. *Architecture of a dually connected model*

Additionally, we can also create another branch to learn the optimal (different) initial hidden state for the second recurrent layer independently (Figure 5-67). This eases expressivity restrictions that may become prohibitive for complex problems.

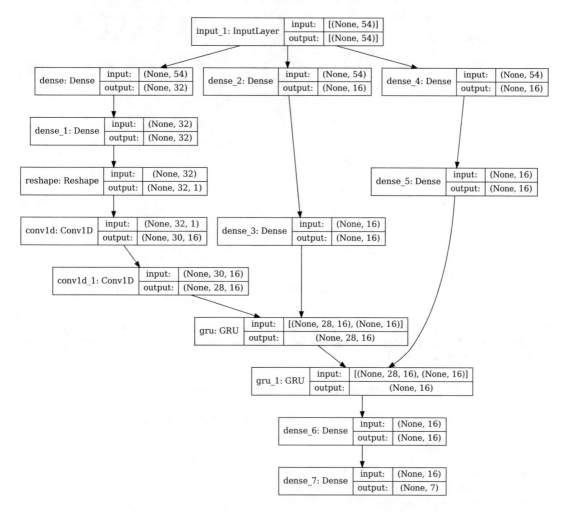

Figure 5-67. *Architecture of an independently learned dually connected model*

LSTMs have both a cell state and a hidden state and therefore lend themselves to even more complex systems where the network simultaneously derives the optimal sequence input, initial hidden state, and initial cell state from the original tabular data and puts all the pieces together in a powerful recurrent layer stack (Figure 5-68).

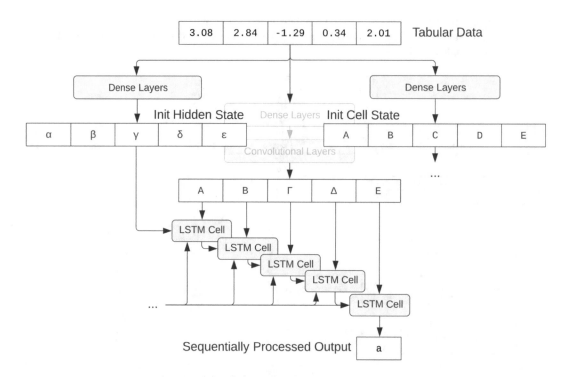

Figure 5-68. *Learning the cell state, hidden state, and input sequence to an LSTM model. The cell state is represented at the bottom as a single channel traveling through the entire sequence for visualization purposes (despite being slightly inaccurate)*

Listing 5-34 and Figure 5-69 demonstrate one such implementation.

Listing 5-34. Implementing an LSTM model in which all relevant inputs are learned

```
init_vec = L.Input((54,))

dense1 = L.Dense(32, activation='relu')(init_vec)
dense2 = L.Dense(32, activation='relu')(dense1)
reshape = L.Reshape((32,1))(dense2)
conv1 = L.Conv1D(16, 3)(reshape)
conv2 = L.Conv1D(16, 3)(conv1)

hidden_dense1 = L.Dense(16, activation='relu')(init_vec)
hidden_dense2 = L.Dense(16, activation='relu')(hidden_dense1)

cell_dense1 = L.Dense(16, activation='relu')(init_vec)
cell_dense2 = L.Dense(16, activation='relu')(cell_dense1)

rnn1 = L.LSTM(16, return_sequences=True)(conv2,
                               initial_state=[hidden_dense2,
                                              cell_dense2])

rnn2 = L.LSTM(16)(rnn1)
```

```
predense = L.Dense(16, activation='relu')(rnn2)
out = L.Dense(7, activation='softmax')(predense)
model = keras.models.Model(inputs=init_vec, outputs=out)

model.compile(optimizer='adam', loss='sparse_categorical_crossentropy',
              metrics=['accuracy'])

model.fit(X_train, y_train, epochs=20,
          validation_data=(X_valid, y_valid),
          batch_size = BATCH_SIZE)
```

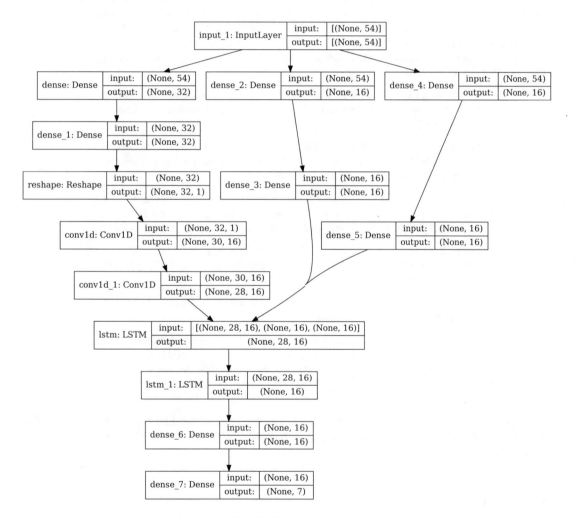

Figure 5-69. *Architecture of an LSTM model with a learned sequence input, cell state, and hidden state*

You can add multiple connections between the learned optimal initial hidden and cell states and different levels of the recurrent layer stack too, as demonstrated previously.

447

We should take a step back to appreciate the architecture that we have arrived at. The primary hesitance toward applying recurrent models directly toward tabular data is, if we think about it, that there doesn't seem to be an effective way for tabular data to pass through the recurrent model in a "natural" way. In this final model, tabular data is used to control all components of the tabular model – the initial states, its transformation through time, and its final interpretation. In this sense, it is as expressive as a feed-forward layer but provides crucial scaffolding to formulate the development of ideas throughout time.

Further Resources

These are all novel approaches, which may or may not work well on any one case. However, they have demonstrated promise in several key areas. Even if you are not convinced of the validity or correctness of the architectures and techniques proposed in this section, hopefully the more philosophical implications encourage greater fluidity and intellectual innovativeness in your modeling.

The following references are applied examples of recurrent models being successfully applied on tabular data and may be informative:

Althubiti, S.A., Nick, W., Mason, J.C., Yuan, X., & Esterline, A.C. (2018). Applying Long Short-Term Memory Recurrent Neural Network for Intrusion Detection. *SoutheastCon 2018*, 1–5.

Kim, J., Kim, J., Thu, H.L., & Kim, H. (2016). Long Short Term Memory Recurrent Neural Network Classifier for Intrusion Detection. *2016 International Conference on Platform Technology and Service (PlatCon)*, 1–5.

Le, T., Kim, J., & Kim, H. (2017). An Effective Intrusion Detection Classifier Using Long Short-Term Memory with Gradient Descent Optimization. *2017 International Conference on Platform Technology and Service (PlatCon)*, 1–6.

Nikolov, D., Kordev, I., & Stefanova, S. (2018). Concept for network intrusion detection system based on recurrent neural network classifier. *2018 IEEE XXVII International Scientific Conference Electronics – ET*, 1–4.

Prajyot, M.A. (2018). Review on Intrusion Detection System Using Recurrent Neural Network with Deep Learning.

Wang, S., Xia, C., & Wang, T. (2019). A Novel Intrusion Detector Based on Deep Learning Hybrid Methods. *2019 IEEE 5th Intl Conference on Big Data Security on Cloud (BigDataSecurity), IEEE Intl Conference on High Performance and Smart Computing, (HPSC) and IEEE Intl Conference on Intelligent Data and Security (IDS)*, 300–305.

For a sample code notebook, see Kaggle user Kouki's solution to the Mechanisms of Action competition, which uses a recurrent tabular model: `www.kaggle.com/code/kokitanisaka/moa-ensemble/notebook?scriptVersionId=48123609`.

Key Points

This chapter discussed three popular forms of recurrent models; demonstrated applications of recurrent models to text, time-series, and multimodal data; and proposed several methods to directly apply recurrent layers to tabular data.

- Recurrent neurons process sequence data iteratively by "snaking" the output from the previous timestamp back into the input as a hidden state, combined with the current timestamp's input. This allows the model to effectively learn patterns through time as an ordered sequence.

- LSTMs improve on standard recurrent neurons by addressing the vanishing gradient problem. They use a gated mechanism, which allows the model to access crucial information from older timestamps, allowing for gradients that still retain meaningful information flowing backward all the way to the beginning timestamp.

- Exploding gradients in recurrent models can be solved using gradient clipping.

- GRU is a modification to LSTM to simplify its training process by scraping the long-term memory unit from LSTMs and introducing update and reset gates as cheaper alternatives.

- To use recurrent layers to model text data: vectorize the text dataset, pass through an embedding layer to obtain embedded vectors, and pass through a stack of recurrent layers.

- There are three general templates of time-series prediction: next-timestep forecasting, time-dependent target prediction, and time-independent time prediction. Deep learning approaches to time-series modeling often work well on high-frequency time-series datasets, such as high-volume stock datasets or audio waves. Such architectures often perform well with convolutional heads that "smooth out" and extract key information-rich sequences for a recurrent layer stack to process.

- By constructing multi-input network architectures, you can create models that simultaneously accept different data modalities. This can be used to process data for which there is both a text component and a tabular component, which is a common occurrence in online platform and business data science contexts.

- To understand the premise, motivation, and justification of direct application of recurrent models (and all other nontraditional mechanisms) to tabular data, we must understand the reversion of the traditional modeling paradigm "if the data has [property], then apply [model built for data with this property]" instead as "if we apply [model built for data with this property] to [data without this property in raw form], then the resulting data will realize [property]." The realization of novel properties as caused by mechanisms assuming such properties opens up the door for new, previously sealed-off techniques and methods.

- By processing a tabular input with fully connected layers before passing into recurrent layers, the tabular input can be mapped to the optimal sequential representation. Applying convolutional layers before inputting to the recurrent layer stack can improve realization of contiguous semantics.

- We can also task full -connected layers to learn the optimal initial hidden state of the network and run it on a dummy stimulus sequence (one vector, transformer-style positional encoding, etc.) to produce a sequentially informed result. This method does not require the "input" to the mechanism (i.e., the learned optimal hidden state) to be assumed as ordered, but nevertheless produces an ordered result that can be processed sequentially by another recurrent layer.

- By combining the two previously discussed methods, we can construct a model that derives the optimal input sequence and the optimal hidden state (and optimal cell state for LSTMs) from the tabular input. This allows the recurrent modeling mechanism to be informed and optimized in heavy connectivity with respect to the input and increases expressivity and topological complexity/nonlinearity.

In the next chapter, we will build upon recurrent models to explore the attention mechanism – both in its original introduction as a supercharger mechanism for recurrent models and as the key player in the transformer architecture – and how it can be applied to tabular data.

CHAPTER 6

■ ■ ■

Applying Attention to Tabular Data

A wealth of information creates a poverty of attention.

—Herbert A. Simon, Political Scientist, Economist, and Early AI Pioneer

Compared with the feed-forward, convolutional, and recurrent mechanisms discussed in Chapters 3, 4, and 5, the attention mechanism has been popular in deep learning for a very short amount of time. Despite the brevity of its existence, it has become the basis of modern natural language processing models. Moreover, it is a very natural mechanism to compute relationships not just between tokens in a language sequence but also between features in a tabular dataset – which is why a significant body of recent work on deep learning tabular data methods centers on attention.

We will begin by contextualizing the attention mechanism in the original context it was introduced and developed in – natural language. Then, we will implement attention in Keras, both "from scratch" and using natively available layers, and demonstrate its behavior on synthetic datasets for the purpose of understanding. Afterward, we will show how attention can be integrated with recurrent language and multimodal models explored in Chapter 5 and be applied directly to tabular datasets. Lastly, we will cover the design behind four tabular deep learning models in recent research – TabTransformer, TabNet, SAINT, and ARM-Net.

Attention Mechanism Theory

In this section, we will track the meteoric rise of the attention mechanism from its role as a sequence aligner situated within a recurrent model to becoming the basis of near all modern language models. Along the way, we'll gain valuable theoretical knowledge about how the attention mechanism operates and why it is a natural idea to apply to tabular data.

© Andre Ye and Zian Wang 2023
A. Ye and Z. Wang, *Modern Deep Learning for Tabular Data*,
https://doi.org/10.1007/978-1-4842-8692-0_6

The Attention Mechanism

The attention mechanism as is now popular in deep learning was introduced by Bahdanau, Cho, and Bengio in 2015[1] to address the problem of dependency forgetting in large language translation tasks. Consider a translation problem of some sequence x to y: $\{x_0, x_1, ..., x_{n-1}\} \rightarrow \{y_0, y_1, ..., y_{n-1}\}$ (Figure 6-1). Say y_{n-1} depends heavily on x_0; that is, the last output relies heavily on the first input. Because the hidden state is pushed through the cell at each timestep in a vanilla recurrent cell, the signal becomes lost and diluted. Long Short-Term Memory networks help to address this problem by adding an additional cell state channel, which allows signals to travel across longer sequences less impeded. But say y_0 depends heavily on x_{n-1}; that is, the first output relies heavily on the last input. (Such long-range dependencies are common in language.) We cannot "look ahead" because the recurrent mechanism proceeds sequentially along the sequence. Bidirectional models address this by reading both forward and backward.

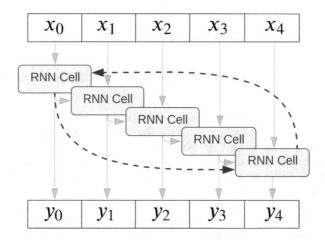

Figure 6-1. *Examples of long-term dependencies in text sequences*

However, it's still read sequentially in either direction: what if some output token y_k depends jointly on x_0 and x_{n-1} and some other output token y_j depends jointly on x_1 and x_{n-2}? Will the signals from x_0 and x_{n-1} be able to "reach" the prediction of y_k while also "carrying" the signals from x_1 and x_{n-2} to y_j? What if we have a dual dependency where the decision for one timestep is dependent on the decision for another timestep, which is itself dependent on the original timestep (Figure 6-2)?

---

[1] Bahdanau, D., Cho, K., & Bengio, Y. (2015). Neural Machine Translation by Jointly Learning to Align and Translate. *CoRR, abs/1409.0473.* https://arxiv.org/abs/1409.0473.

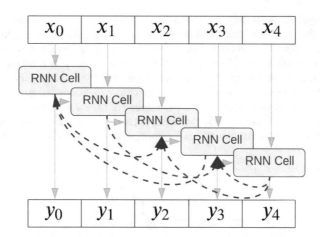

Figure 6-2. *Examples of even more convoluted dependencies in text sequences*

We see that there is an underlying problem of long-range dependencies, which is not quite captured by recurrent models, even with their cell state and bidirectional upgrades. There are always uncovered dependency cases that cannot be reconciled with. Fundamentally, the problem of keeping track of dependencies is still being solved sequentially. This makes it difficult to keep track of the myriad of complex short and long intersequence dependencies that determine the meaning and significance of a sequence. Thus, we often observe dependency forgetting and relatively suboptimal performance on advanced sequence-to-sequence tasks in recurrent models.

The idea behind attention, in a nutshell, is to directly model dependency relationships between timesteps without the hinderance of a necessarily sequential processing directionality (Figure 6-3).

	x_0	x_1	x_2	x_3	x_4
y_0					
y_1					
y_2					
y_3					
y_4					

Figure 6-3. *Visual representation of how an attention mechanism computes attention scores between timesteps of two sequences*

Sequence-to-sequence modeling tasks prior to the introduction of the attention mechanism by Bahdanau et al. used an encoder-decoder structure, in which a recurrent stack encoder encodes the input sequence and a recurrent decoder "interprets" the encoding into the novel output sequence domain (Figure 6-4). Bahdanau et al. apply attention to such an encoder-decoder recurrent structure as follows. The encoder outputs a hidden state hi at every timestep i (think `return_sequences=True`). We can generate a *context vector* for the output at some timestep t as a weighted sum of each of the hidden states (across all the timesteps):

$$c_t = \sum_{i=0}^{n-1} \alpha_{t,i} h_i$$

The weight $\alpha_{t,i}$ is the *alignment score*. This score is learned by another feed-forward neural network with a single hidden layer and represents how important each hidden state is to predicting the output at that timestep. Rephrased, the alignment score measures how much the input at time i represented by the encoder hidden state h_i "matches" or "is relevant to" the output at time t, y_t, represented by the decoder hidden state s_t. The alignment-score-computing network takes in the current decoder hidden state timestep s_t concatenated with the current encoder hidden state timestep h_i to compute the score. This generates a grid-like set of scores (like in Figure 6-3), in which we obtain a dependency score for every combination of input and output timesteps. The context vector, which is jointly informed by all relevant parts of the hidden state sequence, is then passed into the decoder at the appropriate timestep to make a prediction.

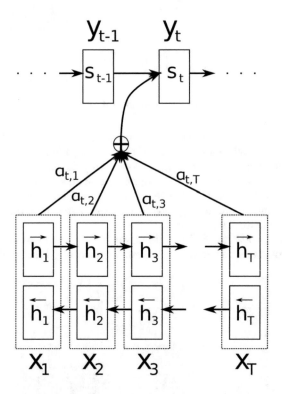

Figure 6-4. *Demonstration of how an attention mechanism can be used to align relevant timesteps with hidden states of the bidirectional recurrent layer. From Bahdanau et al.*

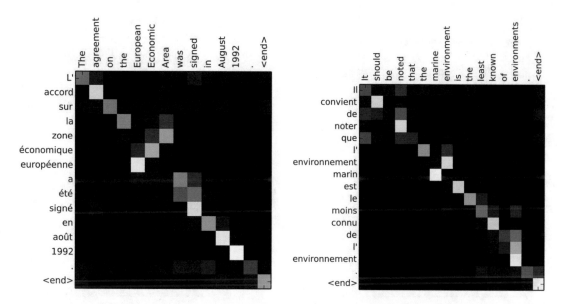

Figure 6-5. *Attention matrices between a French timestep and the English translation. Note the correspondence "alignment" between relevant words – hence "learning alignment." For instance, in French "zone économique europe énne" is aligned with "European Economic Area" in a nondirect way. From Bahdanau et al.*

However, there are alternative modes of deriving alignment scores. Dot product attention (introduced by Luong et al. in 2015) calculates the score simply as the dot product between the decoder hidden state s_t and the encoder hidden state h_i: $\text{score}(s_t, h_i) = s_t^T h_i$. This removes the need to learn alignment scores with another feed-forward network but requires the encoder and decoder hidden states to be already "calibrated" in relation to each other such that the dot product makes "sense." Dot product attention has the same theoretical complexity as additive attention, but the former is faster in practice due to matrix multiplication execution optimization and is more commonly used. The scaled dot product (introduced by Vaswani et al. in 2017) adds a scaling factor: $\text{score}(s_t, h_i) = (s_t^T h_i) / \sqrt{n}$, where n is the length of the hidden state. This scaling is a technical trick that allows for smaller gradients to pass through the softmax function, which is applied to the score set after calculation.

The Google Neural Machine Translation (GNMT) paper, published in 2016 by Wu et al.,[2] uses a recurrent encoder-decoder architecture with Bahdanau-style attention to perform translation (Figure 6-6). The encoder and decoder architectures are composed of eight LSTMs each to "capture subtle irregularities in the source and target languages."

[2] Wu, Y., Schuster, M., Chen, Z., Le, Q.V., Norouzi, M., Macherey, W., Krikun, M., Cao, Y., Gao, Q., Macherey, K., Klingner, J., Shah, A., Johnson, M., Liu, X., Kaiser, L., Gouws, S., Kato, Y., Kudo, T., Kazawa, H., Stevens, K., Kurian, G., Patil, N., Wang, W., Young, C., Smith, J.R., Riesa, J., Rudnick, A., Vinyals, O., Corrado, G.S., Hughes, M., & Dean, J. (2016). Google's Neural Machine Translation System: Bridging the Gap between Human and Machine Translation. *ArXiv, abs/1609.08144*. https://arxiv.org/abs/1609.08144.

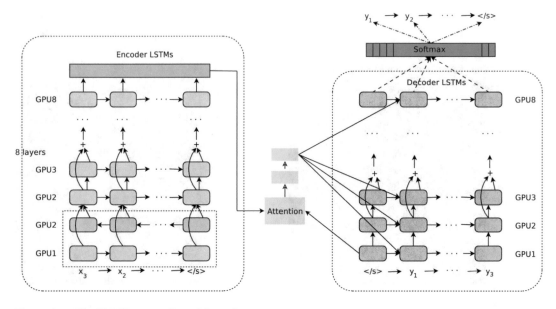

Figure 6-6. *The GNMT system. From Wu et al.*

Wu et al. made some modifications to the LSTM layers. To encourage greater gradient flow, GNMT uses residual LSTMs rather than conventional stacked LSTMs (Figure 6-7). Residual LSTMs add the original input at some timestep to the corresponding hidden state output, such that the hidden state models the difference between the input and the desired output rather than the output itself. (This can be implemented in Keras by applying `keras.layers.Add` to the original input and the hidden state sequence output.)

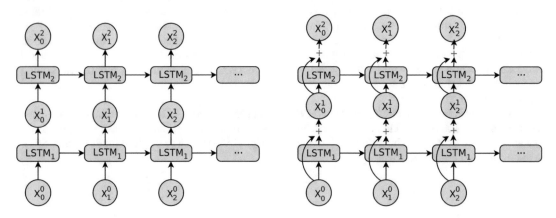

Figure 6-7. *A standard LSTM (left) and an LSTM with residual connections (right). From Wu et al.*

Moreover, Wu et al. used bidirectional LSTMs for the first layer of the encoder to maximize the context given to the later layers (Figure 6-8).

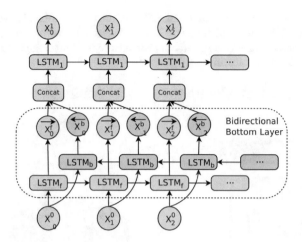

Figure 6-8. *A bidirectional LSTM layer. From Wu et al.*

The Transformer Architecture

In 2017 the infamously titled "Attention Is All You Need" paper by Vaswani et al.[3] introduced the transformer architecture, which dominates work in sequence-to-sequence problems today. While the attention mechanism developed as a method to improve dependency modeling in recurrent models, the transformer model shows that one can use only attention mechanisms to model language, without any usage of recurrent layers. By repeatedly stacking a novel and more powerful spin on attention – multi-head attention – the transformer architecture more freely models cross-text relationships and content.

Rather than accepting a decoder hidden state and an encoder hidden state Bahdanau-style, the transformer interprets the attention mechanism as a query-key-value "lookup":

$$\text{Attention}(Q, K, V) = \text{softmax}\left(\frac{QK^T}{\sqrt{n}}\right)V$$

Think of the key-value pair as elements stored in an abstract database: if the query "matches" with the key, it "unlocks" the desired key for later usage. Of course, this is taking place in continuous space rather than a formal, rigidly segmented database. The query and key interact with each other to determine which areas of the value V to attend to. For some index i in each vector, the attention score is maximized by the product of the ith element of the query and the ith element of the key. This correspondingly controls for how important (how attended to) the ith element of the value is. The interaction between the query and the key is a scaled version of Luong-style dot product attention.

Attention can be reappropriated as *self-attention* by deriving the query, key, and value from the same vector. Self-attention is a method to compute correlations or dependencies between any token in a sequence and other tokens in the same sequence, rather than between tokens in different sequences (like an input and target language for translation tasks). Self-attention is a crucial mechanism in the transformer architecture introduced by Vaswani et al., all successive transformer models, and attention-based deep learning approaches to tabular data.

---

[3] Vaswani, A., Shazeer, N.M., Parmar, N., Uszkoreit, J., Jones, L., Gomez, A.N., Kaiser, L., & Polosukhin, I. (2017). Attention Is All You Need. *ArXiv, abs/1706.03762.* https://arxiv.org/abs/1706.03762.

To allow for multiple different areas and modes of attention, the multi-head attention mechanism allows for multiple different "versions" of the query-key-vector set to be learned via fully connected layers; each "version" is passed through a scaled dot product attention mechanism, concatenated, and "compressed" by a linear layer into an output (Figure 6-9).

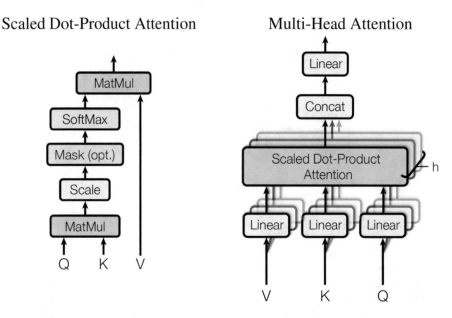

Figure 6-9. Scaled dot product attention (left) vs. multi-head attention (right). From Vaswani et al.

It is quite impressive to observe that the basis of the transformer model really is "just," if we accept a bit of reductionism for cognitive clarity, a complex series of fully connected layers arranged in intricate, self-interactive ways.

■ **Note** Although we may refer to the query, key, and value as "vectors," in practice the vectors are bundled into matrices and computed together. For clarity of understanding, you can think of the operations as happening on vectors, though.

The attention mechanism here is a generic structure for computing an interaction between three inputs and can be used in different ways. The transformer architecture uses multi-head attention three different ways:

- *Encoder-decoder attention*: Queries are derived from the previous decoder layer. The keys and values are derived from the encoder output. The decoder therefore attends to relevant positions in the encoder sequence by manipulating the query.

- *Encoder self-attention*: Keys, values, and queries all come from the output of the previous layer in the encoder; each position in the encoder has "access" to all positions in the previous layer.

- *Decoder self-attention*: Keys, values, and queries all come from all previous positions of the decoder.

Now, we can understand the complete transformer architecture proposed by Vaswani et al. (Figure 6-10). We begin by adding the positional encoding vector to the input encoding, which is calculated as follows for the timestep pos given the model's embedding dimensionality d_{model} at the ith element in the positional encoding vector:

$$PE_{(pos,i)} = \begin{cases} \sin\left(\dfrac{pos}{10,000^{\frac{2i}{d_{model}}}}\right), & \mod(i,2) = 0 \\[4ex] \cos\left(\dfrac{pos}{10,000^{\frac{2i}{d_{model}}}}\right), & \mod(i,2) = 1 \end{cases}$$

This input sequence is then passed through a series of N transformer encoder blocks. Each block is comprised of a multi-head self-attention mechanism followed by a feed-forward layer set with residual connections and layer normalization following each. Layer normalization normalizes all the values in the same layer (as opposed to batch normalization, which normalizes all the values for a particular node across all samples in the batch).

In the decoding component of the model, the current output sequence (beginning with the start token) is passed into the transformer decoder block. The transformer decoder block passes the current output sequence into a masked multi-head self-attention mechanism. The masking here prevents past timesteps to attend to future timesteps by zeroing out the relevant attention timesteps. Another multi-head attention mechanism operates both on the encoder outputs and the output of the masked multi-head attention mechanism (thus performing cross-attention rather than self-attention). The result is processed with feed-forward layers. This decoder block is repeated N times to produce an output. The decoder is autoregressive, meaning that the output is then concatenated to the output sequence as the input for the next step.

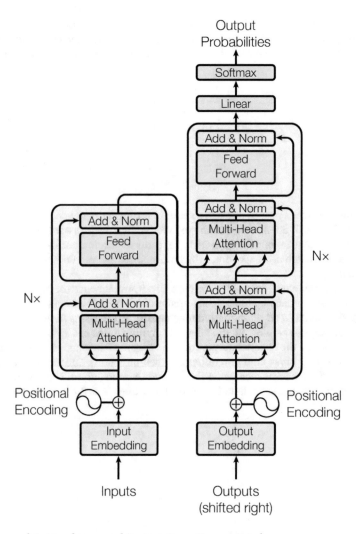

Figure 6-10. *The complete transformer architecture. From Vaswani et al.*

The transformer architecture surpassed the performance of existing dominant models at a fraction of the training cost, which generally used recurrent, convolutional, and "primitive" attention-based designs on a variety of tasks, including language translation, part-of-speech tagging, and other sequence-to-sequence problems (Table 6-1).

Table 6-1. *Performance of the transformer model on a German-English and French-English translation dataset. Lower is better for the BLEU (Bilingual Evaluation Understudy) metric. From Vaswani et al.*

Model	BLEU		Training Cost (FLOPs)	
	EN-DE	EN-FR	EN-DE	EN-FR
ByteNet [18]	23.75			
Deep-Att + PosUnk [39]		39.2		$1.0 \cdot 10^{20}$
GNMT + RL [38]	24.6	39.92	$2.3 \cdot 10^{19}$	$1.4 \cdot 10^{20}$
ConvS2S [9]	25.16	40.46	$9.6 \cdot 10^{18}$	$1.5 \cdot 10^{20}$
MoE [32]	26.03	40.56	$2.0 \cdot 10^{19}$	$1.2 \cdot 10^{20}$
Deep-Att + PosUnk Ensemble [39]		40.4		$8.0 \cdot 10^{20}$
GNMT + RL Ensemble [38]	26.30	41.16	$1.8 \cdot 10^{20}$	$1.1 \cdot 10^{21}$
ConvS2S Ensemble [9]	26.36	**41.29**	$7.7 \cdot 10^{19}$	$1.2 \cdot 10^{21}$
Transformer (base model)	27.3	38.1	$\mathbf{3.3 \cdot 10^{18}}$	
Transformer (big)	**28.4**	**41.8**	$2.3 \cdot 10^{19}$	

BERT and Pretraining Language Models

The next major development in natural language modeling was the BERT model, introduced by Jacob Devlin et al. from Google in the 2019 paper "BERT: Pre-training of Deep Bidirectional Transformers for Language Understanding."[4] The authors use a transformer-like architecture for their base BERT model, with 12 layers, a hidden size of 768, and 12 attention heads totaling 110 million parameters. BERT-Large has double the number of layers, a hidden size of 1024, and 16 attention heads totaling 340 million parameters.

It is worth noting that the BERT architecture uses the GELU (Gaussian Error Linear Unit) activation rather than the standard ReLU activation. Most modern language transformers, including some attention-based deep tabular models we will cover later in this chapter, use GELU. For some input x, GELU is defined as x multiplied by the probability that some value drawn from a unit normal distribution is larger than x[5]:

$$\text{GELU}(x) = xP(X \leq x), X \sim N(0,1) = 0.5x\left(1 + \frac{2}{\sqrt{\pi}} \int_{0}^{\frac{x}{\sqrt{2}}} e^{-t^2} dt\right)$$

$$\approx 0.5x\left(1 + \tanh\left(\sqrt{\frac{2}{\pi}}\left(x + 0.44715x^3\right)\right)\right)$$

In practice, it is a rounded version of ReLU (Figure 6-11).

---

[4] Devlin, J., Chang, M., Lee, K., & Toutanova, K. (2019). BERT: Pre-training of Deep Bidirectional Transformers for Language Understanding. *ArXiv, abs/1810.04805.* https://arxiv.org/abs/1810.04805.
[5] You can read more about the derivation and justification for GELU here: https://arxiv.org/pdf/1606.08415v3.pdf. Hendrycks, D., & Gimpel, K. (2016). Gaussian Error Linear Units (GELUs). *arXiv: Learning.*

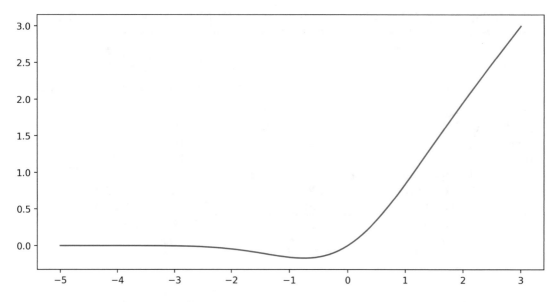

Figure 6-11. *The GELU activation function*

In this sense, it is conceptually similar to the swish activation,[6] which is defined as $x \cdot \sigma(x)$ and also possesses a "dip" slightly left of $x = 0$ on a ReLU "backbone" (Figure 6-12).

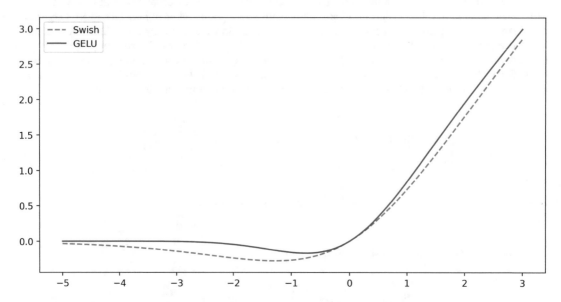

Figure 6-12. *The GELU activation plotted against the similar swish activation*

---

[6] Ramachandran, P., Zoph, B., & Le, Q.V. (2017). Swish: a Self-Gated Activation Function. *arXiv: Neural and Evolutionary Computing.* https://arxiv.org/pdf/1710.05941v1.pdf.

The key contribution of the BERT paper is in introducing pretraining schemes and demonstrating the power of transfer learning in a natural language context (Figure 6-13). While the architectures used for pretraining and fine-tuning toward the actual desired task are very similar or even identical (after adapting I/O sizes), pretraining improves the efficiency and power of the model during fine-tuning.

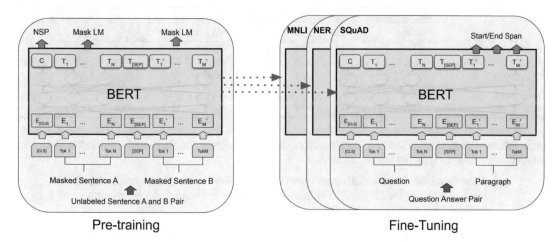

Figure 6-13. *Pretraining with BERT followed by downstream fine-tuning tasks. From Devlin et al.*

Any supervised task must have inputs and labels; it turns out that – in the paradigm of self-supervised learning – we can derive our labels from our inputs by corrupting the input and training the model to undo or correct for said corruption, in the process learning important information about the structure of the input in a label-cheap (free), unsupervised fashion. Devlin et al. introduce two such pretraining tasks: masked language modeling (MLM) and next sentence prediction (NSP).

In masked language modeling, a certain percentage of the input – 15% in the paper – is masked with a [MASK] token, and the model is trained to predict the masked tokens. The objective of such a pretraining task is to encourage the development of deep bidirectional representations since the model must parse the entire structure on both sides of the masked tokens to have any chance at accurately inferring the true token.

In next sentence prediction, the model is presented with two sentences and trained to predict whether the second follows from the first or not in the complete text passage that the sentences were derived from. This forces the model to not only develop cross-token semantic understandings but also learn about semantic continuity across sentences.

Devlin et al. find that both BERT and BERT-Large outperform competitors on the GLUE (General Language Understanding Evaluation) benchmark tasks (Table 6-2).

Table 6-2. *Performance of BERT on various datasets in the GLUE collection. MNLI: Multi-genre Natural Language Inference. QQP: Quora Question Pairs. QNLI: Question Natural Language Inference. SST-2: Stanford Sentiment Treebank. CoLA: Corpus of Linguistic Acceptability. STS-B: Semantic Textual Similarity Benchmark. MRPC: Microsoft Research Paraphrase Corpus. RTE: Recognizing Textual Entailment*

System	MNLI-(m/mm)	QQP	QNLI	SST-2	CoLA	STS-B	MRPC	RTE	Average
	392k	363k	108k	67k	8.5k	5.7k	3.5k	2.5k	-
Pre-OpenAI SOTA	80.6/80.1	66.1	82.3	93.2	35.0	81.0	86.0	61.7	74.0
BiLSTM+ELMo+Attn	76.4/76.1	64.8	79.8	90.4	36.0	73.3	84.9	56.8	71.0
OpenAI GPT	82.1/81.4	70.3	87.4	91.3	45.4	80.0	82.3	56.0	75.1
BERT$_{BASE}$	84.6/83.4	71.2	90.5	93.5	52.1	85.8	88.9	66.4	79.6
BERT$_{LARGE}$	**86.7/85.9**	**72.1**	**92.7**	**94.9**	**60.5**	**86.5**	**89.3**	**70.1**	**82.1**

More or less all modern language models are transformers or take strong inspiration from transformer architectures; it is not within the scope of the book to further discuss them, but some important ones are curated in the following for the interested reader:

- *"Improving Language Understanding by Generative Pre-Training," Alec Radford et al. 2018*: Introduces the GPT architecture and proposes a self-supervised pretraining framework similar to that of BERT.

- *"Improving Language Understanding by Generative Pre-Training," Alec Radford et al. 2019*: Introduces the GPT-2 architecture and demonstrates zero-shot task transfer properties.

- *"Language Models Are Few-Shot Learners," Tom B. Brown et al. 2020*: Introduces the GPT-3 architecture; engages in intense discussion of zero- and few-shot model properties; mentions societal impact, fairness, and bias implications.

- *"Zero-Shot Text to Image Generation," Aditya Ramesh et al. 2021*: Introduces the DALL-E architecture, a modified version of GPT-3 that can be used to generate images form text descriptions.

"LaMDA: Language Models for Dialog Applications," Romal Thoppilan et al. 2022: Introduces the LaMDA model family for conversational dialog. LaMDA has been a recent subject of extreme controversy.

Taking a Step Back

However, we should not neglect that transformers are not the be-all and end-all of language modeling, even though it most certainly seems that way given the present state of research in natural language processing. Stephen Merity's 2019 independent research paper "Single Headed Attention RNN: Stop Thinking With Your Head"[7] is simultaneously an empirical, philosophical, technical, and deeply comedic skepticism of the modern transformers craze. Only Merity's writing can speak for itself as well as it does; the paper's full abstract is reprinted in the following:

> The leading approaches in language modeling are all obsessed with TV shows of my youth – namely Transformers and Sesame Street. Transformers this, Transformers that, and over here a bonfire worth of GPU-TPU-neuromorphic wafer scale silicon. We opt for the lazy path of old and proven techniques with a fancy crypto inspired acronym: the Single Headed Attention RNN (SHA-RNN). The author's lone goal

---

[7] Merity, S. (2019). Single Headed Attention RNN: Stop Thinking With Your Head. *ArXiv, abs/1911.11423*. https://arxiv.org/abs/1911.11423.

is to show that the entire field might have evolved a different direction if we had instead been obsessed with a slightly different acronym and slightly different result. We take a previously strong language model based only on boring LSTMs and get it to within a stone's throw of a stone's throw of state-of-the-art byte level language model results on enwik8. This work has undergone no intensive hyperparameter optimization and lived entirely on a commodity desktop machine that made the author's small studio apartment far too warm in the midst of a San Franciscan summer. The final results are achievable in plus or minus 24 hours on a single GPU as the author is impatient. The attention mechanism is also readily extended to large contexts with minimal computation. Take that Sesame Street.

Driven by the success of the transformer model introduced by Vaswani et al. in 2017, recurrent models have been condemned to a slow death in the research community. Merity argues that modern large language models – which seem to be an arms race to scale up models orders of magnitude larger than the previous state-of-the-art – lack the power of reproducibility and therefore sustainability and potentially efficiency.

To demonstrate the power of small architectures, Merity proposes the Single Headed Attention RNN (SHA-RNN) architecture. SHA-RNN (Figure 6-14) applies an LSTM, followed by a single-headed dot product self-attention mechanism (Figure 6-15) and a "Boom" layer, both with residual connections. A dense layer is applied only to the query key, and all other operations are all unparametrized. The "Boom" layer maps a vector from R^{1024} to R^{4096} and then back down to R^{1024} (boom!). The first mapping is performed with a dense layer, whereas the second is performed parameter-free by summing up adjacent blocks of four elements in a 1D-pooling-style gesture. The SHA-RNN model can be applied to an embedded input as many times as desired and passed through a softmax layer in the last iteration. This architecture's intentional parametrically and computationally conservative design makes it feasible to train relatively quickly on relatively unadvanced computational resources. The author, for instance, writes that he trains the model on a single NVIDIA Titan V GPU.

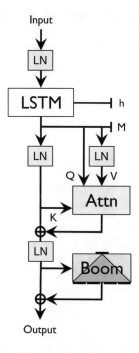

Figure 6-14. *The SHA-RNN block. From Merity*

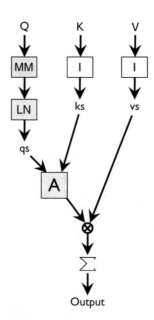

Figure 6-15. *The attention mechanism used in the SHA-RNN block. From Merity*

We see that SHA-RNN outperforms other similarly sized and larger models (Table 6-3).

Table 6-3. *Performance (bits per character) of different-sized SHA-RNN/LSTM models compared with other models on the enwiki dataset. From Merity*

Model	Heads	Valid	Test	Params
Large RHN (Zilly et al., 2016)	0	–	1.27	46M
3 layer AWD-LSTM (Merity et al., 2018b)	0	–	1.232	47M
T12 (12 layer) (Al-Rfou et al., 2019)	24	–	1.11	44M
LSTM (Melis et al., 2019)	0	1.182	1.195	48M
Mogrifier LSTM (Melis et al., 2019)	0	1.135	1.146	48M
4 layer SHA-LSTM ($h = 1024$, no attention head)	0	1.312	1.330	51M
4 layer SHA-LSTM ($h = 1024$, single attention head)	1	1.100	1.076	52M
4 layer SHA-LSTM ($h = 1024$, attention head per layer)	4	1.096	1.068	54M
T64 (64 layer) (Al-Rfou et al., 2019)	128	–	1.06	235M
Transformer-XL (12 layer) (Dai et al., 2019)	160	–	1.06	41M
Transformer-XL (18 layer) (Dai et al., 2019)	160	–	1.03	88M
Adaptive Transformer (12 layer) (Sukhbaatar et al., 2019)	96	1.04	1.02	39M
Sparse Transformer (30 layer) (Child et al., 2019)	240	–	0.99	95M

The paper argues not for the *uselessness* of transformer models, but rather encourages greater healthy skepticism and valuing of efficiency in modern deep learning research culture:

Perhaps we were too quick to throw away the past era of models simply due to a new flurry of progress. Perhaps we're too committed to our existing stepping stones to backtrack and instead find ourselves locked to a given path.

In the next sections, we will focus on implementing attention-based approaches to language, multimodal, and tabular contexts.

Working with Attention

Here, we'll explore different ways to use attention in Keras: a simple custom model of Bahdanau attention, different forms of native Keras attention, and using attention in sequence-to-sequence problems.

Simple Custom Bahdanau Attention

We'll begin by implementing a custom Bahdanau-style attention layer with Keras. The exact code here is adapted with modifications from Jason Brownlee. This layer accepts a set of hidden outputs (the output of a recurrent layer with `return_sequences=True`) and computes a form of self-attention. Because we are not applying this in an encoder-decoder context, it resembles Bahdanau-style self-attention for a single decoder timestep. It is one of the several dozen prominent flavors of attention that can be used in a variety of contexts. This particular flavor abides by the following steps:

1. Take in an input x with shape (s, h) for some sequence length s and hidden state length h.

2. Perform dot product multiplication between x and W, a learned weight matrix with shape $(h, 1)$. The result is a matrix/vector of shape $(s, 1)$ (Figure 6-16).

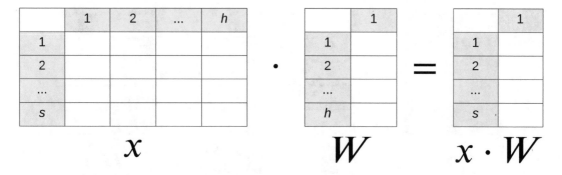

Figure 6-16. *Dot product between the input matrix and the weight matrix*

3.   Add a bias b of shape $(s, 1)$ (Figure 6-17).

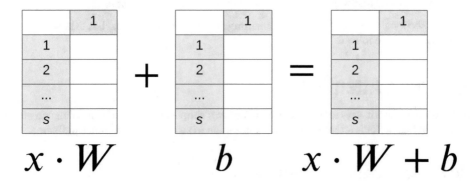

$$x \cdot W \qquad\qquad b \qquad\qquad x \cdot W + b$$

Figure 6-17. *Adding the bias to the dot product between the input matrix and the weight matrix*

4.   Apply the hyperbolic tangent activation to the result. This result represents the result of processing by a neural network with one hidden layer.

5.   Squeeze the second dimension of the matrix such that it becomes a vector of length s.

6.   Apply softmax to the vector such that the elements sum to 1. This s-length vector stores the scores corresponding to each of the hidden states at every timestep.

7.   Expand the squeezed dimension such that the s-length vector becomes a matrix of shape $(s, 1)$. This matrix stores the alpha values, or the scores (Figure 6-18).

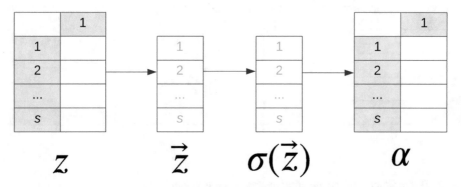

$$z \qquad\qquad \vec{z} \qquad \sigma(\vec{z}) \qquad\qquad \alpha$$

Figure 6-18. *Let $z = tanh\ (x \cdot W + b)$. Dimension squeezing followed by the application of softmax (represented as σ), followed by a dimension expansion. We obtain a matrix of attention scores as a result*

8.   Multiply the score at each timestep s_t by the corresponding hidden state x_t. The result is a hidden state sequence weighted by the derived scores (Figure 6-19).

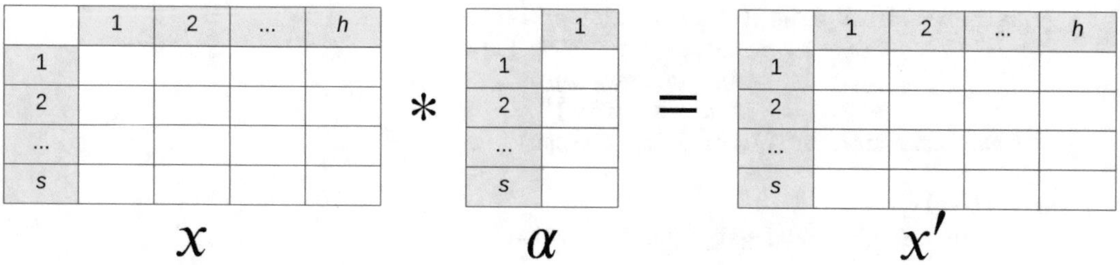

Figure 6-19. Multiplying the inputs by the attention scores to produce the attended scores

9. Sum across the timesteps such that the result is a single "weighted sum hidden state." This "aggregate hidden state" is informed appropriately by the hidden states at all timesteps from the sequence (Figure 6-20).

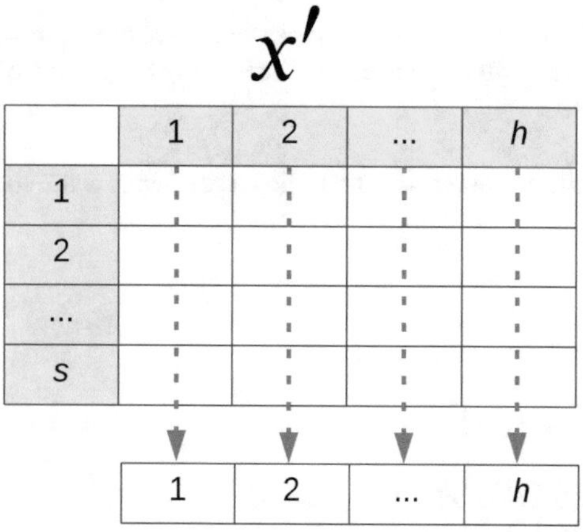

Figure 6-20. Summing to produce the output aggregate attended features

To build a custom layer (Listing 6-1), we inherit from keras.layers.Layer. To provide shape information relevant for constructing a graph before training, we provide a build function that allows Keras to "lazily" build necessary parameters through .add_weight(). When we apply the layer with call(), we simply return the weighted sum of the input. The weights are the alpha values/scores, computed in get_alpha.

Listing 6-1. A custom attention layer

```
class Attention(keras.layers.Layer):
    def __init__(self,**kwargs):
        super(Attention,self).__init__(**kwargs)

    def build(self,input_shape):
        self.W=self.add_weight(name='attention_weight',
                               shape=(input_shape[-1],1),
                               initializer='random_normal',
                               trainable=True)
```

```
        self.b=self.add_weight(name='attention_bias',
                               shape=(input_shape[1],1),
                               initializer='zeros',
                               trainable=True)
        super(Attention, self).build(input_shape)

    def call(self,x):
        return K.sum(x * self.get_alpha(x), axis=1)

    def get_alpha(self,x):
        e = K.tanh(K.dot(x, self.W)+self.b)
        e = K.squeeze(e, axis=-1)
        alpha = K.softmax(e)
        alpha = K.expand_dims(alpha, axis=-1)
        return alpha
```

Let's construct a synthetic task: given a ten-element sequence of length-8 normally distributed random vectors, predict the sum of the seventh and ninth vectors (Listing 6-2). None of the other timesteps are relevant to predicting the label.

Listing 6-2. Generating a synthetic dataset in which the target vector is the sum of the second-to-last and fourth-to-last elements

```
x, y = [], []

NUM_SAMPLES = 10_000

next_element = lambda arr: arr[-2] + arr[-4]

vector_switch = [np.zeros((1,8)), np.ones((1,8))]
for i in tqdm(range(NUM_SAMPLES)):
    seed = np.random.normal(0, 5, size=(10,8))
    x.append(seed)
    y.append(next_element(seed))3

x = np.array(x)
y = np.array(y)

from sklearn.model_selection import train_test_split as tts
X_train, X_valid, y_train, y_valid = tts(x, y, train_size=0.8)
```

One architecture used to model this synthetic problem is a double-GRU stack followed by the custom attention mechanism (Listing 6-3, Figure 6-21). The output is a vector (recall the weighted sum of hidden states); we will simply process it with additional feed-forward layers into an output. Note than an alternative is to use L.RepeatVector to construct a series of these vectors and apply additional recurrent layers.

Listing 6-3. Constructing an architecture to model the synthetic dataset created in Listing 6-2

```
inp = L.Input((10,8))
lstm1 = L.GRU(16, return_sequences=True)(inp)
lstm2 = L.GRU(16, return_sequences=True)(lstm1)
attention = Attention()
```

```
attended = attention(lstm2)
dense = L.Dense(16, activation='relu')(attended)
dense2 = L.Dense(16, activation='relu')(dense)
out = L.Dense(8, activation='linear')(dense2)

model = keras.models.Model(inputs=inp, outputs=out)
```

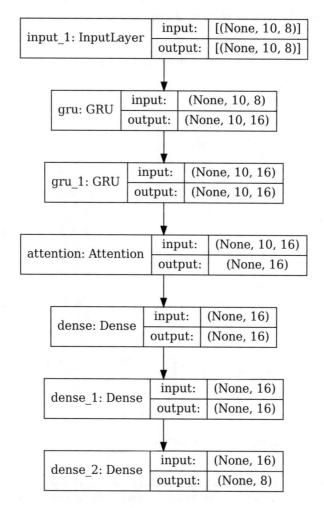

Figure 6-21. *Diagram of the architecture created in Listing 6-3*

The model obtains very good training and validation performance after several hundred epochs of training. We can construct a submodel to derive the output of the second recurrent layer and then pass this output into the custom attention layer's .get_alpha() method to derive the weightage/score for the hidden state at each timestep (Listing 6-4, Figure 6-22).

Listing 6-4. Obtaining and plotting attention scores

```
inp = L.Input((10,8))
rnn1 = model.layers[1](inp)
rnn2 = model.layers[2](rnn1)
submodel = keras.models.Model(inputs=inp, outputs=rnn2)

recurrent_out = tensorflow.constant(submodel.predict(x))

plt.figure(figsize=(10, 5), dpi=400)
plt.bar(range(10), attention.get_alpha(recurrent_out[0,:,0],
color='red')
plt.ylabel('Alpha Values')
plt.xlabel('Time Step')
plt.show()
```

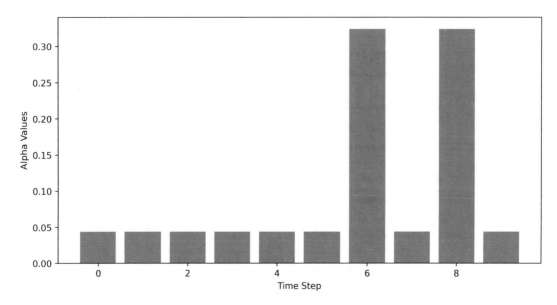

Figure 6-22. *Attention scores averaged across the entire input set*

We see very clearly that the seventh and ninth elements have significantly higher weightage than their counterparts in the input sequence. The attention mechanism allows the network to conveniently extract important components of the timestep without the burden of sequential navigation.

Native Keras Attention

Keras offers two "base" implementations of attention: Luong-style dot product attention (the most commonly used form) and additive attention (Bahdanau-style, less often used).

- `keras.layers.Attention`: Performs Luong-style dot product attention. Parameters: `use_scale=True` creates an additional trainable scalar variable to scale the attention scores. This enables the attention scores to reach larger ranges after being passed through softmax. `score_mode` must be set to 'dot' (default) or 'concat'. The former uses the dot product between the query and key vectors; the latter uses the hyperbolic tangent of the concatenation of the query and key vectors (this resembles Bahdanau-style attention, but without the learned alpha values). dropout must be set to a float between 0 (default) and 1; this indicates the proportion of attention scores to drop. Adding dropout forces the attention mechanism to develop more robust forms of broad attention, which are not overly dependent on specific elements.

- `keras.layers.AdditiveAttention`: Performs Bahdanau-style additive attention. The query and key are added together, passed through a hyperbolic tangent, and summed across the last axis. The Keras implementation does not use trainable weights and biases to learn the alpha values. Parameters: `use_scale=True` creates an additional trainable scalar variable to scale the attention scores. dropout must be set to a float between 0 (default) and 1 and indicates the proportion of attention scores to drop.

To demonstrate usage, let's create a more complex synthetic task. Rather than compositing the output as a sum of individual selected timesteps of the input, we derive it as a weighted sum of all timesteps in the input (Listing 6-5). The weight for some timestep t will be calculated as $4 \cdot \sigma(x - 5) \cdot \sigma(5 - x)$, where σ is the sigmoid function (Figure 6-23). This is a shifted and scaled version of the derivative of the sigmoid function, which is given by $\sigma(x) \cdot \sigma(-x)$.

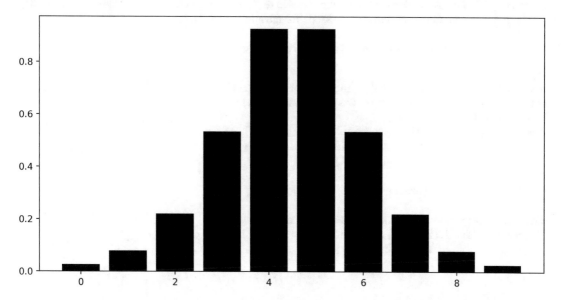

Figure 6-23. *Weightages determined by $4 \cdot \sigma(x - 5) \cdot \sigma(5 - x)$ in the shape of a quasi-normal distribution*

Listing 6-5. Deriving a synthetic dataset with a quasi-normally distributed weighted sum

```
sigmoid = lambda x: 1/(1 + np.exp(-x))
sigmoid_deriv = lambda x: sigmoid(x) * sigmoid(-x)
adjusted_sigmoid_deriv = lambda x: 4 * sigmoid_deriv(x - 5)
weights = adjusted_sigmoid_deriv(np.linspace(0, 10, 10))
```

```
x, y = [], []

NUM_SAMPLES = 10_000

next_element = lambda arr: np.dot(weights, arr)

for i in tqdm(range(NUM_SAMPLES)):
    seed = np.random.normal(0, 1, size=(10,8))
    x.append(seed)
    y.append(next_element(seed))

x = np.array(x)
y = np.array(y)

from sklearn.model_selection import train_test_split as tts
X_train, X_valid, y_train, y_valid = tts(x, y, train_size=0.8)
```

Let us construct the model architecture (Figure 6-24). After extracting relevant features with a bidirectional LSTM, we will perform self-attention with scaled Luong-style dot product attention by passing in the output of the first LSTM as both the query and the key in a list (Listing 6-6). If the value is not provided, it is assumed that the key and the value are the same. In this case, the query, key, and value are all the same. The output of the attention mechanism is passed through another LSTM.

Listing 6-6. Defining the architecture and fitting on the synthetic dataset

```
inp = L.Input((10,8))
lstm1 = L.Bidirectional(L.LSTM(8, return_sequences=True))(inp)
attended = L.Attention(use_scale=True)([lstm1, lstm1])
lstm2 = L.LSTM(16)(attended)
dense = L.Dense(16, activation='relu')(lstm2)
dense2 = L.Dense(16, activation='relu')(dense)
out = L.Dense(8, activation='linear')(dense2)

model = keras.models.Model(inputs=inp, outputs=out)

model.compile(optimizer='adam', loss='mse', metrics=['mae'])
history = model.fit(X_train, y_train, epochs=1000,
                    validation_data=(X_valid, y_valid))
```

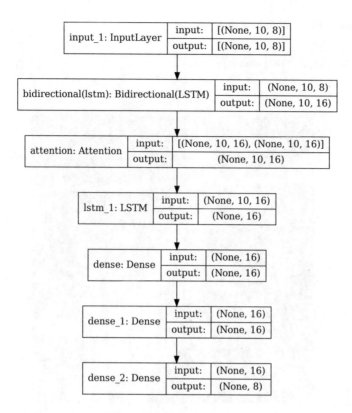

Figure 6-24. *Diagram of the architecture created in Listing 6-6*

When calling an attention layer, we can collect the attention scores by passing `return_attention_scores=True` in addition to the inputs. We can rebuild parts of the model into a submodel to collect the outputted attention scores (Listing 6-7).

Listing 6-7. Obtaining attention scores

```
lstm1_ = model.layers[1](inp)
_, attn = model.layers[2]([lstm1_, lstm1_],
                          return_attention_scores=True)
submodel = keras.models.Model(inputs=inp, outputs=attn)

scores = submodel.predict(X_train)
```

A more convenient way to collect the attention scores if you don't have philosophical issues with leaving "open variables" is to replace `attended = L.Attention(...)` with `attended, scores = L.Attention(return_attention_scores=True, ...)` and construct the submodel as `submodel = keras.models.Model(inputs=inp, outputs=scores)` directly.

The attention scores have shape (number of samples, length of query sequence, length of value/key sequence). The resulting matrix visualizes which timesteps self-attend to other timesteps (Listing 6-8, Figure 6-25).

475

Listing 6-8. Plotting a sample attention score map

```
plt.figure(figsize=(12,12), dpi=400)
sns.heatmap(scores[0,:,:], cbar=False)
plt.show()
```

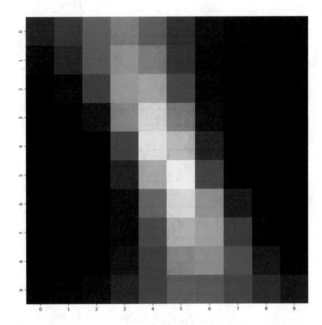

Figure 6-25. *Self-attention matrix from a bidirectional model with dot product attention*

As we expect, the general direction of the self-attention proceeds along the identity diagonal; that is, a timestep t generally attends to timesteps near t. The largest attention values by magnitude (indicated visually by brightness/whiteness) appear at $t \in [4, 5]$ and decay as t increases or decreases, which matches the weight of each timestep on the output vector.

We can change to additive attention by changing L.Attention in Listing 6-6 to L.AdditiveAttention. The resulting self-attention matrix for such a model trained on this dataset looks as follows (Figure 6-26).

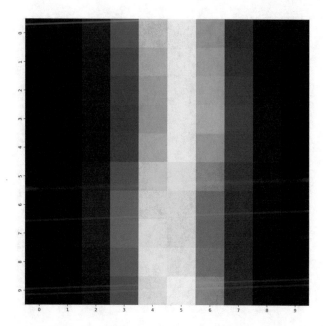

Figure 6-26. *Self-attention matrix from a bidirectional model with additive attention*

Note that the resulting self-attention matrix derived from additive attention is significantly more vertical than that of Luong-style attention. The attention scores are generally independent of the query (the "y-axis") and heavily dependent on the key (the "x-axis").

We see that additive attention has learned attention representations, which are completely independent of the query value. The key alone is enough to determine how the mechanism attends to the value. Given the simplicity of this problem, such behavior is possible. Nevertheless, we note that the attention mechanism still attends to the middle timesteps of the key the highest, with decaying attention scores toward the beginning and ending timesteps. While additive attention and dot product attention obtain similarly near-perfect scores, their attention feature maps are quite different.

As another experiment, let's remove the bidirectionality of the first encoder layer and observe the effect on the attention maps. The dot product attention feature map (Figure 6-27) demonstrates "skew" toward later timesteps, as if wind was blowing southeast and pushing the magnitude of attention values in that direction. The highest attention score is no longer distributed equally in the $[4, 5] \times [4, 5]$ timestep grid, but rather squarely at $(5, 5)$ – the later end of the peak weight timestep region. This makes sense: without bidirectionality, it is still possible for later timesteps to "look back" but not for earlier timesteps to "look forward."

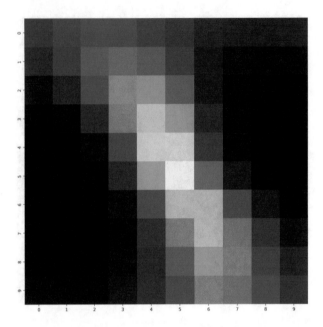

Figure 6-27. *Self-attention matrix from a unidirectional model with dot product attention*

We observe a similar shifted phenomenon when removing bidirectionality from a model fitted with an additive attention mechanism (Figure 6-28).

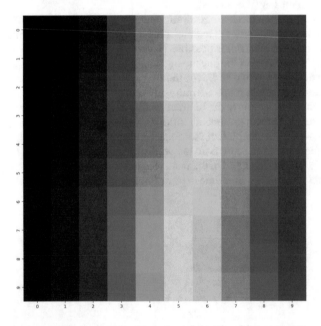

Figure 6-28. *Self-attention matrix from a unidirectional model with additive attention*

Let's adapt our problem once again to experiment with multi-head attention: rather than weighting with a single-humped distribution, we will use a double-humped distribution formed by adding symmetrically shifted single-humped distributions. Let $\sigma'(x) = \sigma(x) \cdot \sigma(-x)$; the weight at timestep x is given by $w(x) = 4 \cdot (\sigma'(x-2) + \sigma'(x-8))$ (Listing 6-9, Figure 6-29).

Listing 6-9. Deriving a bimodal distribution for a weighted sum synthetic dataset

```
sigmoid = lambda x: 1/(1 + np.exp(-x))
sigmoid_deriv = lambda x: sigmoid(x) * sigmoid(-x)
adjusted_sigmoid_deriv1 = lambda x: 4 * sigmoid_deriv(x - 2)
adjusted_sigmoid_deriv2 = lambda x: 4 * sigmoid_deriv(x - 8)
x = np.linspace(0, 10, 10)
weights = adjusted_sigmoid_deriv1(x) + adjusted_sigmoid_deriv2(x)
```

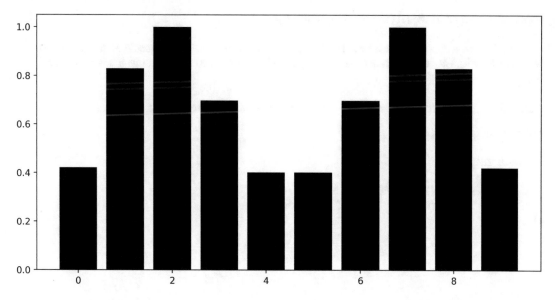

Figure 6-29. *Weightages determined by $4 \cdot (\sigma'(x-2) + \sigma'(x-8))$ in the shape of a quasi-normal distribution*

To use multi-head attention in Keras, specify the number of heads and the dimension of the inputted key (Listing 6-10). When calling the attention layer to link it as part of the graph, pass in the query and key as separate parameters rather than as elements in a bundled list (like with L.AdditiveAttention and L.Attention). value_dim is set to key_dim by default. Key dim is how many dimensions it is projected into by the dense layer. You can specify a key, too, if desired.

Listing 6-10. Deriving a bidirectional recurrent model with multi-head attention to fit on the synthetic dataset derived in Listing 6-9

```
inp = L.Input((10,8))
lstm1 = L.Bidirectional(L.LSTM(8, return_sequences=True))(inp)
attended, scores = L.MultiHeadAttention(num_heads=4,
                                        key_dim=16)(lstm1,
                                                    lstm1,
                                                    return_attention_scores=True)
lstm2 = L.LSTM(16)(attended)
dense = L.Dense(16, activation='relu')(lstm2)
dense2 = L.Dense(16, activation='relu')(dense)
out = L.Dense(8, activation='linear')(dense2)

model = keras.models.Model(inputs=inp, outputs=out)
```

The derived scores in this case have shape (number of samples, number of heads, sequence length, sequence length). We can plot them (Listing 6-11) to interpret how the model attends to the sequences (Figure 6-30).

Listing 6-11. Plotting the attention scores of the multi-head attention mechanism

```
plt.figure(figsize=(24,24), dpi=400)

for i in range(2):
    for j in range(2):
        plt.subplot(2, 2, 2*i + j + 1)
        sns.heatmap(scores[0,2*i + j,:,:], cbar=False)
plt.show()
```

480

Figure 6-30. *Each of the four attention scores for a sample on a multi-head attention model*

Attention in Sequence-to-Sequence Tasks

Lastly, we will also demonstrate the usage of the Keras attention mechanism in its original context of sequence-to-sequence problems, Bahdanau-style. Consider the following seq2seq problem, in which the ith timestep of the target sequence is the sum of the $i + 4$th, $i + 5$th, and $i + 6$th timesteps of the input sequence (Listing 6-12). (The original input sequence is a randomly generated sequence of vectors.)

481

Listing 6-12. Deriving a synthetic sequence-to-sequence dataset

```
x, y = [], []

NUM_SAMPLES = 10_000

next_element = lambda arr: np.stack([arr[(i+4)%10] + arr[(i+5)%10] + arr[(i+6)%10] for i in
range(10)])

for i in tqdm(range(NUM_SAMPLES)):
    seed = np.random.normal(0, 5, size=(10,8))
    x.append(seed)
    y.append(next_element(seed))

x = np.array(x)
y = np.array(y)

from sklearn.model_selection import train_test_split as tts
X_train, X_valid, y_train, y_valid = tts(x, y, train_size=0.8)
```

We will encode the input with two LSTM layers, the first with bidirectionality. The output of the encoder is passed into the decoder. We compute the result of attention between the decoder hidden states (the query) and the encoder output/hidden states (the key *and* value) to determine which elements of the encoder are relevant to attend to. The decoder output is concatenated with the attention mechanism output. For each timestep in the resulting sequence, we project this concatenated vector into an output with a fully connected layer using L.TimeDistributed (Listing 6-13, Figure 6-31). Time-distributed wrappers apply the same layer across multiple time slices, such that we can project the concatenation of the decoder output and the attended encodings into the output "vocabulary."

Listing 6-13. Creating a sequence-to-sequence model with attention

```
inp = L.Input((10,8))
encoder = L.Bidirectional(L.LSTM(16, return_sequences=True))(inp)
encoder2 = L.LSTM(16, return_sequences=True)(encoder)
decoder = L.LSTM(16, return_sequences=True)(encoder2)
attn, scores = L.Attention(use_scale=True)([decoder, encoder2],
                                           return_attention_scores=True)
concat = L.Concatenate()([decoder, attn])
out = L.TimeDistributed(L.Dense(8, activation='linear'))(concat)

model = keras.models.Model(inputs=inp, outputs=out)
```

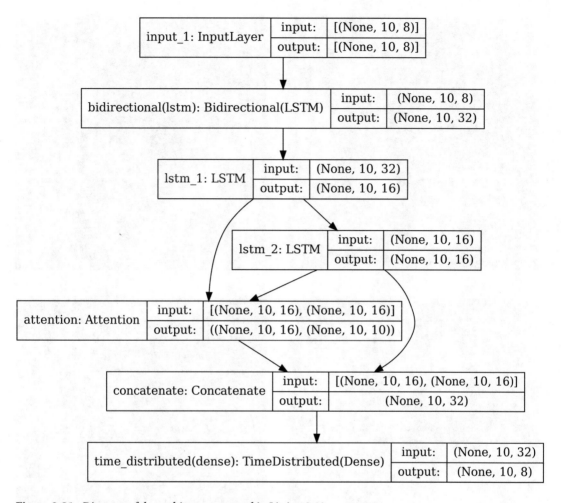

Figure 6-31. *Diagram of the architecture created in Listing 6-13*

We can visualize the learned attention scores for some samples as follows (Listing 6-14, Figure 6-32).

Listing 6-14. Plotting sample attention mechanisms from the sequence-to-sequence model

```
submodel = keras.models.Model(inputs=inp, outputs=scores)
scores = submodel.predict(X_train)

for i in range(4):
    plt.figure(figsize=(12,12), dpi=400)
    sns.heatmap(scores[i,:,:], cbar=False)
    plt.show()
```

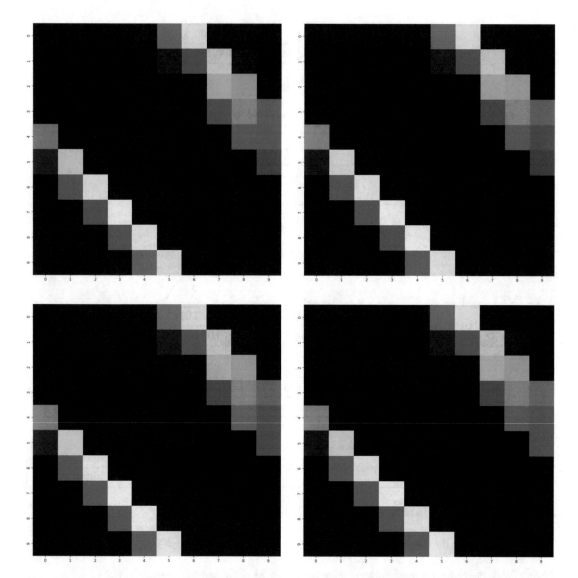

Figure 6-32. *Four attention score matrices for a single sample, derived from different heads of the same attention layer*

Note that our model learns a pretty cool pattern indicative of the true derivation of the output sequence. At the first timestep of the query (representing the output sequence, the first row of the attention grid), the mechanism attends roughly to the region between the fourth and sixth timesteps of the key (representing the input sequence). At the second timestep, the attended region shifts over; as we proceed along the temporal dimension of the query, the attended region wraps around.

You can use this sort of design to solve sequence-to-sequence problems, as well as in creative ways to solve sequence-to-vector or multimodal sequence-and-tabular-to-x problems. For instance, you can build a multitask autoencoder design (see Chapter 8), which both uses a sequence-to-sequence backbone and has an additional output connecting from the encoder output and/or attended encoder output.

Improving Natural Language Models with Attention

Attention was built for and continues to dominate language modeling. In Chapter 5, we built and trained recurrent text models. We can, in some cases, improve these models by adding attention mechanisms. (Note that the success of applying attention directly to non-seq2seq text problems is somewhat limited and case-dependent.)

We will apply attention to a straightforward text-to-vector problem on the TripAdvisor Hotel Reviews dataset. This dataset contains hotel reviews collected from the review platform TripAdvisor with the associated 1–5 rating. The objective is to predict the rating given the review text (Listing 6-15, Figure 6-33).

Listing 6-15. Reading and displaying the TripAdvisor dataset

```
data = pd.read_csv('../input/trip-advisor-hotel-reviews/tripadvisor_hotel_reviews.csv')
data.head()
```

	Review	Rating
0	nice hotel expensive parking got good deal sta...	4
1	ok nothing special charge diamond member hilto...	2
2	nice rooms not 4* experience hotel monaco seat...	3
3	unique, great stay, wonderful time hotel monac...	5
4	great stay great stay, went seahawk game aweso...	5

Figure 6-33. *Head of the TripAdvisor reviews dataset*

Let's begin by creating a model with the custom attention mechanism implemented in the first subsection of the previous section (Listing 6-16, Figure 6-34).

Listing 6-16. A double-LSTM natural language stack

```
inp = L.Input((SEQ_LEN,))
embed = L.Embedding(MAX_TOKENS, EMBEDDING_DIM)(inp)
rnn1 = L.LSTM(16, return_sequences=True)(embed)
rnn2 = L.LSTM(16, return_sequences=True)(rnn1)
attn = Attention()(rnn2)
dense = L.Dense(16, activation='relu')(attn)
dense2 = L.Dense(16, activation='relu')(dense)
out = L.Dense(5, activation='softmax')(dense2)

model = keras.models.Model(inputs=inp, outputs=out)
```

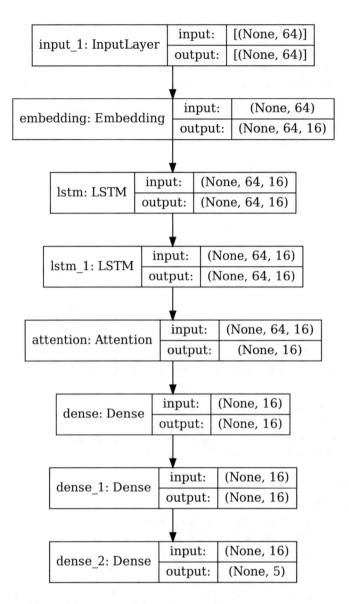

Figure 6-34. *Diagram of the architecture created in Listing 6-16*

We can use a submodel to obtain the attention scores for some input and visualize the scores at each timestep (Listing 6-17, Figures 6-35 to 6-37).

Listing 6-17. Obtaining and plotting attention scores for each word in the sequence

```
inp = L.Input((SEQ_LEN,))
embed = model.layers[1](inp)
rnn1 = model.layers[2](embed)
rnn2 = model.layers[3](rnn1)
```

```
submodel = keras.models.Model(inputs=inp, outputs=rnn2)

for index in range(3):

    fig, ax = plt.subplots(figsize=(10, 5), dpi=400)
    lstm_encodings = tensorflow.constant(submodel.predict(X_train_vec[index:index+1]))
    alpha_values = model.layers[4].get_alpha(lstm_encodings)[0,:,0]
    bars = ax.bar(range(SEQ_LEN), alpha_values, color='red', alpha=0.7)
    text = X_train[X_train.index[index]].split(' ')
    text += ['']*(SEQ_LEN - len(text))
    for i, bar in enumerate(bars):
        height = bar.get_height()
        ax.text(x=bar.get_x() + bar.get_width() / 2 - 0.02, y=height+.0002,
                rotation = 90, size=6,
                s=text[i],
                ha='center')
    ax.set_ylabel('Alpha Values')
    ax.set_xlabel('Time Step')
    ax.axes.yaxis.set_visible(False)
    plt.show()
```

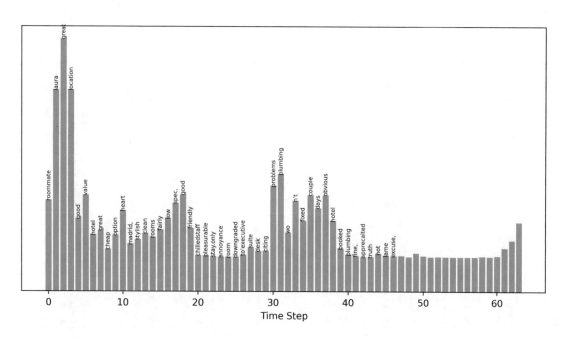

Figure 6-35. *Attention scores for each of the words in a sequence, index 0*

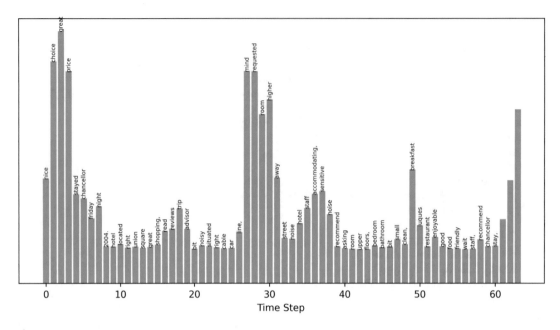

Figure 6-36. *Attention scores for each of the words in a sequence, index 1*

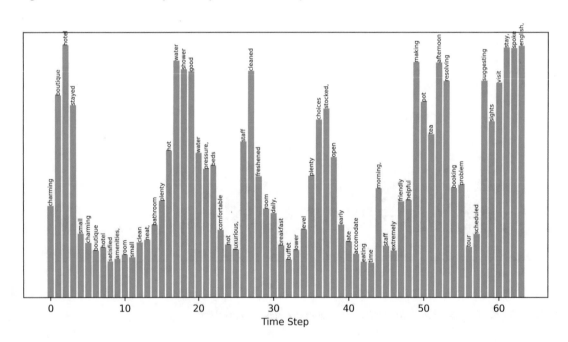

Figure 6-37. *Attention scores for each of the words in a sequence, index 2*

We see that the first few words in the sequence are strongly attended to, along with certain relevant segments in the middle.

We can also use a native Keras multi-head attention approach (Listing 6-18, Figure 6-38).

Listing 6-18. Using a multi-head attention version of the model in Listing 6-16

```
inp = L.Input((SEQ_LEN,))
embed = L.Embedding(MAX_TOKENS, EMBEDDING_DIM)(inp)
rnn1 = L.Bidirectional(L.GRU(16, return_sequences=True))(embed)
attn, scores = L.MultiHeadAttention(num_heads=4, key_dim=4)(rnn1, rnn1,
                                                 return_attention_scores=True)
rnn2 = L.LSTM(16, return_sequences=True)(attn)
rnn3 = L.LSTM(16)(rnn2)
dense = L.Dense(8, activation='relu')(rnn3)
dense2 = L.Dense(8, activation='relu')(dense)
out = L.Dense(5, activation='softmax')(dense2)

model = keras.models.Model(inputs=inp, outputs=out)
```

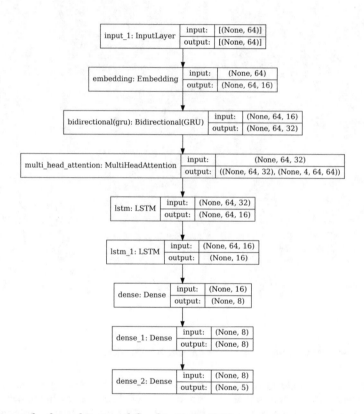

Figure 6-38. Diagram for the architecture defined in Listing 6-18

The attention masks can similarly be visualized (Figure 6-39). We observe that the attention mechanism here is learning a lot of cross-sequence dependency/relationships. Particularly, note that adjectives and the nouns they refer to (e.g., "place," "paradise," "fabulous," etc.) have high self-attention, whereas unrelated components have low self-attention scores. This self-attention matrix demonstrates a lot of verticality/horizontality, meaning that certain words have consistent semantic relevance across the entire sequence.

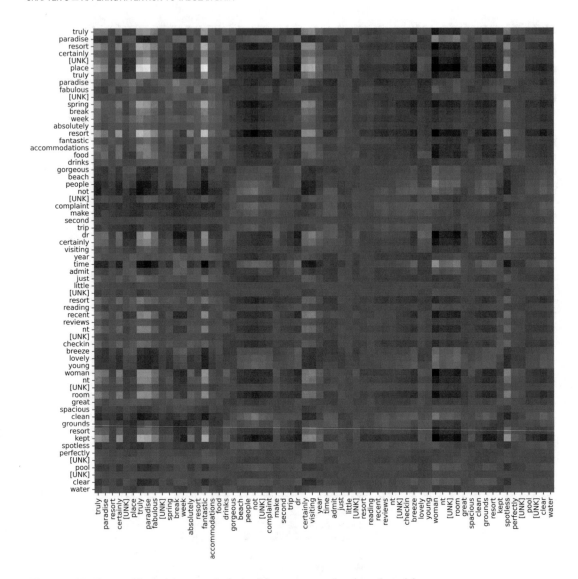

Figure 6-39. *Large self-attention matrix derived from an attention-based model*

By improving recurrent language models, we can also improve the modeling capability of multimodal models. Improved text modeling not only allows us to better model the relationship between the text input and the output for multimodal problems but also allows us to better model the tabular input by interpreting it with improved characterization from the text input.

Let's return to the stock news and forecasting multimodal dataset discussed in Chapter 5. We can modify the text reading component by adding a shared attention mechanism and training appropriately (Listing 6-19, Figure 6-40).

Listing 6-19. Adapting a multi-head multimodal model with attention mechanisms

```
lstm1 = L.Bidirectional(L.LSTM(16, return_sequences=True))
top1_lstm1 = lstm1(top1_embed)
top2_lstm1 = lstm1(top2_embed)
top3_lstm1 = lstm1(top3_embed)

attn = L.Attention(use_scale=True)

lstm2 = L.LSTM(32)
top1_lstm2 = lstm2(attn([top1_lstm1, top1_lstm1]))
top2_lstm2 = lstm2(attn([top2_lstm1, top2_lstm1]))
top3_lstm2 = lstm2(attn([top3_lstm1, top3_lstm1]))
```

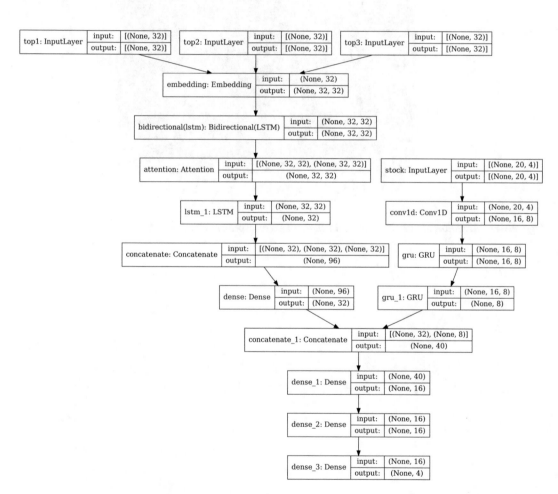

Figure 6-40. *Diagram of the architecture created in Listing 6-19*

Training with attention also provides us strong interpretability into how the model is making its decision. When observing the attention feature maps, we observe that a small set of keywords are strongly relevant for prediction (Figures 6-41 to Figure 6-46).

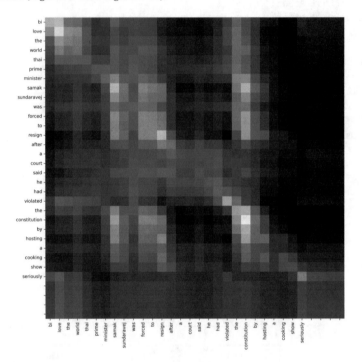

Figure 6-41. *Attention score matrix. Note the high attention score on "constitution"*

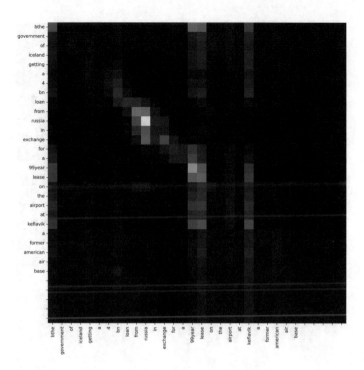

Figure 6-42. *Attention score matrix. Note the high attention score on "russia"*

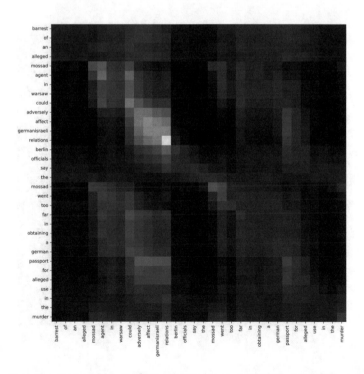

Figure 6-43. *Attention score matrix. Note the high attention score on "relations" and the corresponding high scores in the region surrounding "adversely," "affect," and "german-israeli"*

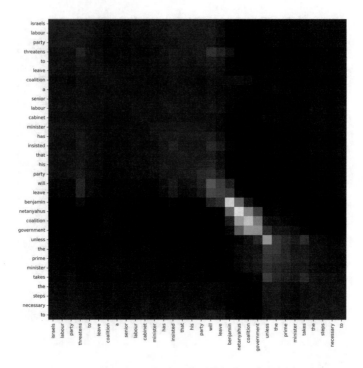

Figure 6-44. *Attention score matrix. Note the high attention scores on "benjamin," "netanyahu," "cocoalition,"*
and "government"

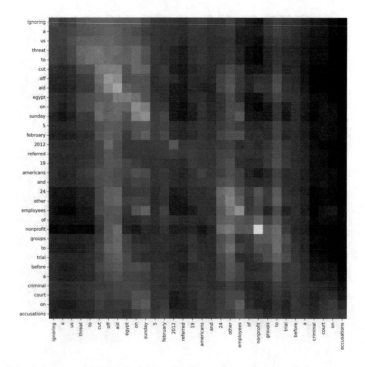

Figure 6-45. *Attention score matrix. Note the high attention score on "nonprofit"*

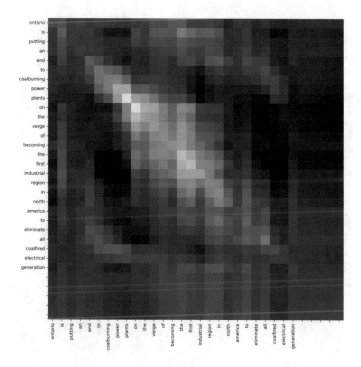

Figure 6-46. *Attention score matrix. Note the high attention scores centering on the middle of the sentence*

We can also replace the single attention mechanism for multi-head attention (Listing 6-20, Figure 6-47).

Listing 6-20. Replacing the single attention mechanism with multi-head attention

```
attn = L.MultiHeadAttention(num_heads=8, key_dim=32,
                            dropout=0.1)
```

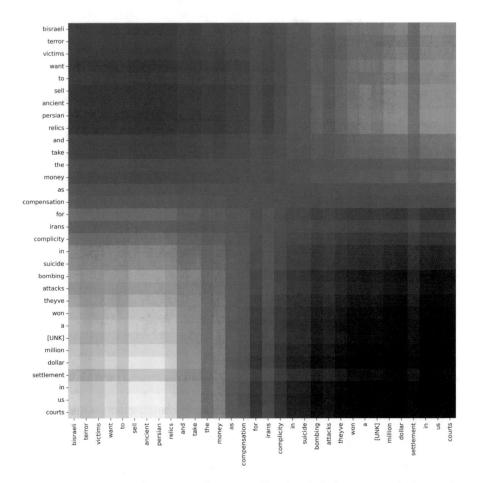

Figure 6-47. *One of the attention feature maps from several heads, which demonstrates high attention between the beginning of the sub-sentence "israeli terror victims want to sell ancient persian relics" and "a million dollar sentiment in us courts"*

Direct Tabular Attention Modeling

Chapters 4 and 5 have introduced methods that, to most readers, seem foreign and unnatural. Even as we, the authors, try to demonstrate their empirical validity and offer conceptual models to understand them, we, too, understand that convolutions and recurrent layers just intuitively or fundamentally don't fit "naturally" with tabular data, and that this is unsettling. Tabular data does not natively possess contiguous semantics, whereas convolutions and recurrent layers operate on the assumption that inputted data possesses contiguous semantics.

Attention, however, shows to be a very natural mechanism for tabular data. Recall from the beginning of this chapter that attention addresses problems in the restrictive sequential reading of natural language data. Rephrased, attention frees natural language data from its contiguous semantics by offering a more direct way to relate between cross-sequence/temporal dependencies in the sequence. Attention operates upon an "anti-contiguous semantics" and can be applied to any input, sequential or not, to model non-contiguous dependencies. (Of course, it can be adapted with appendages like causal masking to support modeling of contiguous dependency.) Due to this, it is no surprise that one dominant trend in deep-learning-for-tabular-data research is dedicated toward attention and transformer architectures. (See the last section of this chapter for a discussion of several such models.)

We'll begin by creating an input head and reshaping the input into two dimensions (Listing 6-21). This is necessary for applying attention layers.

Listing 6-21. The input and reshaping layer for the network

```
inp = L.Input((len(X_train.columns),))
reshape = L.Reshape((len(X_train.columns),1))(inp)
```

Next, we'll build a sample "attention block" (Listing 6-22). We begin by applying two fully connected layers. If a fully connected layer with r nodes is applied to inputs with shape (p, q), the result has shape (p, r). A dense mapping is learned for each "slice" along the first axis. Afterward, we apply Luong-style self-attention with scaling, perform layer normalization, and return.

Listing 6-22. Defining an attention block

```
def attn_block(inp,
               dense_units=8,
               num_heads=4,
               key_dim=4):
    dense = L.Dense(dense_units, activation='relu')(inp)
    dense2 = L.Dense(dense_units, activation='relu')(dense)
    attn_out = L.Attention(use_scale=True)([dense2, dense2])
    layer_norm = L.LayerNormalization()(attn_out)
    return layer_norm
```

We can stack the attention blocks like such (Listing 6-23).

Listing 6-23. Arranging attention blocks into a complete model

```
attn1 = attn_block(reshape)
attn2 = attn_block(attn1)
flatten = L.Flatten()(attn2)
predense = L.Dense(32, activation='relu')(flatten)
out = L.Dense(7, activation='softmax')(predense)

model = keras.models.Model(inputs=inp, outputs=out)
```

Note that the first layer applied to the inputs is a fully connected dense layer; this can be thought of as an embedding that transforms the input of shape $(n_{features}, 1)$ into shape $(n_{features}, d_{embed})$. All following fully connected layers process the vectors corresponding to each feature independently, while each attention mechanism forces cross-relationships between the features. The information resulting from two attention blocks is flattened into a single vector and projected into the output space. The full sample architecture adapted for the Forest Cover dataset is shown in Figure 6-48.

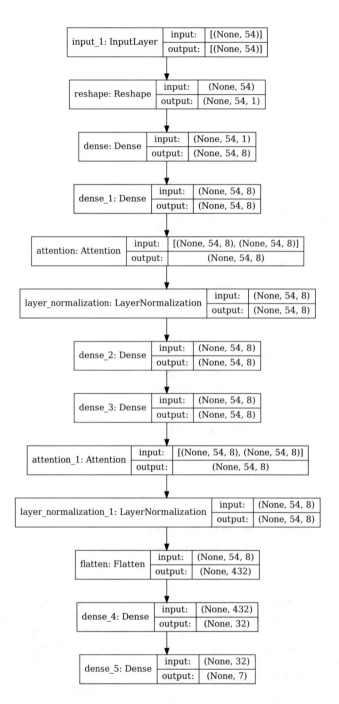

Figure 6-48. *Diagram for the architecture defined in Listing 6-23*

Alternatively, we can use multi-head self-attention, modifying our attention block code appropriately (Listing 6-24).

Listing 6-24. Defining an attention block

```
def attn_block(inp,
               dense_units=8,
               num_heads=4,
               key_dim=4):
    dense = L.Dense(dense_units, activation='relu')(inp)
    dense2 = L.Dense(dense_units, activation='relu')(dense)
    attn_out = L.MultiHeadAttention(num_heads=num_heads,
                                     key_dim=key_dim)(dense2, dense2)
    layer_norm = L.LayerNormalization()(attn_out)
    return layer_norm
```

Note that if we wanted to do something like Luong-style attention but derive different keys and queries (and/or values) from a shared vector, this would just be a case of multi-head attention with one head!

Incorporating attention mechanisms into your tabular model is generally straightforward and effective. You can add, for example, the following features: residual connections, multiple parallel multi-head attention branches, applying convolutions and/or recurrent layers along the per-feature dimension to extract additional features, and/or incorporating attention into direct recurrent modeling techniques proposed in Chapter 5, among others.

For instance, see the AutoInt model (Figure 6-49) proposed by Weiping Song et al. in the paper "AutoInt: Automatic Feature Interaction Learning via Self-Attentive Neural Networks."[8] This centerpiece of this relatively simple architecture is a multi-head self-attention layer with a residual connection and performs exceptionally well on click-through rate (CTR) prediction problems.

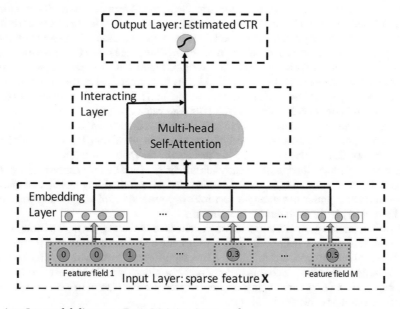

Figure 6-49. *AutoInt model diagram. From Weiping Song et al.*

---

[8] Song, W., Shi, C., Xiao, Z., Duan, Z., Xu, Y., Zhang, M., & Tang, J. (2019). AutoInt: Automatic Feature Interaction Learning via Self-Attentive Neural Networks. *Proceedings of the 28th ACM International Conference on Information and Knowledge Management.*

Attention-Based Tabular Modeling Research

There is a significant body of research applying attention to tabular data. As discussed previously, the attention mechanism is a very natural mechanism to select for different components of an input information stream. This section discusses four such papers from recent research in-depth and provides guidance on implementing the model or using an existing implementation. We will formalize each model using the authors' notation for relevant variables, operations, and functions. This is essential to developing a concrete understanding of how the model operates, but – as a cautionary note – does tend to vary between papers.

TabTransformer

The TabTransformer model, introduced by Xin Huang et al. 2020 in the paper "TabTransformer: Tabular Data Modeling Using Contextual Embeddings,"[9] is a relatively straightforward application of transformer/attention-based blocks to tabular data.

Tabular datasets are generally comprised of just two types of features: categorical and continuous. Following the paper's notation, let m be the number of categorical features, and let c be the number of continuous variables. The set of categorical features is thus $x_{cat} := \{x_1, x_2, ..., x_m\}$, and the set of continuous features is $x_{cont} \in \mathbb{R}^c$. Continuous features are rich with information that can often successfully be mapped to new spaces by neural networks. We can correspondingly think of continuous features as lacking in "potential information," in the same way that a fast-moving ball has less potential energy than a slow-moving ball. Because continuous features span over a wider domain of values, relationships between specific values can be more easily worked out.

On the other hand, categorical features suffer from a dearth of information, but therefore have high potential information. Each specific class in a categorical feature can be associated with some collection of attributes, which becomes useful when interpreted with respect to other features, categorical and continuous. In natural language processing, embeddings map specific class values to a continuous vector; the categorical "feature" at each timestep in the case of language has V total classes, where V is the vocabulary size. Correspondingly, classical machine learning algorithms that perform well on mixed-type tabular datasets construct implicit "embeddings" for classes in a categorical feature too. Suppose a node in a decision tree goes one way if the grade level is ten or higher. This is a rough equivalent to defining some attribute in an embedding vector that maps inputs with values 10, 11, and 12 close in one region and all other inputs close in another region; density in these regions is then "read" and combined with other information from this implicitly constructed embedding to form a prediction. However, such implicit "embeddings" aren't explicit or specific, limited in precision by the node conditions.

We can generate *column embeddings* for each column in x_{cat}. Let d be the dimensionality of the embedding space. For each column in x_{cat}, we maintain a trainable embedding lookup in which each unique value in that column corresponds to a d-length vector. To accommodate missing values, you can also generate an additional embedding to handle the n/a case.

After embedding the categorical features through column embedding, we have a (m, d)-shaped tensor. This is passed through a transformer block N times: the transformer block here is comprised of a standard multi-head attention mechanism followed by a feed-forward layer, with residual connections and layer normalization after each. Each transformer block produces what Huang et al. term "contextual embeddings": that is, embeddings are not just created relative to other classes in a single categorical feature but related across/in context of all other features.

[9] Huang, X., Khetan, A., Cvitkovic, M.W., & Karnin, Z.S. (2020). TabTransformer: Tabular Data Modeling Using Contextual Embeddings. *ArXiv, abs/2012.06678*. https://arxiv.org/pdf/2012.06678.pdf.

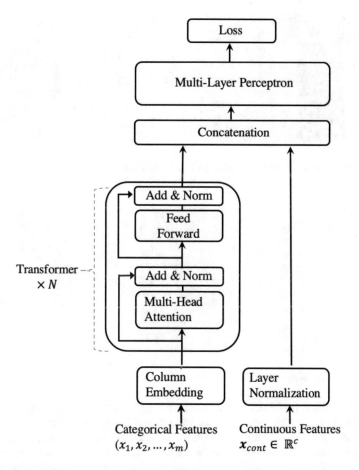

Figure 6-50. *The TabTransformer architecture. From Huang et al.*

After repeated processing by the transformer block stack, the resulting (n, d)-shaped contextual embedding tensor is flattened/lined end-to-end into a vector with length $n \cdot d$ and concatenated with layer-normalized continuous features, such that the resulting concatenated vector has shape $n \cdot d + c$. This vector contains rich computed contextual information, which is passed into a standard feed-forward network/ multilayer perceptron into an output. The TabTransformer's architecture (displayed in full in Figure 6-50) can be summarized as a standard multilayer perceptron model with transformer-based contextualization of the categorical features.

TabTransformer pretrains the categorical embeddings and the transformer stack with two types of self-supervised pretraining: masked language modeling (MLM) (Figure 6-51) and replaced token detection (RTD) (Figure 6-52). In BERT-style masked language modeling pretraining, certain columns in the input are randomly masked, and the objective is to predict the replaced column values. Replaced token detection is a variation in which the values in certain columns are scrambled or otherwise tampered and the objective is to identify which columns have been altered and which have not. Both tasks require embeddings and the contextual processing transformer layers to learn important relationships in an unsupervised manner.

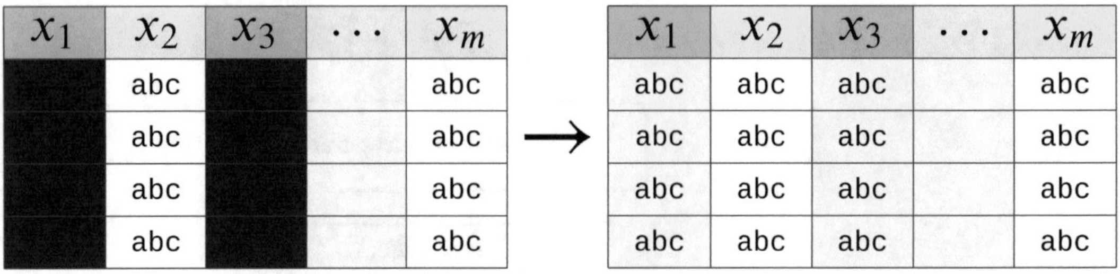

x_1	x_2	x_3	$\cdots$	x_m
	abc			abc
	abc			abc
	abc			abc
	abc			abc

x_1	x_2	x_3	$\cdots$	x_m
abc	abc	abc		abc
abc	abc	abc		abc
abc	abc	abc		abc
abc	abc	abc		abc

Figure 6-51. Visualization of the masked language modeling task

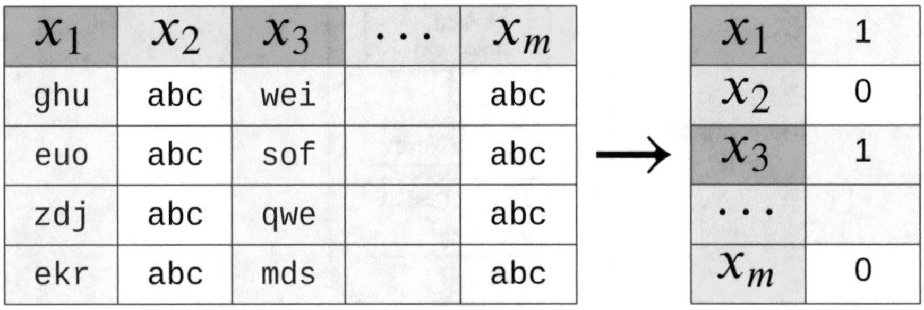

x_1	x_2	x_3	$\cdots$	x_m
ghu	abc	wei		abc
euo	abc	sof		abc
zdj	abc	qwe		abc
ekr	abc	mds		abc

x_1	1
x_2	0
x_3	1
$\cdots$	
x_m	0

Figure 6-52. Visualization of the replaced token detection

The authors of the TabTransformer paper benchmark their model against a collection of 15 datasets. They use a hidden embedding dimension of 32, six transformer blocks, and eight attention heads in each block. The authors find that TabTransformer outperforms a baseline multilayer perceptron in almost all cases, albeit with a generally marginal improvement. Note that TabTransformer is really just an MLP fitted with a transformer-based categorical feature contextual embedding learner, however, so the improvement gains can be attributed just to this mechanism. Moreover, TabTransformer outperforms other deep learning models designed for tabular datasets and approaches the performance of a hyperparameter-optimized Gradient Boosting Decision Tree (GBDT) (Tables 6-4 and 6-5).

Table 6-4. *Performance of TabTransformer against a baseline MLP on several datasets. Percentages in AUC accuracy. From Huang et al.*

Dataset	Baseline MLP	TabTransformer	Gain (%)
albert	74.0	75.7	**1.7**
1995_income	90.5	90.6	**0.1**
dota2games	63.1	63.3	**0.2**
hcdr_main	74.3	75.1	**0.8**
adult	72.5	73.7	**1.2**
bank_marketing	92.9	93.4	**0.5**
blastchar	83.9	83.5	-0.4
insurance_co	69.7	74.4	**4.7**
jasmine	85.1	85.3	**0.2**
online_shoppers	91.9	92.7	**0.8**
philippine	82.1	83.4	**1.3**
qsar_bio	91.0	91.8	**0.8**
seismicbumps	73.5	75.1	**1.6**
shrutime	84.6	85.6	**1.0**
spambase	98.4	98.5	**0.1**

Table 6-5. *Mean dataset performance of TabTransformer against other models. Percentages in AUC accuracy. From Huang et al.*

Model Name	Mean AUC (%)
TabTransformer	**82.8** $\pm$ 0.4
MLP	81.8 $\pm$ 0.4
GBDT	**82.9** $\pm$ 0.4
Sparse MLP	81.4 $\pm$ 0.4
Logistic Regression	80.4 $\pm$ 0.4
TabNet	77.1 $\pm$ 0.5
VIB	80.5 $\pm$ 0.4

One advantage of the TabTransformer model beyond its demonstrated improvement in performance is interpretability. Because the model explicitly learns embeddings associated with each unique class value in each categorical feature, the learned embeddings can be analyzed and interpreted to understand how the model is making decisions. The authors perform a t-SNE reduction on the embeddings derived for the Bank Marketing dataset and find that "semantically similar classes are close with each other and form clusters in the embedding space" (Figure 6-53).

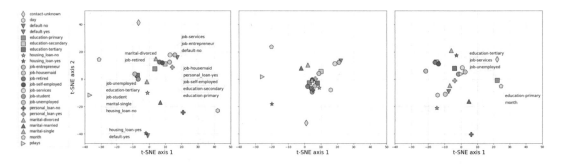

Figure 6-53. *Visualization of category embeddings reduced in low-dimensional space. From Huang et al.*

To demonstrate the richness of the embeddings, the authors train linear models on the contextual embeddings for each categorical feature outputted at each of the successive transformer blocks and show that the derived features, even before the first application of the transformer block, are enough to reach at least 90% of the accuracy obtained by the complete TabTransformer model (Figure 6-54).

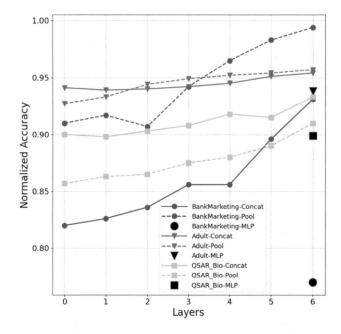

Figure 6-54. *Three different datasets (Bank Marketing, Adult Census, and QSAR) are shown – the performance of a regression model trained on the attended features after the nth layer. From Huang et al.*

Another primary advantage Huang et al. highlight is robustness to noisy and missing data, which tree-based methods often have comparatively more difficulty dealing with. The TabTransformer model performs better than a baseline MLP by a significant margin for both data noise (Figure 6-55) and data deletion (missing data) attacks (Figure 6-56). TabTransformer can withhold an impressively high proportion of the original performance at no corruption at high levels of noise and missing data.

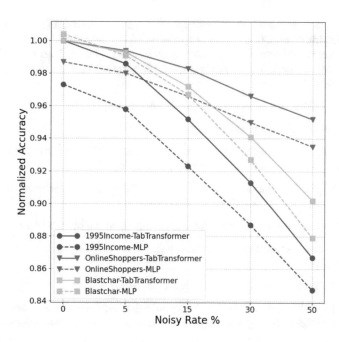

Figure 6-55. *Degradation of performance relative to a model trained on uncorrupted data from different levels of data corruption via data noise. From Huang et al.*

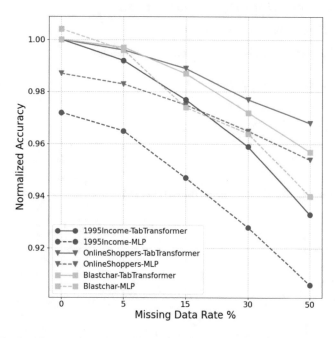

Figure 6-56. *Degradation of performance relative to a model trained on uncorrupted data from different levels of data corruption via missing data. From Huang et al.*

Lastly, TabTransformer also lends itself nicely to self-supervised pretraining and has been shown to be one of the more promising architectures for self-supervised tabular learning.

The TabTransformer architecture is relatively straightforward to build "from scratch" using Keras layers and is a good exercise. We'll begin by defining the following key configuration parameters for our architecture (Listing 6-25):

- NUM_CONT_FEATS: The number of continuous features

- NUM_CAT_FEATS: The number of categorical features

- NUM_UNIQUE_CLASSES: A list of the number of unique classes in each category. Should have the same length as the number of categorical features

- EMBEDDING_DIM: The dimensionality of the embedding (i.e., the vector associated with each unique class in each categorical feature)

- NUM_HEADS: The number of heads in each multi-head attention mechanism

- KEY_DIM: The dimension to project the key and query to in the multi-head attention layer

- NUM_TRANSFORMERS: The number of transformer blocks to stack

- FF_HIDDEN_DIM: The number of hidden units in the feed-forward component of the transformer block before projecting into the dimension of the embedding dimension

- MLP_LAYERS: The number of hidden feed-forward layers in the multilayer perceptron component at the end of the TabTransformer model

- MLP_HIDDEN: The number of units in each hidden layer in the multilayer perceptron component at the end of the TabTransformer model

- OUT_DIM: The dimensionality of the output

- OUT_ACTIVATION: The activation function to use in the output

Listing 6-25. Establishing configuration constants

```
'''
CONFIG
'''

NUM_CONT_FEATS = 8
NUM_CAT_FEATS = 4
NUM_UNIQUE_CLASSES = [32 for i in range(NUM_CAT_FEATS)]

EMBEDDING_DIM = 32

NUM_HEADS = 4
KEY_DIM = 4
NUM_TRANSFORMERS = 6
FF_HIDDEN_DIM = 32

MLP_LAYERS = 4
MLP_HIDDEN = 16

OUT_DIM = 1
OUT_ACTIVATION = 'linear'
```

506

Let's begin by defining the inputs (Listing 6-26). The continuous feature input head is straightforward: we define an input head that accepts a vector of length NUM_CONT_FEATS and apply layer normalization. Because we need to generate a unique embedding scheme for each categorical feature, we create a list of input heads corresponding to each categorical feature. We correspondingly generate embeddings that link from each categorical feature input head and associate with an EMBEDDING_DIM-sized vector. The vocabulary size for each embedding layer is provided to us in NUM_UNIQUE_CLASSES. Each embedding at this point will yield a tensor of shape (batch size, 1, EMBEDDING_DIM). We want to "link" up the embeddings for each of the categorical features, so we concatenate upon the second axis (axis 1 with zero-based indexing) to yield a grouped embedding tensor of shape (batch size, NUM_CAT_FEATS, EMBEDDING_DIM). Note that this tensor shape resembles that of natural language sequences, which have tensor shape (batch size, sequence length, embedding dimension). In this case, we do not assume that the categorical features in axis 1 are arranged sequentially, which works fine with transformer blocks.

Listing 6-26. Defining the input heads

```
cont_inp = L.Input((NUM_CONT_FEATS,), name='Cont Feats')
normalize = L.LayerNormalization()(cont_inp)

cat_inps = [L.Input((1,), name=f'Cat Feats {i}') for i in \
            range(NUM_CAT_FEATS)]
zipped = zip(NUM_UNIQUE_CLASSES, cat_inps)
embeddings = [L.Embedding(uqcls, EMBEDDING_DIM)(cat_inp) for uqcls, cat_inp in zipped]
concat_embed = L.Concatenate(axis=1)(embeddings)
```

Because we will stack several transformer blocks together, it's useful to define a function that performs the block linkage for us (Listing 6-27). We begin by computing multi-head self-attention with tensors from the input layer. To form a residual connection, we add the result of self-attention with the original input. (Note that the result of self-attention is the same as the input shape by default, although you can specify the output to be projected to a different dimension.) We apply layer normalization, followed by two feed-forward layers. Another residual connection is built between the output of layer normalization and the output of the feed-forward component of the transformer block. The result is normalized again and returned.

Listing 6-27. Defining a transformer block

```
def transformer(inp):
    attention = L.MultiHeadAttention(num_heads=NUM_HEADS,
                                     key_dim=KEY_DIM)(inp, inp)
    add = L.Add()([inp, attention])
    norm = L.LayerNormalization()(add)
    dense1 = L.Dense(FF_HIDDEN_DIM, activation='relu')(norm)
    dense2 = L.Dense(EMBEDDING_DIM, activation='relu')(dense1)
    add2 = L.Add()([norm, dense2])
    norm2 = L.LayerNormalization()(add2)
    return norm2
```

We can apply this transformer block several times (Listing 6-28). The output tensor still has shape (batch size, NUM_CAT_FEATS, EMBEDDING_DIM), but each of the embeddings is now contextualized with respect to the other categorical features. We flatten the result into a batch of vectors with shape (batch size, NUM_CAT_FEATS × EMBEDDING_DIM).

Listing 6-28. Stacking transformers together, followed by flattening

```
transformed = concat_embed
for i in range(NUM_TRANSFORMERS):
    transformed = transformer(transformed)
contextual_embeddings = L.Flatten()(transformed)
```

The contextual embeddings can be concatenated with the normalized continuous variables and fed into the multilayer perceptron (Listing 6-29).

Listing 6-29. Defining the MLP that accepts the transformed (flattened) features and outputs the final decision

```
all_feat_concat = L.Concatenate()([normalize, contextual_embeddings])
mlp = all_feat_concat
for i in range(MLP_LAYERS):
    mlp = L.Dense(MLP_HIDDEN, activation='relu')(mlp)
out = L.Dense(OUT_DIM, activation=OUT_ACTIVATION)(mlp)
```

To build the graph into a model, we collect all the inputs and call keras.models.Model to connect the inputs to the output (Listing 6-30, Figure 6-57).

Listing 6-30. Defining the TabTransformer model

```
all_inps = cat_inps + [cont_inp]
model = keras.models.Model(inputs=all_inps, outputs=out)
```

Figure 6-57. *The custom TabTransformer model defined in Listings 6-26 to 6-30*

For the purposes of cleaner visualization and to adhere to good implementation practice, we can use a compartmentalized design to define the transformer block as a separate submodel (Listing 6-31). Rather than returning the output tensor, we construct a new submodel with a unique name (submodels in the same graph must have unique names).

Listing 6-31. Modifying the transformer block with compartmentalized design

```
def build_transformer(inp_shape, id_=0):
    inp = L.Input(inp_shape)
    attention = L.MultiHeadAttention(num_heads=NUM_HEADS,
                                     key_dim=KEY_DIM)(inp, inp)
    add = L.Add()([inp, attention])
    norm = L.LayerNormalization()(add)
    dense1 = L.Dense(FF_HIDDEN_DIM, activation='relu')(norm)
    dense2 = L.Dense(EMBEDDING_DIM, activation='relu')(dense1)
    add2 = L.Add()([norm, dense2])
    norm2 = L.LayerNormalization()(add2)
    return keras.models.Model(inputs=inp, outputs=norm2,
                              name=f'Transformer_Block_{id_}')
```

The code to define the TabTransformer model is largely the same (Listing 6-32; transformer block, Figure 6-58; TabTransformer architecture, Figure 6-59).

Listing 6-32. Defining a model with compartmentalized design

```
cont_inp = L.Input((NUM_CONT_FEATS,), name='Cont Feats')
normalize = L.LayerNormalization()(cont_inp)

cat_inps = [L.Input((1,),
                    name=f'Cat Feats {i}') for i in \
                    range(NUM_CAT_FEATS)]
zipped = zip(NUM_UNIQUE_CLASSES, cat_inps)
embeddings = [L.Embedding(uqcls, EMBEDDING_DIM)(cat_inp) for uqcls, cat_inp in zipped]
concat_embed = L.Concatenate(axis=1)(embeddings)

transformed = concat_embed
for i in range(NUM_TRANSFORMERS):
    transformer = build_transformer((NUM_CAT_FEATS, EMBEDDING_DIM), \
                                    id_=i)
    transformed = transformer(transformed)
contextual_embeddings = L.Flatten()(transformed)

all_feat_concat = L.Concatenate()([normalize, contextual_embeddings])
mlp = all_feat_concat
for i in range(MLP_LAYERS):
    mlp = L.Dense(MLP_HIDDEN, activation='relu')(mlp)
out = L.Dense(OUT_DIM, activation=OUT_ACTIVATION)(mlp)

all_inps = cat_inps + [cont_inp]
model = keras.models.Model(inputs=all_inps, outputs=out)
```

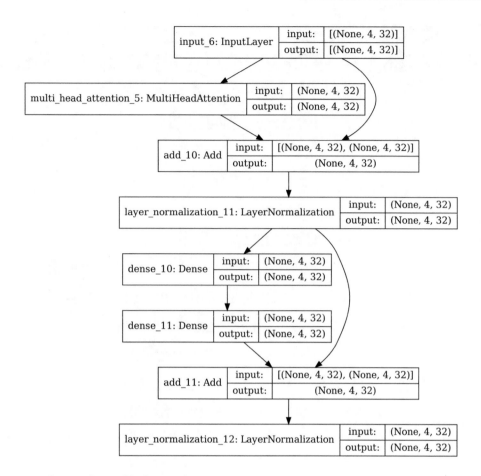

Figure 6-58. *The transformer block*

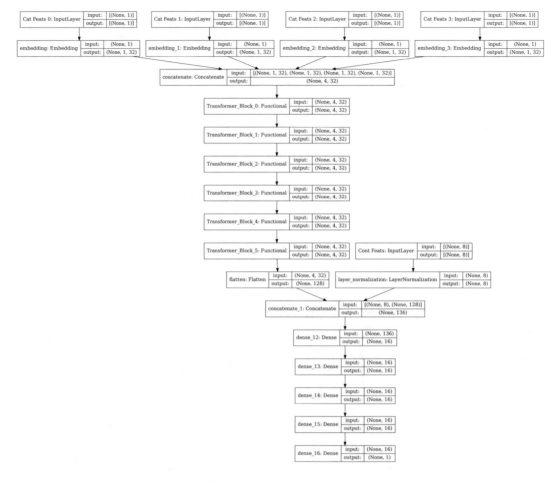

Figure 6-59. *The TabTransformer model in Keras with compartmentalized transformer blocks*

Let's demonstrate the application of such a model to the Ames Housing dataset, which we have encountered previously. The Ames Housing dataset has many categorical features and is a great mixed-type tabular dataset to build models for. We will begin by reading the data, scaling the target, and determining which features are categorical and which are continuous (Listing 6-33).

Listing 6-33. Reading the Ames Housing dataset and identifying categorical features

```
df = pd.read_csv('https://raw.githubusercontent.com/hjhuney/Data/master/AmesHousing/
train.csv')
df = df.dropna(axis=1, how='any').drop('Id', axis=1)
x = df.drop('SalePrice', axis=1)
y = df['SalePrice'] / 1000

cat_features = []
for colIndex, colName in enumerate(x.columns):
    # find categorical variables to process
```

```
    if type(x.iloc[0, colIndex]) == str or len(x[colName].unique()) <= 5:
        cat_features.append(colName)
cont_features = [col for col in x.columns if col not in cat_features]
```

Because the embeddings require all class information to be label-/ordinal-encoded as integers beginning from 0, we will use scikit-learn's OrdinalEncoder to appropriately transform all categorical features (Listing 6-34). We will also cast all continuous features as float32 to avoid type issues later on.

Listing 6-34. Encoding relative features

```
from sklearn.preprocessing import OrdinalEncoder
encoders = {col:OrdinalEncoder() for col in cat_features}
for cat_feature in cat_features:
    encoder = encoders[cat_feature]
    x[cat_feature] = encoder.fit_transform(np.array(x[cat_feature]).reshape(-1, 1))

for cont_feature in cont_features:
    x[cont_feature] = x[cont_feature].astype(np.float32)
```

After preprocessing is completed, we perform train-validation split (Listing 6-35). It is important we perform encoding beforehand such that we do not run the risk of encountering classes that have not been associated with an integer by the encoder because it was not seen during training. Because we are just applying a simple ordinal encoder, this does not warrant data leakage.

Listing 6-35. Train-validation split

```
from sklearn.model_selection import train_test_split as tts
X_train, X_valid, y_train, y_valid = tts(x, y, train_size=0.8)
```

You can also use the from-scratch TabTransformer model we built, but the community also has many great implementations. Cahid Arda has implemented TabTransformer in Keras, which can be directly accessed within-notebook with the following (Listing 6-36).

Listing 6-36. Cloning Cahid Arda's TabTransformer repository. The renaming is to avoid Python syntax issues with hyphens in imports

```
!git clone https://github.com/CahidArda/tab-transformer-keras.git
os.rename('./tab-transformer-keras', './tab_transformer_keras')
```

We can import the TabTransformer model and specify desired configuration parameters and use the get_X_from_features utility function to prepare the inputs for the model (Listing 6-37). This separates the continuous features form the categorical features and dices up each of the categorical features into individual items. The complete set of continuous features and each categorical feature are placed into a list, which is accepted by the multi-head TabTransformer model.

Listing 6-37. Instantiating the TabTransformer model

```
from tab_transformer_keras.tab_transformer_keras.tab_transformer_keras import TabTransformer
from tab_transformer_keras.misc import get_X_from_features

X_train_tt = get_X_from_features(X_train, cont_features, cat_features)
X_valid_tt = get_X_from_features(X_valid, cont_features, cat_features)
```

```
class_counts = [x[col].nunique() for col in cat_features]
model = TabTransformer(
    categories = class_counts,
    num_continuous = len(cat_features),
    dim = 16,
    dim_out = 1,
    depth = 6,
    heads = 8,
    attn_dropout = 0.1,
    ff_dropout = 0.1,
    mlp_hidden = [(32, 'relu'), (16, 'relu')]
)
```

The result is a Keras model that can be compiled and fitted (Listing 6-38). Note that because the model is custom-defined, it lacks certain functionalities like visualization. You can view the source code yourself at https://github.com/CahidArda/tab-transformer-keras.

Listing 6-38. Compiling and fitting the model

```
model.compile(optimizer='adam', loss='mse', metrics=['mae'])
history = model.fit(X_train_tt, y_train, epochs=500,
                    validation_data=(X_valid_tt, y_valid))
```

TabTransformer is an extraordinarily versatile architecture, and there's a lot to explore with. You could use Hyperopt to perform hyperoptimization on the key structural hyperparameters, since there aren't very many. Another idea is to use transformer blocks to jointly process the categorical and continuous features by learning "embeddings" for the continuous features too ("embeddings" in the sense of projecting into a space with the embedding dimension). This could allow contextual embeddings to be contextualized not just with respect to other categorical features but the entire width of the dataset. (Indeed, this is the approach adopted by a later paper we will cover, SAINT.)

TabNet

The TabNet architecture, introduced by Sercan O. Arik and Tomas Pfister from Google Cloud AI in the paper "TabNet: Attentive Interpretable Tabular Learning" in 2019,[10] is another popular deep learning model for tabular data. It is considerably more complex than TabTransformer but shares many fundamental similarities. The fundamental paradigm of TabNet is that decision-making is done in a series of sequential steps; at each step, the model reasons which features to process with input from the progress derived by previous timesteps (Figure 6-60). Each of the steps uses an attention-like mask to select for certain desired features – hence "*attentive* tabular learning." The reasoning from each timestep is then aggregated to produce a final output.

---

[10] Arik, S.Ö., & Pfister, T. (2021). TabNet: Attentive Interpretable Tabular Learning. *ArXiv, abs/1908.07442.* https://arxiv.org/abs/1908.07442.

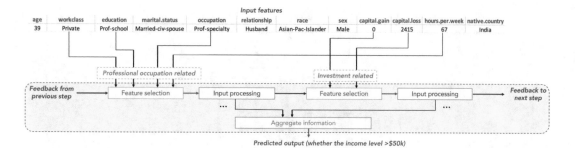

Figure 6-60. *Schematic for feature selection and reasoning from a tabular data row instance across multiple decision steps. From Arik and Pfister*

Let us formalize our understanding of the TabNet model with the authors' notation. There are B samples in each batch, and each row contains D features (i.e., each batch contains B D-dimensional vectors). We thus have that the features $f \in \mathbb{R}^{B \times D}$. Let N_{steps} be the number of decision steps; the outputs of the ith input are passed into the $i + 1$th step. For each timestep i, the model holds a unique learnable mask $M[i] \in \mathbb{R}^{B \times D}$. The mask operates multiplicatively; $M[i] \cdot f$ represents the masked results. Values in $M[i]$ range between 0 and 1, and selection is ensured as *sparse*: most features will have either a comparatively high probability or a near-zero probability such that during multiplication the latter are masked. The masks are calculated as a function of the *prior scale term* $P[i-1]$ and the processed features (i.e., output) of the previous step $a[i-1]$:

$$M[i] = \text{sparsemax}\left(P[i-1] \cdot h_i\left(a[i-1]\right)\right)$$

In this calculation hi is a mapping function parametrized by a feed-forward layer architecture. This function helps to "reinterpret" or "prepare" the previous step's output for interaction with the prior scale term.

The prior scale term $P[i-1]$ represents how much a feature has been previously attended to. Given some relaxation parameter γ, the prior scale term is calculated as follows:

$$P[i] = \prod_{j=1}^{i}\left(\gamma - M[j]\right)$$

Consider $\gamma = 1$. Say that the mask value for some feature at the first timestep value is 1 – that is, the feature was selected for processing. The prior scale term $P[1]$ for that feature then evaluates to $\gamma - M[1] = 1 - 1 = 0$. Note that because the mask $M[i]$ is calculated as a product of the prior scale term, that particular feature will be zeroed out and not selected in the following timestep. Moreover, this prevents usage in any of the later timesteps, since $\prod x = 0$ for some series x if $0 \in x$. Similarly, this forces features that were not previously used to be used. Of course, the softmax mechanism is softer, but the prior scale mechanism still acts as a force to ensure most relevant features "get a shot in the spotlight," so to speak. The first prior scale term $P[0]$ is initialized to all-ones (i.e., $P[0] := 1^{B \times D}$).

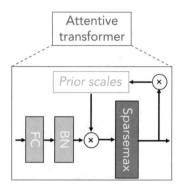

Figure 6-61. *Attentive transformer architecture. From Arik and Pfister*

This component of the TabNet architecture is the *attentive transformer* (Figure 6-61). It generates a mask to determine which features to select in an attention-like manner from the output features of the previous timestep (or, for the first step, from the input features) and prior scale scoring. Arik and Pfister describe such a mask as performing "soft selection of the salient features." "Through sparse selection of the most salient features," the authors write, "the learning capacity of a decision step is not wasted on irrelevant ones, and thus the model becomes more parameter efficient." Such adaptable masks are a philosophical and pragmatic advantage over decision tree–like feature selection, which is "hard" and nonadaptable/static.

At each step i, after feature selection via learned masks, the selected features are passed through a feature transformer F_i (Figure 6-62). We split the output of the feature transformer to collect the decision step output $\boldsymbol{d}[i] \in \mathbb{R}^{B \times N_d}$ for some dimensionality N_d and information for the next step $\boldsymbol{a}[i] \in \mathbb{R}^{B \times N_a}$ for some dimensionality N_a. (Both dimensions are preset.)

$$\left[\boldsymbol{d}[i], \boldsymbol{a}[i]\right] = F_i\left(\boldsymbol{M}[i] \cdot \boldsymbol{f}\right)$$

In terms of specific architecture, the feature transformer contains a shared and independent component. Let a *block* refer to the following stack: fully connected layer, batch normalization, gated linear unit activation (GLU). (The gated linear unit activation was introduced by Dauphin et al. (2015) as GLU(a, b) = $a \otimes \sigma(b)$, which intuitively forces selection of units in a along the gate b. GLU can be used as an activation function on a single vector when a and b represent halves of the vector.) A feature transformer consists of four blocks, with normed residual connections around the latter three. The first two blocks are universally shared across all decision steps, whereas the last two are unique to each decision step. Sharing half of the feature transformer universally across all steps helps to speed up training, improve parameter efficiency, and improve learning robustness.

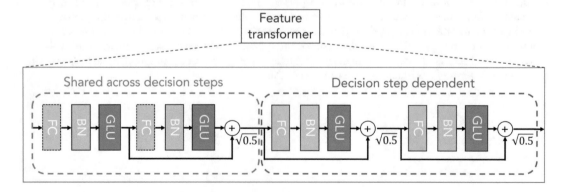

Figure 6-62. *Feature transformer model. From Arik and Pfister*

To aggregate across all decision outputs, TabNet takes the sum of ReLUs across the decision outputs $\boldsymbol{d}$ for all time steps:

$$d_{\text{out}} = \sum_{i=1}^{N_{\text{steps}}} \text{ReLU}\left(\boldsymbol{d}[\mathbf{i}]\right)$$

This joint output is passed through a final linear mapping layer to obtain the true output $\boldsymbol{W}_{\text{final}}\boldsymbol{d}_{\text{out}}$, with final activations (e.g., softmax) applied appropriately on a context-dependent basis. This final aggregation step is reminiscent of *DenseNet*-style residual connections (see Chapter 4), in which each layer is connected to residual connections from all previous anchor points. In a similar way, the final output of the TabNet model is a sum of the outputs of all the steps, such that later steps must modulate/correct for/"keep in mind" the influence of previous steps.

The feature selection masks offer value for interpreting TabNet's decision-making process. If $\boldsymbol{M}_{q,j}[i] = 0$, then the jth feature of the qth sample had no contribution to the decision at step i. However, different steps themselves differ in their contribution to the final output. The authors propose the following formula for the function $\eta_q[i]$, which gives the aggregate decision contribution on the output of the qth sample at the ith decision step:

$$\eta_q[i] = \sum_{c=1}^{N_d} \text{ReLU}\left(\boldsymbol{d}_{\text{q},c}[i]\right)$$

The η function provides us a way to scale the decision mask at each decision step to weight each mask by that step's relevance to the output. Arik and Pfister articulate the following aggregate-level feature importance mask (with an introduced placeholder variable j to iterate over columns):

$$\boldsymbol{M}_{\text{agg}-q,j} = \frac{\sum_{i=1}^{N_{\text{steps}}} \eta_q[i]\boldsymbol{M}_{q,j}[i]}{\sum_{j=1}^{D}\sum_{i=1}^{N_{\text{steps}}} \eta_q[i]\boldsymbol{M}_{q,j}[i]}$$

This formula is quite intuitive: it returns the weighted sum of the masks, normalized such that the sum of the feature importance masks across all features for a sample is 1.

The complete architecture is visualized in Figure 6-63. The input features are passed through an initial feature transformer and then into the first step. At each step, the attentive transformer generates the feature selection mask from the previous step's carry-through output (i.e., $\boldsymbol{a}[i-1]$) and is applied in multiplicative fashion to the original input features. The selected features are passed through the feature transformer; part of the output is passed into the next step (as $\boldsymbol{a}[i]$) and the other as the decision output ($\boldsymbol{d}[i]$).

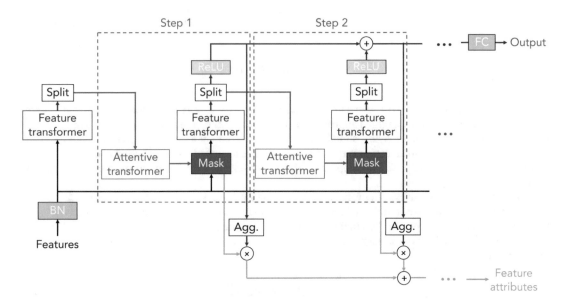

Figure 6-63. *Complete schematic of the TabNet model architecture across several steps. This diagram also shows the collection of the masks for feature attribution at the bottom (leading to "feature attributes"). Don't be confused by these – these are no part of the formal supervised model, which is optimized during training. They represent information flows used for explainability in the prediction phase. From Arik and Pfister*

TabNet, like TabTransformer, also uses an unsupervised/self-supervised pretraining scheme (Figure 6-65). The authors create a *TabNet decoder* (Figure 6-64), which is comprised of several feature transformers arranged in a similar multistep fashion with fully connected layers at each decision step; the summed decision outputs are used to obtain the reconstructed features. The pretraining task is to predict masked feature columns. A binary mask for the batch is generated: $S \in \{0, 1\}^{B \times D}$. The encoder accepts the input $(1 - S) \cdot f$, and the decoder predicts $S \cdot f$. After this pretraining task, the decoder is detached, and the decision-making output is used in replacement for supervised fine-tuning (Figure 6-64).

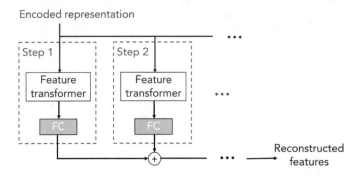

Figure 6-64. *The decoder architecture, which accepts the encoded representation and outputs the reconstructed features. From Arik and Pfister*

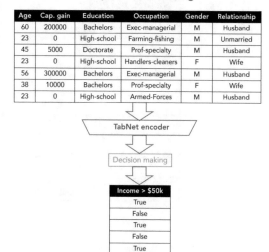

Figure 6-65. *The TabNet two-stage training scheme. From Arik and Pfister*

The authors evaluate TabNet on a wide variety of synthetic and "natural" datasets; they find that TabNet performs competitively and sometimes better than the best tree-based and DNN-based tabular models (Tables 6-6 through 6-11). On synthetic datasets, TabNet performs near the top, with a significantly reduced parameter size (26–31k compared with 101k for INVASE and 43k for other deep learning methods).

Table 6-6. *Performance on a synthetic dataset pack (Chen 2018). From Arik and Pfister*

Model	Test AUC					
	Syn1	Syn2	Syn3	Syn4	Syn5	Syn6
No selection	.578 ± .004	.789 ± .003	.854 ± .004	.558 ± .021	.662 ± .013	.692 ± .015
Tree	.574 ± .101	.872 ± .003	.899 ± .001	.684 ± .017	.741 ± .004	.771 ± .031
Lasso-regularized	.498 ± .006	.555 ± .061	.886 ± .003	.512 ± .031	.691 ± .024	.727 ± .025
L2X	.498 ± .005	.823 ± .029	.862 ± .009	.678 ± .024	.709 ± .008	.827 ± .017
INVASE	**.690 ± .006**	.877 ± .003	**.902 ± .003**	**.787 ± .004**	.784 ± .005	.877 ± .003
Global	.686 ± .005	.873 ± .003	.900 ± .003	.774 ± .006	.784 ± .005	.858 ± .004
TabNet	.682 ± .005	**.892 ± .004**	.897 ± .003	.776 ± .017	**.789 ± .009**	**.878 ± .004**

Table 6-7. *Performance on the Forest Cover dataset (Dua and Graff 2017). From Arik and Pfister*

Model	Test accuracy (%)
XGBoost	89.34
LightGBM	89.28
CatBoost	85.14
AutoML Tables	94.95
TabNet	**96.99**

Table 6-8. *Performance on the Poker Hand dataset (Dua and Graff 2017). From Arik and Pfister*

Model	Test accuracy (%)
DT	50.0
MLP	50.0
Deep neural DT	65.1
XGBoost	71.1
LightGBM	70.0
CatBoost	66.6
TabNet	**99.2**
Rule-based	100.0

Table 6-9. *Performance on the Sarcos dataset (Vijayakumar and Schaal 2000). From Arik and Pfister*

Model	Test MSE	Model size
Random forest	2.39	16.7K
Stochastic DT	2.11	28K
MLP	2.13	0.14M
Adaptive neural tree	1.23	0.60M
Gradient boosted tree	1.44	0.99M
TabNet-S	**1.25**	**6.3K**
TabNet-M	**0.28**	**0.59M**
TabNet-L	**0.14**	**1.75M**

Table 6-10. *Performance on the Higgs Boson dataset (Dua and Graff 2017). From Arik and Pfister*

Model	Test acc. (%)	Model size
Sparse evolutionary MLP	**78.47**	**81K**
Gradient boosted tree-S	74.22	0.12M
Gradient boosted tree-M	75.97	0.69M
MLP	78.44	2.04M
Gradient boosted tree-L	76.98	6.96M
TabNet-S	78.25	81K
TabNet-M	**78.84**	**0.66M**

Table 6-11. *Performance on the Rossmann Store Sales dataset (Kaggle 2019). From Arik and Pfister*

Model	Test MSE
MLP	512.62
XGBoost	490.83
LightGBM	504.76
CatBoost	489.75
TabNet	**485.12**

Moreover, TabNet offers somewhat similar yet different interpretations for feature interpretability (Table 6-12).

Table 6-12. *Comparison of feature importance rankings by different methods. From Arik and Pfister*

Feature	SHAP	Skater	XGBoost	TabNet
Age	1	1	1	1
Capital gain	3	3	4	6
Capital loss	9	9	6	4
Education	5	2	3	2
Gender	8	10	12	8
Hours per week	7	7	2	7
Marital status	2	8	10	9
Native country	11	11	9	12
Occupation	6	5	5	3
Race	12	12	11	11
Relationship	4	4	8	5
Work class	10	8	7	10

It should be noted, moreover, that TabNet shares many architectural and conceptual similarities with tree-based models. The set of sequence operations provides a decision tree–like decision scaffold and is capable of representing decision tree–style feature space separations (Figure 6-66). The authors note that the attention-like mechanism allows for a softer, adaptive version of tree node separation criteria. Moreover, the sequential stacking nature of the TabNet model is conceptually similar to stacking and boosting in tree models, in which units learn with/from/around the previous unit's output.

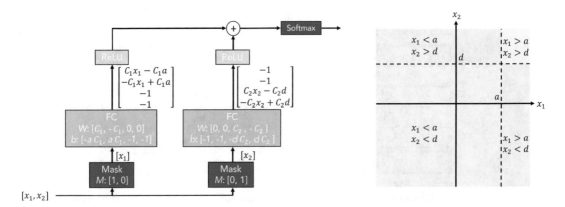

Figure 6-66. *Schematic of how the TabNet architecture can be used to represent decision tree–like logical decision-making. From Arik and Pfister*

Another advantage of TabNet (and all deep learning for tabular data models) is its ability to be trained on unlabeled data. Due to the confluence of several mechanisms like sparsity, masking, and timestep-independent shared weights, TabNet is one of the lightest and most powerful recent tabular deep learning models.

Because TabNet comes from Google Cloud AI, the codebase (model code available at `https://github.com/google-research/google-research/blob/master/tabnet/tabnet_model.py`) is written in TensorFlow and should be somewhat readable. We will use a modified version in proper Keras by Somshubra Majumdar, which comes with additional conveniences and utilities. You can see the implementation here: `https://github.com/titu1994/tf-TabNet/blob/master/tabnet/tabnet.py`. The code is available as a library in pip; its PyPI page is `https://pypi.org/project/tabnet/`, and it can be installed with `pip install tabnet`.

The `tabnet` library comes with two output-fitted models: `TabNetClassifier` and `TabNetRegressor` for classification and regression problems, respectively. Both share the same `TabNet` base model architecture but use different output activations (as one can confirm by looking at the source code). At the minimum, you need to specify the number of input features and the number of output classes. Moreover, you can specify the feature dimension `feature_dim` (this is N_a), the output dimension `output_dim` (this is N_d), the number of decision steps `num_decision_steps` (this is N_{steps}), the relaxation factor `relaxation_factor` (this is γ), and a sparsity coefficient `sparsity_coefficient` to control the severity of adherence toward sparsity, among additional parameters (Listing 6-39.)

Listing 6-39. Instantiating a TabNetClassifier model

```
from tabnet import TabNetClassifier
model = TabNetClassifier(feature_columns=None,
                         num_features=X.shape[-1],
                         num_classes=7,
                         feature_dim=32,
                         output_dim=16,
                         num_decision_steps=8,
                         relaxation_factor=0.7,
                         sparsity_coefficient=1e-6)
```

We can compile and fit as a standard Keras model (Listing 6-40). Use large batch sizes for TabNet – even as high as 10–15% of the total dataset size, if memory permits. Since this implementation does not support self-supervised learning easily, it may take an extended period of training to acclimate to the labels. While self-supervised learning helps, training directly on labels usually yields competitive results too. Implementing self-supervised pretraining is not difficult and can be built from the existing source code building blocks.

Listing 6-40. Compiling and fitting the TabNet model

```
model.compile(optimizer='adam', loss='sparse_categorical_crossentropy',
            metrics=['accuracy'])
model.fit(X_train, y_train, epochs=100,
          validation_data=(X_valid, y_valid),
          batch_size=10_000)
```

For technical TensorFlow reasons, to easily obtain the values for the feature selection masks, we need to pass the desired dataset through the model. We do not need to save the outputs; the point of this command is to force the model to run in eager execution mode. From there, we can collect the masks (a list of tensors in raw original form) and access the data in NumPy arrays (Listing 6-41).

Listing 6-41. Obtaining TabNet feature selection masks on the validation dataset

```
_ = model(X_valid)
fs_masks_orig = model.tabnet.feature_selection_masks
fs_masks = np.stack([mask.numpy()[0,:,:,0] for mask in fs_masks_orig])
```

There are $N_{steps} - 1$ feature selection maps (plotted by Listing 6-42, Figure 6-67). Note that this particular model reasons from individual "key" features in the first few steps and then progressively incorporates input from other features in later steps to further inform the decision-making procedure.

Listing 6-42. Plotting the TabNet feature selection masks

```
for i in range(7):
    plt.figure(figsize=(15, 8), dpi=400)
    sns.heatmap(fs_masks[i,:100,:],
                xticklabels=columns,
                yticklabels=[])
    plt.xlabel('Columns')
    plt.ylabel('Samples')
    plt.title(f'Sample of Mask Values for Layer {i+1}')
    plt.show()
```

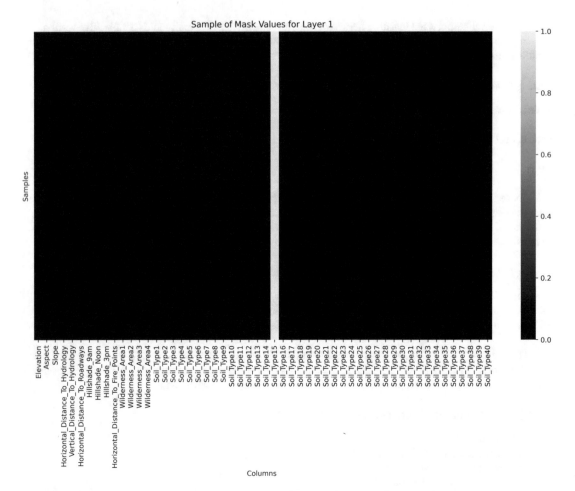

Figure 6-67. *The features attended to at each iteration of a TabNet block (arranged along the x-axis) for several samples (arranged along the y-axis)*

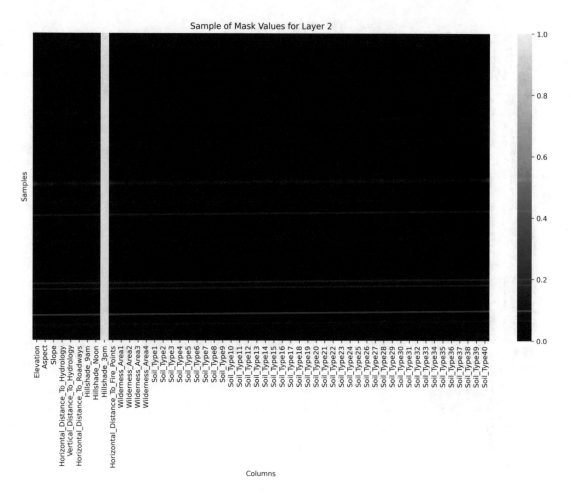

Figure 6-67. *(continued)*

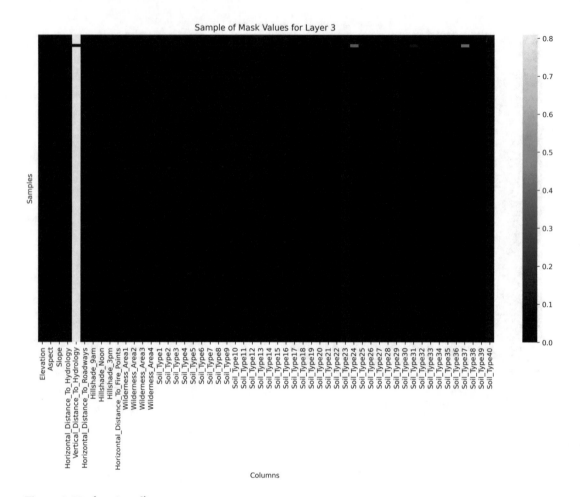

Figure 6-67. *(continued)*

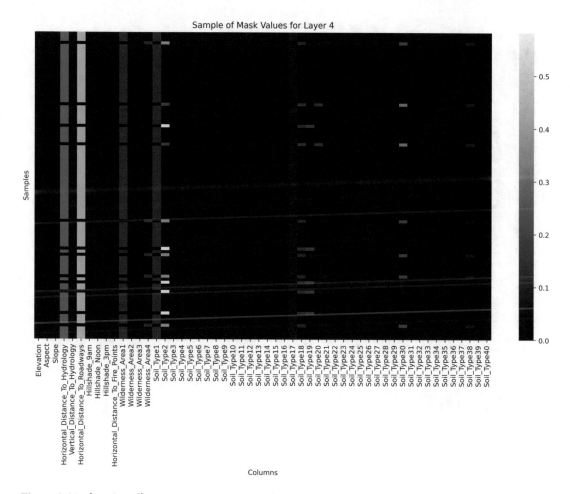

Figure 6-67. *(continued)*

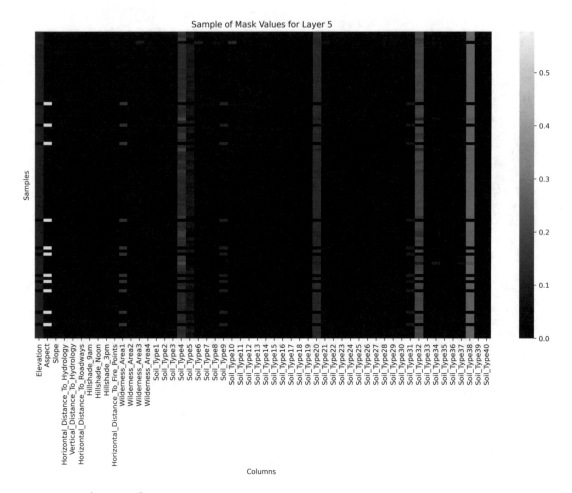

Figure 6-67. *(continued)*

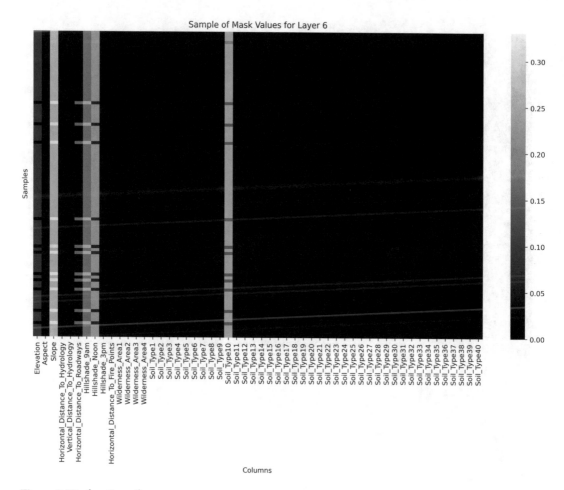

Figure 6-67. *(continued)*

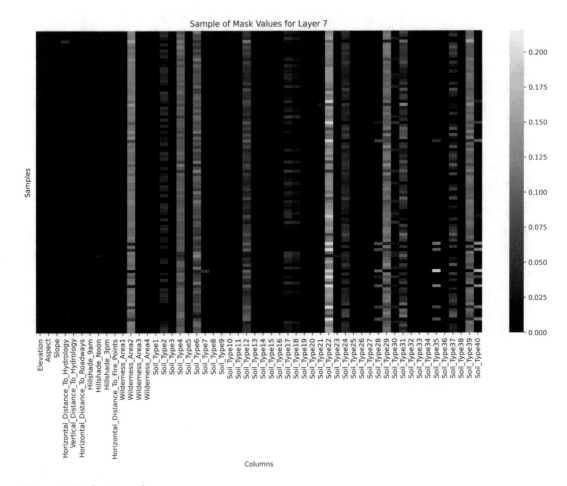

Figure 6-67. *(continued)*

We can similarly access the aggregate feature mask, which informs us of the individual feature contributions to the output weighed across decision steps for all samples in the submitted dataset (Listing 6-43, Figure 6-68).

Listing 6-43. Obtaining aggregate feature masks across all TabNet blocks

```
agg_mask = model.tabnet.aggregate_feature_selection_mask
plt.figure(figsize=(15, 8), dpi=400)
sns.heatmap(agg_mask.numpy()[0,:100,:,0],
            xticklabels=columns,
            yticklabels=[])
plt.xlabel('Columns')
plt.ylabel('Samples')
plt.title(f'Aggregate Feature Mask')
plt.show()
```

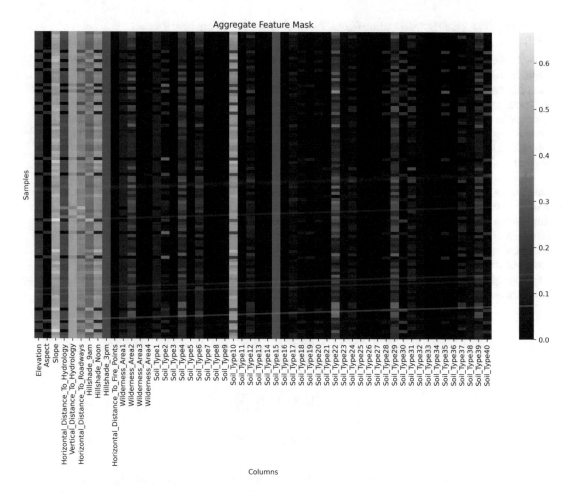

Figure 6-68. *Plotting the aggregate feature mask across several examples*

In sum, TabNet employs an attention-based feature selection mechanism, which forces sequential reasoning and processing of features in similar fashion to tree ensembles, with the advantage of interpretability.

SAINT

The Self-Attention and Intersample Attention Transformer (SAINT) is a recent model, introduced by Gowthami Somepalli et al. from the University of Maryland and Capital One Machine Learning in the 2021 paper "SAINT: Improved Neural Networks for Tabular Data via Row Attention and Contrastive Pre-Training."[11] SAINT's novel contribution is the introduction of *intersample attention*, which allows rows to relate to one another in attentive fashion, as opposed to standard column-wise attention.

---

[11] Somepalli, G., Goldblum, M., Schwarzschild, A., Bruss, C.B., & Goldstein, T. (2021). SAINT: Improved Neural Networks for Tabular Data via Row Attention and Contrastive Pre-Training. *ArXiv, abs/2106.01342.* https://arxiv.org/abs/2106.01342.

Let us conceptualize SAINT using the authors' notation (slightly adapted for clarity). Let $D := \{x_i, y_i\}_{i=1}^m$: the dataset D contains m pairs of an $n - 1$-dimensional feature vector (x_i) and the associated label yi. The true dataset contains n features; a classification $[\text{CLS}]$ token is added as an additional feature: $x_i = \{[\text{CLS}], f_i^1, f_i^2, \ldots, f_i^{n-1}\}$, where f_i^j represents the value of the jth feature of the ith sample. The $[\text{CLS}]$ token serves as a "blank feature." SAINT embeds each feature into d-dimensional space independently, like TabTransformer. Unlike TabTransformer, SAINT embeds all features – categorical and continuous – whereas TabTransformer selectively embeds categorical features. The embedding layer is denoted by E and applies different embedding functions to different categorical features. The $[\text{CLS}]$ token is embedded as if it were a feature. Its relevance will become clearer when the complete architecture is laid out.

Like TabTransformer and TabNet, SAINT's primary architectural body is comprised of L attention-based steps. Each step consists of a self-attention transformer block, followed by an intersample attention transformer block. The self-attention block is identical to the one used in the Vaswani et al. original transformer paper: a multi-head self-attention (MSA) layer followed by feed-forward (FF) layers with a Gaussian Error Linear Unit (GELU) activation. Moreover, let MISA be the multi-head intersample self-attention layer (this mechanism will be explained more in-depth later), LN be the layer normalization layer, and b be the batch size. Both MSA and MISA have residual connections following. A step S_k^q at step q for a sample index k can be accordingly formulated as follows with intermediates $z_k^{q,1}$, $z_k^{q,2}$, and $z_k^{q,3}$ for notational convenience:

$$z_k^{q,1} = \text{LN}\left(\text{MSA}\left(S_{q-1}\right)\right) + S_{q-1}$$

$$z_k^{q,2} = \text{LN}\left(\text{FF}_1\left(z_i^{q,1}\right)\right) + z_k^{q,1}$$

$$z_k^{q,3} = \text{LN}\left(\text{MISA}\left(\{z_i^{q,2}\}_{i=1}^b\right)\right) + z_k^{q,2}$$

$$S_k^q = \text{LN}\left(\text{FF}_2\left(z_k^{q,3}\right)\right) + z_k^{q,3}$$

Moreover, we have $S_k^0 = E(x_k)$ such that the inputs to S_k^1 are the embeddings generated for that sample. Note that to compute the multi-head intersample self-attention layer, we need to compare derived self-attention features across all samples within the batch.

The complete architecture is visualized in Figure 6-69.

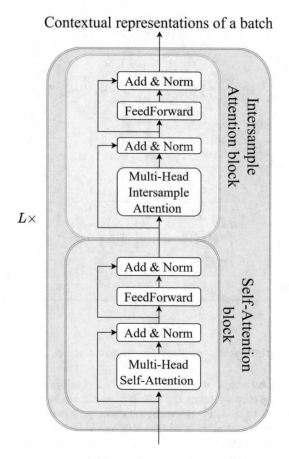

Figure 6-69. *The complete SAINT transformer block. From Somepalli et al.*

To make the final prediction on a supervised problem, the embedding corresponding to the [CLS] token from the last step (S_0^L , assuming that the [CLS] token corresponds to the first feature index) is extracted and passed through a simple multilayer perceptron into the output. After training, the [CLS] feature embedding will have been informed via several steps of cross-feature and cross-sample attending. This is a clever trick to force information into a low-dimensional single embedding (as opposed to concatenating embeddings for every feature TabTransformer-style, which has a significantly larger dimensionality).

To understand the multi-head intersample self-attention mechanism, we will begin by reformalizing the standard multi-head self-attention mechanism (Figure 6-70). Let a be the attention matrix and $a_{i,j}$ indicate the attention score between the query derived from the ith feature and the key derived from the jth feature. a is an $n \times n$ matrix, with self-attention scores computed between embeddings corresponding to elements in $x_i = \{[\text{CLS}] f_i^1 f_i^2 ,...f_i^{n-1})\}$. The ith value of the output is $\sum_{j=0}^{n-1} a_{i,j} * v_i$, where v_i is the value vector derived from the ith feature. This is repeated with multiple heads (i.e., multiple keys, queries, and values are derived from each of the features).

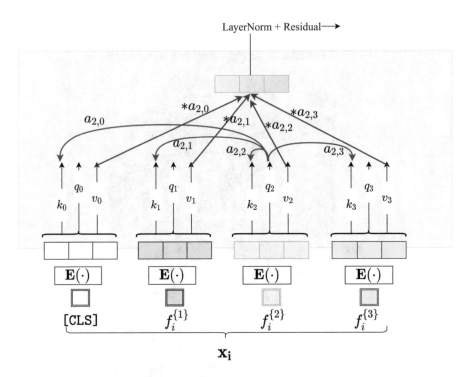

Figure 6-70. *Visualization of inter-feature attention (standard attention). From Somepalli et al.*

Multi-head intersample self-attention (Figure 6-71) is a "mega self-attention" across a batch. Rather than operating on embeddings corresponding to the collection of features $x_i = \left\{ [CLS], f_i^1, f_i^2, \ldots, f_i^{n-1} \right\}$, it operates on a batch of samples $\{x_1, x_2, \ldots, x_b\}$. The embeddings for each of the features within each sample are concatenated with each other, independent of sample. Key, query, and value vectors are derived from these concatenated embeddings such that there are as many (K, Q, V) sets as there are attention heads per sample. Then, we apply attention in standard form: let a be a $b \times b$ attention matrix and $a_{i,j}$ indicate the attention score between the query derived from the concatenated embedding corresponding to x_i and the key derived from the concatenated embedding corresponding to x_j. To obtain the final output, the ith value is given by

$$\sum_1^b a_{i,j} * v_i,$$ where v_i is the value vector derived from the concatenated embedding corresponding to x_i.

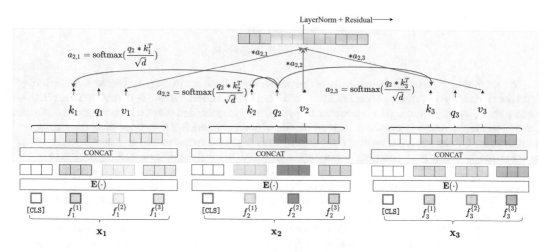

Figure 6-71. Visualization of intersample attention. From Somepalli et al.

Like most deep learning approaches to tabular data, SAINT is pretrained with a self-supervised training task. Rather than using BERT/masked language model–style pretraining tasks, like TabTransformer's replaced token detection pretraining task, SAINT uses contrastive learning. Contrastive learning is a training paradigm in which the goal is not to strictly learn an associated target given an input, but rather to identify shared or different attributes (i.e., to "compare and *contrast*") between a set of inputs.

SAINT's pretraining task is as follows. For each sample x_i, generate a corrupted version x_i'. This is done using the CutMix augmentation, which uses the following calculation given a randomly selected sample x_a and a binary mask vector sampled from a Bernoulli distribution $\boldsymbol{m}$:

$$x_i' = x_i \otimes m + x_a \otimes (1 - \boldsymbol{m})$$

We will embed x_i to obtain $p_i = E(x_i)$. Then, we will generate a corrupted embedding, given the mix-up parameter α and another randomly selected sample xb:

$$p_i' = \alpha * E\left(x_i'\right) + (1 - \alpha) * E\left(x_b'\right)$$

Now, we have four sets of data: x_i, the original untouched sample; x_i', the corrupted sample; p_i, the original untouched embedding; and p_i', the corrupted embedding. We can pass the two embeddings through the SAINT model, which we will denote with a bold S (the compositing of all individual steps), to obtain $S(p_i)$ and $S\left(p_i'\right)$. To reduce the dimensionality of these representations, we pass these through an additional MLP g_1 and g_2 to obtain $g_1(S(p_i))$ and $g_2\left(S(p_i')\right)$. We can use this to calculate the *contrastive loss*, for some temperature parameter τ:

$$\text{Contrastive Loss} = -\sum_{i=1}^{m} \log \left(\frac{\exp\left(\dfrac{g_1\left(S(p_i)\right) \cdot g_2\left(S(p_i')\right)}{\tau}\right)}{\sum_{k=1}^{m} \exp\left(\dfrac{g_1\left(S(p_i)\right) \cdot g_2\left(S(p_k')\right)}{\tau}\right)} \right)$$

Let's break down this formula. The temperature parameter, logarithm, and exponentials can be more or less ignored since they do not affect the principal dynamics of the expression. We can rough-handedly "simplify" it as follows:

$$-\sum_{i=1}^{m} \frac{g_1\left(S(p_i)\right) \cdot g_2\left(S(p_i')\right)}{\sum_{k=1}^{m}\left(g_1\left(S(p_i)\right) \cdot g_2\left(S(p_k')\right)\right)}$$

It becomes more readable in this form. In the numerator, we are comparing the representation derived from the clean input with the representation derived from the corrupted version of that same input. In the denominator, we are summing the interaction between the representation derived from the clean input and the representation derived from the corrupted version of every element in the dataset. (Recall the authors denote an un-bolded m as the length of the dataset.) The dot product of some vector $\vec{a}$ and another vector $\vec{b}$ is maximized when $\vec{a} = \vec{b}$, holding one of the vectors fixed. In ideal conditions, $g_1(S(p_i))$ and $g_2\left(S(p_i')\right)$ will be very close since they both fundamentally derive from the same sample, even if one is corrupted. In this case, the numerator will be large, and the overall term will evaluate to a high value relative to if $g_1(S(p_i))$ and $g_2\left(S(p_i')\right)$ were farther apart. We sum all such values across all items in the dataset. Because we want to minimize the loss, we negate the summation.

We can think of the entire SAINT architecture S as a giant embedding machine; the contrastive loss incentivizes intelligently mapped embeddings, which place close points physically close together in the embedding space.

The authors also introduce a denoising loss, in which the objective is to decode the original input x_i from the representation derived from the corrupted embedding, $g_2\left(S(p_i')\right)$. A unique multilayer perceptron model is constructed for each individual feature to perform the "denoising." The loss $\mathcal{L}_j$, which is binary cross-entropy for categorical features and Mean Squared Error for continuous features, is taken between the original input and the derived representation. This is summed across all features (n total features) and across all samples (m total samples) and then scaled by λ_{pt} such that it is on an appropriate magnitude relative to the contrastive loss:

$$\text{Denoising Loss} = \lambda_{pt} \sum_{i=1}^{m} \sum_{j=1}^{n} \mathcal{L}_j\left(\text{MLP}_j\left(g_2\left(S(p_i')\right)\right), x_i\right)$$

The overall training loss is the sum of the contrastive and denoising losses:

$$\mathcal{L}_{\text{pretraining}} = \text{Contrastive Loss} + \text{Denoising Loss}$$

$$\mathcal{L}_{\text{pretraining}} = -\sum_{i=1}^{m} \log\left(\frac{\exp\left(\frac{g_1\left(S(p_i)\right) \cdot g_2\left(S(p_i')\right)}{\tau}\right)}{\sum_{k=1}^{m} \exp\left(\frac{g_1\left(S(p_i)\right) \cdot g_2\left(S(p_k')\right)}{\tau}\right)}\right) + \lambda_{pt} \sum_{i=1}^{m} \sum_{j=1}^{n} \mathcal{L}_j\left(\text{MLP}_j\left(g_2\left(S(p_i')\right)\right), x_i\right)$$

Then, as previously mentioned, the model is fine-tuned in a supervised learning regime; the embedding corresponding to the [CLS] token at the last step S_0^L is passed into an MLP with a single hidden layer to obtain the output:

$$\mathcal{L}_{\text{fine-tuning}} = \sum_{i=1}^{m} \text{BCE}\left(y_i, \text{MLP}\left(S\left(E(x_i)\right)\right)\right)$$

The complete SAINT training pipeline is visualized in Figure 6-72.

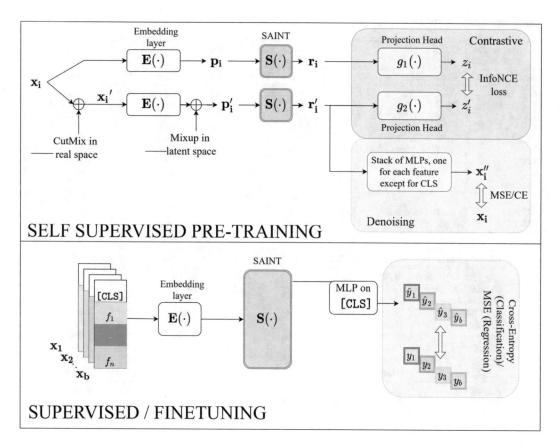

Figure 6-72. *The complete SAINT training pipeline and architectures employed. For notation: a bold $S(...)$ indicates the complete SAINT pipeline (i.e., the compositing of the all steps together), and r_i is the output of $S(...)$. From Somepalli et al.*

Somepalli et al. evaluate three versions of SAINT (standard SAINT; SAINT-s, with only self-attention; SAINT-i, with only intersample attention) across 16 datasets and demonstrate improved performance over other tree-based and deep learning tabular data models (Table 6-13). Note that SAINT-s is more or less identical to the original Vaswani et al. transformer block but applied to tabular data. We see that intersample attention helps provide improvement over just self-attention in a large number of datasets, however.

Table 6-13. *Mean AUROC*

| Dataset size | 45,211 | 7,043 | 452 | 200 | 495,141 | 12,330 | 32,561 | 58,310 | 60,000 | |
| Feature size | 16 | 20 | 226 | 783 | 49 | 17 | 14 | 147 | 784 | |
Model \ Dataset	Bank	Blastchar	Arrhythmia	Arcene	Forest	Shoppers	Income	Volkert†	MNIST†	Mean
Logistic Reg.	90.73	82.34	86.22	91.59	84.79	87.03	92.12	53.87	89.89*	89.25
Random Forest	89.12	80.63	86.96	79.17	98.80	89.87	88.04	66.25	93.75	89.52
XGBoost [4]	92.96	81.78	81.98	81.41	95.53	92.51	92.31	68.95	94.13*	91.06
LightGBM [22]	93.39	83.17	88.73	81.05	93.29	**93.20**	**92.57**	67.91	95.2	90.13
CatBoost [10]	90.47	84.77	87.91	82.48	85.36	93.12	90.80	66.37	96.6	90.73
MLP	91.47	59.63	58.82	90.26	96.81	84.71	92.08	63.02	93.87*	84.59
VIME [49]	76.64	50.08	65.3	61.03	75.06	74.37	88.98	64.28	95.77*	76.07
TabNet [1]	91.76	79.61	52.12	54.10	96.37	91.38	90.72	56.83	96.79	83.88
TabTransf. [18]	91.34	81.67	70.03	86.8	84.96	92.70*	90.60*	57.98	88.74	90.86
SAINT-s	**93.61**	**84.91**	93.46	86.88	99.67	92.92	91.79	62.91	90.52	92.59
SAINT-i	92.83	84.46	**95.8**	**92.75**	99.45	92.29	91.55	**71.27**	**98.06**	93.09
SAINT	93.3	84.67	94.18	91.04	**99.7**	93.06	91.67	70.12	97.67	**93.13**

Model \Dataset	Credit	HTRU2	QSAR Bio	Shrutime	Spambase	Philippine	KDD99
Logistic Regression	96.85	98.23	84.06	83.37	92.77	79.48	99.98
Random Forest	92.66	96.41	91.49	80.87	98.02	81.29	**100.00**
XGBoost	**98.20**	97.81	92.70	83.59	98.91	**85.15**	**100.00**
LightGBM	76.07	98.10	92.97	85.36	**99.01**	84.97	**100.00**
CatBoost	96.83	97.85	93.05	85.44	98.47	83.63	**100.00**
MLP	97.76	98.35	79.66	73.70	66.74	79.70	99.99
VIME	82.63	97.02	81.04	70.24	69.24	73.51	99.89
TabNet	95.24	97.58	67.55	75.24	97.93	74.21	**100.00**
Tab Transformer	97.31	96.56	91.80	85.60	98.50	83.40	**100.00**
SAINT-s	98.08	98.16	92.89	86.40	98.21	79.30	**100.00**
SAINT-i	98.12	**98.36**	**93.48**	85.68	98.40	80.08	**100.00**
SAINT	97.92	98.08	93.21	**86.47**	98.54	81.96	**100.00**

Moreover, the authors find that SAINT is highly robust to significant data corruption and that changing the batch size has little effect on the performance, assuming a minimum batch size of 32. This suggests that just a "critical mass" of samples in a batch is required for effective comparison and cross-sample comparison.

The attention maps of the layers can be interpreted to understand how the model makes decisions (Figure 6-73).

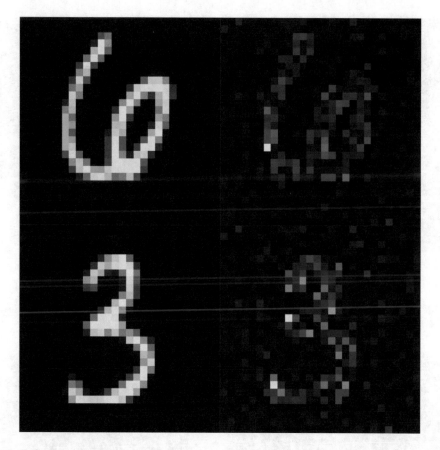

Figure 6-73. *Left column: sample inputs, reshaped into two dimensions. Right: self-attention scores, selected and reshaped into two dimensions. From Somepalli et al.*

However, SAINT is unique because we can also understand how the network makes decisions for a particular sample based on other examples. On MNIST, the network tends to heavily consult a set of exemplars (Figure 6-74), perhaps because they are difficult-to-classify examples with high information value. However, intersample attention grids are more varying on the more complex Volkert dataset (Figure 6-75). The authors speculate that intersample attention density rises with dataset complexity.

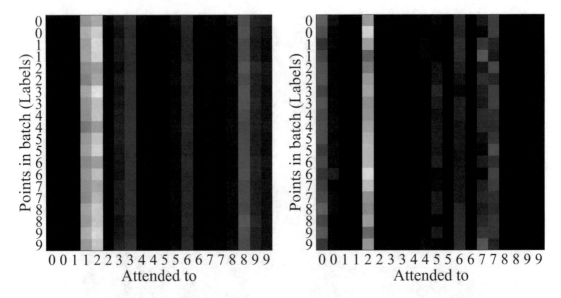

Figure 6-74. *Left: intersample attention for SAINT. Right: intersample attention for SAINT-i. On MNIST dataset. From Somepalli et al.*

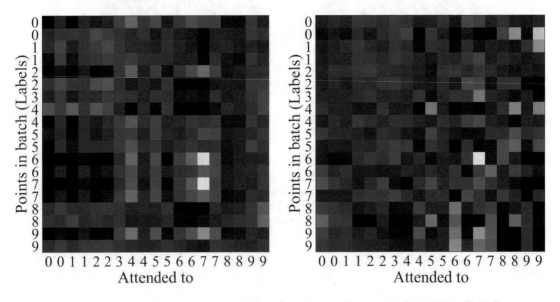

Figure 6-75. *Left: intersample attention for SAINT. Right: intersample attention for SAINT-i. On Volkert dataset. From Somepalli et al.*

The paper's authors have implemented SAINT in PyTorch, which is available on the official repository here: https://github.com/somepago/saint. As of our knowledge, there exists no readily available Keras or TensorFlow implementation. However, Somepalli et al.'s implementation is user-friendly and directly accessible via the command line without knowledge of PyTorch.

Begin by cloning the repository and creating and activating the provided environment (Listing 6-44).

Listing 6-44. Cloning the repository and creating and activating the environment

```
git clone https://github.com/somepago/saint.git
conda env create -f saint_enviornment.yml
conda activate saint_env
```

In the current implementation, the model directly pulls data from OpenML – a data platform with the advantage of concretely defined and standardized (in the organizational, not the statistical, sense) features, labels, and other data attributes. Navigate to www.openml.org/ to browse existing datasets or to upload and create your own. Importantly, each dataset page has a numerical integer ID, which we will provide as a flag to identify the dataset we desire to train on. The Forest Cover dataset, for instance, has the OpenML ID 180 (Figure 6-76).

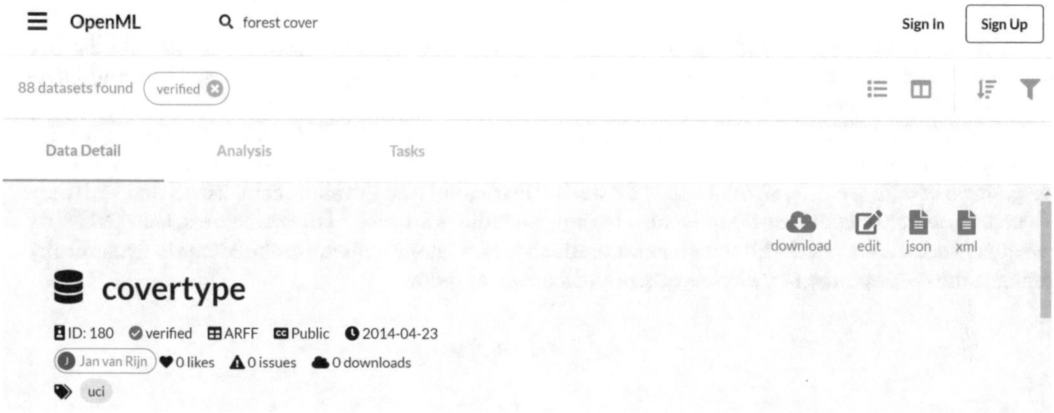

Figure 6-76. The Forest Cover dataset on OpenML

Once you've obtained the OpenML ID, you can initiate the training process: `python train.py --dset_id 180 --task multiclass`. The dataset ID and task are the only two required flags; you can also specify parameters like the number of attention heads, whether to use pretraining, the embedding size, and so on. See the repository's README for more information. Note that the repository maintainers, as of the time of this writing, have verified the code only for Linux. You may run into problems on other operating systems.

ARM-Net

ARM-Net, introduced by Shaofeng Cai et al. in 2021 in the paper "ARM-Net: Adaptive Relation Modeling Network for Structured Data,"[12] employs a unique architecture that could be described as more "involved" than the previously discussed attention-based architectures. Rather than using the standard attention mechanism as a transformation method for abstractly attending to different relevant features (and samples, in the case of SAINT), ARM-Net uses attention to help explicitly compute information-rich cross-features in a tabular dataset, which can be used for a supervised task. ARM-Net is structured in three modules: a preprocessing module, an adaptive relation modeling module, and a prediction module.

Let us begin by formalizing the preprocessing module with the authors' notation. Let m be the number of features in the dataset and let the input vector $x = [x_1, x_2, ..., x_m]$. Each feature is mapped to an embedding: $E = [e_1, e_2, ..., e_m]$. Categorical features are mapped via an embedding lookup, and continuous features are transformed via linear transformation. Let n_e be the embedding dimension.

The key transformation mechanism of the adaptive relation modeling module is the exponential neuron. For some interaction weight matrix w, we can compute the ith element of the exponential neuron output y given the embeddings set e as follows:

$$y_i = \exp\left(\sum_{j=1}^{m} w_{i,j} e_j\right) = \exp(e_1)^{w_{i,1}} \otimes \exp(e_2)^{w_{i,2}} \otimes ... \otimes \exp(e_2)^{w_{i,m}}$$

The interaction weight matrix determines the influence each embedding has on the output – like in a standard artificial neural network neuron – but operates in exponential space, as opposed to the additive-multiplicative dynamics of standard neurons.

To obtain the interaction weight matrix w, the adaptive relation modeling module uses multi-head gated attention. When determining the value of the ith neuron y_i, we need to obtain relevant power term weightings: $w_i = [w_{i,1}, w_{i,2}, ..., w_{i,m}]$. Let $v_i \in \mathbb{R}^m$ be the (learnable) weight value vector associated with the ith neuron, which encodes the attentiveness to the embeddings for each of the m features. Let $q_i \in \mathbb{R}^{n_e}$ be the query vector associated with the ith neuron, which is used along with the embeddings to dynamically generate the *bilinear attention alignment score*, calculated as follows:

$$\phi_{att}(q_i, e_j) = q_i^T W_{att} e_j$$

$$\tilde{z}_{i,j} = \phi_{att}(q_i, e_j)$$

$$z_i = \alpha \text{entmax}(\tilde{z}_i)$$

Here, $W_{att} \in \mathbb{R}^{n_e \times n_e}$ is a weight matrix for bilinear attention. The shared bilinear attention function $\phi_{att}(q_i, e_j)$ is calculated by performing a product between the query vector, the bilinear attention weight matrix, and the embedding. The transposed query has shape $(1, n_e)$; the product between this and the (n_e, n_e)-shaped weight matrix yields a $(1, n_e)$-shaped matrix; the product between this matrix and the $(n_e, 1)$-shaped embedding matrix yields a $(1, 1)$-shaped result (i.e., a scalar). Thus, $\tilde{z}_{i,j}$ stores the attention

[12] Shaofeng Cai, Kaiping Zheng, Gang Chen, H. V. Jagadish, Beng Chin Ooi, and Meihui Zhang. 2021. ARM-Net: Adaptive Relation Modeling Network for Structured Data. In Proceedings of the 2021 International Conference on Management of Data (SIGMOD '21), June 20–25, 2021, Virtual Event, China. ACM, New York, NY, USA, 14 pages. https://doi.org/10.1145/3448016.3457321

score between the query vector of the ith neuron and the embedding corresponding to the jth feature. Thus, $\tilde{z}_i \in \mathbb{R}^m$; that is, $\tilde{z}_i$ has length m – representing the attention scores between the ith exponential neuron and each of the m features. We compute the true embedding scores by applying the αentmax (sparse softmax) function to obtain z_i. The sparse softmax function – as used previously in other tabular attention architectures in various modified forms – encourages sparsity by pushing smaller values to zero while preserving the signature softmax one-sum property.[13]

We can therefore calculate the interaction weights as follows:

$$w_i = z_i \otimes v_i$$

Because $z_i \in \mathbb{R}^m$ and $v_i \in \mathbb{R}^m$, we have that $w_i \in \mathbb{R}^m$. This is the gate-like nature of the mechanism: z_i serves as the "gate" determining which elements in v_i "can pass" (i.e., as relevant for downstream tasks). This weight, informed by relevant features, is then used to "program" the behavior of the exponential neurons.

We can reexpress the calculation for the ith exponential neuron more completely as follows, given the learned "atomics" q_i, W_{att}, and v_i and the embeddings e:

$$y_i = \exp\left(\sum_{j=1}^m \left(\alpha\text{entmax}\left(q_i^{\mathrm{T}} W_{att} e_j \right) \otimes v_i \right)_j e_j \right)$$

The authors adopt a multi-head version of this system. Let K be the number of heads and o be the number of exponential neurons. $q_i^{(k)}$ denotes the kth query key for the ith neuron, and likewise $v_i^{(k)}$ denotes the kth value key for the ith neuron. Note that W_{att} is shared across heads. From each of the heads, we can derive a final output of the adaptive relation modeling module Y, which is the concatenation of the outputs from each of the heads. (In this context, $a \oplus b$ refers to vector concatenation.)

$$y_i^{(k)} = \exp\left(\sum_{j=1}^m \left(\alpha\text{entmax}\left(q_i^{(k)\mathrm{T}} W_{att} e_j \right) \otimes v_i^{(k)} \right)_j e_j \right)$$

$$Y^{(k)} = \left[y_1^{(k)}, y_2^{(k)}, \ldots, y_o^{(k)} \right]$$

$$Y = Y^{(1)} \oplus Y^{(2)} \oplus \ldots \oplus Y^{(K)}\}$$

The complete adaptive relation modeling module is displayed in Figure 6-77.

---

[13] For the mathematically inclined reader, the authors formally define the sparse softmax function as follows:

$$\alpha\text{entmax}(z) = \underset{p \in \Delta^d}{\arg\min}\left\{ p^T z + H_\alpha^T(p) \right\}$$

$$H_\alpha^T(p) = \begin{cases} \dfrac{1}{\alpha(\alpha-1)} \sum_j \left(p_j - p_j^\alpha \right), & \alpha \neq 1 \\ -\sum_j p_j \log p_j & \alpha = 1 \end{cases}$$

$\Delta^d := \{ p \in \mathbb{R}^d : p \geq 0, \|p\|_1 = 1 \}$ (probability simplex)

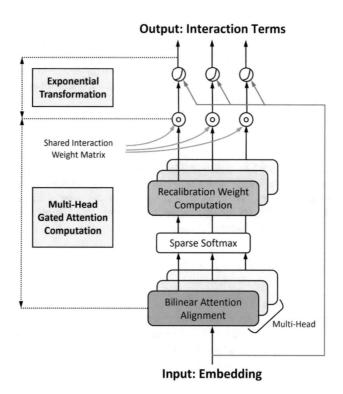

Figure 6-77. *Adaptive relation modeling module architecture. From Cai et al.*

We have $Y \in \mathbb{R}^{K \cdot o \cdot n_e}$; the prediction module uses a multilayer perceptron to project this vector into the final desired output:

$$\hat{y} = \text{MLP}(Y)$$

The complete ARM-Net architecture is displayed in Figure 6-78.

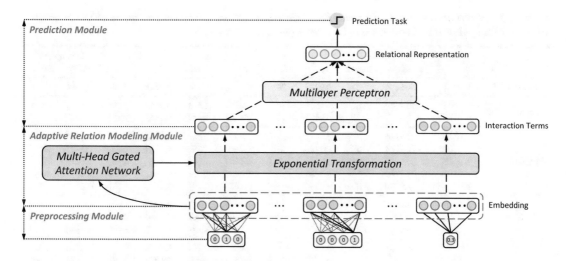

Figure 6-78. *Complete ARM-Net architecture. From Cai et al.*

Overall, the ARM-Net architecture takes a more involved design toward modeling tabular data. Rather than openly applying attention in large-scale blocks and employing strategies like self-supervised pretraining, like previously discussed work, ARM-Net is highly "strict" about how attention mechanisms are utilized. ARM-Net explicitly models cross-features between embeddings using exponential neurons and dynamically determines how such modeling is performed using a multi-head attention mechanism. This "stringent" architecture ensures high parameter efficiency, as information flow is explicitly directed rather than set open and expected to be fully learned (e.g., through expensive self-supervised learning campaigns). Moreover, like previous work, the attention weights can be interpreted to understand how the model makes predictions on any sample.

Cai et al. apply ARM-Net to several large benchmark tabular datasets and find the ARM-Net performs competitively with other deep tabular models with a reasonable parameter size (Table 6-14).

Table 6-14. *Performance of ARM-NET and ARM-Net+ (ARM-Net ensembled with a standard DNN) against other varieties of models on five benchmark models from different contexts*

Model Class	Model	Frappe		MovieLens		Avazu		Criteo		Diabetes130	
		AUC	Param	AUC	Param	AUC	Param	AUC	Param	AUC	Param
First-Order	LR	0.9336	5.4K	0.9215	90K	0.6900	1.5M	0.7741	2.1M	0.6701	370
Second-Order	FM	0.9709	5.4K	0.9384	90K	0.6797	1.5M	0.7663	2.1M	0.6594	370
	AFM	0.9665	5.7K	0.9473	91K	0.6857	1.6M	0.7847	2.1M	0.6774	7.6K
Higher-Order	HOFM	0.9778	170K	0.9435	2.9M	0.6919	18M	0.7788	107M	0.6714	11K
	DCN	0.9583	56K	0.9401	510	0.7460	3.3K	0.7959	6.6K	0.6765	2.2K
	CIN	0.9766	111K	0.9416	153K	0.6859	5.2M	0.7904	4.2M	0.6776	23K
	AFN	0.9779	3.1M	0.9470	242K	0.7456	3.3M	0.8061	7.8M	0.6778	306K
	ARM-Net	**0.9786**	867K	**0.9550**	140K	**0.7651**	147K	**0.8086**	1.5M	**0.6853**	102K
NN-based	DNN	**0.9787**	122K	**0.9540**	101K	0.7513	126K	**0.8082**	449K	0.6753	130K
	GCN	0.9732	1.6M	0.9404	365K	0.7506	964K	0.7984	2.2M	0.6828	4.0M
	GAT	0.9744	404K	0.9420	821K	**0.7525**	302K	0.8047	2.2M	**0.6846**	1.3M
Ensemble	Wide&Deep	0.9762	127K	0.9477	192K	0.6893	1.7M	0.7913	2.5M	0.6626	130K
	KPNN	0.9787	140K	0.9546	102K	0.7514	195K	0.8089	893K	0.6794	582K
	NFM	0.9745	100K	0.9214	186K	0.6874	1.6M	0.7833	2.3M	0.6695	4.6M
	DeepFM	0.9773	127K	0.9481	192K	0.6891	1.7M	0.7899	2.5M	0.6683	131K
	DCN+	0.9786	213K	0.9553	192K	0.7487	168K	0.8079	703K	0.6844	227K
	xDeepFM	0.9775	538K	0.9481	215K	0.6913	1.8M	0.7917	5.0M	0.6659	55M
	AFN+	0.9790	365K	0.9563	343K	0.7524	3.0M	0.8074	8.0M	0.6825	741K
	ARM-Net+	**0.9800**	263K	**0.9592**	217K	**0.7656**	339K	**0.8090**	1.3M	**0.6871**	1.7M

The authors provide the official repository at `https://github.com/nusdbsystem/ARM-Net`. It is, like SAINT, also written in PyTorch with no easily findable Keras or TensorFlow reimplementation. After cloning and installing dependencies, you can view examples of command-line scripts in `ARM-Net/run.sh`, where you can control the model architecture, training parameters, and the dataset used.

Key Points

This chapter discussed the attention mechanism, its usage in transformer models, and the application of attention-based models for language, multimodal, and tabular context data.

- Attention mechanisms allow for direct, explicit modeling of the dependencies between two sequences by associating every timestep pair with an attention score indicating the relevance of the cross-timestep dependency. This helps address signal propagation and dependency forgetting problems in recurrent models, which can hinder performance on advanced sequence-to-sequence tasks even with upgrades like long-term memory states and bidirectionality. Models like the Google Neural Machine Translation model (Wu et al. 2016) employed large LSTM stack encoders and decoders with attention mechanisms.

- The transformer architecture (Vaswani et al. 2017) demonstrated that one could build successful sequence-to-sequence models without recurrent models, relying only upon attention as the core mechanism modeling sequence dependencies. Vaswani et al. use multi-head attention, in which multiple versions of the value, key, and query key are derived via linear layers; attention is computed between each value-key-query combination and concatenated and then mapped linearly to an output. This theoretically allows for multiple representations of the value, key, and query to be learned and processed. The transformer architecture uses self-attention in the encoder and decoder layers and cross-attention between the encoder and the decoder in autoregressive decoding. Later transformer-based models like BERT (Devlin et al. 2019) demonstrate that transformer models can significantly benefit from self-supervised pretraining, like masked language modeling and next sentence recognition tasks. Most all modern natural language models are transformers or heavily inspired by transformers. Some researchers (Merity 2019) express skepticism for the research craze around transformer models, demonstrating that much more sustainable and lightweight models can obtain good performance and advocating for greater valuing of reproducibility.

- Keras offers three native flavors of attention: Luong-style dot product attention with `L.Attention()`, Bahdanau-style additive attention with `L.AdditiveAttention()`, and Vaswani-style multi-head attention with `L.MultiHeadAttention`. By passing the `return_attention_scores=True` argument into the call of any of these layers, you can also collect the attention scores yielded in any one pass. This allows you to interpret how the particular layer attends to certain timesteps.

- You can build attention layers into recurrent models for text and into recurrent heads for multi-head multimodal models to improve how text is modeled in relationship to tabular data.

- You can directly apply attention layers to embedded tabular data. This is a natural way to compute interactions between different features in the dataset.

- There is a significant body of work applying attention-based deep learning models to tabular data. We covered four research papers in this chapter exemplifying work in the area.

  - TabTransformer (Huang et al. 2020) embeds categorical features and applies a transformer block, which consists of a multi-head attention layer followed by a feed-forward layer with residual connections and layer normalization after each, several times. The derived features are concatenated with layer-normalized continuous features and processed with a standard MLP.

  - TabNet (Arik and Pfister 2019) uses a similar multi-step decision model; at each step, the model "selects" a certain subset of features using an attention-like mechanism and then processes the features with several feed-forward layers and residual connections. TabNet employs self-supervised pretraining; certain values in the input are randomly masked, and TabNet must reconstruct the masked values. After the pretraining task, the TabNet decoder is discarded and replaced with a decision-making module, which can be used for fine-tuning.

  - SAINT (Somepalli et al. 2021) uses a multi-step architecture; each step consists of a standard transformer-like block followed by an intersample attention block. The intersample attention block computes relationships between samples, as opposed to computing relationships between features or timestamps of inputs. This allows the model to explicitly reason about information relative to other samples in its batch and in turn is informative from an explainability perspective.

- The ARM-Net model (Cai et al. 2021) does not use a transformer-like architecture, but rather uses attention mechanisms to control the behavior of novel exponential neurons, which allow for very direct learning of explicit interactions between features.

The next chapter will explore work in tree-based deep learning models, the other significant body of modern research on deep learning for tabular data.

CHAPTER 7

■ ■ ■

Tree-Based Deep Learning Approaches

Imitation is the most sincere form of flattery.

—Oscar Wilde, Irish Poet, among many others

In this last chapter of Part 2, we explore deep models that draw upon the success of tree-based models to inform their architectures and training processes. Together with attention-based models, this constitutes the majority of current tabular deep learning research. Tree-based models are well-adapted toward many domains of tabular data because of their sharp and rigidly hierarchical shape-cutting structure, compounded in ensembles. By creating neural network models that simulate these properties, we can hopefully realize these strengths on new scales and levels of efficiency.

This chapter is structured in three parts, each sampling methods from different dominant approaches in the field. We will begin by discussing tree-structured neural networks, which explicitly build tree node–like logic into the structural units of the architecture, followed by boosting and stacking neural networks, which imitate successful tree ensembling paradigms. Lastly, we will explore distillation, in which knowledge from a tree is transferred into a neural network. The chapter adopts a dominantly research-based approach, presenting relevant components of curated papers and the implementation(s) when available.

Tree-Structured Neural Networks

Many deep models take *structural* interpretation from trees by imitating the structure of a decision tree using differentiable "neural" equivalents. This section discusses five sampled research papers adopting this approach:

- *"Deep Neural Decision Trees" by Yongxin Yang et al.*: Training decision trees through deep learning methods by soft binning networks.

- *"SDTR: Soft Decision Tree Regressor for Tabular Data" by Haoran Luo et al.*: A soft decision tree (SDT) architecture is built from perceptron nodes.

- *"Neural Oblivious Decision Ensembles for Deep Learning on Tabular Data" by Sergei Popov et al.*: An ensemble of naïve neural decision trees in which each tree level has the same splitting condition.

© Andre Ye and Zian Wang 2023
A. Ye and Z. Wang, *Modern Deep Learning for Tabular Data*,
https://doi.org/10.1007/978-1-4842-8692-0_7

- *"Deep Neural Network Initialization with Decision Trees" by Kelli Humbird et al.*: Cleverly mapping the structure of a decision tree to neural networks, acting as a warm start to training.

- *"DNF-Net: A Neural Architecture for Tabular Data" by Ami Abutbul*: Differentiable analogs to AND and OR gates are constructed to represent trees using "soft" logical expressions.

Deep Neural Decision Trees

Tree-based models are highly interpretable due to their use of greedy function approximation.[1] Their unique mechanism of splitting data based on a trained threshold allows for state-of-the-art performance on tabular data. Ensemble tree methods such as Random Forests and Gradient Boosting Machines are typically used for tabular data benchmarking as they are competitive, if not better than deep learning methods. Moreover, tree-based models can be easily visualized, thus informative of how and why certain features contribute to certain decisions. Interpretability is critical to many real-world applications in areas such as business, law, and much more. This is not to say that neural networks do not provide any practical advantage over classical machine learning methods. In neural networks, parameters are updated simultaneously. In contrast, tree-based models update their parameters individually through traversing branches and leaves in order; this optimization approach can be suboptimal compared with the simultaneous parameter updates in neural networks. By combining the advantages of both, we would theoretically be able to construct a differentiable tree-based model that is able to excel in tabular data prediction tasks.

Yongxin Yang, Irene Garcia Morillo, and Timothy M. Hospedales propose a mix between tree-based models and neural networks referred to as "deep neural decision trees" (DNDTs).[2] Deep neural decision trees optimize their parameters simultaneously using gradient descent. From the technical perspective, DNDT can be implemented in practically any deep learning framework; thus, it's able to utilize the computational power of accelerators such as GPUs or TPUs. Since the model's parameters can be optimized by gradient descent, the model can also be treated as a building block for a much larger end-to-end modeling scheme.

We can formulate Decision Tree training as adjusting the values of various discrete binning functions. More specifically, we treat each split of a decision node as binning the sample into one of the two branches (which, in the context of a binning function, can be seen as bins) stemming from the node. Training can be seen as optimizing the value of the binning threshold to achieve minimal loss. However, in standard Decision Trees, those metaphoric "binning" functions are undifferentiable due to their discrete nature. Yang et al. propose a soft approximation alternative to the non-differentiable binning procedure. Their binning function is continuous, allowing for optimization through gradient-based methods.

Consider a non-binary decision tree where a differentiable binning function models every branch node decision. Notably, each binning function contains $n + 1$ available bins, where each bin represents a branch in the decision tree. Having $n + 1$ available bins equates to needing n "cut-off" points or thresholds. Using notations adopted from the original paper, the cut-off points can be denoted as a list: $[\beta_1, \beta_2, \beta_3, \beta_4, ..., \beta_n]$. The roles of these values will come into play later as we expand on how the binning functions are optimized.

---

[1] The term "greedy function approximation" is adopted from the Gradient Boosting Machine paper "Greedy Function Approximation: A Gradient Boosting Machine". The term symbolizes the tree-like structure of Gradient Boosting machines and decision trees, indirectly reflecting their high interpretability.

[2] Yang, Y., Morillo, I.G., & Hospedales, T.M. (2018). Deep Neural Decision Trees. *ArXiv, abs/1806.06988.*

A single-layer neural network, f, parameterizes every binning function. We can construct the network as follows:

$$f_{w,b,\tau}(x) = \text{softmax}\big((wx + b)/\tau\big)$$

Let's break the equation down and define each parameter in f:

- The weight of the network, w, is non-trainable and set as a constant before initiating training. Its values will always be set to $w = [1, 2, ..., n + 1]$ regardless of outside conditions.

- The network's bias, b, is also pre-initialized to a set value before training, but its values are adjustable by gradient descent. For every training iteration, the bias is constructed as $b = [0, -\beta_1, -\beta_1 - \beta_2, -\beta_1 - \beta_2 - \beta_3, ..., -\beta_1 - \beta_2 - ... - \beta_n]$ where every unique β value is trainable through backpropagation.

- The softmax activation bounds the output vector between 0 and 1 while keeping the sum across all values to 1. We can interpret this output as a list of probabilities defining which of the $n + 1$ branches the input belongs to. The temperature factor τ controls the output sparsity. As $\tau \to 0$, the output tends to be a one-hot vector indicating the index of the branch to which the input belongs.

To demonstrate the effect of the temperature τ on the output vector, consider an example where $wx + b$ is computed to [1 6 9] and the temperature is set to 10:

$$\text{softmax}\left(\frac{(wx + b)}{\tau}\right)$$

$$\text{softmax}\big([1 \ 6 \ 9]/10\big) \approx [0.205 \ 0.338 \ 0.456]$$

By adjusting τ to 1, we observe a sparser output:

$$\text{softmax}\big([1 \ 6 \ 9]/1\big) \approx [0.000 \ 0.047 \ 0.952]$$

Further lowering the temperature value to 0.1, the output tends to be a complete one-hot encoded vector:

$$\text{softmax}\big([1 \ 6 \ 9]/0.1\big) \approx [0.000 \ 0.000 \ 0.999]$$

Generally, decision trees are built from top to bottom in a greedy manner where each decision node is defined individually and optimized before moving on to the next. Not only is this method of optimization suboptimal but it's also resource intensive when the tree is non-binary. Instead, Yang et al. utilize neural networks' ability to update their parameters simultaneously and train a single binning network for every feature prior to building the tree. We can then build the tree recursively in the following steps:

1. Treat each binning network as a decision node where the number of branches stemming from the node is determined by the number of bins available or the length of the output vector. Reusing the notations introduced previously, every decision node should have $n + 1$ branches.

2. Pick any decision node (binning network) as the root node. Results will be the same regardless of which decision node is selected. Add a note about how the result will be the same.

3. From here, every tree level will be assigned one single decision node, where it will become the child node of every branch from the previous level. Rephrased, it's connected to the previous decision node. Essentially, every level will contain n^l of the same decision node where l is the tree level.

4. Assuming that there are D features, there will be n^D leaf nodes at the final level of the tree. Unlike standard decision trees where leaf nodes directly represent model predictions, leaf nodes in DNDT are mere indicators of "clusters" that samples belong to based on their feature values. One further processing step is required to obtain the final prediction. Usually, a linear model is employed to classify samples arriving at leaf nodes.

Mathematically, we can exhaustively find the final leaf node that one sample leads to using the Kronecker product. For those unfamiliar, the Kronecker product is usually denoted as $\otimes$ and is a special form of matrix/vector multiplication. For two matrices $A \in \mathbb{R}^{m \times n}$ and $B \in \mathbb{R}^{p \times q}$, the Kronecker product is defined as

$$A \otimes B = \begin{bmatrix} a_{1,1}B & \cdots & a_{1,n}B \\ \vdots & \ddots & \vdots \\ a_{m,1}B & \cdots & a_{m,n}B \end{bmatrix}$$

The resulting matrix will be in the shape of $mp \times nq$. We denote D as the number of features present in the dataset and f_i as the ith binning network. The Kronecker product is repeatedly applied to yield an almost one-hot encoded vector (for a low temperature value τ), signifying the index of the leaf node that the input x will lead to:

$$f_1(x_1) \otimes f_2(x_2) \otimes f_3(x_3) \otimes \ldots \otimes f_D(x_D)$$

The resulting "leaf nodes" vector would be in length n^D if every binning network/decision node had n available bins/branches. Note that the value of n can be different for each binning network. The final linear model will receive the "leaf nodes" vector as input and produce a prediction relevant to the dataset classes (or continuous values in the case of regression).

DNDT cleverly avoids the suboptimal training scheme of tree-based models by training neural networks and optimizing their parameters without depending on each other. DNDT provides the advantage of scalability; however, this only applies to the size of samples, not features. Due to the use of the Kronecker product, computation becomes significantly expensive as the number of features increases. The authors, therefore, propose to use Random Forest–style training where several weak learners are each trained on a subset of features. Figure 7-1 is a representation of DNDT as well as its equivalent decision tree. For explanatory purposes, only two features from the Iris Flower dataset are selected in the diagram.

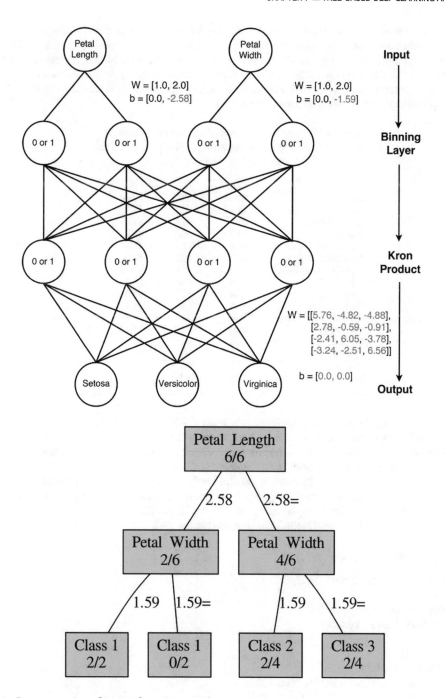

Figure 7-1. *Representation of DNDT from Yang et al.*

The authors of the paper compared DNDT with a decision tree baseline and a shallow two-layer neural network with 50 neurons in each hidden layer. The number of cut points for each feature in DNDT is all set to 1, meaning that there are only two branches for every node. A total of 14 datasets are retrieved from Kaggle and UCI. For datasets with more than 12 features, Random Forest–style training is adapted for DNDT, with each weak learner randomly learning from ten features with a total of ten weak learners. The following diagram shows the result of the comparison (Table 7-1).

Table 7-1. *Comparison of DNDT against various other algorithms. Datasets marked with * indicate the ensemble version of DNDT was used. From Yang et al.*

Dataset	DNDT	DT	NN
Iris	**100.0**	**100.0**	**100.0**
Haberman's Survival	**70.9**	66.1	**70.9**
Car Evaluation	95.1	**96.5**	91.6
Titanic	**80.4**	79.0	76.9
Breast Cancer Wisconsin	94.9	91.9	**95.6**
Pima Indian Diabetes	66.9	**74.7**	64.9
Gime-Me-Some-Credit	98.6	92.2	**100.0**
Poker Hand	50.0	**65.1**	50.0
Flight Delay	**78.4**	67.1	78.3
HR Evaluation	92.1	**97.9**	76.1
German Credit Data (*)	**70.5**	66.5	**70.5**
Connect-4 (*)	66.9	**77.7**	75.7
Image Segmentation (*)	70.6	**96.1**	48.05
Covertype (*)	49.0	**93.9**	49.0
# of wins	5	7	5
Mean Reciprocal Rank	0.65	**0.73**	0.61

Although Decision Trees still are empirically superior to DNDT across this selection of benchmark datasets, DNDT can still match Decision Trees' performance in most scenarios. DNDT also provides flexibility as the number of cut points can be changed for each individual feature. It has been shown that increasing the cut points improves model performance significantly.

DNDT can be implemented in PyTorch or TensorFlow in around 20 lines of code from the official implementation done by the authors of the paper. Since TensorFlow custom training loops can be quite confusing, we will use the PyTorch implementation. The Iris Flower dataset will be used as an example. We can start by importing PyTorch and loading the dataset (Listing 7-1).

Listing 7-1. Imports

```
from sklearn.datasets import load_iris
import torch
# used for implementation later
from functools import reduce

data = load_iris()
X = np.array(data.data)
X = torch.from_numpy(X.astype(np.float32))
y = torch.from_numpy(np.array(data.target))
```

Following, we can define custom functions for each component of DNDT as shown in Listing 7-2.

Listing 7-2. Custom implementation of DNDT components

```python
def torch_kron_prod(a, b):
    res = torch.einsum('ij,ik->ijk', [a, b])
    res = torch.reshape(res, [-1, np.prod(res.shape[1:])])
        return res

def torch_bin(x, cut_points, temperature=0.1):
    # x is a N-by-1 matrix (column vector)
    # cut_points is a D-dim vector (D is the number of cut-points)
    # this function produces a N-by-(D+1) matrix, each row has only one element being one
        and the rest are all zeros
    D = cut_points.shape[0]
    W = torch.reshape(torch.linspace(1.0, D + 1.0, D + 1), [1, -1])
    cut_points, _ = torch.sort(cut_points)  # make sure cut_points is monotonically
        increasing
    b = torch.cumsum(torch.cat([torch.zeros([1]), -cut_points], 0),0)
    h = torch.matmul(x, W) + b
    res = torch.exp(h-torch.max(h))
    res = res/torch.sum(res, dim=-1, keepdim=True)
        return h

def nn_decision_tree(x, cut_points_list, leaf_score, temperature=0.1):
    # cut_points_list contains the cut_points for each dimension of feature
    leaf = reduce(torch_kron_prod,
                  map(lambda z: torch_bin(x[:, z[0]:z[0] + 1], z[1], temperature),
                  enumerate(cut_points_list)))
return torch.matmul(leaf, leaf_score)
```

Before training, a few hyperparameters of the model will be defined (Listing 7-3).

Listing 7-3. Defining parameters for training

```python
num_cut = [2]*4  # 4 features with 2 cut points each
num_leaf = np.prod(np.array(num_cut) + 1) # number of leaf node
num_class = 3
# randomly initialize cutpoints
cut_points_list = [torch.rand([i], requires_grad=True) for i in num_cut]
leaf_score = torch.rand([num_leaf, num_class], requires_grad=True)
loss_function = torch.nn.CrossEntropyLoss()
optimizer = torch.optim.Adam(cut_points_list + [leaf_score], lr=0.001)
```

Finally, we can start the training process with PyTorch's custom training loops (Listing 7-4).

Listing 7-4. Training DNDT with PyTorch

```python
from sklearn.metrics import accuracy_score

for i in range(2000):
    optimizer.zero_grad()
    y_pred = nn_decision_tree(X, cut_points_list, leaf_score, temperature=0.05)
    loss = loss_function(y_pred, y)
```

555

```
loss.backward()
optimizer.step()
if (i+1) % 100 == 0:
    print(f"EPOCH {i} RESULTS")
    print(accuracy_score(np.array(y), np.argmax(y_pred.detach().numpy(), axis=1)))
```

Three main factors can improve or worsen training results: the number of splits for each feature, the temperature, and the learning rate. These values should be carefully selected based on domain knowledge or through hyperparameter tuning, as subtle changes can influence training results significantly.

The core of DNDT shines at being able to simultaneously update parameters through gradient descent while having a tree-based architecture. The scalability of DNDT also provides conveniences that most tree-based models do not possess. Although DNDT may require some hyperparameter tuning to match the performance of current state-of-the-art models, it still stands as an alternative or a mix between deep learning and tree-based models. Finally, DNDT opens the door to building better neural networks that imitate tree-based model logic, as we will see in later sections.

Soft Decision Tree Regressors

Recall that in a standard decision tree, each node represents a binary decision threshold for a particular feature, which is executed against the incoming sample and used to determine which path (left or right) the sample ends up taking. Haoran Luo, Fan Cheng, Heng Yu, and Yuqi Yi propose a soft analog for such a decision tree on regression problems, which can be trained with gradient descent, in the 2021 paper "SDTR: Soft Decision Tree Regressor for Tabular Data."[3]

Consider a full binary tree with nodes denoted by their index i. In the soft decision tree model, each node does not output a binary left or right decision, but instead a soft probability using a perceptron. We define the probability of choosing the left branch at the ith node given the input $\bar{x}$ as follows, for some learned weight $\overline{w}_i$ and bias bi:

$$p_i(\bar{x}) = \sigma\left(\beta_i\left(\overrightarrow{w_i}\,\bar{x} + b_i\right)\right)$$

Note that regardless of how deep the node is, the decision is always made with access to the full set of features. The parameter β_i is a scaling coefficient used to prevent "too soft decisions" – that is, to push the weighted sum away from zero and toward more extreme ends that produce soft decisions closer to zero and one.

Every leaf node is associated with a scalar R_i; this is the predicted output for a sample that travels down the tree to that particular leaf node. Let P_l represent the probability of choosing a leaf l. Then we can calculate P_l as follows, where i is either 1 or 0 depending on whether a left or right step is taken on the path toward l:

$$P_l(x) = \prod_{i\in\text{Path}(l)} p_i(x)^i \left(1 - p_i(x)\right)^{1-i}$$

Rather than just evaluating the difference between the value associated with the leaf node with the highest probability and the true value, we define the loss as the sum of differences between *every leaf* and the true value, weighted by the leaf probabilities:

$$L(x) = \sum_{l\in\text{LeafNodes}} P_l(x)\left(R_l - y\right)^2$$

---

[3] Luo, H., Cheng, F., Yu, H., & Yi, Y. (2021). SDTR: Soft Decision Tree Regressor for Tabular Data. *IEEE Access, 9*, 55999-56011.

While the structure of the tree is technically fixed – there is a specified depth and a static binary structure – the conditions that fill each of the nodes are learnable by the model through the weights and biases, optimized to minimize the loss (Figure 7-2).

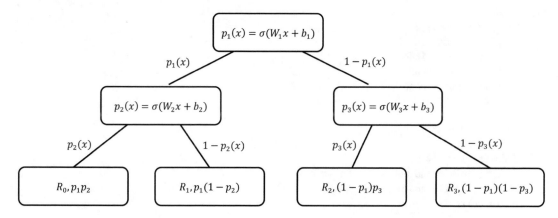

Figure 7-2. *Diagram of the relationship between leaf nodes and connections. From Luo et al.*

The paper describes additional mechanisms to enforce proper learning. Just for the purposes of understanding the key elements of the architecture, however, we will build a very simple, bare-bones version of the model. What is interesting about this particular model design is that the model essentially is a multi-output architecture with one dense layer extending from the input for each node (Listing 7-5). The tree-like architecture is realized in how the different fully connected layers are brought in relation with one another to calculate the loss.

Listing 7-5. Generating all nodes in the tree

```
MAX_DEPTH = 5

inp = L.Input((INPUT_DIM,))
outputs = []
for node in range(sum([2**i for i in range(MAX_DEPTH + 1)])):
    outputs.append(L.Dense(1, activation='sigmoid')(inp))
model = keras.models.Model(inputs=inp, outputs=outputs)
```

Note that we are assuming in this case that the soft decision tree model is being trained on a binary prediction task, such that all layers in the model (both output layers and intermediate probabilistic tree nodes) use the sigmoid activation, for simplicity.

We want to associate each of these layers with a particular position in a binary tree structure. There are many clever ways to go about this, but we will stick with the canonical object-oriented approach, which has the benefit of interpretability and ease of navigation. Each Node object corresponds to an index in the outputs list. Note that it does not matter which nodes correspond to which indices, as long as there is only one node per index and vice versa. We can do this by recursively building linked nodes in a binary tree fashion, with a global index variable incremented upon the instantiation of a Node (Listing 7-6).

Listing 7-6. Defining a node class and generating a binary tree with a specified depth

```
index = 0

class Node():
    def __init__(self):
        global index
        self.index = index
        self.left = None
        self.right = None
        index += 1

def add_nodes(depths_left):
    curr = Node()
    if depths_left != 0:
        curr.left = add_nodes(depths_left - 1)
        curr.right = add_nodes(depths_left - 1)
    return curr

root = add_nodes(MAX_DEPTH)
```

In order to calculate the loss, we need to multiply each leaf node by all the node probabilities that lead toward that leaf node. We can recursively create a collection of outputs by navigating our tree structure (Listing 7-7).

Listing 7-7. Getting all the leaf nodes

```
def get_outs(root, y_pred):
    if not root.left and not root.right: # is a leaf
        return [y_pred[root.index]]
    lefts = get_outs(root.left, y_pred)
    lefts = [y_pred[root.index] * prob for prob in lefts]
    rights = get_outs(root.right, y_pred)
    rights = [(1-y_pred[root.index]) * prob for prob in rights]
    return lefts + rights
```

We need to define a custom loss that evaluates the average loss between the truth and each leaf node value multiplied by the probability series (Listing 7-8).

Listing 7-8. Defining the loss function

```
from tensorflow.keras.losses import binary_crossentropy as bce
NUM_OUT = tf.constant(2**MAX_DEPTH, dtype=tf.float32)
def custom_loss(y_true, y_pred):
    outputs = get_outs(root, y_pred)
    return tf.math.divide(tf.add_n([bce(y_true, out) for out in outputs]), NUM_OUT)
```

Because this loss function aggregates multiple outputs rather than acting independently on a single model output, it is more convenient for us to define a custom model with a specific fit method (Listing 7-9). (With the default compiling and fitting steps, we can only specify losses that act on an output or several losses that each act on a single output in the case of multimodal models. There is no easy way to define a loss that accepts multiple outputs.) We can do this by overriding the default train_step method.

Listing 7-9. Writing a custom training function

```
import tensorflow as tf
avg_loss = tf.keras.metrics.Mean('loss', dtype=tf.float32)
class custom_fit(tf.keras.Model):
    def train_step(self, data):
        images, labels = data
        with tf.GradientTape() as tape:
            outputs = self(images, training=True) # forward pass
            total_loss = custom_loss(labels, outputs)
        gradients = tape.gradient(total_loss, self.trainable_variables)
        self.optimizer.apply_gradients(zip(gradients, self.trainable_variables))
        avg_loss.update_state(total_loss)
        return {"loss": avg_loss.result()}
```

After instantiation, the model can be trained (Listing 7-10).

Listing 7-10. Compiling and fitting the model

```
model = custom_fit(inputs=inp, outputs=outputs)
model.compile(optimizer='adam')
history = model.fit(x, y, epochs=20)
```

Again, this model doesn't do very well by itself, but it illustrates the fundamental idea.

The authors offer a model implemented in PyTorch. Minimal PyTorch is needed to begin using the model. We begin by loading the soft decision tree model from the official repository (Listing 7-11).

Listing 7-11. Obtaining the SDT from the official repository

```
!wget -O SDT.py https://raw.githubusercontent.com/xuyxu/Soft-Decision-Tree/master/SDT.py
import SDT
import importlib
importlib.reload(SDT)
```

The first step is to define a PyTorch dataset (Listing 7-12). The PyTorch dataset format is almost exactly the same as the TensorFlow custom dataset syntax (recall from Chapter 2): we need to define a __len__ and a __getitem__ method.

Listing 7-12. Writing a PyTorch dataset

```
import torch
from torch.utils.data import Dataset, DataLoader
from sklearn.model_selection import train_test_split as tts

class dataset(Dataset):

    def __init__(self, data, seed = 42):
        X_train, X_valid, y_train, y_valid = tts(data.drop('Cover_Type', axis=1),
                                                 data['Cover_Type'],
                                                 random_state = seed)
        self.x_train=torch.tensor(X_train.values,
                                  dtype=torch.float32)
```

559

```
        self.y_train=torch.tensor(pd.get_dummies(y_train).values,
                                      dtype=torch.float32)

    def __len__(self):
        return len(self.y_train)

    def __getitem__(self,idx):
        return self.x_train[idx],self.y_train[idx]
```

We can instantiate the dataset on the Forest Cover dataset, for instance. The DataLoader wraps around the Dataset and provides additional training-level tooling for feeding the data to the model (Listing 7-13).

Listing 7-13. Reading a CSV file into a PyTorch dataset and converting into a DataLoader

```
import pandas as pd, numpy as np
df = pd.read_csv('../input/forest-cover-type-dataset/covtype.csv')
data = dataset(df.astype(np.float32))
dataloader = DataLoader(data, batch_size=64, shuffle=True)
```

The soft decision tree can be instantiated and trained as follows (Listing 7-14).

Listing 7-14. Training the SDT

```
from SDT import SDT
model = SDT(input_dim = len(X_train.columns),
            output_dim = len(np.unique(y_train)))

import torch.optim as optim
import torch.nn as nn
criterion = nn.CrossEntropyLoss()
optimizer = optim.SGD(model.parameters(), lr=0.001, momentum=0.9)

for epoch in range(10):

    running_loss = 0.0
    for i, data in enumerate(dataloader, 0):
        inputs, labels = data
        optimizer.zero_grad()

        outputs = model(inputs)
        loss = criterion(outputs, labels)
        loss.backward()
        optimizer.step()

        print(f'[Epoch: {epoch + 1}; Minibatch: {i + 1:5d}]. Loss: {loss.item():.3f}',
              end='\r')

    print('\n')

print('Finished Training')
```

The syntax used here is similar to that of writing a custom loop in TensorFlow. The primary difference is the requirement to explicitly call steps occurring during the feed-forward and backpropagation stages in the loss and optimizer objects.

NODE

Consider the game of "laggy" 20 Questions. To guess an object that player A is thinking of, player B asks a series of questions to which player A answers "yes" or "no." The twist is that player B receives all the answers to their questions only *after they finish asking all the questions*, rather than immediately after each question.

This is an example of an *oblivious decision tree* – a tree in which each level has the same splitting criteria, as opposed to using different criteria at different levels. For instance, the following "3 Questions" decision tree is oblivious:

- Is it an animal?
    - *If yes*: Does it fly?
        - *If yes*: Is it fast?
            - *If yes*: eagle
            - *If no*: woodcock
        - *If no*: Is it fast?
            - *If yes*: cheetah
            - *If no*: tortoise
    - *If no*: Does it fly?
        - *If yes*: Is it fast?
            - *If yes*: plane
            - *If no*: paraglider
        - *If no*: Is it fast?
            - *If yes*: racecar
            - *If no*: rock

In contrast, the following decision tree, which represents how one might play a more standard game of 20 Questions, is not oblivious:

- Is it an animal?
    - *If yes*: Does it live in the water?
        - *If yes*: Is it a predator?
            - *If yes*: shark
            - *If no*: sardine
        - *If no*: Does it have four legs?
            - *If yes*: lion
            - *If no*: flamingo

- *If no*: Is it a vehicle?
    - *If yes*: Does it have four wheels?
        - If yes: car
        - *If no*: bicycle
    - *If no*: Does it fly?
        - *If yes*: plane
        - *If no*: ketchup

Note that a standard non-oblivious decision tree is more expressive than an oblivious decision tree because it is not beholden to the requirement that each level must operate on the same splitting condition. That means that we can construct downstream splitting conditions that are informed by previous known information; for instance, we ask if the object lives in the water after we know that it is an animal. However, oblivious decision trees have the advantage of being simple in computation and complexity. In fact, oblivious decision trees are not so much trees as they are large binary lookup tables. In an oblivious decision tree, the same tree can be represented with different permutations of splitting conditions on each level because no condition is dependent on another condition.

Sergei Popov et al. introduced the NODE model in the paper "Neural Oblivious Decision Ensembles for Deep Learning on Tabular Data"[4]; it is similar to previous neural network mimics of tree-based models but uses an ensemble of naïve oblivious decision trees rather than optimizing a single complex decision tree.

It also directly chooses features to select rather than employing abstract learned linear combinations like soft decision tree regressors. F_i represents the feature of the data χ selected for splitting at the ith tree level, and b_i represents the required threshold for the feature F_i at the ith tree level. Thus, $F_i - b_i$ will be positive if the feature exceeds the threshold and negative if not.

The αentmax function is applied to "binarize" this result (i.e., to effectively make the decision). Recall the αentmax function from Chapter 6 is a modified version of the softmax function, which is sparser and encourages more extreme values. Each node therefore possesses a more decisive determination. The soft decision tree regressor struggles with this and requires additional mechanisms such as the pre-activation scaling coefficient to reconcile it (Figure 7-3).

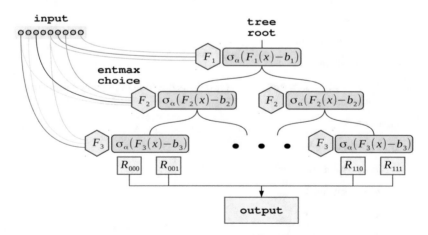

Figure 7-3. *A schematic of a single NODE layer/tree. From Popov et al.*

[4] Popov, S., Morozov, S., & Babenko, A. (2020). Neural Oblivious Decision Ensembles for Deep Learning on Tabular Data. *ArXiv, abs/1909.06312*.

Otherwise, NODE is optimized very similarly to SDTR: the loss is expressed as a sum of all of the leaf nodes weighted by the path probability to that node. This forms a single NODE layer – a neural oblivious decision tree, which is differentiable and can therefore be trained with backpropagation techniques.

The power of NODE comes from stacking multiple individual NODE layers together into a joint ensemble. The authors propose a DenseNet-style stacking (see Chapter 4 on DenseNet), in which each layer is connected to every other layer (Figure 7-4).

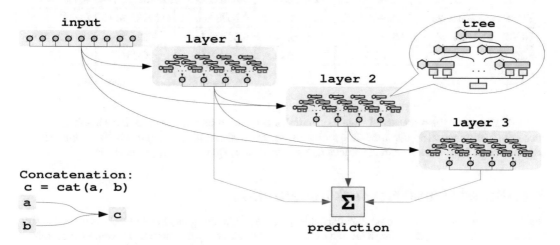

Figure 7-4. *The arrangement of NODE into a multilayer model. From Popov et al.*

The authors find that NODE outperforms CatBoost and XGBoost on several benchmark tabular datasets (Tables 7-2 and 7-3). When these competitors undergo hyperparameter optimization, NODE performs slightly worse on some datasets but still retains overall dominance across the evaluated datasets.

Table 7-2. *Performance of NODE against CatBoost and XGBoost with default hyperparameters. From Popov et al.*

	Epsilon	YearPrediction	Higgs	Microsoft	Yahoo	Click
Default hyperparameters						
CatBoost	0.1119±2e−4	80.68±0.04	0.2434±2e−4	0.5587±2e−4	0.5781±3e−4	0.3438±1e−3
XGBoost	0.1144	81.11	0.2600	0.5637	0.5756	0.3461
NODE	**0.1043±4e−4**	**77.43±0.09**	**0.2412±5e−4**	**0.5584±3e−4**	**0.5666±5e−4**	**0.3309±3e−4**

Table 7-3. *Performance of NODE against competitors with tuned hyperparameters. From Popov et al.*

	Epsilon	YearPrediction	Higgs	Microsoft	Yahoo	Click
	Tuned hyperparameters					
CatBoost	0.1113±4e−4	79.67±0.12	0.2378±1e−4	0.5565±2e−4	0.5632±3e−4	0.3401±2e−3
XGBoost	0.1112±6e−4	78.53±0.09	0.2328±3e−4	**0.5544±1e−4**	**0.5420±4e−4**	0.3334±2e−3
FCNN	0.1041±2e−4	79.99±0.47	0.2140±2e−4	0.5608±4e−4	0.5773±1e−3	0.3325±2e−3
NODE	**0.1034±3e−4**	**76.21±0.12**	**0.2101±5e−4**	0.5570±2e−4	0.5692±2e−4	**0.3312±2e−3**
mGBDT	OOM	80.67	OOM	OOM	OOM	OOM
DeepForest	0.1179	—	0.2391	—	—	0.3333

NODE is one of the most directly tree-like neural network architectures proposed and has found success in many general tabular modeling problems. See the well-written and approachable supplementary code notebook demonstrations in the official repository: `https://github.com/Qwicen/node`.

Tree-Based Neural Network Initialization

Neural networks are known for their flexibility, allowing the user to design and create architectures that suit their needs. Without a doubt, there has been a tremendous amount of work put into developing methods and techniques to search and obtain optimal or suboptimal network architectures without human trial and error; these types of algorithms are usually referred to as Neural Architecture Search (NAS). Refer to Chapter 10 for detailed explanations and implementation regarding NAS. The NAS process is typically time-consuming, and the algorithm does not consider the specific problem type, whether it be image recognition, text analysis, or tabular data in the context of this book. On the other hand, the structure of tree-based models allows them to excel in structured data tasks. Instead of attempting to adapt tree-based models to gradient-based training, we can shift our goal to designing a tree-like architecture for neural networks. K. D. Humbird, J. L. Peterson, and Rand. G. McClarren, in their paper "Deep Neural Network Initialization with Decision Trees," [5] seek to accomplish precisely that.

Unlike neural networks, tree-based models progressively construct their nodes and branches during training; this eliminates the need to manually design model architectures prior to training. Since sophisticated mathematical computations always inform the shape and form of tree-based models, their architectural design is almost always better than human trial and error. One can think of designing a neural network architecture for tabular data as manually designing the placement of each and every branch in a decision tree. Undoubtedly, this shows how difficult it is for human-designed network architectures to outperform tree-based models consistently. K. D. Humbird et al. attempt to "map" the core structure of tree-based models onto deep neural networks. According to the authors, the mapping will create a "boosted start" for the neural network in terms of structure and weight initialization to achieve better training results. The network produced by the mapping algorithm is referred to as "Deep Jointly Informed Neural Network (DJINN)."

---

[5] Humbird, K.D., Peterson, J.L., & McClarren, R.G. (2019). Deep Neural Network Initialization with Decision Trees. *IEEE Transactions on Neural Networks and Learning Systems, 30,* 1286-1295.

The construction and training process of DJINN can be split into three general steps as outlined in the following:

1. Train any tree-based model on the selected dataset. Note that for the purpose of concise explanations, the decision tree will be used as the target tree-based model. However, any kind of tree-based ensemble algorithm would work. The algorithm is repeated n times for each of the n weak learners in the ensemble, essentially mapping out n neural networks.

2. We recursively traverse the trained decision tree to map its structure onto a neural network. This is done by following specific rules defined by the DJINN mapping algorithm.

3. The mapped network is trained like any other ANN. The paper suggests Adam as the optimizer and ReLU as hidden layer activations.

To formalize the algorithm for mapping decision trees to DJINN, we adapt notations used by the authors in the original paper:

– Denote l as the level index for the decision tree and the layer index for the neural network where $l = 0$ represents the first layer of the neural network and the first level of the decision tree.

– The value of l falls within the interval $[0, D_t]$ where D_t is the max depth of the decision tree. This indirectly tells us that the mapped neural network will have $D_t + 1$ total layers (including the input and output layers).

– Denote D_b as the maximum level that branch nodes exist. Rephrased, D_b is the lowest level in the decision tree that it will still be possible for nodes to partition the data further. This indirectly tells us about the relationship between D_t and D_b: $D_b = D_t - 1$.

– Let $N_b(l)$ return the number of branches in the decision tree at level l. The number of neurons in any hidden layer of the network can be computed as $n(l) = n(l-1) + Nb(l)$.

– Let $L_i^{\max}$ be a list containing the furthest level that each feature appeared as a branch node. To clarify, say that "feature 1" was chosen by the tree to partition the data at tree levels 2, 4, and 5. Then $L_i^{\max}$ for "feature 1" would be 5 since it's the deepest level that the feature appeared in the tree.

– Let W^l denote the weight matrix for layer l. For every unique feature in the dataset, $i = 0$, 1, 2, ..., number of features – 1, we set $W_{i,i}^l$ to unity[6] for every value of $l < L_i^{\max}$. For a better understanding, the following visual in Figure 7-5 demonstrates which neurons are initialized to unity for $L_i^{\max} = (2,2,1)$.

---

[6] For those unfamiliar, "unity" simply refers to the number 1.

$$L_i^{max} = (2, 2, 1) \text{ for } x_1, x_2, x_3$$

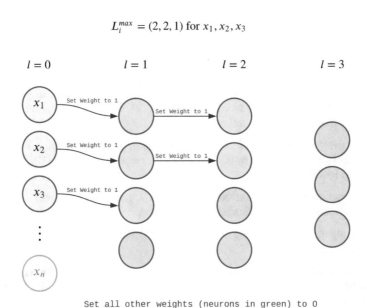

Figure 7-5. Weight initialization of the network

Currently, every weight and bias of the mapped network is set to zero other than those pre-initialized unity weights mentioned previously. After applying the DJINN mapping algorithm, the final network architecture will be determined by pruning neurons with their weight still at zero. Removing zero-weight neurons will be discussed in detail after explaining the mapping algorithm since the process will be much more relevant then. The core idea of the mapping algorithm works by traversing through the decision tree and "reinitializing" neurons corresponding to the positions of nodes and branches in the tree. We can interpret neurons with zero weight as "disconnected neurons" since they cannot pass on information without a bias attached to them. On the other hand, neurons "reinitialized" by the mapping algorithm will possess nonzero values and thus can be interpreted as "connected neurons" for their ability to pass on information without a bias attached to them. Note that the algorithm will not "reinitialize" every neuron in the network. Neurons that weren't "reinitialized" by the mapping algorithm will be selectively pruned according to their bias value. All biases in the network will be randomly initialized from a normal distribution, and neurons with a negative bias value with zero weights will be scrapped. The selective pruning will inject randomness into the network architecture, providing better flexibility and more significant potential than the original pretrained tree-based model.

We start at $l = 1$ since at $l = 0$ the network is the input layer where every weight is set to 1 beforehand and the number of neurons is restricted to the number of features. When neurons are reinitialized, their weight is randomly chosen from the distribution $(0, \sigma^2)$ where

$$\sigma^2 = \frac{3}{\text{sum of previous} + \text{current layer neuron count}}$$

As we traverse through the decision tree recursively, for each node in every level $l \, \epsilon \, [1, D_l]$, we denote the current node as c. There are two possibilities for what c could be:

– Node c is a branch node, meaning that the node further splits into branches or the current level of the decision tree $< D_t$. In this case, we initialize a new neuron, turning it from being disconnected to connected at layer l. Then, we record the feature used in the branch node to split the data further and find the input neuron associated with that feature. We can temporarily denote the input neuron as n sub f e a. t. By using neurons that we initialized earlier to unity, from n_{feat}, we string the input neuron all the way to c. Finally, we connect c to its equivalent "parent node," or the neuron that we initialized for the parent node of c.

– Node c is a leaf node. In the case of a regression task, we simply connect the output neuron to its equivalent "parent node," or the neuron we initialized for the parent node of c given the context of decision trees. In the case of classification, we connect the output neuron outputting the same class as the leaf c to its equivalent "parent node" neuron.

We can visualize this process from the original paper's example depicted in Figure 7-6.

By examining the trained decision tree on the left, we see that $L_i^{max} = (2,1,2)$ for x_1, x_2, x_3. Neurons with a blue cross marked on them are initialized to unity according to the L_i^{max} for each feature (Figure 7-6). Note that neurons can be initialized to unity but never be connected when mapping the tree to the network; thus, some neurons are marked with the blue cross but shaded in gray.

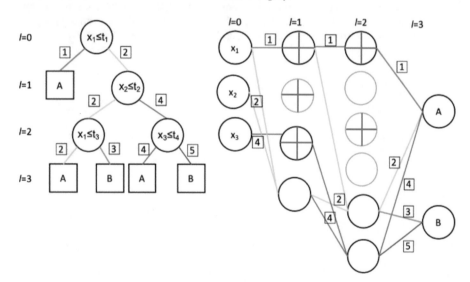

Figure 7-6. *Visualization of mapping decision trees to a neural network, constructing DJINN. From K. D. Humbird et al. with slight modifications*

We iterate through every tree level, mapping every node from that level to its respective neuron in the network, going from left to right. At $l = 1$ of the decision tree, the first node that we iterate to is a leaf node of class A, denoting it as c. The parent node for c is the input node of the decision tree, where the feature x_1 is selected to split the data. To connect the input neuron all the way to the output neuron of class A (corresponding to c), we utilize neurons that we initialized to unity. The red path labeled "1" represents this connection, mapping the leaf c to the network.

We are moving to the right in the tree to the node splitting on x_2. We instantiate a new neuron in the respective layer of the network (at layer $l = 1$). We first connect the new neuron to its parent node, or the neuron mapped from the tree's input node. We then connect the input neuron corresponding to the feature used in the current node, the x_1 input neuron. Both connections are shown in the yellow path labeled as "2".

Moving down to $l = 2$ in the decision tree, the first neuron to map is a node that splits on the feature x_1. Again, we first instantiate a new neuron in the respective network layer ($l = 2$). We then connect the new neuron to the neuron in the previous layer, which was mapped by the parent node of the current node. To clarify, that was the neuron we initialized and connected for the node splitting on x_2. Finally, using neurons that we initialized to unity at the start, we can string the input neuron of x_1 all the way to the new neuron we just initialized. Both connections are drawn by the yellow path between layers 1 and 2 labeled "2".

For the last branch node, the tree chose to split on the feature x_3. The exact process is repeated on this node for mapping the network: connect the neuron created by the current node's parent branch to the current one and connect the current neuron to the x_3 input neuron by using neurons with unity weight. Both connections are shown in the blue path labeled "4".

Finally, moving to the last level of the tree, there are a total of four leaves, with two pointing to class A, while the other two pointing to class B. For the leaf node on the left, connect the class A output neuron to the neuron we created for its parent node. This is shown in the yellow path labeled "2". Moving to the right, for the leaf node with the same parent branch, we simply connect the output neuron for class B to the same neuron in the previous layer, represented by the green path labeled "3". The final two leaf nodes are both child nodes of a branch node splitting on x_3; we connect the respective output neurons to the neuron mapped by that branch node. The two connections are drawn by the blue "4" path and the purple "5" path, respectively. The complete pseudocode for the mapping algorithm is shown in Figure 7-7.

Algorithm 1 DJINN Tree to Neural Network Mapping

1: Recurse through paths of the decision tree:
 · Determine max branch depth (D_b)
 · Count number of branches at each level $N_b(l)$
 · Record max depth each input occurs as a branch:
 $L_i^{\max}$

▷ For a max branch depth D_b, there will be D_b hidden layers, an input layer with N_{in} neurons, and an output layer with N_{out} (regression) or N_{class} (classification) neurons in the neural network. Each hidden layer will have $n(l)$ neurons, where

$$n(l) = n(l-1) + N_b(l) \tag{1}$$

This "copies" the previous hidden layer and adds "new" neurons for each branch in the current level of the tree.

2: Create arrays W^l of dimension $n(l)$ x $n(l-1)$, $l=1,...,D_b$, and $W^{D_{b+1}}$ with dimension $n(D_b)$ x N_{out} (or N_{class}) to store initial weights. Initialize arrays to 0.
3: For each input $i=0,1,...,N_{in}-1$:
 · Set $W_{i,i}^l = 1$ for $l < L_i^{\max}$
 ▷ This ensures input values are passed through hidden layers until the decision tree no longer splits on them.
4: Recurse through decision paths of the tree:
 For levels $l=1,...,D_b$:
  For each node c in level l:
   ·Define p as the neuron created by the parent branch
   If c = branch:
    ▷ According to Eq. 1, a new neuron has been added to layer l
    · Initialize $W_{new,p}^l \sim \mathcal{N}(0,\sigma^2)$, connecting branch p and new neuron
    · Initialize $W_{new,c}^l \sim \mathcal{N}(0,\sigma^2)$, connecting branch c and new neuron
   If c = leaf:
    · Initialize $W_{p,p}^l \sim \mathcal{N}(0,\sigma^2)$, $l=l+1 ...D_b-1$
    · Initialize $W_{p,out}^{D_b} \sim \mathcal{N}(0,\sigma^2)$
    ▷ Classification: out = neuron for the class
    ▷ Regression: out = output neurons

Figure 7-7. Pseudocode for decision tree mapping to neural networks. From K. D. Humbird et al.

As mentioned earlier, those disconnected neurons will be randomly chosen to be included in the final architecture decided by their bias initialization. Biases of all neurons will be randomly selected from a Gaussian distribution. DJINN essentially utilizes the optimal structure created by the trained decision tree while allowing a small amount of freedom to account for inaccuracies. This cleverly avoids the time-consuming process of NAS while creating a dynamic method of designing/producing ANN architecture specialized for tabular data. Furthermore, the interpretability of decision trees also partially carries over to the DJINN. One can observe a highly interpretable network structure through the example of decision trees trained for logic operations shown in Figure 7-8. Note that gray neurons are initialized with the architecture but randomly chosen to be included in the final network by their bias value.

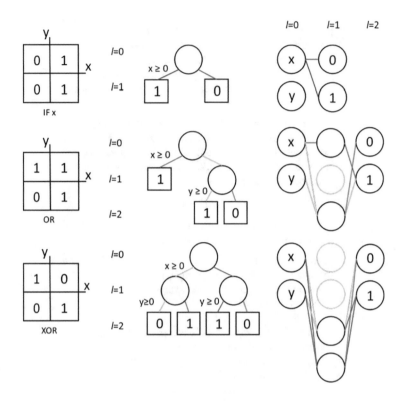

Figure 7-8. *Decision trees trained for logic operations are then mapped to neural networks, providing high structural interpretability. From K. D. Humbird et al.*

The authors performed a series of tests and comparisons for DJINN. Here's a summary of their findings:

1. Using ensemble tree-based models as the pretrained tree-based model consistently outperformed single-tree models. Bagging approaches such as Random Forest can map to multiple weak neural networks. The final prediction is simply the average of all mapped networks. Figure 7-9 are results for DJINN trained on four different tabular datasets (Boston Housing, Diabetes Progression, California Housing, and Inertial Confinement Nuclear Fusion Implosion Simulations) with varying tree counts for the ensemble. An increased number of trees in the ensemble is empirically better than less. The following graph shows the Mean Squared Error as a function of the number of trees in the ensemble (Figure 7-9).

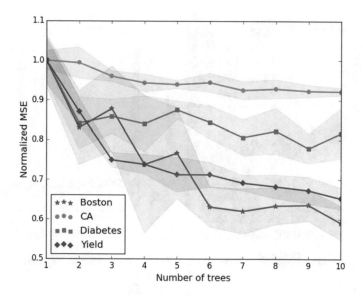

Figure 7-9. *Performance of DJINN with four different tabular datasets compared against the number of trees used in the ensemble. From K. D. Humbird et al.*

2. The tree-based structure of DJINN can be seen as a warm start to model training. Two distinct characteristics distinguish DJINN as a warm start technique from others: its sparsity in nonzero weights and their placements. These advantages are shown in their comparison with other weight initialization methods, including densely connected Xavier initialized weights,[7] randomly initializing the same number of nonzero weights per layer with random placement, and finally a standard two–hidden layer ANN. Again, the MSE metric is plotted against the number of epochs trained (Figure 7-10).

---

[7] Xavier weight initialization randomly sets the lth layer weights from a random uniform distribution with range

$$\left(-\frac{\sqrt{6}}{\sqrt{n_i + n_{i+1}}}, \frac{\sqrt{6}}{\sqrt{n_i + n_{i+1}}} \right)$$

where n_i is the number of incoming connections from layer $l - 1$ and n_{i+1} is the number of outgoing connections to layer $l + 1$.

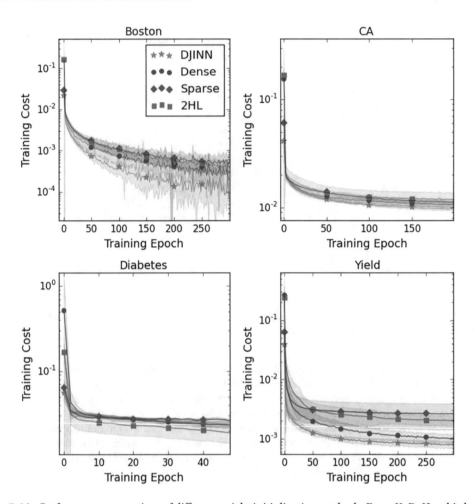

Figure 7-10. *Performance comparison of different weight initialization methods. From K. D. Humbird et al.*

Implementation for DJINN can be found on GitHub at `https://github.com/LLNL/DJINN` for TensorFlow. To install, we can clone the repository to our current directory and download packages mentioned in `requirement.txt` (Listing 7-15).

Listing 7-15. Installing DJINN and importing the package

```
git clone https://github.com/LLNL/DJINN.git
cd DJINN
pip install -r requirements.txt
pip install .

from djinn import djinn
```

To illustrate the simple pipeline that the package provides for selecting various hyperparameters and training, the Breast Cancer dataset will be used as the data for the model (Listing 7-16).

Listing 7-16. Importing the dataset and performing train-test split

```
breast_cancer_data = load_breast_cancer()
X = breast_cancer_data.data
y = breast_cancer_data.target

X_train,X_test,y_train,y_test=train_test_split(X, y, test_size=0.25)
```

Next, we will create a DJINN_Classifier object and specify hyperparameters for the tree. Based on those hyperparameters, the library will search for the optimal training parameters for the mapped network (Listing 7-17).

Listing 7-17. Instantiating a DJINN classifier object and getting the optimal hyperparameters for the mapped network

```
# dropout keep is the probability of keeping a neuron in dropout layers
djinn_model = djinn.DJINN_Classifier(ntrees=1, maxdepth=6, dropout_keep=0.9)

# automatically search for optimal hyperparameters
optimal_params = djinn_model.get_hyperparameters(X_train, y_train)
batch_size = optimal_params['batch_size']
lr = optimal_params['learn_rate']
num_epochs = optimal_params['epochs']
```

Once the training parameters are obtained, we simply call the train method on the model and fill in the optimal hyperparameters (Listing 7-18).

Listing 7-18. Training DJINN using the train method

```
model.train(X_train, y_train,epochs=num_epochs,learn_rate=lr, batch_size=batch_size,
display_step=1,)
```

The prediction can be generated by calling the model's prediction method (Listing 7-19).

Listing 7-19. Predicting using DJINN

```
from sklearn.metrics import auc

preds = djinn_model.predict(X_test)
print(auc(preds, y_test))
```

DJINN can be viewed as an optimal network architecture for tabular data tasks and a brand-new modeling technique for optimizing deep learning performance on structured datasets. Its cleverness in manipulating and utilizing tree-based model architecture also adds another layer of structural interoperability; in other words, we know why the network architecture is structured as it is.

Net-DNF

In classical binary logic, *disjunctive normal form* (DNF) is a logical expression in which two or more conjunctions of literals are joined by a disjunction in the main scope. To provide the relevant vocabulary to understand DNF

- A *variable* holds one of two truth values (True or False) and is represented generally by a capital letter, such as A, B, or C (and so on).

- *Negation* is equivalent to the logical *not* operator and is represented in logic by ¬. For instance, ¬A means "not A" or "the negation of A." If A is True, then ¬A evaluates to False.

- A *literal* is either a variable or the negation of a variable. For instance, the following are literals: A, ¬ B, ¬ C, F.

- *Conjunction* is equivalent to the logical *and* operator and is represented in logic by ∧. For instance, $A \land B$ means "A and B"; if A is True and B is False, then $A \land B$ evaluates to False. Alternatively, $A \land$ ¬ B evaluates to True. A conjunction is not a literal.

- *Disjunction* is equivalent to the logical *or* operator and is represented in logic by ∨. For instance, $A \lor B$ means "A or B"; if A is True and B is False, $A \lor B$ evaluates to True. Alternatively, ¬$A \lor B$ evaluates to False. A disjunction is not a literal.

- Conjunction or disjunction is considered to be in *wide scope* if it is not "wrapped" in any other operators. For instance, conjunction is in wide scope in the expression $(A \lor$ ¬ $B) \land C$ because the conjunction ∧ is the "outermost" operator; nothing is wrapping around it. On the other hand, conjunction is not in wide scope in the expression $(A \land$ ¬ $B) \lor C$ because it is wrapped by the disjunction ∨. In this case, disjunction is in wide scope.

Putting this all together, disjunctive normal form features disjunctions that are all at wide scope; each argument of the disjunction (i.e., the expressions being disjoined together) must be either a literal or a conjunction of literals.[8] The following are examples of expressions in DNF:

- $(¬A \land$ ¬ $B \land C) \lor D$

- $A \lor (¬B \land$ ¬ $C \land D)$

- $A \lor (¬B \land$ ¬ $C \land D) \lor E$

- $A \lor (¬B \land$ ¬ $C \land D) \lor E \lor (A \land$ ¬ $F)$

- $A \lor B$

The following are examples of expressions which are *not* in DNF:

- $A \land B$; conjunction rather than disjunction is in wide scope.

- $A \lor (A \land (¬B \lor C))$; a disjunction is not in wide scope.

- ¬$(A \lor B)$; negation rather than disjunction is in wide scope.

Why is DNF relevant at all? Decision Trees can be represented as DNF formulas over feature split conditions. For instance, consider the following decision tree logic expressed in nested form[9]:

---

[8] Technically speaking, each argument must be in conjunctive normal form (CNF); that is, the conjunctions and negations cannot be nested, and all must be on the same level.

[9] This logic is not medical advice. In fact, it is pretty shoddy advice in general.

- If temperature larger than 80 degrees

  - Pack sunscreen.

- If temperature not larger than 80 degrees

  - If going to shaded region

    - Do not pack sunscreen.

  - If not going to shaded region

    - Pack sunscreen.

Let A be a Boolean variable representing the truth value of the statement "the temperature is larger than 80 degrees." Let B represent the truth value of the statement "you are going to a shaded region." We can express the decision tree intuitively in disjunctive normal form as follows:

$$A \vee \neg B$$

The truth value represents whether we should pack sunscreen (True) or not (False). Say that the temperature is not larger than 80 degrees (A = False) and we are not going to a shaded region (B = False). Then, $A \vee \neg B$ = False $\vee \neg$ False = True. As our tree dictates, we will pack sunscreen.

Ami Abutbul, Gal Elidan, Liran Katzir, and Ran El-Yaniv of the Technion – Israel Institute of Technology and Google propose building a deep learning network out of disjunctive-normal-form "building blocks" to model tabular data in the 2020 ICLR paper "DNF-Net: A Neural Architecture for Tabular Data."[10] The authors propose that using DNF units theoretically capable of expressing decision tree logic could help develop neural emulations of successful decision tree models in tabular data problems:

> …the "universality" of forest methods in handling a wide variety of tabular data suggests that it might be beneficial to emulate, using neural networks, the important elements that are part of the tree ensemble representation and algorithms.

In order to integrate a custom unit design into a neural network architecture, however, it must be differentiable. The traditional disjunction and conjunction operators are not differentiable. Abutbul et al. propose the *disjunctive normal neural form* (DNNF) block, which uses "soft" and therefore differentiable generalization of these logic gates.

DNNF is implemented with a two–hidden layer neural network and replicates the constitution of a formula in disjunctive normal form. The first layer derives "literals," which are passed into a series of soft conjunctions. These soft conjunctions are then passed into a disjunction gate.

The soft disjunction and conjunction gates are defined as follows:

$$\text{or}(\vec{x}) = \tanh\left(\sum_{i=1}^{d} \vec{x}_i + (d - 1.5)\right)$$

$$\text{and}(\vec{x}) = \tanh\left(\sum_{i=1}^{d} \vec{x}_i - (d - 1.5)\right)$$

---

[10] Abutbul, A., Elidan, G., Katzir, L., & El-Yaniv, R. (2020). DNF-Net: A Neural Architecture for Tabular Data. *ArXiv, abs/2006.06465.*

Let's consider the input $\vec{x} := \langle -1, 1, -1 \rangle$. This is representative of the inputs False, True, and False. If we are to apply disjunction (or) to the input, we are computing a representation of False ∨ True ∨ False. We can calculate the disjunction and conjunction of this input as follows:

$$\text{or}\left(\langle -1,1,-1 \rangle\right) = \tanh\left((-1+1-1)+(3-1.5)\right)$$

$$= \tanh(-1+1.5) = \tanh(0.5) \approx 0.46$$

$$\text{and}\left(\langle -1,1,-1 \rangle\right) = \tanh\left((-1+1-1)-(3-1.5)\right)$$

$$= \tanh(-1-1.5) = \tanh(-2.5) \approx -0.99$$

Indeed, False ∨ True ∨ False = True (0.46 being closer to "True" +1 than "False" −1) and False ∧ True ∧ False = False (−0.99 being closer to "False" −1 than "True" +1).

Note, however, that these soft neural gates also do a little bit of quantification as to just "how true" the disjunction or conjunction is. False ∨ True ∨ False can be thought of as "weakly true" because only one of the arguments makes the disjunction true. On the other hand, or(⟨1, 1, 1⟩) obtains a much higher result of 0.9998, since True ∨ True ∨ True is "strongly" true.

The authors use a modified version of conjunction to select certain literals rather than being forced to accept all of them. Recall that in DNF, literals are joined with conjunction; we want to give the network a mechanism to select only a subset of literals to conjoin. Otherwise, given some hypothetical set of literals $A, B, C, \ldots$, the only possible DNF expression uses the following pattern (with the argument of disjunction repeated an arbitrary number of times):

$$(A \wedge B \wedge C \wedge \ldots) \vee (A \wedge B \wedge C \wedge \ldots) \vee \ldots$$

This is not very informative. But say we "mask" out variables depending on the argument of conjunction to form a more expressive formula:

$$(A \wedge B \wedge D) \vee (B \wedge D \wedge F) \vee (A \wedge F)$$

As a technical detail, the authors restrict it such that each literal can only belong in *one* conjunction, for example:

$$(A \wedge C) \vee (D \wedge E) \vee (B \wedge F)$$

In order to enable such masking, we use a *projected conjunction gate*, which accepts some mask vector $\vec{u} \in \{0,1\}^d$ to select for variables in the input $\vec{x}$:

$$\text{and}_{\vec{u}}(\vec{x}) = \tanh\left(\vec{u}^T \vec{x} - |\vec{u}|_1 + 1.5\right)$$

Note that this is a generalization of the originally introduced formula for soft conjunction. We sum only the selected variables, subtract the number of selected variables (given to us by the L1 norm of the mask vector, $|\vec{u}|_1$, because $\vec{u}$ is binary), and add a bias of 1.5.

We can formally define the disjunctive normal neural form block as follows:

$$L(\vec{x}) = \tanh\left(\vec{x}^T W + \vec{b}\right)$$

$$\text{DNNF}(\vec{x}) = \text{or}(\text{and}_{\overrightarrow{c^1}}(L(\vec{x})), \text{and}_{\overrightarrow{c^2}}(L(\vec{x})), \ldots, \text{and}_{\overrightarrow{c^k}}(L(\vec{x}))$$

Note that c^i denotes a d-length mask vector, which determines which variables are selected for conjunction in the ith argument of the wide-scope disjunction. These are learnable, but the technical implementation of how this is done is omitted from the chapter and can be found in the paper. The authors employ gradient tricks to overcome gradient problems caused by learning binary masks in a continuous optimization regime.

$L(\vec{x})$ helps generate "literals," which are processed in subsequent layers *without any learned parameters* through a soft disjunctive normal form expression. This generation process can be thought of as a neural equivalent of creating splitting conditions in a tree context.

A *DNF-Net* is formed by stacking together n DNNF blocks, the outputs of which are linearly transformed and summed with a standard dense layer:

$$\text{DNFNet}(\vec{x}) = \sigma\left(\sum_{i=1}^{n}(w_i \text{DNNF}_i(\vec{x}) + b_i\right)$$

DNF-Net performs competitively against XGBoost and consistently better than a standard fully connected network on a variety of tabular datasets (Table 7-4). While DNF-Net is not a hands-down superior competitor to XGBoost, its differentiable emulation of tree-like logical structures in soft, neural form is promising and may become the basis of improved research.

Table 7-4. *Performance of DNF-Net on several datasets, compared with XGBoost and a fully connected neural network. From Abutbul et al.*

Dataset	Test Metric	DNF-Net	XGBoost	FCN
Otto Group	log-loss	**45.600 ± 0.445**	45.705 ± 0.361	47.898 ± 0.480
Gesture Phase	log-loss	86.798 ± 0.810	**81.408 ± 0.806**	102.070 ± 0.964
Gas Concentrations	log-loss	**1.425 ± 0.104**	2.219 ± 0.219	5.814 ± 1.079
Eye Movements	log-loss	68.037 ± 0.651	**57.447 ± 0.664**	78.797 ± 0.674
Santander Transaction	roc auc	88.668 ± 0.128	**89.682 ± 0.165**	86.722 ± 0.158
House	roc auc	95.451 ± 0.092	**95.525 ± 0.138**	95.164 ± 0.103

We will implement a very simple and incomplete modified version of Net-DNF to concretely illustrate the previously discussed theory. The authors of the paper add additional mechanisms to improve performance and functionality; the repository can be viewed here: `https://github.com/amramabutbul/DisjunctiveNormalFormNet`.

We'll start by defining the following configurations for the network (Listing 7-20):

- *The number of literals generated by* $L(\vec{x})$: This constitutes the "vocabulary" available to each of the DNNF blocks.

- *The number of arguments to the disjunction*: This is the number of conjunction expressions we generate and pass into the disjunction in each DNNF block.

- *Average number of conjunction literals*: This is the mean number of literals selected from the total array of available literals for conjunction.

- The number of DNNF blocks

Listing 7-20. Setting relevant constants

```
NUM_LITERALS = 64
NUM_DISJ_ARGS = 32
AVG_NUM_CONJ_LITS = 16
NUM_DNNF_BLOCKS = 8
```

Let's begin by defining the neural disjunction gate. We set the number of disjunction arguments (this is the length of the vector input to disjunction) as a constant and use it in the neural disjunction calculation (Listing 7-21).

Listing 7-21. Defining a neural OR function

```
NUM_DISJ_ARGS_const = tf.constant(NUM_DISJ_ARGS, dtype=tf.float32)
def neural_or(x):
    return K.tanh(K.sum(x, axis=1) + NUM_DISJ_ARGS_const - 1.5)
neural_or = L.Lambda(neural_or)
```

We pass both the input $\bar{x}$ and the mask vector $\bar{u}$ into the neural gate and use the given calculation (Listing 7-22).

Listing 7-22. Defining a neural AND function

```
def neural_and(inputs):
    x, u = inputs
    u = tf.reshape(u, (NUM_LITERALS,1))
    return K.tanh(K.dot(x, u) - K.sum(u) + 1.5)
neural_and = L.Lambda(neural_and)
```

To simplify things, we'll select literals for conjunction in the following manner: in the creation of each DNNF block, we select a random proportion of literals (with the specified average proportion), which is fixed – it becomes an intrinsic part of the layer.

We can do this by defining a "stimulus" tensor, which has shape (number of disjunction arguments, number of literals) and is filled with samples randomly drawn from a uniform distribution [0, 1]. All elements of the tensor are set to 1 if less than average number of conjunction literals/number of literals and 0 otherwise. This creates a random fixed mask to select literals for conjunction.

Then, for each disjunction argument, we perform conjunction on the selected literals by passing the full literal set (literals) and the corresponding mask vector (masks[i]). The outputs are concatenated together to produce a single vector output, which is passed into a neural disjunction output.

The DNNF function (Listing 7-23) accepts an input and connects it to an output layer, which is returned.

Listing 7-23. Defining a DNNF layer

```
def DNNF(inp_layer):

    stimulus = tf.random.uniform((NUM_DISJ_ARGS, NUM_LITERALS))
    ratio = tf.constant(AVG_NUM_CONJ_LITS / NUM_LITERALS)
    masks = tf.cast(tf.math.less(stimulus, ratio), np.float32)

    literals = L.Dense(NUM_LITERALS, activation='tanh')(inp_layer)
    disj_args = []
    for i in range(NUM_DISJ_ARGS):
        disj_args.append(neural_and([literals, masks[i]]))
    disj_inp = L.Concatenate()(disj_args)
    disj = neural_or(disj_inp)
    return L.Reshape((1,))(disj)
```

The DNF-Net can be built in the following fashion (Listing 7-24).

Listing 7-24. Defining the DNF-Network

```
def DNF_Net(input_dim, output_dim):

    inp = L.Input((input_dim,))
    dnnf_block_outs = []
    for i in range(NUM_DNNF_BLOCKS):
        dnnf_block_outs.append(DNNF(inp))
    concat = L.Concatenate()(dnnf_block_outs)
    out = L.Dense(output_dim, activation='softmax')(concat)

    return keras.models.Model(inputs=inp, outputs=out)
```

The model can then be instantiated, compiled, and fitted on a dataset.

Boosting and Stacking Neural Networks

Other models imitate the success of tree ensembles using boosting and stacking by applying boosting techniques to the neural network domain. This section discusses two sampled research papers adopting this approach:

- *"GrowNet" by Sarkan Badirli et al.*: Training an ensemble of multiple weak NN learners by Gradient Boosting methods.

- *"XBNet" by Tushar Sarkar*: Training an ensemble of networks through the XGBoost algorithm.

Note that Chapter 11 discusses multi-model arrangement techniques, which are a broad category of modeling methods for GrowNet and XBNet.

GrowNet

GrowNet (Gradient Boosting neural networks) perfectly fits under "boosting and stacking neural networks" – it mimics a Gradient Boosting Machine where each weak learner is a shallow ANN – proposed by S. Badirli, X. Liu, Z. Xing, A. Bhowmik, and S.S. Keerthi in 2020. GrowNet utilizes the powerful structural advantage of Gradient Boosting frameworks while letting neural networks discover complex relationships where GBMs cannot fathom.[11]

Intuitively, the proposed method of training neural networks seems like it would perform better than standard ANNs. Simple ensemble techniques such as weighted average and majority voting were proposed as early as the 1990s. Ensemble models almost always outperform single models as they can combine the learning results from different perspectives. Recall from Chapter 1 that Gradient Boosting is considered an ensemble technique, falling under the boosting category. Badirli et al. adapted the structure of GBMs and their learning method; they applied those key characteristics to neural networks. Currently, GBMs dominate the field of tabular data AI along with other tree-based models. Designing neural networks that consistently perform well on structured data still presents an immense challenge. However, typical weak learners of Gradient Boosting Machines are unable to discover complex, nonlinear relationships that might define the correlation between features and targets. GrowNet replaces the usual tree-based weak learners of Gradient Boosting Machines with shallow ANNs in hopes that it can utilize the unique boosting concept while having the learning power of neural networks.

---

[11] Badirli, S., Liu, X., Xing, Z., Bhowmik, A., & Keerthi, S.S. (2020). Gradient Boosting Neural Networks: GrowNet. *ArXiv, abs/2002.07971.*

Following the notation used by the authors, assume a dataset D with n samples in a d-dimensional feature space: $D = (x_i, y_i)_{i=1}^n$. GrowNet uses K additive functions, or weak learners, to predict the output:

$$\hat{y}_i = \sum_{k=0}^{K} \alpha_k f_k(x_i)$$

In the language of GBMs, α_k represents the boosting rate, which controls how much each weak learner contributes to the final prediction. Recall from Chapter 1 that the goal of weak learners is to progressively correct the mistake of the previous learner by predicting pseudo residuals. In GrowNet, instead of tree-based models as weak learners, shallow neural networks, f_k, are tasked with predicting pseudo residuals. We define the loss function as $L = \sum_{i=0}^{n} l(y_i, \hat{y}_i)$. The value of $\hat{y}_i$ is computed additively at each stage during boosting since each learner is sought greedily, and its predictions are calculated using the output from the previous stage. At stage t, the loss can be computed as

$$L^{(t)} = \sum_{i=0}^{n} l\left(y_i, \hat{y}_i^{(t-1)} + \alpha_k f_k(x_i)\right)$$

Second-order optimization techniques are adopted as they can lead to faster convergence and are superior to first-order optimization approaches in the current context. Neural networks in GrowNet are trained with Newton-Raphson steps. Additionally, the second-order Taylor expansion of the loss function is used to ease computational complexity. Following the expansion, the loss function can be simplified as

$$L^{(t)} = \sum_{i=0}^{n} h_i \left(\tilde{y}_i - \alpha_k f_k(x_i)\right)^2$$

where $\tilde{y}_i = -g_i / h_i$ with g_i and h_i being the first- and second-order gradients, respectively, of the loss function at x_i with respect to $\hat{y}_i^{(t-1)}$. Refer to the original paper for a detailed mathematical explanation. The general architecture of GrowNet can be seen in Figure 7-11.

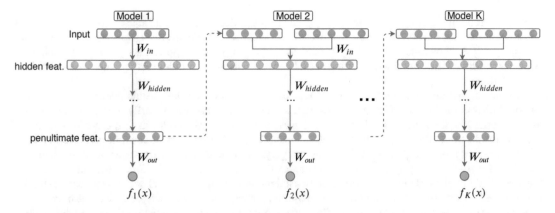

Figure 7-11. *Architecture representation of GrowNet. From Badirli et al.*

One of the significant disadvantages that tree-based models have compared with neural networks is their inability to update their parameters simultaneously (for more details, refer back to the "Deep Neural Decision Trees" section). Instead, they can only optimize parameters one at a time, finding the optimal value for one parameter before moving on to the next. Even though the GrowNet algorithm relies on several small neural networks as its foundation, these "weak learners" aren't able to further update their parameters past their training stage. To address this issue, the paper's authors implemented a corrective step executed after every stage. During the corrective step, all weak learners' parameters are updated with respect to the entire GrowNet. Additionally, the boosting rate at each stage is dynamically adjusted through backpropagation during the corrective step. Details of the corrective step are shown in Figure 7-12.

# Part 2 - Corrective step
for $epoch = 1$ **to** T **do**

    Calculate GrowNet output: $\hat{y}_i^{(k)} = \sum_{m=0}^{k} \alpha_m f_m(x_i), \forall x_i \in \mathcal{D}_{tr}$

    Calculate Loss from GrowNet: $\mathcal{L} = \frac{1}{n} \sum_i^n l(y_i, \hat{y}_i^{(k)})$

    Update model f_m parameters through back-propagation $\forall m \in \{1, ...k\}$

    Update step size α_k through back-propagation

end for

Figure 7-12. *Algorithmic details of the GrowNet corrective step. From Badirli et al.*

Consider a regression example where the Mean Squared Error (MSE), denoted as l, is adopted as the loss function. The general training procedure for GrowNet can be outlined in the following:

1. Instantiate a shallow neural network. This will be the first stage of training out of the total K stages in GrowNet. The network is trained on the complete dataset $\{x, y\}$ where x represents the features and y represents the targets.

2. From the second training stage up until the last training stage, K, the model will be trained on $\{\tilde{x}, \tilde{y}\}$. Let's break each variable down individually. The features, $\tilde{x}$, combine both the original dataset features and the penultimate features obtained from the previous weak learner $K - 1$. Here, penultimate features are obtained from the raw output from the final hidden layer (which is not the output layer). The dimension of $\tilde{x}$ will always be the number of dataset features plus the number of neurons in the final hidden layer of the weak learner. This will remain constant regardless of the training stage since every weak learner has the same network architecture. Recall from the preceding modified loss function, $\tilde{y}$ is computed as the negative quotient of the first- and second-order gradients of the loss function w.r.t. $\hat{y}^{(t-1)} : -g/h$. We can calculate the respective gradients for the MSE loss function as such:

$$g = 2\left(\hat{y}^{(t-1)} - y\right), \quad h = 2$$

Simplifying, we obtain $\tilde{y}$ as

$$\tilde{y} = -\frac{g}{h} = y - \hat{y}^{(t-1)}$$

We discover that this is the exact calculation for pseudo residuals in GBMs.

3. The corrective step is implemented. We treat GrowNet as one giant neural network and compute its output by $\sum_{k=0}^{K} \alpha_k f_k(x)$ and its loss by $L = \sum_{i=0}^{n} l(y_i, \hat{y}_i)$. Next, for every instantiated weak learner f, its parameters are updated through backpropagation and the boosting rate αk at each stage. This corrective step is repeated for a set number of epochs.

4. Steps 2 and 3 are repeated K times for every stage of training.

The complete technical algorithm of GrowNet, along with the corrective step, is depicted in the authors' pseudocode (Figure 7-13).

Algorithm 1 Full GrowNet training

Input: $f_0(x) = log(\frac{n_+}{n_-})$, α_0, Training data $\mathcal{D}_{tr}$
Output: GrowNet $\mathcal{E}$
for $k = 1$ **to** M **do**
    # Part 1 - Individual model training
    Initialize model $f_k(x)$
    Calculate 1st order grad.: $g_i = \partial_{\hat{y}_i^{(k-1)}} l(y_i, \hat{y}_i^{(k-1)})$, $\forall x_i \in \mathcal{D}_{tr}$
    Calculate 2nd order grad.: $h_i = \partial^2_{\hat{y}_i^{(k-1)}} l(y_i, \hat{y}_i^{(k-1)})$, $\forall x_i \in \mathcal{D}_{tr}$
    Train $f_k(\cdot)$ by least square regression on $\{x_i, -g_i/h_i\}$
    Add the model $f_k(x)$ into the GrowNet $\mathcal{E}$
    # Part 2 - Corrective step
    for $epoch = 1$ **to** T **do**
        Calculate GrowNet output: $\hat{y}_i^{(k)} = \sum_{m=0}^{k} \alpha_m f_m(x_i)$, $\forall x_i \in \mathcal{D}_{tr}$
        Calculate Loss from GrowNet: $\mathcal{L} = \frac{1}{n} \sum_i^n l(y_i, \hat{y}_i^{(k)})$
        Update model f_m parameters through back-propagation $\forall m \in \{1, ...k\}$
        Update step size α_k through back-propagation
    end for
end for

Figure 7-13. Full algorithm of GrowNet. From Badirli et al.

GrowNet can be adapted to classification, regression, and learning to rank. Experiment results showed GrowNet's empirical advantage over XGBoost and another similarly structured model, AdaNet. Each weak learner adopted has two standard dense layers with the number of neurons equal to half of the input feature dimension. Boosting rate was initially set to 1 and then later adjusted by the model automatically. For classification, the model is trained on the Higgs Boson dataset, while the Computed Tomography (retrieve the location of CT slices on an axial axis) and Year Prediction MSD (a subset of a million-song dataset) datasets are used for regression. Experiment results are shown in Tables 7-5 and 7-6.

Table 7-5. Experiment results from classification. From Badirli et al.

XGBoost	0.8304
GrowNet (all data)	**0.8510**
GrowNet (data sampling= 10%)	0.8439
GrowNet (data sampling= 1%)	0.8180

Table 7-6. *Experiment results from regression. From Badirli et al.*

	Music Year Pred.	Slice Localz.
XGBoost	8.9301	6.6744
AdaNet	12.1778	5.3824
GrowNet	**8.8156** (0.0061)	**5.3112** (0.3512)

Empirical results do indeed demonstrate that GrowNet outperforms XGBoost and its competitor, AdaNet, which uses a similar fashion of building neural networks. An implementation of GrowNet can be found at the following GitHub gist written by Yam Peleg, `https://gist.github.com/ypeleg/576c9c647` `0e7013ae4b4b7d16736947f`, which we can download by cloning the gist.

The following lines of code in Listing 7-25 will download and import the gist.

Listing 7-25. Downloading and importing the GrowNet package

```
!git clone https://gist.github.com/576c9c6470e7013ae4b4b7d16736947f.git grow_net
import grow_net
```

We can simply create an architecture using Keras and wrap it with the `GradientBoost` object. Calling `fit` will train the network as a GrowNet model. The California Housing dataset will be used in the following example (Listing 7-26).

Listing 7-26. Defining and compiling the GrowNet model

```
from sklearn.datasets import fetch_california_housing
from sklearn.model_selection import train_test_split

import tensorflow as tf
import tensorflow.keras.callbacks as C
import tensorflow.keras.layers as L
import tensorflow.keras.models as M

data = fetch_california_housing()
X = data.data
y = data.target
X_train, X_test, y_train, y_test = train_test_split(X, y, test_size=0.25)

inp = L.Input(X.shape[1])
x = L.BatchNormalization()(inp)
x = L.Dense(64, activation="swish")(x)
x = L.Dense(128, activation="swish")(x)
x = L.BatchNormalization()(x)
x = L.Dropout(0.25)(x)
x = L.Dense(32, activation="swish")(x)
x = L.BatchNormalization()(x)
out = L.Dense(1, activation='linear')(x)
model = M.Model(inp, out)
model.compile(tf.keras.optimizers.Adam(learning_rate=1e-3), 'mse')
model = GradientBoost(model, batch_size=4096, n_boosting_rounds=15, boost_rate = 1, epochs_
per_stage=2)
```

Finally, we can call the train and predict methods as usual (Listing 7-27).

Listing 7-27. Training and predicting using GrowNet

```
model.fit(X_train, y_train,batch_size=4096)
from sklearn.metrics import mean_squared_error
mean_squared_error(y_test, model.predict(X_test))
```

Like most tree-based ensemble models, heavy hyperparameter tuning is needed to squeeze every bit of performance out of GrowNet. GrowNet provides many improvements to both neural networks and tree-based methods; it's a result of the mixture between the two. Its performance can edge or even outperform models previously known to excel in structured data tasks.

XBNet

XBNet, proposed by Tushar Sarkar in 2021, combines the power of XGBoost and neural networks through a simple yet effective approach without much modification to either model.[12] Recall XGBoost from Chapter 1, short for Extreme Gradient Boosting. XGBoost is a highly optimized Gradient Boosting algorithm that combines efficiency, speed, regularization, and performance into one model. XGBoost not only uses a custom-defined splitting criterion but also prunes every tree after its creation to introduce regularization as well as to increase training speed. Most tree-based models and GBMs, including XGBoost, can easily compute feature importance values for the dataset. One numerical value is produced for each feature, which informs relatively how beneficial the feature is to model prediction compared with other features. Most deep learning models lack the ability to interpret how each individual feature contributes to model predictions. Consequently, feature importance is a commonly used tool to better understand the relationship between certain features and labels and is used as a measure for feature selection to improve model performance.

XBNet, like most other networks discussed in this chapter, attempts to combine the advantages of deep learning approaches and tree-based models. Like GrowNet, XBNet uses neural networks as its base model and modifies its training procedure to include characteristics from Gradient Boosting Machines. Specifically, those modifications can be summarized into two main ideas: (a) smart weight initialization based on GBM feature importance values and (b) modified backpropagation to allow weight updates through GBM feature importance values. Without much modification to the structure of the network and the process of parameter updates, we can keep the advantages that neural networks provide while injecting insights that XGBoost has on the data (Figure 7-14).

---

[12] Sarkar, T. (2022). XBNet : An Extremely Boosted Neural Network. *ArXiv, abs/2106.05239.*

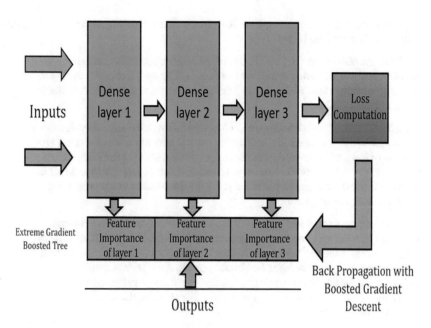

Figure 7-14. *XBNet's architecture. From Tushar Sarkar*

Additionally to incorporating feature importance, Tushar Sarkar also proposed tweaking the loss function to include a L2 regularization parameter to prevent overfitting. XBNet may be more prone to overfitting than standard ANNs since it includes added information from GBM feature importance values. It may be more severe than training a regular network due to the added information from XGBoost.

In certain prediction tasks, neural networks may be initialized with pretrained or precomputed weight to be better fitted for the problem situation, essentially giving the network a "boosted start" (and, yes, this is similar to DJINN in some aspects!). The idea of using pretrained weight to increase performance in domain-specific tasks is usually applied to unstructured datasets such as image, audio, or text data. Having pretrained weights for tabular data is rarely seen as the diversity of tabular datasets is vast. It's a rather daunting task to find a tabular dataset that can be used to pretrain a specific network architecture and is able to generalize to other datasets without difficulty.

Tushar Sarkar's proposed smart weight initialization approach involves using GBMs, specifically, the XGBoost model. An XGBoost model is trained on the entire dataset prior to the initialization of XBNet. Succeeding the network initialization, we set every input neuron's weight to its corresponding feature importance value calculated by XGBoost.[13] To an extent, we assigned each input neuron a weight representing how much each input neuron, or each feature in the dataset, could potentially contribute to the network training/prediction. It's probable that these preassigned values are suboptimal to the network since these are insights obtained from XGBoost: a model that utilizes completely different training schemes in comparison with neural networks. Nevertheless, the "boosted start" XGBoost can provide to XBNet surpasses what random weight initialization can achieve.

---

[13] Typically, feature importance values for each attribute in XGBoost are calculated by averaging the Information Gain/entropy that one feature retains for one weak learner across all weak learners.

As per usual, to explain and demonstrate the XBNet algorithm in detail, we can outline the XBNet training procedure in the following steps:

1. Initialize an XGBoost model and fit it on the entire dataset. We can then obtain feature importance values for the dataset. The vector that stores feature importance values for the dataset should have a length equal to the number of input neurons to XBNet.

2. Initialize XBNet. The architecture of XBNet should look like any other standard ANNs: the number of neurons, hidden layers, and activation functions are all choices of the user. The weights of input neurons are set to feature importance values obtained from step 1. Each input neuron should correspond to a feature in the dataset; the feature importance value for that feature will be the input neuron's initial weight value.

3. We will perform a modified version of the feed-forward operation in neural networks by training a separate XGBoost model from each layer's feed-forward outputs. Let $w^{(l)}$ and $b^{(l)}$ be the weight and bias for layer l, respectively. Denote $z^{(l)}$ as the raw output from layer l without applying the layer's activation function $g^{(l)}(x)$. Lastly, we will let $A^{(l)}$ represent the final layer output after applying the activation function. The following equations can conduct the feed-forward operation for layer l:

$$z^{(l)} = w^{(l)} A^{(l-1)} + b^{(l)}$$

$$A^{(l)} = g^{(l)}\left(z^{(l)}\right)$$

We will instantiate a new XGBoost model for layer l and denote it as $\text{xgb}^{(l)}$; this is not the same model used in step 1. The model will be trained on the raw output from layer l against the ground-truth values for the current batch $y^{(i)}$, and its feature importance values (here, we treat every neuron output in $A^{(l)}$ as a feature) will be stored in $f^{(l)}$:

$$f^{(l)} = \text{xgb}^{(l)}.\text{train}\left(A^{(l)}, y^{(i)}\right).\text{importance}$$

This step in the training procedure is repeated for every layer in the network. The XGBoost model instantiated for each layer will have the same hyperparameters. From a technical perspective, this allows us only to store one XGBoost model in memory and reset its training history every layer.

4. We will implement a modified backpropagation algorithm using $f^{(l)}$. First, the loss is computed with the L2 regularization (for more information, refer to Chapter 1) where $\mathcal{L}$ denotes the loss function and λ is the regularization strength hyperparameter:

$$\text{loss} = \frac{1}{m}\sum \mathcal{L}\left(\hat{y}^{(i)}, y^{(i)}\right) + \frac{\lambda}{2m}\sum\left(\left\|w^{(l)}\right\|\right)^2$$

Next, we can update weights and biases using the standard gradient descent update rule (or any other update rule depending on the optimizer used):

$$w^{(l)} := w^{(l)} - \alpha \nabla \mathbf{w}^{(l)}$$

$$b^{(l)} := b^{(l)} - \alpha \nabla b^{(l)}$$

By utilizing the same intuition that we deduced for initializing the input layer weights, we can incorporate $f^{(l)}$ into the update rule:

$$w^{(l)} := w^{(l)} + f^{(l)}$$

With added feature importance values, we essentially change how significant each weight in layer l contributes to the network by using measures from XGBoost. However, we cannot simply add $f^{(l)}$ to the weight matrix since we cannot ensure that they will be in the same order every epoch. The scale of feature importance values will remain the same by virtue of its definition in the context of XGBoost, which has no relation to the neural network. Due to gradient-based updates, the network weights will not remain in the same order as $f^{(l)}$; thus, we multiply $f^{(l)}$ by a scalar before using it to update $w^{(l)}$. We can replace the previously shown update rule $w^{(l)} := w^{(l)} + f^{(l)}$ with $w^{(l)} := w^{(l)} + f^{(l)} 10^{\log\left(\min\left(w^{(l)}\right)\right)}$.

The min value across the entire weight matrix $w^{(l)}$ is used as a scaling factor to ensure that every value of $f^{(l)}$ would be in the same order as $w^{(l)}$.

Again, this step is repeated for all layers of the network (as backpropagation should).

5. Inference in XBNet is done like any other ANN by a forward pass.

For reference, in Figure 7-15 is the pseudocode for XBNet.

Algorithm 2: Training Algorithm for XBNet using Boosted Gradient
Descent

Result: Cost function is minimized using Boosted Gradient Descent

Initialize w,b, α,tree;

$w^{[1]} = tree.train(X, y)$.importance;

for $t= 1,2,..,m$ **do**

> Forward propagation on X^t;
>
> $z^{[l]} = w^{[l]}A^{[l-1]} + b^{[l]}$;
>
> $A^{[l]} = g^{[l]}(z^{[l]})$;
>
> $f^{[l]} = tree.train(A^{[l]}, y^{(i)})$.importance;
>
> Compute cost J=
>
> $$\frac{1}{m}\sum \mathcal{L}(\hat{y}^{(i)}, y^{(i)}) + \frac{\lambda}{2m}\sum (\||w^{[l]}\||)^2_f$$
>
> Backward propagation on J^t;
>
> $w^{[l]} = w^{[l]} - \alpha\nabla w^{[l]}$;
>
> $f^{[l]} = f^{[l]} \times 10^{log(min(w^{[l]}))}$;
>
> $w^{[l]} = w^{[l]} + f^{[l]}$;
>
> $b^{[l]} = b^{[l]} - \alpha\nabla b^{[l]}$;

end

Figure 7-15. Complete algorithm/pseudocode for XBNet's boosted gradient descent from Tushar Sarkar

The author mentioned that having every layer to be "boosted" by XGBoost is not always helpful. In most cases, having a few hidden layers that are not boosted by XGBoost can achieve optimal performance. Furthermore, to reduce training costs, the n_estimator parameter of all trained XGBoost models is typically fixed to 100. Another rule of thumb for designing the architecture of XBNet is to have fewer layers and neurons compared with vanilla ANNs and boost layers toward the front portion of the network instead of the back.

XBNet was benchmarked against some well-known tabular datasets; although specific hyperparameters weren't mentioned, in most cases, XBNet can edge the performance of XGBoost (Table 7-7).

Table 7-7. *Results on benchmarked datasets in comparison with XGBoost from Tushar Sarkar*

Table 4: Performance Comparision of XBNet and XGBoost

Dataset	XBNet	XGBoost
Iris	100	97.7
Breast Cancer	96.49	96.47
Wine	97.22	97.22
Diabetes	78.78	77.48
Titanic	79.85	80.5
German Credit	71.33	77.66
Digit Completion	85.98	78.24

Although the official implementation for XBNet is written in PyTorch, one does not need any knowledge of PyTorch to utilize the implementation. The code can be found in the following GitHub repository: https://github.com/tusharsarkar3/XBNet. The library is currently not in PyPI; to install it, we need to directly download the relevant code from the Web using the following pip command: `pip install --upgrade git+https://github.com/tusharsarkar3/XBNet.git`. It's worthwhile to mention that it's impossible to adjust the hyperparameters of boosted trees within the official XBNet implementation. For those curious, every XGBoost model in XBNet is set to have 100 estimators, while every other parameter remains to be the default parameter set by the XGBoost library.

Once installed, we can import a few things to help us train and instantiate the model and predict (Listing 7-28).

Listing 7-28. Imports needed for XBNet

```
from XBNet.training_utils import training,predict
from XBNet.models import XBNETClassifier
from XBNet.run import run_XBNET
```

Since the library is still based in PyTorch, we also need to install PyTorch and import Torch as shown in Listing 7-29.

Listing 7-29. Installing and importing PyTorch

```
!pip install torch
import torch
```

The library contains separate models for classification and regression. Both model classes work the same way, the only difference being the prediction task type. We will use the Iris Flower dataset to train an XBNet model (Listing 7-30) for demonstration purposes.

Listing 7-30. Loading the dataset from sklearn and performing train-test split

```
# example dataset using sklearn's iris flower
# dataset
from sklearn.datasets import load_iris
raw_data = load_iris()
X, y = raw_data["data"], raw_data["target"]

from sklearn.model_selection import train_test_split
X_train, X_test, y_train, y_test = train_test_split(X, y, test_size=0.25)
```

The first two parameters passed in during instantiation are the X and y data. Next, we can specify the number of layers and the number of boosted layers. As mentioned previously, the library does not provide much flexibility in the boosted tree itself. The instantiation does use a command prompt window that asks for the number of neurons in each layer and the activation function of the output layer. Again, there is a lack of flexibility in the customization of the network, but it provides easy usage without having to learn the syntax of PyTorch (Listing 7-31).

Listing 7-31. Instantiating an XBNet classifier

```
xbnet_model = XBNETClassifier(X_train,y_train, num_layers=3, num_layers_boosted=2,)
```

To train the model, we need to define our loss function and optimizer using PyTorch syntax. All the options for predefined losses in PyTorch can be found under torch.nn or through their official documentation: https://pytorch.org/docs/stable/nn.html#loss-functions. Optimizers are found under torch.optim or on PyTorch's optimizer documentation for a full list: https://pytorch.org/docs/stable/optim.html. In the following code, the cross-entropy loss is used along with the Adam optimizer with the suggested learning rate of 0.01 from the original paper (Listing 7-32).

Listing 7-32. Defining loss function and the optimizer

```
criterion = torch.nn.CrossEntropyLoss()
optimizer = torch.optim.Adam(xbnet_model.parameters(), lr=0.01)
```

To begin training, we use the function from XBNet.run called run_XBNet. This function takes in ten parameters in the following order: features (X_train), validation features (X_test), targets (y_train), validation targets (y_test), the model object, the criterion or the loss function, the optimizer, the batch size (defaults to 16), the number of epochs, and whether to save the model after training. The run_XBNet returns the model object itself, the training accuracy and loss, as well as the validation accuracy and loss in that order. In Listing 7-33 is an example of running an XBNet model.

Listing 7-33. Training XBNet

```
xbnet_model, accuracy, loss, val_acc, val_loss = run_XBNET(X_train,X_test,y_train,y_test,
                                            xbnet_model,criterion,
                                            optimizer,batch_size=16,
                                            epochs=100, save=False)
```

The XBNetRegressor can be used the exact same way. For inference, we call the predict function that we imported previously from training_utils and pass it in our model along with the features we want predictions from (Listing 7-34).

Listing 7-34. Inference using XBNet

```
predict(xbnet_model, X_test)
```

Overall, XBNet is relatively simple in structure and has a minimal modification to standard ANNs. Still, its implication and the goals that it attempts to achieve provide itself a huge advantage over many other methods. By combining the training and understanding power of ANNs and XGBoost, networks trained on tabular data can obtain insights from both models.

Distillation

Model, or knowledge, distillation describes the process of transferring the learning from one model to another. Usually, distillation is used for downscaling large models to suit the situation better. The following paper proposes a distillation-like approach to train deep learning–backed models for tabular data.

DeepGBM

In recent years, as the popularity of machine learning has increased, the need for diverse model structures that situate each type of scenario increased as well. Algorithms introduced beforehand in this chapter only considered the tabular data category as a whole, ignoring potentially different types of structured data present within. Sparse categorical data, or categorical data with mostly zeros, has proven to be challenging for traditional GBMs and deep learning methods. For most Gradient Boosting approaches (except CatBoost), inputting sparse categorical data that was one-hot encoded yields minimal Information Gain during node splits, thus potentially missing crucial knowledge that these features might contribute to predicting the target.

DeepGBM, proposed by Guolin Ke, Zhenhui Xu, Jia Zhang, Jiang Bian, and Tie-Yan Liu from the Microsoft Research team, can handle a mixture of sparse categorical data and non-sparse data while combining the power of Gradient Boosting and deep learning through distillation.[14] Although not the focus of this book, DeepGBM has the ability to perform online learning by updating model parameters in real time. A great deal of tabular data learning is involved in business forecasting, predicting trends, and diagnosing in the medical industry, all of which can potentially involve online learning. GBM's greedily based learning approach makes it impossible to update its parameters during real-time forecasting continuously. Although it may be an excellent suit for tabular data in general, GBM still lacks usability in real-world applications.

DeepGBM has two main components, CatNN and GBDT2NN, which deal with sparse categorical and dense numerical data. CatNN is a neural network–based model that learns sparse data effectively by embeddings. Trained embeddings can convert high-dimensional sparse vectors into densely packed numerical data, scaling down the difficulty for the model. Along with embeddings, the authors proposed to use FMs, or Factorization Machines. FMs are commonly seen in recommendation systems to determine n-way feature interactions. FMs can handle high-dimensional sparse data, reducing the computational cost from a polynomial scale to a linear one. Using notations defined in the paper, we can obtain the ith feature's embedding as such:

$$E_{V_i}(x_i) = \text{embedding\_lookup}(V_i, x_i)$$

---

[14] Ke, G., Xu, Z., Zhang, J., Bian, J., & Liu, T. (2019). DeepGBM: A Deep Learning Framework Distilled by GBDT for Online Prediction Tasks. *Proceedings of the 25th ACM SIGKDD International Conference on Knowledge Discovery & Data Mining.*

where x_i is the value of the ith feature, while V_i stores all embeddings of x_i. Similar to most embeddings used throughout neural networks, it can be learned by backpropagation to produce an accurate depiction of the sparse feature. Once we obtain dense numerical representation of our sparse features, the FM component can then learn the linear (order 1) and pairwise (order 2) feature interactions as such:

$$y_{FM}(x) = w_0 + \langle w, x \rangle + \sum_{i=1}^{d} \sum_{j=i+1}^{d} \left\langle E_{V_i}(x_i), E_{V_j}(x_j) \right\rangle x_i x_j$$

where $\langle \cdot, \cdot \rangle$ denotes the dot product. The global bias w_0 and weight w can be optimized using common methods such as SGD or the Adam optimizer. FMs are able to learn lower-order feature interactions efficiently, but for higher-order feature relations, a multilayer NN is adapted and can be described by the following equation:

$$y_{Deep}(x) = \mathcal{N}\left(\left[E_{V_1}(x_1)^T, E_{V_2}(x_2)^T, \ldots, E_{V_d}(x_2)^T \right]^T; \theta \right)$$

where $\mathcal{N}(x; \theta)$ is a multilayer NN with parameters θ and input x. We can denote the number of features as d and the number of samples as n. Notice that each extracted embedding of dimension $1 \times n$ is transposed to a column vector of size $n \times 1$, stacked together horizontally to produce a matrix of size $n \times d$, and then transposed again to a matrix of size $d \times n$. Through the series of transpose operations, embedding dimensions were corrected so that the network input would be of the appropriate size. Figure 7-16 visualizes the $y_{Deep}(x)$ component of CatNN (the tedious process of transposing matrices is omitted for clarity).

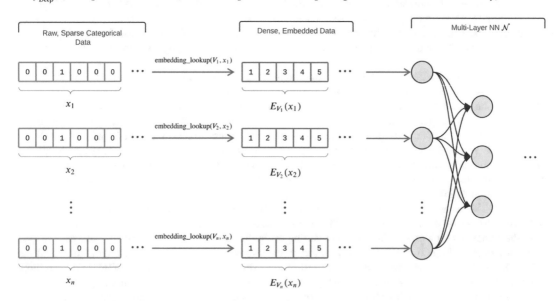

Figure 7-16. *The yDeep(x) component of CatNN*

The outputs of both components are combined to produce the final prediction of CatNN:

$$y_{Cat}(x) = y_{FM}(x) + y_{Deep}(x)$$

The second part of DeepGBM, GBDT2NN, utilizes distillation to transfer knowledge between tree-based models and neural networks. Typical distillation between models refers to transferring knowledge or learned relations from one model to another. Distillation is commonly used for scaling down model size while preserving near-exact performance compared to the larger model. Tree-based models and neural networks are inherently different models. Thus, standard distillation techniques cannot apply here.

Decision trees can be interpreted as clustering functions since they partition the data into different clusters where the number of clusters is equal to the tree's leaf nodes. An arbitrary clustering function can be defined for any decision tree where the function produces an index for the cluster (leaf node) that the input belongs to. Since neural networks can theoretically approximate any function (refer to the Universal Approximation Theorem in Chapter 3), we can, in turn, let NNs approximate the structural function of a tree-based model. Moreover, tree-based models naturally perform feature selection during training as not all features may be used for splitting; the model will scrap those with less Information Gain. Thus, our distilled network will only use features that the tree-based model selects.

From here, we can define some notations:

- Denote $\mathbb{I}^t$ as the indices used by the tree-based model t. We can then deduce the input features for the network: $x\left[\mathbb{I}^t\right]$ where x is the input data containing the entire set of features.

- Denote the structural function of a trained tree-like model as $Ct(x)$ in which the output returns the leaf index to which the input belongs. Note that the output can also be interpreted as the cluster index that the input fits in.

- Denote the network that approximates $Ct(x)$ as $\mathcal{N}(x;\theta)$.

- Denote the output from $Ct(x)$, or the one-hot encoded leaf/cluster indices, as $L^{t,i}$.

- Denote the loss function for the network as $\mathcal{L}'$.

The overall distillation process from the tree-based model to the network can be written as such:

$$\min_{\theta}\frac{1}{n}\sum_{i=1}^{n}\mathcal{L}'\left(\mathcal{N}\left(x^i\left[\mathbb{I}^t\right];\theta\right),L^{t,i}\right)$$

From the one-hot encoded output indicating the leaf index that the input will end up on, we can map the output to the corresponding leaf value (i.e., in classification, the corresponding class of that leaf) from the actual tree-based model. Mathematically, we can represent this as $L^t \times q^t$ where q^t is a vector containing all the leaf node values from the trained tree model. Rephrased, the ith value of q^t is the value of the ith leaf node in the trained tree model t. Finally, the output of the distilled network can be represented by the following product:

$$y^t(x)=\mathcal{N}\left(x^i\left[\mathbb{I}^t\right];\theta\right)\times q^t$$

Shown in Figure 7-17 is a representation of the distillation process described previously. Note that up until now, only a single tree has been distilled into the network; thus, it cannot be called GBDT2NN yet.

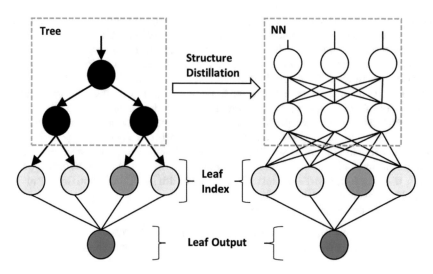

Figure 7-17. *Representation of distilling a single tree into a network. From G. Ke et al.*

The number of trees in GBDT can often be up to the hundreds. To map such a number of decision trees to neural networks will not only be difficult due to the sheer number of NNs trained but also because of the size of L^t. The authors propose two solutions for reducing computational and time complexity when mapping GBDT to NNs: leaf embedding distillation for reducing the size of L^t and tree grouping, reducing the number of NNs trained.

Embedding mechanisms are used again to reduce the dimension of the one-hot encoded leaf index while still retaining the information contained within. By utilizing the bijection relationship between the leaf indices and the actual leaf values, the learned embedding can be mapped to the desired output that corresponds with the original leaf index. More specifically, the conversion from one-hot encoded leaf indices L^t to dense embedding representation H^t by a one-layer fully connected network can be written as $H^{t,i} = \mathcal{H}\left(L^{t,i};\omega^t\right)$ where ω^t is the network parameter. Subsequently, the mapping from the embedding to actual leaf values can be denoted as

$$\min_{w,w_0,\omega^t} \frac{1}{n}\sum_{i=1}^{n}\mathcal{L}''\left(w^T\mathcal{H}\left(L^{t,i};\omega^t\right)+w_0,p^{t,i}\right)$$

where $\mathcal{L}''$ is the same loss function used in the tree model and $p^{t,i}$ is the actual leaf node value for input x^i. The weight term w^T and the bias term w_0 are trained to map the embedding $H^{t,i}$ to $p^{t,i}$. Since we reduced the dimensionality of the sparse one-hot encoded representation L^t, during the distillation learning process, it can be replaced with H^t:

$$\min_{\theta} \frac{1}{n}\sum_{i=1}^{n}\mathcal{L}'\left(\mathcal{N}\left(x^i\left[\mathbb{I}^t\right];\theta\right),H^{t,i}\right)$$

The output for distillation from a single decision tree to a neural network can be described by the following equation:

$$y(x)=w^T \times \mathcal{N}\left(x^i\left[\mathbb{I}^t\right];\theta\right)+w_0$$

The following visual shown in Figure 7-18 strings all the components discussed for embedding distillation from a single decision tree to a neural network. The index superscript i is omitted from the diagram since it only shows the distillation process for one sample.

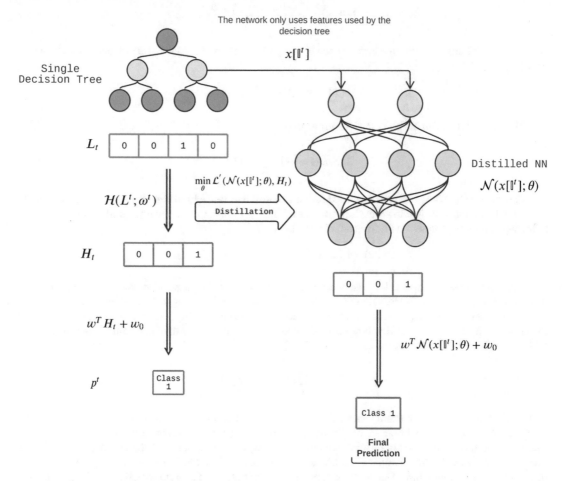

Figure 7-18. *Distillation from a single tree by embeddings into a neural network*

To reduce the number of neural networks, the authors propose to randomly group multiple trees and distill their learning into one neural network. The learning process for converting leaf indices to embeddings can also be extended to process multiple trees' outputs at once. The concatenation operation is imposed between every one-hot encoded vector produced by the group of trees. Trees from the GBDT model are randomly sectioned into k groups with every group having $s = \lceil m/k \rceil$ trees where m denotes the total number of trees present in the GBDT. We denote the entire group of s total trees as $\mathbb{T}$. Intuitively, we can modify the equation for learning the embeddings from one-hot encoded vectors to learn from multi-hot encoded vectors produced from s trees:

$$\min_{w,w_0,\omega^{\mathbb{T}}} \frac{1}{n} \sum_{i=1}^{n} \mathcal{L}'' \left(w^T \mathcal{H} \left(\|_{t \in \mathbb{T}} \left(L^{t,i} \right); \omega^{\mathbb{T}} \right) + w_0, \sum_{t \in \mathbb{T}} p^{t,i} \right)$$

where $\|(\cdot)\|$ denotes the concatenate operation. The embedding generated from $\mathcal{H}\left(\|_{t\in\mathbb{T}}\left(L^{t,i}\right);\omega^{\mathbb{T}}\right)$, $G^{\mathbb{T},i}$, can replace $H^{t,i}$ in the distillation equation for mapping multiple trees to one single network:

$$\min_{\theta^{\mathbb{T}}}\frac{1}{n}\sum_{i=1}^{n}\mathcal{L}'\left(\mathcal{N}\left(x^{i}\left[\mathbb{I}^{\mathbb{T}}\right];\theta\right),G^{\mathbb{T},i}\right)$$

where $\mathbb{I}^{\mathbb{T}}$ represents every unique feature used by the group of trees $\mathbb{T}$. Finally, the prediction for i^{th} tree group $\mathbb{T}_{i}$ is

$$y_{\mathbb{T}_{i}}(x)=w^{T}\times\mathcal{N}\left(x^{i}\left[\mathbb{I}^{t}\right];\theta\right)+w_{0}$$

Then the final output of GBDT2NN becomes

$$y_{GBDT2NN}(x)=\sum_{j=1}^{k}y_{\mathbb{T}_{j}}(x)$$

CatNN and GBDT2NN can be combined for end-to-end learning where the output of each component is assigned a trainable weight to adjust how much each model contributes to the final prediction. The following equation denotes the final output of DeepGBM:

$$\hat{y}(x)=\sigma'\left(w_{1}\times y_{GBDT2NN}(x)\right)+w_{2}\times y_{CatNN}(x)$$

where σ' can be taken as the last-layer activation of a neural network (sigmoid for binary classification, linear for regression, etc.). The loss function used for the end-to-end training of DeepGBM is a mixture of the loss for the current task (classification or regression) and the embedding loss for tree groups with two hyperparameters α and β specifying how much each loss contributes to the final loss value. The combined loss of DeepGBM can be computed as

$$\mathcal{L}=\alpha\mathcal{L}''\left(\hat{y}(x),y\right)+\beta\sum_{j=1}^{k}\mathcal{L}^{\mathbb{T}_{j}}$$

where $\mathcal{L}''$ is the loss for the current task and $\mathcal{L}^{\mathbb{T}_{j}}$ denotes the embedding loss for the tree group $\mathbb{T}_{j}$. Although DeepGBM is also designed for effective online learning, since this is outside the scope of the current section, only comparison results from offline training will be presented. A total of six models – GBDT, Logistic Regression, FM with a linear model, Wide and Deep, DeepFM, and PNN – are compared against three variations of DeepGBM – DeepGBM where GBDT2NN is replaced by a standard GBDT (D1 in the table), DeepGBM without CatNN (D2 in the table), and finally the full version of DeepGBM (Table 7-8).

Table 7-8. *Training results of DeepGBM compared with other models*

Model	Binary Classification						Regression
	Flight	Criteo	Malware	AutoML-1	AutoML-2	AutoML-3	Zillow
LR	0.7234 ±5e-4	0.7839 ±7e-5	0.7048 ±1e-4	0.7278 ±2e-3	0.6524 ±2e-3	0.7366 ±2e-3	0.02268 ±1e-4
FM	0.7381 ±3e-4	0.7875 ±1e-4	0.7147 ±3e-4	0.7310 ±1e-3	0.6546 ±2e-3	0.7425 ±1e-3	0.02315±2e-4
Wide&Deep	0.7353 ±3e-3	0.7962 ±3e-4	0.7339 ±7e-4	0.7409 ±1e-3	0.6615 ±1e-3	0.7503 ±2e-3	0.02304 ±3e-4
DeepFM	0.7469 ±2e-3	0.7932 ±1e-4	0.7307 ±4e-4	0.7400 ±1e-3	0.6577 ±2e-3	0.7482 ±2e-3	0.02346 ±2e-4
PNN	0.7356 ±2e-3	0.7946 ±8e-4	0.7232 ±6e-4	0.7350 ±1e-3	0.6604 ±2e-3	0.7418 ±1e-3	0.02207 ±2e-5
GBDT	0.7605 ±1e-3	0.7982 ±5e-5	0.7374 ±2e-4	0.7525 ±2e-4	0.6844 ±1e-3	0.7644 ±9e-4	0.02193 ±2e-5
DeepGBM (D1)	0.7668 ±5e-4	**0.8038 ±3e-4**	0.7390 ±9e-5	0.7538 ±2e-4	0.6865 ±4e-4	**0.7663 ±3e-4**	0.02204 ±5e-5
DeepGBM (D2)	**0.7816 ±5e-4**	0.8006 ±3e-4	**0.7426 ±5e-5**	**0.7557 ±2e-4**	0.6873 ±3e-4	0.7655 ±2e-4	**0.02190 ±2e-5**
DeepGBM	**0.7943 ±2e-3**	0.8039 ±3e-4	0.7434 ±2e-4	0.7564 ±1e-4	0.6877 ±8e-4	0.7664 ±5e-4	**0.02183 ±3e-5**

One explanation for the dominating performance of DeepGBM other than its unique training method and structure can be reasoned from its multi-model nature. DeepGBM can be considered an ensemble algorithm with the combined power of CatNN and GBDT2NN.

To use DeepGBM, one can clone the official repository from GitHub: `https://github.com/motefly/DeepGBM`. In order to train DeepGBM, three main folders must be present within the project with the data folder containing the dataset, the preprocess folder containing helper functions for feature encoding and manipulation, and the model folder containing code for the actual DeepGBM model. The last two folders are already included in the repository, while the data folder requires the user to create themselves. Before running the model, all data must go through preprocessing by encoders defined in the preprocess folder and then converted into `.npy` format. Inside the `main.py` file, data will be loaded by `dh.load_data` with the pointer to the desired dataset specified through `argparse`. Finally, training can be instantiated by `train_DEEPGBM` where arguments are passed in through `argpase`'s `parse_args()`.

For more detailed training instruction, refer to the repository as the training process may be different depending on the dataset and the entire operation spans across multiple files.

Key Points

In this chapter, we discussed several tree-based deep learning models.

- Tree-structured deep learning approaches attempt to incorporate deep learning components to tree-based models. The following approaches either propose some form of differentiable trees or tree-like neural networks.

  - Deep neural decision trees take on the task of training tree-based models through backpropagation by utilizing soft binning functions and brute-forcing the final leaf nodes by the Kronecker product. The model works well with datasets of smaller dimensions but gets significantly more expensive as the number of features present in the dataset increases.

  - Soft decision tree regressors imitate a decision tree with "soft" probabilistic decision nodes learned by fully connected layers attached to the input. The model is trained to optimize the loss between the true value and each leaf node output multiplied by the probability series leading to said node.

  - Neural oblivious decision ensemble models use ensembles of neural oblivious trees, in which the condition at each level is the same.

  - Tree-based neural network initialization maps decision tree structure to a neural network while initializing its weights based on the mapping.

  - Decision trees can be represented as logical statements in disjunctive normal form (DNF), with node conditions as variables. DNF-Net imitates a DNF expression using soft OR and AND gate analogs.

- Boosting and stacking networks utilize the unique yet effective structure of a Gradient Boosting Machine and attempt to apply it to ANNs.

  - GrowNet constructs and trains an ensemble of neural networks in Gradient Boosting style.

  - XBNet mimics the training techniques of XGBoost and adapts it to neural networks.

- Distillation approaches transfer knowledge from one model to another. In the current context with DeepGBM as the only approach under this category, distillation refers to distilling from a Gradient Boosting Machine to neural networks.

  - DeepGBM involves several components, which can process sparse categorical features along with dense numerical features. It has the ability to perform online learning while incorporating elements from GBDT by distillation.

This marks the last chapter in Part 2. We have developed an expansive knowledge of deep learning in this part. In the final part, we will explore different methods to supercharge deep learning modeling. In the next chapter, we will discuss how autoencoder architectures can be utilized to pretrain, speed up, "sparsify," and denoise neural networks.

## PART III

Deep Learning Design and Tools

CHAPTER 8

■ ■ ■

Autoencoders

Weak encoding means mistakes and weak decoding means illiteracy.

—Rajesh Walecha, Author

An autoencoder is a very simple model: a model that predicts its own input. In fact, it may seem deceivingly simple to the point of being worthless. (After all, what use is a model that predicts what we already know?) Autoencoders are extraordinarily valuable and versatile architectures not because of their functionality to reproduce the input, but because of the internal capabilities developed in order to obtain said functionality. Autoencoders can be chopped up into desirable parts and stuck onto other neural networks, like playing with Legos or performing surgery (take your pick of analogy), with incredible success, or can be used to perform other useful tasks, such as denoising.

This chapter begins by explaining the intuition of the autoencoder concept, followed by a demonstration of how one would implement a simple "vanilla" autoencoder. Afterward, four applications of autoencoders – pretraining, denoising, sparse learning for robust representations, and denoising – are discussed and implemented.

The Concept of the Autoencoder

The operations of encoding and decoding are fundamental to information. Some hypothesize that all transformation and evolution of information results from these two abstracted actions of encoding and decoding (Figures 8-1 and 8-2). Say Alice sees Humpty Dumpty hit his head on the ground after some precarious wall-sitting and tells Bob, "Humpty Dumpty hit his head badly on the ground!" Upon hearing this information, Bob *encodes* the information from a language representation into thoughts and opinions – what we might call *latent* representations.

Say Bob is a chef, and so his encoding "specializes" in features relating to food. Bob then *decodes* the latent representations back into a language representation when he tells Carol, "Humpty Dumpty cracked his shell! We can use the innards to make an omelet." Carol, in turn, encodes the information.

Say Carol is an egg activist and cares deeply about the well-being of Humpty Dumpty. Her latent representations will encode the information in a way that reflects her priorities and interests as a thinker. When she decodes her latent representations into language, she tells Drew that "People are trying to eat Humpty Dumpty after he has suffered a serious injury! It is horrible."

So on and so forth. The conversation continues and evolves, information passed and transformed from thinker to thinker. Because each thinker *encodes* the information represented in language in a semantic system relevant to their experiences, priorities, and interests, they correspondingly *decode* information in a fashion colored through these lens.

© Andre Ye and Zian Wang 2023
A. Ye and Z. Wang, *Modern Deep Learning for Tabular Data*,
https://doi.org/10.1007/978-1-4842-8692-0_8

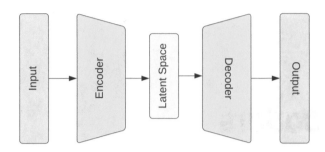

Figure 8-1. *A high-level autoencoder architecture*

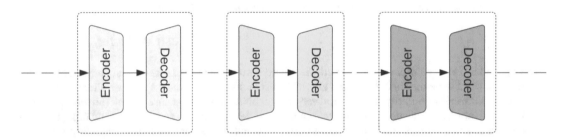

Figure 8-2. *Transformation of information as a series of encoding and decoding operations*

Of course, this interpretation of encoding and decoding is very broad and more psychological than anything else. In the strict context of computer science, encoding is an operation to represent some information in another form, usually with smaller information content (there are few applications for encoding techniques that make the storage size larger). Decoding, in turn, "undoes" the encoding operation to recover the original information. Encoding and decoding are commonly used terms in the context of compression (Figure 8-3). Various computer scientists throughout the decades have proposed very clever algorithms to map information to smaller storage sizes with lossless and lossy reconstruction of the original information, making the transmission of large data like long text, images, and videos across limited information transfer connections feasible.

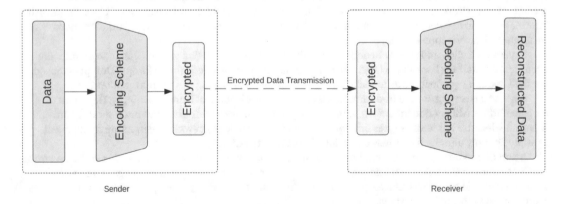

Figure 8-3. *Interpretation of autoencoders as sending and receiving encrypted data*

Encoding and decoding in deep learning are a bit of a fusion of these two understandings. *Autoencoders* are versatile neural network structures consisting of an encoder and a decoder component. The encoder maps the input into a smaller latent/encoded space, and the decoder maps the encoded representation back to the original input. The goal of the autoencoder is to reconstruct the original input as faithfully as possible, that is, to minimize the *reconstruction loss*. In order to do so, the encoder and decoder need to "work together" to develop an encoding and decoding scheme.

Autoencoders reduce the representation size of the original input information, which can be thought of as a form of lossy compression. However, numerous studies have demonstrated that autoencoders are generally quite bad at compression when put into contrast with human-designed compression schemes. Rather, when we build autoencoders, it is almost always to *extract meaningful features at the "core" of the data*. The smaller representation size of the latent space compared with the original input emerges only because we need to impose an information bottleneck to force the network to learn meaningful latent features. The architectures in Figures 8-4 and 8-5 demonstrate constant and expanded information representations compared with the input; the autoencoder can trivially learn weights that simply pass/carry the input from the input to the output. On the other hand, the architecture in Figure 8-6 must learn nontrivial patterns to compress and reconstruct the original input. The information bottleneck and information compression, therefore, is the means, not the end, of the autoencoder.

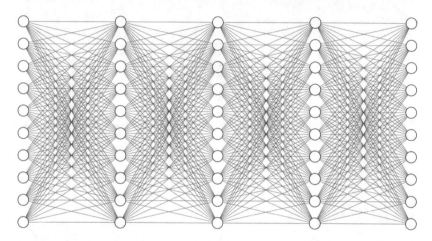

Figure 8-4. *A bad autoencoder architecture (latent space representation size is equal to input representation size)*

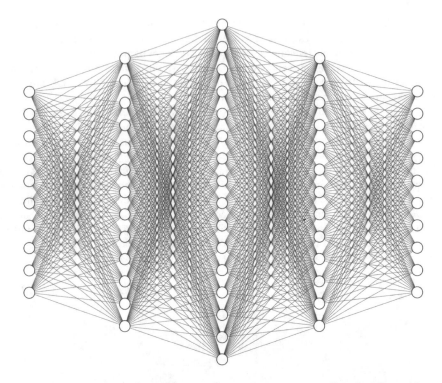

Figure 8-5. *An even worse autoencoder architecture (latent space representation size is larger than input representation size)*

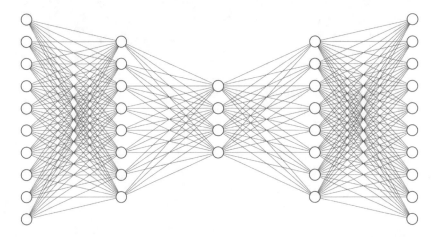

Figure 8-6. *A good autoencoder architecture (latent space representation size is smaller than input representation size)*

Autoencoders are very good at finding higher-level abstractions and features. In order to reliably reconstruct the original input from a much smaller latent space representation, the encoder and decoder must develop a system of mapping that most meaningfully characterizes and distinguishes each input or set of inputs from others. This is no trivial task!

Consider the following autoencoder design scheme adapted for humans (Figure 8-7): person A is the encoder and attempts to "encode" a high-resolution image of a sketch as a natural language description, restricted to N words or less; person B is the decoder and attempts to "decode" the original image person A was looking at by drawing a reconstructed image based on person A's natural language description. Person A and person B must work together to develop a system to reliably reconstruct the original image.

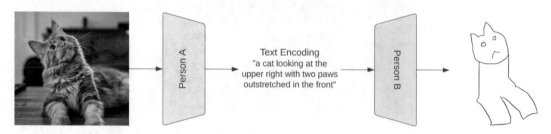

Figure 8-7. Image-to-text encoding guessing game

Say that you are person B and you are given the following natural language description by person A: "a black pug dressed in a black and white scarf looks at the upper-left region of the camera among an orange background." For the sake of intuition, it is a worthwhile exercise to try actually playing the role of person B in this game by sketching out/"decoding" the original input.

Figure 8-8 shows the (hypothetical) image that person A encoded into the given natural language description. Chances are that your sketch is very different from the actual image. By performing this exercise, you will have experienced first-hand two key low-level challenges in autoencoding: reconstructing a complex output from a comparatively simpler encoding requires a lot of thinking and conceptual reasoning about the encoding, and the encoding scheme itself needs to effectively communicate both key concepts and precision/positioning information.

Figure 8-8. *What person A was hypothetically looking at when they provided you the natural language encoding. Taken by Charles Deluvio*

In this example, the latent space is in the form of language – which is discrete, sequential, and variable-length. Most autoencoders for tabular data use latent spaces that satisfy none of these attributes: they are (quasi-) continuous, read and generated all at once rather than sequentially, and fixed-length. These general autoencoders can reliably find effective encoding and decoding schemes with lifted restrictions, but the two-player game is still good intuition for thinking through challenges associated with autoencoder training.

Although autoencoders are relatively simple neural network architectures, they are incredibly versatile. In this section, we will begin with the plain "vanilla" autoencoder and move to more complex forms and applications of autoencoders.

Vanilla Autoencoders

Let's begin with the traditional understanding of an autoencoder, which merely consists of an encoder and a decoder component working together to translate the input into a latent representation and then back into original form. The value of autoencoders will become clearer in following sections, in which we will use autoencoders to substantively improve model training.

The goal of this subsection is not only to demonstrate and implement autoencoder architectures but also to understand implementation best practices and to perform technical investigations and explorations into how and why autoencoders work.

Autoencoders are traditionally applied to image and text-based datasets, because this sort of data often features semantic concepts that should take a smaller amount of space to represent than is used in raw form. For instance, consider the following approximately 3000-by-3000 pixel image of a line (Figure 8-9).

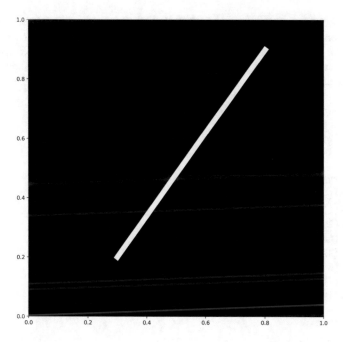

Figure 8-9. *An image of a line*

This image contains nine million pixels, meaning we are representing the concept of this line with nine million data values. However, in actuality we can express any line with just four numbers: a slope, a *y*-intercept, a lower *x* bound, and a higher *x* bound (or a starting *x* point, a starting *y* point, an ending *x* point, and an ending *y* point). If we were to design an encoding and decoding scheme set, the encoder would identify these four parameters – yielding a very compact four-dimensional latent space – and the decoder would redraw the line given those four parameters. By collecting higher-level abstract *latent* features from the semantics represented in the images, we are able to represent the dataset more compactly. We'll revisit this example later in the subsection.

Notice, however, that the autoencoder's reconstruction capability is conditional on the existence of structural similarities (and differences) within the dataset. An autoencoder cannot reliably reconstruct an image of random noise, for instance.

The MNIST dataset is a particularly useful demonstration of autoencoders. It is technically visual/image-based, which is useful for understanding various autoencoder forms and applications (given that autoencoders are most well developed for images). However, it spans a small enough number of features and is structurally simple enough such that we can model it without any convolutional layers. Thus, the MNIST dataset serves as a nice link between the image and tabular data worlds. Throughout this section, we'll use the MNIST dataset as an introduction to autoencoder techniques before demonstrating applications to "real" tabular/structured datasets.

Let's begin by loading the MNIST dataset from Keras datasets (Listing 8-1).

Listing 8-1. Loading the MNIST dataset

```
from keras.datasets.mnist import load_data
(x_train, y_train), (x_valid, y_valid) = load_data()
x_train = x_train.reshape(len(x_train),784)/255
x_valid = x_valid.reshape(len(x_valid),784)/255
```

Recall that the key feature of an autoencoder is an information bottleneck. We want to begin from the original representation size, progressively force the information flow into smaller vector sizes, and then progressively force the information back into the original size. Such a design is simple to quickly implement in Keras, where we can successively decrease and increase the number of nodes in a sequence of fully connected layers (Listing 8-2).

Listing 8-2. Building an autoencoder sequentially

```
import keras.layers as L
from keras.models import Sequential

# define architecture
model = Sequential()
model.add(L.Input((784,)))
model.add(L.Dense(256, activation='relu'))
model.add(L.Dense(64, activation='relu'))
model.add(L.Dense(32, activation='relu'))
model.add(L.Dense(64, activation='relu'))
model.add(L.Dense(256, activation='relu'))
model.add(L.Dense(784, activation='sigmoid'))

# compile
model.compile(optimizer='adam',
              loss='binary_crossentropy')

# fit
model.fit(x_train, x_train, epochs=1,
          validation_data=(x_valid, x_valid))
```

The architecture is visualized in Figure 8-10.

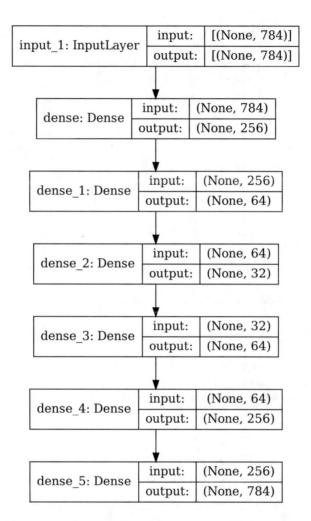

Figure 8-10. *A sequential autoencoder architecture*

There are a few features of this autoencoder architecture to note. Firstly, the output activation of the autoencoder is a sigmoid function, but this is only because the input vector has values ranging from 0 to 1 (recall that we scaled the dataset upon loading in Listing 8-1). If we had not scaled the dataset as such, we would need to change the activation function such that the network could feasibly predict in the entire domain of possible values. If the input values consist of values larger than 0, ReLU may be a good activation output choice. If the inputs contain both positive and negative values, using a plain linear activation may be the easiest possible option. Moreover, the loss function chosen must be reflective of the output activation. Since our particular example contains outputs between 0 and 1 and the distribution of values is more or less binary (i.e., most values are very close to 0 or 1, as shown in Figure 8-11), binary cross-entropy is a suitable loss to apply. We can treat reconstruction as a series of binary classification problems for each pixel in the original input.

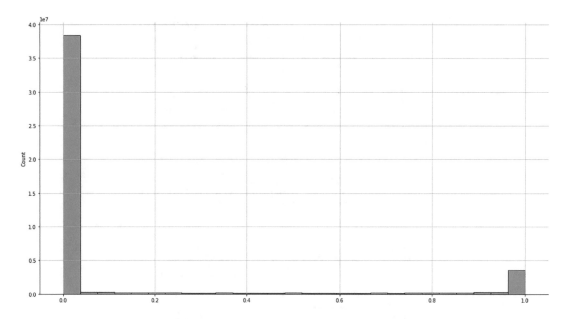

Figure 8-11. *Distribution of pixel values (scaled between 0 and 1) in the MNIST dataset*

However, in other cases, reconstruction is more of a regression problem in which the distribution of possible values is not binarized toward the ends of the domains but rather more spread out. This is common in more complex image datasets (Figure 8-12) and in many tabular datasets (Figure 8-13).

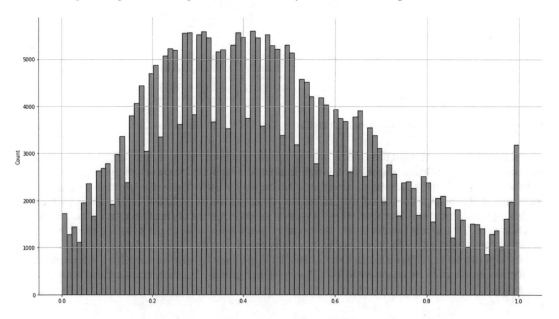

Figure 8-12. *Distribution of pixel values (scaled between 0 and 1) from a set of images in CIFAR-10*

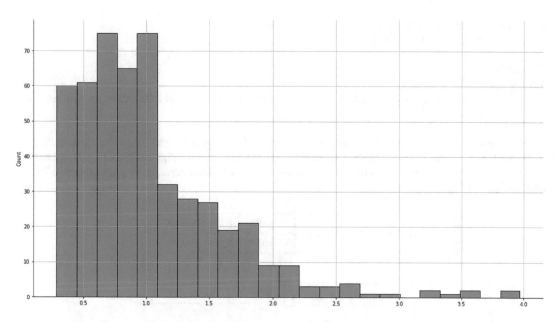

Figure 8-13. *Distribution of values for a feature in the Higgs Boson dataset (we will work with this dataset later in the chapter)*

In these cases, it is more suitable to use a regression loss, like the generic Mean Squared Error or a more specialized alternative (e.g., Huber). Refer to Chapter 1 for a review of regression losses.

Autoencoders are generally easier to work with when implemented in *compartmentalized form*. Rather than simply constructing the autoencoder as a continuous stack of layers with a bottleneck, we can build encoder and decoder models/components and chain them together to form a complete autoencoder (Listing 8-3).

Listing 8-3. Building an autoencoder with compartmentalized design

```
from keras.models import Model

# define architecture components
encoder = Sequential(name='encoder')
encoder.add(L.Input((784,)))
encoder.add(L.Dense(256, activation='relu'))
encoder.add(L.Dense(64, activation='relu'))
encoder.add(L.Dense(32, activation='relu'))

decoder = Sequential(name='decoder')
decoder.add(L.Input((32,)))
decoder.add(L.Dense(64, activation='relu'))
decoder.add(L.Dense(256, activation='relu'))
decoder.add(L.Dense(784, activation='sigmoid'))

# define model architecture from components
ae_input = L.Input((784,), name='input')
ae_encoder = encoder(ae_input)
ae_decoder = decoder(ae_encoder)
```

```
ae = Model(inputs = ae_input,
           outputs = ae_decoder)

# compile
ae.compile(optimizer='adam',
           loss='binary_crossentropy') # note that in other situations other losses may be
           more suitable
```

This method of construction is philosophically more desirable because it reflects our understanding of the autoencoder structure as meaningfully composed of a separate encoding and decoding component. When we visualize our architecture, we obtain a much cleaner high-level breakdown of the autoencoder model (Figure 8-14).

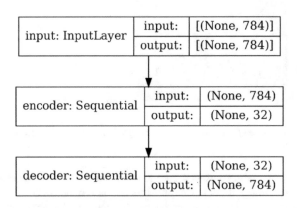

Figure 8-14. *Visualization of the compartmentalized model*

However, using compartmentalized design is incredibly helpful because we can reference the encoder and decoder components separately from the autoencoder. For instance, if we desire to obtain the encoded representation for an input, we can simply call encoder.predict(…) on our input. The encoder and decoder are used to build the autoencoder; after the autoencoder is trained, the encoder and decoder still exist as references to components of that (now trained) autoencoder. The alternative would be to go searching for the latent space layer of the model and create a temporary model to run predictions, in a similar approach to the demonstration in Chapter 4 used to visualize learned convolutional transformations in CNNs. Similarly, if we desire to decode a latent space vector, we can simply call decoder.predict(…) on our sample latent vector.

For instance, Listing 8-4 demonstrates visualization (Figures 8-15 through 8-18) of the internal state and reconstruction of the autoencoder created in Listing 8-3 after training.

Listing 8-4. Visualizing the input, latent space, and reconstruction of an autoencoder

```
for i in range(10):
    plt.figure(figsize=(10, 5), dpi=400)
    plt.subplot(1, 3, 1)
    plt.imshow(x_valid[i].reshape((28, 28)))
    plt.axis('off')
    plt.title('Original Input')
    plt.subplot(1, 3, 2)
    plt.imshow(encoder.predict(x_valid[i:i+1]).reshape((8, 4)))
    plt.axis('off')
```

```
plt.title('Latent Space (Reshaped)')
plt.subplot(1, 3, 3)
plt.imshow(ae.predict(x_valid[i:i+1]).reshape((28, 28)))
plt.axis('off')
plt.title('Reconstructed')
plt.show()
```

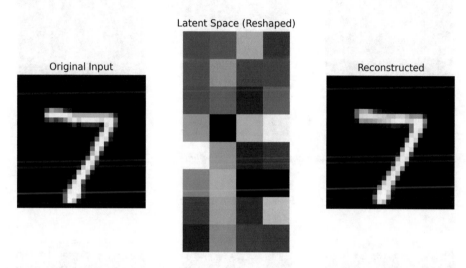

Figure 8-15. *Sample latent shape and reconstruction for the digit "7"*

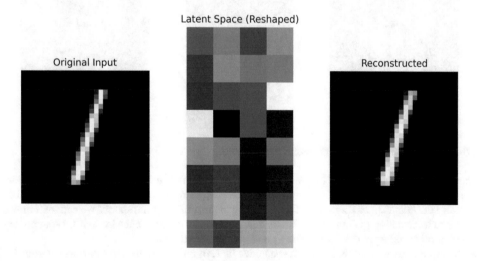

Figure 8-16. *Sample latent shape and reconstruction for the digit "1"*

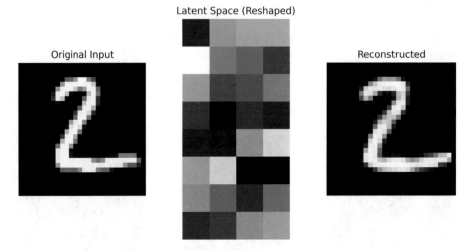

Figure 8-17. *Sample latent shape and reconstruction for the digit "2"*

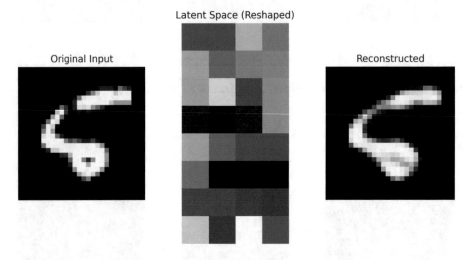

Figure 8-18. *Sample latent shape and reconstruction for the digit "5"*

When we build standard neural networks that we may want multiple models of with small differences, it is often useful to create a "builder" or "constructor." The two key parameters of a neural network are the input size and the latent space size. Given these two key "determining' parameters," we can infer how we generally want information to flow. For instance, halving the information space in each subsequent layer in the encoder (and doubling in the decoder) is a good generic update rule.

Let the input size be I, and let the latent space size be L. In order to maintain this rule, we want all intermediate layers to use nodes as multiples of L. Consider the case in which $I = 4L$, for instance (Figure 8-19).

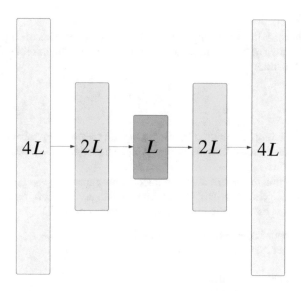

Figure 8-19. *Visualization of a "halving" autoencoder architecture logic*

We see that the number of layers needed to either reduce the input to the latent space or to expand the latent space to the output is

$$\log_2 \frac{I}{L}$$

This simple expression measures how many times we need to multiply L by 2 in order to reach I.

However, it will often be the case that $\frac{I}{L} \notin \mathbb{Z}$ (i.e., I does not divide cleanly into L), in which case our earlier logarithmic expression will not be integer. In these cases, we have a simple fix: we can cast the input to a layer with N nodes, where $N = 2^k \cdot L$ for the largest integer k such that $N < I$. For instance, if $I = 4L + 8$, we first "cast" down to $4L$ and execute our standard halving policy from that point (Figure 8-20).

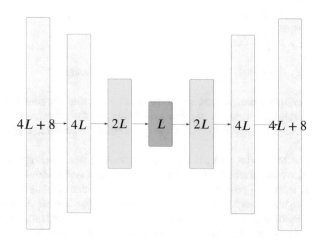

Figure 8-20. *Adapting the halving autoencoder logic to inputs that are not powers of 2*

To accommodate for cases in which $\log_2 \frac{I}{L} \notin \mathbb{Z}$ (i.e., we cannot express the input size in relationship to the layer size as an exponent of 2), we can modify our expression for the number of layers required by wrapping with the floor function:

$$\left\lfloor \log_2 \frac{I}{L} \right\rfloor$$

Using this halving/doubling information flow logic, we can create a generalized buildAutoencoder function that constructs a feed-forward autoencoder given an input size and a latent size (Listing 8-5).

Listing 8-5. A general function to construct an autoencoder architecture given an input size and a desired latent space, constructed using halving/doubling architectural logic. Note this implementation also has an outActivation parameter in cases where our output is not between 0 and 1

```
def buildAutoencoder(inputSize=784, latentSize=32,
                     outActivation='sigmoid'):

    # define architecture components
    encoder = Sequential(name='encoder')
    encoder.add(L.Input((inputSize,)))
    for i in range(int(np.floor(np.log2(inputSize/latentSize))), -1, -1):
        encoder.add(L.Dense(latentSize * 2**i, activation='relu'))

    decoder = Sequential(name='decoder')
    decoder.add(L.Input((latentSize,)))
    for i in range(1,int(np.floor(np.log2(inputSize/latentSize)))+1):
        decoder.add(L.Dense(latentSize * 2**i, activation='relu'))
    decoder.add(L.Dense(inputSize, activation=outActivation))

    # define model architecture from components
    ae_input = L.Input((inputSize,), name='input')
    ae_encoder = encoder(ae_input)
    ae_decoder = decoder(ae_encoder)
    ae = Model(inputs = ae_input,
               outputs = ae_decoder)

    return {'model': ae, 'encoder': encoder, 'decoder': decoder}
```

Rather than just returning the model, we also return the encoder and decoder. Recall from earlier discussion of compartmentalized design that retaining a reference to the encoder and decoder components of the autoencoder can be helpful. If not returned, these references – created internally inside the function – will be lost and irretrievable.

Having a generalized autoencoder creation function allows us to perform larger-scale autoencoder experiments. One particularly important phenomenon to understand is the trade-off between model performance and the latent size. As previously mentioned, the latent size must be configured properly such that the task is challenging enough to force the autoencoder to develop meaningful and nontrivial representations, but also feasible enough such that the autoencoder can gain traction at solving the problem (rather than stagnating and not learning anything at all due to the difficulty of the reconstruction problem). Let's train several autoencoders on the MNIST dataset with bottleneck sizes 2^n where $n \in [1, 2, ..., \lfloor \log_2 I \rfloor]$ (the last value of n being the largest power of 2 less than the original input size) and obtain each one's validation performance (Listing 8-6, Figure 8-21).

Listing 8-6. Training autoencoders with varying latent space sizes and observing the performance trend

```
inputSize = 784

earlyStopping = keras.callbacks.EarlyStopping(monitor='loss',
                                              patience=5)

latentSizes = list(range(1, int(np.floor(np.log2(inputSize)))))
validPerf = []
for latentSize in tqdm(latentSizes):
    model = buildAutoencoder(inputSize, 2**latentSize)['model']
    model.compile(optimizer='adam', loss='binary_crossentropy')
history = model.fit(x_train, x_train, epochs=50,
                    callbacks=[earlyStopping], verbose=0)
    score = keras.metrics.MeanAbsoluteError()
    score.update_state(model.predict(x_valid), x_valid)
    validPerf.append(score.result().numpy())

plt.figure(figsize=(15, 7.5), dpi=400)
plt.plot(latentSizes, validPerf, color='red')
plt.ylabel('Validation Performance')
plt.xlabel('Latent Size (power of 2)')
plt.grid()
plt.show()
```

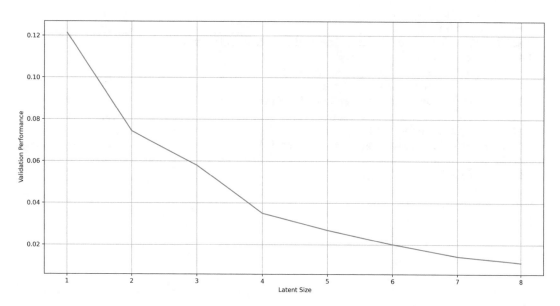

Figure 8-21. *Relationship between the latent size of a tabular autoencoder (2x neurons) and the validation performance. Note the diminishing returns*

The diminishing returns for larger latent sizes are very apparently clear. As the latent size increases, the benefit we can reap from it decreases. This phenomenon is true generally in deep learning models (recall "Deep Double Descent" from Chapter 1, which similarly compared model size vs. performance in a supervised domain with CNNs).

We can do one better and visualize the differences in the learned latent representations for different bottleneck sizes. The latent representations for the training set after the autoencoder has been trained can be obtained via encoder.predict(x_train). Of course, the latent representations will be in different dimensions for each autoencoder. We can use the t-SNE method (introduced in Chapter 2) to visualize these latent spaces (Listing 8-7, Figures 8-22 through 8-30).

Listing 8-7. Plotting a t-SNE representation of the latent space of autoencoders with varying latent space sizes

```
from sklearn.manifold import TSNE

inputSize = 784

earlyStopping = keras.callbacks.EarlyStopping(monitor='loss',
                                              patience=5)

latentSizes = list(range(1, int(np.floor(np.log2(inputSize))) + 1))

for latentSize in tqdm(latentSizes):

    modelSet = buildAutoencoder(inputSize, 2**latentSize)
    model = modelSet['model']
    encoder = modelSet['encoder']
    model.compile(optimizer='adam', loss='binary_crossentropy')
model.fit(x_train, x_train, epochs=50,
        callbacks=[earlyStopping], verbose=0)
    transformed = encoder.predict(x_train)
    tsne_ = TSNE(n_components=2).fit_transform(transformed)

    plt.figure(figsize=(10, 10), dpi=400)
    plt.scatter(tsne_[:,0], tsne_[:,1], c=y_train)
    plt.show()
    plt.close()
```

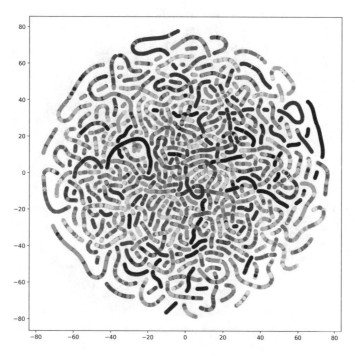

Figure 8-22. *t-SNE projection of a latent space for an autoencoder with a bottleneck size of two nodes trained on MNIST. Note that in this case, we are projecting into a number of dimensions (2) equal to the dimensionality of the original dataset (2), hence the pretty snake-like arrangements*

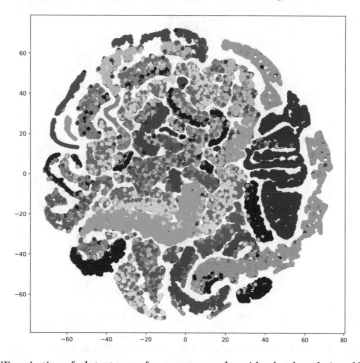

Figure 8-23. *t-SNE projection of a latent space for an autoencoder with a bottleneck size of four nodes trained on MNIST*

619

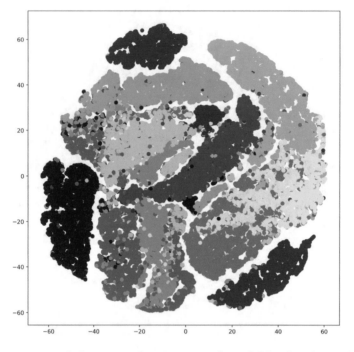

Figure 8-24. *t-SNE projection of a latent space for an autoencoder with a bottleneck size of eight nodes trained on MNIST*

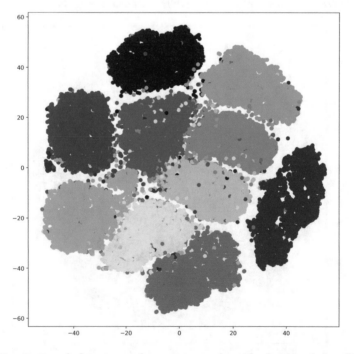

Figure 8-25. *t-SNE projection of a latent space for an autoencoder with a bottleneck size of 16 nodes trained on MNIST*

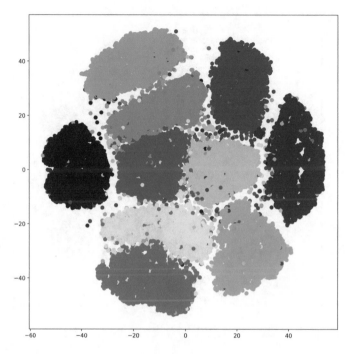

Figure 8-26. *t-SNE projection of a latent space for an autoencoder with a bottleneck size of 32 nodes trained on MNIST*

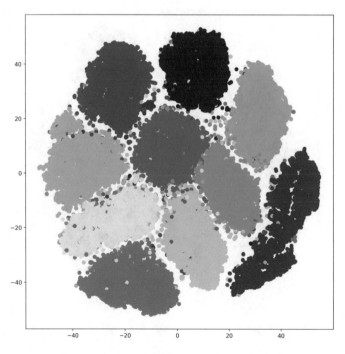

Figure 8-27. *t-SNE projection of a latent space for an autoencoder with a bottleneck size of 64 nodes trained on MNIST*

621

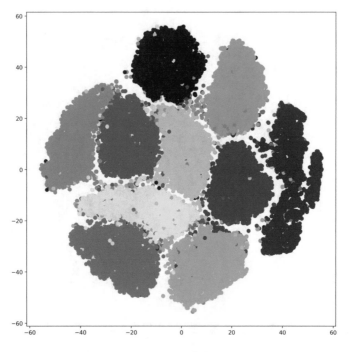

Figure 8-28. *t-SNE projection of a latent space for an autoencoder with a bottleneck size of 128 nodes trained on MNIST*

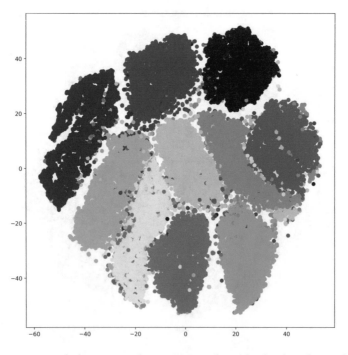

Figure 8-29. *t-SNE projection of a latent space for an autoencoder with a bottleneck size of 256 nodes trained on MNIST*

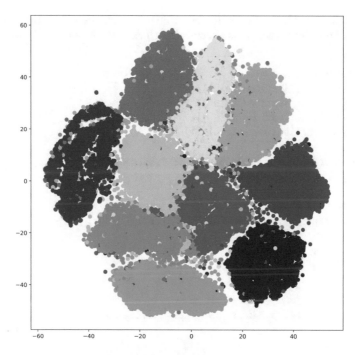

Figure 8-30. *t-SNE projection of a latent space for an autoencoder with a bottleneck size of 512 nodes trained on MNIST*

■ **Note**   If we had loaded the model as model = buildAutoencoder(784, 32)['model'] and the encoder as encoder = buildAutoencoder(784, 32)['encoder'], we indeed would obtain a model architecture and an encoder architecture – but they wouldn't be "linked." The stored model would be associated with an encoder that we haven't captured, and the stored encoder would be part of an overarching model that we haven't captured. Thus, we make sure to store the entire set of model components into modelSet first.

Each individual point is colored by the target label (i.e., the digit associated with the data point) for the purpose of exploring the autoencoder's ability to implicitly "cluster" points of the same digit together or separate them, even though the autoencoder was never exposed to the labels. Observe that as the dimensionality of the latent space increases, the overlap between data samples of different digits decreases until there is functionally complete separation between digits of different classes.

If we build an architecture in which the input is expanded rather than compressed and visualize a dimensionality reduction of the latent space (Listing 8-8), we find that the learned representations are significantly less meaningful (Figure 8-31) – despite this architecture obtaining very high performance (i.e., low training error).

Listing 8-8. Training and visualizing the latent space of an overcomplete, architecturally redundant autoencoder architecture. This particular architecture has slightly over 5.8 million parameters!

```
model = Sequential()
model.add(L.Input((784,)))
model.add(L.Dense(1024, activation='relu'))
model.add(L.Dense(2048, activation='relu'))
model.add(L.Dense(1024, activation='relu'))
model.add(L.Dense(784, activation='sigmoid'))

model.compile(optimizer='adam', loss='binary_crossentropy')
model.fit(x_train, x_train, epochs=50)

transformed = encoder.predict(x_train)
tsne_ = TSNE(n_components=2).fit_transform(transformed)

plt.figure(figsize=(10, 10), dpi=400)
plt.scatter(tsne_[:,0], tsne_[:,1], c=y_train)
plt.show()
```

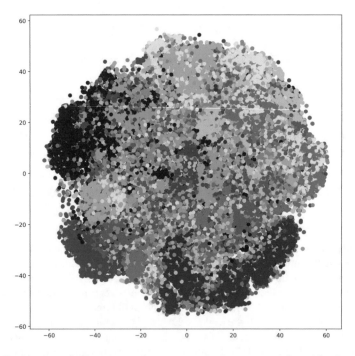

Figure 8-31. *t-SNE projection of a latent space for an overcomplete autoencoder with a bottleneck size of 2048 trained on MNIST*

Let's revisit the example given at the beginning of this subsection: reconstruction of an image of a line. Listing 8-9 generates a dataset of 50-by-50 images with randomly placed line segments using the image processing library cv2 (Listing 8-9).

Listing 8-9. Generating a dataset of 50-by-50 images of lines

```
x = np.zeros((1024, 50, 50))
for i in range(1024):
    start = [np.random.randint(0, 50), np.random.randint(0, 50)]
    end = [np.random.randint(0, 50), np.random.randint(0, 50)]
    x[i,:,:] = cv2.line(x[i,:,:], start, end, color=1, thickness=4)
x = x.reshape((1024, 50 * 50))
```

Since theoretically we can intuitively represent each line segment with four values, we'll build and train an autoencoder with four neurons in the latent space on the dataset (Listing 8-10).

Listing 8-10. Fitting a simple autoencoder on the synthetic toy line dataset

```
modelSet = buildAutoencoder(50 * 50, 4)
model = modelSet['model']
encoder = modelSet['encoder']
model.compile(optimizer='adam', loss='binary_crossentropy')
model.fit(x, x, epochs=400, validation_split=0.2)
```

The model reaches near 0.03 binary cross-entropy, which is quite good. Its reconstructions are very accurate (Figure 8-32).

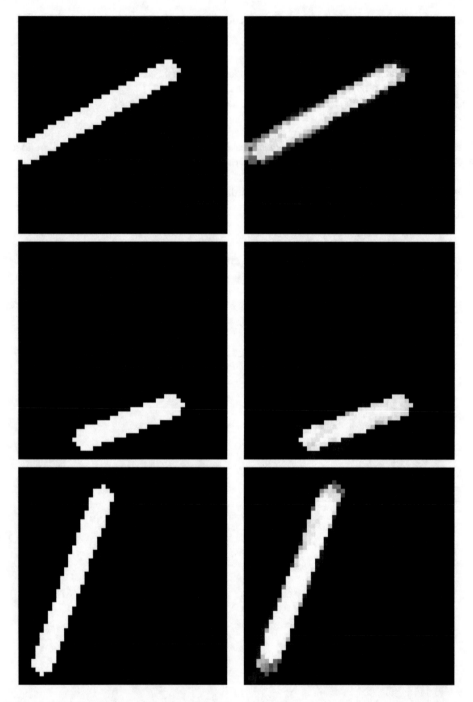

Figure 8-32. *Left column: original input images of lines. Right column: reconstructions via an autoencoder with a latent space dimensionality of 4*

In fact, an autoencoder trained with only two neurons does a decent job at identifying the general shape of the line marked in the input (Figure 8-33). If you look closely, you will identify the silhouette of other lines. There are many hypotheses to explain their presence. One possibility is that the autoencoder has "memorized"/"internalized" a set of generally useful "landmark" samples that are then mapped to during prediction and that with a larger latent space increased information for precise placement could be passed through.

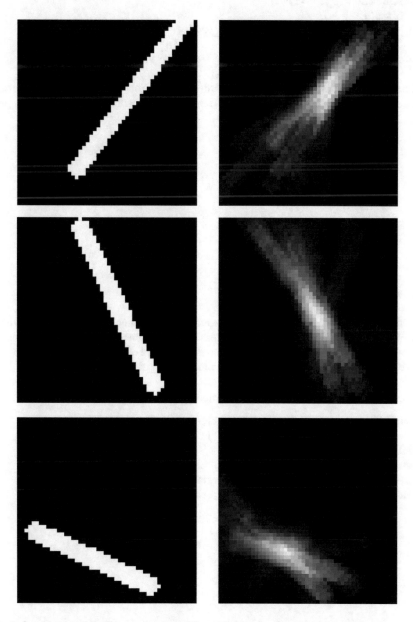

Figure 8-33. *Left column: original input images of lines. Right column: reconstructions via an autoencoder with a latent space dimensionality of 2*

Finally, let's explore how we can apply autoencoders to a strictly tabular dataset – the Mice Protein Expression dataset, used in previous chapters (Listing 8-11).

Listing 8-11. Splitting the dataset into training and validation sets

```
from sklearn.model_selection import train_test_split as tts
mpe_x = df.drop('class', axis=1)
mpe_y = df['class']
mpe_x_train, mpe_x_valid, mpe_y_train, mpe_y_vaid = tts(mpe_x, mpe_y,
                                      train_size=0.8,
                                      random_state=42)
```

Recall that we need to look at the input data in order to gauge how to deal with the model output in autoencoders. If we call mpe_x_train.min(), Pandas returns a series with the minimum value per column.

```
DYRK1A_N      0.156849
ITSN1_N       0.261185
BDNF_N        0.115181
NR1_N         1.330831
NR2A_N        1.737540
                ...
H3MeK4_N      0.101787
CaNA_N        0.586479
Genotype      1.000000
Treatment     1.000000
Behavior      1.000000
Length: 80, dtype: float64
```

Calling .min() again takes the minimum of the minimums across columns. We find that the smallest value across the entire dataset is –0.062007874, whereas the maximum is 8.482553422. Since values can theoretically be negative, we'll use a linear output activation instead of a ReLU and optimize using the standard Mean Squared Error loss for regression problems (Listing 8-12).

Listing 8-12. Fitting an autoencoder on the Mice Protein Expression dataset

```
modelSet = buildAutoencoder(len(mpe_x.columns), 8,
                      outActivation='linear')
model = modelSet['model']
encoder = modelSet['encoder']
model.compile(optimizer='adam', loss='mse', metrics=['mae'])
history = model.fit(mpe_x_train, mpe_x_train, epochs=150)
```

After 150 epochs of training, which progresses very quickly (this is a comparatively small dataset), the autoencoder obtains good training and validation performance (Table 8-1, Figure 8-34).

Table 8-1. *Performance of the autoencoder trained on the Mice Protein Expression dataset*

	Train	Validation
Mean Squared Error	0.0117	0.0118
Mean Absolute Error	0.0626	0.0625

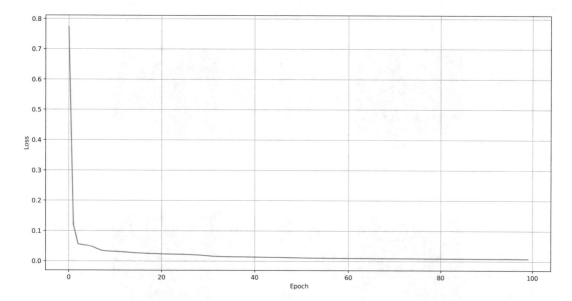

Figure 8-34. *Training history of an autoencoder trained on the Mice Protein Expression dataset*

Figures 8-35 demonstrates some sample latent vectors and reconstructions made by our autoencoder, with the input and reconstructed vectors reshaped into 8-by-10 grids for more convenient viewing.

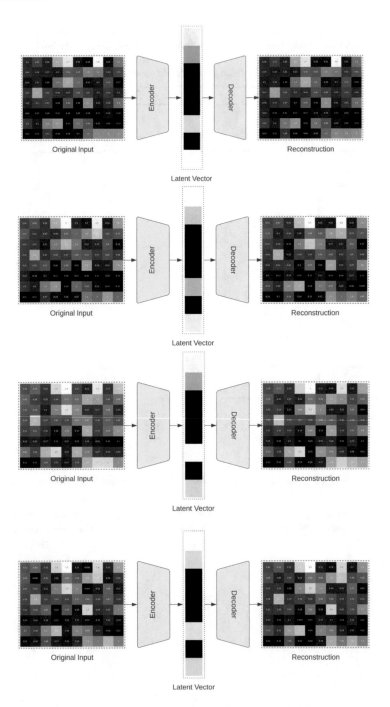

Figure 8-35. *Samples and the associated latent vector and reconstruction by an autoencoder trained on the Mice Protein Expression dataset. Samples and reconstructions are represented in two spatial dimensions for convenience of viewing*

We can employ a similar technique as previously employed on the MNIST dataset – visualizing the latent space of an autoencoder using t-SNE. Each data point in Figure 8-36 is colored by one of the eight classes each row in the Mice Protein Expression dataset falls into. This tabular autoencoder obtains pretty good separation between classes without any exposure to the labels.

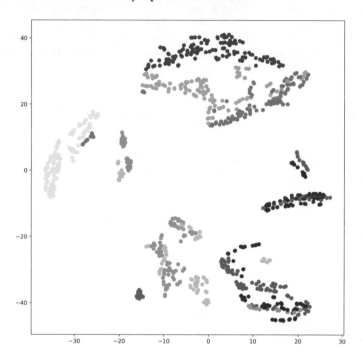

Figure 8-36. *t-SNE projection of a latent space for an autoencoder trained on the Mice Protein Expression dataset*

Note that a more formal/rigorous tabular autoencoder design would require us to standardize or normalize all columns to within the same domain. Tabular datasets often contain features that operate on different scales; for instance, say feature A represents a proportion (i.e., between 0 and 1, inclusive), whereas feature B measures years (i.e., likely larger than 1000). Regression losses simply take the mean error across all columns, which means that the reward for correctly reconstructing A is negligible compared to reconstructing feature B. In this case, however, all columns are in roughly the same range, so skipping this step is tolerable.

In the next subsection, we will explore a direct application of autoencoders to concretely improve the performance of supervised models.

Autoencoders for Pretraining

Vanilla autoencoders, as we have already seen, can do some pretty cool things. We see that a vanilla autoencoder trained on various datasets can perform implicit clustering and classification of digits, without being exposed to the labels themselves as well. Rather, natural differences in the input resulting from differences in labels are independently observed and implicitly recognized by the autoencoder.

This sort of impressive feature extraction capability is valuable in the context of training neural networks to perform supervised tasks. Say we want a neural network to classify digits from the MNIST dataset. If we start from scratch, we are asking the neural network to learn both how to extract the optimal set of features and how to interpret them – all at once, with no prior information. However, we see that the encoder of an autoencoder trained on the MNIST dataset has developed an impressive feature extraction and class separation scheme. We can use the encoder of the autoencoder as a *pretraining* instrument; rather than building and training a new network that learns both extraction and interpretation from scratch, we can simply append a model component to the output of the encoder to interpret the already-learned feature extractor (i.e., the encoder) (Figure 8-37).

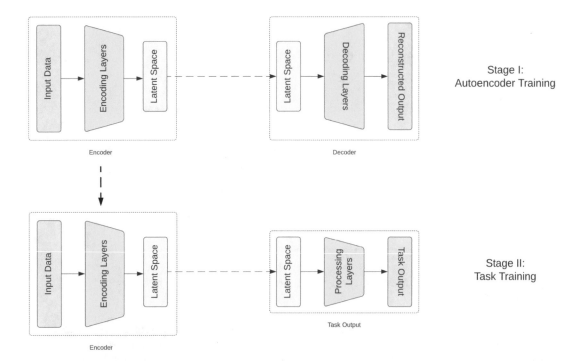

Figure 8-37. *Schematic of multistage pretraining*

In the first stage of training, we train the autoencoder on the standard input reconstruction task. After sufficient training, we can extract the encoder and append an "interpretation"-focused model component that assembles and arranges the features extracted by the encoder into the desired output.

During stage 2, we impose *layer freezing* upon the encoder, meaning that we prevent its weights from being trained. This is to retain the learned structures of the encoder. We spent a significant amount of effort obtaining a good feature extractor; if we do not impose layer freezing, we will find that optimizing a good feature extractor connected to a very poor (randomly initialized) feature interpreter degrades the feature extractor.

However, once good performance is obtained on training with a frozen feature extractor and a trainable feature interpreter, the entire model can be trained for a few epochs for the purposes of fine-tuning (Figure 8-38). The idea here is that the feature interpreter has developed a good relationship with the static feature extractor, but now both can be jointly optimized to improve the relationship. (Just like couples in relationships, it's not healthy if one partner is always static!)

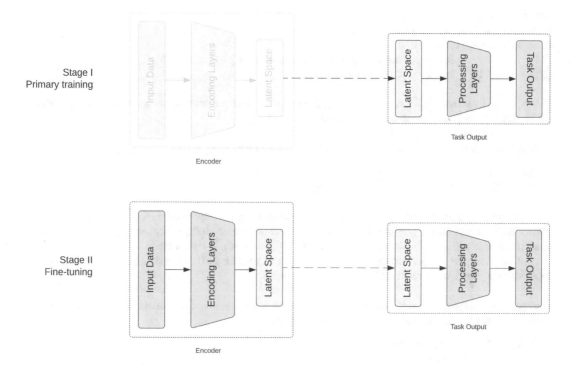

Figure 8-38. *Freezing followed by fine-tuning can be an effective way to perform autoencoder pretraining.*

Let's begin by demonstrating autoencoder pretraining on MNIST. We'll use the buildAutoencoder function defined previously to fit an autoencoder, making sure to retain references to both the original model and the encoder (Listing 8-13).

Listing 8-13. Training an autoencoder on MNIST

```
modelSet = buildAutoencoder(784, 32)
model = modelSet['model']
encoder = modelSet['encoder']
model.compile(optimizer='adam', loss='binary_crossentropy')
model.fit(x_train, x_train, epochs=20)
```

After the model has been sufficiently trained, we can extract the encoder and stack it as the feature extraction unit/component of our task model (Listing 8-14). The outputs of the encoder (named encoded in the following script) are further interpreted via several fully connected layers. The encoder is set not to be trainable (i.e., layer freezing). The task model is trained on the original supervised task.

Listing 8-14. Repurposing the encoder of the autoencoder as the frozen encoder/feature extractor of a supervised network

```
inp = L.Input((784,))
encoded = encoder(inp)
dense1 = L.Dense(16, activation='relu')(encoded)
dense2 = L.Dense(16, activation='relu')(dense1)
dense3 = L.Dense(10, activation='softmax')(dense2)
```

```
encoded.trainable = False
task_model = Model(inputs=inp, outputs=dense3)
task_model.compile(optimizer='adam',
                   loss='sparse_categorical_crossentropy')
task_model.fit(x_train, y_train, epochs=50)
```

After sufficient training, it is common practice to make the encoder trainable again and fine-tune the entire architecture in an end-to-end fashion (Listing 8-15).

Listing 8-15. Fine-tuning the whole supervised network by unfreezing the encoder

```
encoded.trainable = True
task_model.fit(x_train, y_train, epochs=5)
```

We often reduce the learning rate on fine-tuning tasks to prevent destruction/"overwriting" of information learned during the pretraining process. This can be accomplished by recompiling the model after pretraining with an optimizer configured with a different initial learning rate.

We can compare the performance of this model to one with no pretraining (i.e., begins learning in a supervised fashion from scratch) (Listing 8-16, Figure 8-39).

Listing 8-16. Training a supervised model with the same architecture as the model with pretraining, but without pretraining the encoder via an autoencoding task

```
modelSet = buildAutoencoder(784, 32)
model = modelSet['model']
encoder = modelSet['encoder']

inp = L.Input((784,))
encoded = encoder(inp)
dense1 = L.Dense(16, activation='relu')(encoded)
dense2 = L.Dense(16, activation='relu')(dense1)
dense3 = L.Dense(10, activation='softmax')(dense2)
task_model = Model(inputs=inp, outputs=dense3)
model.compile(optimizer='adam', loss='sparse_categorical_crossentropy')
history2 = model.fit(x_train, y_train, epochs=20)

plt.figure(figsize=(15, 7.5), dpi=400)
plt.plot(history.history['loss'], color='red',
         label='With AE Pretraining')
plt.plot(history2.history['loss'], color='blue',
         label='Without AE Pretraining')
plt.grid()
plt.xlabel('Epoch')
plt.ylabel('Loss')
plt.legend()
plt.show()
```

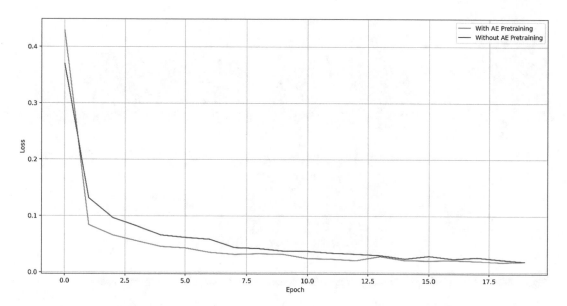

Figure 8-39. *Comparing the training curves for a classifier trained on the MNIST dataset with and without autoencoder pretraining*

The MNIST dataset is relatively simple, so both models converge relatively quickly to good weights. However, the model with pretraining is noticeably "ahead" of the other. By taking the difference between the epoch at which a model with and without pretraining obtains some performance value, we can estimate how "far ahead" a model with autoencoder pretraining is. For any loss p (at least one epoch in training), the model with pretraining reaches p two to four epochs before the model without pretraining.

This process seems and is superfluous on the MNIST dataset, which has a comparatively simple set of rules in a comparatively small number of dimensions. However, this advantage manifests more significantly for more complex datasets, as has been shown with more advanced computer vision and natural language processing tasks. Neural networks trained to perform large-scale image classification (e.g., ImageNet), for instance, benefit significantly from performing an autoencoder pretraining task that learns useful latent features that are later interpreted and fine-tuned. Similarly, it has been shown that language models learn important fundamental structures of language by performing reconstruction tasks, which can be later used as the basis for a supervised task like text classification or generation (Figure 8-40).

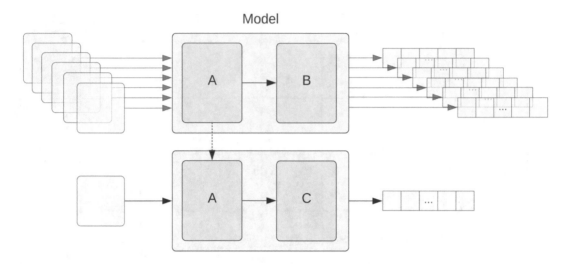

Figure 8-40. *General transfer learning/pretraining design used dominantly in computer vision*

Recall, for instance, the Inception and EfficientNet models discussed in Chapter 4. Keras allows users to load weights from a model trained on ImageNet because the feature extraction "skills" required to perform well on a wide-ranging task like ImageNet are valuable or can be adapted to become valuable in most computer vision tasks.

However, as we have previously seen in Chapters 4 and 5, the success of a deep learning method on complex image and natural language data does not necessarily bar it from being useful to tabular data applications too.

Let's consider the Mice Protein Expression dataset. We can begin by instantiating and training a sample autoencoder (Listing 8-17).

Listing 8-17. Building and training an autoencoder on the Mice Protein Expression dataset

```
modelSet = buildAutoencoder(len(mpe_x_train.columns), 32,
                            outActivation='linear')
model = modelSet['model']
encoder = modelSet['encoder']
model.compile(optimizer='adam', loss='mse')
history = model.fit(mpe_x_train, mpe_x_train, epochs=50)
```

We can now create and fit a task model using the trained encoder in two phases, the first in which the encoder is frozen and the second in which the encoder is trainable (Listing 8-18, Figure 8-41).

Listing 8-18. Using the pretrained encoder in a supervised task

```
inp = L.Input((len(mpe_x_train.columns),))
encoded = encoder(inp)
dense1 = L.Dense(32, activation='relu')(encoded)
dense2 = L.Dense(32, activation='relu')(dense1)
dense3 = L.Dense(32, activation='relu')(dense2)
dense4 = L.Dense(8, activation='softmax')(dense2)
encoded.trainable = False
task_model = Model(inputs=inp, outputs=dense4)
```

636

```
task_model.compile(optimizer='adam', loss='sparse_categorical_crossentropy',
                metrics=['accuracy'])
history_i = task_model.fit(mpe_x_train, mpe_y_train-1, epochs=30,
                        validation_data=(mpe_x_valid, mpe_y_valid-1))

encoded.trainable = True
history_ii = task_model.fit(mpe_x_train, mpe_y_train-1, epochs=10,
                        validation_data=(mpe_x_valid,
                                        mpe_y_valid-1))
```

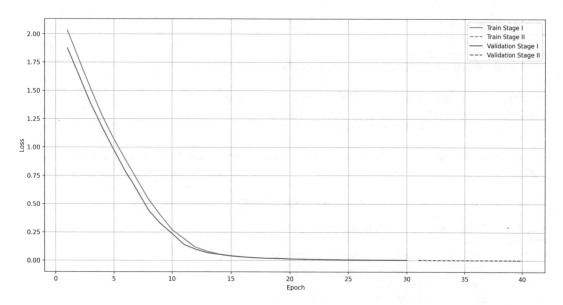

Figure 8-41. *Validation and training curves for stages 1 and 2*

Alternatively, consider the Higgs Boson dataset. This dataset only has 28 features. If we use our standard autoencoder logic, which halves the number of nodes in each encoder layer and doubles the number of nodes in each decoder layer, we will either need to have a very smaller number of layers to use a reasonable latent space size or a very small latent space to use a reasonable number of layers. For instance, if our latent space has only eight features, the autoencoder logic would build only two layers ($28 \rightarrow 16 \rightarrow 8$). On the other hand, if we want a larger number of layers (e.g., five), we would need a very small latent space (e.g., an autoencoder with $28 \rightarrow 16 \rightarrow 8 \rightarrow 4 \rightarrow 2 \rightarrow 1$). In this case, it's most beneficial to design a custom autoencoder with a sufficiently large latent space and a sufficient number of layers. We could design an autoencoder, for instance, with six layers in the encoder and decoder each and a latent space of 16 dimensions (Listing 8-19).

Listing 8-19. Defining a custom autoencoder architecture for the Higgs Boson dataset

```
encoder = Sequential()
encoder.add(L.Input((len(X_train.columns),)))
encoder.add(L.Dense(28, activation='relu'))
encoder.add(L.Dense(28, activation='relu'))
encoder.add(L.Dense(28, activation='relu'))
encoder.add(L.Dense(16, activation='relu'))
encoder.add(L.Dense(16, activation='relu'))
```

```
encoder.add(L.Dense(16, activation='relu'))

decoder = Sequential()
decoder.add(L.Input((16,)))
decoder.add(L.Dense(16, activation='relu'))
decoder.add(L.Dense(16, activation='relu'))
decoder.add(L.Dense(16, activation='relu'))
decoder.add(L.Dense(28, activation='relu'))
decoder.add(L.Dense(28, activation='relu'))
decoder.add(L.Dense(28, activation='linear'))

inp = L.Input((28,))
encoded = encoder(inp)
decoded = decoder(encoded)
ae = keras.models.Model(inputs=inp, outputs=decoded)

ae.compile(optimizer='adam', loss='mse', metrics=['mae'])
history = ae.fit(X_train, X_train, epochs=100,
                 validation_data=(X_valid, X_valid))
```

We can treat a static encoder as a feature extractor for our task model (Listing 8-20, Figures 8-42 and Figure 8-43).

Listing 8-20. Using the pretrained encoder as a feature extractor for a supervised task

```
inp = L.Input((len(X_train.columns),))
encoded = encoder(inp)
dense1 = L.Dense(16, activation='relu')(encoded)
dense2 = L.Dense(16, activation='relu')(dense1)
dense3 = L.Dense(16, activation='relu')(dense2)
dense4 = L.Dense(1, activation='sigmoid')(dense3)
encoded.trainable = False
task_model = keras.models.Model(inputs=inp, outputs=dense4)
task_model.compile(optimizer='adam', loss='binary_crossentropy',
                   metrics=['accuracy'])

history_i = task_model.fit(X_train, y_train, epochs=70,
                           validation_data=(X_valid, y_valid))

encoded.trainable = True
history_ii = task_model.fit(X_train, y_train, epochs=30,
                            validation_data=(X_valid, y_valid))
```

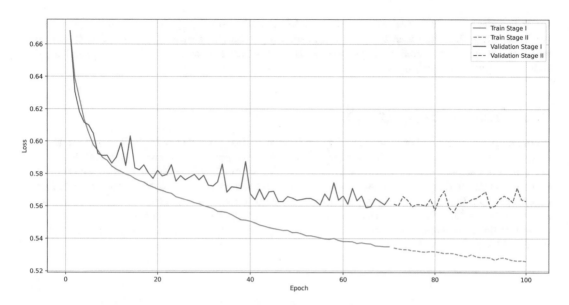

Figure 8-42. *Validation and training loss curves for stages 1 and 2*

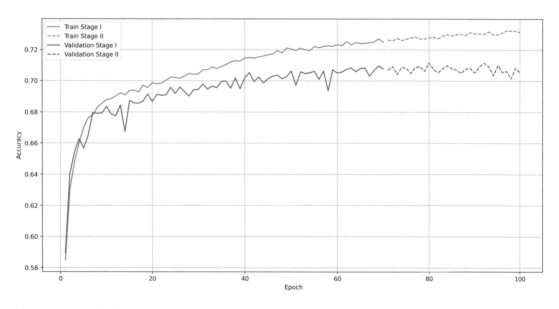

Figure 8-43. *Validation and training accuracy curves for stages 1 and 2*

We can observe a significant amount of overfitting in this particular case. We can attempt to improve generalization by employing best practices such as adding dropout or batch normalization.

Lastly, it should be noted that using autoencoders for pretraining is a great *semi-supervised* method. Semi-supervised methods make use of data with and without labels (and are used most often in cases where labeled data is scarce and unlabeled data is abundant). Say you possess three sets of data: $X_{unlabeled}$, $X_{labeled}$, and y (which corresponds to $X_{labeled}$). You can train an autoencoder to reconstruct $X_{unlabeled}$ and then use the

639

frozen encoder as the feature extractor in a task model to predict y from $X_{labeled}$. This technique generally works well even when the size of $X_{unlabeled}$ is significantly larger than the size of $X_{labeled}$; the autoencoding task learns meaningful representations that should be significantly easier to associate with a supervised target than beginning from initialization.

Multitask Autoencoders

Pretraining with autoencoders is often an effective strategy to take advantage of quality learned latent features. However, one criticism of the system is that it proceeds *sequentially* – autoencoder training takes place at a separate stage than the task training. Multitask autoencoders train the network on the autoencoder task and the intended task *simultaneously* (hence the name multitask). These autoencoders accept one input that is encoded by the encoder into a latent space. This one set of latent features is decoded separately by two "decoders" into two outputs; one output is dedicated to the autoencoder task, while the other is dedicated to the intended task. The network learns both of these tasks at the same time during training (Figures 8-44 and 8-45).

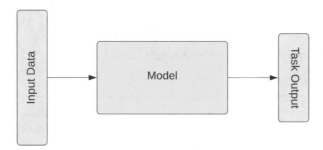

Figure 8-44. *Original task model*

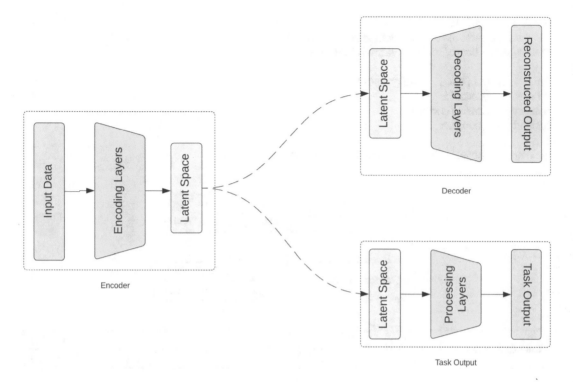

Figure 8-45. *Multitask learning*

By training the autoencoder simultaneously along the task network, we can theoretically experience the benefits of the autoencoder in a dynamic fashion. Say the encoder has "difficulty" encoding features in a way relevant to the task output, which can be difficult. However, the encoder component of the model can still decrease the overall loss by learning features relevant to the autoencoder reconstruction task. These features may provide continuous support for the task output by providing the optimizer a viable path to loss minimization – it is "another way out," so to speak. Using multitask autoencoders is often an effective technique to avoid or minimize difficult local minimum problems, in which the model makes mediocre to negligible progress in the first few moments of training and then plateaus (i.e., is stuck in a poor local minimum).

In order to construct a multitask autoencoder, we begin by initializing an autoencoder and extracting the encoder and decoder components. We create a "tasker" model that accepts latent features (i.e., data of the shape of the encoder output) and processes them into the task output (i.e., one of ten digits, in the case of MNIST). Each of these components can be linked using functional API syntax to form a complete multitask autoencoder architecture (Listing 8-21, Figure 8-46).

Listing 8-21. Building a multitask autoencoder for the MNIST dataset

```
modelSet = buildAutoencoder(784, 32)
model = modelSet['model']
encoder = modelSet['encoder']
decoder = modelSet['decoder']

tasker = keras.models.Sequential(name='taskOut')
tasker.add(L.Input((32,)))
```

```
for i in range(3):
    tasker.add(L.Dense(16, activation='relu'))
tasker.add(L.Dense(10, activation='softmax'))

inp = L.Input((784,), name='input')
encoded = encoder(inp)
decoded = decoder(encoded)
taskOut = tasker(encoded)

taskModel = Model(inputs=inp, outputs=[decoded, taskOut])
```

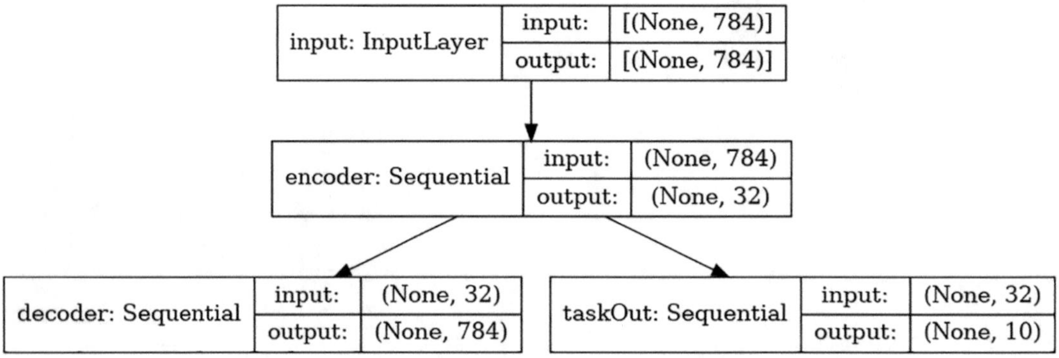

Figure 8-46. *Visualization of a multitask autoencoder architecture*

Because the multitask autoencoder has multiple outputs, we need to specify losses and labels for each of the outputs by referencing a particular output's name. In this case, the two outputs have been named "decoder" and "taskOut." The decoder output will be given the original input (i.e., x_train) and optimized with binary cross-entropy, since its objective is to perform pixel-wise reconstruction. The task output will be given the image labels (i.e., y_train) and optimized with categorical cross-entropy, since its objective is to perform multiclass classification (Listing 8-22).

Listing 8-22. Compiling and fitting the task model

```
taskModel.compile(optimizer='adam',
                loss = {'decoder':'binary_crossentropy',
                        'taskOut':'sparse_categorical_crossentropy'})

history = taskModel.fit(x_train, {'decoder':x_train,
                                'taskOut': y_train},
                        epochs=100)
```

We can observe from the training history that the model is able to reach both a fairly good task loss and a reconstruction loss within just a few dozen epochs (Listing 8-23, Figure 8-47).

Listing 8-23. Plotting out different dimensions of the performance over time

```
plt.figure(figsize=(15, 7.5), dpi=400)
plt.plot(history.history['decoder_loss'], color='red', linestyle='--',
label='Reconstruction Loss')
plt.plot(history.history['taskOut_loss'], color='blue', label='Task Loss')
plt.plot(history.history['loss'], color='green', linestyle='-.', label='Overall Loss')
plt.grid()
plt.xlabel('Epoch')
plt.ylabel('Loss')
plt.legend()
plt.show()
```

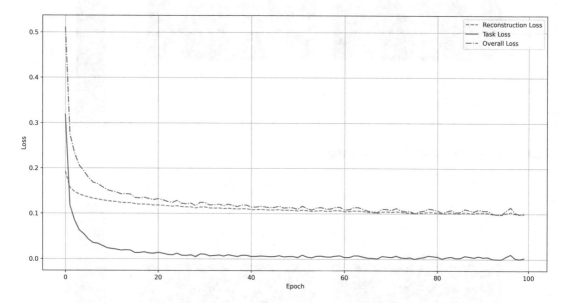

Figure 8-47. Different dimensions of performance (reconstruction loss, task loss, overall loss)

Figures 8-48 to 8-51 visualize how the state of the multitask autoencoder progresses throughout each epoch.

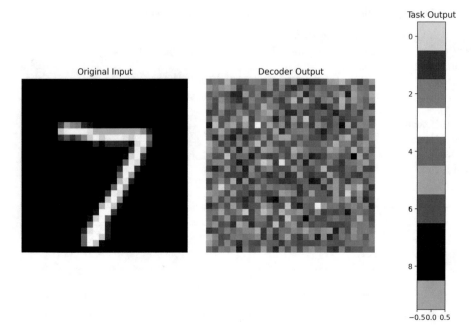

Figure 8-48. *Multitask autoencoder at zero epochs*

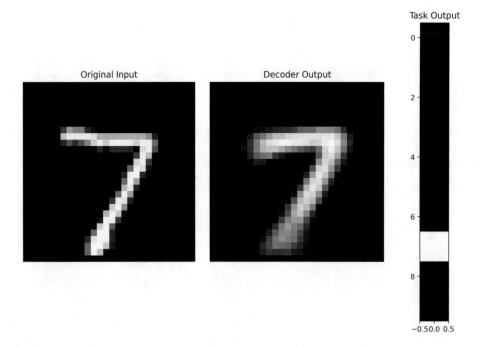

Figure 8-49. *Multitask autoencoder at one epoch*

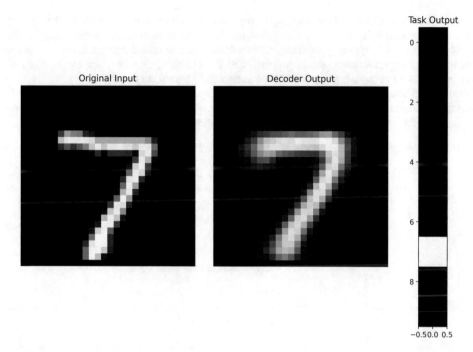

Figure 8-50. *Multitask autoencoder at two epochs*

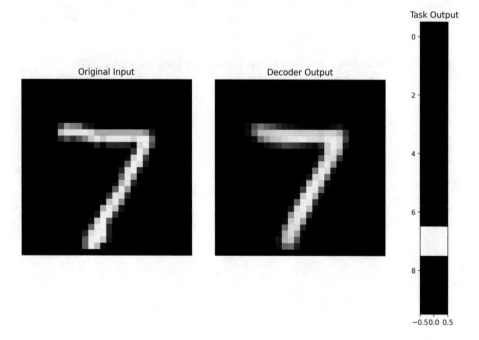

Figure 8-51. *Several more epochs*

From these visualizations and the training history, we see that the multitask autoencoder obtains better performance on the task than the autoencoding task that is intended to assist task performance! In this case, the MNIST dataset's task output is more straightforward than the autoencoding task, which makes sense. In this case, using a multitask autoencoder is not beneficial. It probably is more beneficial to directly train or use an autoencoder for pretraining when multitask autoencoders perform poorly.

We can use an adapted approach on the Mice Protein Expression dataset, in which we see that autoencoding is a more approachable problem than the task of classification itself, both from the training history (Figures 8-52) and output state progression visualizations (Figures 8-53 through 8-56).

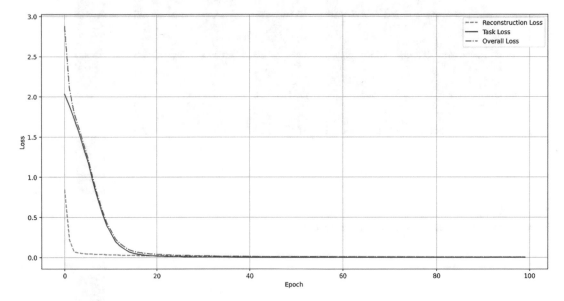

Figure 8-52. *Different dimensions of performance on the Mice Protein Expression dataset*

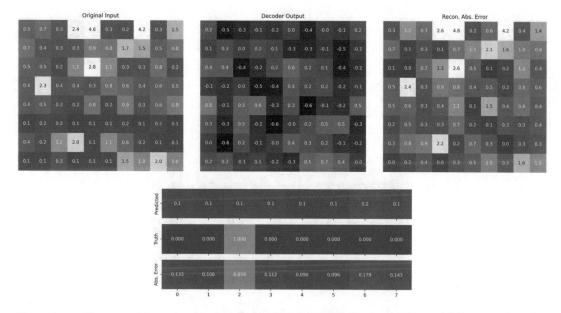

Figure 8-53. *The state of the multitask autoencoder after zero epochs (i.e., upon initialization). Top: displays the original set of 80 features in the Mice Protein Expression dataset (arranged in a grid for more convenient visual viewing), the output of the decoder (of which the goal is to reconstruct the input), and the absolute error of the reconstruction. Bottom: the predicted and true classes (eight in total) and the absolute probability error*

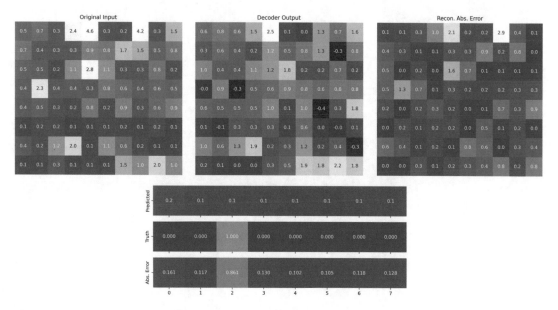

Figure 8-54. *The state of the multitask autoencoder after one epoch*

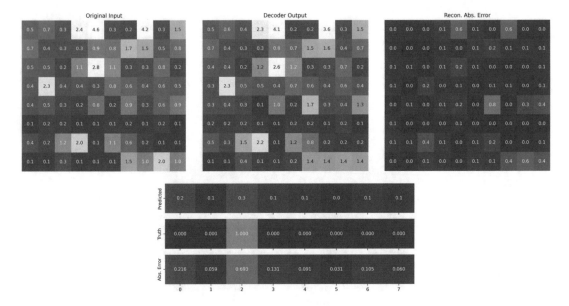

Figure 8-55. *The state of the multitask autoencoder after five epochs*

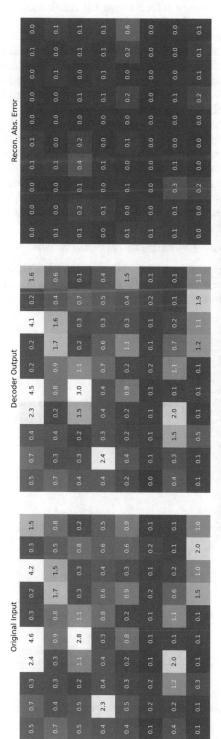

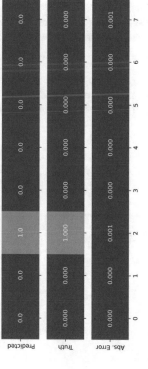

Figure 8-56. *The state of the multitask autoencoder after 50 epochs*

Figures 8-53 through 8-56 demonstrate the performance of the reconstruction task alongside the classification task at various stages in training. Notice that the reconstruction error converges near zero quickly and helps "pull"/"guide" the task error to zero over time.

In many cases, simultaneous execution of the autoencoder task and the original desired task can help provide stimulus to "push" progress on the desired task. However, you may make the valid objection that once the desired task reaches sufficiently good performance, it becomes limited by the autoencoding task.

One method to reconcile with this is simply to detach the autoencoder output from the model by creating a new model connecting the input to the task output and fine-tuning on the dataset.

Another more sophisticated technique is to change the loss weights between the original desired task and the autoencoding task. While Keras weighs multiple losses equally by default, we can provide different weights to reflect different levels of priority or importance delegated to each of the tasks. At the beginning of training, we can give a high weight to the autoencoding task, since we want the model to develop useful representations through a (ideally somewhat easier) task of autoencoding. Throughout the training duration, the weight on the original task model loss can be successively increased and the weight on the autoencoder model loss decreased. To formalize this, let α be the weight on the task output loss, and let $1 - \alpha$ be the weight on the decoder output loss (with $0 < \alpha < 1$).

The sigmoid equation $\sigma(x) = \dfrac{1}{1+e^{-x}}$ is a pretty good way to get from a value very close to some minimum bound to another value very close to an upper bound. Over the span of 100 epochs, we can employ a simple (arbitrarily set but functional) transformation on the sigmoid function to obtain a smooth transition from a slow to high value of α (visualized by Listing 8-24 in Figure 8-57), where t represents the epoch number:

$$\alpha = \sigma\left(\frac{t-50}{10}\right) = \frac{1}{1+e^{-\left(\frac{t-50}{10}\right)}}$$

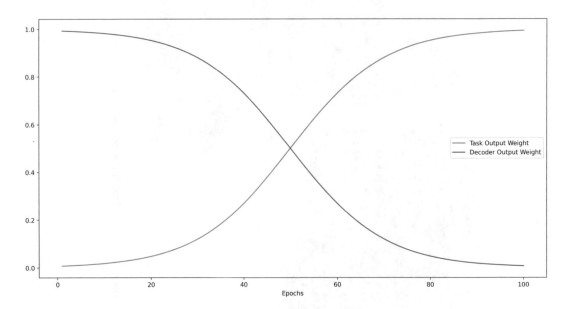

Figure 8-57. *Plot of the task output loss weight and the decoder weight across each epoch*

Listing 8-24. Plotting out our custom α-adjusting curve

```
plt.figure(figsize=(15, 7.5), dpi=400)
epochs = np.linspace(1, 100, 100)
alpha = 1/(1 + np.exp(-(epochs-50)/10))
plt.plot(epochs, alpha, color='red', label='Task Output Weight')
plt.plot(epochs, 1-alpha, color='blue', label='Decoder Output Weight')
plt.xlabel('Epochs')
plt.legend()
plt.show()
```

A generalized equation to scale α through t_{max} using a transformation of the sigmoid function is as follows:

$$\alpha = \sigma \left(\frac{t - \dfrac{t_{max}}{2}}{\dfrac{t_{max}}{10}} \right) = \frac{1}{1 + e^{-\left(\frac{t - \frac{t_{max}}{2}}{\frac{t_{max}}{10}} \right)}}$$

At initial conditions, we have α at a very small value. (For the purposes of this calculation, we use $t = 1$ for simplification of calculation.)

$$\alpha @ \{t \approx 0\} \rightarrow \frac{1}{1 + e^{-\left(\frac{t_{max}}{2} \Big/ \frac{t_{max}}{10} \right)}} = \frac{1}{1 + e^{5}} \approx 0.006692$$

The training regime completes at $t = t_{max}$, at which α is very close to 1:

$$\alpha @ \{t = t_{max}\} \rightarrow \frac{1}{1 + e^{-\left(\frac{t_{max} - \frac{t_{max}}{2}}{\frac{t_{max}}{10}} \right)}} = \frac{e^{5}}{1 + e^{5}} \approx 0.993307$$

Moreover, we observe by taking the derivative and solving for the maximum that the largest change for some t_{max} is $\dfrac{5}{2 \cdot t_{max}}$. As t_{max} increases, analysis of the derivative reveals that the overall change becomes more uniformly spread out. For large values of t_{max}, this functionally becomes a horizontal line (i.e., the derivative becomes near 0). A simple linear transformation of α also suffices in most cases in which t_{max} is reasonably large.

Loss weighting is conveyed in the compiling stage. This means that we'll have to recompile and fit every epoch. This is not difficult to do; we can write a for loop that loops through every epoch, calculates the α value for that epoch, compiles the model with that loss weighting, and fits for one epoch. Collecting the training history is slightly more manual; we need to collect the metrics for the single epoch and append them to user-created lists (Listing 8-25).

Listing 8-25. Recompiling and fitting a multitask autoencoder with varied loss weighting

```
total_epochs = 100

lossParams = {'decoder':'binary_crossentropy',
              'taskOut':'sparse_categorical_crossentropy'}
```

```
loss, decoderLoss, taskOutLoss = [], [], []

for epoch in range(1, total_epochs+1):
    alpha = 1/(1 + np.exp(-(epoch-50)/10))
    taskModel.compile(optimizer='adam',
                      loss = lossParams,
                      loss_weights = {'taskOut': alpha,
                                      'decoder': 1-alpha})
    history = taskModel.fit(x_train, {'decoder':x_train,
                                      'taskOut': y_train},
                            epochs = 1)
    loss.extend(history.history['loss'])
    decoderLoss.extend(history.history['decoder_loss'])
    taskOutLoss.extend(history.history['taskOut_loss'])
```

For another higher-code but perhaps smoother approach to dynamically adjusting the loss calculation weights of multi-output models, which does not require repeated refitting, see Anuj Arora's well-written post on adaptive loss weighting in Keras using callbacks: https://medium.com/dive-into-ml-ai/adaptive-weighing-of-loss-functions-for-multiple-output-keras-models-71a1b0aca66e.

Figure 8-58 demonstrates the history of the reconstruction, task, and overall losses throughout training of the multitask autoencoder, with the background shaded by the value of α used at that epoch. Note that the reconstruction task is more trivial than the original intended task (hence the faster decline in loss) and the logistically shaped overall loss function that makes major changes in α in epochs 40–60 and "switches" bounds from the reconstruction to the original task loss.

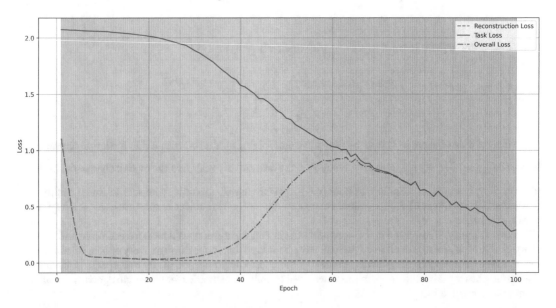

Figure 8-58. *Diagram of reconstruction loss, task loss, and overall loss (now a dynamically weighted sum) with the weighting gradient shaded in the background*

Multitask autoencoders perform best in difficult supervised classification tasks that benefit from rich latent features, which can be learned well by autoencoders.

Sparse Autoencoders

Standard autoencoders are given the limitation of size representation – autoencoder architectures are built with a "physical" bottleneck through which information must be compressed. The autoencoder attempts to maximize the amount of information it can squeeze through a significantly compressed latent space such that the information can reliably be decoded into the original output (Figure 8-59).

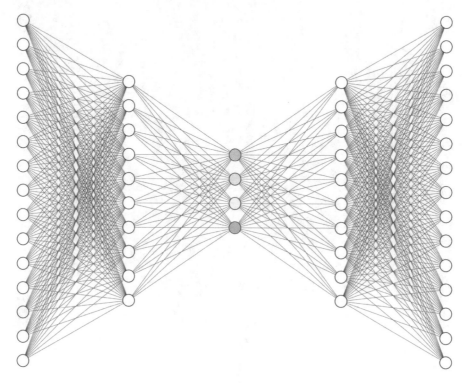

Figure 8-59. *A standard autoencoder, which encodes information into a densely packed and quasi-continuous latent space*

However, this is not the only limit we can impose. Another information bottleneck tool is *sparsity*. We can make the bottleneck layer very large, but force only a few nodes to be active at any one pass. While this still forces a limitation on the amount of *information* that can pass through the bottleneck layer, the network is given more freedom and control to "choose" *which* nodes information passes through, which itself is an additional medium of information expression (Figure 8-60).

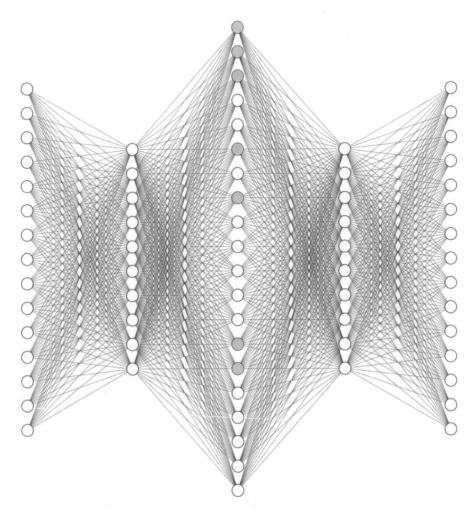

Figure 8-60. *A sparse autoencoder, in which a much larger latent size is accessible but only a few nodes can be used at any one time*

To maintain sparsity, we generally impose L1 regularization on the layer's activity. (Recall the discussion of regularization in Chapter 3, "Regularization Learning Networks.") L1 regularization penalizes the output activity of the bottleneck layer by being too large. Assuming a network uses binary cross-entropy loss to minimize task output and λ represents the overall activity/output of the bottleneck layer, the joint loss of an L1-regularized network is as follows:

$$\text{loss} = \text{BCE}\left(y_{\text{pred}}, y_{\text{true}}\right) + \alpha \cdot |\lambda|$$

The parameter α is user-defined and controls the "importance" of the L1 regularization term relative to the task loss. Setting the correct value of α is important for correct behavior. If α is too small, the network ignores the sparsity restriction in favor of completing the task, which is now made quasi-trivial by the overcomplete bottleneck layer. If α is too large, the network ignores the task by learning the "ultimate sparsity" – predicting all zeros in the bottleneck layer, which entirely minimizes λ but performs poorly on the actual task we want it to learn.

An alternative commonly used penalty is L2 regularization, in which the square rather than the absolute value is penalized:

$$\text{loss} = \text{BCE}\left(y_{\text{pred}}, y_{\text{true}}\right) + \alpha \cdot \lambda^2$$

This is a common machine learning paradigm. L2 regularization tends to produce sets of values generally near zero but not at zero, whereas L1 regularization tends to produce values solidly at zero. An intuitive explanation is that L2 regularization significantly discounts the need to decrease values that are already somewhat near zero. The decrease from 3 to 2, for instance, is rewarded with a penalty decrease of $3^2 - 2^2 = 5$. The decrease from 1 to 0, on the other hand, is rewarded with a measly penalty decrease of $1^2 - 0^2 = 1$. On the other hand, L1 regularization rewards the decrease from 3 to 2 identically as the decrease from 1 to 0. We generally use L1 regularization to impose sparsity constraints because of this property.

To implement this, we need to make a slight modification to our original buildAutoencoder function. We can build the autoencoder as if we were leading up to and from a certain *implicit* latent size, but replace the *implicit* latent size with the *real* (expanded) latent size. For instance, consider an autoencoder build with an input of 64 dimensions and an implicit latent space of 8 dimensions. The node count progression in each layer of a standard autoencoder using our prebuilt autoencoder logic would be $64 \to 32 \to 16 \to 8 \to 16 \to 32 \to 64$. However, because we are planning to impose a sparsity constraint on the bottleneck layer, we need to provide an expanded set of nodes to pass information through. Say the real bottleneck size is 128 nodes. The node count progression in each layer of this sparse autoencoder would be $64 \to 32 \to 16 \to 128 \to 16 \to 32 \to 64$.

To actually implement the sparsity constraint, note that almost all layers in Keras have an activity_ regularizer parameter, set upon initialization. This parameter penalizes the *activity*, or the output, of the layer (Listing 8-26). Note that you can also set the weight_regularizer or bias_regularizer parameter if you desire to penalize the learned weights or biases. In this case, we don't care about how the encoder *arrives* at a sparse encoding, only that the encoder *creates* a sparse encoding. Hence, we perform regularization on the layer activity. The arguments accept a keras.regularizers object. We will use the L1 regularization object, which accepts the specific weighting of the penalty as a parameter. Setting the weight is important and should be given thought and experimentation, considering the model power, difficulty of autoencoding, and latent space size. As discussed previously, setting an improper weight in either direction (too large or too small) yields adverse outcomes.

Listing 8-26. Defining a sparse autoencoder with L1 regularization

```
from keras.regularizers import L1

def buildSparseAutoencoder(inputSize=784,
                           impLatentSize=32,
                           realLatentSize=128,
                           outActivation='sigmoid'):

    # define architecture components
    encoder = Sequential(name='encoder')
    encoder.add(L.Input((inputSize,)))
    for i in range(int(np.floor(np.log2(inputSize/impLatentSize))), -1, -1):
        encoder.add(L.Dense(impLatentSize * 2**i, activation='relu'))
        encoder.add(L.Dense(impLatentSize * 2**i, activation='relu'))
    encoder.add(L.Dense(realLatentSize, activation='relu',
                        activity_regularizer = L1(0.001)))

    decoder = Sequential(name='decoder')
    decoder.add(L.Input((realLatentSize,)))
```

```
for i in range(1,int(np.floor(np.log2(inputSize/impLatentSize)))+1):
    decoder.add(L.Dense(impLatentSize * 2**i, activation='relu'))
    decoder.add(L.Dense(impLatentSize * 2**i, activation='relu'))
decoder.add(L.Dense(inputSize, activation=outActivation))

# define model architecture from components
ae_input = L.Input((inputSize,), name='input')
ae_encoder = encoder(ae_input)
ae_decoder = decoder(ae_encoder)
ae = Model(inputs = ae_input,
           outputs = ae_decoder)

return {'model': ae, 'encoder': encoder, 'decoder': decoder}
```

Figure 8-61 demonstrates performance of the sparse autoencoder on the MNIST dataset, in which a 64-dimensional latent space vector is reshaped into an 8-by-8 grid for convenient viewing. The reconstruction is not visibly worse than a standard autoencoder without a sparsity constraint. Notice that only two to five of the 64 nodes are active at any one pass (and that which node(s) are active vary for each image). A standard autoencoder trained even with five nodes in the bottleneck layer (no sparsity requirement) would obtain poor performance on reconstruction, demonstrating the informational richness of "choosing" which nodes are active.

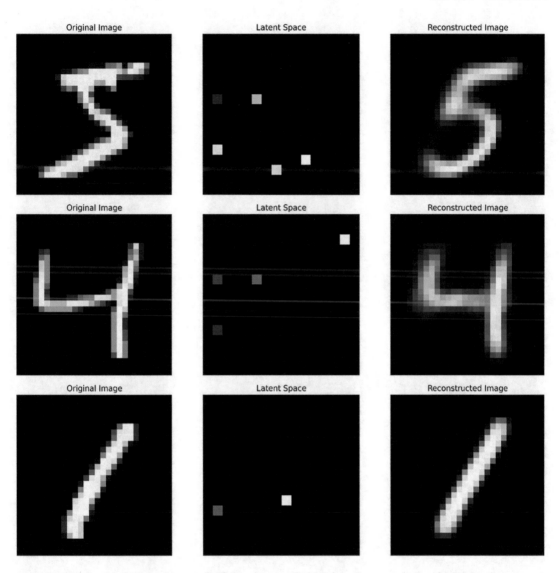

Figure 8-61. *Sampled original inputs (left), latent space (middle), and reconstruction (right) for a sparse autoencoder trained on MNIST. The latent space is 256 neurons reshaped into a 16-by-16 grid for viewing. The actual latent space is not arranged in two spatial directions*

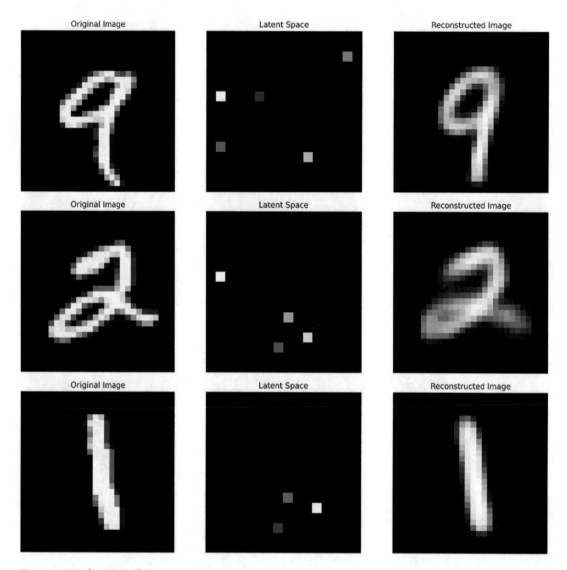

Figure 8-61. *(continued)*

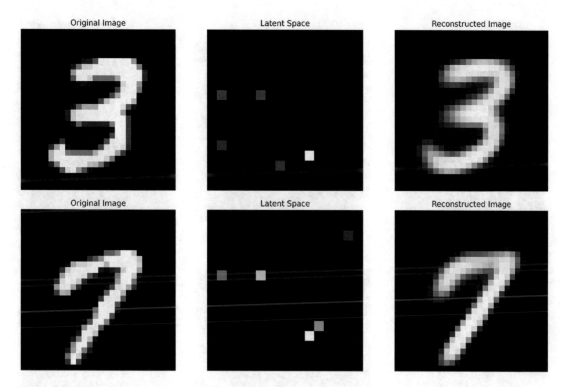

Figure 8-61. (continued)

If we decreased the regularization alpha value (i.e., the L1 penalty would be weighted less relative to the loss), the network would obtain better overall loss at the cost of decreased sparsity (i.e., more nodes would be active at any one pass). If we increased the regularization alpha, the network would obtain worse overall loss at the benefit of increased sparsity (i.e., even fewer nodes would be active at any one pass).

We can apply the same sparse autoencoding scheme to the Higgs Boson dataset, encoding a 28-dimensional input vector into a 64-dimensional latent space. At each pass, about one-fourth to one-third of the latent space is active, although many bottleneck nodes are "quasi-active" – they are not zero, but very close. Figure 8-62 demonstrates the internal state and reconstruction of the sparse autoencoder on different inputs, with 28-dimensional input vectors reshaped into 7-by-4 grids for more convenient viewing.

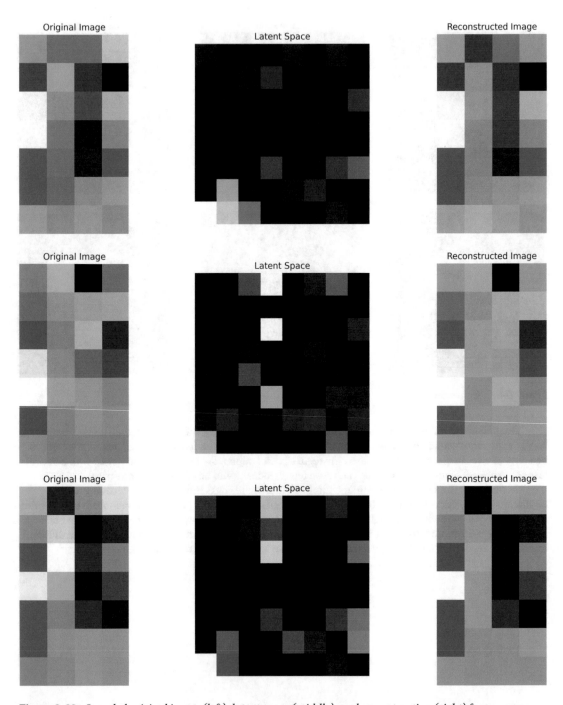

Figure 8-62. *Sampled original inputs (left), latent space (middle), and reconstruction (right) for a sparse autoencoder trained on the Higgs Boson dataset. The latent space is 256 neurons reshaped into a 16-by-16 grid for viewing; the input and reconstruction are 28 dimensions arranged into 7-by-4 grids*

Original Image
Latent Space
Reconstructed Image

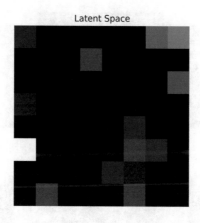

Figure 8-62. *(continued)*

Similarly, Figure 8-63 shows the application of a trained sparse autoencoder to various elements of the Mice Protein Expression dataset.

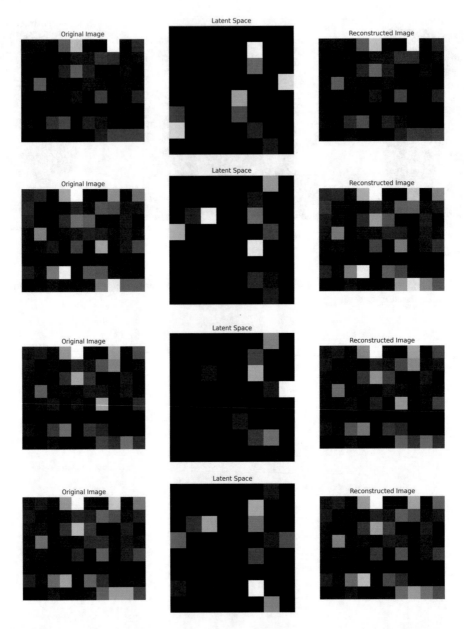

Figure 8-63. *Sampled original inputs (left), latent space (middle), and reconstruction (right) for a sparse autoencoder trained on the Mice Protein Expression dataset. The latent space is 256 neurons reshaped into a 16-by-16 grid for viewing; the input and reconstruction are 80 dimensions arranged into 8-by-10 grids*

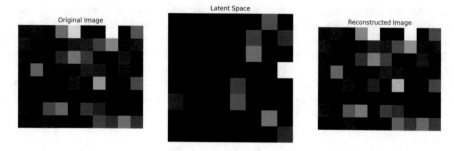

Figure 8-63. *(continued)*

Why would you want to use sparse autoencoders? The primary reason is to take advantage of sparse encoders' robustness properties. Adversarial examples are instances generated to deliberately fool a neural network into some image originally correctly classified as class *A* into class *B* with high confidence simply by making miniscule, barely visible changes to the input. The canonical example in the field is a diagram created by Ian Goodfellow et al. in the paper "Explaining and Harnessing Adversarial Examples." The Fast Signed Gradient Method (FSGM) generates a permutation matrix that adjusts every pixel in the input in a way that significantly changes the network's final prediction (Figure 8-64).

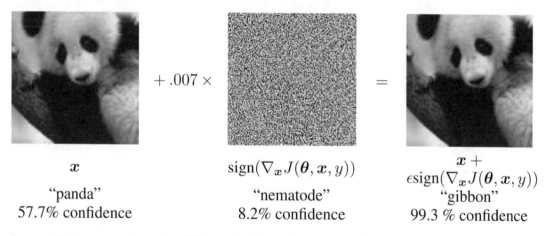

Figure 8-64. *Demonstration of the FSGM method. From "Explaining and Harnessing Adversarial Examples," Goodfellow et al.*

Adversarial example finders profit from continuity and gradients. Because neural networks operate in very large continuous spaces, adversarial examples can be found by "sneaking" through smooth channels and ridges in the surface of the landscape. Adversarial examples can be security threats (some instances of naturally occurring adversarial examples, like tape placed onto a traffic sign in a particular orientation causing egregious misidentification), as well as potential symptoms of poor generalization.[1] However, sparse encoders impose a discreteness upon the encoded space. It becomes significantly more difficult to generate successful adversarial examples when a frozen encoder is used as the feature extractor for a network.

Sparse autoencoders can also be useful for the purposes of interpretability. We'll talk more about specialized interpretability techniques later in this chapter, but sparse autoencoders can be easily interpreted without additional complex theoretical tools. Because only a few neurons are active at any one time, understanding which neurons are activated for any one input is relatively simple, especially compared with the latent vectors generated by standard autoencoders.

Denoising and Reparative Autoencoders

So far, we've only considered applications of autoencoder training in which the desired output is identical to the input. However, autoencoders can perform another function: to repair or restore a damaged or noisy input.

Here's the clever way we go about it – we artificially add realistic noise or corruption to a "pure"/"clean" dataset and then train the model to recover the cleaned image from its artificially corrupted version (Figure 8-65).

---

[1] This is an ongoing research topic in the field and a debated position. The paper "Adversarial Examples Are Not Bugs, They Are Features" is worth reading as complementary to the landmark Goodfellow et al. FGSM paper. A commonly raised criticism of neural networks vulnerable to adversarial examples is that they do not reflect generalization in the same way that humans can (i.e., humans can look at an image without an adversarial mask applied and an image with an adversarial mask applied but identify both as of the same class, as in Figure 8-64). However, some researchers, such as Alec Bunn at the University of Washington, have suggested that humans may also be prone to adversarial examples – examples deliberately designed to trick the system given the system's behavior – by somehow tracing perception and thought patterns throughout the brain, but that we simply don't currently possess the knowledge to neurologically generate adversarial examples for humans.

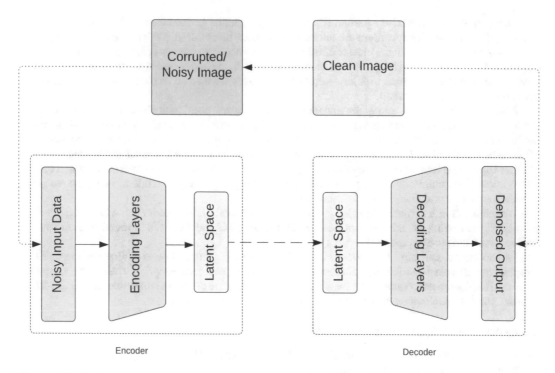

Figure 8-65. *Deriving a noisy image as input and the original clean image as the desired output of a denoising autoencoder*

There are many applications for such a model. We can use it, most obviously, to denoise a noisy input; the "cleaned" input can then be used for other purposes. Alternatively, if we are developing a model that we know will operate in a domain with lots of noisy data, we can use the encoder of a denoising autoencoder as a robust or resilient feature extractor (similarly to in autoencoder pretraining), exploiting the encoder's "denoised" latent representations (Figure 8-66).

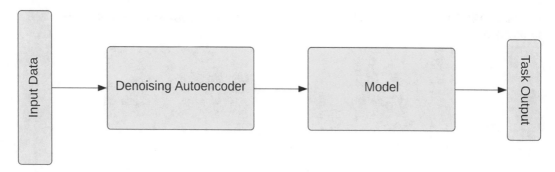

Figure 8-66. *A potential application of denoising autoencoders as a structure that learns to clean up the input before it is actually used in a model for a task*

These reparative models have particularly exciting applications for intelligent or deep graphics processing. Many graphics operations are not *trivially two-way invertible* in that it is trivial to go from one state to another but not in the inverse direction. For instance, if I convert a color image or movie into grayscale (for instance, using the pixel-wise methods covered in the image case study in Chapter 2), there is no simple way to invert it back to color. Alternatively, if you spill coffee on an old family photo, there is no trivial process to "erase" the stain.

Autoencoders, however, exploit the triviality of going from the "pure" to the "corrupted" state by artificially imposing corruption upon pure data and forcing powerful autoencoder architectures to learn the "undoing." Researchers have used denoising autoencoder architectures to generate color versions of historical black-and-white film and to repair photos that have been ripped, stained, or streaked. Another application is in biological/medical imaging, where an imaging operation can be disrupted by environmental conditions; replicating this noise/image damage artificially and training an autoencoder to become robust to it can make the model more resilient to noise.

We will begin with demonstrating the application of a denoising autoencoder to the MNIST dataset by successively increasing the amount of noise in the image and observing how well the denoising autoencoder performs (similarly to exercises in Chapter 4).

We can use a simple but effective technique to introduce noise into an image: adding random noise sampled from a normal distribution with mean 0 and a specified standard deviation. The result is clipped to ensure that the resulting value is still between 0 and 1, the feasible domain of pixel values. Listing 8-27 implements and visualizes artificial noise for a given standard deviation std.

Listing 8-27. Displaying data corrupted by random noise

```
modified = x_train + np.random.normal(0, std, size=x_train.shape)
modified_clipped = np.clip(modified, 0, 1)

plt.set_cmap('gray')
plt.figure(figsize=(20, 20), dpi=400)
for i in range(25):
    plt.subplot(5, 5, i+1)
    plt.imshow(modified_clipped[i].reshape((28, 28)))
    plt.axis('off')
    plt.show()
```

Figure 8-67 demonstrates a grid of sample images with no artificial noise added as reference for comparison.

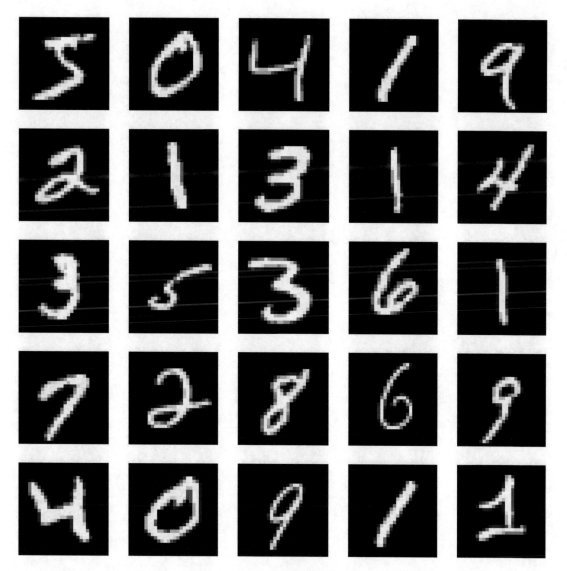

Figure 8-67. *A grid of untampered clean images from MNIST for reference*

Figure 8-68 demonstrates the same set of images with random noise sampled from a normal distribution with standard deviation of 0.1. We can observe marginal noise, especially in affecting the consistency of the digit outlines.

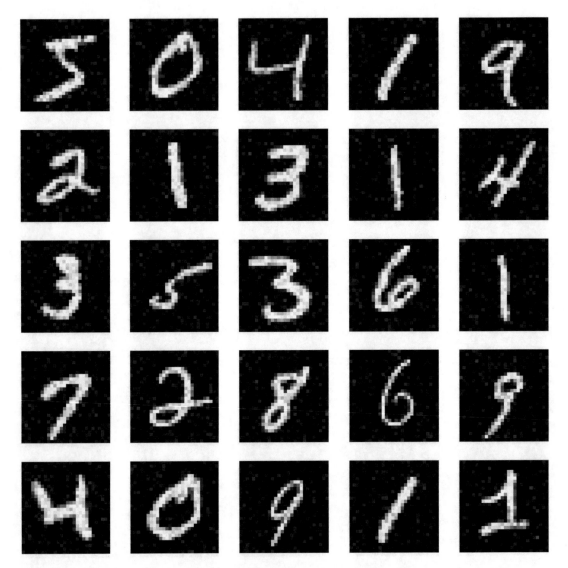

Figure 8-68. *A sample of MNIST images with added normally distributed random noise using standard deviation 0.1*

Let's build an autoencoder to denoise this data (Listing 8-28). There is no difference between the architecture of the autoencoder used here and in previous applications; the difference rather is in the data that we pass in (namely, that the input should have artificial noise applied). In this implementation, we compute new noise in each epoch. This is desired because it provides "fresh" noise that the denoising autoencoder must learn to denoise rather than to "accept"/"memorize."

Listing 8-28. Training the denoising autoencoder on novel corrupted MNIST data each epoch

```
models = buildAutoencoder(784, 32)
model = models['model']
encoder = models['encoder']

model.compile(optimizer='adam', loss='mse')
TOTAL_EPOCHS = 100
loss = []
for i in tqdm(range(TOTAL_EPOCHS)):
    modified = x_train + np.random.normal(0, std, size=x_train.shape)
    modified_clipped = np.clip(modified, 0, 1)
    history = model.fit(modified_clipped, x_train, epochs=1, verbose=0)
    loss.append(history.history['loss'])
```

After training, we can evaluate the Mean Absolute Error on a fresh validation set of noisy images (Listing 8-29).

Listing 8-29. Evaluating the performance of the denoising autoencoder on a fresh set of noisy images

```
modified = x_valid + np.random.normal(0, std, size=x_valid.shape)
modified_clipped = np.clip(modified, 0, 1)

from sklearn.metrics import mean_absolute_error as mae
mae(model.predict(modified_clipped), x_valid)
```

Listing 8-30 and Figure 8-69, respectively, implement and demonstrate a sampling of images with normally distributed random noise, using a standard deviation of 0.1. A denoising autoencoder trained to recover the original version given a noisy image generated using this procedure obtains a validation Mean Squared Error of 0.0266.

Listing 8-30. Displaying the corrupted image, the reconstruction, and the desired reconstruction (i.e., the original uncorrupted image)

```
plt.set_cmap('gray')

for i in range(3):
    plt.figure(figsize=(15, 5), dpi=400)

    plt.subplot(1, 3, 1)
    plt.imshow(modified_clipped[i].reshape((28, 28)))
    plt.axis('off')
    plt.title('Noisy Input')

    plt.subplot(1, 3, 2)
    plt.imshow(x_valid[i].reshape((28, 28)))
    plt.axis('off')
    plt.title('True Denoised')

    plt.subplot(1, 3, 3)
    plt.imshow(model.predict(x_valid[i:i+1]).reshape((28, 28)))
    plt.axis('off')
```

669

```
plt.title('Predicted Denoised')

plt.show()
```

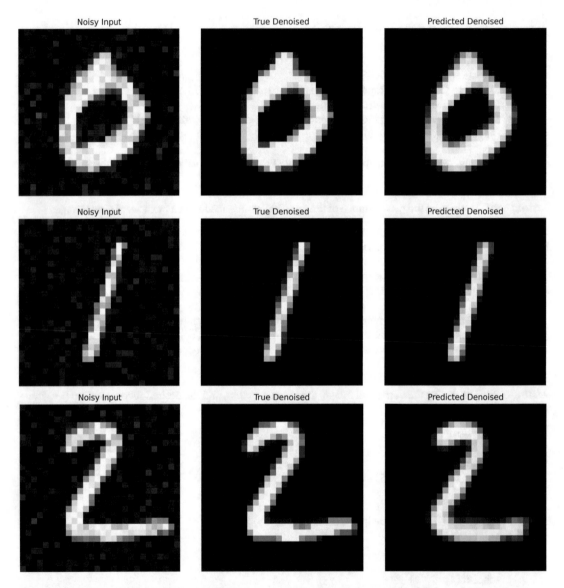

Figure 8-69. *The noisy/perturbed input (left), the unperturbed desired output (middle), and the predicted output (right) for a denoising autoencoder trained on MNIST with a noisy normal distribution of standard deviation 0.1*

Let's increase the standard deviation to 0.2. Figure 8-70 demonstrates the noise effect on the images, and Figure 8-71 demonstrates the reconstruction performance on a set of images. The denoising autoencoder obtains a validation Mean Absolute Error of 0.0289, slightly more than that of the denoising autoencoder trained on noise drawn from a normal distribution with standard deviation 0.1.

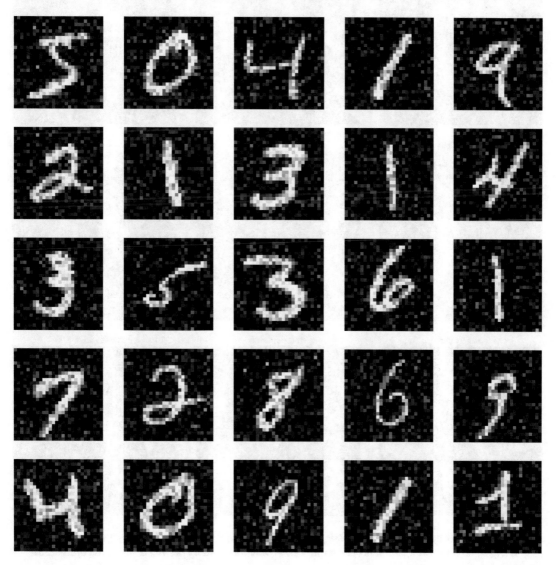

Figure 8-70. *A sample of MNIST images with added normally distributed random noise using standard deviation 0.2*

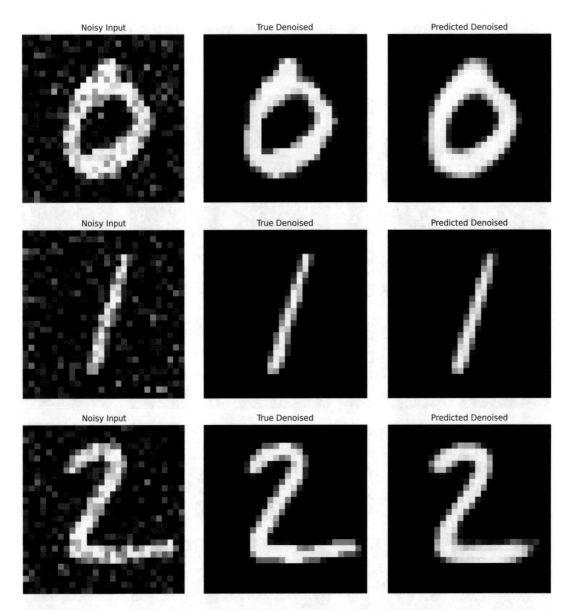

Figure 8-71. *The noisy/perturbed input (left), the unperturbed desired output (middle), and the predicted output (right) for a denoising autoencoder trained on MNIST with a noisy normal distribution of standard deviation 0.2*

Figures 8-72 and 8-73 demonstrate a sample of images and the denoising autoencoder performance on images corrupted using noise drawn from a normal distribution with standard deviation 0.3. The denoising autoencoder obtains a validation Mean Absolute Error of about 0.0343.

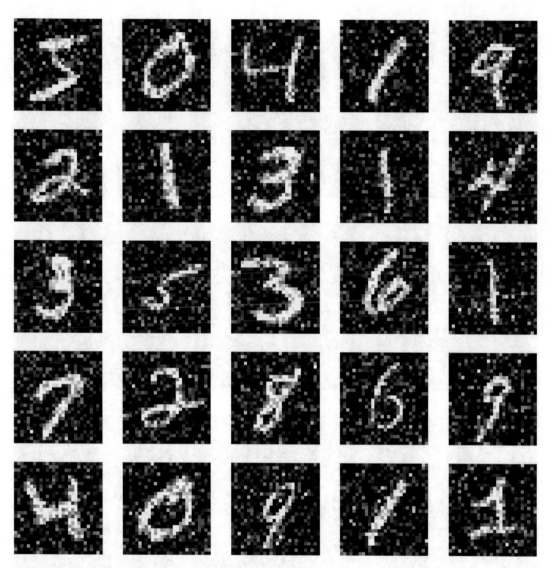

Figure 8-72. *A sample of MNIST images with added normally distributed random noise using standard deviation 0.3*

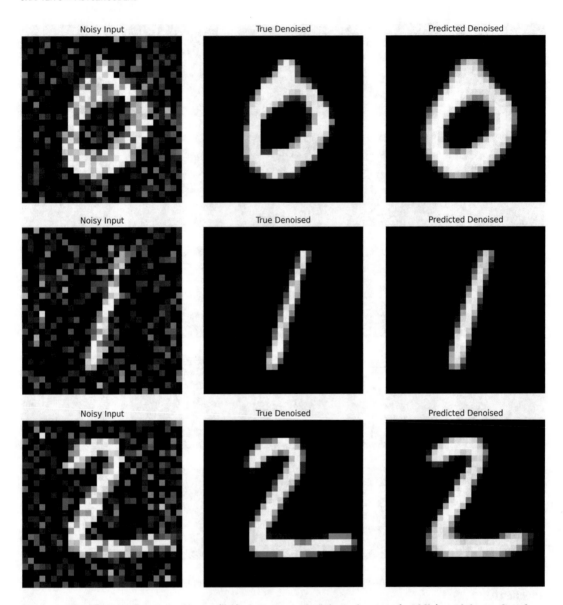

Figure 8-73. *The noisy/perturbed input (left), the unperturbed desired output (middle), and the predicted output (right) for a denoising autoencoder trained on MNIST with a noisy normal distribution of standard deviation 0.3*

Figures 8-74 and 8-75 demonstrate a sample of images and the denoising autoencoder performance on images corrupted using noise drawn from a normal distribution with standard deviation 0.5. The denoising autoencoder obtains a validation Mean Absolute Error of about 0.0427.

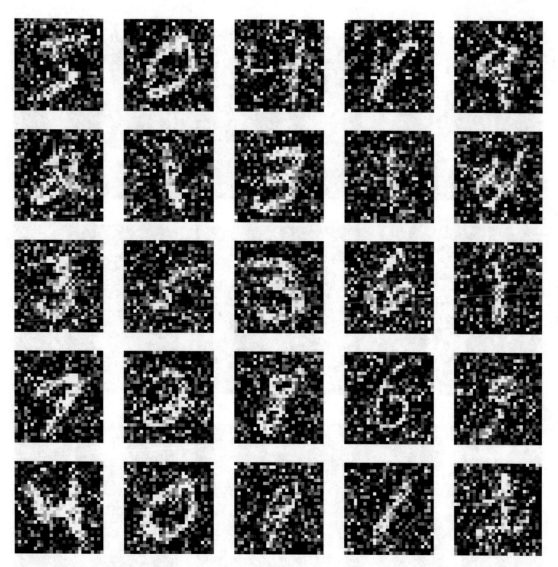

Figure 8-74. *A sample of MNIST images with added normally distributed random noise using standard deviation 0.5*

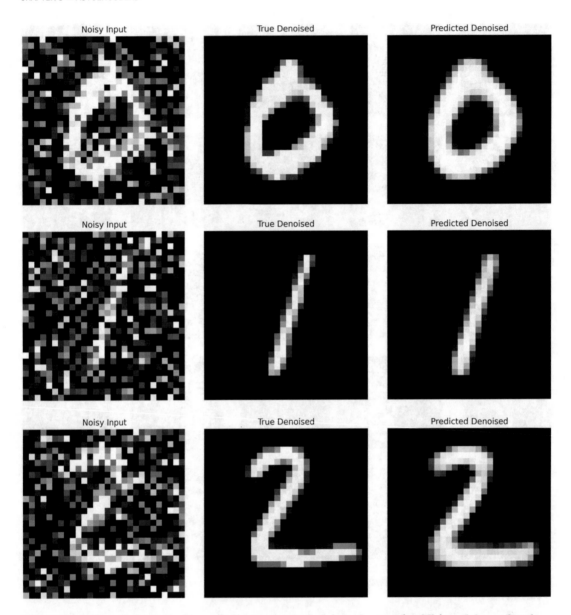

Figure 8-75. *The noisy/perturbed input (left), the unperturbed desired output (middle), and the predicted output (right) for a denoising autoencoder trained on MNIST with a noisy normal distribution of standard deviation 0.5*

Figures 8-76 and 8-77 demonstrate a sample of images and the denoising autoencoder performance on images corrupted using noise drawn from a normal distribution with standard deviation 0.9. The denoising autoencoder obtains a validation Mean Absolute Error of about 0.0683. Note that this is an exceedingly nontrivial task – even humans would have some difficulty denoising many of the shown samples! The autoencoder reconstructions are more abstracted – there is no physical way to exactly reconstruct all the details due to the significant amount of information corruption, so the autoencoder instead performs implicit digit recognition and reconstructs the image as a "generalized" digit with specific positional and orientational characteristics.

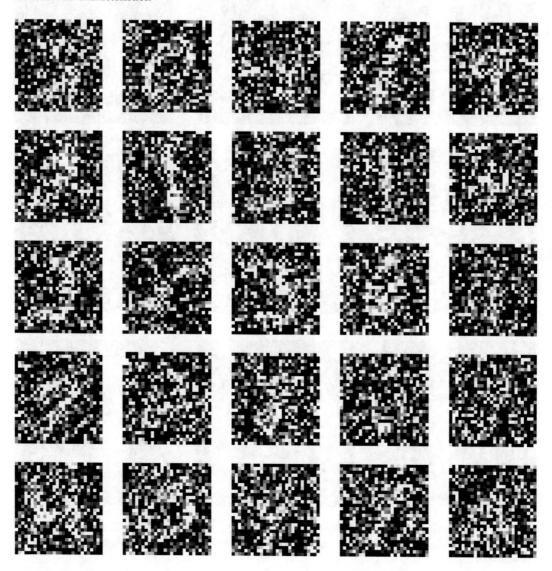

Figure 8-76. *A sample of MNIST images with added normally distributed random noise using standard deviation 0.9*

Figure 8-77. *The noisy/perturbed input (left), the unperturbed desired output (middle), and the predicted output (right) for a denoising autoencoder trained on MNIST with a noisy normal distribution of standard deviation 0.9*

We can see that denoising autoencoders can perform reconstruction to a pretty impressive degree. In practice, however, we want to keep our noise level somewhat low; increasing the noise level can destroy information and cause the network to develop incorrect and/or overly simplified representations of decisions.

A similar logic can be applied to tabular data. There are many situations in which you find that a tabular dataset is particularly noisy. This is especially common in scientific datasets recording variable physical activity, like low-level physics dynamics or biological system data.

Let's build a denoising autoencoder for the Mice Protein Expression dataset. Listing 8-31 loads the dataset and splits it into a training and a validation set.

Listing 8-31. Loading and splitting the Mice Protein Expression dataset

```
data = pd.read_csv('../input/mpempe/mouse-protein-expression.csv').drop(['Unnamed: 0',
'class'], axis=1)
train_indices = np.random.choice(data.index, replace=False,
                                 size = round(0.8 * len(data)))
valid_indices = np.array([ind for ind in data.index if ind\
                          not in train_indices])
x_train, x_valid = data.loc[train_indices], data.loc[valid_indices]
```

Listing 8-32 builds a standard autoencoder architecture.

Listing 8-32. Building an autoencoder architecture to fit the Mice Protein Expression dataset

```
models = buildAutoencoder(len(data.columns), 16)
model = models['model']
encoder = models['encoder']

model.compile(optimizer='adam', loss='mse')
```

To train, we generate noise to the input and train the model to reconstruct the original input from the noisy input. In tabular datasets, we generally cannot add randomly distributed noise to the entire set of data in blanket fashion, because different features operate on different scales. Instead, the noise should be dependent on the standard deviation of each feature itself. In this implementation, we add noise randomly sampled from a normal distribution with a standard deviation equal to one-fifth of the actual feature's standard deviation (Listing 8-33).

Listing 8-33. Adding noise to each column of the Mice Protein Expression dataset with a reflective standard deviation

```
TOTAL_EPOCHS = 100
loss = []
stds = x_train.std()
for i in tqdm(range(TOTAL_EPOCHS)):
    noise = pd.DataFrame(index=x_train.index, columns=x_train.columns)
    for col in noise.columns:
        noise[col] = np.random.normal(0, stds[col]/5,
                                      size=(len(x_train),))
    history = model.fit(x_train + noise, x_train, epochs=1, verbose=0)
    loss.append(history.history['loss'])
```

Listing 8-34 demonstrates the evaluation of such a model on novel validation noisy data.

Listing 8-34. Evaluating the performance of the denoising tabular autoencoder on novel noisy data

```
noise = pd.DataFrame(index=x_valid.index, columns=x_valid.columns)
for col in noise.columns:
    noise[col] = np.random.normal(0, np.sqrt(stds[col]),
                                  size=(len(x_valid),))
```

```
from sklearn.metrics import mean_absolute_error as mae
mae(model.predict(x_valid + noise), x_valid)
```

After training, the encoder of the denoising autoencoder can be used for pretraining or other previously described applications.

Key Points

In this chapter, we discussed the autoencoder architecture and how it can be used in four different contexts – pretraining, multitask training, sparse autoencoders, and denoising autoencoders.

- Autoencoders are neural network architectures trained to encode an input into a latent space with a representation size smaller than the original input and then to reconstruct the input from the latent space. Autoencoders are forced to learn meaningful latent representations of the data because of this imposed information bottleneck.

- The encoder of a trained autoencoder can be detached and built as the feature extractor of a supervised network; that is, the autoencoder serves the purpose of pretraining.

- In cases where supervised learning is difficult to get started with, creating a multitask autoencoder that can optimize its loss by performing both the supervised task and an auxiliary autoencoding task can help overcome initial learning hurdles.

- Sparse autoencoders use a significantly expanded latent space size, but are trained with restrictions on latent space activity, such that only a few nodes/neurons can be active at any one pass. Sparse autoencoders are thought to be more robust.

- Denoising autoencoders are trained to reconstruct clean data from an artificially corrupted, noisy data. In the process, the encoder learns to look for key patterns and "denoises" data, which can be a useful component for supervised models.

In the next chapter, we will look into deep generative models – including a particular type of autoencoder, the Variational Autoencoder (VAE) – which can notably be used to reconcile unbalanced datasets, improve model robustness, and train models on sensitive/private data, in addition to other applications.

CHAPTER 9

■ ■ ■

Data Generation

It is the real, and not the map, whose vestiges persist here and there in the deserts that are
no longer those of the Empire, but ours: The desert of the real itself.

—Jean Baudrillard, Philosopher

Data generation can be defined as creating synthetic data samples based on a selected, existing dataset that
resembles the original dataset. To an extent, the term "resemble" is vague since there's no universal metric
to define one sample's similarity to another without being indifferent. Evaluation of synthetic image data can
be purely judged by human eyes, while tabular data requires calculating the bivariate relationship between
each feature and comparing it with the original dataset.

Now comes the question: Why do we need artificially created data samples when there are data
collected from the real world that may be more representative of the situation? The most straightforward
answer is that we don't have enough data. There can be a vast number of reasons this is so, such as a lack of
resources available for further data collection or the data collection process being too time-consuming, to
name a few.

Machine learning models, especially those associated with deep learning, absolutely require a large
number of data points to achieve decent performance, thus the need for synthetic data to add to potentially
small datasets. Additionally, many classification datasets are imbalanced – their labels are heavily skewed
to one or more classes, leaving few samples for the remaining classes. Many of these cases occur in medical
diagnosis datasets, with significantly fewer positive cases than negative ones. By utilizing conditional data
generation, we can insert positive cases into the dataset until it's balanced, where the model would be able
to classify both classes with equal confidence and accuracy.

Another issue with datasets collected from the real world is sensitive information. Some information
in certain datasets is private to others and protected by law. To train on these datasets, we can generate new
samples by synthesizing these sensitive fields, effectively protecting the victim of the subject. The following
sections will introduce you to various data generation algorithms, including Variational Autoencoders and
Generative Adversarial Neural Networks.

Variational Autoencoders

Variational Autoencoders (VAEs) are one of the most exciting applications of autoencoders. VAEs allow us
to generate new data, which has been used to create realistic images resembling those in the dataset it was
trained on. An advantage of VAEs compared with other data generation techniques (i.e., primarily GANs or
Generative Adversarial Networks) is that of fine control: we can somewhat control what we want to see in
the generated output. VAEs can be used, for instance, to draw a realistic mustache on an image of someone

without a mustache. In the context of tabular data, VAEs allow us to generate synthetic data to increase the size and/or representation of minority samples within tabular datasets. This dataset can then be used to train classical machine learning or deep learning models with better validation or real-world deployment performance.

Theory

To understand the Variational Autoencoder, we need to begin with reunderstanding the logic of the autoencoder itself and the relationship between the encoder and the decoder. The encoder encodes inputs into a latent space, whereas the decoder learns to decode the sample from the encoded latent space vector. However, the decoder cannot simply memorize the association between every latent space vector and the corresponding input (not in a well-designed autoencoder, anyway). The encoder must work to structure the latent space in a way that allows for the decoder to generalize the task of reconstruction. For instance, two very similar images of the digit "3" should have latent space vectors very close to each other, since they are structurally quite similar – the decoder should be able to apply very similar faculties when decoding those inputs. In order for autoencoders to succeed, the latent space must be structured such that similar items are closer and dissimilar items are farther. We have been exploiting this property of autoencoder latent spaces in many of the previous applications and demonstrations; we saw that vanilla autoencoders could perform implicit clustering/separation, that the learned latent space was useful for pretraining, and that the latent space could be simultaneously structured for autoencoding and task training in multitask learning.

In all these applications and demonstrations, the decoder is only allowed to decode vectors given to it by the encoder. However, the specific operations of the encoder and decoder and the individual input encoding pairs are not quite as relevant as the concept of the latent space. These "mappings" help to define the latent space, but ultimately it is the *space*, rather than the specific *points* in the space, that matters. In concrete terms, we should be able to obtain a realistic output by decoding any vector in the latent space, not just the ones that correspond to items in the dataset.

Assume we have an autoencoder that has already been trained on the MNIST dataset (with standard halving/doubling architectural logic). The decoder has learned, to an extent, the *space* around the latent space vector, rather than just the latent vector itself. Holding that an autoencoder has been trained on MNIST, let us demonstrate what happens when we progressively "step away" from a latent encoding directly corresponding to a known sample item. We see that the decoding remains constant for the first few steps and then quickly deforms into something else entirely (Listing 9-1, Figure 9-1).

Listing 9-1. Sampling from the latent space

```
encoded = encoder(X_train[0:1])0

plt.figure(figsize=(10, 10), dpi=400)
for i in range(5):
    for j in range(5):
        plt.subplot(5, 5, i*5+j+1)
        modified_encoded = encoded + 0.5 * (i*5+j+1)
        decoded = decoder(modified_encoded).numpy()
        plt.imshow(decoded.reshape((28,28)))
        plt.axis('off')
plt.show()
```

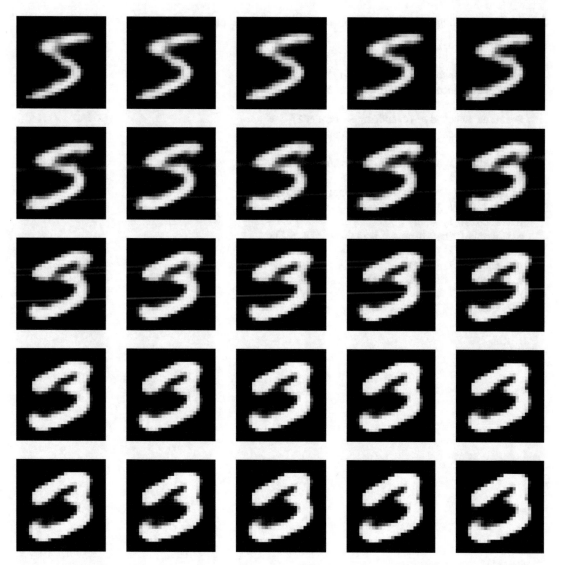

Figure 9-1. *Sample digits decoded by "walking away" from a "real" latent space vector, arranged in a grid for convenient viewing*

We might characterize such a space as "discrete" – notice that for a very large distance everything is constant, but at a critical threshold the decoding suddenly changes and remains constant, too, thereafter. It is not continuous; the change in the shape of the decoded digit is not proportional to how far away we step from the original point in the latent space.

Alternatively, we can try to decode a linear interpolation between two relatively significantly different samples – averaging the latent space vectors to obtain a vector representing a point that is "in between" the latent space vectors of those two samples. Using our hypothesis, which suggests that autoencoders learn the space rather than merely the set of points, the output should be valid and ideally some sort of "in-between" mesh between the two real samples. Let's sample a few (Listing 9-2, Figure 9-2).

Listing 9-2. Sampling using linear interpolation from the latent space

```python
for i in range(10):

    encoded1 = encoder(X_train[i:i+1])
    encoded2 = encoder(X_train[i+1:i+2])

    modified_encoded = (encoded1 + encoded2) / 2
    decoded = decoder(modified_encoded)

    plt.figure(figsize=(10, 3), dpi=400)
    plt.subplot(1, 3, 1)
    plt.imshow(X_train[i:i+1].reshape((28,28)))
    plt.axis('off')
    plt.subplot(1, 3, 2)
    plt.imshow(decoded.numpy().reshape((28,28)))
    plt.axis('off')
    plt.subplot(1, 3, 3)
    plt.imshow(X_train[i+1:i+2].reshape((28,28)))
    plt.axis('off')
    plt.show()
```

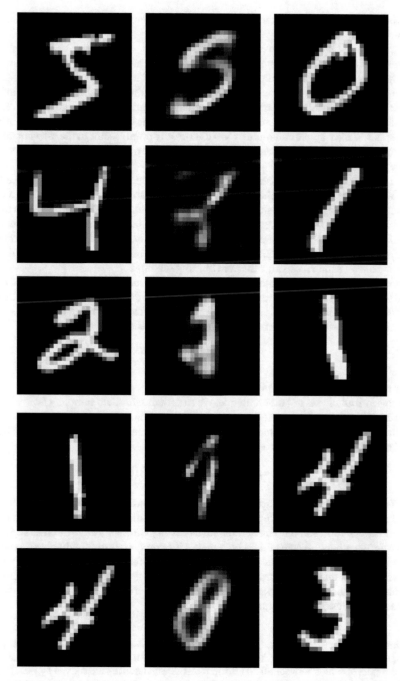

Figure 9-2. *Displaying the decoded results (center column) of linear interpolation between latent vectors of the two images shown on the left and right columns*

The decoded linear interpolations are not valid digits nor meaningful "in-between" meshes of the two digits from which they were derived.

What's going on here? Is our hypothesis/logic incorrect?

The answer is partially correct and partially incorrect. Vanilla autoencoders have no need to learn *most* of the relevant latent space; they only need to learn *some* of the latent space around similar examples. Instead, autoencoders make use of discrete separation of items in the latent space to help "bin" samples in discrete fashion (as we observed previously), which the decoder can coordinate and decode. We want a way to impose continuity upon the latent space such that the decoder is forced to learn most, if not all, of the relevant latent space.[1] This way, we can more or less choose any point/vector in the relevant latent space and reasonably decode it into a realistic-looking digit. The latent space must be structured to be continuously and realistically interpolatable.

Variational Autoencoders accomplish this by forcing the encoder to predict a latent space *distribution* in lieu of each latent space dimension. The encoder predicts a mean and a standard deviation to define a normal distribution rather than a single scalar value (for each dimension). The decoder then decodes a latent space vector randomly sampled from this set of distributions and attempts to reconstruct the original input as faithfully as possible. By representing the latent space probabilistically rather than explicitly, the Variational Autoencoder is forced to learn feasible intermediate representations throughout the entire relevant latent space and cannot construct discrete binning schemes.

We will formalize the Variational Autoencoder: the encoder produces two outputs, μ and σ; both are n-dimensional vectors representing the means and standard deviations of n normal distributions, where n is the dimensionality of the latent space. We would like to sample a latent vector z from this n-dimensional normal distribution, where the ith element of z is sampled as follows:

$$z^{(i)} \sim \mathcal{N}\left(\mu^{(i)}, \sigma^{(i)}\right)$$

The latent vector z is then decoded by the decoder.

However, this formulation is not differentiable. We cannot differentiate sampling from a normal distribution defined by a learned mean and standard deviation. Thus, we use a *reparameterization trick*, articulated in the original variational autoencoding paper "Auto-Encoding Variational Bayes" by Diederik Kingma and Max Welling approximately as follows[2]:

$$z^{(i)} = \mu^{(i)} + \sigma^{(i)} \odot \epsilon^{(i)}, \quad \epsilon \sim \mathcal{N}(0, c)$$

Here, $\odot$ refers to element-wise multiplication, and c is an arbitrary constant. This formulation provably has the same character as directly sampling from a normal distribution, but it re-expresses the sampling with respect to the means and standard deviations such that all parameters are written as additions and multiplications of each other and constants, which makes it a completely differentiable scheme.

Thus, the objective of the Variational Autoencoder can be expressed as follows, where y represents the ground truth, z represents the vector sampled from the output of the encoder, x represents the input to the encoder (and the network, technically speaking), E represents the encoder, D represents the decoder, and $L(y, \hat{y})$ represents some loss/error function:

---

[1] The qualification for "relevant," of course, is blurry. A reader who is disturbed by this ambiguity may find the following notes helpful. The intuitive definition is "space that is between samples"; if a one-dimensional latent space only maps inputs to values between –5 and 5, then the relevant latent space is given by [–5, 5] and perhaps a bit more on the sides – [–6, 6]. However, as demonstrated by Balestriero, Pesenti, and LeCun in the 2021 paper "Learning in High Dimension Always Amounts to Extrapolation," this intuition does not hold for higher-dimensional spaces. For the purposes of this book, a convex hull of the latent space suffices to qualify as the "relevant latent space."

Balestriero, R., Pesenti, J., & LeCun, Y. (2021). Learning in High Dimension Always Amounts to Extrapolation. *ArXiv, abs/2110.09485.*

[2] Kingma, D.P., & Welling, M. (2014). Auto-Encoding Variational Bayes. *CoRR, abs/1312.6114.*

$$\min_{E,D} - \mathbb{E}_{z \sim \mathcal{N}\left(E(x)_{\mu}, E(x)_{\sigma}\right)} L\big(x, D(z)\big)$$

Here, we want to find the parameters of the functions E and D, which minimize the average difference between the input x and the decoding of a random latent space vector, which is sampled from $\mathcal{N}\left(E(x)_{\mu}, E(x)_{\sigma}\right)$, a normal distribution defined in the latent space with the mean and standard deviation provided by the encoder.

However, given that autoencoders tend to learn discrete representations of the latent space, we have a problem: it is advantageous for the network to learn small standard deviations $\sigma \to 0$ and to maximize the distance between the mean vectors, or $\sum_{i=1}^{n}\sum_{j=i+1}^{n}\left\| \mu^{(i)} - \mu^{(j)} \right\|$, such that it is essentially identical to a standard autoencoder, which outputs only scalars (a distribution with zero standard deviation is a single point) in a discrete space.

As such, we can formulate the following penalty term:

$$\sigma - \log(\sigma) + \mu^2$$

Firstly, we want to minimize the means; thus, we add L2-style regularization in which the network is penalized if the means are too large. Secondly, we want to maximize the standard deviations, but we also want them to be broadly uniform to ensure continuous homogeneity across distributions in the latent space; thus, the $-\log(\sigma)$ term severely punishes standard deviations that are too small, but the $+\sigma$ term also tends standard deviations toward smaller values if they are too large. The minimum of $x - \log x$ is about 0.797 and occurs approximately at $x = 0.434$. Some variations of the penalty will square the standard deviations (which is unnecessary but may improve convergence in some cases) and subtract a constant for aesthetic purposes (such that this penalty term can theoretically be zero):

$$\sigma^2 - \log(\sigma) + \mu^2 - 1$$

The formal optimization problem, then, can be expressed in one form as follows:

$$\min_{E,D} - \mathbb{E}_{z \sim \mathcal{N}\left(E(x)_{\mu}, E(x)_{\sigma}\right)} L\big(x, D(z)\big) + \mathbb{E}\left(E(x)_{\sigma} - \log E(x)_{\sigma} + E(x)_{\mu}^2\right)$$

Implementation

Let's begin by demonstrating a canonical introduction of the Variational Autoencoder – interpolating between digits from the MNIST dataset. We'll begin by creating the encoder architecture (Listing 9-3). Rather than only having one latent space output, we'll have two outputs – one defining the mean of the latent space and the other the standard deviation. Note that we technically learn the logarithm of the standard deviation (which will be exponentiated later) such that the network can more easily grapple with larger scales of standard deviations. (The larger the standard deviations predicted, the more strictly and widely continuity is imposed upon the space.) This allows us to have an open linear activation rather than a zero-bounded one too.

Listing 9-3. Building the encoder of the Variational Autoencoder, which outputs both the means and the log-standard deviations of the latent variables

```
# encoder
enc_inputs = L.Input((784,), name='input')
enc_dense1 = L.Dense(256, activation='relu',
                     name='dense1')(enc_inputs)
```

```
enc_dense2 = L.Dense(128, activation='relu',
                        name='dense2')(enc_dense1)
means = L.Dense(32, name='means')(enc_dense2)
log_stds = L.Dense(32, name='log_stds')(enc_dense2)
```

Note that the output dimensionalities of the means and log_stds layers are the same because they must correspond with each other.

Next, we will define a sampling layer that takes in the derived means and log-standard deviations (Listing 9-4). In order to maintain the ability to propagate information through the means and standard deviations, we sample a small normally distributed noise vector centered at zero, multiply it by the "standard deviation," and add it to the mean (using the reparameterization trick rather than simply sampling from a normal distribution with the specified mean and standard deviation).

Listing 9-4. Defining a custom sampling layer, which samples a random vector from the outputted latent space

```
def sampling(args):
    means, log_stds = args
    eps = tf.random.normal(shape=(tf.shape(means)[0], 32),
                            mean=0, stddev=0.1)
    return means + tf.exp(log_stds) * eps

x = L.Lambda(sampling, name='sampling')([means, log_stds])
```

We can complete the encoder (Listing 9-5). The encoder technically outputs only the sampled latent space vector x, but in order to compute the penalty term, we will also output means and log_stds (which we will use to calculate the loss but will not pass to the decoder).

Listing 9-5. Defining the encoder

```
encoder = keras.Model(inputs=enc_inputs,
                        outputs=[means, log_stds, x],
                        name='encoder')
```

The decoder model is fairly standard; it simply takes in the sampled latent space vector and decodes it back into the original output (Listing 9-6).

Listing 9-6. Defining the decoder

```
# decoder
dec_inputs = L.Input((32,), name='input')
dec_dense1 = L.Dense(128, activation='relu',
                        name='dense1')(dec_inputs)
dec_dense2 = L.Dense(256, activation='relu',
                        name='dense2')(dec_dense1)
output = L.Dense(784, activation='sigmoid',
                    name='output')(dec_dense2)
decoder = keras.Model(inputs=dec_inputs,
                        outputs=output,
                        name='decoder')
```

To construct the complete model, we pass the input through the encoder and the encoded latent space vector through the decoder (Listing 9-7).

Listing 9-7. Constructing the entire Variational Autoencoder

```
# construct vae
vae_inputs = enc_inputs
encoded = encoder(vae_inputs)
decoded = decoder(encoded[2])
vae = keras.Model(inputs=vae_inputs,
                  outputs=decoded,
                  name='vae')
```

Now, we can compute the loss as a relationship between the outputs of the models; since this custom loss is dependent on layers that are not strictly the output of the primary model (i.e., the means and log-standard deviations), we add the loss separately from compilation (Listing 9-8).

Listing 9-8. Adding the custom loss and fitting

```
from keras.losses import binary_crossentropy
reconst_loss = binary_crossentropy(vae_inputs, decoded)
kl_loss = log_stds - tf.exp(log_stds) + tf.square(means)
kl_loss = tf.square(tf.reduce_sum(kl_loss, axis=-1))
vae_loss = tf.reduce_mean(reconst_loss + kl_loss)

# compile model
vae.add_loss(vae_loss)
vae.compile(optimizer='adam')

# fit
vae.fit(x_train, x_train, epochs=20)
```

After training, the Variational Autoencoder should have learned the *entire* relevant latent space (by relevant, we mean the entire space enclosed by a theoretical boundary surface). We can visualize an example traversal of this latent space. We will begin with a base latent space vector that represents the encoder-learned representation for a sample that does exist in the dataset. Then, we'll "cut" a planar cross-section to visualize a "slice" of the learned latent space. There are many ways to do so, but in this case, we simply create a linear addition (or subtraction) to the base vector dependent on a grid of row and column values (Listing 9-9).

Listing 9-9. Obtaining a grid of spatially interpolated and "continuous" images by taking a cross-section of the learned latent space

```
i = 0

base = encoder.predict(x_train[i:i+1])[2]

plt.figure(figsize=(10, 10), dpi=400)
for row in range(10):
    for col in range(10):
        plt.subplot(10, 10, (row) * 10 + col + 1)
        add = np.zeros(base.shape)
        add[:, [0, 2, 4, 6]] = 0.25 * (row - 5)
        add[:, [1, 3, 5, 7]] = 0.25 * (col - 5)
        decoded = decoder.predict(base + add)
```

```
        plt.imshow(decoded.reshape((28, 28)))
        plt.axis('off')
plt.show()
```

The result, shown in Figure 9-3, demonstrates such a cross-section of the learned latent space. Note how image representations spatially situated closer together in the latent space cross-section are also more similar in digit morphology. We can select any two identifiable digits, like the 7 in the top-left corner and the 1 in the top-right corner, draw a line between the two, and trace an "interpolation" from one digit to the other. Functionally, all these digits, except for the base case in the top-left corner, are synthetically generated, yet they look – for the most part – realistic. Forcing the autoencoder to learn the entire latent space, then interpolating within the latent space, allows us to generate realistic synthetic data.

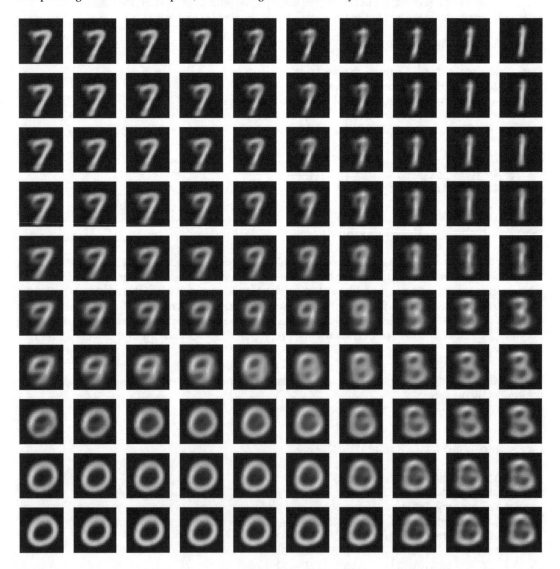

Figure 9-3. *A visualization of a cross-section of the latent space learned by a Variational Autoencoder trained on MNIST*

Variational Autoencoders can just as easily be applied to generate synthetic tabular data. Let's adapt our VAE code to fit the Higgs Boson dataset (Listing 9-10).

Listing 9-10. Building a Variational Autoencoder for the Higgs Boson dataset

```
# encoder
enc_inputs = L.Input((28,), name='input')
enc_dense1 = L.Dense(16, activation='relu',
                        name='dense1')(enc_inputs)
enc_dense2 = L.Dense(16, activation='relu',
                        name='dense2')(enc_dense1)
means = L.Dense(8, name='means')(enc_dense2)
log_stds = L.Dense(8, name='log-stds')(enc_dense2)

def sampling(args):
    means, log_stds = args
    eps = tf.random.normal(shape=(tf.shape(means)[0], 8),
                           mean=0, stddev=0.15)
    return means + tf.exp(log_stds) * eps

x = L.Lambda(sampling, name='sampling')([means, log_stds])

encoder = keras.Model(inputs=enc_inputs,
                        outputs=[means, log_stds, x],
                        name='encoder')

# decoder
dec_inputs = L.Input((8,), name='input')
dec_dense1 = L.Dense(16, activation='relu',
                        name='dense1')(dec_inputs)
dec_dense2 = L.Dense(16, activation='relu',
                        name='dense2')(dec_dense1)
output = L.Dense(28, activation='linear',
                    name='output')(dec_dense2)
decoder = keras.Model(inputs=dec_inputs,
                        outputs=output,
                        name='decoder')

# construct vae
vae_inputs = enc_inputs
encoded = encoder(vae_inputs)
decoded = decoder(encoded[2])
vae = keras.Model(inputs=vae_inputs,
                    outputs=decoded,
                    name='vae')

# build loss function
from keras.losses import mean_squared_error
reconst_loss = mean_squared_error(vae_inputs, decoded)
kl_loss = 1 + log_stds - tf.square(means) - tf.exp(log_stds)
kl_loss = tf.square(tf.reduce_sum(kl_loss, axis=-1))
vae_loss = tf.reduce_mean(reconst_loss + kl_loss)
```

```
# compile model
vae.add_loss(vae_loss)
vae.compile(optimizer='adam')

# fit
vae.fit(X_train, X_train, epochs=20)
```

There are many ways to generate data from the trained Variational Autoencoder. One way is to select several latent space "bases" by obtaining the learned encodings and randomly moving "around" those samples (Listing 9-11).

Listing 9-11. Generating novel tabular data samples by randomly moving around known latent space encodings

```
NUM_BASES = 40
NUM_PER_SAMPLE = 20
samples = []

for i in tqdm(range(NUM_BASES)):
    base = encoder.predict(X_train[i:i+1])[2]
    for i in range(NUM_PER_SAMPLE):
        add = np.random.normal(0, 1, size=base.shape)
        generated = decoder.predict(base + add)
        samples.append(generated[0])

samples = np.array(samples)
generated = pd.DataFrame(samples, columns=X.columns)
```

Let's visualize some representations of the dataset structure of the original dataset compared to the generated dataset. We can get a decent perspective of a dataset's structure using pairplots. Listing 9-12 demonstrates the code used to generate Figures 9-4 and 9-5, sample pairplots for the generated and real datasets, respectively. Note that many of the bivariate relationships are distributed quite similarly.

Listing 9-12. Plotting bivariate relationships between two sets of five variables in the true and generated datasets

```
plt.figure(figsize=(50, 50), dpi=400)
sns.pairplot(generated,
             x_vars = X.columns[:5],
             y_vars = X.columns[5:10],
             kind='kde')
plt.show()

plt.figure(figsize=(50, 50), dpi=400)
sns.pairplot(X.iloc[np.random.choice(len(X), size=800, replace=False)],
             x_vars = X.columns[:5],
             y_vars = X.columns[5:10],
             kind='kde')
plt.show()
```

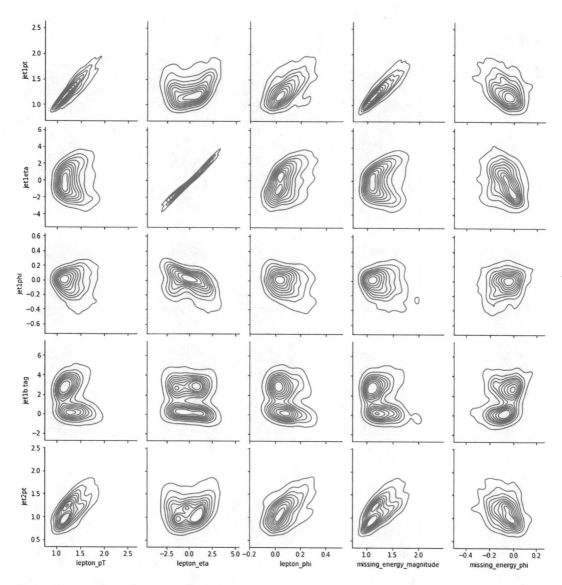

Figure 9-4. *Bivariate relationships/interactions between two sets of five features generated by a Variational Autoencoder trained on the Higgs Boson dataset*

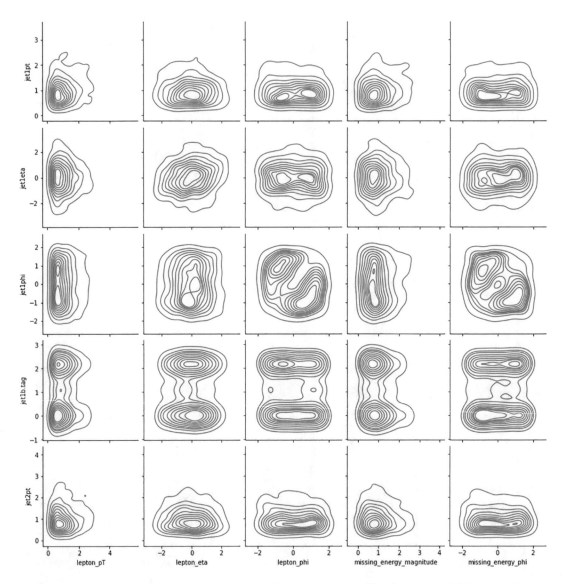

Figure 9-5. *True bivariate relationships/interactions between two sets of five features pulled from the Higgs Boson dataset*

Figures 9-6 and 9-7 demonstrate pairplots of the generated and real data from the Mice Protein Expression dataset. Here, too, the distribution and shape of bivariate relationships within the generated dataset is quite similar to that of the real dataset.

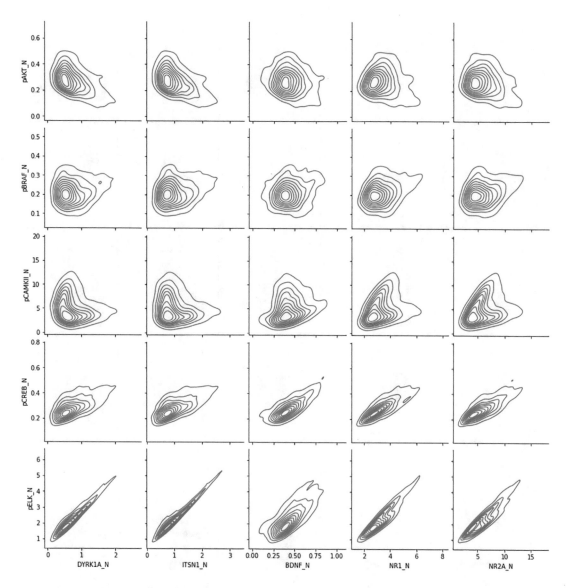

Figure 9-6. *Bivariate relationships/interactions between two sets of five features generated by a Variational Autoencoder trained on the Mice Protein Expression dataset*

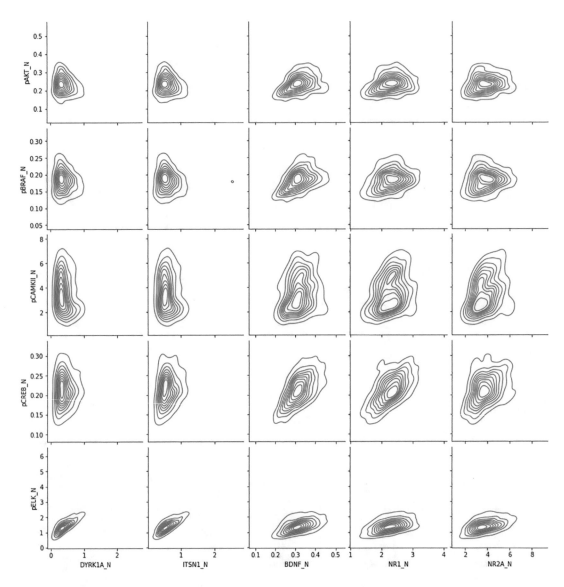

Figure 9-7. *True bivariate relationships/interactions between two sets of five features pulled from the Mice Protein Expression dataset*

Variational Autoencoders offer sophisticated data generation capabilities, which can help build more successful machine learning models (neural networks or classical models) on otherwise weak or lacking data.

Generative Adversarial Networks

The painting *Edmond de Belamy* looks like a standard seventeenth- and eighteenth-century painting. A hunched man, suit painted in broad strokes of black ink blending into a dark background, looks blankly at the viewer with shaded eyes. But in the bottom right, in lieu of the artistic signature, is in elegant script

$$\min_{G} \max_{D} E_{x} \left[\log(D(x)) \right] + E_{y}[\log(1 - D(G(y)))]$$

This is a simplified expression representing the objective of the Generative Adversarial Network (GAN) model, which constitutes the other major deep generative model family besides Variational Autoencoders. Notably, GANs have been the underlying technology behind impressively realistic art and image generation, but they have also been applied in a less common but no less promising capacity to generate other forms of data, including tabular data.

Theory

Let us begin by formalizing the equation articulated in *Edmond de Belamy*:

- Let z represent a random noise vector.

- Let x represent a data sample drawn from the data.

- Let G be a neural network that accepts y and transforms it into an output that mimics a sample drawn from the data.

- Let D be a neural network that outputs the probability that the given sample is drawn from the data (as opposed to being generated by G).

- Let $Exf(x)$ represent the average value of a function f that depends on some variable x.

The expression represents the discriminator's average log-loss[3] and consists of two terms: $E_{x}[\log(D(x))]$ and $E_{z}[\log(1 - D(G(z)))]$. The discriminator can be thought of as having two jobs: classifying samples drawn from the data (i.e., x) as being drawn from the data and classifying synthetic samples created by the generator from a random noise vector (i.e., $G(z)$) as not being drawn from the data. These two objectives are respectively represented by the expression.

The first term represents the average logarithm of the discriminator's prediction on samples drawn from the data. If the discriminator is working perfectly, then this value will be maximized because $D(x)$ is predicting 1, $\log(D(x)) = 0$, and $E_{x}[\log(D(x))] = 0$. The other term represents the average logarithm of the discriminator's inverted prediction on generated samples. If the discriminator is working perfectly, this value will also be maximized because $D(x)$ is predicting 0, $1 - D(G(z)) = 1$, $\log(1 - D(G(z)) = 0$, and $E_{z}[\log(1 - D(G(z)))] = 0$. Thus, the sum of these two terms is maximized to the value zero when the discriminator performs perfectly.

The objective of the discriminator is to maximize this expression, whereas the objective of the generator is to minimize it. Putting the parts together, we obtain the following system in which the generator and discriminator play against each other:

$$\min_{G} \max_{D} E_{x} \left[\log(D(x)) \right] + E_{z}[\log(1 - D(G(z)))]$$

Figure 9-8 presents a visual representation of the previous equation.

---

[3] Assuming the number of synthetic samples and real samples the model is exposed to is the same, which is generally the case.

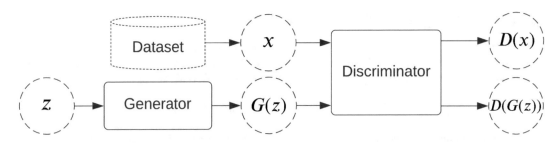

Figure 9-8. *Schematic of the Generative Adversarial Network system*

The more complete articulation presented by Ian Goodfellow et al. in the original paper[4] is as follows, with z a random vector drawn from the distribution $pz(z)$ and x a data sample drawn from the distribution $p_{\text{data}}(x)$:

$$\min_{G} \max_{D} \mathrm{E}_{x \sim p_{\text{data}}(x)}\left[\log(D(x))\right] + \mathrm{E}_{z \sim p_z(z)}\left[\log\left(1 - D\left(G(z)\right)\right)\right]$$

To update the discriminator, we use the gradient of its loss with respect to the discriminator parameters θ_d:

$$\nabla_{\theta_d} E\left[\log D(x) + \log\left(1 - D\left(G(z)\right)\right)\right]$$

The gradients are used for *ascending* because the discriminator aims to maximize rather than to minimize its performance. Recall from earlier that this expression is maximized when the discriminator performs perfectly. The only difference with gradient descent is to move in the direction of greatest increase rather than the direction opposite to it (i.e., of greatest decrease).

After updating the discriminator, we update the generator in descending fashion using the gradient of the second term with respect to the generator parameters θ_g:

$$\nabla_{\theta_g} E\left[\log\left(1 - D\left(G(z)\right)\right)\right]$$

We do not need to bother with the first term because the generator is not involved in it; it has no contribution to the gradient calculation. Note that the generator is explicitly updated to "fool" the discriminator by minimizing the discriminator's performance classifying generated items as generated.

By updating the discriminator followed by the generator repeatedly, the two models play an adversarial game.

The formal training algorithm articulated in the original paper is as follows:

1. For k steps, do

   a. Sample a minibatch of m noise samples $\{z^{(1)}, z^{(2)}, ..., z^{(m)}\}$ from the noise prior $pg(z)$.

   b. Sample a minibatch of m training samples $\{x^{(1)}, x^{(2)}, ..., x^{(m)}\}$ from the true data distribution $p_{\text{data}}(x)$.

---

[4] Goodfellow, I.J., Pouget-Abadie, J., Mirza, M., Xu, B., Warde-Farley, D., Ozair, S., Courville, A.C., & Bengio, Y. (2014). Generative Adversarial Nets. *NIPS*.

c. Update the discriminator by stochastic gradient ascent (or another optimization method using the same loss):

$$\nabla_{\theta_d} \frac{1}{m} \sum_{i=1}^{m} \left[\log D\left(x^{(i)}\right) + \log\left(1 - D\left(G\left(z^{(i)}\right)\right)\right) \right]$$

2. Sample a new minibatch of m noise samples $\{z^{(1)}, z^{(2)}, ..., z^{(m)}\}$ from the noise prior $pg(z)$.

3. Update the generator by stochastic gradient descent (or another optimization method using the same loss):

$$\nabla_{\theta_g} \frac{1}{m} \sum_{i=1}^{m} \left[\log\left(1 - D\left(G\left(z^{(i)}\right)\right)\right) \right]$$

4. Repeat until the specified number of training iterations is completed.

Note the following features of the optimization procedure:

- By changing k, you control the ratio of the number of "moves" the discriminator has relative to the generator. The authors of the original GAN paper use $k = 1$ to maximize computational efficiency.

- From the perspective of the computational graph, we are technically optimizing the same collection of operations – just different parts of it at any one time. Observe how the respective expressions for gradient ascent and descent in the discriminator and generator update formulas involve the same nested $D(G(...))$ expression, but only one part of it is updated, whereas the other is held static.

- When we update the generator, we select a new random minibatch of noise samples rather than using the same set of samples the discriminator was updated on.

GANs are, however, notoriously difficult to train. The system is often unstable and fails to converge. Tim Salimans and other researchers from OpenAI suggest several actionable training modifications for improving convergence and performance of GAN systems in the paper "Improved Techniques for Training GANs,"[5] some of which are summarized in the following:

- *Feature matching*: It can be difficult for the generator to produce samples given the explicit objective to maximize the discriminator's prediction on generated samples. Instead, the generator can be trained to match the activations of an intermediate layer of the discriminator. If the generator produces an output that, on aggregate, yields discriminator activations for some feature, which are very different from the activations derived for real items, then the generator will likely not succeed in fooling the discriminator. Feature matching is a technique to directly optimize the generator using the "internal thinking processes" of the discriminator.

- *Historical averaging*: Each player is penalized by the magnitude of its weight updates relative to a historical time period of length t: $\left\| \theta[0] - \frac{1}{t} \sum_{i=1}^{t} \theta[i] \right\|$, where θ is an array of the model parameters and $\theta[0]$ represents the most recent, current set of parameters. This provides a constant force, which pushes players toward convergence: any large updates will be penalized by this quantity, which therefore can constrain erratic nonconverging behavior.

---

[5] Salimans, T., Goodfellow, I., Zaremba, W., Cheung, V., Radford, A., & Chen, X. (2016). Improved Techniques for Training GANs. *ArXiV, abs/1606.03498*.

- *One-sided label smoothing*: Label smoothing, in which discrete binary labels are replaced with probabilistic approximations (e.g., replacing 0 with 0.1, 1 with 0.9). We can replace positive discriminator labels (i.e., real data samples that have a label of 1) with a smooth approximation (e.g., 0.9 or even 0.8) such that the generator can more easily achieve equivalence with "real" samples in the "eye" of the discriminator even if it cannot produce a high discriminator output. For instance, let x be a real data sample and $D(x) = 1$. Let z be a random vector; say $D(G(z)) = 0.9$. In this case the discriminator was fooled but not completely. If we apply one-sided label smoothing, however, the discriminator will learn $D(x) = 0.9$. Therefore, $G(z)$ and x are comparable in that the discriminator yields the same determination for both. This can make generation an easier task.

Many subsequent methods have been proposed to expand GAN capabilities. The conditional GAN, introduced by Medhi Mirza and Simon Osindero shortly after the original GAN paper in the 2014 paper entitled "Conditional Generative Adversarial Nets,"[6] allows for the generation of outputs conditioned on some property. For instance, rather than just generating arbitrary digits, a conditional generative model can generate a particular type of digit (0, 1, 2, etc.).

This is accomplished with a simple modification to the original GAN system optimization objective: the discriminator accepts both the original input (either a sample from the dataset or synthesized from the generator) and conditional information y. The output of the generator is generated from both the random vector z and the conditional information y:

$$\min_{G} \max_{D} \mathrm{E}_{x \sim p_{\text{data}}(x)}\left[\log(D(x|y))\right] + \mathrm{E}_{z \sim p_z(z)}\left[\log(1 - D(G(z|y)))\right]$$

To be concrete, consider applying a conditional GAN to generate images of digits. We begin by selecting n images of digits x and their corresponding class y (e.g., the first image has class 2, the second has class 9, ..., the nth has class 0). Next, we randomly sample noise vectors z and pass both z and y into the generator. Now, we have generated samples: $G(z|y)$. The discriminator predicts on both the original and generated samples with access to the corresponding classes. Even if the generator generates realistic images, the discriminator can theoretically detect if a sample is generated if the given input, say, is an image of the digit 8 but the associated class (i.e., of y) is 2. Thus, the generator must generate realistic images conditional on the given attributes (Figure 9-9).

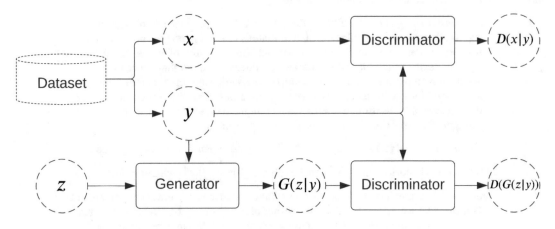

Figure 9-9. *Schematic of the conditional Generative Adversarial Network system*

---

[6] Mirza, M., & Osindero, S. (2014). Conditional Generative Adversarial Nets. *ArXiv, abs/1411.1784.*

While GANs have found incredible success in image generation, particularly in text-conditioned generation, tabular data generation remains a difficult problem. The cross-column heterogeneity of tabular data – that is, the variation across columns in properties like value range, sparsity, distribution, discreteness/continuity, value imbalance, and so on – makes it difficult to easily transfer scales and rules across the span of the dataset.

Lei Xu, Maria Skoularidou, Alfredo Cuesta-Infante, and Kalyan Veeramachaneni proposed the very successful Conditional Tabular GAN (CTGAN) to address the deep adversarial generation of tabular data in the 2019 paper "Modeling Tabular Data Using Conditional GAN."[7]

The Conditional Tabular GAN model is quite complex and has many moving parts. Here, we summarize important elements of the architecture and training process. View the original paper for details.

Each row is represented by concatenating continuous and discrete columns. Discrete columns are one-hot encoded, and continuous features with complex multimodal distributions (in this context, multimodal refers to distributions with multiple "humps," or modes, as opposed to a singular mass – like in a normal distribution) are normalized using a randomly sampled mode. Let the row be denoted as $\hat{r}$.

In order to ensure that the GAN system has roughly equal exposure to different values in discrete columns – even highly imbalanced or sparse ones – the authors propose a conditional vector, which, in rough terms, represents which types of values in which discrete columns must be reproduced by the generator. Let m represent this conditional vector. Then, we have the generator output as $G(z|m)$, given a randomly sampled vector z and the conditional vector m. This is a similar training paradigm to masked language modeling pretraining discussed in Chapter 6, on attention and transformers. In masked language modeling, the model is presented with part of the words in a sentence and must fill in the masked tokens. The generator in the CTGAN is presented with "part of the sentence" (i.e., the selected values in the selected columns) and must "fill in" the rest of the row (i.e., which continuous column and non-selected discrete column values make sense given the provided conditional vector). Given an equal conditional vector generation procedure, the generator and discriminator are exposed to a much wider range of the feature space and acclimate to more complex data forms.

The generator model consists of two hidden layers using ReLU activations, batch normalization, and residual connections. There are two relevant outputs – the continuous features c and the discrete features d. It can be approximately formally articulated as follows, where $\oplus$ represents vector concatenation and Gumbel represents the softmax Gumbel function[8]:

$$h_0 = z \oplus m$$

$$h_1 = h_0 \oplus \text{ReLU}\left(\text{BN}\left(\text{FC}(h_0)\right)\right)$$

$$h_2 = h_0 \oplus \text{ReLU}\left(\text{BN}\left(\text{FC}(h_0)\right)\right)$$

---

[7] Xu, L., Skoularidou, M., Cuesta-Infante, A., & Veeramachaneni, K. (2019). Modeling Tabular Data Using Conditional GAN. *ArXiv, abs/1907.00503.*

[8] The softmax Gumbel function, introduced by Eric Jang, Shixiang Gu, and Ben Poole in the paper "Categorical Reparametrization with Gumbel_Softmax", is a clever trick to discretely sample from a differentiable distribution. Given class probabilities π, a temperature parameter τ, and independently and identically distributed samples g sampled from the Gumbel distribution (PDF: $f(x) = e^{-\left(x+e^{-x}\right)}$), we can sample from π a 'discrete' vector y where the ith element is given by the formula (assuming k represents the number of unique classes):

$$y_i = \frac{\exp\left(\dfrac{\log \pi_i + g_i}{\tau}\right)}{\sum_{j=1}^{k} \exp\left(\dfrac{\log \pi_j + g_j}{\tau}\right)}$$

Jang, E., Gu, S.S., & Poole, B. (2017). Categorical Reparameterization with Gumbel-Softmax. *ArXiv, abs/1611.01144.*

$$c = \tanh\left(FC\left(h_2\right)\right)$$

$$d = \text{Gumbel}\left(FC\left(h_2\right)\right)$$

$$G : z, m \mapsto c, d$$

The CTGAN model performs well on tabular data generation across a wide range of tabular datasets, including heterogenous tabular data, which has challenged previous tabular generation attempts. It has shown promise on addressing imbalanced datasets and providing data augmentation.

In the next subsections, we will demonstrate the implementation of a sample GAN model from scratch and using the CTGAN.

Simple GAN in TensorFlow

To demonstrate the flow of modeling and training a simple GAN, the MNIST dataset will be adopted as our target generated images. The dataset is available through `tensorflow.keras.datasets` as shown in Listing 9-13. To ensure our generation produces a variety of images, the number "1" is removed from the dataset to avoid mode collapse as it's probable that the generator can fool the discriminator by simply producing slanted images of a line.

Listing 9-13. Imports and retrieving the dataset

```python
import numpy as np
import pandas as pd
import tensorflow as tf
import os
import matplotlib.pyplot as plt

import tensorflow.keras.layers as L
import tensorflow.keras.models as M
import tensorflow.keras.callbacks as C
from tensorflow.keras.datasets import mnist

(X_train, y_train), (X_test, y_test) = mnist.load_data()
# remove 1
X_train = X_train[y_train!=1]
# reshape to (28, 28, 1)
X_train = np.expand_dims(X_train, axis=3)
# for operations later
X_train = X_train.astype("float32")
# normalize
X_train /= 255.0
```

Next, we're going to build the discriminator. Recall from earlier that regardless of the generation task, the discriminator will always be a classifier that distinguishes between real and generated images. To make things simple, the discriminator model architecture shown in the following will be a five-layer fully connected network (Listing 9-14). You can further improve the discriminator's performance by using convolutional layers.

Listing 9-14. Defining the discriminator

```
# discriminator
# simple fully-connected NN, can be modified to CNN to improve performance
# flatten 2D images
inp = L.Input(shape=(28, 28, 1))
x = L.Flatten(input_shape=[28, 28])(inp)
x = L.Dense(512, activation=L.LeakyReLU(alpha=0.25))(x)
x = L.Dropout(0.3)(x)
x = L.Dense(1024, activation=L.LeakyReLU(alpha=0.25))(x)
x = L.Dropout(0.3)(x)
x = L.Dense(256, activation=L.LeakyReLU(alpha=0.25))(x)
x = L.Dropout(0.3)(x)
x = L.Dense(512, activation=L.LeakyReLU(alpha=0.25))(x)
x = L.Dropout(0.3)(x)
x = L.Dense(64, activation="swish")(x)
out = L.Dense(1, activation="sigmoid")(x)

# beta_1 is set to 0.5 in the adam optimizer for more stable training
discriminator = M.Model(inputs=inp, outputs=out)
discriminator.compile(loss="binary_crossentropy",
                      optimizer=tf.keras.optimizers.Adam(lr=0.0002, beta_1=0.5),
                      metrics=["acc"])
```

The generator model will receive an arbitrary number of points in latent space as random Gaussian noise. It's responsible for creating fake images based on the gradients flowing back from the discriminator to produce images that the discriminator cannot correctly classify as fake. The dimension of the latent space is irrelevant to model performance – in the following example, 128 was chosen as the input dimension. The model's architecture is defined in an expanding manner, increasing the number of neurons from 128 to 1024. The final layer will be a reshape to the correct image size of $(28, 28)$. Note that the final layer activation will be sigmoid since we normalized the image pixel values to be between 1 and 0.

Add comments (Listing 9-15) wherever you want.

Listing 9-15. Defining the generator

```
# generator

# 128 as latent dimension
inp_gen = L.Input(shape=(128))
y = L.Dense(224)(inp_gen)
y = L.LeakyReLU(alpha=0.2)(y)
y = L.Dense(256)(inp_gen)
y = L.LeakyReLU(alpha=0.2)(y)
y = L.Dense(512)(y)
y = L.LeakyReLU(alpha=0.2)(y)
y = L.Dense(664)(y)
y = L.LeakyReLU(alpha=0.2)(y)
y = L.Dense(1024)(y)
y = L.LeakyReLU(alpha=0.2)(y)
# shape of mnist image
y = L.Dense(784, activation="sigmoid")(y)
# reshape to dimensions of an image
```

```
out_gen = L.Reshape([28, 28, 1])(y)

# do not compile since the generator will never be trained alone
generator = M.Model(inputs=inp_gen, outputs=out_gen)
```

The entire GAN model consists of the generator and the discriminator in that order (Listing 9-16). To initiate training, the trainable attribute of the discriminator is set to false. We freeze the discriminator's weights prior to training the entire GAN model since the gradients flowing back from the discriminator at each generated sample must remain unchanged (in terms of learning) across the entire iteration. The classification aspect of the discriminator is trained as a standalone model.

Listing 9-16. Defining the GAN

```
# combine model and make discriminator untrainable
gan_model = M.Sequential([generator, discriminator])
discriminator.trainable=False
gan_model.compile(loss="binary_crossentropy", optimizer=tf.keras.optimizers.Adam(lr=0.0002,
beta_1=0.5), metrics=["acc"])
```

As per usual for image-related tasks, TensorFlow datasets will be used to batch samples into a generator-like format (Listing 9-17).

Listing 9-17. Defining Tensorflow Datasets

```
def build_dataset(data, batch_size=32):
    AUTO = tf.data.experimental.AUTOTUNE
    dset = tf.data.Dataset.from_tensor_slices(data).shuffle(1024)
return dset.batch(batch_size, drop_remainder=True).prefetch(AUTO)
batch_size = 256
real_img_dataset = build_dataset(X_train, batch_size=batch_size)
```

Training is accomplished mainly through two nested loops, one looping through the desired number of epochs while, for each epoch, every batch of the dataset is looped through. For each batch of data, the generator produces one batch of fake images along with one batch of real images. Their labels are assigned through tf.constant and then concatenated with the images to create the training dataset for the discriminator. To clarify, one batch of fake images with 0s as their label is joined with one batch of real images labeled 1 and fed into the discriminator using the train_on_batch method (essentially training two batches worth of data). The generator is trained subsequently with the discriminator's weights frozen and calling train_on_batch on the entire GAN (Listing 9-18).

Listing 9-18. Training GAN

```
# retrieve each individual model
generator, discriminator = gan_model.layers

epochs = 150

for epo in range(epochs):

    print(f"TRAINING EPOCH {epo+1}")

    for idx, cur_batch in enumerate(real_img_dataset):
```

704

```
# random noise for generating fake img
noise = tf.random.normal(shape=[batch_size, 128])
# generate fake img and label
fake_img, fake_label = generator(noise), tf.constant([[0.0]]*batch_size)
# extract one batch of real img and label
real_img, real_label = tf.dtypes.cast(cur_batch, dtype=tf.float32),
tf.constant([[1.0]]*batch_size)

# the X of discriminator, consists of half fake img, half real img
discriminator_X = tf.concat([real_img, fake_img], axis=0)
# the y of discriminator, 1s and 0s
discriminator_y = tf.concat([real_label, fake_label], axis=0)
# set to trainable
discriminator.trainable = True
# train discriminator as standalone classification model
d_loss = discriminator.train_on_batch(discriminator_X, discriminator_y)

# X of generator, noise
gan_x = tf.random.normal(shape=[batch_size, 128])
# y of generator, set to "real"
gan_y = tf.constant([[1.0]]*batch_size)
# set discriminator to untraibable
gan_model.layers[1].trainable = False
gan_loss = gan_model.train_on_batch(gan_x, gan_y)

# avoid OOM
del fake_img, real_img, fake_label, real_label,
del discriminator_X, discriminator_y

if (idx+1) % 100 == 0:
    print(f"\t On batch {idx+1}/{len(real_img_dataset)}   Discriminator Acc:
    {d_loss[1]}  GAN Acc {gan_loss[1]}")

if (epo+1)%10==0:
    # plot results every 10 epochs
    print(f"RESULTS FOR EPOCH {epo}")
    gen_img = generator(tf.random.normal(shape=[5, 128]))
    columns = 5
    rows = 1

    fig = plt.figure(figsize=(12, 2))
    for i in range(rows*columns):
        fig.add_subplot(rows, columns, i+1)
        plt.imshow(gen_img[i], interpolation='nearest', cmap='gray_r')
    plt.tight_layout()
    plt.show()
```

After several dozen epochs with a batch size of 256, the GAN can generate somewhat convincing handwritten digits (Figure 9-10). You can further improve the performance by modifying the discriminator or the generator to include more layers or different activation functions or switching to a convolutional-based design. Due to the instability of GAN, one minor change to the model can influence results significantly.

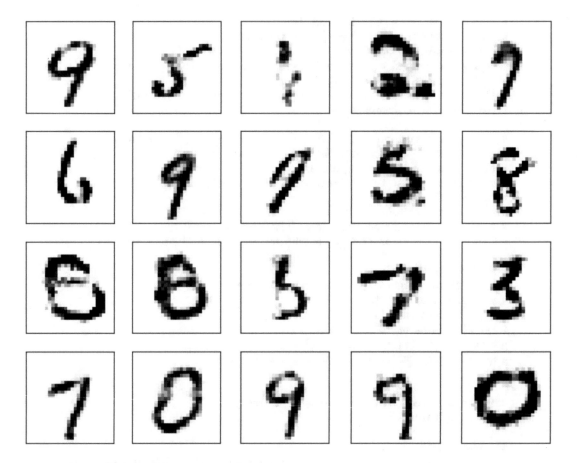

Figure 9-10. *Preliminary GAN training results*

CTGAN

The official implementation of CTGAN comes in a convenient, easy-to-use package. To install from pip, type the following command: pip install sdv. For consistency and comparison purposes, the Higgs Boson dataset will be reused here for CTGAN to produced artificial samples. The following shows a bare-bones example of training and then sampling fake data from CTGAN (Listing 9-19).

Listing 9-19. Simple CTGAN demo

```
# using test dataset since it has more samples
data = pd.read_csv("../input/higgsb/test.csv")
from sdv.tabular import CTGAN
ctgan_model = CTGAN(verbose=True)
ctgan_model.fit(data)
new_data = ctgan_model.sample(num_rows=800)
```

With absolutely no hyperparameter tuning, the performance of CTGAN is beyond impressive compared with VAE discussed previously. As shown in Figures 9-11 and 9-12, the pairplots generated by both the synthetic data and the actual dataset for the bivariate relationships between features are almost indistinguishable. Although feature correlations are not the only method or the best method perse to measure performance of generative models, the results that we can produce after merely five lines of code are quite astounding.

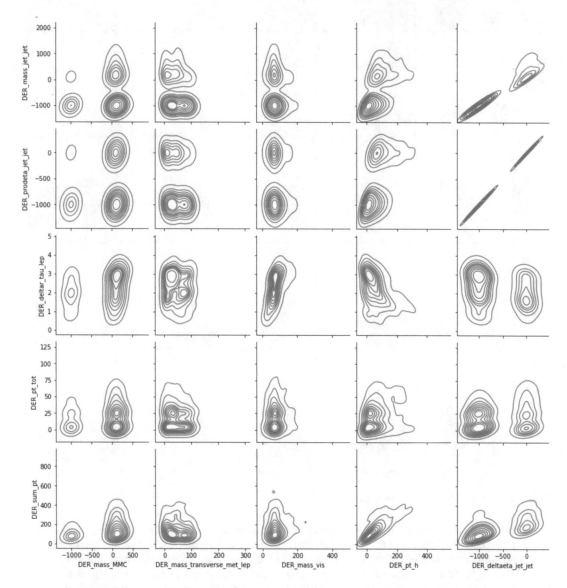

Figure 9-11. *Side-by-side comparison of synthetic data generated by CTGAN (top) and 800 randomly sampled data points from the original source (bottom)*

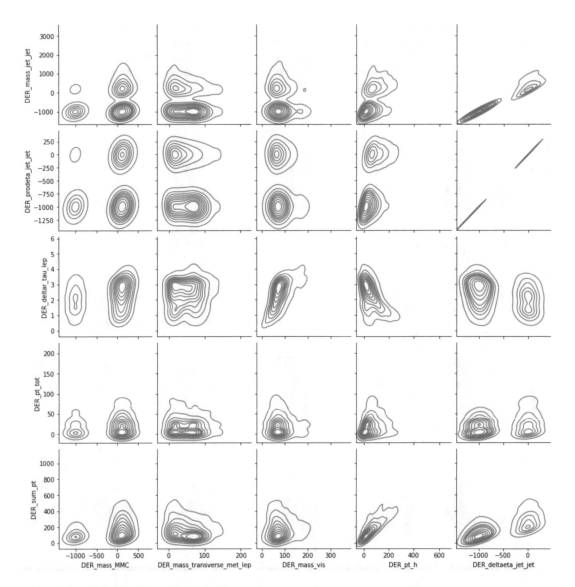

Figure 9-12. *Pairplot of the actual dataset*

The "C" in CTGAN (conditional) really comes into play with the flexibility that the model has. We can specify a "primary_key" to generate unique data for that specific feature along with the "anonymize_field" option to produce purely artificial samples that aren't included in the training data (Listing 9-20). For the anonymize_field option, there are predefined potential data categories that could be anonymized. Depending on what gets passed in, CTGAN will retrieve data from a set of pre-generated data points.

Listing 9-20. Conditional generation

```
# an example out of the dataset's context
# the generated column of 'DER_mass_MMC' will be treated as names and anonymized accordingly
# the full list of categories can be found at
# https://sdv.dev/SDV/user_guides/single_table/ctgan.html#anonymizing-personally-
identifiable-information-pii
ctgan_model = CTGAN(
    primary_key='EventId',
    anonymize_fields={
        'DER_mass_MMC': 'name'
    }
)
```

Sampling of synthetic data can be conditional as well. There are two general approaches to setting constraints during generation:

1. By initializing a Condition object with a dictionary specifying the column/columns that will be constrained. The dictionary's keys will represent the constrained columns, while its value will be the only value that the model will produce. Note that for continuous features, only values within the range of training data can be produced.

2. By directly calling sample_remaining_columns on the CTGAN model. It's exactly what it sounds like: a Pandas DataFrame is passed in containing columns already set; then the model will generate the remaining columns.

In Listing 9-21 is a demonstration of both approaches.

Listing 9-21. Conditional generation

```
from sdv.sampling import Condition
condition = Condition({
    'DER_deltar_tau_lep': 2.0,
    # categotical features' values can be passed in as a string
})
constrained_sample = ctgan_model.sample_conditions(condition)

given_colums = pd.DataFrame({
    # arbitrary values
    "DER_mass_MMC": [120.2, 117.3, -988, 189.9]
})
constrained_sample = ctgan_model.sample_remaining_columns(given_columns)
```

Finally, the low-level parameters of the GAN, such as the number of epochs, batch sizes, latent space dimension, learning rates, and decays, can be adjusted. More details can be found at CTGAN's official documentation: https://sdv.dev/SDV/user_guides/single_table/ctgan.html.

Key Points

In this chapter, we discussed various data generation algorithms, from a simpler and quicker approach through VAEs to more complex, sophisticated GANs. The application of tabular data generation methods is wider than what most people might assume.

- Variational Autoencoders predict probability distributions rather than explicit scalars for each dimension of the latent space, which allows them to learn the entire latent space in a continuous, rather than discrete, fashion. Correspondingly, vectors in the latent space can be interpolated between and decoded into realistic outputs. These can be used for data generation.

- Generative Adversarial Network systems consist of a discriminator model and a generator model; the generator model accepts a random vector and synthesizes artificial samples, and the discriminator determines whether the inputted sample is real (drawn from the dataset) or not (generated by the generator). The generator is trained to minimize the discriminator's performance, whereas the discriminator is trained to maximize it.

  - Conditional GANs allow for samples to be conditioned on a particular class or property by passing sample attribute information both to the generator and the discriminator.

  - The CTGAN (Conditional Tabular GAN) model uses conditional generation to synthesize new robust and representative tabular data, even in complex heterogeneous and unbalanced environments.

In the next chapter, we will explore meta-optimization. Meta-optimization involves tuning hyperparameters across the training pipeline, as we will realize that this is especially important for tabular data modeling.

CHAPTER 10

■ ■ ■

Meta-optimization

If you optimize everything, you will always be unhappy.

—Donald Knuth, Computer Scientist

The legendary Donald Knuth was right about many things, but meta-optimization is indeed strong evidence that most people can be pretty happy optimizing most things (not quite *everything*, admittedly). Optimization in its standard form concerns the reconciliation of ambiguities and uncertainties on the direct level of parameter-to-loss model. But when we construct said models, we also run into ambiguities and uncertainties that are generally overlooked and forgotten. Meta-optimization concerns the optimization of these *meta-parameters* (alternatively, "hyperparameters"), which can not only improve performance but also potentially be more efficient than manual tuning.

This chapter begins with a brief discussion of key concepts and motivations of meta-optimization, followed by an exploration of no-gradient optimization methods often used for meta-optimization procedures. The next two sections demonstrate how meta-optimization with the library Hyperopt can be used to optimize the model and the data pipeline. Lastly, the chapter concludes with a brief section on Neural Architecture Search.

Meta-optimization: Concepts and Motivations

If you have been following along with the scripts outlined throughout the book, you inevitably will sometimes encounter points at which you must decide how to build a system, but which specific possibility you choose feels arbitrary or equally unsure – for instance, whether to use the GradientBoostingClassifier or the AdaBoostClassifier, whether the maximum tree depth should be 128 or 64, how many layers should be in a particular neural network cell, how many nodes to put in a particular layer, the exact proportion used for a dropout layer, and so on.

The truth is that for most of these decisions, it doesn't really matter which path you choose. Whether a specific layer has 128 or 256 nodes likely will not noticeably impact the final performance because deep learning systems are so large and one single decision is comparatively insignificant. Which optimizer you use might have an impact, but not one that is incredible (unless you are switching from a primitive caveman optimizer, of course). We can call these decisions *meta-parameters* because they are decisions determining the "form" and "shape" of the model rather than the specific weights determining its predictive behavior.

However, if we somehow make decisions for all these meta-parameters simultaneously – that is, we set each meta-parameter such that it is optimal with respect to all other meta-parameters – we can end up significantly altering the system's performance overall.

Meta-optimization, also referred to sometimes as meta-learning or auto-ML, is the process of "learning how to learn" – a meta-model (or "*controller* model") finds the best meta-parameters for a *candidate* model's

© Andre Ye and Zian Wang 2023
A. Ye and Z. Wang, *Modern Deep Learning for Tabular Data*,
https://doi.org/10.1007/978-1-4842-8692-0_10

validation performance (Figure 10-1). We can use meta-optimization to find the best type of model to use, the best meta-parameters for that best model, and the best data preprocessing and postprocessing schemes for that best model and best set of meta-parameters.

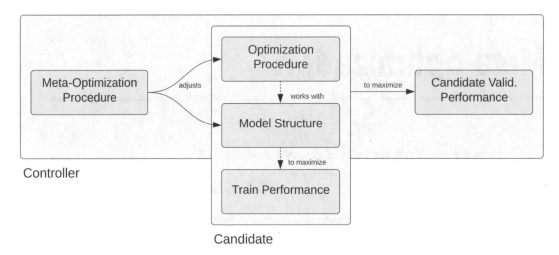

Figure 10-1. *Relationship between controller model and controlled model in meta-optimization*

There are three key components of a meta-optimization campaign: the meta-parameter space, the objective function, and the optimization procedure.

- The *meta-parameter space* describes (a) which meta-parameters are being optimized and (b) how the optimization procedure should go about sampling the meta-parameters being optimized. Some meta-parameters must be integers (e.g., number of nodes), whereas others must be proportions (e.g., dropout rate). Moreover, we may have more confidence in the success of certain domains of the meta-parameter space over others. For instance, a low dropout rate (e.g., 0.1–0.3) is almost surely more effective at regularization than a high dropout rate (e.g., 0.7–0.9), because high dropout rates significantly hinder information transfer. This distribution of confidence must be specified in the construction of the meta-parameter space.

- The *objective function* describes how the meta-parameters being sampled come together into a model and returns the performance of a model built and trained using the sampled meta-parameters. For instance, if we want to optimize the number of hidden nodes in a shallow neural network with only one hidden layer, the objective function will accept the sampled number of hidden nodes, construct a neural network with that many hidden nodes, fit until convergence, and return performance on the validation dataset. The goal of meta-optimization is to minimize this objective function.

- The *optimization procedure* is the algorithm by which new combinations of meta-parameters are sampled from the search space given the performance of previous meta-parameter combinations. Ideally, such an algorithm would be adaptive – if a set of meta-parameters performs very poorly, it should sample meta-parameters quite far away; if a set performs exceptionally well, it should sample nearby meta-parameters to "hang on to" and exploit the good performance. The user generally does not need to implement the optimization procedure.

Hyperparameter optimization in deep learning is generally considered to be an expensive task, because (a) neural networks have so many more possible directions for hyperparameter variation compared with classical machine learning algorithms and (b) neural networks generally take significantly longer than classical machine learning algorithms to train. Research in fields like Neural Architecture Search, a subfield of hyperparameter optimization in which a meta-model/controller (often another neural network) attempts to find the optimal neural network architecture for a task, can involve experiments lasting several days on supercharged hardware.

However, hyperparameter optimization in deep learning for structured data is comparatively more accessible than optimizing deep learning pipelines for computer vision and natural language processing tasks. Neural network models for structured data are often faster to train and smaller; this allows the optimization procedure to repeatedly sample a much wider array of neural network meta-parameter combinations without specialized hardware in a feasible amount of time.

In this section, we'll explore hyperparameter optimization for a variety of applications, including optimizing classical machine learning models, deep learning training procedures, deep learning architectures, data pipelines, and more.

No-Gradient Optimization

The hyperparameter optimization tools we will be using are more broadly tools for *no-gradient optimization*.

Consider some function $f(x)$. You only have access to its output given a certain input, and you know that it is expensive to calculate (i.e., it takes a nontrivial amount of time to process one query). Your task is to find the set of inputs that minimizes the output of the function as much as possible.

This sort of setup is known as a black box optimization problem (Figure 10-2), because the algorithm or entity attempting to find a solution to the problem has access to very little or no information about the function. You have access only to the output of any input passed into the function, but not the derivative. This bars the usage of gradient-based methods that have proved successful in the domain of neural networks.

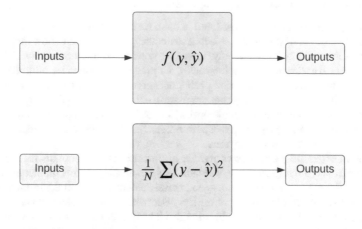

Figure 10-2. *Objective functions to minimize. Top: black box function with no explicit information given. Bottom: explicitly defined loss function (in this case MSE)*

Black box optimization is, generally speaking, the name of the game in feasible meta-parameter optimization. While there are several methods for differentiating meta-parameter optimization,[1] black box optimization has been a long-studied problem that has strong applications to expensive black box problems. The meta-parameter optimization procedure is given no information other than the loss incurred by a candidate model trained with the sampled meta-parameters.

So-called "naïve" meta-optimization algorithms/procedures use the following general structure to solve black box optimization problems:

1. Select structural parameters for a proposed controlled model.

2. Obtain the performance of a controlled model trained under those selected structural parameters.

3. Repeat.

There are two generally recognized naïve meta-optimization algorithms used as a baseline against more sophisticated meta-optimization methods:

- *Grid search*: In a grid search, every combination of a user-specified list of values for each parameter is tried and evaluated. Consider a hypothetical model with two structural parameters we would like to optimize, A and B. The user may specify the search space for A to be [1, 2, 3] and the search space for B to be [0.5, 1.2]. Here, "search space" indicates the values for each parameter that will be tested. A grid search would train six models for every combination of these parameters – $A = 1$ and $B = 0.5$, $A = 1$ and $B = 1.2$, $A = 2$ and $B = 0.5$, and so on.

- *Random search*: In a random search, the user provides information about a feasible distribution of potential values that each structural parameter could take on. For instance, the search space for A may be a normal distribution with mean 2 and standard deviation 1, and the search space for B might be a uniform choice from the list of values [0.5, 1.2]. A random search would then randomly sample parameter values and return the best-performing set of values.

Grid search and random search are considered to be naïve search algorithms because they do not incorporate the results from their previously selected structural parameters into how they select the next set of structural parameters; they simply repeatedly "query" structural parameters blindly and return the best-performing set. While grid and random searches have their place in certain meta-optimization problems – grid search suffices for small meta-optimization problems, and random search proves to be a surprisingly strong strategy for models that are relatively cheaper to train – they cannot produce consistently strong results for more complex models, like neural networks. The problem is not necessarily that these naïve methods inherently cannot produce good sets of parameters, but that they take too long to do so due to the large search space and slow black box query time.

A key component of the unique character of meta-optimization that distinguishes it from other fields of optimization problems is the impact of the evaluation step in amplifying any inefficiencies in the meta-optimization system. Generally, to quantify how good certain selected structural parameters are, a model is fully trained under those structural parameters, and its performance on the test set is used as the evaluation. In the context of neural networks, this evaluation step can take hours. Thus, an effective meta-optimization system should attempt to require as few models to be built and trained as possible before arriving at a good

---

[1] An intuitive example is DARTS, or Differentiable Architecture Search.
Liu, H., Simonyan, K., & Yang, Y. (2019). DARTS: Differentiable Architecture Search. *ArXiv, abs/1806.09055.*

solution. (Compare this with standard neural network optimization, in which the model queries the loss function and updates its weights accordingly anywhere from hundreds of thousands to millions of times in the span of hours.)

To prevent inefficiency in the selection of new structural parameters to evaluate, successful meta-optimization methods used for models like neural networks include another step – incorporating knowledge from previous "experiments" into determining the next best set of parameters to select:

1. Select structural parameters for a proposed controlled model.

2. Obtain the performance of a controlled model trained under those selected structural parameters.

3. Incorporate knowledge about the relationship between selected structural parameters and the performance of a model trained under such parameters into the next selection.

4. Repeat.

Hyperopt, the popular hyperparameter optimization framework we will use for meta-optimization, uses Bayesian optimization to address the black box optimization problem. Provided with a well-designed search space, Hyperopt can generally find solutions that perform better than manual designs.

Bayesian optimization is often used in black box optimization problems because it succeeds in obtaining reliably good results with a relatively small number of required queries to the objective function. The spirit of Bayesian modeling is to begin with a set of *prior* beliefs and continually update that set of beliefs with new information to form *posterior* beliefs. It is this spirit of continuous update – searching for new information in places where it is needed – that makes Bayesian optimization a robust and versatile tool in black box optimization problems.

Consider a hypothetical objective function, represented in Figure 10-3. In the context of meta-optimization, this function represents the loss or cost of a certain model (y-axis) incurred by a model trained using the sample parameter(s) (x-axis). The goal is to find the value(s) of x that minimizes the cost (Figure 10-3).

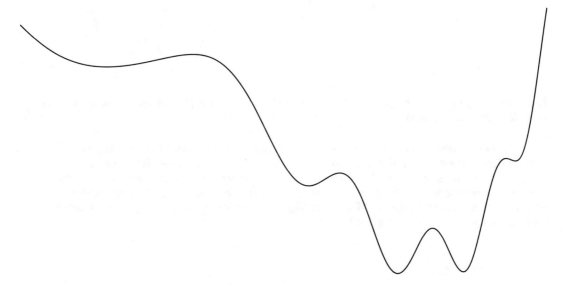

Figure 10-3. *A hypothetical cost function, showing the loss incurred by a model with some parameter x. For the sake of convenient visualization, in this case we are optimizing a one-parameter model or optimizing the loss with respect to only one parameter (i.e., looking at a cross-section of the full loss landscape)*

However, to the meta-optimization algorithm, the cost function is a black box. It cannot "see" the entire function – if it could, solving the minimization problem would be trivial. The function is displayed for the user for convenience in understanding Bayesian optimization, but be sure to distinguish between what the optimization process does and doesn't know!

All the optimization procedure has access to is the set of sampled points. Using these sampled points, it develops a *hypothesis* about the shape of the true cost function. This is known formally as a *surrogate function*. The surrogate function approximates the objective/cost function and represents the current set of beliefs on how the objective function behaves with respect to sampled parameters/independent variables.

Figure 10-4 demonstrates how two sampled points are used as the basis for a surrogate function (red, dotted). While the model isn't able to "see" the complete objective function, it *can* "see" the complete surrogate function. The surrogate function is a mathematically represented function whose properties can be easily accessed through established techniques. The optimization procedure can determine which points are promising or risky based on the surrogate model. If a certain proposed point x has a high surrogate function value, then it is a riskier decision. If a certain proposed point x has a low surrogate function value, then it is more promising (Figure 10-4).

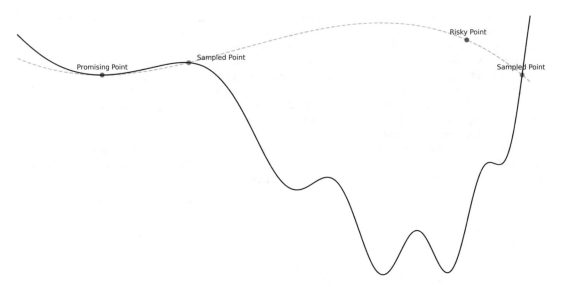

Figure 10-4. *Fitting a surrogate function to two sampled points and using the surrogate function to develop estimations for how promising and risky certain unsampled points are*

Because our surrogate function here is defined only by two points, it is unwise for the optimization procedure to enact a strictly greedy sampling policy. Suppose the procedure decides to evaluate the true value of both of these points (by evaluating them through the objective function). We find that the risky point actually minimized the objective function more than the promising point, which means that taking the risk paid off! We can now update the surrogate function to reflect these sampled points and identify new promising points to sample (Figure 10-5).

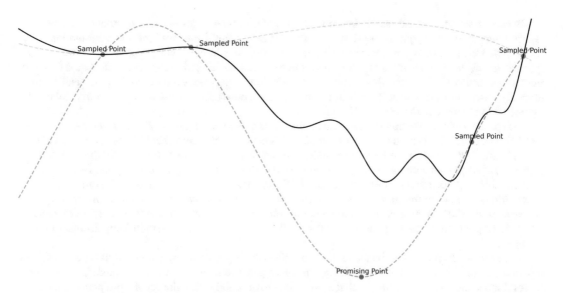

Figure 10-5. *A refitting of the surrogate function to newly sampled points and an update of estimations for how promising or risky unsampled points are*

This process repeats: increased "intelligent"/informed sampling helps to define the surrogate function such that it becomes a more and more accurate representation of the true objective function (Figure 10-6).

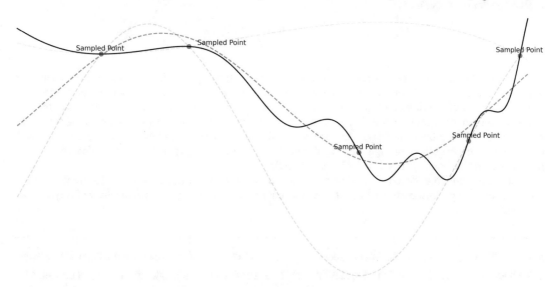

Figure 10-6. *A second refitting of the surrogate function to newly sampled points and an update of estimations for how promising or risky unsampled points are*

Note that the visualization of the surrogate models here is deterministic, but in practice the surrogate functions used are probabilistic. The functions return $p(y|x)$, or the probability that the objective function's output is y given x. Probabilistic surrogate functions are more natural to update and sample from in a Bayesian manner.

Because a random or grid search does not take any of the previous results into consideration when determining the next sampled set of parameters, these "naïve" algorithms save time and computation in calculating the next set of parameters to sample. However, the additional computation Bayesian optimization algorithms use to determine the next point to sample is used to construct a surrogate function more intelligently with fewer queries. Net-wise, the reduction in necessary queries to the objective function generally outweighs the increase in time and computation to determine the next sampled point, making the Bayesian optimization method more efficient.

This process of optimization is known more abstractly as *Sequential Model–Based Optimization (SMBO)*. It operates as a central concept or template against which various model optimization strategies can be formulated and compared. SMBO contains one key feature: a surrogate function for the objective function that is updated with new information and used to determine new points to sample. Two key attributes differentiate different SMBO methods: the design of the acquisition function (the process determining how new points are sampled given the surrogate models) and the method of constructing the surrogate model (how to incorporate sampled points into an approximate representation of the objective function). Hyperopt uses the Tree-Structured Parzen Estimator (TPE) surrogate model and acquisition strategy.

The Expected Improvement measurement quantifies the expected improvement with respect to the set of parameters to be optimized, x. For instance, if the surrogate model $p(y|x)$ evaluates to zero for all values of y less than some threshold value y^* – that is, there is zero probability that the set of input parameters x could yield an output of the objective function less than y^* – there is probably no improvement to be found by sampling x.

The Tree-Structured Parzen Estimator is built to work toward a set of parameters x that maximizes Expected Improvement. Like all surrogate functions used in Bayesian optimization, it returns $p(y|x)$ – the probability that the objective function's output is y given an input x. Instead of directly obtaining this probability, it uses Bayes's rule:

$$p(y|x) = \frac{p(x|y) \cdot p(y)}{p(x)}$$

The $p(x|y)$ term represents the probability that the input to the objective function was x given an output y. To calculate this, two distribution functions are used: $l(x)$ if the output y is less than some threshold y^* and $g(x)$ if the output y is less than some threshold y^*. To sample values of x that yield objective function outputs less than the threshold, the strategy is to draw from $l(x)$ rather than $g(x)$. (The other terms, $p(x)$ and $p(y)$, can be easily calculated as they do not involve conditionals.) Sampled values with the highest expected improvement are evaluated through the objective function. The resulting value is used to update the probability distributions $l(x)$ and $g(x)$ for better prediction.

Ultimately, the Tree-Structured Parzen Estimator strategy attempts to find the best objective function inputs to sample by continually updating its two internal probability distributions to maximize the quality of prediction.

■ **Note** You may be wondering: What is tree-structured about the Tree-Structured Parzen Estimator strategy? In the original TPE paper, the authors suggest that the "tree" component of the algorithm's name is derived from the tree-like nature of the hyperparameter space: the value chosen for one hyperparameter determines the set of possible values for other parameters. For instance, if we are optimizing the architecture of a neural network, we first determine the number of layers before determining the number of nodes in the third layer.

Let's begin exploring the implementation of no-gradient optimization by finding the minima of simple scalar-to-scalar functions.

Consider the following problem: find the value of x that minimizes

$$f(x) = (x-1)^2$$

(Recall a similar exercise in Chapter 1, used to demonstrate the gradient descent algorithm.)

Some simple math will tell you that the global minimum is at $x = 1$. However, the model does not have access to the analytical expression – it must find the minimum by iteratively sampling from the function (Listing 10-1, Figure 10-7).

Listing 10-1. Plotting the graph $y = (x - 1)^2$

```
plt.figure(figsize=(10, 5))
x = np.linspace(-5, 5, 100)
y = (x - 1)**2
plt.plot(x, y, color='red')
plt.scatter([1], [0], color='red')

plt.grid()
plt.show()
```

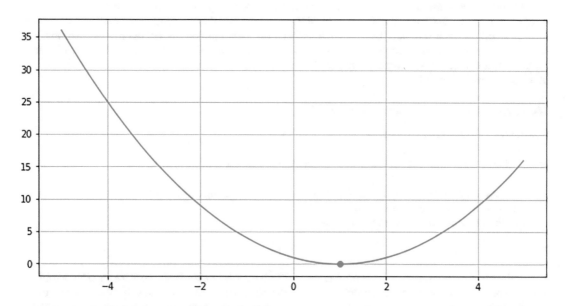

Figure 10-7. *The graph of $y = (x - 1)^2$ and its minimum at $x = 1$*

The first step is to define the search space. Say that we are most confident that the true value of x that minimizes $y = (x - 1)^2$ is somewhere near 0 and that the farther x is from 0, the less confidence that it will minimize the function. This reflects a *normal distribution* with a center of 0. The standard deviation represents how quickly our confidence drops in x's merits to minimize the objective function as x moves farther from the mean.

To define spaces in Hyperopt, you must create a dictionary in which keys are parameter identifiers/references/names and the values are Hyperopt hyperparameter search space objects (Listing 10-2). In this case, we will use hp.normal, which defines a parameter that will be sampled in a Gaussian fashion.

All hp.type spaces accept a name as the first argument. For hp.normal, we can specify the mean (mu) and standard deviation (sigma) afterward.

Listing 10-2. Defining a search space for *x*

```
# define the search space
from hyperopt import hp
space = {'x':hp.normal('x', mu=0, sigma=10)}
```

Our objective function takes in a dictionary params. params is structured in the same way as our search space, except each of the keys has been replaced with the sampled value. params['x'] will return a float sampled from the normal distribution, rather than a hp.normal object. We can evaluate the sampled value through our sample function (Listing 10-3).

Listing 10-3. Defining the objective function

```
# define objective function
def obj_func(params):
    return (params['x']-1)**2
```

To perform optimization, use the function minimization function fmin, which accepts the objective function and search space arguments. You can additionally provide the algorithm (the previously discussed Tree-Structured Parzen Estimator is recommended) and the maximum number of queries/evaluations that can be made to the objective function. Note that Hyperopt may stop earlier than the maximum number of evaluations if it reaches a satisfactory solution and determines that it is unlikely there are other unexplored better solutions (Listing 10-4).

Listing 10-4. Running the minimization campaign

```
# perform minimization procedure
from hyperopt import fmin, tpe
best = fmin(obj_func, space, algo=tpe.suggest, max_evals=500)
```

After several hundred evaluations, Hyperopt finds the minimizing value of *x* to be 1.0058378772703804, which is very close to the true value of 1. Success!

Consider another function defined as follows:

$$f(x) = \sin^2 x$$

Multiples of π minimize this function, since $\sin(k\pi) = 0$ for any integer k. Suppose we want Hyperopt to find a solution that minimizes the function and resides between 2 and 4. We know that $2 < \pi < 4$ and that $\text{argmin}(\sin^2 x) = k\pi$; thus, $\text{argmin}(\sin^2 x\{2 < x < 4\}) = \pi$. Tackling this optimization problem leads us nicely to an approximation for π (Listing 10-5, Figure 10-8)!

Listing 10-5. Plotting the graph of $y = sin^2x$ and a minimum at $x = \pi$

```
plt.figure(figsize=(10, 5), dpi=400)
x = np.linspace(2, 4, 1000)
y = np.sin(x)**2
plt.plot(x, y, color='red')
plt.scatter([np.pi], [0], color='red')
```

```
x = np.linspace(-2, 8, 1000)
y = np.sin(x)**2
plt.plot(x, y, color='red', alpha=0.3, linestyle='--')
plt.scatter([np.pi], [0], color='red')

plt.grid()
plt.show()
```

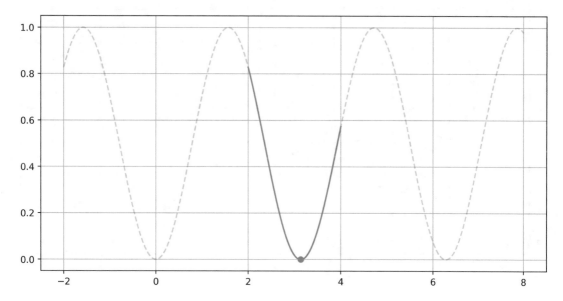

Figure 10-8. *The graph of sin²x and one of its minima at x = π marked*

Rather than using a normal distribution, we use a uniform distribution with minimum 2 and maximum 4 to reflect our knowledge of the space (Listing 10-6).

Listing 10-6. Defining a search space and objective function

```
# define the search space
from hyperopt import hp
space = {'x':hp.uniform('x', 2, 4)}

# define objective function
def obj_func(params):
    return np.sin(params['x'])**2
```

After ten trials, Hyperopt's approximation of π is 3.254149; after 100 trials, it is 3.139025; after 1000, it is 3.141426. For reference, the actual value of pi rounded to six digits is 3.141593 – our no-gradient approximation was only about 0.00017 off!

Often, however, functions are not defined for all values in the domain we want to search on. Hyperopt allows us to specify when a parameter or set of parameters is invalid using *statuses*. Consider minimizing the following function, which is undefined at $x = 0$ (Listing 10-7, Figure 10-9):

$$y = \frac{1}{x^2} + x^2$$

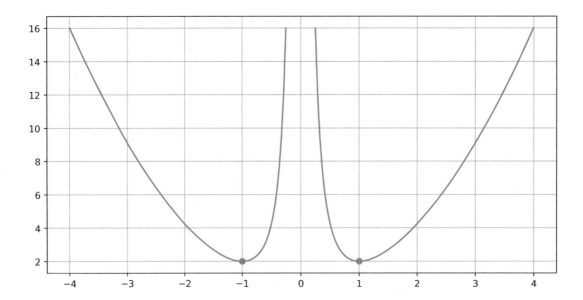

Figure 10-9. *The graph of $\frac{1}{x^2} + x^2$, with both minimas at $x \in \{-1, 1\}$*

Listing 10-7. Plotting the graph of $y = \frac{1}{x^2} + x^2$

```
plt.figure(figsize=(10, 5), dpi=400)
x = np.linspace(-3.99214, -0.25049, 100)
y = 1/(x**2) + (x)**2
plt.plot(x, y, color='red')

x = np.linspace(0.25049, 3.99214, 100)
y = 1/(x**2) + (x)**2
plt.plot(x, y, color='red')
plt.scatter([-1, 1], [2, 2], color='red')

plt.grid()
plt.show()
```

It's not easy to specify in the search space that x should not be 0; instead, we return `{'status':'fail'}` from the objective function if it receives an invalid value. If the sampled parameters are valid, we return `{'status':'ok', 'loss': ...}` – filling in the "loss" with whatever loss is incurred by the given set of sampled parameters (Listing 10-8).

Listing 10-8. Defining the search space and objective function, with statuses built in for invalid sampled inputs

```
# define the search space
from hyperopt import hp
space = {'x':hp.normal('x', mu=0, sigma=10)}

# define objective function
def obj_func(params):
    if params['x']==0:
        return {'status':'fail'}
    return {'loss':1/(params['x']**2) + params['x']**2,
            'status':'ok'}
```

Notice that this particular function has two global minima, both of which are equally valid. Given that the search space is symmetric and centered directly in between the minimizing values of x, Hyperopt will arrive at a particular solution entirely random about half of the time.

Sometimes, we can't even describe precisely which sampled parameters or parameter sets will yield invalid results. For instance, consider the following function, which is invalid on the real domain both for values of $x < 0$ and $\sin(x^2) < 0$ (the latter condition occurring more frequently as $x \to \infty$) (Listing 10-9, Figure 10-10):

$$y = \frac{1}{\sqrt{x}} - \sqrt{\sin(x^2)}$$

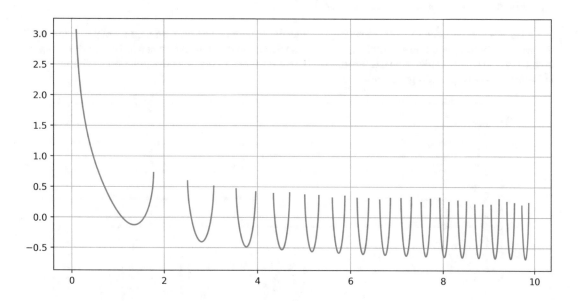

Figure 10-10. *The graph of* $y = \frac{1}{\sqrt{x}} - \sqrt{\sin(x^2)}$

Listing 10-9. Plotting the graph of $y = \dfrac{1}{\sqrt{x}} - \sqrt{\sin\left(x^2\right)}$

```
plt.figure(figsize=(10, 5), dpi=400)
x = np.linspace(0.1, 10, 10_000)
y = 1/(np.sqrt(x)) - np.sqrt(np.sin(x**2))
plt.plot(x, y, color='red')

plt.grid()
plt.show()
```

Say we want to minimize this function between 0 and 10. Rather than specifying which sampled parameter values will yield an invalid result – which is a pretty arduous algebraic task – we can carry out the calculation first and then deal with the fallout afterward if it blows up. In this case, if we perform an invalid operation with NumPy, the result will be np.nan. We can execute the objective function evaluation and return a failed status if it is nan (Listing 10-10).

Listing 10-10. Defining the objective function, with a general fail-safe for invalid inputs using statuses

```
# define objective function
def obj_func(params):
    result = 1/np.sqrt(params['x']) - np.sqrt(np.sin(params['x']**2))
    if result == np.nan:
        return {'status':'fail'}
    return {'loss': result, 'status':'ok'}
```

In alternative situations where executing the objective function given the sampled set of parameters throws an error (e.g., building a neural network with an invalid architecture), you can also use a try/except architecture in which a failed status is returned if there is any error in the objective function evaluation. Hyperopt finds a reasonably good minimum (Figure 10-11).

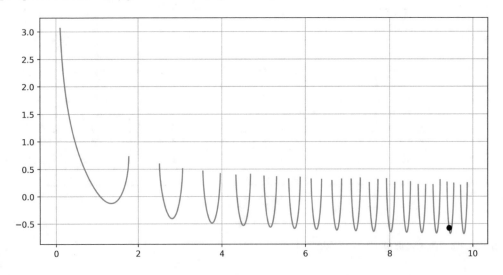

Figure 10-11. The graph of $y = \dfrac{1}{\sqrt{x}} - \sqrt{\sin\left(x^2\right)}$ with the minimum discovered by Hyperopt marked as a point

Optimizing Model Meta-parameters

In this section, we'll apply our knowledge of Hyperopt's syntax to begin meta-optimization of classical machine learning models and deep learning models. Rather than dealing with simple $f: x \rightarrow y$ minimization problems, we will be constructing large possible search spaces that are used to build and train a model in the objective function.

Let's consider the Higgs Boson dataset, which we have approached from multiple different directions previously in the book. This dataset is a generally difficult modeling problem, but perhaps we will be able to find a better solution using meta-optimization (Figure 10-12).

	class	lepton_pT	lepton_eta	lepton_phi	missing_energy_magnitude	missing_energy_phi	jet1pt	jet1eta	jet1phi	jet1b.tag	...	jet4eta	jet4phi	jet4b.tag	m_jj	m_jjj	m_lv	m_jlv	m_bb	m_wbb	m_wwbb
0	1	0.907542	0.329147	0.359412	1.497970		-0.313010	1.095531	-0.557525	-1.588230	2.173076	...	-1.138930	-0.000819	0.000000	0.302220	0.833048	0.985700	0.978098	0.779732	0.992356	0.798343
1	1	0.798835	1.470639	-1.635975	0.453773		0.425629	1.104875	1.282322	1.381664	0.000000	...	1.128848	0.900461	0.000000	0.909753	1.108330	0.985692	0.951331	0.803252	0.865924	0.780118
2	0	1.344385	-0.876626	0.935913	1.992050		0.882454	1.786066	-1.646778	-0.942383	0.000000	...	-0.678379	-1.360356	0.000000	0.946652	1.028704	0.998656	0.728281	0.869200	1.026736	0.957904
3	0	1.595839	-0.607811	0.007075	1.818450		-0.111906	0.847550	-0.566437	1.581239	2.173076	...	-0.654227	-1.274345	3.101961	0.823761	0.938191	0.971768	0.789176	0.430553	0.961357	0.957818
4	1	0.409391	-1.884684	-1.027292	1.672452		-1.604598	1.338015	0.055427	0.013466	2.173076	...	0.069496	1.377130	3.101961	0.869418	1.222083	1.000627	0.545045	0.698653	0.977314	0.828786
...	...	...	...	...	...		...	...	...	...	...	...	...	...	...	...	...	...	...	...	...	...
68631	0	0.631747	-1.732745	0.734498	0.647226		-0.795422	0.768035	0.966439	0.950915	0.000000	...	0.079490	-0.501350	3.101961	1.843042	1.384108	0.989323	0.978744	0.467165	1.412571	2.054788
68632	0	0.967752	-0.317568	-1.579379	0.512029		0.764061	0.684489	-0.537720	-0.855346	0.000000	...	0.173599	0.239561	3.101961	2.000598	1.331450	0.985605	0.880876	0.899824	0.978647	0.915264
68633	1	0.908091	-0.825006	-0.830871	0.736298		1.512713	0.881811	-0.363440	0.006813	1.086538	...	-0.187013	0.716785	0.000000	1.317681	1.010795	0.985962	0.957878	1.454672	0.903937	0.786069
68634	0	0.903699	0.261943	-0.429149	1.892855		0.313687	0.493397	-1.494282	-1.458506	0.000000	...	-0.283621	1.110772	3.101961	0.527038	0.607263	1.125286	0.634107	0.115543	0.425828	0.692506
68635	0	0.566047	-0.317568	0.062561	0.358186		-1.315823	0.691176	1.154583	-0.242759	2.173076	...	-0.844943	-0.294922	0.000000	0.747239	1.008975	0.989497	1.203147	0.892492	1.090807	0.888965

68636 rows × 29 columns

Figure 10-12. *The Higgs Boson dataset*

We will begin by trying to find the best classical machine learning model: one of Logistic Regression, Decision Tree, Random Forest, Gradient Boosting, AdaBoost, and multilayer perceptron (Figure 10-13). This is the most overarching implicit meta-parameter in the modeling pipeline: which model you even choose to model a dataset with is itself a meta-parameter! Beyond this, however, there are meta-parameters within each model that need to be optimized. For instance, there are different depth values and criteria used in constructing a Decision Tree; these can significantly alter the behavior of one Decision Tree compared with the other. The Random Forest and Gradient Boosting models have an additional meta-parameter – the number of estimators in the ensemble. We not only want to choose the best model but also the best set of meta-parameters for that model.

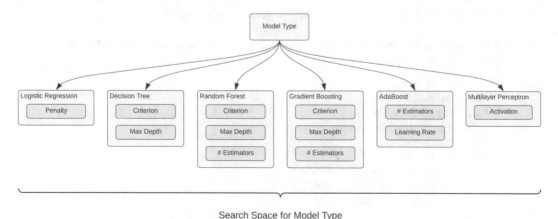

Figure 10-13. *Search space for which model to use and parameters relevant to that model*

Let's begin by importing each of our relevant models (Listing 10-11).

Listing 10-11. Importing relevant models

```
from sklearn.linear_model import LogisticRegression
from sklearn.tree import DecisionTreeClassifier
from sklearn.ensemble import RandomForestClassifier,
                             GradientBoostingClassifier,
                             AdaBoostClassifier
from sklearn.neural_network import MLPClassifier
```

We will use the F1 score to evaluate the best model. Recall from Chapter 1 that the F1 score is a more balanced and informed metric for binary classification problems than accuracy. Additionally, we need to import our space definition (`hyperopt.hp`), function minimization (`hyperopt.fmin`), and Tree-Structured Parzen Estimator (`hp.tpe`) utilities (Listing 10-12).

Listing 10-12. Importing the F1 score metric from scikit-learn and relevant Hyperopt functions

```
from sklearn.metrics import f1_score
from hyperopt import hp, fmin, tpe
```

We can begin to construct our space. We want to make a decision about which model to use; the key-value pair `'model': hp.choice('model', models)` tells Hyperopt that this meta-parameter called 'model' must be sampled as one choice out of the given list of models (which are uninstantiated model objects). If the selected model is Linear Regression, we want to choose whether to use no regularization or L2 regularization; if the selected model is a decision tree, we want to choose whether to use Gini or entropy criterion (see Chapter 1) and the maximum depth. To define the sampling space of a meta-parameter that is roughly continuous but only takes on quantized values, we use `hp.qtype`; in this case, `quniform('name', 1, 30, q=1)` defines a uniform distribution between 1 and 30 quantized with factor 1 (quantization defined as $x \rightarrow \dfrac{\text{round}(q \cdot x)}{q}$ for some quantization factor q). If we wanted multiples of two, we could set the quantization factor to q=2. We can continue listing out each of the meta-parameters in a similar fashion (Listing 10-13).

Listing 10-13. Setting up a potential search space to optimize which model to use and the best of hyperparameters for that model

```
models = [LogisticRegression,
          DecisionTreeClassifier,
          RandomForestClassifier,
          GradientBoostingClassifier,
          AdaBoostClassifier,
          MLPClassifier]
space = {'model': hp.choice('model', models),
         'lr_penalty': hp.choice('lr_penalty', ['none', 'l2']),
         'dtc_criterion': hp.choice('dtc_criterion', ['gini',
                                                      'entropy']),
         'dtc_max_depth': hp.quniform('dtc_max_depth', 1, 30, q=1),
         'rfc_criterion': hp.choice('rfc_criterion', ['gini',
                                                      'entropy']),
         'rfc_max_depth': hp.choice('rfc_max_depth', 1, 30, q=1),
```

```
'rfc_n_estimators': hp.qnormal('rfc_n_estimators', 100, 30,
                                q=1)
...}
```

However, this is a very poor way to structure our search space for a number of reasons. Philosophically, we are not structuring the space in a way consistent with the inherent nature of the meta-parameters. Everything is placed linearly, whereas they should be somehow nested (as shown in the hierarchical diagram in Figure 10-13). From a technical-theoretical perspective, most of the points we end up sampling have no effect on the output. For instance, if the selected model is Linear Regression, then only one other meta-parameter is relevant (lr_penalty), and all other 11 are irrelevant. We are structuring the space in a way that makes optimization difficult, because the majority of sampled parameters are guaranteed to not affect the outcome at all. From an implementation perspective, this sort of search space structuring forces us to write a horrendously long objective function. We need to duplicate everything and implicitly structure relationships anyway, as shown in Listing 10-14.

Listing 10-14. The objective function that one would have to write if they defined the search space in a similar manner as in Listing 10-13

def objective(params):

```
if params['model'] == LogisticRegression:
    model = LogisticRegression(lr_penalty = params['lr_penalty'])
elif params['model'] == DecisionTreeClassifier:
    model = DecisionTreeClassifier(criterion = params['dtc_criterion'],
                                    max_depth = params['dtc_max_depth'])
elif params['model'] == RandomForestClassifier:
    model = RandomForestClassifier(criterion = params['rfc_criterion'],
                                    max_depth = params['rfc_max_depth'],
                                    n_estimators = params['rfc_n_estimators'])
...
model.fit(x_train, y_train)
return -f1_score(model.predict(x_valid), y_valid)
```

To avoid this, we will use a *nested search space*. Rather than just choosing a model objective, we choose a model "bundle." This "bundle" includes not only the actual model itself but a *sub–search space* for meta-parameters specific to that selected model. By nesting dictionaries within each other, we can more accurately capture the relationship between meta-parameters. Technically speaking, we are creating a list of dictionaries, in which each dictionary represents a subspace with potentially more search spaces and each hp.choice() samples a dictionary during exploration (Listing 10-15).

Listing 10-15. A better, nested search space for model hyperparameter optimization

space = {}

```
models = [{'model': LogisticRegression,
           'parameters':{'penalty':hp.choice('lr_penalty', ['none', 'l2'])}},
          {'model': DecisionTreeClassifier,
           'parameters': {'criterion': hp.choice('dtc_criterion', ['gini', 'entropy']),
                          'max_depth': hp.quniform('dtc_max_depth', 1, 30, 1)}},
          {'model': RandomForestClassifier,
           'parameters': {
               'criterion': hp.choice('rfc_criterion', ['gini', 'entropy']),
               'max_depth': hp.quniform('rfc_max_depth', 1, 30, q=1),
```

```
            'n_estimators': hp.qnormal('rfc_n_estimators', 100, 30, 1)}},
        {'model': GradientBoostingClassifier,
         'parameters':
         {'criterion': hp.choice('gbc_criterion', ['friedman_mse',
                                                   'squared_error',
                                                   'mse', 'mae']),
          'n_estimators': hp.qnormal('gbc_n_estimators', 100, 30, 1),
          'max_depth': hp.quniform('gbc_max_depth', 1, 30, q=1)}},
        {'model': AdaBoostClassifier,
         'parameters': {
             'n_estimators': hp.qnormal('abc_n_estimators', 50, 15, 1),
             'learning_rate': hp.uniform('abc_learning_rate', 1e-3, 10)}
        },
        {'model': MLPClassifier,
         'parameters':{'activation': hp.choice('mlp_activation', ['logistic', 'tanh',
         'relu'])}
        }]
```

```
space['models'] = hp.choice('models', models)
```

When we define our search space this way, our objective function becomes exceptionally clean (Listing 10-16).

Listing 10-16. A clean objective function following from a well-designed nested search space in Listing 10-15

def objective(params):

```
    model = params['models']['model'](**params['models']['parameters'])
    model.fit(X_train, y_train)
    return -f1_score(model.predict(X_valid), y_valid)
```

What is Listing 10-16 doing? Recall that the model is stored as an uninstantiated object, which can be instantiated using the parenthesis notation (e.g., model = DecisionTreeClassifier()). Additionally, we want to construct the model with the sampled set of parameters. From the search space defined in Listing 10-15, we see that params['models']['parameters'] yields a dictionary in which keys are parameter names and values will be the sampled values for those parameters. The unpacking ** command in Python can be used to "translate" this dictionary into object construction such that each key is a parameter and each value is the parameter value. We conveniently have named each of the keys in our parameter dictionaries in the search space the same as the initialization arguments in the model constructor, and thus we can initialize our model and all the sampled parameters in one line of code.

Practically speaking, however, we have to make one adjustment. Hyperopt samples quantized inputs (e.g., for tree depth of a Decision Tree) as floats, even if the actual values are mathematically integers (e.g., 2.0, 42.0). sklearn throws an error when we pass a float value into an argument that accepts only integers. Therefore, we can add some "cleaning" code in our objective function to make sure that the n_estimators parameter is cast to an integer and also is at least 1 (it is possible for Hyperopt to sample values less than 1, since we technically defined a quantized *normal* distribution). We can initialize our model as previously with this set of cleaned parameters (Listing 10-17).

Listing 10-17. Adding additional catches for potential invalid sampling that are easier to address in the objective function than elsewhere

```
def objective(params):
    cleanedParams = {}
    for param in params['models']['parameters']:
        value = params['models']['parameters'][param]
        if param == 'n_estimators':
            if value < 1:
                value = 1
            value = int(value)
        cleanedParams[param] = value

    model = params['models']['model'](**cleanedParams)
    model.fit(X_train, y_train)
    return -f1_score(model.predict(X_valid), y_valid)

best = fmin(objective, space, algo=tpe.suggest, max_evals=30)
```

Note additionally that we return the negative F1 score because Hyperopt *minimizes* the given objective function. If we were to omit the negation, Hyperopt would in fact be finding the worst model. (Purely theoretically speaking, the worst model is very close to the best model – just flip the labels.)

After 30 evaluations, Hyperopt determines that the best set of parameters (stored in best) is as follows:

```
{'models': 2,
 'rfc_criterion': 1,
 'rfc_max_depth': 16.0,
 'rfc_n_estimators': 69.0}
```

The selected value for the parameter 'models' is 2; the model at index 2 in the list of models in the search space is the Random Forest classifier. The optimal model is a Random Forest classifier using the entropy criterion with a maximum tree depth of 16 and 69 estimators. After training separately, this model obtains an F score of about 0.73, which is quite high for the Higgs Boson dataset.

We can also use meta-optimization to optimize how we train neural networks. For instance, we may want to choose the best optimizer, optimizer parameters, and the decay factor and patience of a learning rate manager. Note that we need to structure our optimizer in a nested/hierarchical fashion because some parameters are only relevant if another parameter is sampled at a certain value (i.e., we want to capture parameter dependencies). However, we will reliably need to sample the patience and decay factor of a learning rate manager; thus, we can add these to the space as parameters in a sub-dictionary for the purposes of organization but don't "choose" whether they are relevant or not via nested choice (Listing 10-18).

Listing 10-18. Defining a search space for the optimal neural network training hyperparameters

```
from tensorflow.keras.optimizers import Adam, SGD, RMSprop, Adagrad

from sklearn.metrics import f1_score

from hyperopt import hp
from hyperopt import fmin, tpe

space = {}
```

```
optimizers = [{'optimizer':SGD,
               'parameters':{
                   'learning_rate': hp.uniform('sgd_lr', 1e-5, 1),
                   'momentum': hp.uniform('sgd_mom', 0, 1),
                   'nesterov': hp.choice('sgd_nest', [False, True])
               }},
              {'optimizer':RMSprop,
               'parameters':{
                   'learning_rate': hp.uniform('rms_lr', 1e-5, 1),
                   'momentum': hp.uniform('rms_mom', 0, 1),
                   'rho': hp.normal('rms_rho', 1.0, 0.3),
                   'centered': hp.choice('rms_cent', [False, True])
               }},
              {'optimizer':Adam,
               'parameters':{
                   'learning_rate': hp.uniform('adam_lr', 1e-5, 1),
                   'beta_1': hp.uniform('adam_beta1', 0.3, 0.9999999999),
                   'beta_2': hp.uniform('adam_beta2', 0.3, 0.9999999999),
                   'amsgrad': hp.choice('amsgrad', [False, True])
               }},
              {'optimizer':Adagrad,
               'parameters':{
                   'learning_rate': hp.uniform('adagrad_lr', 1e-5, 1),
                   'initial_accumulator_value': hp.uniform('adagrad_iav', 0.0, 1.0)
               }}]
space['optimizers'] = hp.choice('optimizers', optimizers)

from keras.callbacks import ReduceLROnPlateau
space['lr_manage'] = {'factor': hp.uniform('lr_factor', 0.01, 0.95),
                      'patience': hp.quniform('lr_patience', 3, 20, q=1)}
```

Since creating a neural network requires a nontrivial amount of code, it's good practice to create a separate method that constructs a neural network architecture. In this case, we will employ a simple static seven-layer network (Listing 10-19).

Listing 10-19. A function that builds a simple static seven-layer network

```
def build_NN(input_dim = len(X_train.columns)):
    model = keras.models.Sequential()
    model.add(L.Input((input_dim,)))
    model.add(L.Dense(input_dim, activation='relu'))
    model.add(L.Dense(input_dim, activation='relu'))
    model.add(L.Dense(input_dim, activation='relu'))
    model.add(L.BatchNormalization())
    model.add(L.Dense(16, activation='relu'))
    model.add(L.Dense(16, activation='relu'))
    model.add(L.Dense(16, activation='relu'))
    model.add(L.Dense(1, activation='sigmoid'))
    return model
```

Our objective is to minimize the validation binary cross-entropy. The objective function takes in the sampled parameters, initializes a neural network, and fits it using the given optimizer parameters and learning rate manager (Listing 10-20).

Listing 10-20. Defining an objective function to optimize training parameters

```
from keras.callbacks import EarlyStopping

bce = tf.keras.losses.BinaryCrossentropy(from_logits=True)

def objective(params):
    model = build_NN()
    es = EarlyStopping(patience=5)
    rlrop = ReduceLROnPlateau(**params['lr_manage'])

    optimizer = params['optimizers']['optimizer']
    optimizer_params = params['optimizers']['parameters']
    model.compile(loss='binary_crossentropy',
                  optimizer=optimizer(**optimizer_params))
    model.fit(X_train, y_train, callbacks=[es, rlrop],
              epochs = 50, verbose = 0)

    pred = model.predict(np.array(X_valid))
    truth = np.array(y_valid).reshape((len(y_valid),1))
    valid_loss = bce(pred.astype(np.float16)),
                     truth.astype(np.float16)).numpy()
    return valid_loss

best = fmin(objective, space, algo=tpe.suggest, max_evals=100);
```

Note that because we are using a relatively lightweight neural network architecture and tabular datasets are comparatively small, training and evaluating 100 neural networks (or more) is not computationally unfeasible on ordinary hardware.

The best solution yielded in one run is as follows:

```
{'lr_factor': 0.7749528095685804,
 'lr_patience': 14.0,
 'optimizers': 0,
 'sgd_lr': 0.23311588639160802,
 'sgd_mom': 0.5967800410439047,
 'sgd_nest': 1}
```

This particular neural network model reaches an F1 score of almost 0.74.

Beyond just optimizing the model's training parameters, we can also try to optimize the neural network's architecture. Let's parametrize the build_NN function by allowing for various dimensions of the network (Listing 10-21). In an effort to build more complex nonlinear network topologies, we will use multiple branches (see Figure 10-14). Each branch is composed of a certain number of 28-neuron fully connected layers, followed by a certain number of 16-neuron fully connected layers. These branches are merged together through a certain joining method (either adding or concatenating) and processed by a certain number of 16-neuron layers. This gives us five dimensions of parametrization.

Listing 10-21. A function that builds a neural network given five adjustable architectural parameters

```python
def build_NN(num_branches,
             num28repeats, num16repeats,
             join_method, numOutRepeats):
    inp = L.Input((28,))
    out_tensors = []
    for i in range(int(num_branches)):
        x = L.Dense(28, activation='relu')(inp)
        for i in range(int(num28repeats-1)):
            x = L.Dense(28, activation='relu')(x)
        for i in range(int(num16repeats)):
            x = L.Dense(16, activation='relu')(x)
        out_tensors.append(x)
    if num_branches == 1:
        join = out_tensors[0]
    elif join_method == 'concat':
        join = L.Concatenate()(out_tensors)
    else:
        join = L.Add()(out_tensors)
    x = L.Dense(16, activation='relu')(join)
    for i in range(int(numOutRepeats-1)):
        x = L.Dense(16, activation='relu')(x)
    out = L.Dense(1, activation='sigmoid')(x)
    return keras.models.Model(inputs=inp, outputs=out)
```

We can correspondingly design a search space (Listing 10-22). In addition to optimizing for the optimizer and learning rate management, we will also provide five fields for Hyperopt to optimize the neural network architecture. quniform works fine for all the fields except for the join method, which is a choice between adding and concatenating.

Listing 10-22. Defining a search space to optimize the architectural parameters of the neural network constructor defined in Listing 10-21

```python
space = {}

space['optimizers'] = hp.choice('optimizers', optimizers)
space['lr_manage'] = {'factor': hp.uniform('lr_factor', 0.01, 0.95),
                      'patience': hp.quniform('lr_patience', 3, 20, q=1)}
space['architecture'] = {'num_branches': hp.quniform('num_branches', 1, 5, q=1),
                         'num28repeats': hp.quniform('num28repeats', 1, 5, q=1),
                         'num16repeats': hp.quniform('num16repeats', 1, 5, q=1),
                         'join_method': hp.choice('join_method', ['add', 'concat']),
                         'numOutRepeats': hp.quniform('numOutRepeats', 1, 5, q=1)}
```

To understand how different branch sizes manifest, see Listing 10-23 and Figure 10-14.

Listing 10-23. Visualizing variation across the num_branches architectural dimension

```
for i in range(1, 7):
    model = build_NN(num_branches=i,
                     num28repeats=2,
                     num16repeats=2,
                     join_method='add',
                     numOutRepeats=2)

    tensorflow.keras.utils.plot_model(model, dpi=400, to_file=f'branches{i}.png')

plt.figure(figsize=(7, 5), dpi=400)
for i in range(1, 7):
    plt.subplot(2, 3, i)
    plt.title(f'{i} branches')
    plt.axis('off')
    plt.imshow(plt.imread(f'branches{i}.png'))
plt.show()
```

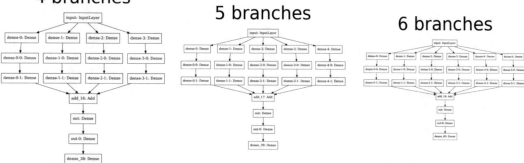

Figure 10-14. *Visualization of architectures with varying numbers of branches (holding other architectural features constant)*

733

Moreover, to illustrate the architectural diversity of topologies generated from our five dimensions of architecture parametrization, see Listing 10-24 and Figure 10-15. You can think of each of the displayed architectures as occupying some point in the search space.

Listing 10-24. Randomly sampling different possible architectures from the search space

```
for i in range(35):
    model = build_NN(num_branches=np.random.choice([1, 2, 3, 4, 5, 6]),
                    num28repeats=np.random.choice([1, 2, 3, 4, 5, 6]),
                    num16repeats=np.random.choice([1, 2, 3, 4, 5, 6]),
                    join_method=np.random.choice(['add', 'concat']),
                    numOutRepeats=np.random.choice([1, 2, 3, 4, 5, 6]))

    tensorflow.keras.utils.plot_model(model, dpi=400, to_file=f'{i}.png')

plt.figure(figsize=(15, 21), dpi=400)
for i in range(35):
    plt.subplot(5, 7, i+1)
    plt.axis('off')
    plt.imshow(plt.imread(f'{i}.png'))
plt.show()
```

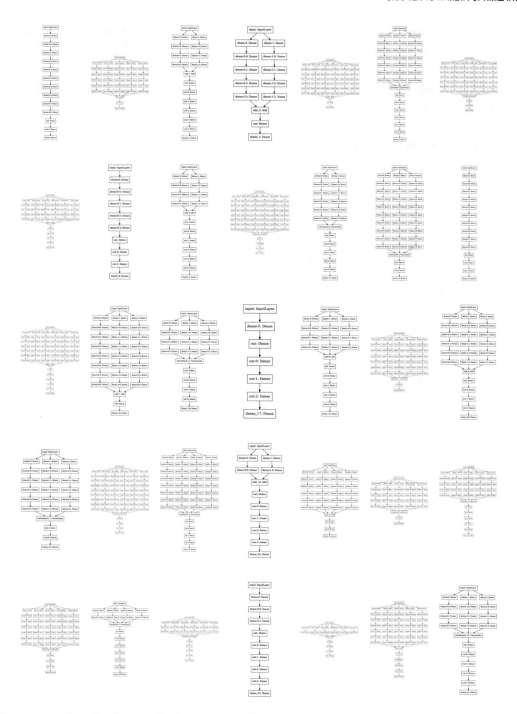

Figure 10-15. *Sample selection of architectures possible in our search space*

The objective function is not difficult to write, due to our well-designed implementation of the search space. We can simply unpack relevant architectural parameters and pass them into the build_NN function, which instantiates the sampled architecture (Listing 10-25).

Listing 10-25. Writing the objective function for the best-architecture search problem and executing the optimization operation

```
def objective(params):
    model = build_NN(**params['architecture'])
    es = EarlyStopping(patience=5)
    rlrop = ReduceLROnPlateau(**params['lr_manage'])

    optimizer = params['optimizers']['optimizer']
    optimizer_params = params['optimizers']['parameters']
    model.compile(loss='binary_crossentropy',
                  metrics=['accuracy'],
                  optimizer=optimizer(**optimizer_params))
    model.fit(X_train, y_train, callbacks=[es, rlrop],
              epochs = 50, verbose = 0)

    pred = model.predict(np.array(X_valid)
    truth = np.array(y_valid).reshape((len(y_valid),1))
    valid_loss = bce(pred.astype(np.float16)),
                     truth.astype(np.float16)).numpy()
    return valid_loss

best = fmin(objective, space, algo=tpe.suggest, max_evals=100);
```

The best-discovered architecture for a sample run had two branches, three 28-node layers followed by three 110-node layers in each branch and four 110-node layers after an add-based joining. The discovered architecture seems reasonable: it exploits the nonlinearity of multiple branches, but is overall a well-balanced topology (Figure 10-16).

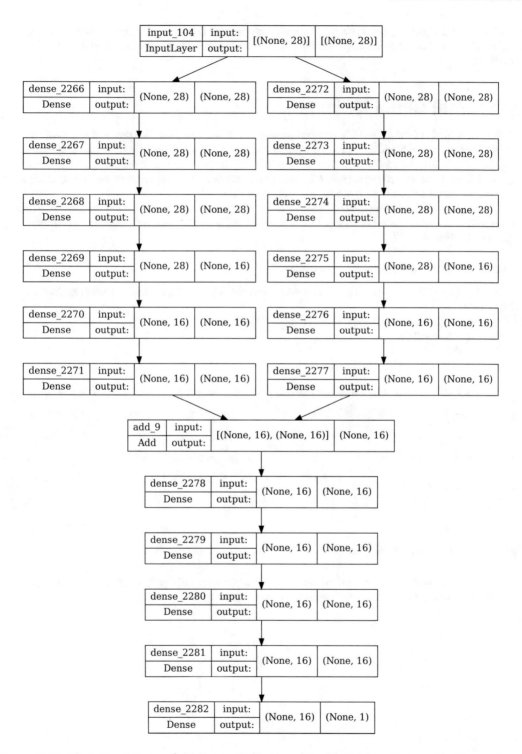

Figure 10-16. *The best architecture (within our problem space) for this problem, discovered by Hyperopt*

However, note that Hyperopt was not designed for intensive architecture search. Later in this chapter, we'll explore how we can use more specialized frameworks to run Neural Architecture Search (NAS) operations.

Optimizing Data Pipelines

Building the model itself is a relatively small component of the overall modeling process! Any data scientist who has worked on real-world data problems understands that tasks like data cleaning and preparation comprise a large part of the labor and time required to work one's way through the data pipeline. Along the way, you will make many decisions on how to manipulate the data that may seem arbitrary or unoptimized. We can optimize these decisions, too, with meta-optimization tools.

Let's use the Ames Housing dataset, which has many categorical features that require intensive data encoding. Recall we used this dataset in Chapter 2 to try out different categorical encoding techniques (Listing 10-26, Figure 10-17).

Listing 10-26. Reading the Ames Housing dataset

```
df = pd.read_csv('https://raw.githubusercontent.com/hjhuney/Data/master/AmesHousing/
train.csv')
df = df.dropna(axis=1, how='any').drop('Id', axis=1)
x = df.drop('SalePrice', axis=1)
y = df['SalePrice']
```

	MSSubClass	MSZoning	LotArea	Street	LotShape	LandContour	Utilities	LotConfig	LandSlope	Neighborhood	...	OpenPorchSF	EnclosedPorch	3SsnPorch	ScreenPorch	PoolArea	MiscVal	MoSold	YrSold	SaleType	SaleCondition
0	60	RL	8450	Pave	Reg	Lvl	AllPub	Inside	Gtl	CollgCr	...	61	0	0	0	0	0	2	2008	WD	Normal
1	20	RL	9600	Pave	Reg	Lvl	AllPub	FR2	Gtl	Veenker	...	0	0	0	0	0	0	5	2007	WD	Normal
2	60	RL	11250	Pave	IR1	Lvl	AllPub	Inside	Gtl	CollgCr	...	42	0	0	0	0	0	9	2008	WD	Normal
3	70	RL	9550	Pave	IR1	Lvl	AllPub	Corner	Gtl	Crawfor	...	35	272	0	0	0	0	2	2006	WD	Abnorml
4	60	RL	14260	Pave	IR1	Lvl	AllPub	FR2	Gtl	NoRidge	...	84	0	0	0	0	0	12	2008	WD	Normal
...	...	...	...	...	...	...	...	...	...	...	...	...	...	...	...	...	...	...	...	...	...
1455	60	RL	7917	Pave	Reg	Lvl	AllPub	Inside	Gtl	Gilbert	...	40	0	0	0	0	0	8	2007	WD	Normal
1456	20	RL	13175	Pave	Reg	Lvl	AllPub	Inside	Gtl	NWAmes	...	0	0	0	0	0	0	2	2010	WD	Normal
1457	70	RL	9042	Pave	Reg	Lvl	AllPub	Inside	Gtl	Crawfor	...	60	0	0	0	2500	5	2010	WD	Normal
1458	20	RL	9717	Pave	Reg	Lvl	AllPub	Inside	Gtl	NAmes	...	0	112	0	0	0	0	4	2010	WD	Normal
1459	20	RL	9937	Pave	Reg	Lvl	AllPub	Inside	Gtl	Edwards	...	68	0	0	0	0	0	6	2008	WD	Normal

1460 rows × 60 columns

Figure 10-17. The Ames Housing dataset

Plotting out a univariate distribution of house prices demonstrates a pretty significant range of prices, spanning the range of hundreds of thousands of dollars (Figure 10-18).

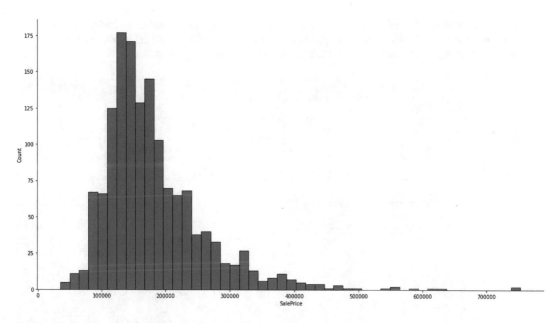

Figure 10-18. *Distribution of house prices in the Ames Housing dataset*

We mainly care about categorical features here, since these are ones that need preprocessing. We want to find which type of categorical encoding is optimal for each feature.

Categorical features need to be encoded somehow. Instead of guessing which categorical encoding technique to use for each feature (or across all features), we can define a search space in which we want to choose the best categorical encoder for each categorical column in the Ames Housing dataset (Figure 10-19).

Street	Bathrooms	Area	City	Has Patio?
Paved	3	2301	Seattle	Yes
Gravel	2	1580	Portland	No
Gravel	5	2560	San Francisco	Yes
...	...	...	...	...

Search Space

Label Encoder	Label Encoder		Label Encoder	Label Encoder
One-Hot Encoder	One-Hot Encoder		One-Hot Encoder	One-Hot Encoder
Binary Encoder	Binary Encoder		Binary Encoder	Binary Encoder
Target Encoder	Target Encoder		Target Encoder	Target Encoder
Count Encoder	Count Encoder		Count Encoder	Count Encoder
Leave 1 Out Encoder	Leave 1 Out Encoder		Leave 1 Out Encoder	Leave 1 Out Encoder
James-Stein Encoder	James-Stein Encoder		James-Stein Encoder	James-Stein Encoder
CatBoost Encoder	CatBoost Encoder		CatBoost Encoder	CatBoost Encoder

Figure 10-19. *Demonstration of mapping a set of possible categorical encoders to each categorical feature*

Ideally, at the conclusion of such a search, the meta-optimization procedure would select the best combination of categorical encoding techniques for each column such that a model trained on the categorically encoded dataset yields the best validation performance (Figure 10-20).

Street	Bathrooms	Area	City	Has Patio?
Paved	3	2301	Seattle	Yes
Gravel	2	1580	Portland	No
Gravel	5	2560	San Francisco	Yes
...	...	...	...	...

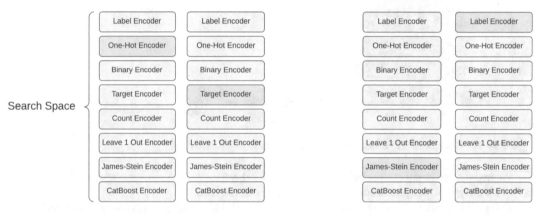

Figure 10-20. *Demonstration of selecting a single best categorical encoder from the set of categorical encoders for each categorical feature in the dataset*

First, we need to determine which columns in the dataset are categorical or not. This is not a trivial task, because there are several dozen columns in the dataset and some categorical features are numbers. Moreover, some features can be reasonably considered both to be categorical and continuous, like the number of bathrooms in a house. This quantity is technically continuous, but is categorical in nature because there are only really four to five unique values that this feature could feasibly occupy. We'll define a categorical feature as a feature that satisfies the following condition: it either is filled with strings or contains five or less unique values (Listing 10-27). This isn't perfect, but it gets the job done.

Listing 10-27. Acquiring a list of which columns are considered categorical

```
cat_features = []
for colIndex, colName in enumerate(x.columns):
    # find categorical variables to process
    if type(x.iloc[0, colIndex]) == str or len(x[colName].unique()) <= 5:
        cat_features.append(colName)
```

Now, we need to actually construct our space. To begin with, we want a set of categorical encoders to consider for each feature (Listing 10-28). We'll choose several earlier discussed in Chapter 2 and create a list of possible encoders.

Listing 10-28. Creating a list of possible categorical encoders

```
from category_encoders.ordinal import OrdinalEncoder
from category_encoders.one_hot import OneHotEncoder
from category_encoders.binary import BinaryEncoder
from category_encoders.target_encoder import TargetEncoder
from category_encoders.count import CountEncoder
from category_encoders.leave_one_out import LeaveOneOutEncoder
from category_encoders.james_stein import JamesSteinEncoder
from category_encoders.cat_boost import CatBoostEncoder

encoders = [OrdinalEncoder(),
            OneHotEncoder(),
            BinaryEncoder(),
            TargetEncoder(),
            CountEncoder(),
            LeaveOneOutEncoder(),
            JamesSteinEncoder(),
            CatBoostEncoder()]
```

To create the space, we simply need to loop through each column in the dataset and determine if it is a categorical feature or not. If it is, we take note of it in `cat_features` and create a parameter in the search space that allows Hyperopt to choose between any of the encoders to apply for that particular feature (Listing 10-29).

Listing 10-29. Programmatically generating a categorical encoder search space

```
space = {}

cat_features = []
for colIndex, colName in enumerate(x.columns):
    # find categorical variables to process
    if type(x.iloc[0, colIndex]) == str or len(x[colName].unique()) <= 5:
        cat_features.append(colName)
        space[f'{colName}_cat_enc'] = hp.choice(f'{colName}_cat_enc', encoders)
```

Now, our space has a parameter for each categorical variable. You can probably already imagine the structure of the objective function from our previous exercises. The objective function will accept a dictionary of parameters, and we will apply the selected encoders to each feature (using something like `encoder.fit_transform(data)`).

However, there's a problem. Some of the categorical encoders (e.g., count encoder, target encoder) used require a target feature y in addition to the categorical feature x, whereas others (e.g., one-hot encoder, ordinal encoder) require only the categorical feature x. We cannot train all encoders the same way. To reconcile with this, we can associate each encoder in the search space with metadata (similar to the practice of "bundling" introduced earlier with nested search spaces). In this case, we can attach a Boolean to each encoder indicating if it requires a target feature during training or not. When we read in the set of selected parameters, we can determine exactly how to use the encoder with respect to the categorical feature by observing that Boolean status.

To implement this, we can restructure our list of encoders to instead a list of encoder "bundles," in which each bundle contains the actual encoder object as its first element and the Boolean value as the second element (Listing 10-30). Note that Hyperopt doesn't care about how you structure the choice search space, as long as you give it a list of some objects (e.g., sets, tuples, lists) to choose from.

Listing 10-30. Using a modified collection of encoders that bundles an encoder with important information about how to instantiate it

```
encoders = [[OrdinalEncoder(), False],
            [OneHotEncoder(), False],
            [BinaryEncoder(), True],
            [TargetEncoder(), True],
            [CountEncoder(), True],
            [LeaveOneOutEncoder(), True],
            [JamesSteinEncoder(), True],
            [CatBoostEncoder(), True]]
```

Now we can build our objective function. We will loop through each of the categorical features and encode them using the sampled encoder and then add them to a dataset (Listing 10-31). We'll run train-test split with the same seed and train a Random Forest regressor model on that dataset. The performance of the set is the mean absolute error of that model on the validation dataset.

Listing 10-31. Defining the objective function

```
from sklearn.ensemble import RandomForestRegressor
from sklearn.metrics import mean_absolute_error as mae
from sklearn.model_selection import train_test_split as tts

def objective(params):
    x_ = pd.DataFrame()
    for colName in cat_features:
        colValues = np.array(x[colName])
        encoder = params[f'{colName}_cat_enc'][0]
        if params[f'{colName}_cat_enc'][1]:
            transformed = encoder.fit_transform(colValues, y)
        else:
            transformed = encoder.fit_transform(colValues)
        x_ = pd.concat([x_, transformed], axis=1)
    nonCatCols = [col for col in x.columns if (col not in cat_features)]
    x_ = pd.concat([x_, x[nonCatCols]], axis=1)

    X_train, X_valid, y_train, y_valid = tts(x_, y, train_size = 0.8, random_state = 42)

    model = RandomForestRegressor(random_state = 42)
    model.fit(X_train, y_train)
    return mae(model.predict(X_valid), y_valid)
```

We can fit it using standard objectives (Listing 10-32).

Listing 10-32. Executing the optimization algorithm

```
best = fmin(objective, space, algo=tpe.suggest, max_evals=1000);
```

The best model obtains a validation loss of 825. This is a good result – that's 825 dollars off on predicting house price with a much larger target output range.

Moreover, we can print out the contents of the best-discovered set of encoders. Hyperopt displays the indices of the best encoders for a particular column. The results are interesting to think about; you can do a lot of important analyses about the nature of the feature based on which type of encoder is most beneficial to prediction:

```
{'BldgType_cat_enc': 0,        # ordinal encoder
 'BsmtFullBath_cat_enc': 2,    # binary encoder
 'BsmtHalfBath_cat_enc': 0,    # ordinal encoder
 'CentralAir_cat_enc': 5,      # leave-one-out encoder
 'Condition1_cat_enc': 6,      # james-stein encoder
 'Condition2_cat_enc': 2,      # binary encoder
 'ExterCond_cat_enc': 0,       # ordinal encoder
 'ExterQual_cat_enc': 3,       # target encoder
 'Exterior1st_cat_enc': 0,     # ordinal encoder
 'Exterior2nd_cat_enc': 5,     # leave-one-out encoder
 'Fireplaces_cat_enc': 2,      # binary encoder
 'Foundation_cat_enc': 3,      # target encoder
 'FullBath_cat_enc': 7,        # catboost encoder
 'Functional_cat_enc': 5,      # leave-one-out encoder
 'GarageCars_cat_enc': 7,      # catboost encoder
 'HalfBath_cat_enc': 0,        # ordinal encoder
 'HeatingQC_cat_enc': 4,       # count encoder
 'Heating_cat_enc': 5,         # leave-one-out encoder
 'HouseStyle_cat_enc': 1,      # one-hot encoder
 'KitchenAbvGr_cat_enc': 5,    # leave-one-out encoder
 'KitchenQual_cat_enc': 6,     # james-stein encoder
 'LandContour_cat_enc': 0,     # ordinal encoder
 'LandSlope_cat_enc': 6,       # james-stein encoder
 'LotConfig_cat_enc': 3,       # target encoder
 'LotShape_cat_enc': 5,        # leave-one-out encoder
 'MSZoning_cat_enc': 0,        # ordinal encoder
 'Neighborhood_cat_enc': 6,    # james-stein encoder
 'PavedDrive_cat_enc': 3,      # target encoder
 'RoofMatl_cat_enc': 2,        # binary encoder
 'RoofStyle_cat_enc': 1,       # one-hot encoder
 'SaleCondition_cat_enc': 4,   # count encoder
 'SaleType_cat_enc': 2,        # binary encoder
 'Street_cat_enc': 1,          # one-hot encoder
 'Utilities_cat_enc': 5,       # leave-one-out encoder
 'YrSold_cat_enc': 3}          # target encoder
```

We can also optimize which model is used. It may be the case that the best set of encoders for the Random Forest regressor is not the best set of encoders for a different regressor that performs better overall. This follows similar logic as previously discussed; we can define a search space of possible models that will be instantiated and trained inside the objective function (Listing 10-33).

Listing 10-33. Importing various relevant models

```
from sklearn.linear_model import LinearRegression, Lasso
from sklearn.tree import DecisionTreeRegressor
from sklearn.ensemble import RandomForestRegressor, GradientBoostingRegressor,
AdaBoostRegressor
from sklearn.neural_network import MLPRegressor
```

```
...

space['model'] = hp.choice('model',
                            [LinearRegression, Lasso,
                             DecisionTreeRegressor,
                             RandomForestRegressor,
                             GradientBoostingRegressor,
                             AdaBoostRegressor,
                             MLPRegressor])
```

Within the objective function, we instantiate the model with ( ) and fit on the encoded set of features, returning the validation Mean Absolute Error as a measure of badness to be minimized (Listing 10-34).

Listing 10-34. Defining the objective function

```
def objective(params):
    x_ = pd.DataFrame()
    for colName in cat_features:
        colValues = np.array(x[colName])
        encoder = params[f'{colName}_cat_enc'][0]
        if params[f'{colName}_cat_enc'][1]:
            transformed = encoder.fit_transform(colValues, y)
        else:
            transformed = encoder.fit_transform(colValues)
        x_ = pd.concat([x_, transformed], axis=1)
    nonCatCols = [col for col in x.columns if (col not in cat_features)]
    x_ = pd.concat([x_, x[nonCatCols]], axis=1)

    X_train, X_valid, y_train, y_valid = tts(x_, y, train_size = 0.8,
                                             random_state = 42)

    model = params['model']()
    model.fit(X_train, y_train)
    return mae(model.predict(X_valid), y_valid)
```

This model performs much better, obtaining a Mean Absolute Error of 373.589. The best selected model uses LASSO regression, which you likely would not have expected to be the best-performing algorithm!

Moreover, we can optimize the hyperparameters of the model. This brings everything previously discussed together: we are optimizing, more or less, the entire data pipeline from encoding to model training. The definition of the model hyperparameter search space is very similar to the previously discussed sample, but in this case we use regression rather than classification models and therefore optimize a somewhat different set of hyperparameters (Listing 10-35).

Listing 10-35. Defining an "ultimate" search space in which the data encodings, the type of model, and the hyperparameters of the model are all optimized simultaneously

```
space = {}
encoders = [[OrdinalEncoder(), False],
            [OneHotEncoder(), False],
            [BinaryEncoder(), True],
            [TargetEncoder(), True],
            [CountEncoder(), True],
```

```
                        [LeaveOneOutEncoder(), True],
                        [JamesSteinEncoder(), True],
                        [CatBoostEncoder(), True]]

models = [{'model': LinearRegression,
           'parameters':{}},
          {'model': Lasso,
           'parameters': {'alpha': hp.uniform('lr_alpha', 0, 5),
                          'normalize': hp.choice('lr_normalize', [True, False])}},
          {'model': DecisionTreeRegressor,
           'parameters': {'criterion': hp.choice('dtr_criterion', ['squared_error', 'friedman_mse'
                                                 'absolute_error', 'poisson']),
                          'max_depth': hp.quniform('dtr_max_depth', 1, 30, 1)}},
          {'model': RandomForestRegressor,
           'parameters': {
               'criterion': hp.choice('rfr_criterion', ['squared_error', 'friedman_mse',
                                                 'absolute_error', 'poisson']),
               'max_depth': hp.quniform('rfr_max_depth', 1, 30, q=1),
               'n_estimators': hp.qnormal('rfr_n_estimators', 100, 30, 1)}},
          {'model': GradientBoostingRegressor,
           'parameters': {'criterion': hp.choice('gbr_criterion', ['squared_error', 'absolute_erro
                                                 'huber', 'quantile']),
           'n_estimators': hp.qnormal('gbr_n_estimators', 100, 30, 1),
           'criterion': hp.choice('gbr_criterion', ['squared_error', 'friedman_mse',
                                                 'absolute_error', 'poisson']),
           'max_depth': hp.quniform('gbr_max_depth', 1, 30, q=1)}},
          {'model': AdaBoostRegressor,
           'parameters': {'n_estimators': hp.qnormal('abr_n_estimators', 50, 15, 1),
           'loss': hp.choice('abr_loss', ['linear', 'square', 'exponential'])}},
          {'model': MLPRegressor,
           'parameters':{'activation': hp.choice('mlp_activation', ['logistic', 'tanh', 'relu'])}}]

cat_features = []
for colIndex, colName in enumerate(x.columns):
    # find categorical variables to process
    if type(x.iloc[0, colIndex]) == str or len(x[colName].unique()) <= 5:
        cat_features.append(colName)
        space[f'{colName}_cat_enc'] = hp.choice(f'{colName}_cat_enc', encoders)
space['models'] = hp.choice('models', models)

def objective(params):
    x_ = pd.DataFrame()
    for colName in cat_features:
        colValues = np.array(x[colName])
        encoder = params[f'{colName}_cat_enc'][0]
        if params[f'{colName}_cat_enc'][1]:
            transformed = encoder.fit_transform(colValues, y)
        else:
            transformed = encoder.fit_transform(colValues)
        x_ = pd.concat([x_, transformed], axis=1)
    nonCatCols = [col for col in x.columns if (col not in cat_features)]
```

```
    x_ = pd.concat([x_, x[nonCatCols]], axis=1)

    X_train, X_valid, y_train, y_valid = tts(x_, y, train_size = 0.8, random_state = 42)

    cleanedParams = {}
    for param in params['models']['parameters']:
        value = params['models']['parameters'][param]
        if param == 'n_estimators':
            value = int(value)
        cleanedParams[param] = value

    model = params['models']['model'](**cleanedParams)
    model.fit(X_train, y_train)
    return mae(model.predict(X_valid), y_valid)

best = fmin(objective, space, algo=tpe.suggest, max_evals=1000)
```

We get a validation error of about 145 with the following parameter dictionary. This is almost a fivefold decrease in error from our first hyperparameter optimization procedure on this dataset and likely an even larger decrease in error compared with a manually designed model:

```
{'BldgType_cat_enc': 6,
 'BsmtFullBath_cat_enc': 3,
 'BsmtHalfBath_cat_enc': 0,
 'CentralAir_cat_enc': 2,
 'Condition1_cat_enc': 5,
 'Condition2_cat_enc': 1,
 'ExterCond_cat_enc': 3,
 'ExterQual_cat_enc': 6,
 'Exterior1st_cat_enc': 5,
 'Exterior2nd_cat_enc': 5,
 'Fireplaces_cat_enc': 5,
 'Foundation_cat_enc': 0,
 'FullBath_cat_enc': 5,
 'Functional_cat_enc': 5,
 'GarageCars_cat_enc': 5,
 'HalfBath_cat_enc': 4,
 'HeatingQC_cat_enc': 4,
 'Heating_cat_enc': 6,
 'HouseStyle_cat_enc': 6,
 'KitchenAbvGr_cat_enc': 3,
 'KitchenQual_cat_enc': 6,
 'LandContour_cat_enc': 2,
 'LandSlope_cat_enc': 3,
 'LotConfig_cat_enc': 4,
 'LotShape_cat_enc': 6,
 'MSZoning_cat_enc': 3,
 'Neighborhood_cat_enc': 5,
 'PavedDrive_cat_enc': 2,
 'RoofMatl_cat_enc': 7,
 'RoofStyle_cat_enc': 2,
```

```
'SaleCondition_cat_enc': 6,
'SaleType_cat_enc': 3,
'Street_cat_enc': 0,
'Utilities_cat_enc': 5,
'YrSold_cat_enc': 6,
'lr_alpha': 3.305989653851934,
'lr_normalize': 0,
'models': 1}
```

However, it should be noted that meta-optimization can – in a minority of cases – be prone to *meta-overfitting*, in which we overly parametrize our search space and end up creating a highly specialized model that performs very well on the validation dataset, but not on new data. Meta-overfitting is only relevant when the search space size drastically overshadows the size of the dataset, because it is quite difficult to overfit on a "secondary" level through the outputs of a model. This can occur when optimizing a very large set of neural network meta-parameters on a comparatively much smaller dataset, but seldom in other cases. Imagine, for instance, that your hands were replaced with monkeys that had their own will and actions. Performing intricate, fine-grained, and highly coordinated movements like grabbing a cup of coffee (i.e., having the monkey's hand grab the coffee and bring it to your lips) is very difficult because you only have "meta-level" control over the monkeys, but not direct control over the monkeys' hands. In cases where meta-overfitting may be a risk, either reduce the search space, increase the dataset (perhaps through data generation techniques like VAEs), and/or separate the validation dataset itself into two parts such that you can evaluate the "true" performance of the hyper-optimized best model.

A common criticism of meta-optimization is that it takes a long time. However, the evaluation metric for whether to use meta-optimization or not is to evaluate *its efficiency relative to your efficiency*. We can define efficiency as follows:

$$\text{efficiency} = \frac{\text{success of model}}{\text{time and labor exerted creating model}}$$

More often than not, using meta-optimization techniques will have higher efficiency than manually creating a model, because manually designing models often takes a nontrivial amount of time and labor yet on average yields average performance. Even if the meta-optimization space is even just reasonably set up, chances are that you will get something out of it.

In many ways, meta-optimization allows you to integrate many of the deep learning tools learned before (and tools we will learn later). For instance, you could use meta-optimization to determine whether to and how to use artificial, convolutional, and/or recurrent processing layers in your neural network design. The design is in your hands!

Neural Architecture Search

Previously, we demonstrated how general Bayesian hyperparameter optimization methods implemented in Hyperopt could be used to accomplish tasks like optimizing a neural network architecture. However, neural network architecture optimization is a quite specific task, and attempting to apply general tools to solve specific tasks generally causes inefficiency due to overhead (like using a generic tool like a shovel to move large mounds of rock when a specialized tool like a bulldozer is needed).

The specific theory behind the subfield of Neural Architecture Search is beyond the scope of the book. In this section, we will demonstrate client-end usage of the friendly, high-level NAS library AutoKeras. True to its name, AutoKeras is a Keras/TensorFlow-based Neural Architecture Search library that uses a variation of Bayesian optimization designed particularly for NAS.

AutoKeras's syntax reflects that of Keras: different "blocks" or "layers" can be chained together in functional API–like syntax. The blocks in AutoKeras can be thought of as "super-layers" – the use can specify the general nature of the block, but the specific composition of the block is optimized using the AutoKeras NAS algorithm.

Say we want to find the optimal neural network architecture for the Higgs Boson dataset. We will use the ak.StructuredDataInput block to accept structured data input (i.e., tabular data). Afterward, we attach ak.StructuredDataBlock, which is an abstracted block object that processes structured data. The output of the structured data block is passed into ak.ClassificationHead, because we want to perform a classification task (Listing 10-36).

Listing 10-36. Creating a neural network search structure in AutoKeras

```
import autokeras as ak
input_node = ak.StructuredDataInput()
output_node = ak.StructuredDataBlock()(input_node)
output_node = ak.ClassificationHead()(output_node)
```

The layers can then be compiled into ak.AutoModel. The overwrite parameter allows new better-performing models to replace older poorer-performing models in storage and is recommended for large-scale trials. The max_trials parameter determines the maximum number of sample neural network architectures AutoKeras will run before experiment termination. The search can be fitted using standard Keras syntax. Note that AutoKeras further splits the X_train and y_train into "subtraining" and "subvalidation" sets. After training, the best model can be exported using .export_model(). Since the output is a Keras model, we can perform operations like saving the model or plotting its architecture (Listing 10-37, Figure 10-21).

Listing 10-37. Arranging the search structure into an AutoModel, fitting, and plotting the best-performing model

```
clf = ak.AutoModel(
    inputs=input_node, outputs=output_node, overwrite=True,
    max_trials=100
)
clf.fit(X_train, y_train, epochs=100)
keras.utils.plot_model(clf.export_model(), show_shapes=True, dpi=400)
```

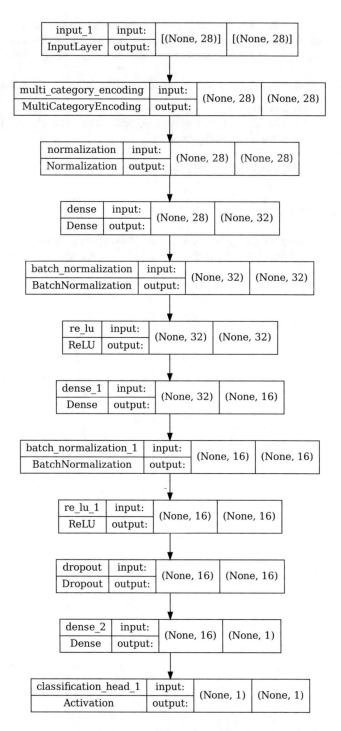

Figure 10-21. The best deep learning model architecture for the Higgs Boson dataset, discovered by AutoKeras

We can also adapt our search to a regression problem – the Ames Housing dataset – by simply swapping the output ak.ClassificationHead with ak.RegressionHead. In this case, we can also pass in categorical_encoding = True to have AutoKeras learn optimal categorical encoding techniques too (Listing 10-38).

Listing 10-38. Fitting a Neural Architecture Search campaign on the Ames Housing dataset

```
input_node = ak.StructuredDataInput()
output_node = ak.StructuredDataBlock(categorical_encoding=True)(input_node)
output_node = ak.RegressionHead()(output_node)
clf = ak.AutoModel(
    inputs=input_node, outputs=output_node, overwrite=True,
    max_trials=100
)
clf.fit(x, y, epochs=100)
```

AutoKeras's best solution performs quite poorly (although acceptably, speaking on the absolute) relative to our more manual design implemented earlier in the "Optimizing Data Pipelines" subsection (Figure 10-22).

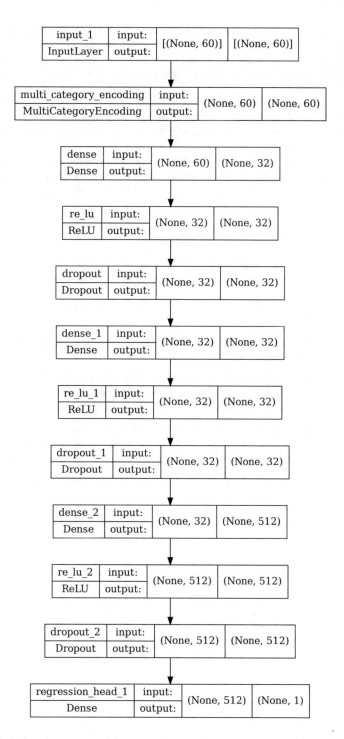

Figure 10-22. *The best deep learning model architecture for the Ames Housing dataset, discovered by AutoKeras*

The poor performance of AutoKeras's search for the optimal Ames Housing dataset neural network can demonstrate one or multiple of three takeaways: (a) neural networks are not the best for every problem (and can actually do very poorly in some contexts), (b) custom-optimized categorical encoding can do wonders for categorical feature–heavy datasets, and (c) Neural Architecture Search – for all its shiny glitz and glamour – doesn't always work out.

AutoKeras has many additional adaptations for multi-input and multi-output heads, which can be useful if you want to build multimodal models like in Chapter 4 or 5. It is especially helpful for processing text data, which can require a significant labor investment; AutoKeras's text heads and embedding layers require little manual work.

Generally speaking, AutoKeras and NAS are neither tools to be wholly relied upon nor wholly dismissed. In many cases, a NAS-discovered architecture may be the inspiration for a high-performing somewhat manually designed network, as is often the case in deep learning research.

Key Points

In this chapter, we discussed meta-optimization and its application to various components of the model development process.

- In meta-optimization, a meta-model searches for the set of hyperparameters that optimizes a model's performance.

- In Bayesian optimization, a probabilistic surrogate model is both updated by sampled points and used to inform next points to sample.

- Meta-optimization can be used to optimize which model is used, model hyperparameters (including the architecture and training parameters of models), and data encodings (among other components of the modeling process).

- Meta-optimization is especially feasible with tabular data models, relatively speaking, since these tend to be smaller than computer vision or natural language processing models.

In the next chapter, we will explore how models can be effectively arranged to form powerful ensembles and "self-aware" systems.

CHAPTER 11

■ ■ ■

Multi-model Arrangement

Democracy cannot succeed unless those who express their choice are prepared to choose wisely. The real safeguard of democracy, therefore, is education.

—Franklin D. Roosevelt, US President

A single model calling all the shots on the final decisions is somewhat of a dictatorship. A dictator can be efficient, but often successful decisions are informed by a collective group of thinkers. *Multi-model arrangement* concerns the construction of model systems in which models interact in different ways to produce a theoretically more informed output.

Different models offer different perspectives when modeling any datasets. Even though two models may obtain the same performance metric value, their *performance distributions* – that is, how their strengths and weaknesses are distributed across the dataset and the space of new data – are likely different. One particular model may be better suited toward certain types of samples or approaches prediction in broad strokes rather than thin slices. One model might be more aggressive and volatile with predictions, whereas a different one may be more conservative and reserved across samples.

Rather than just choosing one particular model, we can choose several different models to form a multi-model arrangement. We pass the input into the entire system, rather than a singular model, and aggregate the output to form a prediction informed by multiple different perspectives. Aggregating/ensembling models in well-designed ways can be incredibly effective, as has been demonstrated in previous chapters (notably, in the "Boosting and Stacking Neural Networks" section of Chapter 7, on tree-Based models).

In this brief chapter, we will cover three key multi-model arrangement techniques: average weighting, input-informed weighting, and meta-evaluation. We will also discuss meta-evaluation, a multi-model arrangement technique that can be used to estimate model confidence/error in contexts where a probability is not easily available.

Average Weighting

Average weighting is a fairly intuitive and simple method to aggregate the outputs of multiple models together. Given a set of model outputs $\{\hat{y}_1, \hat{y}_2, \ldots, \hat{y}_n\}$, we can simply take the average of the outputs to obtain an aggregate result $\bar{\hat{y}}$ (Figure 11-1):

$$\bar{\hat{y}} = \frac{1}{n}\sum_{i=1}^{n}\hat{y}_i$$

A. Ye and Z. Wang, *Modern Deep Learning for Tabular Data*,
https://doi.org/10.1007/978-1-4842-8692-0_11

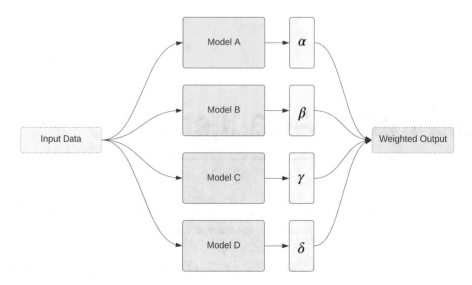

Figure 11-1. *Averaging ensemble*

For instance, say we have a Random Forest classifier, decision tree classifier, Gradient Boosting classifier, and neural network model trained on the Higgs Boson dataset. We can take the average of their outputs and hopefully obtain improved performance.

However, it is almost always true that some models are better than others. Models that are better should have a higher weight in the output. Rather than taking a simple average, we can take a *weighted average* by associating each model with a weighting coefficient representing how much of an impact it has on the final system output. Given a set of model weights $\{w_1, w_2, ..., w_n\}$ and model outputs $\{\hat{y}_1, \hat{y}_2, ..., \hat{y}_n\}$, we can simply obtain a weighted average/linear combination $\bar{\hat{y}}$ (Figure 11-2):

$$\bar{\hat{y}} = \sum_{i=1}^{n} \hat{y}_i \cdot w_i$$

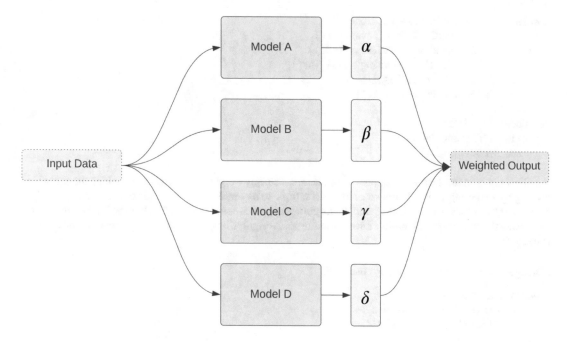

Figure 11-2. *Weighted averaging ensemble*

There are many approaches we can use to find the optimal model weighting set w. If the models are performing a regression or binary classification problem, we can train a Linear Regression model to predict the ground truth given a dataset of model outputs $\{\hat{y}_1, \hat{y}_2, \dots, \hat{y}_n\}$ for each sample. This is quick to implement and train.

Note that it is considered good practice to train meta-models on the validation dataset, although the subject is up for debate. The logic is as such: when we actually deploy the model in a real-world setting, we want the meta-model to incorporate the predictions of models operating in a real-world setting (in which models may make mistakes), rather than in a training regime. However, this can require infeasible data splitting, especially in the context of small datasets. It is acceptable to train an aggregatory Linear Regression model to predict the training set labels given model predictions informed by the training set inputs, given that the models are not overfitting.

Using Linear Regression is less suited for multiclass problems, because the model outputs one of many possible classes for each sample. One cannot simply average the predictions in a trivial way. Instead, we can use Bayesian optimization frameworks to find the optimal weighting for each model prediction.

To demonstrate, let's consider the NASA Wildfire dataset. The objective is to categorize a satellite-observed wildfire into one of four types – presumed vegetation fire, active volcano, other static land source, or offshore – given collected satellite reading data.

Let's begin by creating an ensemble of classifiers and training each on the dataset (Listing 11-1).

Listing 11-1. Creating and individually fitting an ensemble of models

```
from sklearn.linear_model import LogisticRegression
from sklearn.tree import DecisionTreeClassifier
from sklearn.ensemble import RandomForestClassifier, GradientBoostingClassifier,
AdaBoostClassifier
from sklearn.neural_network import MLPClassifier
```

```python
models = {'lr': LogisticRegression(),
          'dtc': DecisionTreeClassifier(),
          'rfc': RandomForestClassifier(),
          'gbc': GradientBoostingClassifier(),
          'abc': AdaBoostClassifier(),
          'mlpc': MLPClassifier()}

for model in models:
    print(f'Fitting {model}')
    models[model].fit(X_train, y_train)
```

For convenience, let's write an ensemble class that simply weights all models' predictions the same. We begin by creating a set of blank votes for each sample to be predicted upon. Each model adds a "vote" for which class it predicts a certain sample falls into. We then choose the class that has the highest number of votes as the final class. All models have the same vote weight; this gives it the average ensembling quality (Listing 11-2).

Listing 11-2. Defining an averaging ensemble

```python
class AverageEnsemble:
    def __init__(self, modeldic):
        self.modeldic = modeldic

    def predict(self, x, num_classes = 4):
        votes = np.zeros((len(x), num_classes))
        for model in self.modeldic:
            predictions = self.modeldic[model].predict(x)
            for item, vote in enumerate(predictions):
                votes[item, vote] += 1
        return np.argmax(votes, axis=1)

ensemble = AverageEnsemble(models)
```

As one may expect, the performance of such an ensemble is mediocre and worse than several of the top-performing models in isolation.

We can modify our AverageEnsemble class to accept a set of model weights; rather than incrementing a vote for a class with +1, we increment it by the assigned vote weightage. This allows some models to have more general influence than others in determining the final outcome (Listing 11-3).

Listing 11-3. Defining a weighted average ensemble

```python
class WeightedAverageEnsemble:

    def __init__(self, modeldic, modelweights):
        self.modeldic = modeldic
        self.modelweights = modelweights

    def predict(self, x, num_classes = 4):
        votes = np.zeros((len(x), num_classes))
        for model in self.modeldic:
            predictions = self.modeldic[model].predict(x)
```

```
        for item, vote in enumerate(predictions):
            votes[item, vote] += self.modelweights[model]
    return np.argmax(votes, axis=1)
```

We can now use Hyperopt to optimize the set of modelweights. Our objective function creates a WeightedAverageEnsemble with the given set of models and the sampled set of parameters. We'll consider a weighted ensemble to be better if the F1 score is higher. Because this is a multiclass problem but the F1 score is defined in original form for binary classification, we need to pass a mechanism by which to apply the F1 score to multiple classes. Specifying average='macro' aggregates the F1 score intuitively: it is simply the average of the F1 scores for each of the individual classes (Listing 11-4).

Listing 11-4. Using hyperoptimization to find the best set of model weights

```
from hyperopt import hp, tpe, fmin
# define the search space
from hyperopt import hp

space = {model:hp.normal(model, mu = 1, sigma = 0.75) for model in models}

# define objective function
def obj_func(params):
    ensemble = WeightedAverageEnsemble(models, params)
    return -f1_score(ensemble.predict(X_valid), y_valid,
                    average='macro')

# perform minimization procedure
from hyperopt import fmin, tpe
best = fmin(obj_func, space, algo=tpe.suggest, max_evals=500)
```

Let's demonstrate how we might do something similar with neural networks. Consider the following set of five different neural network model architectures: modelA, modelB, modelC, modelD, and modelE (Listing 11-5).

Listing 11-5. Defining five uniquely different neural network model architectures

```
modelA = keras.models.Sequential(name='modelA')
modelA.add(L.Input((len(X_train.columns),)))
modelA.add(L.Dense(16, activation='relu'))
modelA.add(L.Dense(16, activation='relu'))
modelA.add(L.Dense(4, activation='softmax'))

modelB = keras.models.Sequential(name='modelB')
modelB.add(L.Input((len(X_train.columns),)))
modelB.add(L.Dense(16, activation='relu'))
modelB.add(L.Dense(16, activation='relu'))
modelB.add(L.Dense(16, activation='relu'))
modelB.add(L.Dense(16, activation='relu'))
modelB.add(L.Dense(4, activation='softmax'))

inp = L.Input((len(X_train.columns),))
dense = L.Dense(16, activation='relu')(inp)
branch1a = L.Dense(16, activation='relu')(dense)
```

```
branch1b = L.Dense(16, activation='relu')(branch1a)
branch1c = L.Dense(8, activation='relu')(branch1b)
branch2a = L.Dense(8, activation='relu')(dense)
branch2b = L.Dense(8, activation='relu')(branch2a)
concat = L.Concatenate()([branch1c, branch2b])
out = L.Dense(4, activation='softmax')(concat)
modelC = keras.models.Model(inputs=inp, outputs=out, name='modelC')

modelD = keras.models.Sequential(name='modelD')
modelD.add(L.Input((len(X_train.columns),)))
modelD.add(L.Dense(64, activation='relu'))
modelD.add(L.Reshape((8, 8, 1)))
modelD.add(L.Conv2D(8, (3, 3), padding='same', activation='relu'))
modelD.add(L.Conv2D(8, (3, 3), padding='same', activation='relu'))
modelD.add(L.MaxPooling2D(2, 2))
modelD.add(L.Conv2D(16, (3, 3), padding='same', activation='relu'))
modelD.add(L.Conv2D(16, (3, 3), padding='same', activation='relu'))
modelD.add(L.Flatten())
modelD.add(L.Dense(16, activation='relu'))
modelD.add(L.Dense(4, activation='softmax'))

modelE = keras.models.Sequential(name='modelE')
modelE.add(L.Input((len(X_train.columns),)))
modelE.add(L.Dense(64, activation='relu'))
modelE.add(L.Reshape((64, 1)))
modelE.add(L.Conv1D(8, 3, padding='same', activation='relu'))
modelE.add(L.Conv1D(8, 3, padding='same', activation='relu'))
modelE.add(L.MaxPooling1D(2))
modelE.add(L.Conv1D(16, 3, padding='same', activation='relu'))
modelE.add(L.Conv1D(16, 3, padding='same', activation='relu'))
modelE.add(L.Flatten())
modelE.add(L.Dense(16, activation='relu'))
modelE.add(L.Dense(4, activation='softmax'))
```

Each model has a different "spirit" or "character." Some process inputs more bluntly and others more thoroughly. Some use one-dimensional convolutions, some use two-dimensional convolutions, and some use nonlinearity. Diversity makes for a strong ensemble (Figures 11-3 through 11-7).

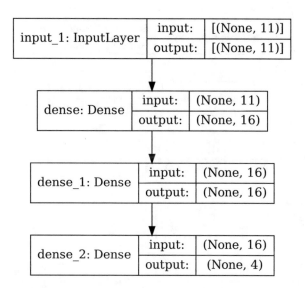

Figure 11-3. *Model A architecture*

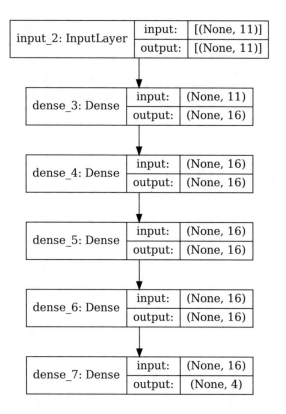

Figure 11-4. *Model B architecture*

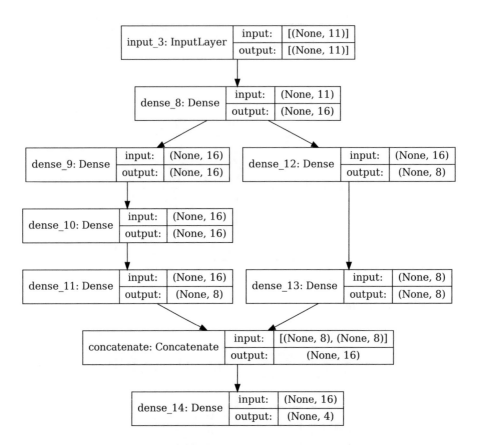

Figure 11-5. *Model C architecture*

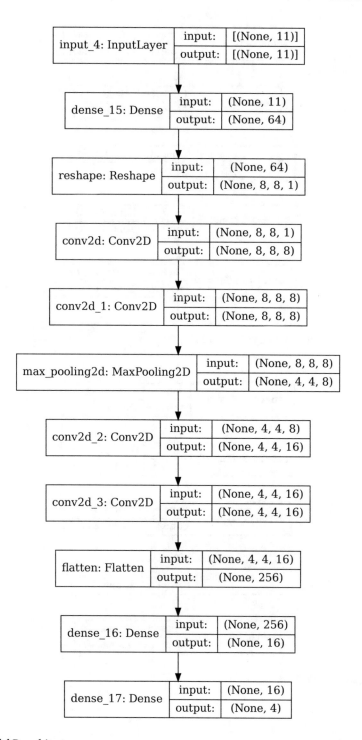

Figure 11-6. *Model D architecture*

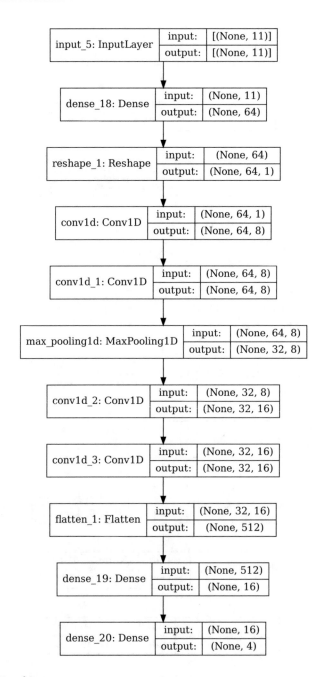

Figure 11-7. *Model E architecture*

Listing 11-6 demonstrates the process of training each of these models.

Listing 11-6. Fitting each model in the ensemble

```
models = {'modelA': modelA,
          'modelB': modelB,
          'modelC': modelC,
          'modelD': modelD,
          'modelE': modelE}
for model in models:
    models[model].compile(optimizer='adam',
                          loss='sparse_categorical_crossentropy',
                          metrics='accuracy')
    models[model].fit(X_train, y_train, epochs=30)
```

Since neural networks output probabilities (as opposed to ordinal integers representing classes used in scikit-learn models), we simply scale the probabilities by a weighting factor and add the predicted probabilities, choosing the class with the highest "continuous vote" as the final class (Listing 11-7).

Listing 11-7. Defining a weighted average ensemble adapted for probabilistic neural network outputs

```
class WeightedAverageEnsemble:
    def __init__(self, modeldic, modelweights):
        self.modeldic = modeldic
        self.modelweights = modelweights
    def predict(self, x, num_classes = 4):
        votes = np.zeros((len(x), num_classes))
        for model in self.modeldic:
            predictions = self.modeldic[model].predict(x)
            votes += self.modelweights[model] * predictions
        return np.argmax(votes, axis=1)
```

Our optimization procedure looks pretty similar as before (Listing 11-8).

Listing 11-8. Optimizing the best weighting for a neural network ensemble

```
# define the search space
from hyperopt import hp

space = {model:hp.normal(model, mu = 1, sigma = 0.75) for model in models}

# define objective function
def obj_func(params):
    ensemble = WeightedAverageEnsemble(models, params)
    return -f1_score(ensemble.predict(X_valid), y_valid,
                     average='macro')

# perform minimization procedure
from hyperopt import fmin, tpe
best = fmin(obj_func, space, algo=tpe.suggest, max_evals=500)
```

763

However, because we are working with neural networks – highly versatile computational structures – we can also use other methods. One, for instance, is to integrate all these models into one large piece. In order to weight/multiply each model's prediction by a certain quantity, we'll use a bit of a "cheat" and apply a 1-length convolution, which simply multiplies each element in a given sequence by the same value (assuming the bias is disabled). Afterward, we will simply add each of the scaled model predictions together and then pass through softmax such that the output of the complete linked model architecture remains a set of valid class probabilities (Listing 11-9, Figure 11-8).

Listing 11-9. An alternative: rearranging each of the models as a fixed subcomponent of a larger meta-model architecture

```
for model in models:
    models[model].trainable = False
inp = L.Input((len(X_train.columns),))
mergeList = []
for model in models:
    modelOut = models[model](inp)
    reshape = L.Reshape((4, 1))(modelOut)
    scale = L.Conv1D(1, 1, use_bias=False)(reshape)
    flatten = L.Flatten()(scale)
    mergeList.append(flatten)
concat = L.Add()(mergeList)
softmax = L.Softmax()(concat)
metaModel = keras.models.Model(inputs=inp, outputs=softmax)
```

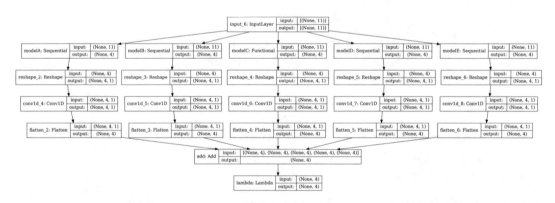

Figure 11-8. *A meta-model architecture combining each of the models together and automatically determining weightings*

Another approach would be to concatenate each of the outputs along a joint axis and to apply a time-distributed dense layer across each dimension of the output.

If you call `model.summary()`, you'll find that the total number of parameters is 22,326; 22,311 are untrainable because they belong to `modelA`, `modelB`, `modelC`, `modelD`, or `modelE`. The remaining five trainable parameters are simply the output weights for each of the models.

Fitting is standard (Listing 11-10).

Listing 11-10. Compiling and fitting the meta-model

```
metaModel.compile(optimizer='adam',
                  loss='sparse_categorical_crossentropy',
                  metrics=['accuracy'])
metaModel.fit(X_train, y_train, epochs=10,
        validation_data=(X_valid, y_valid))
```

An advantage of this is design is that we can do some more fine-tuning overall by unfreezing each of the models and training the entire architecture to further optimize each of the models' weights in relationship to each other (Listing 11-11).

Listing 11-11. Unfreezing each component and fine-tuning

```
for model in models:
    models[model].trainable = True
metaModel.fit(X_train, y_train, epochs=3,
        validation_data=(X_valid, y_valid))
```

This can be a good strategy to build effective large networks from a wide, diverse ensemble of smaller candidate models.

Input-Informed Weighting

Average weighting allows us to understand which models are generally more trustworthy and which ones are less trustworthy. The more trustworthy models are ideally given a higher weight, whereas the less trustworthy models are weighted lower. However, some models may specialize in predicting a minority case; because they perform poorer overall, however, average weighting will mark these "specialist" models as generally untrustworthy and weight their prediction contributions down *for all samples* (even when it is predicting on a sample case it "specializes" in).

To reconcile this problem, we need to use a more complex form of weighting: *input-informed weighting*. In input-informed weighting, the specific weighting configuration is not static but rather dependent on the input received. This way, if a particular model or set of models specializes on a certain input, its contributions will be ideally weighted higher (and lower for other inputs it performs poorly on).

This sort of scheme works best for neural networks, which are able to capture complex relationships between the "specialization" or performance distribution of model performance and certain types of inputs. Let's create a weighting model that takes in an input and outputs several outputs, one for each model (Listing 11-12).

Listing 11-12. An input-informed weighter

```
inp = L.Input((len(X_train.columns),))
dense1 = L.Dense(8, activation='relu')(inp)
dense2 = L.Dense(8, activation='relu')(dense1)
dense3 = L.Dense(8, activation='relu')(dense2)
outLayers = []
for model in models:
    outLayers.append(L.Dense(1, activation='sigmoid')(dense3))
weightingModel = keras.models.Model(inputs=inp, outputs=outLayers)
```

This weighting model can be incorporated into the meta-model: the output of each model is multiplied by the corresponding weighting of the weighting model (which itself predicts weights informed by the input) (Listing 11-13, Figure 11-9).

Listing 11-13. Building the weighting model into the meta-model architecture

```
inp = L.Input((len(X_train.columns),))
weights = weightingModel(inp)
finalVotes = []
for weight, model in zip(weights, models):
    models[model].trainable = False
    modelOut = models[model](inp)
    expand = L.Flatten()(L.RepeatVector(num_classes)(weight))
    scale = L.Multiply()([modelOut, expand])
    finalVotes.append(scale)
out = L.Add()(finalVotes)

metaModel = keras.models.Model(inputs=inp, outputs=out)
```

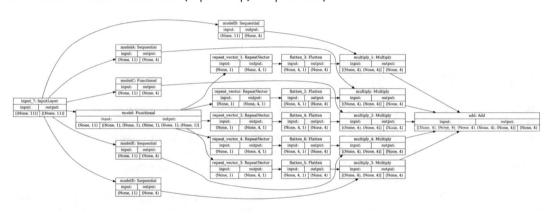

Figure 11-9. *An input-informed ensemble model architecture.*

A good example of similar techniques at work in deep learning research literature can be found in the paper "Who Said What: Modeling Individual Labelers Improves Classification"[1] by Melody Y. Guan, Varun Gulshan, Andrew M. Dai, and Geoffrey E. Hinton. Guan et al. tackle the problem of modeling expert-crowdsourced annotations. In this data collection scheme, a set of specialized data is labeled by multiple expert annotators. This scheme occurs often, for instance, in modeling medical datasets, in which several medical professionals may annotate a single sample. However, we may sometimes encounter disagreement or discrepancy in the labels of different annotators on the same sample.

The simple and dominant method to deal with disagreement in crowdsourced annotations is to somehow aggregate the different annotations and model the aggregated annotation. For instance, if five doctors rate the severity of a diagnosis from 0 to 100 as {88, 97, 73, 84, 86}, the overall annotation for the sample would be listed as 85.6. The model would then be trained to predict the diagnosis severity of the sample as 85.6.

---

[1] Guan, M.Y., Gulshan, V., Dai, A.M., & Hinton, G.E. (2018). Who Said What: Modeling Individual Labelers Improves Classification. *AAAI.*

Guan et al. instead predict each expert's annotations for a particular sample. In continuation of our five-doctor example, the model would be trained to jointly predict each of the diagnoses {88, 97, 73, 84, 86}. Then, the model learns the optimal averaging weights for each model, similar to our previous discussion of weighted averaging of model predictions. Guan et al. also design and train a more complex model in which the weighting is input-informed. Ultimately, the research finds – to paraphrase from the paper itself – that aggregating the modeled annotations beats modeling the aggregated annotation. This is the power of meta-modeling at work (Figure 11-10)!

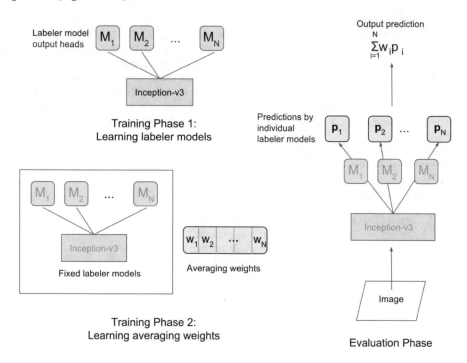

Figure 11-10. *Different learning phases/types of a multi-modeling ensemble. From Guan et al., "Who Said What: Modeling Individual Labelers Improves Classification"*

Meta-evaluation

We can use principles and ideas from multi-model arrangement to understand a useful technique for real-world deployment of machine learning and/or deep learning models: meta-evaluation.

For a certain set of problems, model confidence is trivially obtainable because it is inherently embedded in the algorithm design. Logistic Regression, for instance, directly gives us an estimate of the model's "confidence" that a particular input falls into a particular class. Neural networks trained on classification problems directly return the probability that an input belongs to a certain class. However, in many other cases, gauging model confidence and/or error is not as clear.

■ **Note** In many cases, even the output class probabilities of a classification network don't provide meaningful information about the confidence/trustworthiness of the network's predictions, due to the softmax layer used in classification output layers. This forces all output probabilities to sum to 1. One technique used specifically for classification networks to reconcile this problem is to introduce an additional class representing "none of the above," which can be trained on images from none of the classes in the original dataset.

During model development, we generally care only about aggregate-level model error. We want to minimize the aggregate error (e.g., Mean Squared Error, mean binary cross-entropy, etc.) of the model across the dataset. However, when models are deployed, they operate on a sample-by-sample basis. It becomes important to know when a model's predictions are likely to be correct or incorrect for each sample being used for inference and by how much. This can be used to gauge how to act based on a model's predictions. For instance, if a model gives a diagnosis that a patient does indeed have cancer, we would like to know the model's specific confidence on this sample case more than the general knowledge that the model has such-and-such error rate.

Listing 11-14 defines an architecture (visualized in Figure 11-11) that estimates the error of the prediction given the original input and the predicted output.

Listing 11-14. A meta-model error estimator

```
dataInput = L.Input((len(X_train.columns),), name='data')
dataDense1 = L.Dense(16, activation='relu')(dataInput)
dataDense2 = L.Dense(16, activation='relu')(dataDense1)
dataDense3 = L.Dense(8, activation='relu')(dataDense2)

predInput = L.Input((1,), name='pred')
predDense1 = L.Dense(1, activation='relu')(predInput)

combine = L.Concatenate()([predDense1, dataDense3])
preOut = L.Dense(16, activation='relu')(combine)
out = L.Dense(1, activation='relu')(preOut)

metaEval = keras.models.Model(inputs = [dataInput, predInput],
                              outputs = out)
```

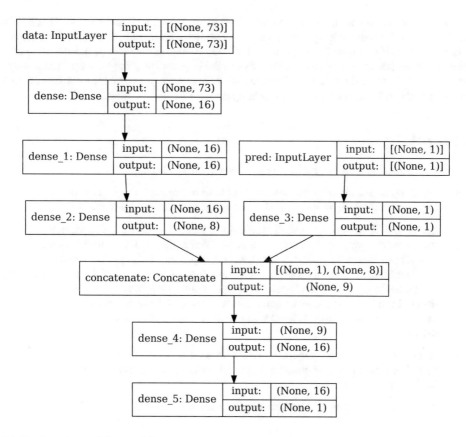

Figure 11-11. *A meta-modeling architecture*

Now, we can train on the error. In this case, we want to give the model both the original inputs and the predicted values; the target is the absolute error of the prediction (Listing 11-15).

Listing 11-15. Training the meta-evaluation model on the prediction residuals

```
preds = model.predict(X_valid)

metaEval.compile(optimizer='adam', loss='mse')
history = metaEval.fit([X_valid, preds],
                        np.abs(preds - y_valid),
                        epochs=1000,
                        verbose=0)
```

During the prediction phase, you can then pass in the prediction (along with the original input) through the meta-evaluation model to get an understanding for how much you can trust the individual prediction.

This technique is similar to that of residual learning employed in boosting ensembles (see Chapter 1, on Gradient Boosting algorithms, and Chapter 7, on GrowNet).

Another technique used for large neural networks is to force the neural network to both make the formal prediction and simultaneously estimate the error of such a prediction. By this method, the error estimation is made "self-consciously" by the model with knowledge of the internal states used to form a prediction. However, it is important to severely discount the importance of the error estimation output early on in training (like changing the value of α for multitask autoencoders) to ensure the network reaches a point of satisfactory performance such that estimating error is feasible and meaningful.

Key Points

In this chapter, we discussed various multi-model arrangement methods.

- Multi-model arrangement provides a set of tools to composite multiple models together. This can make systems more robust, jointly informed, and self-aware.

- In simplest form, the output of all models in an ensemble can be averaged. However, some models may perform better than others. Weighted averaging, in which each model's prediction is associated with a weight, captures relative performance/ trustworthiness across an ensemble of models. Input-informed weighting makes this weighting dependent on the type of input, to accommodate model specialization in certain types of inputs. All weightings can be learned through meta-optimization or, if all models are neural networks, can be fixed and arranged into a meta-neural network.

- Meta-evaluation is a framework to predict a model's error on a per-sample basis. This allows for the characterization of model trustworthiness and confidence in live deployment.

In this book's final chapter, we will look at deep learning interpretability.

CHAPTER 12

■ ■ ■

Neural Network Interpretability

I have always thought the actions of men the best interpreters of their thoughts.

—John Locke, Political Theorist

Neural networks are powerful tools that can be used to solve a host of difficult tabular data modeling challenges. However, they're also less obviously interpretable than other alternatives to modeling tabular data, like Linear Regression or decision trees – from which the model's processing of the data can be more or less directly read off of the learned parameters. The same is not true for neural network architectures, which are significantly more complex and therefore more difficult to interpret. At the same time, it is important to interpret any model used in production to verify that it is not using cheap tricks or other exploitative measures, which can lead to poor behavior in production.

In this chapter, we will briefly explore three techniques for interpreting neural networks: SHAP, LIME, and Activation Maximization. The first two techniques are *model agnostic*, meaning that they can be used to explain the decisions of any model, not just neural networks. The last – Activation Maximization – can only be used on neural networks. A well-informed interpretability analysis considers multiple different techniques.

SHAP

SHAP, which stands for SHapley Additive exPlanations,[1] is a popular framework for model explainability. The framework is based on the game theory concept of a Shapley value, which quantifies how much each player contributes to the overall value in a collaborative game.

For instance, let's say you're trying to analyze players in a three-person basketball team (perhaps for some 3-on-3 ball). The overall reward of the three-person team is trivial to obtain – it's simply the difference between the team's score and the other team's score, or how many more balls they scored in the hoop than the other team. However, the contribution of individual players is less simple to quantify. We could calculate the individual contribution of each player simply by how many balls they scored, but this obscures the role of a defensive player that blocks the other team from making shots. Alternatively, perhaps player B performs well only when player C is present, or player A and player C have a complementary performance relationship (only one of them can perform well; the other performs poorly). Ultimately, there is no easy way to estimate each individual player's contributions from just looking at the raw game data itself because of the complex interdependencies that often manifest in real-world collaborative games (Figure 12-1).

---

[1] Lundberg, S.M., & Lee, S. (2017). A Unified Approach to Interpreting Model Predictions. *ArXiv, abs/1705.07874.*

© Andre Ye and Zian Wang 2023
A. Ye and Z. Wang, *Modern Deep Learning for Tabular Data*,
https://doi.org/10.1007/978-1-4842-8692-0_12

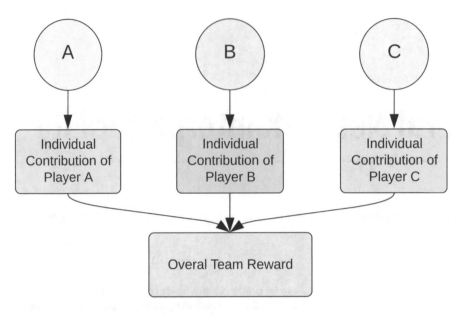

Figure 12-1. *A game schematic with individual players each contributing an unknown individual contribution to a known overall team reward*

The key idea of a Shapley value is that we must consider each possible combination of players in order to determine how important any one individual player is. In the case of three players, there are seven combinations:

- One player
  - {Player A}
  - {Player B}
  - {Player C}
- Two players
  - {Players A, B}
  - {Players B, C}
  - {Players A, C}
- Three players
  - {Players A, B, C}

For reference, if we had five players, we would need 31 combinations:

- One player
  - {Player A}
  - {Player B}
  - {Player C}
  - {Player D}
  - {Player E}

- Two players
  - {Players A, B}
  - {Players A, C}
  - {Players A, D}
  - {Players A, E}
  - {Players B, C}
  - {Players B, D}
  - {Players B, E}
  - {Players C, D}
  - {Players C, E}
  - {Players D, E}
- Three players
  - {Players C, D, E}
  - {Players B, D, E}
  - {Players B, C, E}
  - {Players B, C D}
  - {Players A, D, E}
  - {Players A, C, E}
  - {Players A, C, D}
  - {Players A, B, E}
  - {Players A, B, C}
- Four players
  - {Players A, B, C, D}
  - {Players A, B, C, E}
  - {Players A, B, D, E}
  - {Players A, C, D, E}
  - {Players B, C, D, E}
- Five players
  - {Players A, B, C, D, E}

(The mathematically inclined reader will notice that the number of nonempty subsets in a set with n elements is $2^n - 1$.)

Then, we evaluate the performance of each of these teams. We have player A play against the other team and evaluate its performance and then player B, followed by player C, followed by a two-person team consisting of players A and B, and so on. At the end, we have collected a significant amount of data on the rewards obtained by different subsets of players.

The important attribute of this dataset is that we can compare the difference obtained in reward when some player is or is not present across several different contexts. For instance, if we want to evaluate the individual contribution or importance of player A, we can aggregate the increase in overall team reward in the following contexts:

- Between the set {Player B} and the set {Players A, B}

- Between the set {Player C} and the set {Players A, C}

- Between the set {Players B, C} and the set {Players A, B, C}

These constitute the *marginal contributions* of player A to the game. Each of these marginal contributions is then weighted by relevance to the overall problem and summed to form a Shapley value for player A. We can repeat the same with other players to obtain Shapley values for the other players.

In machine learning, the features are the players, and the game reward is the score of the model! We want to understand which features influence the performance of the model in accurately modeling the dataset by shuffling out combinations of features and measuring the marginal contribution of each feature. Note that this means computing SHAP values can be expensive, since in their pure form $2^n - 1$ models need to be trained, where n is the number of features. Luckily, the shap python library has tricks to improve the efficiency of computing the relevance of a feature.

We can begin by importing shap and initializing the JavaScript back end required for visualization with shap.initjs(); this is required to produce many of the visualizations (these rely upon JavaScript inside the Jupyter Notebook). shap can be installed with pip install shap (Listing 12-1).

Listing 12-1. Importing and initializing SHAP

```
import shap
shap.initjs()
```

For demonstration, we'll load the Adult Census dataset directly from SHAP, which is a preprocessed quasi-dummy dataset containing demographic information (which is usually ripe for interesting interpretability analysis) (Listing 12-2).

Listing 12-2. Splitting the dataset into training and validation datasets

```
x, y = shap.datasets.adult()
y = y.astype(np.int32)

from sklearn.model_selection import train_test_split as tts
X_train, X_valid, y_train, y_valid = tts(x, y, train_size = 0.8,
                                          random_state = 42)
```

We'll use AutoKeras to quickly find a decent model on this dataset and export the Keras neural network to a variable appropriately named model (Listing 12-3).

Listing 12-3. Finding a good neural network architecture automatically with AutoKeras

```
import autokeras as ak
input_node = ak.StructuredDataInput()
output_node = ak.StructuredDataBlock(categorical_encoding=True)(input_node)
output_node = ak.ClassificationHead()(output_node)
```

```
clf = ak.AutoModel(
    inputs=input_node, outputs=output_node,
    overwrite=True, max_trials=20
)
clf.fit(X_train, y_train, epochs=50)
model = clf.export_model()
```

In order to generate feature relevance values, we will use shap.KernelExplainer. This is the most standard model-agnostic object in shap. It accepts two mandatory arguments: a function returning the model predictions given some input and a set of data to evaluate the model's performance on. In Listing 12-4, we initialize the kernel explainer on the first 100 rows of the dataset.

Listing 12-4. Using the generic agnostic kernel explainer

```
def f(x):
    return model.predict(x).flatten()
explainer = shap.KernelExplainer(f, X_valid.iloc[:100,:])
```

Once the SHAP kernel explainer object has been initialized, we can use it to explain how a model arrives at a prediction for a single sample. Listing 12-5 demonstrates a *force plot* visualization (shown in Figure 12-2), in which various features are shown as vectors demonstrating the direction and magnitude effect on the target (Listing 12-5, Figure 12-2).

Listing 12-5. Using the generic kernel explainer to generate force plots for single samples

```
i = 100
shap_values = explainer.shap_values(X_valid.iloc[i], nsamples=500)
shap.force_plot(explainer.expected_value,
                shap_values,
                X_valid.iloc[i,:])
```

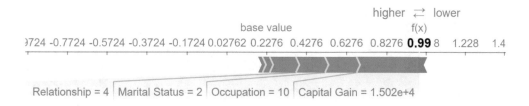

Figure 12-2. *A force plot generated with the sample at* i = 100

By changing the index i of the selected sample to explain, we can obtain force plots to explain how the model approaches other samples. The force plot shows us how different features are weighted and in which direction they act. Figure 12-3 demonstrates an example in which the model output is a probability very close to 0. In this case, there are no factors acting in the positive direction. Figure 12-4 demonstrates a case in which features have values that act both positively and negatively on the target output.

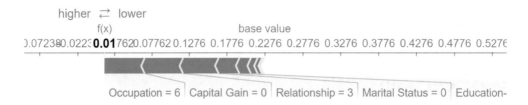

Figure 12-3. A force plot generated with the sample at i = 101

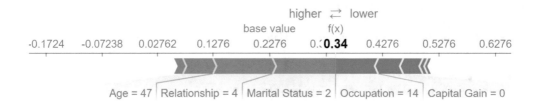

Figure 12-4. A force plot generated with the sample at i = 102

Moreover, SHAP allows you to view multiple force plots for several samples altogether in one convenient interactive visualization. Listing 12-6 demonstrates the code to produce multiple force plots (Figure 12-5).

Listing 12-6. Generating a view of multiple force plots

```
shap.force_plot(explainer.expected_value, shap_values,
                X_valid.iloc[:100])
```

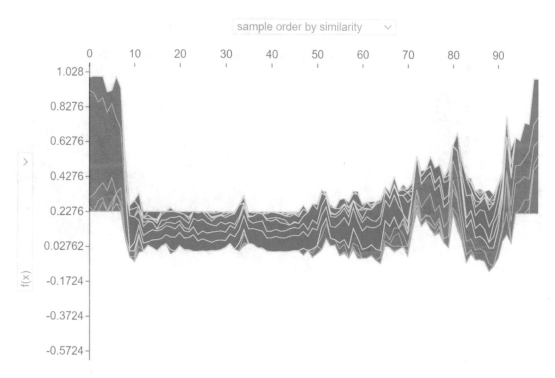

Figure 12-5. *An aggregated force plot visualization*

This is an interactive application: you can view the force plot and responsible features for a particular sample by hovering over a horizontal rule (Figure 12-6).

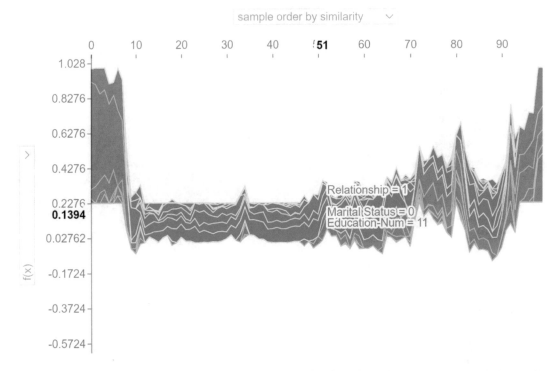

Figure 12-6. *Since SHAP produces JS visualization embedded in the notebook, you can hover over and see the most significant impacting factors for each sample*

Another clean visualization is to plot out the SHAP values for each column across multiple different samples. This allows us to understand the distribution of effects each feature has on the predicted output (Listing 12-7, Figure 12-7).

Listing 12-7. Plotting out the importances/values of each column

```
shap.summary_plot(shap_values, X_valid.iloc[:100])
```

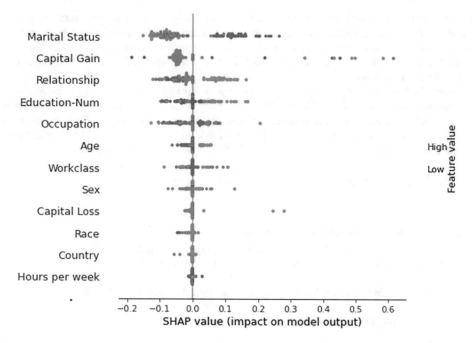

Figure 12-7. *Another visualization of SHAP values to help us understand the distribution of feature importance and influence*

The kernel explainer is model agnostic, which means that it only interprets the model's behavior based on the inputs and the outputs. That is, it treats the model as a black box function and thus can be applied to any model.

SHAP also implements model-specific explainers that look "inside" the model to calculate SHAP values informed not only by the predicted output but also the internal parameters and processes used to derive the prediction. The `TreeExplainer`, for instance, can be applied to tree-based models (e.g., `sklearn.tree.DecisionTree`) by "reading" the learned criteria. SHAP provides the `GradientExplainer` to provide better interpretations of neural networks by exploiting their differentiability.

Listing 12-8 demonstrates the construction of a simple neural network trained on the Mice Protein Expression dataset.

Listing 12-8. Training and fitting a model on the Mice Protein Expression dataset

```
model = keras.models.Sequential()
model.add(L.Input((len(X_train.columns),)))
for i in range(3):
    model.add(L.Dense(32, activation='relu'))
for i in range(2):
    model.add(L.Dense(16, activation='relu'))
model.add(L.Dense(8, activation='softmax'))

model.compile(optimizer='adam', loss='sparse_categorical_
crossentropy',                 metrics=['accuracy'])
model.fit(X_train, y_train-1, epochs=100, verbose=0)
```

Since neural networks are pretty complicated, it is often valuable not just to explain the output of the entire model but also that of intermediate layers. Recall a similar analysis we performed in Chapter 4, when we visualized how learned convolutional kernels convolved feature maps to understand which features were being amplified and which were being dimmed. In order to link an input to the output of a certain layer, we need to construct a submodel.

Let's begin by just explaining the last output layer, which has eight nodes for classification into one of the ten classes. We index the desired layer, create a submodel linking the input to that layer, and pass the model in addition to a dataset of features to the GradientExplainer (Listing 12-9). (In this case, the submodel construction is somewhat redundant because it is identical to the model. However, such a formulation allows us to generalize gradient explanations to other intermediate layers.)

Listing 12-9. Using the SHAP gradient explainer to explain a subnetwork

```
layer = -1

submodel = keras.models.Model(inputs=model.input,
                              outputs=model.layers[layer].output)
explainer = shap.GradientExplainer(submodel, np.array(X_train))
```

We can obtain the SHAP values from the explainer on a certain data subset (Listing 12-10).

Listing 12-10. Obtaining SHAP values from the gradient explainer

```
values = explainer.shap_values(np.array(X_train)[:200])
values = np.array(values)
```

Calling values.shape yields (8, 200, 80). You can understand this as taking the format (# neurons in layer, # samples evaluated across, # features). SHAP tells us how much impact each feature has on each output for each sample. That's a lot of information!

You can derive a lot of meaningful insights from these SHAP values. An obvious aggregation method is to measure the mean SHAP value for each feature across all samples and outputs (Listing 12-11, Figure 12-8).

Listing 12-11. Taking the mean SHAP value of each sample across all samples and outputs

```
mean_importance = np.mean(values.reshape(-1, values.shape[-1]), axis=0)
plt.figure(figsize=(8, 8), dpi=400)
sns.heatmap(mean_importance.reshape((10, 8)), annot=True,
        xticklabels=[], yticklabels=[])
plt.show()
```

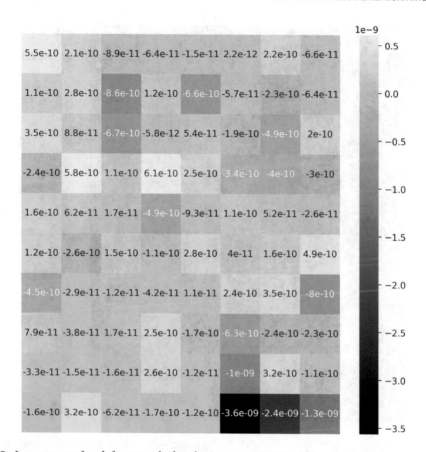

Figure 12-8. *Importance of each feature calculated using mean SHAP with respect to the last output layer*

Alternatively, we can visualize the overall impact of each feature on the output of the second-to-last layer (Figure 12-9) and third-to-last layer (Figure 12-10) to understand the dynamics of internal network computation.

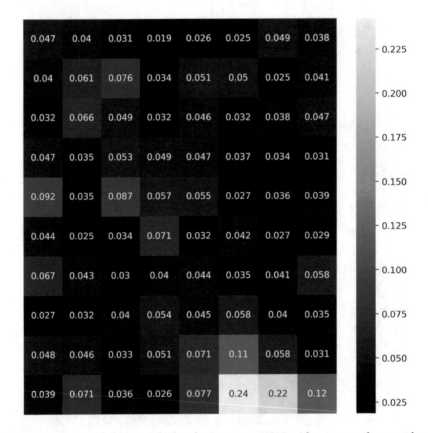

Figure 12-9. *Importance of each feature calculated using mean SHAP with respect to the second-to-last output layer*

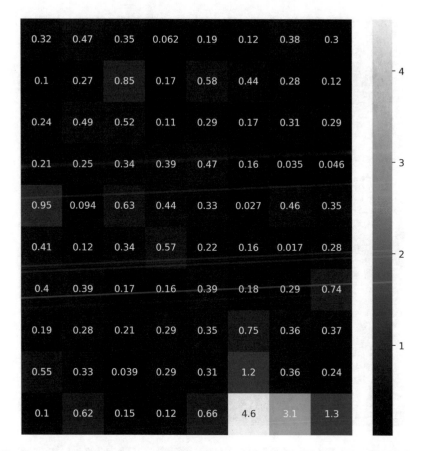

Figure 12-10. *Importance of each feature calculated using mean SHAP with respect to the third-to-last output layer*

We can also take the standard deviation of SHAP values across all outputs and samples to understand the variability in feature impact (Figure 12-11).

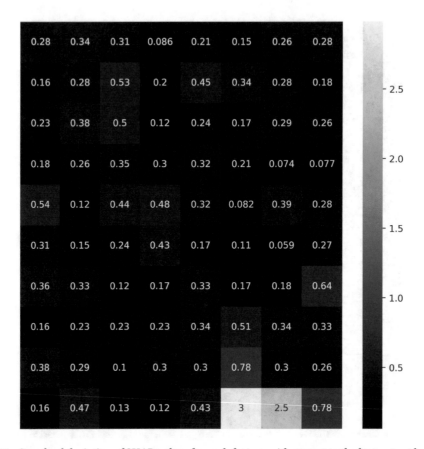

Figure 12-11. *Standard deviation of SHAP values for each feature with respect to the last output layer*

Of course, one can perform much more fine-grained and detailed analyses by isolating a particular feature or set of features of interest and tracking how the impact varies and/or trends throughout samples and outputs. SHAP is an extraordinarily useful tool, and lots of new research are continually pushing at the boundaries of how useful, efficient, and informative SHAP-based methods can be (especially for deep learning!).

LIME

LIME, an acronym for Local Interpretable Model-Agnostic Explanations,[2] is another model interpretation method. LIME relies and operates on the same idea as SHAP as it is also model agnostic. LIME interprets or approximates any machine learning model as a black box function and then explains its behavior and predictions with a local, interpretable model.

The idea of LIME is simple and intuitive. Unlike SHAP, which uses different sets of features, LIME inputs different sets of *samples* into the black box model that it's trying to interpret. However, instead of simply calculating the importance of features from the outputs of the altered dataset, LIME generates a brand-new dataset that's comprised of the perturbed samples along with their prediction from the black box model.

---

[2]Ribeiro, M., Singh, S., & Guestrin, C. (2016). "Why Should I Trust You?": Explaining the Predictions of Any Classifier. *Proceedings of the 22nd ACM SIGKDD International Conference on Knowledge Discovery and Data Mining.*

When LIME perturbs the dataset, continuous samples are randomly chosen from a normal distribution; then the inverse operation of mean-centering and scaling is performed on the selected samples. In contrast, for categorical features, they're perturbed by training distributions. Then, on the generated dataset, LIME trains an interpretable model weighted by the proximity of sampled instances to the instance of interest. As the name describes, the trained model on the modified dataset won't be a good approximation of the black box function globally, but it should be accurate within a local scope. Here, the instance of interest refers to the sample that you want LIME to interpret as it can explain how each feature contributes to the sample prediction.

The key point that differs LIME from SHAP is that LIME focuses on local interpretability, explaining the impact that one prediction has, while SHAP is more global, interpreting the model with respect to the entire model and dataset. Additionally, LIME has a feature called SP-LIME, which lets the model produce an explained subset of the original dataset that would produce a global understanding of the model. Here is a basic overview of LIME's explanation process:

1. Choose one sample of interest that you would like to be analyzed and its prediction explained by LIME.

2. Generate a perturbed dataset based on different sets of features that create a representative subset of the data. The target of these features will be the black box model's prediction of these samples.

3. The samples are assigned a weight according to their distance or proximity to the instance of interest.

4. Train a model that's interpretable, such as LASSO or a decision tree, on the perturbed dataset. LIME uses Ridge as a default model, but any sklearn-like model objects can be passed in.

5. Explain predictions by analyzing the model output.

Although LIME shines in text analysis and image recognition, for the context of this book, we will only explore its application to tabular data.

LIME comes implemented in its own library. For consistency and comparison purposes, we're going to utilize LIME on the same dataset used previously in the "SHAP" section. We can retrieve the dataset and define an AutoKeras model (Listing 12-12) as described previously.

Listing 12-12. Retrieving the dataset and fitting an AutoKeras model

```
import shap
shap.initjs()
x, y = shap.datasets.adult()
y = y.astype(np.int32)

from sklearn.model_selection import train_test_split as tts
X_train, X_valid, y_train, y_valid = tts(x, y, train_size = 0.8,
                                          random_state = 42)
import autokeras as ak
input_node = ak.StructuredDataInput()
output_node = ak.StructuredDataBlock(categorical_encoding=True)(input_node)
output_node = ak.ClassificationHead()(output_node)
clf = ak.AutoModel(
inputs=input_node, outputs=output_node,
overwrite=True, max_trials=20
)
clf.fit(X_train, y_train, epochs=50)
model = clf.export_model()
```

The following is the basic syntax for initializing a LIME explainable object in Listing 12-13.

Listing 12-13. Initializing the LIME explainable object

```
import lime
from lime import lime_tabular

explainer = lime_tabular.LimeTabularExplainer(
    training_data=np.array(X_train),
    feature_names=X_train.columns,
    class_names=["Income < 50k", "Income > 50k"],
    mode='classification'
)
```

Most of the parameters are self-explanatory, but one might get confused with the parameter "class_name". Shown previously is a list of strings passed in as the name of each class that LIME will use during its explanation. According to our dataset, for a classification result of 0, it represents the person with those demographic features has an income less than $50,000 per year and vice versa for class 1.

After initialization, we can simply call the method explain_instance shown in Listing 12-14 to let LIME do all the heavy computation and explanation work.

Listing 12-14. Explaining the instance of interest

```
exp = explainer.explain_instance(
    data_row=X_valid.iloc[4],
    predict_fn=model.predict,
    num_features=8,
    num_samples=1000,
    labels=(0,)
)

exp.show_in_notebook(show_table=True)
```

Here, the data_row parameter specifies which "sample of interest" LIME will attempt to explain, the model_fn parameter provides the model prediction function that LIME will use, and finally both num_features and num_samples give the number of features and samples that LIME will consider during the explanation, respectively.

Since the Keras model.predict function only returns the probability of the predicted class in binary classification, LIME only displays the influence that the various features had on that specific classified class. For a better visual and interpretation, we can create a custom prediction function, which returns both the probability for the predicted class and the probability for the other class by subtracting the returned value from 1 (Listing 12-15). Then, the function is simply passed into the predict_fn parameter. As mentioned previously, the base, interpretable model is set to Ridge, but it can be changed with setting the model_regressor parameter in the explain_instance method to a sklearn model object.

Listing 12-15. Custom prediction function

```
def pred_proba(x):
    p = model.predict(x)
        return np.concatenate([p, 1-p], axis=1)
```

```
exp = explainer.explain_instance(
    data_row=X_valid.iloc[4],
    predict_fn=pred_proba,
    num_features=8,
    num_samples=1000,
    labels=(0,)
)

exp.show_in_notebook(show_table=True)
```

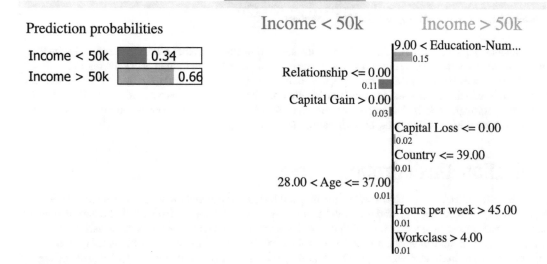

Figure 12-12. *LIME explanation for the fourth sample in the validation dataset*

In this prediction sample (Figure 12-12), the confidence, or the probability that the model has for "Income > 50k", is 66% or 0.66. Each feature on the bar chart to the left is color-coded according to the color of each class. In the bar graph, those orange "bars" that are extended to the left represent how much those features contribute to classifying the target as "Income > 50k". The preceding table is colored-coded as well, with the value column depicting the value of the feature and the color representing which feature contributes to which class's predictions.

Furthermore, there's a little trick in LIME that can improve the bar plot's quality. By calling as_pyploy_ figure on the explanation, the object returned will plot the bar plot as a matplotlib figure (Listing 12-16).

Listing 12-16. Plotting the bar plot using matplotlib

```
plot = exp.as_pyplot_figure(label=0)
```

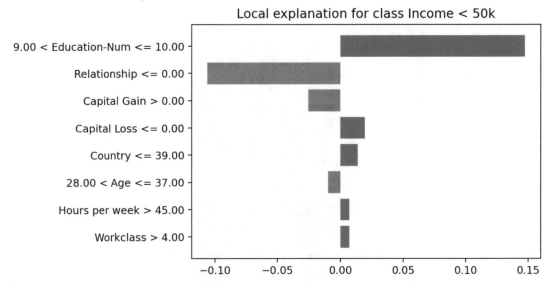

Figure 12-13. *Bar plot plotted with the pyplot method*

The plot (Figure 12-13) essentially displays the same information as the smaller plot shown in the first figure but with a better formatting and a clearer comparison between the influence of each feature.

Compared with SHAP, LIME is more popular in terms of presentation and used by people who may be less familiar with the inner workings surrounding the model; SHAP is more popular among professional data scientists. Although LIME isn't necessarily better than SHAP, but for quick interpretation of the model that comes with intuitive visualizations, LIME is definitely the tool to go.

Activation Maximization

Activation Maximization[3] is a clever technique that can be used to understand what components a network is specifically "looking for." In standard model training, we optimize the set of weights in the neural network such that the difference between the prediction and the true output is minimized (or some variation of this difference). In Activation Maximization, we hold the weights fixed and optimize the input such that the activations in some selected weights are maximized. That is, we perform gradient *ascent* to artificially construct the optimal input to maximize a set of model activations.

---

[3]Often attributed to the following paper in application to convolutional neural networks: Mahendran, A., & Vedaldi, A. (2016). Visualizing Deep Convolutional Neural Networks Using Natural Pre-images. *International Journal of Computer Vision, 120*, 233–255. We apply it to tabular contexts here.

Why is this valuable? Model activations are representations of what factors the network considers during feed-forward prediction. Using Activation Maximization, you can see which extreme cases "most satisfy" "conditions" or features the network is "looking" for – and therefore implicitly understand how the network is interpreting different features and patterns in your dataset broadly.

The library `tf_keras_vis`, which can be installed with `pip install tf_keras_vis`, can be used to execute Activation Maximization on Keras/TensorFlow models. We begin by defining a *loss* – this allows us to specify the optimization objective. We can return a tuple traveling along the output matrix diagonal, representing the model outputs for each of the eight classes. Next, we define a `model_modifier` that changes the last activation to a linear rather than softmax layer (Listing 12-17). This makes the Activation Maximization procedure easier, since it is more difficult to navigate the softmax layer than a set of independent linear activations. Since the weights of the network itself are fixed, this does not change the network's fundamental predictive properties.

Listing 12-17. Defining utility function for Activation Maximization

```
def loss(output):
    return (output[0, 0], output[1, 1], output[2, 2], output[3, 3],
            output[4, 4], output[5, 5], output[6, 6], output[7, 7])

def model_modifier(model):
    model.layers[-1].activation = tensorflow.keras.activations.linear
```

Additionally, `tf_keras_vis` implements Activation Maximization for image-based neural networks (this is the dominant application of Activation Maximization), but our data is tabular and therefore of a different spatial dimensionality. We can address this by creating a new model that accepts data that is technically in the shape of an image, reshaping it into tabular form, and then passing it into our model (Listing 12-18).

Listing 12-18. Creating a model that converts image shaped data to tabular form

```
inp = L.Input((80,1,1))
reshape = L.Reshape((80,))(inp)
modelOut = model(reshape)
act = keras.models.Model(inputs=inp, outputs=modelOut)
```

To run the Activation Maximization procedure, we pass in the model and the model modifier function into the `ActivationMaximization` object (Listing 12-19). Then, we define a `seed_input` that constitutes the initial "guess." The Activation Maximization procedure then iteratively updates this initial input across `steps`, the number of steps, to maximize the loss defined by `loss`. In this case, we are trying to find the inputs that are the "most defining" of each of the eight classes.

Listing 12-19. The Keras Activation Maximization procedure

```
from tf_keras_vis.activation_maximization import ActivationMaximization
visualize_activation = ActivationMaximization(act, model_modifier)
seed_input = tensorflow.random.uniform((7, 80, 1, 1), 0, 1)
activations = visualize_activation(loss,
                                   seed_input=seed_input,
                                   steps=256)
```

Once we have obtained the maps, we can visualize the inputs for each class (shown in Listing 12-20 and Figure 12-14). We have obtained synthetically constructed inputs that the network considers to be "the most representative" of each class!

Listing 12-20. Visualizing the results of Activation Maximization

```
images = [activation.astype(np.float32) for activation in activations]
plt.set_cmap('gray')
plt.figure(figsize=(9,9), dpi=400)
for i in range(0, len(images)):
    plt.subplot(3, 3, i+1)
    visualization = images[i].reshape(8, 10)
    plt.imshow(visualization)
    plt.title(f'Target: {i}')
    plt.axis('off')
 plt.show()
```

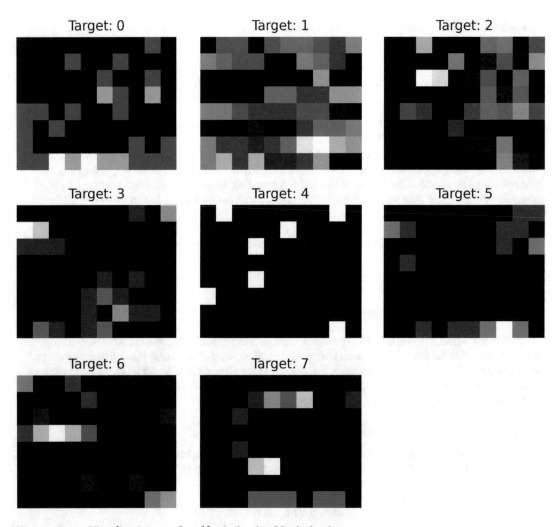

Figure 12-14. *Visualizations produced by Activation Maximization*

Of course, you'll want to run this procedure multiple times, since the possible space is very complex and the results are highly dependent upon the initialization. The resulting synthetically constructed inputs can be analyzed to understand which column values and/or column value groups are most characteristic of each class. Moreover, you can run Activation Maximization on intermediate layers by replacing the complete model with a submodel (as demonstrated previously).

Key Points

In this brief chapter, we discussed three deep learning interpretability methods:

- SHAP is a model-agnostic suite of interpretability methods that estimate the individual contribution of each feature to the output of the neural network by considering large numbers of feature subsets. SHAP also has adaptations to exploit the gradients of neural networks.

- LIME is also a model-agnostic tool targeted toward model interpretation, especially for tabular and text data. LIME utilizes a representative perturbed dataset and produces explainable results by training an additional interpretable model. LIME is generally used for quicker and more intuitive representation.

- Activation Maximization is a neural network–specific technique, in which the input is optimized to maximize the activations in a certain layer of the neural network. Multiple "optimal inputs" can be generated and compared to form an analysis of what certain activations are "looked for" or are "triggered" by.

This chapter has just scraped the surface of deep learning interpretability, to demonstrate where one might start to interpret their tabular deep learning model. In addition to these three methods, we encourage you to explore other deep learning interpretability methods developed for tabular contexts. A sample of interesting papers is presented in the following:

Agarwal, R., Frosst, N., Zhang, X., Caruana, R., & Hinton, G.E. (2021). Neural Additive Models: Interpretable Machine Learning with Neural Nets. *ArXiv, abs/2004.13912.*

Chang, C., Caruana, R., & Goldenberg, A. (2021). NODE-GAM: Neural Generalized Additive Model for Interpretable Deep Learning. *ArXiv, abs/2106.01613.*

Liu, X., Wang, X., & Matwin, S. (2018). Improving the Interpretability of Deep Neural Networks with Knowledge Distillation. *2018 IEEE International Conference on Data Mining Workshops (ICDMW)*, 905–912.

Novakovsky, G., Fornes, O., Saraswat, M., Mostafavi, S., & Wasserman, W.W. (2022). ExplaiNN: Interpretable and transparent neural networks for genomics. *bioRxiv.*

Radenović, F., Dubey, A., & Mahajan, D.K. (2022). Neural Basis Models for Interpretability. *ArXiv, abs/2205.14120.*

Ranjbar, N., & Safabakhsh, R. (2022). Using Decision Tree as Local Interpretable Model in Autoencoder-based LIME. *2022 27th International Computer Conference, Computer Society of Iran (CSICC)*, 1–7.

Richman, R., & Wüthrich, M.V. (2021). LocalGLMnet: Interpretable deep learning for tabular data. *DecisionSciRN: Methods of Forecasting (Sub-Topic).*

This concludes the last part and chapter of the book.

Closing Remarks

We have covered a tremendous amount of ground throughout this book. We began in Part 1 with a thorough overview of machine learning and the data preprocessing pipeline; key machine learning concepts and principles; several classical machine learning algorithms, including Gradient Boosting models – deep learning's dominant competitor on tabular data problems; different data storage and delivery structures; and a variety of data encoding and transformation techniques. In Part 2, we built an expansive knowledge of deep learning across five chapters – artificial neural networks, convolutional neural networks, recurrent neural networks, attention and transformations, and tree-based neural networks. This part covered dozens of deep learning modeling approaches, applied to a variety of data types and contexts, and explored almost 20 research papers. In Part 3, we further expanded our toolkit not only to model tabular data but to pretrain models, develop noise- and perturbation-robust models, generate tabular data, optimize the modeling pipeline, link together models into ensembles and self-aware systems, and interpret models.

We hope that this book has been thought-provoking to those looking to better understand the role of deep learning in tabular data and a helpful resource to those looking to use deep learning in their respective domains. The guiding principle at the forefront of this book was accessibility, and we hope the explanations, diagrams, code, and research walk-throughs supported this.

As the problems we face become more advanced and the data we collect from them become correspondingly more intricate and complex, we find a greater divergence – rather than convergence – in problems and the models that we use to approach them. Thus, the question of "which model performs the best on tabular data" – while admittedly being an efficient question – is a question that already shapes its own obscured answer. We should always keep our minds open and our fingers agile, ready to acquire, test, and synthesize the old and the new. We must strive to be driven less wholly by necessary but restrictive benchmarks and more by the diversity of the problem multiverse.

■ ■ ■

NumPy and Pandas

In this appendix, we discuss the NumPy and Pandas libraries. NumPy is one of the most important libraries, if not arguably *the most important* library, in the Python data science ecosystem. By wrapping data in custom-built NumPy objects optimized for manipulations commonly used in data science, NumPy provides the "atomics" of the "molecular data science ecosystem." Pandas is another important data science library that provides quick tabular data functionality, such as storing, querying, manipulation, and more. Both of these libraries are important to understand to follow through the implementation and exercises in the book.

We will take a thorough look at both NumPy and Pandas in this chapter, with the objective of serving as an effective ramp-up from someone who is totally unfamiliar with these libraries to someone who has a solid grasp of the concepts and syntax behind important operations.

NumPy Arrays

NumPy arrays are perhaps the most important and ubiquitous non-native data storage object in the Python data science ecosystem. In this section, we will learn how to construct and manipulate NumPy arrays.

NumPy Array Construction

We can construct a NumPy array by passing a list into the np.array() constructor function. For instance, arr = np.array([0, 1, 2, 3, 4, 5]) creates a NumPy array under arr holding the values 0, 1, 2, 3, 4, 5.

There are many instances in which we want a NumPy array of elements organized by some sort of pattern or identity, but don't desire to type out manually – like a 1000-element-long array of zeros or an array that counts from 0 to 10^6. NumPy offers several helpful functions for common pattern-based arrays you may want to generate:

- np.arange(start, stop, step) acts like range() in native Python, taking in two parameters indicating the start and end, as well as an optional step size (1 by default). For instance, np.arange(1, 10, 2) creates an array with values [1, 3, 5, 7, 9]. Recall that the end/stop value is not an inclusive bound (i.e., it is not included in the resulting list). Using a negative step value allows for counting backward, that is, start > stop.

- np.linspace(start, end, num) returns an array of length num elements, equally spaced from the first number start to the end number end (inclusive). For instance, np.linspace(1, 10, 5) creates an array with values [1., 3.25, 5.5, 7.75, 10.].

© Andre Ye and Zian Wang 2023
A. Ye and Z. Wang, *Modern Deep Learning for Tabular Data*,
https://doi.org/10.1007/978-1-4842-8692-0

- `np.zeros(shape)` takes in a tuple and initializes an array of that shape with all zeros. For instance, `np.zeros((2, 2, 2))` returns the NumPy array with contents `[[[0, 0], [0, 0]], [[0, 0], [0, 0]]]`.

- `np.ones(shape)` takes in a tuple and initializes an array of that shape with all ones. For instance, `np.ones ((2, 2, 2))` returns the NumPy array with contents `[[[1, 1], [1, 1]], [[1, 1], [1, 1]]]`.

- `np.random.uniform(low, high, shape)` takes in a low bound and a high bound and fills an array with the given shape with uniform-randomly sampled values from that range. If no tuple is provided for the shape parameter, the function returns a single value instead of an array.

- `np.random.normal(mean, std, shape)` takes in a mean and a standard deviation and fills an array with the given shape with values sampled from a normal distribution with that shape. If no tuple is provided for the shape parameter, the function returns a single value instead of an array.

NumPy arrays are of type `numpy.ndarray` – here, "*n*-d" indicates that the array can be of any integer *n* dimensions. The examples we have explored so far are one-dimensional, but we can also build arrays of higher dimensions. A two-dimensional array is an array in which each element itself holds another list/array of elements. A three-dimensional array is an array in which each element holds another list/array of elements and each element of *that* array holds a *third* level of list/array – and so on. The shape of an array indicates its dimensionality and the length/size of each dimension. For instance, the shape (128, 64, 32) indicates that the corresponding array is four-dimensional and has 128 lists of 64 lists each of 32 elements each.

NumPy arrays can be reshaped into any desired shape, as long as the total number of elements is the same in the resulting array. For instance, `np.arange(100)` returns an array with values `[0, 1, 2, ..., 98, 99]`, but `np.arange(100).reshape((10,10))` organizes the 100 elements in 10 arrays of 10 elements each, like `[[0, 1, ..., 8, 9], [10, 11, ..., 18, 19], ..., [90, 91, ..., 98, 99]]` (Figure A-1).

Note   `.reshape()` also accepts negative values: you can specify an "unknown" dimension for reshaping provided you specify all other known dimensions. For instance, one could reshape a 120-element array into shape (3, 4, 2, 5) with .reshape(-1, 4, 2, 5), .reshape(3, 4, -1, 5), and any other permutation. You may not use more than one missing dimension, since this creates ambiguity in the resulting shape. This is often useful both for convenience and in cases where you are dealing with variable-length "lists" of arrays, which each have their own unique structure, where −1 is used in lieu of the dimension expressing the number of arrays in the list.

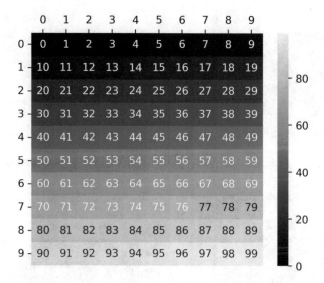

Figure A-1. *A 10-by-10 array of numbers from 0 to 99 (inclusive) generated by the arange and reshape functions*

Simple NumPy Indexing

Indexing one-dimensional NumPy arrays is the same as with native Python. For instance, arr[1:4] can be used to select the second through the fourth elements of the array arr = [0, 1, 2, 3, 4]. NumPy also works with Python's negative indexing syntax. The index arr[1:-1] yields the same result as arr[1:4].

Indexing *n*-dimensional NumPy arrays follows a similar structure, in which the syntax for indexing an individual dimension is defined for each dimension of the *n*-d array. The array np.arange(100). reshape((10, 10))[:5, :5] indexes the first five rows and the first five columns of the 10-by-10 array (Figure A-2). Note that the indexing specification for each dimension is separated by a comma.

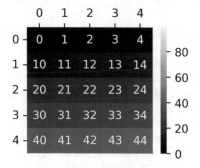

Figure A-2. *The top-left 5-by-5 quadrant of the array visualized in Figure A-1*

If you wish to set indexing specifications for certain dimensions but not others, indicate the lack of an indexing range for a certain dimension by simply typing a colon ':'. For instance, Figure A-3 demonstrates the 10-by-10 array indexed via [:5, :] (left) and [5:, :] (right).

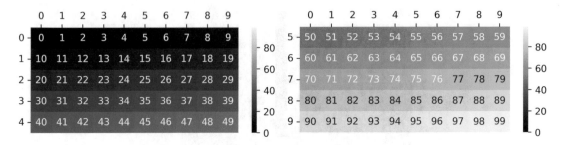

Figure A-3. *Indexing top (left) and bottom (right) halves of the array visualized in Figure A-1*

Figure A-4 demonstrates the result obtained by indexing via [:, :5] and [:, 5:].

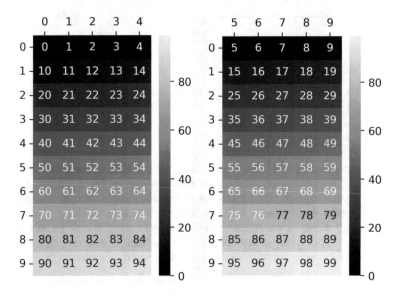

Figure A-4. *Indexing left (left) and right (right) halves of the array visualized in Figure A-1*

Another important concept to understand is the difference between the indexing commands [i] and [i:i+1]. Functionally, the two index the same information: calling np.array([0, 1, 2, 3])[1] indexes the second element (which has value 1), and calling np.array([0, 1, 2, 3])[1:2] begins indexing at the second element and stops at the third element (noninclusve) – this also only indexes the second element. The difference in actual result, however, is that the former index syntax indexes a single element and the latter index syntax indexes a range of elements, even if such a range includes only one method. Thus, np.array([0, 1, 2, 3])[1] returns 1, whereas np.array([0, 1, 2, 3])[1:2] returns np.array([1]).

As another exercise, consider the array initialized by np.zeros((5, 5, 5, 5, 5)) – what is the shape of the index command [:, 0, 3:, 1:2, 2:4]? We can follow through how the indexing specification for each dimension affects the shape of the resulting array, which has shape (5, 1, 3, 1, 2). You can verify this yourself by calling .shape on the indexed array.

Quantitative Manipulation

NumPy offers many functions to manipulate both Python quantitative objects (native integer and float types, for instance) and NumPy arrays (Table A-1). Familiar Python mathematical operations like addition, subtraction, multiplication, division, modulus, exponentiation, binary, and comparative relationships can be applied *element-wise* (i.e., the operation is applied to the *i*th index of the first array and the *i*th index of the second array if two arrays are involved).

Table A-1. *Sample operations performed on NumPy arrays*

Array 1 Contents	Operation	Array 2 Contents	Yields	Result
0, 1, 2	+	2, 1, 0		2, 2, 2
5, 4, 3	-	3, 2, 1		2, 2, 2
4, 2, 1	*	0.5, 1, 2		2, 2, 2
4, 2, 1	/	2, 1, 0.5		2, 2, 2
6, 8, 11	%	4, 3, 3		2, 2, 2
1, 2, 3	**	2, 2, 2		1, 4, 9
2, 3, 4	^	2, 2, 2		0, 1, 6
2, 3, 4	&	2, 2, 2		2, 2, 0
2, 3, 4	|	2, 2, 2		2, 3, 6
-1, 0, 1	<	0, 0, 0		True, False, False
-1, 0, 1	==	0, 0, 0		False, True, False

Note A common mistake for beginners is to confuse the caret operator (^) with exponentiation. Instead, Python uses ** for exponentiation and ^ for XOR. Fortran denoted exponentiation with the double asterisk because most computers at that time used 6-bit encodings and thus did not support a caret character. According to Ken Thompson, co-creator of C, the association of the caret character and the XOR operation was arbitrary – a random choice of the remaining characters! ^ is universally associated with exponentiation in most other contexts likely from its origination as the superscript notation for TeX. The syntax was introduced in algebra systems and graphing calculators in the late 1980s and early 1990s.

This is different from Python syntax. For instance, [0, 1, 2] + [3, 4, 5] will not return [3, 5, 7] but rather [0, 1, 2, 3, 4, 5] if we use standard lists and not NumPy arrays.

The two elements involved in an operation must be the same length, unless one of the arrays is a repetition of the same value; in this case, that array can be replaced with a one-element array containing that value or just that value by itself. For instance, np.arange(100) * np.array([2, 2, ..., 2]) can be replaced with np.arange(100) * np.array([2]) or np.arange(100) * 2.

Otherwise, applying relationships between two NumPy arrays of different lengths that do not fall in the previously mentioned category will yield a ValueError: operands could not be broadcast together.

NumPy also offers multiple mathematical functions that can be applied to a single value or array (in which case the function is applied element-wise and returns an array of the same length) (Table A-2).

Table A-2. *Example NumPy functions*

Function	Usage	Function	Usage
Sine	np.sin(0) -> 0.0	Floor	np.floor(2.4) -> 2
Cosine	np.cos(0) -> 1.0	Ceiling	np.ceil(2.4) -> 3
Tangent	np.tan(0) -> 0.0	Round	np.round(2.4) -> 2
Arcsine	np.arcsin(0) -> 0.0	Exponential	np.exp(0) -> 1.0
Arccosine	np.arccos(1) -> 0.0	Natural Log	np.log(np.e) -> 1.0
Arctangent	np.arctan(0) -> 0.0	Base 10 Log	np.log10(100) -> 2.0
Maximum	np.max([1,2]) -> 2	Square Root	np.sqrt(9) -> 3.0
Minimum	np.min([1,2]) -> 1	Absolute Value	np.abs(-2.5) -> 2.5
Mean	np.mean([1,2]) -> 1.5	Median	np.mean([1,2,3]) -> 2

These functions are efficient and very helpful in obtaining mathematical derivations from arrays. For instance, we may implement the sigmoid function ($\sigma(x) = \dfrac{1}{1+e^{-x}}$) as sigmoid = lambda x: 1/(1 + np.exp(-x)). This function can work with both single scalar values and NumPy arrays.

Advanced NumPy Indexing

This set of simple NumPy indexing should satisfy most important and common operations you'll manipulate arrays with. However, if desired, it can be incredibly helpful to learn the syntax of more advanced NumPy indexing methods, which allow you to express more complex desired outcomes in a syntactically short amount of space.

NumPy colon-and-bracket indexing accepts a third parameter (in addition to the start and stop indexing) indicating the step size. For instance, np.arange(10)[2:6:2] indexes every other element from the array: [2, 4]. As expected, leaving the start and end indices unspecified while providing a step size indexes the entire array with the provided step size: np.arange(10)[::2] yields [0, 2, 4, 6, 8].

If working with arrays containing a large number of axes, NumPy offers the ellipsis (...) to represent lack of an indexing specification for certain dimensions. For instance, if we want to index only the first and last dimensions of an array z with shape (5, 5, 5, 5, 5, 5) and leave the rest untouched, we could write something like z[1:4, :, :, :, :, 1:4] without ellipsis notation or equivalently write (in much more compact form) z[1:4, ..., 1:4].

NumPy arrays also support reassignment. Individual elements can be changed via arr[index] = new_ value. Multiple elements can be reassigned: consider an array defined by arr = np.arange(6); replacing the third through fifth elements with arr[2:5] = np.arange(3) yields arr as [0, 1, 0, 1, 2, 5]. Moreover, the indices do not need to be consecutive: arr[::2] = np.arange(3) yields arr as [0, 1, 1, 3, 2, 5].

Reassignment can be dangerous if you're not paying attention. Consider the following series of array manipulations (Listing A-1): we initialize an array of numbers from 0 to 9 (inclusive), set a new array copy to that array, and then reassign the first element of the first array to 10.

Listing A-1. Danger of reassignment

```
arr = np.arange(10)
copy = arr
arr[0] = 10
```

As expected, the contents of arr are [10, 1, 2, 3, 4, 5, 6, 7, 8, 9]. However, the contents of copy are also [10, 1, 2, 3, 4, 5, 6, 7, 8, 9]! When we set copy to arr, we're not actually copying the contents of arr: we're creating another reference to the original array's location in memory. Thus, when a reassignment is made to arr, it also appears in copy. In order to prevent this linking, we must physically copy an array; this can be done with copy = np.copy(arr) or with copy = arr[:]. The latter method indexes the entire array, but physically copies it in memory such that reassignments and manipulations are not linked.

Note that the indices used in colon-bracket syntax (start:stop:step) are only specifying a set of indices generated by a given set of rules, which means there is no reason we can't specify our own custom indices. If we want the second, fourth, and sixth elements of an array, we can index it with subset = arr[[1, 3, 5]]. The double brackets may feel unnecessary at first, but think of the command as a shorthand for two lines of code: indices = [1, 3, 5] and subset = arr[indices]. For specialized indexing, you can programmatically generate your own index lists.

For certain specialized indexing operations, however, NumPy can help us with *conditional indexing*. For instance, if we want to retrieve all items in an array that are larger than 3 in value, we can call arr[arr > 3]. Recall that arr > 3 returns a Boolean array in which each element is either True if the corresponding index element in arr satisfies the condition of being larger than 3 and False otherwise. When we index an array with these element-wise Boolean specifications, NumPy includes an element of arr if the corresponding Boolean is True and does not if it is False.

NumPy Data Types

NumPy offers for values in NumPy arrays to be stored in multiple different forms, or data types. Here are some of the most important:

- Boolean (np.bool_)

- Unsigned integers (np.uint8, np.uint16, np.uint32, np.uint64)

- Signed integers (np.int8, np.int16, np.int32, np.int64)

- Floats (np.float16, np.float32, np.float64)

When you initialize an array, you can pass in the desired data type with the dtype parameter, for instance, np.array([-1, 0, 1, 2, 3], dtype=np.int8).

You can *cast* ("convert") one data type into another using arr.astype(np.datatype). Consider the following arrays (Listing A-2).

Listing A-2. Casting data types

```
arr1 = np.array([1,2,3])
arr2 = arr1.astype(np.uint8)
```

Calling arr1.dtype yields dtype('int64'); calling arr2.dtype yields dtype('uint8'). When we first construct arr1, integers are represented by the np.int64 type; they are then cast as unsigned integers into arr2 with no effect on the contents. However, note that casting to a lower representation size can alter the values of the array; for instance, casting an array with the values [-1, -2, -3] to np.uint8 yields [255, 254, 253] (calculated by subtracting from 2^8). As another example, casting an array with value [1.123456789] to np.float16 yields the value [1.123].

Generally, you'll need to cast to a lower representation size than a higher one, usually in response to memory/storage problems. Casting is also often utilized to prepare images for image processing libraries, which may require image data to be stored in np.uint8 type to guarantee values consist only of integers from [0, 255].

Function Application and Vectorization

Often, we would like to apply a function element-wise to a NumPy array and that this function is not already supported. (You always want to use default functions if they are implemented for potential efficiency gains.) For instance, say we would like to graph the piecewise function

$$f(x) = \begin{cases} \dfrac{x^2}{25}, & x < 0 \\ \sin x \cdot x^2, & x \geq 0 \end{cases}$$

from $-5 \leq x \leq 5$. The array containing the x-axis values can be generated with inputs = np.linspace (-5, 5, 100). In this case, we are sampling 100 points from the function, which is high enough precision for our visualization purposes.

We can implement the function as follows (Listing A-3).

Listing A-3. A custom piecewise function

```
def f(x):
    if x < 0: return x**2/25
    else: return np.sin(x) * x**2
```

However, simply applying f(inputs) yields a ValueError:

```
---------------------------------------------------------------------------
ValueError                              Traceback (most recent call last)
/tmp/ipykernel_33/829457706.py in <module>
----> 1 f(np.linspace(-5, 5, 100))

/tmp/ipykernel_33/3998949136.py in f(x)
      1 def f(x):
----> 2     if x < 0: return x**2/25
      3     else: return np.sin(x) * x**2

ValueError: The truth value of an array with more than one element is ambiguous. Use a.any()
or a.all()
```

Our multipart function involves some relatively complex logic (i.e., if statements and comparisons), and thus applying the function blankly fails. In this case, the value error stems from using if with multiple Boolean values; since some elements of the Boolean array formed by x < 0 are True and some are False, Python cannot decide whether to execute the code within the if or not – the truth value of the array is *ambiguous*. In this case, Python cannot quite tell that we want it to apply the function element-wise; we must communicate this explicitly.

One manual method is to use list comprehension and create a new array formed by applying the function individually to each element of inputs: outputs = np.array([f(element) for element in inputs]). The shorter but functionally (get it?) equivalent alternative is to use *function vectorization*. np. vectorize takes in a Python function and returns another function that applies the original function element-wise: outputs = np.vectorize(f)(inputs) or alternatively vectorized = np.vectorize(f) and outputs = vectorized(inputs) for a longer but perhaps more readable representation.

Function vectorization is also convenient when we want to apply element-wise operations to multiple inputs. For instance, we may want to return True when the sum of elements across three input arrays is larger than 10 and False otherwise (Listing A-4).

Listing A-4. A sample multi-input function

```
def f(x, y, z):
    if x + y + z > 10: return True
    return False
```

We can apply the function as follows (Listing A-5).

Listing A-5. Using function vectorization on a function with multiple inputs

```
x = np.arange(0, 5)
y = np.arange(7, 2, -1)
z = np.arange(-1, 9, 2)
np.vectorize(f)(x, y, z)
# array([False, False, False, True, True])
```

Note that even though some have observed minor speedups by using `np.vectorize`, the function is "provided primarily for convenience, not for performance" (from the NumPy documentation website).

NumPy Array Application: Image Manipulation

Let's use our knowledge of NumPy arrays to have some fun with image manipulation. The `skimage.io.imread` function can take in a URL of an image and return it as a NumPy array. Our sample image will be a landscape view of the New York City skyline (Listing A-6, Figure A-5).

Listing A-6. Loading a sample image

```
from skimage import io
import matplotlib.pyplot as plt

url = 'https://upload.wikimedia.org/wikipedia/commons/ thumb/2/2b/NYC_Downtown_Manhattan_
Skyline_seen_from_Paulus_Hook_2019-12-20_IMG_7347_FRD_%28cropped%29.jpg/1920px-NYC_Downtown_
Manhattan_Skyline_seen_from_Paulus_Hook_2019-12-20_IMG_7347_FRD_%28cropped%29.jpg'
image = io.imread(url)

plt.figure(figsize=(10, 5), dpi=400)
plt.imshow(image)
plt.show()
```

Figure A-5. *A sample image of the New York skyline*

Calling image.shape yields the tuple (770, 1920, 3). This means the image is 770 pixels high and 1920 pixels wide. The image is in color, and so by standard it has three channels corresponding to red, green, and blue (RGB). We can separate the image into its "color composition" by indexing each of the channels independently and displaying the two-dimensional slice in the corresponding color (Listing A-7, Figure A-6).

Listing A-7. Separating and visualizing the individual red, green, and blue color maps of the single image from Figure A-5

```python
for i, color in enumerate(['Reds', 'Blues', 'Greens']):
    plt.figure(figsize=(10, 5), dpi=400)
    plt.imshow(image[:,:,i], cmap=color)
    plt.show()
```

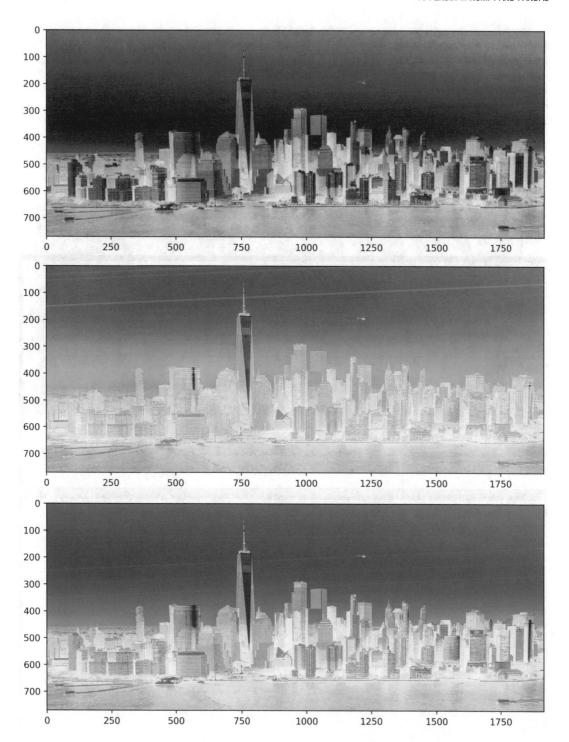

Figure A-6. *Viewing the individual red, green, and blue color maps of the sample image from Figure A-5*

Say we want to "collapse" the three-dimensional image into a two-dimensional one by converting it from color to grayscale. One natural approach is to take the mean of channels for each pixel, which can be implemented with np.mean(image, axis=2) (Listing A-8, Figure A-7). Here, the axis parameter indicates that we are taking the mean along the third axis, indicated by 2 just as the third element of a tuple is indexed with 2.

Listing A-8. Visualizing a mean-based grayscale representation

```
plt.figure(figsize=(10, 5), dpi=400)
plt.imshow(np.mean(image, axis=2), cmap='gray')
plt.show()
```

Figure A-7. *Obtaining a grayscale representation of the image by taking the mean of the image across the color depth axis (i.e., replacing each pixel with the average of the RGB values)*

The result is a pretty good grayscale representation! We can also produce a similar effect by taking the maximum value of each channel per pixel with np.max(image, axis=2), which produces a vintage-looking "overexposed" grayscale representation (Figure A-8).

Figure A-8. *Obtaining a grayscale representation of the image by taking the max of the image across the color depth axis*

And taking the minimum value for each pixel instead with np.min(image, axis=2) yields – as one may expect – a generally darker grayscale representation (Figure A-9).

Figure A-9. *Obtaining a grayscale representation of the image by taking the min of the image across the color depth axis*

We can augment the original image by adding noise (Listing A-9, Figures A-10 and A-11). An array of normally distributed noise with the same shape as the original image can be generated with noise = np.random.normal(0, 40, (770, 1920, 3)). In this case, we center the mean at 0 and use standard deviation 40. Recall that images are generally stored to have numerical pixel values between 0 and 255. A larger standard deviation will yield larger visual noise, whereas a smaller one will yield less visible noise. Additionally, note that we need to cast the noisy image as an unsigned 8-bit integer (between 0 and $2^8 - 1 = 255$) because the noise vector is drawn from a continuous distribution, yielding pixel values that are not within the valid set of all integers from 0 to 255 inclusive that are accepted for image display. Trying to display the image without casting the array as uint8 type will yield a bizarre, mostly blank canvas.

Listing A-9. Generating a noisy image by adding noise randomly drawn from a normal distribution to the image from Figure A-5

```
noise_vector = np.random.normal(0, 40, (770, 1920, 3))
altered_image = image + noise_vector
display_image = altered_image.astype(np.uint8)
```

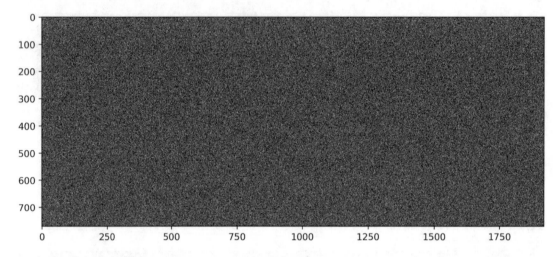

Figure A-10. *A visual representation of the random noise array*

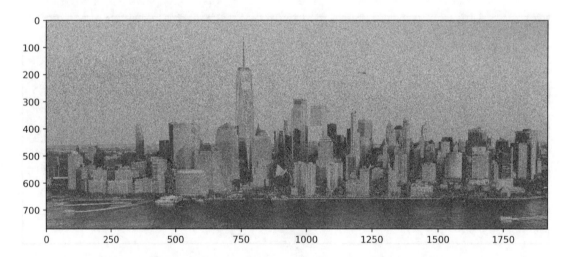

Figure A-11. *Visualizing the combination of the noise array to the original array*

We can also adjust the mean value of the normal distribution from which values for the noise matrix are sampled to generally influence the "feel" of the image overall (Figure A-12, Figure A-13).

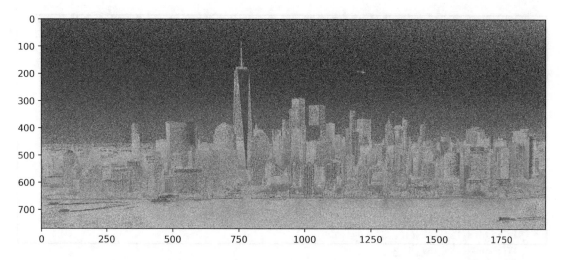

Figure A-12. *Visualizing the combination of the noise array to the original array, with the noise array generated from a normal distribution with mean 100*

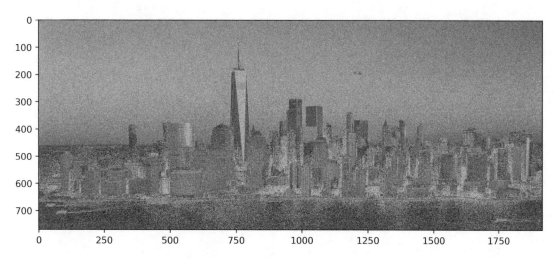

Figure A-13. *Visualizing the combination of the noise array to the original array, with the noise array generated from a normal distribution with mean 200*

The features of the image can also be enhanced or dimmed by multiplying all the values in the image by some constant – $0 \leq k < 1$ to dim the image and $k > 1$ to enhance it. This is known as *contrast*. Note that we need to similarly cast the altered image as an unsigned 8-bit integer because multiplying each value by a non-integer value does not guarantee an integer outcome required for picture display (Listing A-10). Observe that minute differences form harsh, colorful boundaries in high-contrast images due to the exacerbated/exaggerated quantitative value difference (Figure A-14).

Listing A-10. Generating and visualizing different levels of contrast by multiplying values in the sample image by varying factors

```
for factor in [0.2, 0.6, 1.5, 3, 8]:

    altered_image = image * factor
    display_image = altered_image.astype(np.uint8)

    plt.figure(figsize=(10, 5), dpi=400)
    plt.imshow(display_image)
     plt.show()
```

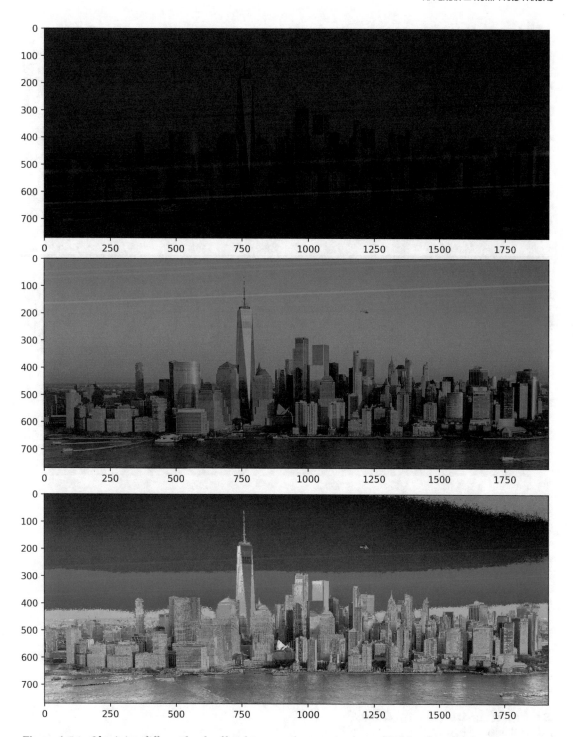

Figure A-14. *Obtaining different levels of brightness and saturation by multiplying the array by a factor*

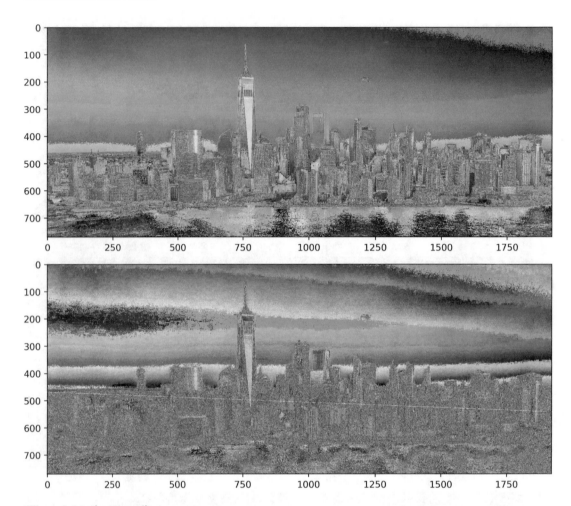

Figure A-14. *(continued)*

Another familiar parameter in image editing tools is brightness, which can be adjusted by adding or subtracting the same value to or from all pixels in an image, thus uniformly increasing or decreasing the array values.

It's common knowledge that the New York City skyline just isn't complete without King Kong and Godzilla battling it out. Let's load in a sample image of the scene (Listing A-11, Figure A-15).

Listing A-11. Loading and displaying a sample image of King Kong battling Godzilla

```
url = 'https://upload.wikimedia.org/wikipedia/commons/ thumb/f/f4/KK_v_G_trailer_%281962%29.
png/440px-KK_v_G_trailer_%281962%29.png'
beasts = io.imread(url)

plt.figure(figsize=(10, 5), dpi=400)
plt.imshow(beasts)
plt.show()
```

Figure A-15. *A sample image of King Kong battling Godzilla*

We'll use a simple bitwise AND operation between two sample images to merge them. In order to do this, we first need to make sure that the images are the same size. One way to ensure equivalent array shapes is to resize the higher-resolution image (the New York City skyline in this case) to the shape of the lower-resolution image. This can be accomplished with the Python `cv2` computer vision library, which offers the helpful function `cv2.resize`: `resized = cv2.resize(original, desired_shape)`.

■ **Note** Do not confuse resizing with reshaping! Resizing is an image operation that reduces the resolution/ dimensions of an image, whereas reshaping is a generalized array operation that shifts how elements in the array are arranged/distributed across dimensions while keeping the number of elements constant.

The resulting merge (Listing A-12, Figure A-16) isn't bad!

Listing A-12. Generating and visualizing a "merger" by using the bitwise OR operation

```
merged = cv2.resize(image, beasts.shape) & beasts
plt.figure(figsize=(10, 5), dpi=400)
plt.imshow(merged)
plt.show()
```

Figure A-16. *A decent merging of two images using a simple bitwise OR operation*

(See more fun image manipulation with convolutions in Chapter 4!)

We were able to do a lot simply by manipulating NumPy arrays with minimal help from additional libraries! Having a strong grasp of NumPy will prove not only helpful for manipulating data stored in array form, but almost any data type in the Python data science ecosystem due to NumPy's ubiquitous nature.

Pandas DataFrames

While NumPy arrays allow for the efficient storage of raw data from images to tables to text in array form, its generality can limit how efficiently we work with specific types of data. Perhaps the best-developed library to work with table-based data is Pandas, which is built upon the *DataFrame*, a two-dimensional container for tabular data. (In fact, Pandas is built upon NumPy!) With Pandas, you can read and write from and to files, select data, filter data, and transform data. Pandas is an essential tool in the context of tabular data: there simply is no other library like Pandas in Python at the time of this book's writing that is as well-maintained and appropriate for effective tabular data manipulation.

Constructing Pandas DataFrames

To construct a Pandas dictionary from scratch, pass in a dictionary to the `pd.DataFrame()` constructor in which each key is a string representing the column name and the value is a list or array representing its values (Listing A-13, Figure A-17).

Listing A-13. Generating a dummy DataFrame

```
df = pd.DataFrame({'a':[1, 2, 3],
                   'b':[4, 5, 6],
                   'c':[7, 8, 9]})
```

	a	b	c
0	1	4	7
1	2	5	8
2	3	6	9

Figure A-17. A simple dummy Pandas DataFrame

If the lists provided for each column are not the same length, you will encounter a `ValueError: All arrays must be of the same length` error.

This method of constructing DataFrames is especially helpful when attempting to create small DataFrames, for instance, as a dummy table to test out manipulations or record and collect data for visualizations.

You can accomplish the same outcome by first initializing a blank DataFrame by passing in no information into the constructor and then creating the columns one by one (Listing A-14).

Listing A-14. Initializing a dummy DataFrame via column creation/assignment

```
df = pd.DataFrame()
df['a'] = [1, 2, 3]
df['b'] = [4, 5, 6]
df['c'] = [7, 8, 9]
```

It should be noted here that this operation is very similar to NumPy array reassignment mechanics (Listing A-15). The bracket notation allows for the indexing of an element or collections of elements along a given object's axes. Note, however, that certain element or dimension of an array needs to already exist to be reassigned in NumPy, whereas in Pandas the DataFrame can be empty before assignment.

Listing A-15. Analog in NumPy to the column assignment operation in Listing A-14

```
arr = np.zeros((3, 3))

arr[0] = [1, 2, 3]
arr[1] = [4, 5, 6]
arr[2] = [7, 8, 9]
```

DataFrame columns can be indexed using brackets and the column's name (Listing A-16). This returns a *Series* object, which can be thought of as a dictionary. In a dictionary, each key is associated with a value; in a Series, each *index* is associated with a value.

Listing A-16. Result of indexing a single column in a Pandas DataFrame

```
df['a']

'''
Returns:
0    1
1    2
2    3
Name: a, dtype: int64
'''
```

Thus, we can obtain the first item of the series indexed in Listing A-16 by df['a'][0], which returns 1.

DataFrames are more like a dictionary than a list because of the explicit and possible modification of the index, even if it is ordered like a list by default.

You'll more commonly be reading data from a file. For instance, if you want to read the data in a comma-separated value file, use data = pd.read_csv(path). Depending on the organization of your .csv file, you may need to specify certain delimiters. Pandas has corresponding reading functions for Excel spreadsheets (pd.read_excel), JSON (pd.read_json), HTML tables (pd.read_html), SQL data (pd.read_sql), and many other file types. Correspondingly, you can export Pandas DataFrames in a desired supported form – for example, data.to_csv(path) or data.to_excel(path). See the IO tools page on the Pandas documentation for a full and up-to-date list of Pandas file reading and processing functionality: https://pandas.pydata.org/pandas-docs/stable/user_guide/io.html.

Simple Pandas Mechanics

We can write a function to construct a multiplication table, represented in Pandas with a DataFrame (Listing A-17, Figure A-18). The multiplication table is a square DataFrame with both indices and columns containing the integers from [1, n] (inclusive); each element within the table is the product of the corresponding index and column coordinates. The function initializes a blank DataFrame with the desired index and column values and then iteratively fills in the desired elements using standard array logic.

Listing A-17. A function that generates a multiplication table using Pandas value reassignment of an arbitrary $n \times n$ size

```
def makeTable(n = 10):
    table = pd.DataFrame(index=range(1, n+1),
                         columns=range(1, n+1))
    for num1 in table.columns:
        for num2 in table.index:
            table[num1][num2] = num1 * num2
    return table

table = makeTable(n=100)
```

	1	2	3	4	5	6	7	8	9	10	...	91	92	93	94	95	96	97	98	99	100
1	1	2	3	4	5	6	7	8	9	10	...	91	92	93	94	95	96	97	98	99	100
2	2	4	6	8	10	12	14	16	18	20	...	182	184	186	188	190	192	194	196	198	200
3	3	6	9	12	15	18	21	24	27	30	...	273	276	279	282	285	288	291	294	297	300
4	4	8	12	16	20	24	28	32	36	40	...	364	368	372	376	380	384	388	392	396	400
5	5	10	15	20	25	30	35	40	45	50	...	455	460	465	470	475	480	485	490	495	500
...	...	...	...	...	...	...	...	...	...	...	...	...	...	...	...	...	...	...	...	...	...
96	96	192	288	384	480	576	672	768	864	960	...	8736	8832	8928	9024	9120	9216	9312	9408	9504	9600
97	97	194	291	388	485	582	679	776	873	970	...	8827	8924	9021	9118	9215	9312	9409	9506	9603	9700
98	98	196	294	392	490	588	686	784	882	980	...	8918	9016	9114	9212	9310	9408	9506	9604	9702	9800
99	99	198	297	396	495	594	693	792	891	990	...	9009	9108	9207	9306	9405	9504	9603	9702	9801	9900
100	100	200	300	400	500	600	700	800	900	1000	...	9100	9200	9300	9400	9500	9600	9700	9800	9900	10000

100 rows × 100 columns

Figure A-18. *Sample generated 100-by-100 multiplication table*

Recall that you can index a column in a DataFrame with a bracket. When we call table[5], we obtain the following displayed Series. In our context, this returns all the multiples of 5 from 1 · 5 to 100 · 5:

```
1          5
2         10
3         15
4         20
5         25
         ...
96        480
97        485
98        490
99        495
100       500
Name: 5, Length: 100, dtype: object
```

Say we want to view multiples of 5, 10, and 15 all at once. Rather than passing in a single reference to a column, we pass in a list of column references: table[[5, 10, 15]] returns the DataFrame shown in Figure A-19.

	5	10	15
1	5	10	15
2	10	20	30
3	15	30	45
4	20	40	60
5	25	50	75
...	...	...	...
96	480	960	1440
97	485	970	1455
98	490	980	1470
99	495	990	1485
100	500	1000	1500

100 rows × 3 columns

Figure A-19. *Indexing a subset of the columns in the 100-by-100 multiplication table visualized in Figure A-18*

We can also index rows with .loc. You can pass in a single row to index or a list of rows to index. table.loc[[5, 10, 15]] returns the DataFrame shown in Figure A-20.

	1	2	3	4	5	6	7	8	9	10	...	91	92	93	94	95	96	97	98	99	100
5	5	10	15	20	25	30	35	40	45	50	...	455	460	465	470	475	480	485	490	495	500
10	10	20	30	40	50	60	70	80	90	100	...	910	920	930	940	950	960	970	980	990	1000
15	15	30	45	60	75	90	105	120	135	150	...	1365	1380	1395	1410	1425	1440	1455	1470	1485	1500

Figure A-20. *Indexing a subset of the rows in the 100-by-100 multiplication table visualized in Figure A-18*

■ **Note** In our discussion of indexing DataFrames, note that *indexing* refers to the process of selecting a subset of the data, whereas the *indices* refer to the row-level references in the DataFrame. Sometimes, "indices" refers to the parameters used in indexing a DataFrame, like the 3:5 in the sample Python list indexing list1[3:5].

Naturally, if you want to specify indices for both columns and rows, you can chain individual index commands together: table[[5,10,15]].loc[[5, 10, 15]]. However, the preferred way is to take advantage of .loc, which supports simultaneous row and column indexing in .loc[row, col] format and therefore is more efficient than chaining separate calls. The equivalent command to index both indices and rows with [10, 15] would be table.loc[[5, 10, 15], [5, 10, 15]] (Figure A-21).

	5	10	15
5	25	50	75
10	50	100	150
15	75	150	225

Figure A-21. *Specifying indices along both the column and row axes of a DataFrame*

Note that by selecting certain indices, our new table does not have "standard" indices 0, 1, 2, ... The `data.reset_index()` method pops out the original index and replaces it with a fresh "standard" index axis (Figure A-22).

	index	5	10	15
0	5	25	50	75
1	10	50	100	150
2	15	75	150	225

Figure A-22. *Resetting an index with the old index "popped out" as a new column*

To prevent the popping out of the old axis as a new column, specify `drop = True` as an argument in the `reset_index()` method (Figure A-23).

	5	10	15
0	25	50	75
1	50	100	150
2	75	150	225

Figure A-23. *Resetting an index without the old index "popped out" as a new column*

In general, to drop a column or set of columns, call `data.drop(col, axis=1, inplace=True)` or `data.drop([col1, col2, ...], axis=1, inplace=True)`. The 1 axis represents the columns, whereas the 0 axis represents the rows. If you want to drop certain rows instead, set axis = 0. The inplace argument determines whether to execute the drop command on the current object or on a copy. If `inplace` is set to `False`, the original DataFrame will not be altered, but another DataFrame with the dropped data will be returned.

■ **Note** A common mistake for beginners to Pandas is to simply call `data.drop(col, axis=1)`. This command has no effect on the data, because the default value for inplace is False. To fix this, you can call `data = data.drop(col, axis=1)`, which reassigns the `data` variable to the returned dropped DataFrame generated from the original DataFrame referenced by `data`. The preferred solution, however, is to use `inplace = True` if you do not need to preserve the original data; this handles all operations internally in a more efficient manner than generating copies and reassigning.

To index a range of columns or indices, Pandas also supports native Python indexing. However, unlike Python, Pandas includes the end index. For instance, the command `table.loc[90:100, 5:100:3]` includes all rows from indices 90 to 100 (inclusive) and all rows from 5 to 100 with step size 3 (Figure A-24).

	5	8	11	14	17	20	23	26	29	32	...	71	74	77	80	83	86	89	92	95	98
90	450	720	990	1260	1530	1800	2070	2340	2610	2880	...	6390	6660	6930	7200	7470	7740	8010	8280	8550	8820
91	455	728	1001	1274	1547	1820	2093	2366	2639	2912	...	6461	6734	7007	7280	7553	7826	8099	8372	8645	8918
92	460	736	1012	1288	1564	1840	2116	2392	2668	2944	...	6532	6808	7084	7360	7636	7912	8188	8464	8740	9016
93	465	744	1023	1302	1581	1860	2139	2418	2697	2976	...	6603	6882	7161	7440	7719	7998	8277	8556	8835	9114
94	470	752	1034	1316	1598	1880	2162	2444	2726	3008	...	6674	6956	7238	7520	7802	8084	8366	8648	8930	9212
95	475	760	1045	1330	1615	1900	2185	2470	2755	3040	...	6745	7030	7315	7600	7885	8170	8455	8740	9025	9310
96	480	768	1056	1344	1632	1920	2208	2496	2784	3072	...	6816	7104	7392	7680	7968	8256	8544	8832	9120	9408
97	485	776	1067	1358	1649	1940	2231	2522	2813	3104	...	6887	7178	7469	7760	8051	8342	8633	8924	9215	9506
98	490	784	1078	1372	1666	1960	2254	2548	2842	3136	...	6958	7252	7546	7840	8134	8428	8722	9016	9310	9604
99	495	792	1089	1386	1683	1980	2277	2574	2871	3168	...	7029	7326	7623	7920	8217	8514	8811	9108	9405	9702
100	500	800	1100	1400	1700	2000	2300	2600	2900	3200	...	7100	7400	7700	8000	8300	8600	8900	9200	9500	9800

Figure A-24. *Indexing columns and rows of a DataFrame via slicing*

Say we want to be a bit mischievous and mess up the multiplication table. We can reassign, for instance, the name of each column to a different random name. In order to do this, we need to give Pandas a mapping between an original column name and a new column name in the form of a dictionary (Listing A-18, Figure A-25). Then, call the `.rename()` method from the DataFrame and specify `columns = dictionary_mapping`. The `.rename()` method has a similar `in_place` parameter as the `.drop()` command.

Listing A-18. A "sabotage" renaming operation that randomly assigns each original column to a new column name. This is done by shuffling the column names and setting each column name to be renamed as the column name after it. The mod 100 operation allows for "wrapping around" (i.e., the very last element is renamed to the name of the very first element)

```
newCol = {}
nums = list(range(1, 101))
np.random.shuffle(nums)
for i in range(len(nums)):
    newCol[nums[i]] = nums[(i+1) % 100]
table = table.rename(columns=newCol)
```

	65	23	81	34	93	30	63	11	74	100	...	28	24	67	1	33	21	76	90	54	42
1	1	2	3	4	5	6	7	8	9	10	...	91	92	93	94	95	96	97	98	99	100
2	2	4	6	8	10	12	14	16	18	20	...	182	184	186	188	190	192	194	196	198	200
3	3	6	9	12	15	18	21	24	27	30	...	273	276	279	282	285	288	291	294	297	300
4	4	8	12	16	20	24	28	32	36	40	...	364	368	372	376	380	384	388	392	396	400
5	5	10	15	20	25	30	35	40	45	50	...	455	460	465	470	475	480	485	490	495	500
...	...	...	...	...	...	...	...	...	...	...		...	...	...	...	...	...	...	...	...	...
96	96	192	288	384	480	576	672	768	864	960	...	8736	8832	8928	9024	9120	9216	9312	9408	9504	9600
97	97	194	291	388	485	582	679	776	873	970	...	8827	8924	9021	9118	9215	9312	9409	9506	9603	9700
98	98	196	294	392	490	588	686	784	882	980	...	8918	9016	9114	9212	9310	9408	9506	9604	9702	9800
99	99	198	297	396	495	594	693	792	891	990	...	9009	9108	9207	9306	9405	9504	9603	9702	9801	9900
100	100	200	300	400	500	600	700	800	900	1000	...	9100	9200	9300	9400	9500	9600	9700	9800	9900	10000

Figure A-25. *Result of randomly renaming columns in a DataFrame*

In machine learning and deep learning, we are often interested in the scaling property of computational structures. (For a deep learning example, see Chapter 4, "Why Do We Need Convolutions?") In the case of generating multiplication tables, we may wonder how the storage needed to store the Pandas DataFrame in memory scales as we increase the dimension of the table *n* (Listing A-19, Figure A-26).

Listing A-19. Plotting out the storage scaling of the current multiplication table function

```
import sys

x = [1, 2, 4, 8, 16, 32, 64, 128, 256, 512, 1024]
y = [sys.getsizeof(makeTable(n=n))/1000 for n in tqdm(x)]

plt.figure(figsize=(10, 5), dpi=400)
plt.plot(x, y, color='black')
plt.grid()
plt.xlabel('$n$')
plt.ylabel('KB')
plt.title('Storage Size for Pandas $n \cdot n$ Multiplication DataFrames')
plt.show()
```

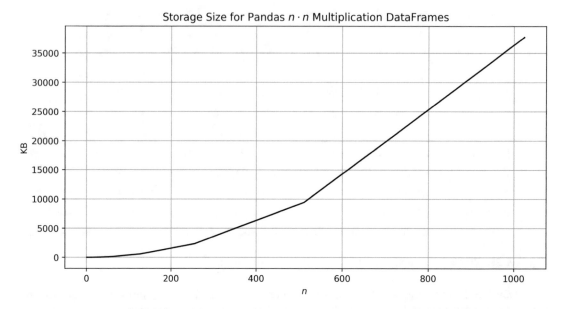

Figure A-26. *Plotting the storage size of a Pandas DataFrame used to store an n-by-n table scaling across values of n*

The storage size grows roughly quadratically, as expected. Even so, the storage size for large values of *n* becomes very large. While we cannot alter the computational complexity of the multiplication-table-building algorithm, we can generally improve the scaling by recognizing and cutting redundancies in the table.

For one, the table is symmetric about the diagonal extending from the top-left corner to the bottom-right corner (i.e., $a \cdot b = b \cdot a$). Thus, slightly less than half of the table contains duplicate information. Let's alter our table function to only fill in unique multiplication equations by only filling index-wise starting from the current column value (Listing A-20).

Listing A-20. Adapting the multiplication table function to fill in only half of the multiplication table

```
def makeHalfTable(n = 10):
    table = pd.DataFrame(index=range(1, n+1),
                         columns=range(1, n+1))
    for num1 in table.columns:
        for num2 in table.index[num1-1:]:
            table[num1][num2] = num1 * num2
    return table
```

We see that all the values that were not filled in contain NaN (Figure A-27), which are left there from initialization (created in line 2 of Listing A-20).

	1	2	3	4	5	6	7	8	9	10	...	91	92	93	94	95	96	97	98	99	100
1	1	NaN	NaN	NaN	NaN	NaN	NaN	NaN	NaN	NaN	...	NaN	NaN	NaN	NaN	NaN	NaN	NaN	NaN	NaN	NaN
2	2	4	NaN	NaN	NaN	NaN	NaN	NaN	NaN	NaN	...	NaN	NaN	NaN	NaN	NaN	NaN	NaN	NaN	NaN	NaN
3	3	6	9	NaN	NaN	NaN	NaN	NaN	NaN	NaN	...	NaN	NaN	NaN	NaN	NaN	NaN	NaN	NaN	NaN	NaN
4	4	8	12	16	NaN	NaN	NaN	NaN	NaN	NaN	...	NaN	NaN	NaN	NaN	NaN	NaN	NaN	NaN	NaN	NaN
5	5	10	15	20	25	NaN	NaN	NaN	NaN	NaN	...	NaN	NaN	NaN	NaN	NaN	NaN	NaN	NaN	NaN	NaN
...	...	...	...	...	...	...	...	...	...	...	...	...	...	...	...	...	...	...	...	...	...
96	96	192	288	384	480	576	672	768	864	960	...	8736	8832	8928	9024	9120	9216	NaN	NaN	NaN	NaN
97	97	194	291	388	485	582	679	776	873	970	...	8827	8924	9021	9118	9215	9312	9409	NaN	NaN	NaN
98	98	196	294	392	490	588	686	784	882	980	...	8918	9016	9114	9212	9310	9408	9506	9604	NaN	NaN
99	99	198	297	396	495	594	693	792	891	990	...	9009	9108	9207	9306	9405	9504	9603	9702	9801	NaN
100	100	200	300	400	500	600	700	800	900	1000	...	9100	9200	9300	9400	9500	9600	9700	9800	9900	10000

Figure A-27. *Result of filling in only half of the multiplication table*

Perhaps surprisingly, the scaling of this half-table method is only negligibly better than the complete table (Figure A-28).

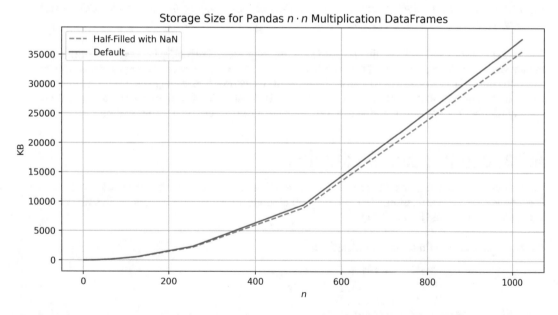

Figure A-28. *Plotting the storage size of a Pandas DataFrame used to store an n-by-n table scaling across values of n, default method vs. half-filling method*

However, if we replace all the np.nan's with 0s, we obtain pretty substantial storage savings. By returning table.fillna(0) – which, as the syntax naming suggests, fills all NA/null/NaN values with 0 – we get a much more lightweight scaling multiplication table generator (Figure A-29).

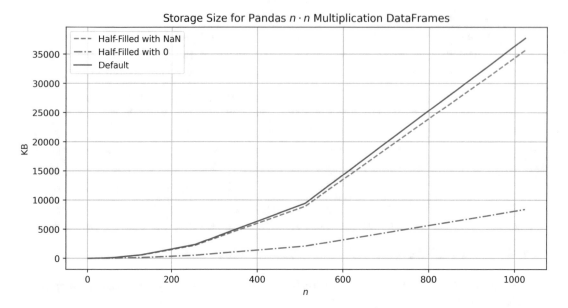

Figure A-29. *Plotting the storage size of a Pandas DataFrame used to store an n-by-n table scaling across values of n, default method vs. half-filling method*

One simplistic explanation is that np.nan is a "bulky higher-level object," whereas 0 is a "primitive Python value"; thus, it makes sense intuitively that Python can handle storage of a large number of 0s more efficiently than a large number of np.nan's. There are, of course, many more neglected complex low-level details that contribute to the optimization of storage and efficiency. This demonstration, however, shows some quick high-level approaches that can be used to cut redundancy and improve scaling.

Advanced Pandas Mechanics

Pandas offers several functions that offer advanced functionality for manipulating the contents of DataFrames. Let's create a dummy DataFrame to manipulate throughout this section to illustrate the various manipulation functions (Listing A-21).

Listing A-21. Constructing a dummy DataFrame

```
construct_dict = {'foo': ['A']*3 + ['B']*3,
                  'bar': ['I', 'II', 'III']*2,
                  'baz': range(1, 7)}
dummy_df = pd.DataFrame(construct_dict)
```

The contents of such a DataFrame are displayed in Table A-3.

Table A-3. *Contents of the sample DataFrame created in Listing A-21*

	foo	bar	baz
0	A	I	1
1	A	II	2
2	A	III	3
3	B	I	4
4	B	II	5
5	B	III	6

Pivot

A pivot operation projects two existing columns in the data as the axes of a new pivoted table (the index and the column). Then, it fills in the pivoted table's elements with values from a third or more columns that correspond to a particular combination of the first two columns. Consider the scheme in Figure A-30: we set "foo" and "bar" to be the *index* and *column* of the new pivoted table (respectively) and use "baz" to fill in *values* of the table. Because "baz" is 1 when "foo" is "A" and "bar" is "I", the element at index "A" and column "I" is 1.

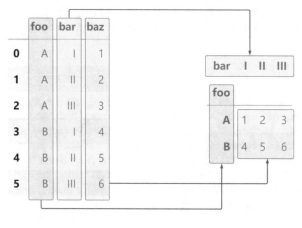

Original DataFrame   Pivoted DataFrame

Figure A-30. *Visualizing the pivot operation*

Pivoting can be implemented as df.pivot(index=..., columns=..., values=...). In Figure A-30, we use the command dummy_df.pivot(index='foo', columns='bar', values='baz').

If you pass in more than one column to the values parameter via a list (Listing A-22), Pandas creates a multilevel column to accommodate different values for combinations of the index and column features (Figure A-31).

Listing A-22. Code to perform a pivot

```
mod_dummy_df = dummy_df.copy()
mod_dummy_df['baz2'] = range(101, 107)
mod_dummy_df.pivot(index='foo', columns='bar',
                     values=['baz','baz2'])
```

	foo	bar	baz	baz2
0	A	I	1	101
1	A	II	2	102
2	A	III	3	103
3	B	I	4	104
4	B	II	5	105
5	B	III	6	106

	baz			baz2		
bar	I	II	III	I	II	III
foo						
A	1	2	3	101	102	103
B	4	5	6	104	105	106

Figure A-31. *Visualizing the pivot operation on Pandas DataFrames*

If there are multiple entries for the same combination of index and column features, Pandas will (as one may expect) throw an error: "Index complains duplicate entries, cannot reshape".

Pivoting is a convenient operation for automatically finding values at the combination of two features.

Melt

Melting can be thought of as an "unpivot." It converts matrix-based data (i.e., the index and column are *both* "significant," instead of the index serving as a counter) into list-based data, whereas pivots do the opposite. You can think of the operation as "melting" away the rigid, organized structure of a matrix into a primitive stream of data, like complex ice sculptures melt into elementary puddles of water.

Consider a melting operation with an *ID feature* "baz" and *value features* "foo" and "bar" (Figure A-32). Two columns in the melted DataFrame are created: "variable" and "value". "value" holds the value stored by the feature name referenced in the "variable" column in the original dataset. The "baz" ID column is used to keep track of which row the melted variable-value pairs belong to.

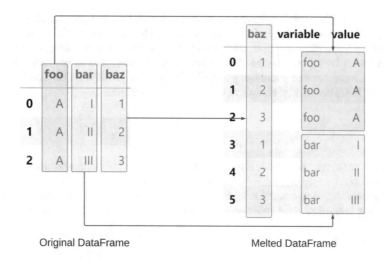

Figure A-32. *Visualizing the melting operation*

Melting can be implemented as `df.melt(id_vars=["baz"], value_vars=["foo", "bar"])`.

Explode

Exploding is an operation that separates columns containing lists into a melted-style form. Consider the schematic in Figure A-33: the original DataFrame contains a list [1, 2] in index 0, where "foo" has value "A", "bar" has value "I", and "baz" has value 1. After the list is exploded, the exploded DataFrame will contain two entries at index 0, with the same values for non-exploded columns ("foo", "bar", "baz") but separate values corresponding to items in the list (1 in one row, 2 in the other).

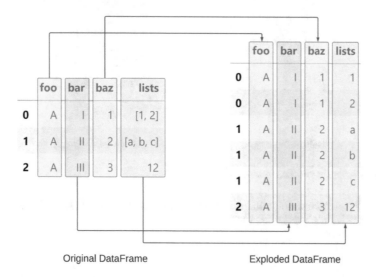

Figure A-33. *Visualizing the exploding operation*

It's generally unlikely that you'll encounter a raw dataset with lists as a column. However, knowing that the explode function exists can be helpful when you're artificially constructing DataFrames; rather than writing code to create a specific organization of elements, just create a column with relevant list values and explode it. While the end result may be the same, it becomes much simpler for you to implement.

Exploding can be implemented as df.explode('column_name'). You can also pass in a list of column names to explode if there are multiple columns containing list elements.

Stack

The stack operation converts a two-dimensional Pandas DataFrame into a DataFrame with multilevel indices and one column by rearranging the elements into "vertical" form.

In the stacking operation visualized in Figure A-34, the first row value in the column "foo" is rearranged as being the first row in the first-level index and the "foo" row in the second-level index under the column 0.

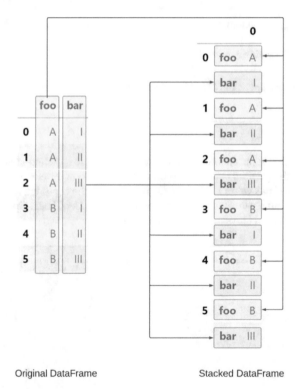

Figure A-34. *Visualizing the stacking operation*

To access multilevel data, simply call index-based retrieval twice, like df.loc[0].loc['foo']. The code to stack is simply df.stack().

Unstack

True to its name, the unstacking operation acts as an inverse to the stack operation, converting a stacked-style DataFrame with multilevel indices into a standard two-dimensional DataFrame. Performing a stack followed by an unstack operation produces no change in the DataFrame, except for the existence of a "0" introduced during stacking (Figure A-35).

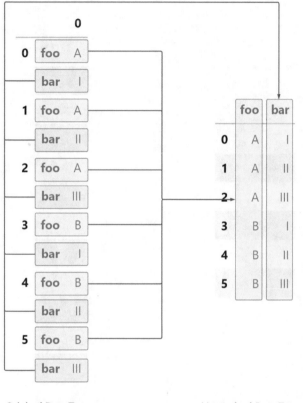

Original DataFrame   Unstacked DataFrame

Figure A-35. *Visualizing the unstacking operation*

The code to unstack is simply `df.unstack()`.

Conclusion

In this appendix, we took a tour through the NumPy and Pandas libraries. As you will see or have already seen, these objects constitute the basis through which we allow data to interact with models, such as in the scikit-learn and Keras/TensorFlow libraries. Happy coding!

Index

A

Accuracy, 43, 56, 164, 167, 222, 227, 325, 344, 503, 590, 639
Activation functions, 609
    gradient descent, 203
    hyperbolic tangent (tanh), 204
    key improvements, 203
    LeakyReLU, 205, 206
    linear functions, 203
    nonlinear functions, 203
    nonlinearity/variability, 208–212
    rectified linear unit, 205
    scaled linear activation, 207
    SELU, 253, 254
    sigmoid function, 204
    swish function, 206–208
Activation Maximization, 788–791
AdaBoostClassifier, 711
Adaptive Gradient Boosting (AdaBoost), 85, 86
Adaptive Moment Estimation (Adam), 219, 220
Adaptive Relation Modeling Network (ARM-Net)
    adaptive relation modeling module, 542
    benchmark models, 546
    definition, 542
    module, 543, 544
    PyTorch, 546
Adult Census dataset, 774
Aggregating/ensembling models, 753
ak.StructuredDataBlock, 748
AlexNet, 307–310
Alignment-score-computing network, 454
Alpha value, 468, 469, 473, 659
Ames Housing dataset, 104, 116, 147, 512, 738, 739, 751
applyKernel function, 277
Artificial neural network (ANN), 184, 185, 192, 257, 281, 542, 792
Asymmetric convolutions, 319
Attention-based tabular modeling, 500

Attention mechanism
    Bahdanau-style attention, 467–472
    BERT, 461, 463, 464
    context vector, 453
    deep learning models, 547, 548
    definition, 452
    GELU, 461, 462
    Keras
        attention scores, 475
        bidirectional model, 476, 477
        bidirectional recurrent model, 480
        implementation, 472
        multi-head attention model, 481
        synthetic dataset, 474
        unidirectional model, 478
    LSTM, 456, 457, 465, 466
    multimodal, 546
    natural language models, 485
    sequence-to-sequence dataset, 481–484
    SHA-RNN model, 465, 466
    text sequences, 452, 453
    transform architecture, 457, 458, 460
Audio files, 411–413
Audio model, 414
Autoencoders, 606, *See also* Vanilla
    abstractions, 605
    architecture, 602, 603
    decoding, 602
    denoising, 664–680
    encoding, 602
    encrypted data, 602
    image-to-text encoding, 605
    internal capabilities, 601
    language representation, 601
    natural language, 605, 606
    nontrivial patterns, 603
    pretraining, 631–640
    reparative, 664–680
    sending/receiving, 602
    sparse, 653–665
    tabular data, 606

AutoInt model, 499
AutoKeras model, 747–752, 785
AutoKeras NAS algorithm, 748
Auto-ML, 711
AverageEnsemble class, 756
Average weighting, 753–765
Averaging ensemble, 754–756

B

Backpropagation process, 183, 188, 385
Backpropagation Through Time (BPTT), 384–388
Bag of Words (BoW) model, 130–132, 138, 178
Bahdanau-style attention layer, 455, 467, 473
Batches, 98, 216, 225, 247, 398, 429, 704
Batch normalization, 225–227, 388, 516, 639
Bayesian optimization, 715, 718, 747, 752, 755
BERT architecture, 461, 501, 535, 547
Bidirectionality, 396, 397, 477, 482
Binary classification problem, 360, 609, 726, 755
Binary cross-entropy (BCE), 68, 195, 200, 536, 609, 625, 642, 731, 768
Black box, 713–716, 779, 784, 785
Blank canvas, 436, 805
Boosting gradient, 83
    AdaBoost, 85, 86
    LightGBM, 88–90
    regression/classification, 83
    sigmoid function, 85
    theory/intuition, 83–85
    visualization, 84
    XGBoost, 86–88
buildAutoencoder function, 616, 633, 655
Builder, 614

C

CatNN, 591, 592, 596, 597
Cell state, 391, 402, 445, 452
Classical machine learning model, 126, 134, 713, 725
Classification and Regression Trees (CART), 73
Click-through rate (CTR) prediction problems, 499
Cluster, 7, 71, 159, 503, 552, 623
Code snippet, 398
Colon-bracket syntax, 799
Compartmentalized form, 611, 612
Compressed Sparse Row (CSR), 101
Conditional Tabular GAN (CTGAN), 701, 710
    anonymize_field option, 708
    approaches, 709
    conditional generation, 709
    official documentation, 709
    official implementation, 706
    performance, 707

simple CTGAN demo, 706
synthetic data generated, 707
tabular data generation, 702
Confusion Matrix, 42
Constructor, 97, 256, 614, 728, 813
Contiguous semantics, 348, 356, 424–426, 439, 449
Continuous vote, 763
Conveyor belt, 389
Convolutional component, 286, 307, 334, 357
Convolutional neural networks (CNNs)
    ANN, 268
    architectures, 259, 267
    build architecture, 266
    EfficientNet, 323–326
    flattened MNIST dataset, 262
    heatmap, 260, 261
    Inception v3
        definition, 318
        expanded filter bank module, 320
        ImageNet, 322
        Keras, 322, 323
        module A/B, 319
        modules, 321
    MNIST digits dataset, 260
    parameters, 263, 264, 266
    pooling operation, 292
    reshaping MNIST dataset, 261
    ResNet, 311
    submodels, 378
    sweep, 378
    validation performance modeling, 265
Convolution operation
    8-by-8 uniform blurring kernel, 275
    first/second layer kernel, 290
    flattening layer, 284
    fourth layer kernel, 292
    grayscale images, 282
    identity kernel, 272
    implementation, 281
    input shape, 282
    kernel, 269, 271
    loss/accuracy history, 287
    low dimensional image, 270
    MNIST dataset, 286
    operation, kernel, 268
    parameters, 283, 285
    peek.predict, 288
    pixel-by-pixel image, 276
    stacking, 282
    3-by-3 blurring kernel, 277–280
    3-by-3 sharpening kernel, 276, 281
    training/validation performance, 286
    2-by-2 uniform blurring kernel, 274
Convolutions, 260–268, 281, 288, 293, 339, 356, 414
Convolution windows, 432

Corrupted image, 669
Cross-validation, 380
Custom loss, 689
cv2 function, 271

■ D

Data generation algorithms, 681
Data leakage, 380, 421
Data pipeline, 738
Data points, 380
Data preparation/preprocessing
    bigrams, 132
    BoW model, 130–132
    categorical encoding methods, 104
    components, 96
    continuous quantitative data, 119
    deep sentiment analysis extraction, 137
    elementary operations, 149
    encoding process, 104–119
    engineering (*see* Engineering techniques)
    entropy, 163
    extraction/engineering methods, 179
    feature extraction, 145
    feature selection, 163
    few-shot learning scheme, 136
    geographical data, 144
    high-correlation method, 167–170
    Information Gain, 163–165
    key components, 177–179
    keyword search function, 127
    LASSO coefficient selection, 175–177
    linear discriminant analysis, 159–161
    min-max scaling, 120, 121
    natural language processing framework, 136
    n-gram data, 133, 134
    non-invertible function, 148
    normalized distribution, 124
    one-shot learning scheme, 136
    original *vs.* robust-scaled distribution, 123
    partial data destruction, 148
    PCA (*see* Principal Component Analysis (PCA))
    permutation importance, 173–175
    quantitative representation, 126
    raw vectorization, 128–130
    recursive feature elimination, 170–173
    robust scaling, 121–125
    selection methods, 179
    sentiment extraction implementation, 134–137
    single/multi-feature transformations, 149–155
    sparsity, 109
    standardization, 125, 126
    storage/manipulation (*see* TensorFlow
        datasets)
    supervised learning scheme, 136

    TARS model, 137
    text data, 126, 127
    TF-IDF encoding, 133, 134
    time/temporal data, 141–144
    train/validation splitting, 127
    transformations, 95
    t-SNE, 156–159
    vaderSentiment library, 135
    variances, 165–167
    Word2Vec, 138–141
    zero-shot learning scheme, 137
Decision trees, 725
    advantages, 74
    CART, 73
    features, 74, 75
    Gini Impurity/Entropy, 75
    high-variance intuition, 82
    implementation, 76–83
    leaf nodes, 73, 74
    mitigate variance intuition, 82
    node class, 78
    prediction functions, 79
    random forest, 80–83
    scikit-learn implementation, 79
decoder.predict(…), 612
Decoding, 459, 547, 602, 603
Deep Double Descent theory, 91, 92
DeepGBM, 591, 598
DeepInsight, 358–367, 377
Deep Jointly Informed Neural Network
    (DJINN), 564
Deep learning (DL), 3, 49, 91, 95, 183, 419, 713
    Keras, 189
    wide model, 250–253
Deep neural decision trees (DNDTs), 550
Default method *vs.* half-filling method, 821
Denoising autoencoders
    application, 665
    blanket fashion, 679
    MNIST data, 666, 669, 670
    MNIST images, 673, 675, 677
    noise level, 678
    noisy images, 665, 669
    noisy input, 665
    noisy/perturbed input, 670, 672
    performance, 674
    predicted output, 670, 672
    random noise, 666, 671
    reflective standard deviation, 679, 680
    reparative models, 666
    sample of images, 672, 676
    tabular data, 678
    triviality, 666
    unperturbed desired output, 670, 672
    validation, 669

Denoising loss, 536
Dense-convolution-recurrent model, 434
DenseNet, 317
Digit recognition, 677
Direct tabular attention modeling, 496–499
Direct tabular recurrent modeling, 424
Discriminator, 697, 703
Disjunctive normal form (DNF), 574, 575, 577, 597
Disjunctive normal neural form (DNNF) block, 575
Distillation
    DeepGBM
        components, 591, 592, 595
        definition, 591
        embeddings, 595
        GBDT, 594
        notations, 593
        training, 596
    definition, 591
Dow Jones Industrial Average (DJIA), 419
.drop() command, 818

E

EfficientNetB1 model, 336
EfficientNet models, 325, 326, 334, 636
Element-wise function, 797
Embedding mechanisms, 334, 594
Embeddings, 405, 407, 418
Embedding vector, 406, 407, 500
encoder.predict(...), 612
Encoding process, 602, 603
    Ames Housing dataset, 105
    binary representation, 109–111
    discrete data, 104, 105
    enumerate() function, 106
    frequency, 111–113
    James-Stein encoding, 116–118
    label encoding models, 105–107
    leave-one-out encoding, 115, 116
    multicollinearity, 108
    numerical representation, 104
    one-hot representation, 107–109
    sparsity/multicollinearity, 109
    target representation, 113–115
    TF-IDF representation, 133, 134
    WoE technique, 118, 119
Engineering techniques
    distribution, 162
    extraction methods, 179
    feature extraction, 161, 162
    homogenous, 162
    PCA, 151
    researches, 149
    statistics, 162
Entire GAN model, 704

Epidemic forecasting, 411
Epochs, 629
Exclusive Feature Bundling (EFB), 88
explain_instance method, 786
Exploding, 825, 826
Extreme Gradient Boosting (XGBoost), 49

F

False positive Rate (FPR), 46, 48
False validation scores, 380
Fast Signed Gradient Method (FSGM), 663
Feature matching, 699
Feed-forward (FF) layers, 532
Feed-forward operation, 186–188, 257, 384
    activation (*see* Activation functions)
    feature feeds, 201
    hidden layer, 201
    notations, 202
First-order optimization approaches, 580
Flexibility, 45, 170, 177, 397, 554, 564
Flow-based malware detection, 356
Force plot visualization, 775, 776
Forget gate, 389, 390
Fully connected component, 286
Function vectorization, 800, 801

G

Gated linear unit activation (GLU), 516
Gated recurrent unit (GRU) network
    components, 396
    forget and input gates, 392
    hidden state, 393
    hyperbolic tangent activation, 395
    LSTMs, 392
    representation, 393
    structure/complexity, 395
    update gate, 394
Gauging model, 767
Gaussian Error Linear Unit (GELU) activation,
    461, 532
Generative Adversarial Network (GAN), 697, 710
    conditional, 700, 701
    formal training algorithm, 698
    generator model, 701
    gradients, 698
    image generation, 701
    optimization procedure, 699
    schematic, 698
    simple GAN, TensorFlow, 702–705
    techniques, 699
    theory, 697–702
Generator, 140, 399, 698
Geographical data, 144

get_alpha() method, 471
Global pooling, 305–307
Gradient-based methods, 550, 713
Gradient-Based One-Side Sampling (GOSS), 88
Gradient Boosting, 83–90, 725
GradientBoostingClassifier, 711
Gradient Boosting Decision Tree (GBDT), 502
Gradient Boosting Machines (GBMs), 83, 550, 579, 597
Gradient Boosting neural networks (GrowNet)
    ANNs, 579
    architecture, 580
    definition, 579
    GBMs, 579
    package, 583
    training procedures, 581
    train/predict methods, 584
    XGBoost, 582
Gradient explanations, 780
Gradients
    exploding, 388–392
    vanishing, 384–391, 393
Grayscale representation, 804, 805
Grid of values, 431
Grid search, 714
GrowNet, 579–584, 597

H

Hadamard product, 395
Halving, 614–616
Helper function, 412, 597
Hidden states, 384, 385, 388, 390, 392, 393, 395, 396,
    401, 407, 436, 437, 445, 447
Hierarchical Data Format (h5), 102
Higgs Boson dataset, 611, 637, 638, 691, 725
High-correlation method, 168–171
High precision, 20, 96, 100, 132, 432
Historical averaging, 699
Horizontal rule, 777
Human-designed compression, 603
Human voice audio, 412
Hyperbolic tangent (tanh), 204
Hyperopt, 711, 715, 718–726, 728, 729, 732, 738, 741,
    742, 747, 757
Hyperoptimization, 514, 757
Hyperparameter optimization tools, 713
Hypothetical objective function, 715

I, J

Identity kernel, 272
Image Generation for Tabular Data (IGTD), 357, 367
Inception v3 model, 318
Indexing commands, 796
Information compression factor, 292

Information gain, 163–165
Information-rich signals, 435
Initial memory states
    architecture, 443, 444
    dual linkage, 438
    feed-forward layer, 448
    hidden state, 441
    hidden-state-learned tabular recurrent
        model, 438
    LSTMs, 445–447
    optimal transformation, 436
    recurrent layer, 436
    schematic diagram, 442
    semantics, 439
    sequence, 441
    sinusoidal position encodings, 436, 440
    tabular data, 448
    tabular input, 436, 437
    timesteps, 436
Input-informed ensemble model architecture, 766
Input-informed weighting, 765–767
Interactive visualization, 776
Internal thinking processes, 699
Interpretation, 27, 145, 200, 315, 632
Intertwined loops, 385

K

Keras, 183, 607, 608, 636, 650, 652, 655
    batch sizes, 397
    bidirectional LSTM, 400
    hidden layer, 409
    input shape, 397
    multiple layers, 408
    preprocessing data, 397
    real-time forecasting, 400
    regularization learning networks, 255
    syntax, 408
    tabular data prediction, 397
Keras application models, 316
Keras library
    customization, 189
    deeper dive
        batch normalization, 226–231
        callbacks, 223–226
        concatenation, 233
        dropouts, 229
        early stopping, 223
        embedding layer, 239
        evaluation phase, 226
        functional API, 230, 231
        hidden layers, 222, 229
        inserting activation, 221
        model architecture, 221, 238
        model checkpoint, 223

Keras library (*cont.*)
    multi-input/multi-output models, 235–238
    nonlinear topology models, 231–235
    parameters, 234
    plotted model diagram, 235
    summary() method, 234
    validation, 222–225
    weight sharing intuition, 240–243
  modeling
    architecture, 192–195
    dense layers, 193
    Fashion MNIST dataset, 189–192
    imshow() function, 190
    model compilation, 195
    sequential model, 192
    softmax, 194
    source code, 194
    training process, 196, 197
    2D array, 192
    workflow, 192
  ONEIROS project, 189
  TensorFlow installation, 189
Keras neural network, 197, 774
keras.utils.plot_model, 263
Kernel-PCA, 363
K-Nearest Neighbors (KNN), 50
  advantages/disadvantages, 90
  algorithm concept, 50
  boosting gradient, 83–89
  decision trees, 73
  Elbow method, 57
  high-correlation method, 168
  implementation, 54–57
  linear regression, 57–66
  logistic regression, 66–73
  Minkowski distance, 55
  NumPy array, 54
  one-hot encoding, 108
  Scikit-learn implementation, 56
  single test data, 54, 55
  standardization, 53
  theory/intuition
    Chebyshev distance, 53
    Euclidian distance, 52, 53
    features, 51
    Manhattan distance, 51
    Minkowski distance, 51
    normalization, 53
    visualization, 51

L

Labels, 6, 44, 156, 254, 397, 412, 584
Label smoothing, 700
Layer freezing, 632, 633

Leaky gates, 248–250
Least Absolute Shrinkage and Selection Operator
    (LASSO), 65, 175–177
Least Square Theory, 57
__len__ and a __getitem__ method, 559
Light Gradient Boosting Machines (LightGBM),
    49, 88–90
Linear Discriminant Analysis (LDA), 159–161
Linear mappings, 431
Linear output activation, 330, 628
Linear regression models, 57–66, 186, 727, 755
  ElasticNet regression, 65
  gradient descent, 64
  implementation
    deterministic results, 62
    gradient descent, 62
    Scikit-learn's implementation, 63
    source code, 61
  LASSO regression, 65
  regularization, 66
  ridge regression, 65
  theory/intuition
    chain rule/derivatives, 60
    composite function, 60
    equation, 60
    estimation line/data point, 59
    explanatory variable, 58
    gradient descent, 59–61
    slope-intercept form, 58
    variables, 58
    weight values, 59
  variations, 63–66
Local Interpretable Model-Agnostic Explanations
    (LIME), 784
  initialize, 786
  matplotlib, 788
  validation dataset, 787
Logistic regression models, 250, 725, 767
  binary classification, 67
  binary cross-entropy (BCE), 68
  classification, 66
  derivatives, 69
  gradient descent, 68, 70, 71
  implementation, 70–72
  scikit-learn, 72
  sigmoid function, 67, 68
  softmax function, 73
  theory/intuition, 66–70
  variations, 72
Long short-term memory (LSTM), 388–392, 452
  cell, 389
  cell state, 390, 446
  diagram, 400
  gate handles, 393
  memories, 390

timestamp, 395
vanishing gradient, 392
Long-term memories, 388, 390, 391
Loss functions, 197–201

M

Machine learning (ML), 3, 681
  accuracy, 43
  algorithms, 49
  applications, 91
  confusion matrix, 42
  Deep Double Descent phenomenon, 92
  F1 score/F-beta implementation, 45
  KNN (see K-Nearest Neighbors (KNN))
  MAE, 38–40
  MSE, 40, 41
  metrics/evaluation, 38
  modeling (see Modeling)
  neural networks, 90, 183
  overfitting-underfitting paradigm, 91
  precision, 43, 44
  recall implementation, 45
  ROC-AUC implementation, 48–51
  Scikit-learn implementation, 41
Manual method, 800
Many-to-many prediction, 386, 392
Mapping, 405, 413
Marginal contributions, 774
Masked language modeling (MLM), 138, 463, 501,
    535, 547, 701
Mean Absolute Error (MAE), 38–40, 209, 671, 674
Mean-based grayscale
    representation, 804
Mean squared error (MSE), 35, 40, 41, 59, 399, 581,
    611, 669
Medical diagnosis datasets, 681
Melting, 824, 825
Memory cells, 381–384
Memory-equipped recurrent models, 411
Memory neuron, 383
Meta-evaluation model, 769
Meta-learning, 711
Meta-model architecture, 764, 766, 769
Meta-model error estimator, 768
Meta-nonlinearities, 424
Meta-optimization, 711, 714, 715, 740, 747, 752
  components, 712
  controller model/controlled model, 712
  meta-parameters, 712
  meta-parameter space, 712
  objective function, 712
  optimization procedure, 712
Meta-overfitting, 747
Meta-parameters, 264, 342, 351, 711, 725

Mice Protein Expression dataset, 628, 629, 636, 646,
    661, 679, 779
min(), 628
Mini-batch stochastic gradient descent
    (SGD), 216–218
Min-max scaling, 120, 121
MNIST dataset, 607, 635
Model accuracy, 397
Model agnostic, 771
Modeling
  approximations/representations, 4
  automated modeling, 5
  bias-variance trade-off
    approaches, 19
    bias/variance errors, 22
    data points, 19, 20
    decomposition, 21
    high/low-variance, 21
    representation, 21
    training data, 20
    underfitting vs. overfitting, 22
  clustering algorithm, 7
  dimensionality reduction algorithms, 7
  domains and levels, 6
  equivalent binary representations, 9
  facial recognition, 6
  fashion model's, 4
  feature space representation
    Chebyshev distance, 30
    count proportion, 28
    distance, 30
    Euclidean distance, 28
    geometric shapes, 29
    hypercubes, 24, 25
    hyperplane separation, 23, 24
    one dimensional code, 25
    plotting code, 29
    proportion, 27
    ratio calculation, 27
    three-dimensional code, 22, 26
    two-dimensional code, 25
  fundamental principles, 4
  learning, 5
  ML data cycle
    alignment/misalignment, 10
    DataFrame, 13
    data leakage, 16, 17
    dataset process, 15
    data type hierarchy, 11
    dominant definitions, 11
    dummy dataset construction, 15
    ecommerce platform, 11, 12
    feature set and a label set, 14
    k-fold evaluation, 17–19
    logical structures, 15

Modeling (*cont.*)
    manual implementation, 16
    memorization/genuine learning, 11
    phenomena/concepts, 10
    random seeding, 17
    train/validation dataset, 10–14
  modes, 5
  optimization/gradient descent
    beta parameter, 36
    global minimum, 32
    gradient descent, 33, 34
    landscape model, 32
    linear regression, 35
    loss functions, 31
    mean squared error, 35
    optimization process, 37
    optimizers, 33
    parameters, 31
    prediction and gradient functions, 36
    sequences, 31
  phenomena, 4
  quantitative conversion/representation, 8, 9
  regression/classification problems, 9
  scientific model, 4
  steady-state approximation, 4
  supervised learning, 5
  unsupervised learning, 6–8
Modern language models, 451, 464
Multi-head attention mechanism, 480, 545
Multi-head attention model, 484
Multi-head intersample self-attention
    mechanism, 532–534
Multi-head self-attention (MSA), 459, 499, 507, 532
Multi-input function, 801
Multimodal
  applications, 326
  components, 331
  definition, 326
  EfficientNetB1, 336, 339
  embedding mechanism, 334
  improvements, 334
  Keras visualization, 332, 337
  requests data, 328
  structure, 327
  tabular dataset, 331
  TensorFlow Sequence dataset, 328–330, 335–336
  training performance, 338
Multimodal recurrent modeling, 418, 419
  compiling and fitting the model, 423
  convolutional model, 433
  convolutions, 430
  embeddings, 418
  high precision/nonlinearity, 432
  multimodal stock data, 420
  recurrent layers, 431

  relevant components, 417
  sequential input, 417
  software reviews dataset, 417
  standard feature vector, 432
  standard feed-forward fashion, 416
  tabular inputs, 417, 435
  temporal relationship, 431
  time series, 419
  training and validation, 421
  usage, 417
  vectorizing, 418
Multimodal text, 421–423
Multi-model arrangement, 753–770
Multi-modeling ensemble, 767
Multipart function, 800
Multiple aggregations, 382
Multi-stack recurrent model, 437
Multitask autoencoders, 770
  absolute error, 647
  α-adjusting curve, 651
  autoencoding task, 650
  compiling stage, 651
  decoders, 640, 641
  dimensions of performance, 643
  dynamic fashion, 641
  encoder, 641
  epochs, 643–646, 648, 650
  learning, 641
  MNIST dataset, 641, 642, 646
  original task model, 640
  outputs, 642
  recompiling and fitting, 651
  reconstruction loss, 652
  reconstruction task, 650
  refitting, 652
  sigmoid equation, 650
  sigmoid function, 651
  training regime, 651
  visualization, 642

■ N

"Naïve" meta-optimization algorithms/
    procedures, 714
NAS-discovered architecture, 326, 752
Natural language models, 403–409
  architecture, 491
  double-LSTM, 485
  feature maps, 496
  multi-head attention approach, 488, 489
  multimodal dataset, 490
  recurrent language models, 490
  score matrix, 492, 493, 495
  submodel, 486, 487
  TripAdvisor dataset, 485

Natural language processing, 136, 266, 464, 500, 635
Nested search space, 727, 741
Nesterov accelerated gradient (NAG), 218, 219
Neural Architecture Search (NAS), 564, 711, 713,
    738, 747, 750, 752
Neural network model, 663, 730, 732, 757, 771
    backpropagation, 257
    backpropagation algorithm, 213–216
    backpropagation process, 183, 188
    feed-forward (see Feed-forward operation)
    gradient descent, 212, 213
    hard-coded algorithms, 183
    Keras (see Keras library)
    loss functions, 197–201
    multilayer perceptron model, 186–188
    optimizers
        Adam optimizer, 219, 220
        adaptive methods, 220
        backpropagation, 216
        gradient descent, 216
        mini-batch SGD, 216–218
        Nesterov accelerated gradient, 218, 219
    perceptron model, 184–186
    regularization, 254–257
    research papers, 246
    self-normalizing model, 253
    SELU activation function, 253, 254
    simple illustration, 184
    single perceptron, 185, 186
    tabular data (see Tabular neural networks)
    tabular deep learning, 258
    Universal Approximation Theorem, 242–245
    variations, 185
Neural network search structure, 748
Next sentence prediction (NSP), 463
No-gradient optimization, 713–425
Noisy image, 665, 805, 806
Noisy normal distribution, 670, 674, 676, 678
Nonlinearity, 202–205, 208–212, 432
Nonlinear topology models, 231–235, 331
Nontemporal relationship, 431
Nontraditional usage, 424
Nontrivial task, 676
Normal distribution, 671, 719
Novel modeling paradigm, 424, 425
NumPy arrays, 398, 399, 402, 412
    advanced indexing, 798, 799
    construction, 793
    data types, 799
    functions, 800, 801
    image manipulation, 801–803, 806–808,
        810, 811
    indexing, 795, 796
    manipulate, 797
    reassignment, 798, 799

    reshape, 794, 795
NumPy functions, 798

O

Oblivious decision tree, 561–563
One-dimensional array, 794
One-dimensional convolutions, tabular data, 414,
    424, 432, 758
    architectural components, 352
    contiguous semantics, 348
    custom function, 340
    cybersecurity, 356
    encoding component, 349
    identification synthetic dataset, 342
    kernel, 339, 340
    loss/accuracy, 344
    measurements, 350
    meta-parameters, 342, 351
    noise standard deviation, 345–346
    numElements elements, 340
    performance history, 355
    powerful/sophisticated soft ordering, 352–353
    training/validation datasets, 341
Open-ended Neuroelectronic Intelligent Robot
    Operating System (ONEIROS), 189
Optimal neural network architecture, 713, 748
Ordinary Least Squares (OLS), 63, 212
OR operation, 811, 812

P, Q

Pandas DataFrame
    advanced mechanics, 822
    dummy, 812, 813
    indexing single column, 813
    mechanics
        indices, 817
        multiplication, 814–816, 819, 820
        random renaming, 819
        resetting index, 817
        slicing, 818
        storage size, 820, 822
    NumPy array, 813
Pattern-based arrays, 793
Pearson's Correlation Coefficient, 168–171, 375
Peek model, 289
Performance distributions, 753, 765
Performance value, 635
Physics dynamics, 678
Piecewise function, 800
Pivot operation, 823, 824
Pixel values, 8, 260, 268, 610, 666, 805
Pixel-wise reconstruction, 642
Plain linear activation, 609

Political sentiment, 411
Pooling operation
    AlexNet architecture, 307–309
    continual convolution stacking, 294–295
    global pooling, 305, 306
    information compression factor, 292
    neural network, 302
    parameter count scales, 294
    parameter scaling, 303–305
    pooled matrix, 298
    scaling capability, 293
    self-extending, 295–297
    two dimensional max, 299–301
Predicted class, 786
Prediction task, 141, 385, 397, 407, 557
Preprocessing pipeline, see Data preparation/
        preprocessing
Pretraining
    autoencoder logic, 637
    computer vision, 635, 636
    datasets, 631
    feature extractor, 638
    fine-tuning, 632–634
    frozen encoder/feature extractor, 633, 634
    large-scale image classification, 635
    latent space size, 637
    layer freezing, 632
    MNIST dataset, 632
    multistage, 632
    multitask, 640–652
    natural language processing, 635
    semi-supervised method, 639
    supervised model training, 634, 635
    supervised target, 640
    supervised task, 636, 637
    task model, 633
    validation and training loss curves, 639
Principal Component Analysis (PCA)
    advantages, 151
    approaches, 154
    components, 152, 155
    diagonal axis, 153
    disadvantages, 154
    dummy dataset, 152
    engineering/extraction, 154
    features extraction, 155
    real-world representation, 152
    rotated data relative, 153
    scree plot, 154
    source code, 155
    visualization, 156
Probabilistic surrogate functions, 717
pyplot method, 788
Python code, 421
PyTorch, 408, 554

R

Random Forests, 132, 133, 550, 725
    classifier, 729
    regressor, 743
    style training, 552
Random Gaussian noise, 703
Random noise, 341, 666–668
Random search, 714
Raw vectorization, 128–130
Realistic synthetic data, 690
Reassignments, 799
Receiver Operating Characteristics (ROC), 46–49
Reconstruction loss, 603, 642, 643, 652
Rectified Linear Unit (ReLU), 205, 207, 208, 211, 330
Recurrent layers, 397, 408, 416
Recurrent layers tabular data, 425
Recurrent models
    customer satisfaction, 403
    dataset, 404
    natural language, 403–409
    probability prediction, 407
    vocabulary size, 405
Recurrent models theory
    bidirectionality, 396, 397
    BPTT, 384–388
    exploding gradients, 388–392
    GRUs, 392–396
    LSTMs, 388–392
    memory cells, 381–384
    recurrent neurons, 381–384
    RNNs (see Recurrent neural networks (RNNs))
    vanishing gradient, 384–388
Recurrent neural networks (RNNs)
    ANNs, 379, 384
    and BPTT, 384–388
    GRUs, 392–396
    LSTMs, 388–392
    memory cells, 381–384
    tabular data, 380
    vanishing gradients, 384–388
Recurrent neurons, 381–388
Recursive Feature Elimination (RFE), 170–173
Red, green, and blue (RGB), 282, 802, 803
Regression, 8, 50, 175, 195, 200, 403, 556, 631
Regression problem, 113, 522, 610, 628, 755
Regularization learning networks (RLN), 254–257
.rename() method, 818
Reparative autoencoders, 664–680
Replaced token detection (RTD), 501, 502, 535
Reshape functions, 795
ResNet
    architecture, 314
    branching operations, 315
    definition, 311

DenseNet-style residual connections, 318
hybrid, 317
Keras, 316, 317
residual connection, 312
style, 313
vanishing gradient problem, 316
Return sequences, 400–403
Return state, 400–403
Robust scaling, 123–127
Root Mean Squared Error (RMSE), 40, 41

■ S

Sampled latent space vector, 688
Sample sequence, 406
Scaled exponential linear unit (SELU), 253, 254
Second element, 129, 382, 407, 741
Second-order optimization techniques, 580
Self-Attention and Intersample Attention
        Transformer (SAINT)
    attention maps, 538
    authors' notation, 532
    cloning repository, 541
    definition, 531
    intersample attention density, 539, 540
    mean AUROC, 538
    OpenML, 541
    pretraining mask, 535
    training pipeline, 536, 537
    transformer block, 533
Self-consciously, 770
Semantic system, 601
Sentiment extraction implementation, 134–137
Sequence data, 382, 407
Sequence length, 386, 414
Sequence-to-sequence modeling, 453
Sequential autoencoder, 609, 610
Sequential Model–Based Optimization
        (SMBO), 718
SHapley Additive exPlanations (SHAP),
        771–774, 780
    gradient explainer, 780
    mean, 781–783
    standard deviation, 784
    subnetwork, 780
Short-term memory, 389
Shuffles, 380
Sigmoid equation, 650
Sigmoid function, 473, 651
SimpleRNN, 398, 400
Single Headed Attention RNN (SHA-RNN)
        architecture, 465
Sinusoidal curves, 439, 440
Skip connections, 230, 311, 387, 388
Sklearn model object, 786

Sneaking, 663
Soft decision tree (SDT), 549
Soft Decision Tree Regressor for Tabular
        Data (SDTR)
    binary tree, 558
    custom training function, 559
    DataLoader, 560
    leaf node, 556
    node probabilities, 558
    nodes/connections, 557
    PyTorch, 559
    training SDT, 560, 561
Softmax function, 455, 764
Softmax Gumbel function, 701
Softmax layer, 407, 465, 789
Softmax output, 73, 405, 425
Soft ordering techniques, 378
Sparse autoencoders
    alpha value, 659
    interpretability, 664
    latent size, 654
    latent space, 657–660
    layer's activity, 654
    MNIST dataset, 656
    parameters, 654
    quasi-active, 659
    quasi-continuous latent space, 653
    reconstruction, 657–660
    regularizations, 655, 656
    sampled original inputs, 657–660
    size representation, 653
Sparsity, 101, 255, 522, 653, 701
Speech Accent Archive dataset, 411
SP-LIME, 785
Stack additional layers, 409
Stack operation, 826, 827
Standard deviation, 53, 120, 162, 226, 345, 688
Standard mean squared error, 628
Standard RNN neuron, 388
Stock data, 420, 421
Stock forecasting, 411, 423
Stock prediction, 381, 419
"Stringent" architecture, 545
Supervised learning, 5, 31
Surrogate function, 716, 717
Symmetric factorization module, 319, 320

■ T

TabNet architecture
    attentive tabular learning, 514
    dataset pack, 519–521
    decision-making procedure, 523
    definition, 514
    feature transformer model, 516, 518, 519

TabNet architecture (*cont.*)
    implementation, 523
    prior scale term, 515, 516
    selection masks, 523, 524, 531
    TabNetClassifier model, 522
tabnet library, 522
TabNet model, 356
TabTransformer model
    architecture, 501
    categorical features, 512, 513
    compartmentalized design, 510–512
    configuration constants, 506
    configuration parameters, 506
    definition, 500
    features, 500
    input heads, 507
    Keras model, 514
    keras.models.Model, 508
    MLP, 504, 508
    replaced token detection, 502
    submodel, 510
    train-validation split, 513
    transformer block, 507
    t-SNE reduction, 503
Tabular data, 631, 636
    generation methods, 710
    models, 397, 421–423
Tabular neural networks
    batch normalization, 247
    distribution calculation, 248
    element-wise linear transformation, 249
    ghost batch normalization, 247–249
    interaction modeling methods, 252
    leaky gates, 248–250
    modifications, 246
    synthetic task, 249
    wide/deep learning, 250–253
Tanh activation function, 204, 391
Task model, 633, 636
taskOut, 642
t-distributed Stochastic Neighbor Embedding
    (t-SNE), 156–159
Tensorflow, 436, 554, 702, 704
TensorFlow database
    Keras library, 189
    neural networks, 257
TensorFlow datasets
    biomedical datasets, 100
    custom dataset, 100
    data formats, 96
    features, 99
    filler code, 99
    handling datasets
        memory, 101
        NumPy memory maps, 104

        Pandas chunking, 102
        pickle files, 101
        Python library h5py, 102
        SciPy/TensorFlow sparse matrix, 101
    key components, 177
    multidimensional array, 97
    NumPy array, 97
    sequence dataset, 98–100
    TensorSliceDataset, 97, 98
    tf.keras.utils.Sequence class, 98, 99
Term Frequency–Inverse Document Frequency
    (TF-IDF), 133, 134
TextVectorization layer, 404
Three-dimensional array, 191, 794
Time-independent target prediction, 411
Time series
    architecture, 415
    audio files, 412
    feature map representation, 414
    high-frequency and high-complexity, 416
    high-frequency data, 411
    next-timestep forecasting, 410
    one-dimensional convolutional layers, 411
    one mode, 410
    prediction, 380, 399
    prices of stocks, 416
    recurrent layers, 411
    sequential data, 380
    sequentially ordered features, 411
    target prediction, 409
    time-dependent prediction, 410
    time-series problems, 411
Time-series data, 381, 421–423
Time-series dataset, 380, 397, 398
TimeseriesGenerator class, 398
Timestamps, 380–382, 384–386, 397, 398, 400, 402
Timestep, 408, 409
Timestep-by-timestep, 411
Training, 421, 704
Training curves, 635, 637
Training GAN, 704
train_step method, 558
Train-validation split, 421
Transformer architecture, 458, 546, 602, 650
Transformer model, 439, 457, 458, 461, 465, 516
Transformer-style positional encoding, 439, 449
Transformer-style sinusoidal position
    encodings, 440
Transporters, 388
Tree-structured deep learning
    approaches, 597
Tree-structured neural networks
    branch node, 568
    configurations, 577
    decision tress, 566, 567

DJINN, 565, 569–573
DNDT, 549
    datasets, 554
    GPUs, 550
    leaf nodes, 552
    non-binary decision tree, 550
    parameters, 555, 556
    PyTorch, 554
    single binning network, 551, 552
    single-layer neural network, 551
    training, 550
DNNF, 578, 579
Grownet, 579
NAS, 564
net-DNF
    literals, 575, 577
    logical expressions, 574, 575
neural AND function, 578
node, 561–564
OR function, 578
SDT, 549
weight initialization, 566
XGBoost, 577
Tree-Structured Parzen Estimator, 718, 726
True positive Rate (TPR), 44, 46, 48
t-SNE method
    latent space sizes, 618–623
    nodes, 619–623
Two-dimensional array, 192, 414, 794
Two-dimensional convolutions, tabular data, 758
    contiguous semantics, 356
    DeepInsight, 358–367
    IGTD, 367–377
    soft ordering, 357

■ U

Unfreezing, 634, 765
Universal Approximation Theorem, 242–245
Unrolled RNN neuron, 383, 384, 386
Unstacking operation, 827
Unsupervised learning, 6–8
Untampered clean images, 667
Update gate, 389, 394
Upvotes, 403
US Amazon Reviews dataset, 403

■ V

Valence Aware Dictionary for Sentiment Reasoning
    (VADER), 135
Validation, 421
Vanilla, 405
    adapting, 615
    autoencoder, 608, 612

compartmentalized model, 612
encoder/decoder, 612
hypotheses, 627
images, 607
image and text-based datasets, 606, 610
image processing library, 624
information flow, 608
larger-scale autoencoder experiments, 616
latent space, 612, 617
lines and reconstructions, 626, 627
logarithmic expression, 615
Mice Protein Expression dataset, 629
MNIST dataset, 607, 631
outActivation parameter, 616
overcomplete autoencoder, 624
parameters, 607
pixel values, 610
reconstruction, 607, 612, 613
regression loss, 611
sample latent shape, 613
samples, 630
sequentially, 608, 609
sigmoid function, 609
standard neural networks, 614
synthetic toy line dataset, 625
tabular autoencoder, 617
training/validation, 628
training/visualizing, 624
training history, 629
t-SNE method, 618–623
Vanilla soft ordering, 357
Vanishing gradient problem, 315, 316
Variance threshold, 165–167
Variational Autoencoders (VAEs), 680, 681, 710
    advantage, 681
    applications/demonstrations, 682
    bivariate relationships, 692
    decoded linear interpolations, 686
    formal optimization problem, 687
    Higgs Boson dataset, 691, 693
    implementation, 687–696
    latent space distribution, 686
    latent space vector, 683
    L2-style regularization, 687
    Mice Protein Expression dataset, 694, 695
    MNIST dataset, 682
    n-dimensional normal
        distribution, 686
    reparameterization trick, 686
    standard autoencoder, 687
    theory, 682
Vectorization methods, 126, 132
Vectorizer, 404
Vectorizing, 421
Voice accent, 411

W

Weighted average, 76, 391, 396, 579, 754, 756, 763
Weighted averaging, 755, 767
Weight of evidence (WoE) technique, 118, 119
Weight sharing model, 242–245
Wide model, 250–253
Window size, 397, 419
Word2Vec representations, 139–142

X, Y

XBNet, 597
    architecture, 585

classifier, 590
definition, 584
instantiate model, 589
pseudocode, 588
training, 589, 590
training procedures, 586, 587
XGBoost, 585
XGBoost, 49, 86–88, 563, 577, 582,
    584, 585, 588

Z

Zero vector, 220, 436

Printed in the United States
by Baker & Taylor Publisher Services